D1013110

ASTROPHYSICAL AGES AND TIME SCALES

Cover by C. Simpson from an original design by N. Manset.

A SERIES OF BOOKS ON RECENT DEVELOPMENTS IN
ASTRONOMY AND ASTROPHYSICS

Publisher

THE ASTRONOMICAL SOCIETY OF THE PACIFIC
390 Ashton Avenue, San Francisco, California, USA   94112-1722
Phone: (415) 337-1100   Fax: (415) 337-5205
E-Mail: catalog@aspsky.org   Web Site: www.aspsky.org

ASP CONFERENCE SERIES - EDITORIAL STAFF
Managing Editor: D. H. McNamara   LaTeX-Computer Consultant: T. J. Mahoney
Associate Managing Editor: J. W. Moody   Production Manager: Enid L. Livingston

PO Box 24453, 211 KMB, Brigham Young University, Provo, Utah, 84602-4463
Phone: (801) 378-2111   Fax: (801) 378-4049   E-Mail: pasp@byu.edu

ASP CONFERENCE SERIES PUBLICATION COMMITTEE:
Alexei V. Filippenko   Geoffrey Marcy
Ray Norris   Donald Terndrup
Frank X. Timmes   C. Megan Urry

A listing of all other ASP Conference Series Volumes and IAU Volumes
published by the ASP is cited at the back of this volume

ASTRONOMICAL SOCIETY OF THE PACIFIC
CONFERENCE SERIES

Volume 245

ASTROPHYSICAL AGES AND TIME SCALES

Proceedings of a conference held in
Hilo, Hawai'i, USA
5-9 February 2001

Edited by

Ted von Hippel
Gemini Observatory, 670 N. A'ohōkū Place, Hilo, Hawai'i, USA

Chris Simpson
Subaru Telescope, 650 N. A'ohōkū Place, Hilo, Hawai'i, USA

and

Nadine Manset
Canada-France-Hawai'i Telescope Corporation
65-1238 Mamalahoa Highway, Kamuela, Hawai'i, USA

© 2001 by Astronomical Society of the Pacific. All Rights Reserved

No part of the material protected by this copyright notice may be reproduced or utilized in any form or by any means – graphic, electronic, or mechanical including photocopying, taping, recording or by any information storage and retrieval system, without written permission from the publisher.

Library of Congress Cataloging in Publication Data
Main entry under title

Card Number: 2001093954
ISBN: 1-58381-083-8

ASP Conference Series - First Edition

Printed in United States of America by Sheridan Books, Chelsea, Michigan

Contents

Session 1. The Nature of Time

Session 2. Precision Timing in Astronomy

Session 3. Planetary and Debris Disk Systems

Contents vii

Preface

During the last ten years alone, astrophysics has made tremendous advances. Brown dwarfs and extrasolar planets, essentially hypothetical objects until six years ago, have not only been discovered but are now turning up regularly. Stellar structure and evolution, although a mature subfield of astrophysics, suffered from numerous tunable parameters masking unknown or hidden physics along with irreducible inaccuracies in key observables. Cosmology, despite providing a clear and robust paradigm for the rest of astrophysics, suffered from tremendous uncertainties in key parameters, especially the expansion rate and the matter and energy densities. While these problems are far from solved, advances have been so quick that many excellent papers on these subjects that were penned merely a decade ago now hold primarily pedagogical relevance. Whether advances come in planet formation or starbursting galaxies, the questions always arise, "When did this happen?" and "How long does this process take?" These are the questions of this conference, a conference dedicated to astrophysical ages and time scales. To the best of our knowledge, there has been no meeting singularly devoted to the topics of ages and time scales in astrophysics. Many recent advances have made this an excellent time to hold such a conference.

The conference started with the broad theme of the Nature of Time (Session 1). From there we focused on astrophysical ages and time scales in the following thematic order: Precision Timing in Astronomy (Session 2), Planetary and Debris Disk Systems (Session 3), the Galaxy and the Local Group (Session 4), Galaxies (Session 5), Galaxy Clusters (Session 6), and the Universe (Session 7). Substantial and new results are presented throughout these proceedings. We hope our readers share the excitement of the conference participants as they delve into the tremendously increased precision in astrophysical ages from the Sun to the Universe.

The conference featured 55 talks and 48 posters, and the following pages contain 97 papers based on these presentations. For completeness, we have included abstracts for the 6 remaining contributions. Despite the wide range of topics, most of the 165 conference attendees from 20 countries interacted throughout the entire week and ensured that the conference worked as a thematic whole, rather than a series of mini-conferences. In order to capture this interaction, we have included much of the discussion that followed each talk, although in the interests of space and readability we have edited the questions and asked speakers to clarify their responses. We wish to thank the conference participants for their enthusiasm, insights, and shared knowledge. We particularly want to thank our speakers and poster contributors for their excellent contributions both at the conference and subsequently in this volume. We also thank Douglas Gough, Alan Guth, and Typhoon Lee for presenting well-attended and well-received public lectures. We were very pleased to see 12 high school science teachers attend the conference as well. In addition, we appreciate Governor Benjamin Cayetano and Mayor Harry Kim for opening the conference and demonstrating their support for astronomy in Hawai'i.

Many thanks are owed to the 15 Local Organizing Committee members (Nancy Bucy, Tom Geballe, Janice Harvey, Martin Houde, Nadine Manset, Peter Michaud, Andrew Pickles, Kalena Quiñones, Antony Schinckel, Ian Shelton, Chris Simpson, Marianne Takamiya, Remo Tilanus, Ted von Hippel, and

Michael West) for their time, ideas, and enthusiasm. They created this conference. Thanks also to the 15 members of the Scientific Organizing Committee (Robert Carswell, Stéphane Charlot, Pierre Demarque, Arjun Dey, Tim de
Zeeuw, Douglas Gough, Steve Kawaler, Ofer Lahav, Typhoon Lee, Joan Najita,
Jean-René Roy, François Schweizer, Kazuhiro Sekiguchi, Jim Truran, and Ted
von Hippel) for their wisdom and guiding hand. We were particularly pleased
to see 9 members of the Scientific Organizing Committee attend, despite the
distance between Hilo and their home institutions. Nearly every astronomical
institution in Hawai'i participated in the conference by contributing staff time
(Gemini Observatory[1], Subaru Telescope[2], Submillimeter Array, Joint Astronomy Centre, University of Hawai'i at Hilo, University of Hawai'i at Manoa, Caltech Submillimeter Observatory[3], and Canada–France–Hawaii Telescope) and
funds (Subaru Telescope, Gemini Observatory, and Caltech Submillimeter Observatory). We also thank Fujitsu America, Inc. and Mitsubishi Electric &
Electronics USA, Inc. for their generous financial support. Likewise, thanks to
our technical editor and proofreader, Pamela Krones, for performing an excellent
job.

One of us (TvH) wishes to acknowledge four wonderful people who died
in the few months since the conference was held in February 2001. Fred Baas,
an excellent observer at the Joint Astronomy Centre, managed to attend the
conference during one of the waning weeks of his life. Fred Gillett, the guiding
hand and visionary behind Gemini, died in April. Helen Husek, a beloved aunt
and brilliant chemist, died in March. Alvin Moe, essentially a second father to
TvH, died in May.

We hope these pages provide you with insight into astrophysical ages and
time scales and with ideas for your research, teaching, or public outreach.

Mahalo nui kākou.

Ted von Hippel
Chris Simpson
Nadine Manset

Hilo, Hawai'i
June 28, 2001

---

[1]Gemini Observatory is operated by AURA, Inc., under a cooperative agreement with the Na-
tional Science Foundation on behalf of the Gemini partnership: the NSF (United States),
PPARC (United Kingdom), NRC (Canada), CONICYT (Chile), ARC (Australia), CNPq
(Brazil), and CONICET (Argentina).

[2]Subaru Telescope is a division of the National Astronomical Observatory of Japan (NAOJ),
which is an Inter-university Research Institute operated under the jurisdiction of the Ministry of
Education, Culture, Sports, Science, and Technology (MEXT; Monbu-Kagaku Sho) of Japan.

[3]The Caltech Submillimeter Observatory is funded by the NSF through contract AST 9615025.

Conference Participants

Alessandra Aloisi — aloisi@stsci.edu
Space Telescope Science Institute, USA
Hiroyasu Ando — ando@naoj.org
Subaru Telescope, USA
Tom Andrews — tom-andrews@msn.com

Antonio Aparicio — aaj@iac.es
Instituto de Astrofísica de Canarias, Spain
Marcel Arnould — marnould@astro.ulb.ac.be
Université Libre de Bruxelles, Belgium
Colin Aspin — caa@gemini.edu
Gemini Observatory, USA
Fred Baas — baas@jach.hawaii.edu
Joint Astronomy Centre, USA
Josh Barnes — barnes@galileo.ifa.hawaii.edu
University of Hawai'i at Manoa, USA
Timothy C. Beers — beers@pa.msu.edu
Michigan State University, USA
Emile Biémont — E.Biemont@ulg.ac.be
Université de Liège, Belgium
Bill Brevoort — billbre@earthlink.net

Moshe Carmeli — carmelim@bgumail.bgu.ac.il
Ben-Gurion University, Israel
Bob Carswell — rfc@ast.cam.ac.uk
University of Cambridge, UK
Roger Cayrel — roger.cayrel@obspm.fr
Observatoire de Paris, France
Brian Chaboyer — brian.chaboyer@dartmouth.edu
Dartmouth College, USA
Norbert Christlieb — nchristlieb@hs.uni-hamburg.de
Hamburger Sternwarte, Germany
Mark Chun — mchun@gemini.edu
Gemini Observatory, USA
Iain Coulson — imc@jach.hawaii.edu
Joint Astronomy Centre, USA
Stéphane Courteau — courteau@astro.ubc.ca
University of British Columbia, Canada
Richard Crowe — rcrowe@maxwell.uhh.hawaii.edu
University of Hawai'i at Hilo, USA
Gavin Dalton — gbd@astro.ox.ac.uk
University of Oxford, UK
Romeel Davé — rad@as.arizona.edu
Steward Observatory, USA
G. Paolo di Benedetto — pdibene@ifctr.mi.cnr.it
Istituto di Fisica Cosmica, Italy
Matthias Dietrich — dietrich@astro.ufl.edu
University of Florida, USA
Ed Duckworth — LeCanard@aol.com
City College of San Francisco, USA
Doug Duncan — duncan@oddjob.uchicago.edu
University of Chicago, USA
Martin Duncan — duncan@astro.queensu.ca
Queen's University, Canada
Michael Edwards — miedward@du.edu
University of Denver, USA

Stefano Ettori   settori@ast.cam.ac.uk
    University of Cambridge, UK
Gary Ferland   gary@pa.uky.edu
    University of Kentucky, USA
Roger Ferlet   ferlet@iap.fr
    Institut d'Astrophysique de Paris, France
Ralf Flicker   rflicker@gemini.edu
    Gemini Observatory, USA
Andre Fludra   fludra@cdso8.nascom.nasa.gov
    Rutherford Appleton Laboratory, UK
Gilles Fontaine   fontaine@astro.umontreal.ca
    Université de Montréal, Canada
Wendy Freedman   wendy@ociw.edu
    Carnegie Observatories, USA
Klaus Fuhrmann   fuhrmann@usm.uni-muenchen.edu
    Universitäts-Sternwarte München, Germany
Gary Fujihara   fujihara@naoj.org
    Subaru Telescope, USA
Tetsuharu Fuse   tetsu@naoj.org
    Subaru Telescope, USA
Wolfgang Gaessler   wgaessle@naoj.org
    Subaru Telescope, USA
Tom Geballe   tgeballe@gemini.edu
    Gemini Observatory, USA
Alvaro Giménez   ag@laeff.esa.es
    Laboratorio de Astrofísica Espacial y Física Fundamental, Spain
Douglas Gough   douglas@ast.cam.ac.uk
    University of Cambridge, UK
Fabio Governato   fabio@merate.mi.astro.it
    Osservatorio Astronomico di Brera, Italy
Evgeny Griv   griv@bgumail.bgu.ac.il
    Ben-Gurion University, Israel
Bengt Gustafsson   bengt.gustaffson@sigtunastiftelsen.se
    Uppsala Astronomical Observatory, Sweden
Alan Guth   guth@ctp.mit.edu
    Massachusetts Institute of Technology, USA
John Hamilton   jhamilton@gemini.edu
    Gemini Observatory, USA
Michael W. Hannawald   Michael.Hannawald@uni-mainz.de
    Universität Mainz, Germany
Hugh Harris   hch@nofa.navy.mil
    US Naval Observatory, USA
Bill Heacox   heacox@hawaii.edu
    University of Hawai'i at Hilo, USA
Gerhard Hensler   hensler@astrophysik.uni-kiel.de
    Universität Kiel, Germany
Klaus W. Hodapp   hodapp@ifa.hawaii.edu
    Univeristy of Hawai'i at Manoa, USA
Jim Hoge   jhoge@jach.hawaii.edu
    Joint Astronomy Centre, USA
Satoshi Honda   honda@optik.mtk.nao.ac.jp
    Graduate University for Advanced Studies, Japan
Martin Houde   houde@ulu.submm.caltech.edu
    Caltech Submillimeter Observatory, USA
James Hough   jhh@star.herts.ac.uk
    University of Hertfordshire, UK

Joanne Hughes Clark   hughes@astro.washington.edu
   Everett Community College and University of Washington, USA
Denise Hurley-Keller   denise@smaug.astr.cwru.edu
   Case Western Reserve University, USA
Hajime Inoue   inoue@astro.isas.ac.jp
   Institute of Space and Astronautical Sciences, Japan
George Isaak   gri@star.sr.bham.ac.uk
   University of Birmingham, UK
Jordi Isern   isern@ieec.fcr.es
   Institut d'Estudis Espacials de Catalunya, Spain
Nob Iwamoto   iwamoto@th.nao.ac.jp
   National Astronomical Observatory of Japan, Japan
Masanori Iye   iye@optik.mtk.nao.ac.jp
   National Astronomical Observatory of Japan, Japan
Joe Jensen   jjensen@gemini.edu
   Gemini Observatory, USA
Heath Jones   hjones@eso.org
   European Southern Observatory, Chile
Inger Jørgensen   ijorgensen@gemini.edu
   Gemini Observatory, USA
Taka Kajino   kajino@nao.ac.jp
   National Astronomical Observatory of Japan, Japan
Victoria Kaspi   vkaspi@physics.mcgill.ca
   McGill University, Canada
Guinevere Kauffmann   gamk@mpa-garching.mpg.de
   Max-Planck Institut für Astrophysik, Germany
Tuba Koktay   tuba@naoj.org
   University of Istanbul, Turkey
Eiichiro Kokubo   kokubo@th.nao.ac.jp
   National Astronomical Observatory of Japan, Japan
Yuri Kolesnik   kolesnik@inasan.rssi.ru
   Institute of Astronomy of the Russian Academy of Sciences, Russia
Yutaka Komiyama   komiyama@naoj.org
   Subaru Telescope, USA
Andreas Korn   akorn@usm.uni-muenchen.edu
   Universitäts-Sternwarte München, Germany
Ofer Lahav   lahav@ast.cam.ac.uk
   University of Cambridge, UK
Bryan E. Laubscher   blaubscher@lanl.gov
   Los Alamos National Laboratory, USA
Dave Leckrone   dleckrone@hst.nasa.gov
   Goddard Space Flight Center, USA
Hyun-chul Lee   hclee@csa.yonsei.ac.kr
   Yonsei University, Korea
Jae-Woo Lee   jaewoo@astro.unc.edu
   University of North Carolina, USA
Typhoon Lee   typhoon@asiaa.sinica.edu.tw
   Academia Sinica, Taiwan
Young-Wook Lee   ywlee@csa.yonsei.ac.kr
   Yonsei University, Korea
Sandy Leggett   skl@jach.hawaii.edu
   Joint Astronomy Centre, USA
Claus Leitherer   leitherer@stsci.edu
   Space Telescope Science Institute, USA
Douglas Lin   lin@ucolick.org
   University of California, Santa Cruz, USA

Harvey Liszt   hliszt@nrao.edu
    National Radio Astronomy Observatory, USA
Thomas Lowe   lowe@jach.hawaii.edu
    Joint Astronomy Centre, USA
Lori Lubin   lml@pha.jhu.edu
    Johns Hopkins University, USA
Lauren MacArthur   lauren@physics.ubc.ca
    University of British Columbia, Canada
Peregrine M. McGehee   peregrine@lanl.gov
    Los Alamos National Laboratory, USA
Nadine Manset   manset@cfht.hawaii.edu
    Canada-France-Hawaii Telescope Corporation, USA
Eduardo Martin   ege@ifa.hawaii.edu
    University of Hawai'i at Manoa, USA
Frank Masci   fmasci@ipac.caltech.edu
    Infrared Processing and Analysis Center, USA
C. J. Masreliez   jmasreliez@estfound.org
    EST Foundation, USA
Lucio Mayer   Lucio.Mayer@uni.mi.astro.it
    University of Washington, USA
Peter Michaud   pmichaud@gemini.edu
    Gemini Observatory, USA
Michelle Mizuno-Wiedner   mizuno@astro.uu.se
    Uppsala Astronomical Observatory, Sweden
J. Ward Moody   jmoody@byu.edu
    Brigham Young University, USA
Gerald Moriarty-Schieven   gms@jach.hawaii.edu
    Joint Astronomy Centre, USA
Kentaro Motohara   motohara@naoj.org
    Subaru Telescope, USA
Matt Mountain   mmountain@gemini.edu
    Gemini Observatory, USA
Anjum Mukadam   anjum@bullwinkle.as.utexas.edu
    University of Texas at Austin, USA
Ralph Neuhaeuser   rne@ifa.hawaii.edu
    University of Hawai'i at Manoa, USA
Kumi Nishimura   KumiN@aol.com

Tetsuo Nishimura   nishimra@naoj.org
    Subaru Telescope, USA
Ryusuke Ogasawara   ryu@naoj.org
    Subaru Telescope, USA
Yoji Osaki   osaki@net.nagasaki-u.ac.jp
    Nagasaki University, Japan
Ruisheng Peng   peng@submm.caltech.edu
    Caltech Submillimeter Observatory, USA
Andrew Pickles   pickles@ifa.hawaii.edu
    University of Hawai'i at Manoa, USA
Piotr Popowski   popowski@igpp.ucllnl.org
    Lawrence Livermore National Laboratory, USA
Phil Puxley   ppuxley@gemini.edu
    Gemini Observatory, USA
Michael Rauch   mr@ociw.edu
    Carnegie Observatories, USA
R. Michael Rich   rmr@astro.ucla.edu
    University of California, Los Angeles, USA

François Rigaut   frigaut@gemini.edu
Gemini Observatory, USA
Helio Rocha-Pinto   helio@iagusp.usp.br
Universidade de São Paulo, Brazil
Marcello Rodonò   mrodono@ct.astro.it
Universitàdi Catania, Italy
Kathy Roth   kroth@gemini.edu
Gemini Observatory, USA
Jean-René Roy   jrroy@gemini.edu
Gemini Observatory, USA
Maurizio Salaris   ms@astro.livjm.ac.uk
Liverpool John Moores University, UK
Toshiyuki Sasaki   sasaki@naoj.org
Subaru Telescope, USA
Steven Savitt   savitt@interchange.ubc.ca
University of British Columbia, Canada
Anthony Schinckel   aschinck@cfa.harvard.edu
Submillimeter Array, USA
Glenn Schneider   gschneider@as.arizona.edu
Steward Observatory, USA
Bob Schommer   bschommer@noao.edu
Cerro Tololo Inter-American Observatory, Chile
François Schweizer   schweizer@ociw.edu
Carnegie Observatories, USA
Kaz Sekiguchi   kaz@naoj.org
Subaru Telescope, USA
Ian Shelton   shelton@naoj.org
Subaru Telescope, USA
Wendy Shook   wshook@gemini.edu
Gemini Observatory, USA
Frank Shu   shu@mars.berkeley.edu
University of California, Berkeley, USA
Chris Simpson   chris@naoj.org
Subaru Telescope, USA
Chris Sneden   chris@verdi.as.utexas.edu
University of Texas at Austin, USA
Alan Stockton   stockton@ifa.hawaii.edu
University of Hawai'i at Manoa, USA
Oscar Straniero   straniero@astrte.te.astro.it
Osservatorio Astronomico di Collurania, Italy
Naoshi Sugiyama   naoshi@th.nao.ac.jp
National Astronomical Observatory of Japan, Japan
Robert Szabó   rszabo@konkoly.hu
Konkoly Observatory, Hungary
Istvan Szapudi   szapudi@ifa.hawaii.edu
University of Hawai'i at Manoa, USA
Marianne Takamiya   mtakamiya@gemini.edu
Gemini Observatory, USA
Tadafumi Takata   takata@naoj.org
Subaru Telescope, USA
Naruhisa Takato   takato@naoj.org
Subaru Telescope, USA
Remo Tilanus   r.tilanus@jach.hawaii.edu
Joint Astronomy Centre, USA
John Tonry   jt@ifa.hawaii.edu
University of Hawai'i at Manoa, USA

Scott Trager   sctrager@ociw.edu
  Carnegie Observatories, USA
Jim Truran   truran@oddjob.uchicago.edu
  University of Chicago, USA
Brent Tully   tully@ifa.hawaii.edu
  University of Hawai'i at Manoa, USA
Michael Turner   mturner@oddjob.uchicago.edu
  University of Chicago, USA
David Tytell   dtytell@skypub.com
  Sky & Telescope, USA
William Unruh   unruh@physics.ubc.ca
  University of British Columbia, Canada
Tomonori Usuda   usuda@naoj.org
  Subaru Telescope, USA
David Valls-Gabaud   dvg@ast.obs-mip.fr
  Observatoire Midi-Pyrénées, France
Gerardo A. Vazquez   gerar@astroscu.unam.mx
  Universidad Nacional Autónoma de México, Mexico
Alfred Vidal-Madjar   alfred@iap.fr
  Institut d'Astrophysique de Paris, France
Gabrielle E. Villa   gev@ifctr.mi.cnr.it
  Istituto di Fisica Cosmica, Italy
Yanick Villedieu   yvilledi@mlink.net
  EnRoute Magazine, Canada
Ted von Hippel   ted@gemini.edu
  Gemini Observatory, USA
Dolores Walther   dwalther@gemini.edu
  Gemini Observatory, USA
William R. Ward   ward@boulder.swri.edu
  Southwest Research Institute, USA
Achim Weiss   aweiss@mpa-garching.mpg.de
  Max-Planck Institut für Astrophysik, Germany
Werner Weiss   weiss@astro.univie.ac.at
  Universität Wien, Austria
Michael West   west@bohr.uhh.hawaii.edu
  University of Hawai'i at Hilo, USA
Brad Whitmore   whitmore@stsci.edu
  Space Telescope Science Institute, USA
Mavourneen Wilcox   mavourneen.wilcox@gte.net
  University of Hawai'i at Hilo, USA
Peter Williams   williams@astr.tohoku.ac.jp
  Tohoku University, Japan
George Wolf   GeorgeWolf@smsu.edu
  Southwest Missouri State University, USA
Hiroshige Yoshida   hiro@submm.caltech.edu
  Caltech Submillimeter Observatory, USA

Conference Photograph

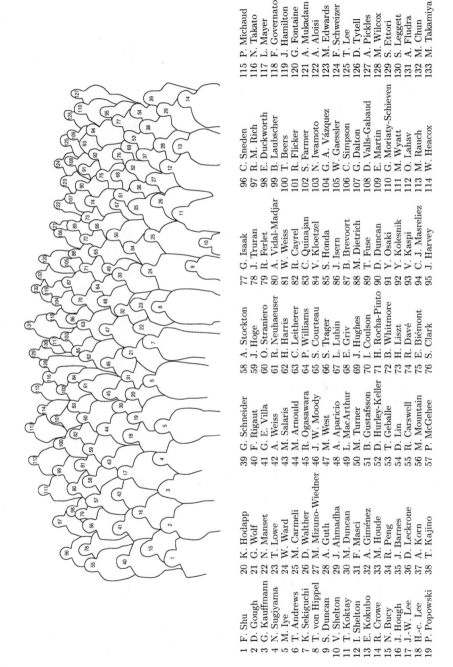

1 F. Shu
2 D. Gough
3 G. Kauffmann
4 N. Sugiyama
5 M. Iye
6 T. Andrews
7 K. Sekiguchi
8 T. von Hippel
9 S. Duncan
10 V. Shelton
11 C. Koktay
12 I. Shelton
13 E. Kokubo
14 R. Crowe
15 N. Bucy
16 J. Hough
17 J-W. Lee
18 H.-c. Lee
19 P. Popowski
20 K. Hodapp
21 G. Wolf
22 N. Manset
23 T. Lowe
24 W. Ward
25 M. Carmeli
26 D. Walther
27 M. Mizuno-Wiedner
28 A. Guth
29 J. Ahmadha
30 M. Duncan
31 F. Masci
32 A. Giménez
33 M. Houde
34 R. Peng
35 J. Barnes
36 D. Leckrone
37 A. Korn
38 T. Kajino
39 G. Schneider
40 F. Rigaut
41 G. E. Villa
42 A. Weiss
43 M. Salaris
44 M. Arnould
45 R. Ogasawara
46 J. W. Moody
47 M. West
48 A. Aparicio
49 L. MacArthur
50 M. Turner
51 B. Gustafsson
52 D. Hurley-Keller
53 T. Geballe
54 D. Lin
55 R. Carswell
56 M. Mountain
57 P. McGehee
58 A. Stockton
59 J. Hoge
60 O. Straniero
61 R. Neuhaeuser
62 H. Harris
63 C. Leitherer
64 P. Williams
65 S. Courteau
66 S. Trager
67 L. Lubin
68 E. Griv
69 J. Hughes
70 I. Coulson
71 H. Rocha-Pinto
72 B. Whitmore
73 H. Liszt
74 R. Davé
75 E. Biémont
76 S. Clark
77 G. Isaak
78 J. Truran
79 R. Ferlet
80 A. Vidal-Madjar
81 W. Weiss
82 R. Cayrel
83 C. Quinajan
84 V. Kloetzel
85 S. Honda
86 J. Isern
87 B. Brevoort
88 M. Dietrich
89 T. Fuse
90 D. Duncan
91 Y. Osaki
92 Y. Kolesnik
93 V. Kaspi
94 C. J. Masreliez
95 J. Harvey
96 C. Sneden
97 R. M. Rich
98 E. Duckworth
99 B. Laubscher
100 T. Beers
101 R. Flicker
102 S. Farmer
103 N. Iwamoto
104 G. A. Vázquez
105 W. Gaessler
106 C. Simpson
107 G. Dalton
108 D. Valls-Gabaud
109 E. Martin
110 E. Moriaty-Schieven
111 M. Wyatt
112 O. Lahav
113 M. Rauch
114 W. Heacox
115 P. Michaud
116 N. Takato
117 L. Mayer
118 F. Governato
119 J. Hamilton
120 G. Fontaine
121 A. Mukadam
122 A. Aloisi
123 M. Edwards
124 F. Schweizer
125 T. Lee
126 D. Tytell
127 A. Pickles
128 M. Wilcox
129 S. Ettori
130 S. Leggett
131 A. Fludra
132 M. Chun
133 M. Takamiya

Session 1

The Nature of Time

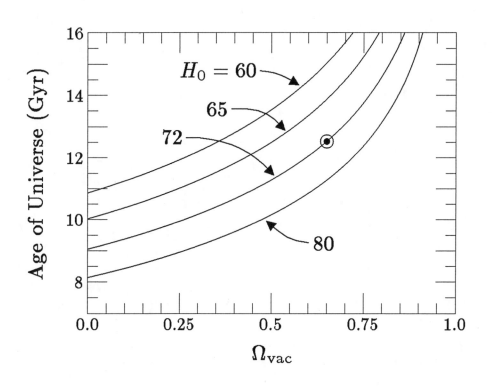

Astrophysical Ages and Time Scales
ASP Conference Series, Vol. 245, 2001
T. von Hippel, C. Simpson, N. Manset

Time Since the Beginning

Alan H. Guth

*Center for Theoretical Physics, Laboratory for Nuclear Science and
Department of Physics, Massachusetts Institute of Technology,
Cambridge, MA 02139, USA*

Abstract. While there is no consensus about the history of time since
the beginning, in this paper I will discuss some possibilities. We have a
pretty clear picture of cosmic history from the electroweak phase transition through the time of recombination, a period that includes the quantum chromodynamics phase transition and Big Bang nucleosynthesis.
This paper includes a quantitative discussion of the age of the Universe,
the radiation–matter transition, and hydrogen recombination. There is
much evidence that at earlier times the Universe underwent inflation, but
the details of how and when inflation happened are still far from certain.
There is even more uncertainty about what happened before inflation,
and how inflation began. I will describe the possibility of eternal inflation, which proposes that our Universe evolved from an infinite tree of
inflationary spacetime. Most likely, however, inflation can be eternal only
into the future but still must have a beginning.

1. Introduction

In this paper I will attempt to discuss the history of time from the beginning,
even though no complete description exists. In Section 2 I will lay out the basic
equations, and in Section 3 I will discuss the time period from about 10^{-12} s to
300 000 yr. In Section 4 I will discuss what happened earlier, suggesting that
inflation is the answer. Section 5 will deal with the question of what happened
before inflation, to which I will argue that the answer is more inflation—i.e.,
eternal inflation. In the final section I will summarize.

2. Fundamentals of Early Universe Physics

The time-evolution of the early Universe seems to be well-described by a remarkably simple theory, known alternatively as the hot Big Bang Theory or
the standard cosmological model. The model assumes that the Universe is well-
approximated as being homogeneous and isotropic, which implies that the metric
can be written in the Robertson–Walker form,

$$ds^2 = -dt^2 + a^2(t) \left\{ \frac{dr^2}{1 - kr^2} + r^2 \left[d\theta^2 + \sin^2 \theta \, d\phi^2 \right] \right\}, \tag{1}$$

where k denotes a constant which indicates whether the Universe is open $(k < 0)$, closed $(k > 0)$, or flat $(k = 0)$, and throughout this article I will use units for which $\hbar \equiv c \equiv k_B \equiv 1$. The Einstein equations imply that the scale factor $a(t)$ evolves according to

$$\left(\frac{\dot{a}}{a}\right)^2 = \frac{8\pi}{3}G\rho - \frac{k}{a^2}, \tag{2}$$

$$\ddot{a} = -\frac{4\pi}{3}G(\rho + 3p)a, \text{ and} \tag{3}$$

$$\frac{d}{dt}\left(a^3\rho\right) = -p\frac{d}{dt}\left(a^3\right), \tag{4}$$

where ρ is the mass density, p is the pressure, G is Newton's constant, and the overdot denotes a derivative with respect to t. These equations are not independent, since any one of them can be derived from the other two. Assuming that the mass density consists of matter ($\rho_m \propto a^{-3}$), radiation ($\rho_r \propto a^{-4}$), and vacuum mass density (ρ_{vac} = constant), then Equation (2) can be integrated to give the relationship between the scale factor a and the time t. Denoting the present value of the Hubble parameter $H \equiv \dot{a}/a$ by H_0, and normalizing the scale factor so that its present value is 1, one finds:

$$t = H_0^{-1} \int_0^a \frac{a'\, da'}{\sqrt{\Omega_m a' + \Omega_r + \Omega_{vac} a'^4 + \Omega_k a'^2}}, \tag{5}$$

where $\Omega_X \equiv \rho_{X0}/\rho_c$, the subscript 0 denotes the present time, ρ_c denotes the critical density $3H_0^2/(8\pi G)$, and $\Omega_k = 1 - \Omega_m - \Omega_r - \Omega_{vac}$.

There is now much evidence that the Universe is flat, coming predominantly from studies of the cosmic microwave background (CMB). Wang, Tegmark, & Zaldarriaga (2001) have carried out a comprehensive study in which they combined the measurements of the CMB (most importantly the results from BOOMERANG (Netterfield et al. 2001), DASI (Halverson et al. 2001), Maxima (Lee et al. 2001), and CBI (Padin et al. 2001)) with measurements of large scale structure (IRAS PSCz survey (Saunders et al. 2000; Hamilton, Tegmark, & Padmanabhan 2000)) and the Hubble parameter (Freedman et al. 2001) to find that $\Omega_k = 0.0 \pm 0.06$ at the 95% confidence level. Since this result is also in agreement with the prediction of the simplest inflationary models, for the remainder of this paper I will consider only models that are exactly flat ($k = 0$).

To apply Equation (5) for times near the present, it is sufficient to neglect $\Omega_r \approx 10^{-4}$, in which case Equation (5) can integrated analytically:

$$t = \frac{2H_0^{-1}}{3\sqrt{\Omega_{vac}}} \tanh^{-1} \sqrt{\frac{\Omega_{vac} a^4}{a^4 \Omega_{vac} + a(1 - \Omega_{vac})}}. \tag{6}$$

To determine the present age we set $a = 1$ in the above equation, finding $t_0 = (2H_0^{-1}/(3\sqrt{\Omega_{vac}})) \tanh^{-1} \sqrt{\Omega_{vac}}$. Numerical evaluations of this formula are shown in Figure 1. The final value for the Hubble parameter obtained by the Hubble Key Project (Freedman et al. 2001) was $H_0 = 72 \pm 8$ km s^{-1} Mpc^{-1}, so I take $H_0 = 72$ as the central value in Figure 1. There is more uncertainty in Ω_{vac}, but I will take $\Omega_{vac} \approx 0.65$ as the central value for the graphs.

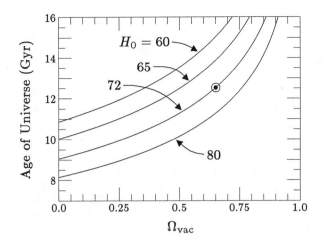

Figure 1.   Calculated age of the Universe as a function of Ω_{vac}, for various values of the Hubble parameter H_0, measured in km s^{-1} Mpc^{-1}. The models are flat and assumed to have negligible radiation density. The circled dot shows the currently popular model with $H_0 = 72$, $\Omega_{\text{vac}} = 0.65$, and an age of 12.5 Gyr.

While radiation is a negligible contribution to the total mass density today, the Universe is believed to have been radiation-dominated from just after inflation until some tens of thousands of years after the Big Bang. The energy density of such thermal radiation is given by

$$\rho = g \frac{\pi^2}{30} T^4, \tag{7}$$

where g denotes the number of effectively massless bosonic spin states, plus 7/8 times the number of effectively massless fermionic spin states. The entropy density is given by

$$s = g \frac{2\pi^2}{45} T^3. \tag{8}$$

Except for inflation the entropy of the early Universe is believed to have remained essentially constant, so that the relationship between time t and temperature T can be found from $a^3 s = constant$ and the dynamical equations (2)–(4). If g can be treated as a constant, as it can for various time intervals, this relation becomes

$$T = \left(\frac{45}{16\pi^3 gG} \right)^{1/4} \frac{1}{\sqrt{t}}. \tag{9}$$

After the disappearance of the muons at about 10^{-4} s, the contributions to g consist of photons ($g = 2$), electron–positron pairs ($g = 7/2$), and three species of neutrinos ($g = 21/4$), for a total of $g = 10\frac{3}{4}$. During this interval Equation (9) reduces to

$$T = \frac{0.8592 \text{ MeV}}{\sqrt{t/(1 \text{ s})}} = \frac{9.971 \times 10^9 \text{ K}}{\sqrt{t/(1 \text{ s})}}. \tag{10}$$

As an order-of-magnitude estimate, one can use the above formula for all times between about 10^{-12} s and 10^3 yr. As a precise formula, Equation (10) begins to fail at about $t = 1$ s, when the e^+e^- pairs start to disappear from the thermal equilibrium mix. By this time the neutrinos have effectively decoupled, so all the entropy of the e^+e^- pairs (with $g = 7/2$) is given to the photons (with $g = 2$) and not the neutrinos. As a result, the entropy density of photons is increased by a factor of $(\frac{7}{2}+2)/2 = 11/4$, so the temperature of the photons relative to the neutrinos is increased by a factor of $(11/4)^{1/3}$. This ratio is believed to persist to the present.

For the period after the disappearance of the e^+e^- pairs, one conventionally uses T to denote the temperature of the photons, while the neutrinos have a temperature $T_\nu = (4/11)^{1/3} T$. The COBE FIRAS measurements (Mather et al. 1999) determined that $T_0 = 2.725 \pm 0.002$ K, which implies a present mass density in photons and neutrinos of 7.804×10^{-34} g cm^{-3}. Using $H_0 = 72$ km s^{-1} Mpc^{-1}, one finds $\Omega_r = 8.013 \times 10^{-5}$.

3. Cosmic Events from 10^{-12} Seconds to 300 000 Years

The first key event of this period is the electroweak phase transition, at which the SU(2)×U(1) symmetry of the Glashow–Weinberg–Salam electroweak theory is broken to the familiar U(1) symmetry of quantum electrodynamics. At this phase transition a Higgs field is believed to acquire a nonzero expectation value. The interaction of the Higgs field with other fields is then responsible for masses of the corresponding particles, which include the W, the Z, the leptons (e, μ, and τ), and the quarks. It is worth noting, however, that the masses acquired by the u and d quarks through the electroweak symmetry-breaking are called the current-quark masses, and have values under 10 MeV. They have very little influence on the masses of protons and neutrons, which are associated with the constituent quark masses that arise from the strong interactions of the quarks.

The details of the electroweak phase transition remain unknown, since the Higgs particle and its detailed properties remain out of reach. The lack of attractive alternatives has convinced most particle physicists that the Higgs particle almost certainly exists, but it remains possible that nature is more complicated than the simple models with a single Higgs field. The energy scale of electroweak symmetry breaking is certainly, however, on the order of 1 TeV, so the time of the electroweak phase transition can be estimated from Equation (10) at about 10^{-12} s.

The next important event was the quantum chromodynamics (QCD) phase transition, which has an energy scale of approximately 1 GeV, and therefore took place at about 10^{-6} s. At this phase transition the quark-gluon plasma, with its essentially free quarks, disappeared in favor of a phase in which the quarks are permanently bound inside mesons and baryons. At about the same time the overwhelming majority of quarks and antiquarks annihilated in pairs. A tiny excess of quarks over antiquarks, of about one part in 10^9, resulted in the survival of a tiny fraction of the hadronic matter, and this tiny excess is responsible for the existence of the protons and neutrons that populate the current Universe. We believe that the excess was generated by a process, known as baryogenesis, which may have occurred anytime from the grand unified theory era through the electroweak phase transition.

At $t \sim 1$ s, when the temperature fell to about 1 MeV, the processes that led to Big Bang nucleosynthesis began. The first step was the decoupling of the neutrinos, which cut off the reactions that had until this time maintained a thermal equilibrium balance between protons and neutrons. Since the neutron mass exceeds the proton mass by 1.29 MeV, the number of neutrons was suppressed relative to the number of protons, but not by a large amount.

It is often pointed out that it appears to be an important coincidence that the temperature at which the neutrinos decouple, determined by the strength of the weak interactions and various cosmological parameters, is very nearly equal to the neutron-proton mass difference, which is presumably the result of an interplay between the strong and electromagnetic interactions. If the neutrinos remained coupled for much longer, the thermal equilibrium between protons and neutrons would be maintained down to lower temperatures, resulting in the almost complete disappearance of neutrons from the Universe. If the neutrinos decoupled earlier, then the Universe would be left with a nearly 50%/50% mix of protons and neutrons, which would result in an almost total conversion to He^4 in Big Bang nucleosynthesis.

After the neutrinos decoupled at $t \sim 1$ s, the only relevant reaction that could interchange protons and neutrons was the free decay of the neutron, with a mean life of about 15 minutes. After about 3 to 4 minutes, however, the temperature fell to $T \sim 0.1$ MeV, which is cool enough for the deuteron to become stable. At this point nuclear reactions proceeded quickly, converting almost all the neutrons that remained into He^4, which today has an abundance of 23–25% by mass (see for example, Burles, Nollett, & Turner 2001b). In addition, detectable amounts of deuterium, He^3, and Li^7 were produced. Note that 0.1 MeV is far below the deuteron binding energy of 2.2 MeV, but that such low temperatures are needed for stability because of the huge ratio of photons to baryons, about $10^9 : 1$. Thus each deuteron that formed must have survived a huge number of photon collisions before it had the chance to proceed with further nuclear reactions.

At $t \sim 30\,000$ years, the mass density of the Universe gradually changed from radiation-dominated to matter-dominated, where "matter" refers to both dark matter and baryonic matter. This change is described by the equations presented in Section 2, and is actually a very gradual transition. The results of a numerical integration of these equations is shown in Figure 2.

Finally, the last important event of this period is known as hydrogen "recombination," although "combination" would be a more accurate term. In the context of the standard cosmological model, the electrons and protons had never been combined at any point in the past. Recombination is often said to take place at a temperature of 4000 K and at a time of 300 000 yr. These numbers are in fact reasonable estimates, but the actual process of recombination, like that of matter-domination, is gradual. Note that 4000 K ~ 0.34 eV, so like the deuteron during nucleosynthesis, atomic hydrogen in the early Universe did not become stable until $k_{\mathrm{B}}T$ was far below its binding energy.

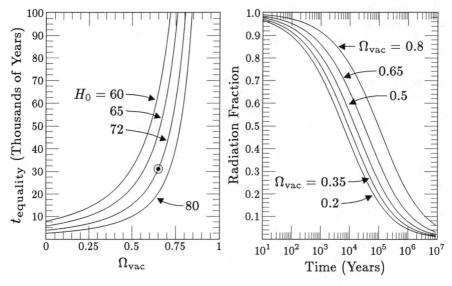

Figure 2.   *Left:* Time of matter-radiation equality as a function of the present value of Ω_{vac}, for various values of H_0 (in km s^{-1} Mpc^{-1}). The circled dot shows the currently popular model with $H_0 = 72$ and $\Omega_{\text{vac}} = 0.65$, with $t_{\text{equality}} = 31\,070$ yr. *Right:* Fraction of the total mass density of the Universe in radiation as a function of time, for various values of Ω_{vac}. Both graphs represent the same flat cosmological models, which are assumed to have three species of massless neutrinos and a present radiation temperature of 2.725 K. The dark energy component is taken to be a cosmological constant, with a fixed vacuum mass density.

If one assumes thermal equilibrium, then the fraction x of protons or electrons that remain ionized is given by the Saha equation,

$$\frac{x^2}{1-x} = \frac{(2\pi m_e k_{\text{B}} T)^{3/2}}{(2\pi \hbar)^3 \, n} \, e^{-B/k_{\text{B}}T}. \tag{11}$$

Here m_e is the electron mass, k_{B} is the Boltzmann constant, and B is the binding energy of hydrogen, 13.60 eV. (For a pedagogical treatment of the Saha equation, see Peebles 1993, pp. 165–167.) The Saha equation provides a reasonable approximation for the onset of recombination, but the process soon departs significantly from thermal equilibrium, as was shown by Peebles (1968). The reasons for the departure from thermal equilibrium are a bit subtle, since the reaction rates for ionization and recombination are much faster than the expansion rate of the Universe. The problem, however, is that almost every decay to the ground state of hydrogen emits a Lyman alpha photon which then has a high probability of ionizing another hydrogen atom. Thus, the sum of the number of ground state hydrogen atoms plus the number of Lyman alpha photons changes slowly, and lags behind thermal equilibrium as the Universe cools. The dominant mechanisms for changing this sum are the rare two-photon decay of the 2s level of hydrogen to the ground state, and the gradual redshifting of the Lyman alpha photons out of the relevant range of frequencies.

Numerical results for recombination are shown in Figure 3, using the currently indicated values of the parameters. In particular, the calculations use a flat model with $T_0 = 2.725$ K, $\Omega_{\text{vac}} = 0.65$, and $\Omega_B h^2 = 0.020$, following Burles, Nollett, & Turner (2001a, 2001b), who found $\Omega_B h^2 = 0.020 \pm 0.002$ (95% confidence level). (Note that h is defined by $H_0 = 100h$ km s^{-1} Mpc^{-1}.) The results were obtained by numerical integrations carried out by me, using the equations of Peebles (1968) and Peebles (1993, pp. 165–173).

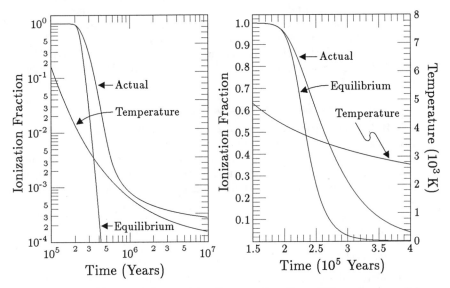

Figure 3.   Process of recombination as a function of time, for a model described in the text. The ionization fraction is the fraction of all protons or electrons that are ionized at any given time. The left graph is logarithmic and the right graph is linear. The line labeled "Equilibrium" shows the result of solving the Saha equation, and the line labeled "Actual" shows the result of integrating the rate equations, revealing the nonequilibrium effects. For reference, the temperature is also shown, keyed to the scale on the right, which applies to both graphs.

4.   Before 10^{-12} Second: Inflation

At times before 10^{-12} s, some of us believe that the Universe almost certainly underwent a period of inflation. The reason is that cosmic inflation can explain a number of features of our Universe that would otherwise be unexplained. In particular, inflation can explain the following questions:

1. *How the Universe acquired $> 10^{90}$ particles*

Starting from the general and then moving toward the specific, one salient feature of the Universe is its enormous size. The visible part of the Universe contains about 10^{90} particles. It is easy to take this for granted, and many cosmologists are not bothered by the fact that the standard FRW cosmology, without inflation, simply postulates that about 10^{90} or more particles were here from the start. However, in the present context many of us hope that even the creation of the Universe can be described in scientific terms, and thus the number of particles would have to be the result of some calculation. The easiest way by far to get a huge number, with presumably only modest numbers as input, is for the calculation to involve an exponential. The exponential expansion of inflation reduces the problem of explaining 10^{90} particles to the problem of explaining 60 to 70 e-foldings of inflation. In fact it is easy to construct underlying particle theories that will give far more than 70 e-foldings of inflation, so inflationary cosmology suggests that the observed Universe is only an infinitesimal fraction of the entire Universe.

2. *Why the Universe is uniformly expanding*

The Hubble expansion is also easy to take for granted, and in the standard FRW cosmology the Hubble expansion is accepted as a postulate about the initial conditions. But inflation offers the possibility of actually explaining how the Hubble expansion began. The repulsive gravity associated with the inflaton field—the scalar field that drives the inflation—is exactly the kind of force needed to propel the Universe into a pattern of motion in which each pair of particles is moving apart with a velocity proportional to their separation.

3. *How the CMB can be uniform to 1 part in 10^5*

The degree of uniformity in the Universe is startling. The intensity of the cosmic background radiation is the same in all directions, after it is corrected for the motion of the Earth, to the incredible precision of one part in 100 000.

The cosmic background radiation was released at the time of recombination, about 300 000 years after the Big Bang, when the Universe cooled enough so that the opaque plasma neutralized into a transparent gas. The cosmic background radiation photons have mostly been traveling on straight lines since then, so they provide an image of what the Universe looked like at 300 000 years after the Big Bang. The observed uniformity of the radiation therefore implies that the observed Universe had become uniform in temperature by that time. In standard FRW cosmology, a simple calculation shows that the uniformity could be established so quickly only if signals could propagate at 100 times the speed of light, a proposition clearly in contradiction with the known laws of physics. In inflationary cosmology, however, the uniformity is easily explained. The uniformity is created initially on microscopic scales, by normal thermal-equilibrium processes, and then inflation takes over and stretches the regions of uniformity to become large enough to encompass the observed Universe.

4. *Why the early Universe was so close to critical density*

I find this issue particularly impressive, because of the extraordinary numbers that it involves. This flatness problem concerns the value of the ratio

$$\Omega_{\text{tot}} \equiv \frac{\rho_{\text{tot}}}{\rho_{\text{c}}}, \tag{12}$$

where ρ_{tot} is the average total mass density of the Universe and $\rho_{\text{c}} = 3H^2/(8\pi G)$ is the critical density, the density that would make the Universe spatially flat. (In the definition of total mass density, I am including the vacuum energy $\rho_{\text{vac}} = \Lambda/8\pi G$ associated with the cosmological constant Λ, if it is nonzero.)

There is now strong evidence that Ω is very near to 1, but the flatness problem is much older and does not require us to believe the most recent results. We have believed for a long time that

$$0.1 \lesssim \Omega_0 \lesssim 2, \tag{13}$$

and this is all that is needed to motivate the flatness problem. Despite the breadth of this range, the value of Ω at early times is highly constrained, since $\Omega = 1$ is an unstable equilibrium point of the standard model evolution. Thus, if Ω was ever *exactly* equal to one, it would remain exactly one forever. However, if Ω differed slightly from one in the early Universe, that difference—whether positive or negative—would be amplified with time. In particular, it can be shown that $\Omega - 1$ grows as

$$\Omega - 1 \propto \begin{cases} t & \text{(during the radiation-dominated era)} \\ t^{2/3} & \text{(during the matter-dominated era)} \end{cases} \tag{14}$$

It was shown by Dicke & Peebles (1979), for example, that as the processes of Big Bang nucleosynthesis were just beginning at $t = 1$ s, Ω must have equaled one to an accuracy of one part in 10^{15}. Classical cosmology provides no explanation for this fact—it is simply assumed as part of the initial conditions. In the context of modern particle theory, where we try to push things all the way back to the Planck time, 10^{-43} s, the problem becomes even more extreme. If one specifies the value of Ω at the Planck time, it has to equal one to 58 decimal places in order to be anywhere in the allowed range today.

While this extraordinary flatness of the early Universe has no explanation in classical FRW cosmology, it is a natural prediction for inflationary cosmology. During the inflationary period, instead of Ω being driven away from one as described by Equation (14), Ω is driven toward one with exponential swiftness:

$$\Omega - 1 \propto e^{-2H_{\text{inf}}t}, \tag{15}$$

where H_{inf} is the Hubble parameter during inflation. Thus, as long as there is a long enough period of inflation, Ω can start at almost any value, and it will be driven to one by the exponential expansion.

5. *Why the inhomogeneities have a nearly flat (Harrison-Zeldovich) spectrum*

The process of inflation smooths the Universe essentially completely, but density fluctuations are generated as inflation ends by the quantum fluctuations of the inflaton field, the scalar field that drives the inflationary expansion. Generically these are adiabatic Gaussian fluctuations with a nearly scale-invariant spectrum (Starobinsky 1982; Guth & Pi 1982; Hawking 1982; Bardeen, Steinhardt, & Turner 1983; Mukhanov, Feldman, & Brandenberger 1992). New data are arriving quickly, but so far the observations are in excellent agreement with the predictions of the simplest inflationary models. For a review, see for example Bond and Jaffe (1999), who find that the combined data give a slope of the primordial power spectrum within 5% of the preferred scale-invariant value. See also Wang et al. (2001), which includes a review of the most current data.

Since the theme here is time and time scales, it is natural to ask *when* inflation occurred. The answer is that we do not really know. Originally inflation was proposed to take place at the scale of grand unified theories, at a characteristic energy scale of 10^{16} GeV (Guth 1981; Linde 1982; Albrecht & Steinhardt 1982). Applying Equation (5) with $g \sim 200$, typical of grand unified theories, one finds a starting time for inflation of about 10^{-39} s. This is extraordinarily early, but it is still late compared to the Planck time, $\sqrt{G\hbar/c^5} \sim 5 \times 10^{-44}$ s, the time scale at which quantum gravity is believed to become important. Thus, it is plausible that the field theoretic formalism that is used to describe inflation is valid at the appropriate energy scale.

It is possible that inflation did occur at the grand unified theory scale, but it might very well have occurred later. The only known restriction on the lateness of inflation is the requirement that baryogenesis occur after inflation, since any net density of baryon number generated before inflation would be diluted to a negligible level. It is now believed that baryogenesis might happen as late as the electroweak scale, operating through the mechanism of electroweak current conservation anomalies (Kuzmin, Rubakov, & Shaposhnikov 1985).

Observationally it is difficult to determine the energy scale and hence the time scale of inflation, since the consequences are very insensitive. The only known way to determine the energy scale of inflation is to directly or indirectly measure the gravitational wave background, which is more intense if the energy scale of inflation was high. In fact the energy scale could not have been significantly higher than the grand unified scale, or else the gravity waves would be so strong that they should have already been detected.

5.   Before Inflation: (Eternal) Inflation

The question of what happened before inflation is an open one, and different cosmologists would venture different ideas. In my opinion, the most plausible answer to what happened before inflation is—more inflation.

Specifically, it appears that essentially all working models of inflation are eternal, in the sense that once inflation starts, it never stops. Instead inflation goes on forever, with pieces of the inflating region breaking off and producing a never-ending stream of pocket universes (Vilenkin 1983; Steinhardt 1983; Linde 1986a, 1986b; Goncharov, Linde, & Mukhanov 1987).

The mechanism that leads to eternal inflation is rather straightforward to understand. Normally one expects inflation to end because the false vacuum—the state of the inflaton field that is responsible for the repulsive gravity driving the inflation—is unstable, so it decays like a radioactive substance. As with familiar radioactive materials, the decay of the false vacuum is generally exponential: during any period of one half-life, on average half of it will decay. This case is nonetheless very different from familiar radioactive decays, however, because the false vacuum is also expanding exponentially. Furthermore, it turns out that the expansion is generally much faster than the decay. Thus, if one waits for one half-life of the decay, half of the false vacuum region would on average convert to ordinary matter. But meanwhile the part that remains would have undergone many doublings, so it would be much larger than the region was at the start. Even though the false vacuum is decaying, the volume of the false vacuum would actually grow with time. The volume of the false vacuum would continue to grow, without limit and without end. Meanwhile pieces of the false vacuum region decay, producing an infinite number of what I call pocket universes.

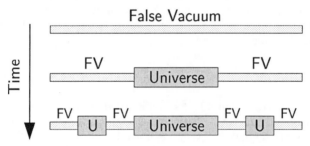

Figure 4. Illustration of eternal inflation, as described in the text.

In Figure 4 I show a schematic illustration of how this works. The top row shows a region of false vacuum, shown very schematically as a horizontal bar. After a certain length of time, a little less than a half-life, the situation looks like the second bar, in which about a third of the region has decayed. The energy released by that decay produces a pocket universe, which will inflate to become much larger than the presently observed Universe.

On the second bar, in addition to the pocket universe, there are two regions of false vacuum. On the diagram I have not tried to show the expansion, so the diagram can fit on the page. So, you are expected to remember that each bar is actually bigger than the previous bar, but drawn on a smaller scale so that it looks the same size. To discuss a definite example, let us assume that each bar represents three times the volume of the previous bar. In that case, each region of false vacuum on the second bar is just as big as the entire bar on the top line.

The process can then repeat. If we wait the same length of time again, the situation will be as illustrated on the third bar of the diagram, which represents a region 3 times larger than the second bar, and 9 times larger than the top bar. For each region of false vacuum on the second bar, about a third of the region decays and becomes a pocket universe, leaving regions of false vacuum in between. Each region of false vacuum shown on the diagram is as large as the original region in the top bar. The process goes on literally forever, producing pocket universes and regions of false vacuum between them, ad infinitum. The Universe on the very large scale acquires a fractal structure.

The illustration of Figure 4 is of course oversimplified in a number of ways: it is one-dimensional instead of three-dimensional, and the decays are shown as if they were very systematic, while in fact they are random. But the qualitative nature of the evolution is nonetheless accurate: eternal inflation really leads to a fractal structure of the Universe, and once inflation begins, an infinite number of pocket universes are produced.

Since inflation is eternal into the future, it is natural to ask if it might also be eternal into the past. The explicit models that have been constructed are eternal only into the future and not into the past, but that does not show whether or not is possible for inflation to be eternal into the past. Borde & Vilenkin (1994) presented a proof that an eternally inflating spacetimes must start from an initial singularity, and hence must have a beginning, but later they pointed out (Borde & Vilenkin 1997) that their proof assumed a condition that is true classically but is violated by quantum field theories. Today the issue is undecided. My own suspicion is that eternally inflating spacetimes must have initial singularities, because it seems significant that no one has been able to construct a model which does not.

6. Summary

For the period between about 10^{-12} s and 300 000 yr, we have a rather detailed description of cosmology that I believe has a good chance of being correct. I believe that inflation played a very significant role at earlier times, but the details are unclear. As one might expect, our view of the earliest moments of the Universe is still clouded with uncertainties.

References

Albrecht, A., & Steinhardt, P. J. 1982, Phys. Rev. Lett, 48, 1220

Bardeen, J. M., Steinhardt, P. J., & Turner, M. S. 1983, Phys. Rev. D, 28, 679

Bond, J. R., & Jaffe, A. H. 1999, Phil. Trans. Roy. Soc. Lond. A, 357, 57

Borde, A., & Vilenkin, A. 1994, Phys. Rev. Lett, 72, 3305

Borde, A., & Vilenkin, A. 1997, Phys. Rev. D, 56, 717

Burles, S., Nollett, K. M., & Turner, M. S. 2001a, Phys. Rev. D, 63, 063512

Burles, S., Nollett, K. M., & Turner, M. S. 2001b (astro-ph/0010171)

Dicke, R. H., & Peebles, P. J. E. 1979, in General Relativity: An Einstein Centenary Survey, ed. S. W. Hawking & W. Israel (Cambridge: Cambridge University Press), 504

Freedman, W. L., et al. 2001, ApJ, 553, 47

Goncharov, A. S., Linde, A. D., & Mukhanov, V. F. 1987, Int. J. Mod. Phys., A2, 561

Guth, A. H. 1981, Phys. Rev. D, 23, 347

Guth, A. H., & Pi, S.-Y. 1982, Phys. Rev. Lett, 49, 1110

Halverson, N. W., et al. 2001, ApJ, submitted (astro-ph/0104489)

Hamilton, A. J. S., Tegmark, M., & Padmanabhan, N. 2000, MNRAS, 317, L23

Hawking, S. W. 1982, Phys. Lett., 115B, 295

Kuzmin, V. A., Rubakov, V. A., & Shaposhnikov, M. E. 1985, Phys. Lett., 155B, 36

Lee, A. T., et al. 2001 (astro-ph/0104459)

Linde, A. D. 1982, Phys. Lett., 108B, 389

Linde, A. D. 1986a, Mod. Phys. Lett., A1, 81

Linde, A. D. 1986b, Phys. Lett., 175B, 395

Mather, J. C., Fixsen, D. J., Shafer, R. A., Mosier, C., & Wilkinson, D. T. 1999, ApJ, 512, 511

Mukhanov, V. F., Feldman, H. A., & Brandenberger, R. H. 1992, Phys. Rep., 215, 203

Netterfield, C. B., et al. 2001 (astro-ph/0104460)

Padin, S., et al. 2001, ApJ, 549, L1

Peebles, P. J. E. 1968, ApJ, 153, 1

Peebles, P. J. E. 1993, Principles of Physical Cosmology (Princeton: Princeton University Press)

Saunders, W., et al. 2000, MNRAS, 317, 55

Starobinsky, A. A. 1982, Phys. Lett., 117B, 175

Steinhardt, P. J. 1983, in Proc. of the Nuffield Workshop, The Very Early Universe, ed. G. W. Gibbons, S. W. Hawking, & S. T. C. Siklos (Cambridge: Cambridge University Press), 251

Vilenkin, A. 1983, Phys. Rev. D, 27, 2848

Wang, X., Tegmark, M., & Zaldarriaga, M. 2001 (astro-ph/0105091)

Acknowledgments. This work is supported in part by funds provided by the U.S. Department of Energy (D.O.E.) under cooperative research agreement #DF-FC02-94ER40818.

Discussion

Moshe Carmeli: Since the Universe now is assumed to be accelerating, does that mean before the time when Ω was equal to 1 the Universe was decelerating and Ω was larger than 1 and after that Ω was smaller than 1?

Alan Guth: By Ω, let me assume that we are talking about the total Ω, which includes not only ordinary matter and dark matter, but also the so-called dark energy. This dark energy could be true vacuum energy, also known as a cosmological constant, or it could be something like a very slowly varying scalar field, also known as quintessence. According to general relativity, if we assume homogeneity and isotropy, then the total Ω is directly related to the geometry of the Universe: $\Omega = 1$ is a geometrically flat universe, $\Omega > 1$ is a closed universe, and $\Omega < 1$ is an open universe. Within the context of FRW cosmology, Ω can never cross one. If the geometry is closed it stays closed, if it is open it stays open, and if it is exactly flat it stays exactly flat. Until recently many of us were taught that Ω is related to the ultimate fate of the Universe, with $\Omega > 1$ corresponding to eventual collapse. This relationship, however, disappears if we allow the possibility of a cosmological constant. It seems very likely that the Universe has a positive cosmological constant, corresponding to a positive vacuum energy. If so, then the Universe can accelerate forever even if it is closed, with $\Omega > 1$.

Inflation drives Ω very sharply towards one. Therefore generic inflationary models—models with a very large amount of inflation—predict that Ω was driven so closely to one by the end of inflation that even today, it should be very close to one. In such models the primary deviations of Ω from one are local deviations, the long wavelength tail of the nearly scale-invariant spectrum of density perturbations that are generated by quantum fluctuations during inflation. We would therefore expect that on the scale of our local Hubble volume, the deviation of Ω from one should be about 10^{-5}, and could go in either direction.

Brian Chaboyer: I just wondered how does the temperature go during inflation?

Alan Guth: It's a hard question to answer. Inflation is a very non-thermal equilibrium process, so probably the best answer is to say that during inflation the scalar field is essentially frozen near the top of the hill in the potential energy diagram, and starts to gradually fall off of it. Or, in the case of chaotic inflation, the scalar field is not near the top of the hill, but is still rolling gradually down it. The dominant effect during inflation, except perhaps at the very beginning of the era of inflation, is presumably the quantum fluctuations. Now it's often pointed out, correctly, that the quantum fluctuations in an exponentially expanding space simulate thermal fluctuations, at a temperature called the Gibbons-Hawking temperature. It is closely analogous to the Hawking temperature of a black hole. The Gibbons-Hawking temperature is $H/2\pi$, where H is the Hubble constant during inflation. So, during inflation the effective temperature is frozen at the constant value, $H/2\pi$.

Taka Kajino: Lepton asymmetry is essential to create baryon asymmetry. So what is the situation with the production of lepton asymmetry? How is this observable in the present Universe?

Alan Guth: Many of the theories we deal with in fact conserve the quantity $B-L$, baryon number minus lepton number, which guarantees that whenever you produce a baryon excess you're also producing a lepton excess of exactly the same magnitude. Even in theories for which $B - L$ is not conserved, one still expects that one would produce a lepton asymmetry by the same kinds of mechanisms that produce baryon asymmetries, and therefore the lepton asymmetry would be of a magnitude comparable to that of baryons. Such a magnitude for a lepton asymmetry, however, is completely unnoticeable—it would correspond, for example, to having an imbalance between the number of neutrinos and the number of antineutrinos by 1 part in 10^9. In fact we have no hope at the present time of even seeing the cosmic background of neutrinos, let alone measuring the density of neutrinos and the density of antineutrinos to an accuracy of 1 part in 10^9 to see the expected asymmetry.

Now it is possible to imagine models in which the lepton asymmetry is huge, say of order 1 rather than order 10^{-9}. In such models the lepton asymmetry can have observable effects. For example, a large imbalance between neutrinos and antineutrinos can shift the balance between protons and neutrons at the time of nucleosynthesis, strongly affecting the predictions of Big Bang Nucleosynthesis. But for conventional models the lepton asymmetry is comparable to the baryon asymmetry, which makes it immeasurably small.

Mark Kloetzel: When you spoke of time at the beginning in terms of seconds and milliseconds, are these seconds as we view seconds, or given the incredible accelerations that are going on are these relativistic seconds, and do you see any cause to differentiate?

Alan Guth: This question concerns the nature of time, and perhaps the next two speakers might want to comment on this, too. Nevertheless I can try to give you my answer, which might be completely different from the other answers. First of all, there are no relativistic corrections to the measurement of time in the early Universe. The time variable that is used in discussing the early Universe is in principle the time that would be measured on a clock that was present in the early Universe, and at rest relative to the matter in the early Universe. From my point of view, this notion of time is simply a straightforward extrapolation of the way we measure time in conventional physics. That is, time appears in our equations as a parameter with respect to which things change. In fact, the equations of motion that we use to describe the early Universe are exactly the same equations of motion that we use to describe the Universe today, with time playing exactly the same role. The only difference is that by extrapolating back to these extraordinarily early times, and consequently to extraordinarily high energy densities, we are discussing events that happen much faster than those to which we are accustomed. By describing time in this way, we are really not making any commitment to any fundamental theory about what time is, at its most basic level. We are really just saying that we can characterize change by thinking of physical quantities varying as a function of a variable that we call time. We have equations that tell us how things vary with time, and we can use the same equations for the early Universe that we use for the Universe today.

Astrophysical Ages and Time Scales
ASP Conference Series, Vol. 245, 2001
T. von Hippel, C. Simpson, N. Manset

Time and Modern Physics

W. G. Unruh

CIAR Cosmology and Gravity Program, Department of Physics,
University of British Columbia, Vancouver V6T 2A6, Canada

Abstract. The role of time in physics has changed substantially in this century. This paper describes some of those changes and discusses the impact they have on the nature of time. I will also review the link between the initial conditions of the Universe and the present.

1. Introduction

The role of time in a theory of science tends to be one of the most fundamental aspects of that theory. We are reminded of the importance of the discovery of Deep Time (as Gould 1988 calls it) to the existence of a theory of both geology and of the existence and development of life on Earth—a theory at odds with physics at the time. Going to the origins of physics, we have the image of Galileo watching the swinging of the chandeliers in the Cathedral, and noticing that the timing of their oscillation was independent of their amplitude. The oft told story is that he used his pulse as a clock to test this notion. Since the pulse is a notoriously unreliable clock, especially when one is excited about a possible scientific breakthrough, I suspect that it was actually the synchronization of the oscillation of the chandeliers with the choir's singing which brought the phenomenon to his attention. It has, for example, been argued by Drake (1975) that Galileo used song to time the rolling of balls down inclined planes in his famous experiments on the falling of bodies under gravity. It is also known that a singer's sense of time in music is not only accurate and reproducible but is reproducible after long periods of time. That Galileo does not reference this use of music may say something about the relations between Galileo Galilei and his then more famous father, Vincenzo Galilei, the top music theorist of the day.

We also read in Newton (1687) the strange paragraph in which he both states that he does not define time

I do not define time ... as being well known to all ...

and then proceeds to define it as something different than everyone else's notion thereof:

I. Absolute, true, and mathematical time, of itself and from its own nature, flows equably without relation to anything external, and by another name is called duration: relative, apparent, and common time, is some sensible and external (whether accurate or inequable) measure of the duration by the means of motion, which is commonly used instead of true time; such as an hour, a day, a month, a year.

This definition is also one which we have problems with understanding the importance of, as our own concepts of time have been so thoroughly formed from a very early age as being precisely Newton's abstract, all-encompassing mathematical time.

That this Newtonian concept is not necessary became clear in this century, with the advent of both of Einstein's theories of Relativity. In special relativity, time, at least time as experienced by any physical system, became identified with the concept of distance. It produced, what would under the Newtonian concept of time be nonsensical statements, e.g., that the lapse of time at a point could differ for different people. Traveling from here now to there at a different time could require different amounts of time, depending on the journey. Under the Newtonian concept, time is simply a property of when you are. It is an absolute and common property. Under Special Relativity however, time becomes like the distance from Vancouver to Hilo. This distance depends not on any absolute property (the whereness say) of Vancouver and Hilo, but on the details of the trip taken to get from one to the other. Thus, Savitt and I could have traveled very different distances to get from Vancouver to Hilo, especially since one took a side trip to Japan. Similarly getting from then to now could take very different times, depending on the details of the trip taken.

But it was in General Relativity that the concept of time was altered in the most radical manner. Time was no longer something which simply exists, but is something which can be altered by the material of the world. In particular, he showed that gravity is intimately linked with time.

It is often stated that gravity can alter time. With the gravitational redshift (e.g., Will 1979) the fact that a clock held stationary far from the surface of the earth ticks at a different, slower rate than a clock on the surface of the earth is taken to demonstrate that gravity affects time. But the General Theory is much more radical than that. Instead of gravity affecting time, it is the fact that time flows differently from place to place that *is* gravity. Gravity and time are not different things which can affect each other. Instead, gravity is just time, the concept of gravity is simply an aspect of time.

I will refer the reader to my description elsewhere of the "beach ball" analogy (Unruh 1995), which demonstrates how, by manufacturing a space-time out of pieces which have a variable amount of time from place to place, and by assuming that objects follow "straight lines" (the shortest distance between two points), one can show how many of the features we usually ascribe to gravity can instead be ascribed to the inequable flow of time from place to place. (In technical terms, this is simply the statement that it is the time-time component of the metric—g_{00}—the rate of flow of time from place to place, which, together with the geodesic equations, explains all of the features of gravity usually ascribed to some mysterious gravitational force.

This ability of matter to change time, and thus for the whole notion of time to loose its absolute qualities and to be conditional on other attributes or events in the world, has presented modern physics with some of its most puzzling problems. If time is conditional, if time can cease to exist, either into the past or into the future, what are the fundamental bases for us to use in describing the world? Let me use this space to outline a few types of questions which this conditional nature of time raises.

2. Inflation

One of the standard arguments in favour of inflation is that the theory leads inevitably to a universe similar to the one we now have about us. Given a few assumptions—that there exists some scalar field which finds itself to have a large value at some initial time over a small region of space, then the dynamics of the Universe is such that that small region will expand exponentially rapidly to a size many many times the size of the currently observable Universe. In the process various inhomogeneities, whether in the matter distribution or in the curvature of space, will be spread out, diluted, so that our current observable Universe is both spatially flat and homogeneous in its matter distribution. In addition, due to the amplification of quantum fluctuations by this expansion, there will also be primordial fluctuations about this flat, homogeneous background, fluctuations for which the theory can be selected to explain the inhomogeneities (galaxies, clusters, stars, people) we now see about us.

However, given the temporal conditionality of General Relativity, this explanation has difficulties. It assumes that the way to explain the situation now, is to do so in terms of the situation in the Universe at a time far in the past. However, in terms of the theory that past may not exist. The Universe itself may not exist at the time at which one is attempting the explanation. I.e., within all standard, Newtonian theories, there is an equivalence between times. The world is as it is because it was as it was. Past and present are equivalent, and if one can find a time in the past for which the explanation of the state of the system becomes simple, then that simple explanation also suffices for the present. However, how can such a simple explanation be considered sufficient, if many of the universes which one is comparing the present Universe to did not even exist at that time, but rather came into being between now and then?

To make this argument (Unruh 1997) perhaps clearer, consider our present state, and consider an inflationary theory which explains our present Universe in all its details. Now, alter the present Universe by some small amount, by say increasing the fluctuations in the microwave background by 10%. To say that inflation explains the present state must also mean that it rules out this alternative (within the context of this definite theory which predicts the present). However, General Relativity and the theories of matter which make up such an inflationary theory are time reversal invariant. (The non-adiabatic reheating of the Universe at the end of inflation makes no contribution to my argument). But if we run back these alternative final conditions we find not that they produce some illegal conditions at that early time. Instead, the Universe comes to an end before (going backwards as we are doing) we get to that initial time at which the inflationary explanation is supposed to have reasonable conditions. I.e., there are a huge number of potential universes which do not even exists at the time that inflation is supposed to provide the explanation for the present. The smaller the structures now which disagree with the inflationary predictions of a particular theory, the closer to the present the Universe ceases to exist within exactly the same inflationary theory. Thus, inflation does not rule out alternatives to the present by demonstrating that the conditions at that early time are unreasonable by some criterion, but rather it rules them out by demanding by fiat that the Universe must have existed, that time must have come into existence, before that time.

When one breaks the relation between the past and the present, as one does in General Relativity, when initial and final conditions are no longer equivalent, what remains of the status of explanation of the present in terms of the past, or a historical explanation for the present?

Another aspect of time which becomes problematic is the very existence of the concept. Barbour (2000) in a recent book has delved into this problem in some detail (though not with answers which I find convincing). The problem arises because General Relativity (and any theory which contains the essential aspects of General Relativity—like string theory) states that there exist no globally preferred sets of coordinates, and no globally preferred notion of time. Time is conditional and dependent. While the theory manages to still be able to describe physics by making use of an arbitrarily defined "coordinate" time. However, the description of all physical attributes of the Universe must be independent of this completely arbitrary choice of coordinate. In the classical theory this is conceptually unproblematic by assuming the existence of an abstract space-time (a manifold) for which the coordinates are just labels for the pre-existing points of this space-time. In a quantum theory, however, this does not appear to be a viable option. Instead, the demand that the physics be independent of the coordinates chosen seems to imply that the only physical observables of the theory are constants of the motion. I.e., it would appear to be impossible to describe precisely that aspect of nature which we believe it is the duty of physics to describe, namely the changes which occur in the world about us.

Far from being *a-priori* necessary for our description of the world, time has become one of the key problems to be overcome in generating a description of the world. Even such self-evident features of the theory, that one describes the world in terms of a theory whose structure is the same everywhere, seem to have become problematic. When the Universe is small, one would expect there to be fewer possibilities, that the total number of possible degrees of freedom should be fewer, than when the Universe is large. The solution to these conundrums and the role which time would play in any such a solution seems very obscure at present, but promises to have immense consequences for both physics and philosophy in the future.

References

Barbour, J. 2000, The End of Time (Oxford: Oxford University Press)

Drake, S. 1975, Scientific American, 232, 98

Gould, S. J. 1988, Time's Arrow/Time's Cycle (Cambridge: Harvard University Press)

Newton, I. 1687, Philosophif naturalis Principia Mathematica, translated A. Motte, 1729, revised F. Cajori, 1934 (University of California Press)

Unruh, W. G. 1995, in Time's Arrow Today, ed. S. Savitt (Cambridge: Cambridge University Press), 1

Unruh, W. G. 1997, in Critical Dialogs in Cosmology, ed. N. Turok (Singapore: World Scientific), 249

Will, C. M. 1979, in General Relativity, An Einstein Centenary Survey, ed. S. W. Hawking & W. Israel (Cambridge: Cambridge University Press)

Discussion

Douglas Gough: You argued from the product of an uncertainty in energy and an uncertainty in time and deduced that only time has a limited range of validity. Could you not have equally come to the conclusion that it was energy that only has a limited range of validity or would you say it's actually both?

Bill Unruh: Look at quantum gravity for example. It turns out that there the energy is equal to zero. So in that sense energy has lost its validity because if you have something which only ever has one single value, at that point it's become totally boring. It certainly doesn't capture what we usually want to have energy explain. Within the context of quantum gravity, energy is starting to get rather amorphous. We knew that already, even in classical General Relativity despite what Alan said this morning about energy of gravitational fields being negative et cetera, his arguments as you noticed were Newtonian arguments. And one of the big reasons for that is that when you try and do that in General Relativity, it all falls apart. And it turns out that you can't really define energy, even within the context of classical gravity. It turns out to depend totally on which coordinate system you happen to choose to define the energy, as to what the energy is in any local region. So yes, energy is also becoming a very uncertain process, but in a sense one wouldn't worry about that too much—after all, energy is really a concept that has been part of our description of the world only for about a hundred years, and the whole idea of the conservation of energy only arose in physics in the late 1800s. Whereas the notion of time as being essential for describing physics certainly went back to Galileo and has gone back philosophically far far longer than that. And losing time I think is something that we would feel much more strongly than losing energy.

Alan Guth: In the course of your talk you took a potshot at eternal inflation by making the unjustified claim that all realizations violate Einstein's equations. I'd just like you to explain what you meant by that.

Bill Unruh: If you look at the technical details then one basically gets a violation of Einstein's equations in the course of this evolution. And I think it's not surprising because what you're really trying to do is to bring in this notion of creation, if you will, of continuous creation of *something*, it's not clear what. And that process is something that's going to have to violate Einstein's equations, because, just as with all of our equations in physics, Einstein's equations are really equations that say the present is as it is because the past was as it was. In that case you push it back to the super-Planck regime and say well, there's stuff going on at these scales which are much much smaller than the Planck scale and they expand out and produce the Universe that we know as we know it now. If you actually think about it in any detail, you realize that talking about super-Planck regimes makes no sense; you're really talking about singularities within the structure of the theory and you're taking about singularities producing what we now see in the world which, within the context of the original theory, makes no sense and yet in the context of a theory where you're trying to grapple with creation—new things coming into being—it seems to make a little bit of sense. So those are the kind of hand-waving thoughts that were going through my mind in making that statement. So in fact it's actually a very positive thing rather than a negative thing; it's trying to go beyond what we currently have to try to see what kind of a theory we can build up which allows for the possibility of creation.

Frank Shu: You made the suggestion that time perhaps was a variable like temperature—a macroscopic rather than essentially a microscopic variable, if I understood you correctly. If you'll indulge me, think of energy in the First Law of Thermodynamics; one basically says heat is a form of energy and, when you take this into account in a closed system, energy is conserved. That's not a terribly interesting statement—the interesting statement in thermodynamics is the Second Law which says there's not only a quantity of energy, but there's a quality to energy and heat is very different from work. In that sense, of course, the Second Law intrinsically contains the concept of time and a flow to time. Would you care to comment on that?

Bill Unruh: This connection between time and thermodymics is one of the most problematic connections because the Second Law does seem to play into the idea of an ordering of time, of time going from past into future, despite the fact that all of our fundamental theories, as Steve Savitt talked about, have no such ordering to them. The future and the past are entirely equivalent as far as all of our fundamental theories are concerned, but here is a place in thermodynamics where one seems to introduce ordering. So thermodynamics does seem to play some role in the ordering of time but not in the usual way of looking at it; it doesn't really address the nature of time in a fundamental sense. One can say that the fundamental thing is not the Second Law of Thermodynamics, the fundamental things are our so-called fundamental equations which say nothing about the nature of time. So one could always fall back on the position that the ordering of time is just our imagination and it doesn't really matter and we don't need it in order to do physics. This connection is harder to get out of and suggest to me at least, although not to everyone, that perhaps we're going to have to set up a theory which doesn't have time in it at all except as a derived quantity just as temperature is a derived quantity from random motions of molecules within thermodynamics. So it was that analogy that I was using rather than anything more specific.

Astrophysical Ages and Time Scales
ASP Conference Series, Vol. 245, 2001
T. von Hippel, C. Simpson, N. Manset

Implications of Spacetime Structure for Transience

Steven F. Savitt

Department of Philosophy, The University of British Columbia, Vancouver V6T 1Z1, Canada

Abstract. This paper describes briefly an historical episode in which philosophical concerns about the nature of time and developments in the general theory of relativity interacted fruitfully. The logician Kurt Gödel developed a novel solution to the Einstein field equations in order to vindicate what he called an "idealistic" view of time. Gödel's novel solution, and others like it, have in turn stimulated further reflection on the nature of time.

Does time pass or flow? Is there objective temporal becoming or transience? This question has occasioned philosophical controversy at least since antiquity when Heraclitus wrote:

> Everything flows and nothing abides; everything gives way and nothing stays fixed.
> You cannot step twice into the same river, for other waters and yet others, go flowing on.[1]

and Parmenides responded:

> There remains, then, but one word by which to express the [true] road: Is. And on this road there are many signs that What Is has no beginning and never will be destroyed: it is whole, still, and without end. It neither was nor will be, it simply is-now, altogether, one, continuous...[2]

The aim of this note is to illustrate how this ancient controversy has been sharpened, although not settled, by setting it in the context of some spacetime structures articulated in the last century. The focus will be arguments found in or inspired by Gödel (1949).

Gödel wrote that the existence of transience or, as he called it, an *objective lapsing of time,* "means (or, at least, is equivalent to the fact) that reality consists of an infinity of layers of 'now' which come into existence successively." (p. 558) While objectively lapsing time seems a natural notion in classical or Newtonian spacetime, in Minkowski spacetime the partition of the spacetime into hyperplanes of simultaneity is arbitrary, dependent on an observer's state

---

[1] *The Fragments of Heraclitus* (circa 500 B. C.) As translated by Philip Wheelwright.

[2] *The Fragments of Parmenides of Elea* (c. 470–460 B. C.) As translated by Philip Wheelwright.

of motion. But Gödel claimed that *existence*, unlike layers of 'now,' can not be so relativized. "The concept of existence," he wrote, "cannot be relativized without destroying its meaning completely." (p 558) From the absolute character of existence and the observer-dependent character of hyperplanes of simultaneity he inferred that Minkowski spacetime is inhospitable to the *objective* lapsing of time and that the special theory of relativity vindicates the view of, amongst others, Parmenides.

While this argument is quite persuasive, Jeans (1935) had argued that in the general theory of relativity the presence of matter can break the special relativistic symmetries and non-arbitrarily select a particular foliation of the spacetime, which may serve as a relativistic surrogate for Newtonian absolute time. Of course, Jeans' argument would side step the argument given immediately above that the relativity of simultaneity rules out objectively lapsing time.

Gödel's response to Jeans' argument was remarkable. After grumbling that even in general relativistic spacetimes that admitted foliations there would likely be an irreducible arbitrariness in choosing one, he produced a novel solution to the Einstein Field Equations that admitted of *no* foliations. He argued that (1) in that spacetime, the so-called *Gödel universe,* which I shall indicate by $\mathcal{M}$, there could be no objectively lapsing time and that therefore (2) there could be no objectively lapsing time in our universe either.

Initially, the feature of $\mathcal{M}$ that attracted comment was that it contained closed timelike curves and so afforded the possibility of a certain sort of time travel. Prompted by Yourgrau (1991), however, philosophers began to think seriously about the argument denying temporal becoming. Earman (1995, Appendix to Chapter 6) subjects Gödel's argument to searching criticism, finding step (2) invalid. At about the same time I (Savitt 1994), aware of Earman's criticism, tried to recast Gödel's argument in a way that would make it a valid argument for a weaker, but nevertheless still interesting, conclusion.

Part (1) I presented as follows:

(A1) The existence of an infinity of layers of 'now' is a necessary condition for the existence of an objective lapse of time in any spacetime.

(A2) A layer of 'now' (in a model of GTR) is a global time slice.

(A3) There are no global time slices in the Gödel universe, $\mathcal{M}$.

(A4) So, there are no layers of 'now' in $\mathcal{M}$.

(A) Therefore, there is no objective lapse of time in $\mathcal{M}$.

This argument I believe is valid—that is, I believe that its conclusion follows logically from its premises. Part (2) of Gödel's argument I presented in the following form:

(A) There is no objective lapse of time in $\mathcal{M}$.

(B1) $\mathcal{M}$ is a mathematical model of a physically possible universe.

(B2) It is physically possible that some galaxy in $\mathcal{M}$, a physically possible universe, is inhabited by creatures just like us.

(B3) A local time function t_L can be defined for the world line of that galaxy such that, whenever x and y are spacetime points in or near that galaxy, $t_L(x) < t_L(y)$ whenever $x \ll y$.[3]

(B4) The direct experience of time of the creatures just like us will be just like our direct experience of time.

(B5) It is therefore possible (by (A) and (B4)) to have direct experience of time just like ours in a universe in which there is no objective lapse of time.

(B6) So, from (B5) it follows that our direct experience of time provides no reason to suppose that there is an objective lapse of time.

(B7) Our direct experience of time provides the only reason to suppose that there is an objective lapse of time in our universe.

(B') Since there is no objective lapse of time in $\mathcal{M}$, (B6) and (B7) imply that there is no reason to suppose that there is an objective lapse of time in our universe either.

This argument is also, I believe, valid. It rests on many premises, however, and may not be sound. The entire argument is formulated using the conceptual apparatus of general relativity, apparatus that may or may not survive the eventual unification of general relativity with quantum theory. And one (or more) of the specific premises may be false. I had expected (B7) to be challenged, but to the best of my knowledge no objections to it have yet been raised. Questions were raised with respect to (B2) in discussion at the conference. Are there realistic solutions to the Einstein field equations containing closed timelike curves?

But there may be another premise in the argument that deserves closer scrutiny. The point of the argument, as I construe it, is that our experience (as) of passing time or transience is consistent with our existing in a universe in which there is no lapsing or passing time, a Parmenidean "block universe." One subtle advantage of the argument is that, at first blush at least, it provides the resources needed to handle a reasonable objection to the Parmenidean view. The objection has been elegantly stated by the physicist-philosopher Abner Shimony (1993, p. 278–9).

> There is a very important principle linking epistemology and ontology, one that is pervasive in the literature of empiricism from Berkeley to the sense-data theorists of the early twentieth century and implicit in other philosophical writings: that *even though the*

---

[3] $x \ll y$ if and only if there is a smooth, non-degenerate, future-directed timelike curve from x to y. Intuitively, if $x \ll y$, then there is a curve in the manifold, M, that represents the trajectory or 'possible life history' of a massive body that travels at subluminal speeds from spacetime point x to spacetime point y, moving into the future at every point along its journey. The most peculiar feature of the Gödel spacetime, $\mathcal{M}$ is that for every spacetime point $x \in \mathcal{M}$, $x \ll x$.

distinction between appearance and reality is maintained, a minimal condition on ontology is to recognize a sufficient set of realities to account for appearances qua appearances. I cannot believe that no name has been given to this principle, but since I do not recall reading one, I propose "the Phenomenological Principle."

The challenge of the Phenomenological Principle, in this context, is to account for the appearance of transience in a static universe. That challenge is met by noting that, given the existence of local time functions, the local appearance of transience can be accommodated even if the global condition (A2) for the existence of transience is not met. But as soon as one's attention is directed to (A2), one notes that in requiring a *global* time-slice (A2) makes a metaphysically-motivated demand that outstrips normal human experience. Why, one might be led to wonder, does transience require a global foliation of the spacetime rather than a local time ordering? Why, in other words, can transience not be a local phenomenon? How, on the other hand, can a plurality of locally passing times be understood? Perhaps what we learn from Gödel's argument is that it may be of interest to consider the plausibility and intelligibility of relaxing (A2).

References

Earman, J. 1995, Bangs, Crunches, Whimpers, and Shrieks: Singularities and Acausalities in Relativistic Spacetimes (Oxford: Oxford University Press)

Gödel, K. 1949, in Albert Einstein: Philosopher-Scientist, ed. P. A. Schilpp (La Salle, Illinois: Open Court), 557

Jeans, J. 1935, in Scientific Progress, ed. J. Jeans, W. Bragg, E. V. Appleton, E. Mellanby, J. B. S. Haldane, & J. Huxley (New York: Macmillan), 13

Savitt, S. 1994, Australasian J. of Philosophy, 72, 463

Shimony, A. 1993, in Search for a Naturalistic World View, Volume II (Cambridge: Cambridge University Press)

Yourgrau, P. 1991, The Disappearance of Time: Kurt Gödel and the Idealistic Tradition in Philosophy (Cambridge: Cambridge University Press)

Acknowledgments. I am grateful to the organizers of the Astrophysical Ages and Time Scales conference for inviting me to participate and to the Social Sciences and Humanities Research Council of Canada for their support of my research.

Discussion

Alan Stockton: As I recall, St. Augustine had an argument that God did not exist in time or did not perceive time as a flow and saw it all essentially as once, which sounds like His view of time would be similar to your eternalist view.

Steven Savitt: There are many theological arguments about the relation between God and time. It's very curious, it seems to me, that on a presentist or possibilist view, it would be very difficult for God, an unchanging entity that's all-knowing and omniscient, to know what time it was actually. But of course that could be countered by the claim that maybe that isn't real knowledge. Anyway, a God's-eye view, as it were, is exactly what eternalism represents.

Alan Guth: You said the Gödel Universe was physically possible but, although it satisfies Einstein's equations, as far as I know, people who have tried to build models of universes with closed time-like curves with realistic forms of matter have failed to do so. I think it's conceivable that physical laws in fact prevent closed time-like curves, and there are conjectures about this, so I think it's an open question at this time whether or not the laws of physics allow closed time-like curves.

Steven Savitt: I was trying to be careful and maybe a little weaselly with that second premise. When I use the term "physically possible" I did use it the way philosophers tend to use it, meaning just compatible with natural laws. It may well be that there's overwhelming evidence, for instance, that this Universe could not be a Gödel universe. I think a Gödel universe is rotating and our Universe is presumably not rotating. It would make the argument considerably less interesting if there were no solutions to the Einstein field equations with closed time-like curves that contain matter. If they were all empty solutions then it would be very hard to see how creatures like us could exist in an empty universe. I'm perfectly happy to be instructed on this score by those who work on that area. You believe that there are now no solutions with matter that contain closed time-like curves?

Alan Guth: It's a little hard to give a definitive answer to that question actually. If one wants to have a solution in which closed time-like curves are created after some finite time, the only way that one knows how to do that is to violate what's usually called the weak energy condition, essentially the positivity of energies of matter. On the other hand, you could, if you're willing to start with a universe which in the beginning has closed time-like curves, then you can do that with any kind of matter you want. In fact, you don't need anything as fancy as the Gödel universe; you could for example just have a static universe where you identify some point in the future with some point in the past and it becomes cyclic.

Steven Savitt: Like the rolled-up picture I gave you.

Session 2

Precision Timing in Astronomy

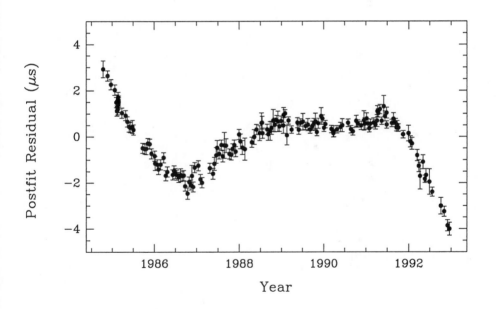

Astrophysical Ages and Time Scales
ASP Conference Series, Vol. 245, 2001
T. von Hippel, C. Simpson, N. Manset

Lessons Learned From Solar Oscillations

D. O. Gough

Institute of Astronomy, Madingley Road, Cambridge CB3 0HA and Department of Applied Mathematics and Theoretical Physics, Silver Street, Cambridge CB3 9EW, UK

Abstract. Helioseismology has demonstrated that standard solar models can be brought into broad agreement with observation, and for many people this provides a high degree of confidence in the theory. If one accepts that theory, one can carry out a calibration of the models against the observational data to obtain a formal value, together with an error estimate, of the age of the Sun. The error estimate is necessarily a lower bound to the uncertainty in the true age, for there may be properties of the Sun that are not incorporated into the theoretical models which affect the age calibration yet which are not detectable by current observations. Moreover, standard solar models are unstable, and therefore, they do not obviously fit into a coherent self-consistent theoretical framework. The additional uncertainties so engendered are difficult to quantify, but one must not forget them when applying solar inferences to broader issues in physics and astronomy.

1. Introduction

More than a decade ago there were several attempts to address the determination of the age of the Sun by helioseismology using the eigenfrequencies of low-degree acoustic modes (Christensen-Dalsgaard 1986; Ulrich 1986; Guenther 1989). Gough & Novotny (1990) compared the attempts and concluded that if the standard errors in the observed frequencies were 0.1 μHz the age of standard solar models could be calibrated within an uncertainty of about 3×10^8 yr by a procedure that does not depend on the reliability of modeling the uncertain outer layers of the Sun. At about that time, frequency errors were typically 0.5 μHz or greater (e.g., Elsworth et al. 1991), so the uncertainty in the determination of the Sun's age would have been 1.5 Gyr or worse.

Since then, the precision of the observed frequencies has increased more than tenfold. Stimulated by this conference I have therefore revisited the matter. In this preliminary reinvestigation, using some of the latest data, I find that the precision of a straightforward calibration is improved as expected. The outcome is not inconsistent with other measures of the solar age, such as that inferred from the much more accurate radioactive dating of meteorites. However, it is not yet precise enough to address the issue of the difference in age between the Sun and the meteorites. Estimates of values of this difference range from several times 10^5 yr to a few times 10^7 yr. A solar calibration with a precision

31

comparable with the lower end of this range is out of sight, but in the foreseeable future a precision capable of resolving timescales near the upper end is perhaps feasible. At present, however, I shall achieve no better than $\pm 10^8$ yr, which I shall use as my fiducial time interval in the discussion that follows.

In this discussion I measure the age from when the Sun effectively arrived on the zero-age main sequence when it had ceased gravitational contraction and when nuclear reactions had begun to provide the light output, yet when essentially no hydrogen had yet been converted into helium and the chemical composition can be considered to have been uniform. According to theory, there was a short but smooth transition from the epoch of Kelvin–Helmholtz contraction (which lasted about 0.05 Gyr) when the source of energy was gravitational to the main-sequence hydrogen-burning phase, a transition which involved burning some ^{3}He and deuterium. It is likely, however, that one can define with reasonable precision a fiducial zero age as the time from which the main sequence star appears to have evolved (by backward extrapolation) from a uniform helium abundance, just as one can define an effective main-sequence end time by forward extrapolation to the time when hydrogen is exhausted from the centre (Gough 1995).

2. Main Sequence Evolution of the Sun

According to standard theory, the Sun is in hydrostatic and thermal balance throughout its main-sequence lifetime. The outer 30 per cent or so by radius is in a state of turbulent convection, interior to which the material is considered to be more-or-less quiescent. In particular, the products of the nuclear reactions are presumed to remain in situ. Although solar models are found to be unstable to grave oscillatory gravity modes driven by the nuclear reactions in the core (e.g., Christensen-Dalsgaard, Dilke, & Gough 1974), it seems not unlikely that the instability does not develop to an amplitude sufficiently large to modify the structure of the central core significantly (Dziembowski 1983; Jordinson & Gough 2000). Consequently, at any given radius r, the hydrogen abundance $X(r, t)$ decreases monotonically with time t due to nuclear processing. The hydrogen abundance $X(r, t_\odot)$ at the presumed current age $t_\odot$ is illustrated in Figure 1. There is also an extremely small contribution to the decrease in X in the core due to displacement by helium and other heavy elements that are gradually settling under gravity, coupled with a corresponding and rather larger (but also small compared with the change in the core due to nuclear reactions) increase in X in the outer layers. A measure of the X profile in the central regions of the Sun is therefore an indicator of the Sun's age.

To a first approximation (at least in a qualitative discussion such as this), the solar material can be thought of as a perfect gas, so that a decrease in the central hydrogen abundance produces a decrease in the number of particles per unit mass and hence in the pressure (at a given density and temperature). The outcome is that the central regions of the Sun gradually contract and heat up further, causing the nuclear energy generation rate, and hence the luminosity L at the surface, to increase with time. The variation of L with time is not very sensitive to the details of the model, and therefore neither are the total energy produced by the Sun over a given time interval on the main sequence and the

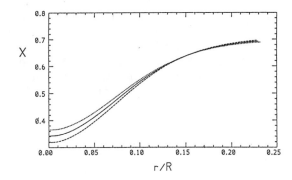

Figure 1. Fractional hydrogen abundance X (by mass) in three solar models. The curves refer to age 4.15 Gyr (dotted), 4.60 Gyr (solid), and 5.10 Gyr (dashed), all with $Z_0 = 0.020$ (Gough & Novotny 1990).

total amount of hydrogen consumed. It follows that the relation between the hydrogen-abundance profile in Figure 1 and the time $t_\odot$ the Sun has spent on the main sequence is robust. At this point one might note that the central hydrogen abundance decreases by about 0.0045 for an increase of 10^8 yr in $t_\odot$.

It is important to emphasize that the hydrogen profiles depicted in Figure 1 are not an evolutionary sequence of a single model, but are a set of three models of different ages, calibrated such that they all have the same luminosity, namely $L_\odot = 3.845 \times 10^{33}$ erg s^{-1}. The calibration has been carried out by adjusting the initial hydrogen-to-helium abundance ratio $X_0{:}Y_0$. The older the model, the higher is X_0 and the lower is the initial luminosity L_0, and indeed the luminosity at any given time t. Therefore the decline of X with present age $t_\odot$ is offset to some degree by the lower L, which implies a lower rate of consumption of hydrogen fuel. According to the computations of Ulrich (1986), the effective main-sequence lifetime of a solar model of age 4.56 Gyr today is 9.23 Gyr (Gough 1995); the central hydrogen abundance declines linearly with time from $X_0 = 0.726$, and therefore decreases by 0.0079 in 10^8 yr, nearly twice as much as the corresponding difference between models calibrated to the same luminosity $L_\odot$. The models in Figure 1 were all computed with the same heavy-element abundance Z, ignoring gravitational settling; models with different Z have somewhat different X.

The intention is to calibrate the profile $X(r, t_\odot)$ by seismology of acoustic modes of oscillation. The frequencies of these modes depend principally on the sound speed $c(r, t_\odot)$ in the star. It is therefore of interest to estimate by how much that might change with age. Using the perfect-gas approximation for the pressure, assuming the gas to be fully ionized and $Z \ll 1$, one obtains for the sound speed $c^2 = \gamma p / \rho \simeq \gamma \mathcal{R}(5X + 3)T/4$, where $\gamma \simeq 5/3$ is the (first) adiabatic exponent and $\mathcal{R}$ is the gas constant. Because nuclear reaction rates are sensitive functions of T, and the change is evaluated at constant L, one might expect as a first approximation that most of the change Δc in c comes from the change ΔX in X, which implies that in 10^8 yr, $\Delta c/c \simeq 2.5\Delta X/(5X + 3) \simeq -0.0025$ at the centre of the model. In fact this must be corrected for the change in temperature: if the nuclear reaction rates at the centre were fixed, T would increase with the age $t_\odot$ of the model to compensate for the decrease in X, thereby tending to augment Δc above the first estimate; but the increase in T tends to increase the region in which significant thermonuclear energy is generated, and therefore the energy generation rate at the centre must be lowered in order to maintain the total rate equal to the observed luminosity $L_\odot$, and that acts on c in the opposite sense. The latter effect actually dominates, and $\Delta c/c \simeq -0.0036$, as is evident in Figure 2a.

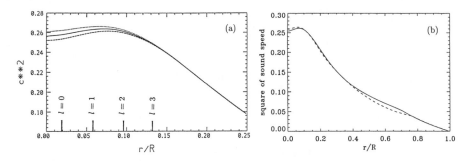

Figure 2.   Square of sound speed (in Mm s^{-1}) in: (a) the models of Figure 1, with the same linestyles, (b) the standard model S (continuous), with $Z_0/X_0 = 0.028$ and $Y_0 = 0.271$, and a model of the same age with $Z_0/X_0 = 0.017$ and $Y_0 = 0.234$ (dashed). Lower turning points r_b of modes with $\nu = 3.0$ mHz are marked on the abscissa scale of (a).

As I mentioned earlier, the sound-speed profile also depends on chemical composition. How does this influence the structure of the solar model? One can simply argue as follows: a higher heavy-element abundance Z produces a higher opacity, hence a steeper temperature gradient, hence a greater central temperature, and so the model would require a lower X to produce the observed luminosity. Thus, associated with an increase in Z must be an increase in the helium abundance Y. That argument is basically correct for the radiative interior of the Sun; the response of the outer layers to a change in Z is complicated by the existence of convection, which I shall not discuss in detail here. Suffice it to say that the model with higher Z, and steeper temperature gradient, is more susceptible to convective instability, and indeed has a deeper convection zone. There are compensating influences on the sound speed in the core: the increase in temperature tends to raise c, the decrease in X tends to lower it. The effect of X dominates near the centre, but not in the rest of the radiative zone; the sound speed is lower for $r \lesssim 0.1R$ in the model with the higher Z, but it is higher further out. This is illustrated in Figure 2b which compares a model not very dissimilar to the Sun with a model that is comparatively deficient in helium and heavy elements. The nuclear reactions in the helium-deficient model, with the lower temperature gradient in the core, are less concentrated near the centre; consequently there is a smaller gradient of $X(r,t)$, and a smaller tendency to decrease c at the very centre of the model, leading to a somewhat smoother sound-speed profile in the central regions. There is evidently some similarity in the effect on the core of small changes in $t_\odot$ and changes in Z_0; the structure of the outer layers, however, is more sensitive to changes in Z_0.

3. Seismic Acoustic Modes in a Nutshell

Acoustic modes (called p modes) are standing sound waves that resonate in the Sun's acoustical cavity. In a stratified medium they satisfy approximately the local dispersion relation between total wavenumber k and frequency ω:

$$\omega^2 = k^2 c^2 + \omega_c^2, \tag{1}$$

where $c(r)$ is the local sound speed and $\omega_c(r)$ is the critical cutoff frequency, i.e., the minimum frequency at which it is possible for a sound wave to propagate in a stratified environment. In an isothermal atmosphere it is given by

$$\omega_c = c/2H \tag{2}$$

(Lamb 1908), where H is the (pressure) scale height, and this formula provides a good guide in a more realistic stellar envelope too. It is plotted in Figure 3a. The eigenfunctions of standing waves in a spherically symmetrical star separate into the product of a function of radius r, associated with which is a vertical wavenumber k_v, and a spherical harmonic of degree l, associated with which we may consider there to be a horizontal component of wavenumber $k_h = (l + \frac{1}{2})/r$. Hence, Equation 1 may be rewritten

$$k_v^2 = \frac{\omega^2 - \omega_c^2}{c^2} - \frac{(l + \frac{1}{2})^2}{r^2}. \tag{3}$$

The acoustic cutoff frequency is significant only in the surface layers of the star. Roughly speaking, H is proportional to the depth z beneath the surface, as is temperature T, at least in the outer layers. Therefore $c \propto T^{1/2} \propto z^{1/2}$, and hence $\omega_c \propto z^{-1/2}$, although it does not actually diverge as $z \to 0$, but becomes roughly constant, $\omega \simeq \omega_{ca}$, in the atmosphere, which to a first approximation can be regarded as being isothermal. I leave aside discussion of the manner in which the level $z = 0$ is defined (but see Lopes & Gough 2001).

Upward propagating waves of given frequency ω travel more and more slowly as they approach the surface, and therefore the wavelength decreases, and when it becomes comparable with the scale H of variation of the background state the waves can no longer propagate. Near the surface, the second term on the right-hand side of Equation 3 is negligible, and $k^2 \simeq (\omega^2 - \omega_c^2)/c^2$; thus k_v becomes imaginary above the upper turning point r_t where $\omega_c(r_t) \simeq \omega$, and the wave is therefore reflected at $r = r_t$. Deep in the star ω_c is negligible; at great depths k_v again becomes imaginary, this time beneath the level r_b at which $(l + \frac{1}{2})c(r_b)/r_b \equiv \Lambda(r_b) \simeq \omega$—downward propagating waves are refracted back upwards at the lower turning point $r = r_b$. The critical frequency $\Lambda(r)$ is sometimes called the Lamb frequency. It is plotted in Figure 3b.

The upper turning points r_t are close to the region where the turbulence in the upper convective boundary layer is in some sense most vigorous, where the convective timescales are comparable with the characteristic timescales of resonant acoustic modes (Figure 3a). The interaction generates acoustic noise, which drives the modes. Most of the power in the modes is in the frequency range 2 mHz $< \nu <$ 4 mHz, where $\nu = \omega/2\pi$ is cyclic frequency. Modes with lower frequency are reflected so far beneath the region of intense turbulence that their evanescent tails are too weak to couple strongly with the acoustic source; modes with higher frequency are reflected very near the surface of the star, and some of their energy leaks out of the star (particularly when ω exceeds ω_{ca}) or is scattered into other modes by inhomogeneity created by the turbulence.

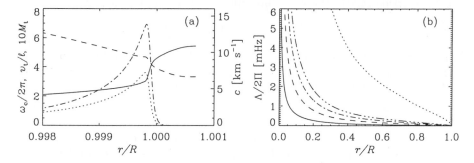

Figure 3.   (a) Acoustic cutoff frequency $\omega_c/2\pi$ (mHz) in the outer layers of a solar model (continuous curve); also the Mach number (multiplied by 10; dotted) and the inverse characteristic time v_t/ℓ of the turbulence (dot-dashed), where v_t and ℓ are characteristic eddy velocity and length scales, and the sound speed (in km s^{-1}; dashed). (b) Critical frequencies $\Lambda/2\pi = (l + \frac{1}{2})c/2\pi r$ (mHz) for $l = 0$ (continuous), $l = 1$ (dashed), $l = 2$ (dot-dashed), $l = 3$ (dash-triple-dotted), and, for comparison, $l = 20$ (dotted).

For the purpose of elementary discussion, we may regard all the modes to have roughly the same frequency (ω varies amongst the dominant modes by only a factor 2), and their character is therefore determined predominantly by their degree l, which varies from 0 to more than 10^3. It is evident, therefore, from the formula for Λ and from Figure 3b, that it is only low-l modes that penetrate to the core of the Sun where the energy is being generated. It is principally these modes that carry the information about the age of the Sun, and the calibration I discuss in this paper is carried out solely with such low-l modes.

It can be shown that when the dispersion condition (Equation 3) is substituted into the quantization condition for resonant waves (e.g., Gough 1993), the eigenvalue equation for the frequency $\nu = \nu_{n,l}$ of degree l in a spherically symmetrical star becomes

$$\nu_{n,l} \sim (n + \tfrac{1}{2}l + \varepsilon)\nu_0 - [A(l + \tfrac{1}{2})^2 - B]\nu_0^2\nu_{n,l}^{-1}, \tag{4}$$

where the integer n is the order of the mode and ε, ν_0, A, B are functionals of the equilibrium state of the star, and do not depend on n or l. This formula was first derived by Tassoul (1980) from the full linearized equations of adiabatic pulsation, rather than from the simple wave dispersion relation (Equation 3), as an asymptotic expression valid for large n. Of particular interest to this discussion are the constants ν_0 and A. They are given by

$$\nu_0^{-1} = 2 \int_0^R \frac{\mathrm{d}r}{c} \tag{5}$$

and

$$A = \frac{1}{4\pi^2 \nu_0} \left[\frac{c(r_{\mathrm{t}})}{r_{\mathrm{t}}} - \int_{r_{\mathrm{b}}}^{r_{\mathrm{t}}} \frac{1}{r} \frac{dc}{dr} dr \right], \tag{6}$$

where R is the radius of the Sun. The quantity ν_0 is the sound travel time from centre to surface of the Sun, and is dominated by conditions in the outer layers where c is small, while A is sensitive to conditions near the centre of the star, owing to the factor r^{-1} in the integrand. The integrand is plotted in Figure 4 for the trio of models illustrated also in Figures 1 and 2a; its integral is much larger than the associated surface term $c(r_{\mathrm{t}})/r_{\mathrm{t}}$. The quantitives ε and B both depend on conditions predominantly near the surface. It is therefore only the quantity A that is sensitive to conditions in the energy-generating core of the Sun, and which can serve as a measure of the solar age. According to Equation 4 it is related to the raw frequencies by the formula

$$d_{n,l} \equiv \frac{3}{2l+3} (\nu_{n,l} - \nu_{n-1,l+2}) \simeq 6A\nu_0^2 \nu_{n,l}^{-1}. \tag{7}$$

The quantity $d_{n,l}$, with or without the factor $3/(2l+3)$, or an average of it such as d_l defined by Equation 8 below, is called the small frequency separation.

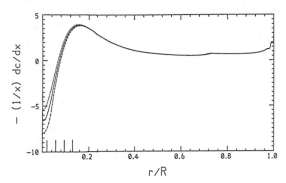

Figure 4.   Integrands in Equation 6 for A, for the models of Figure 1, with the same linestyles: $x = r/R$ and units of c are Mm s^{-1}. The ticks on the abscissa indicate r_{b} for 3 mHz modes with $l = 0, 1, 2$, and 3.

It has been common practice to calibrate (or discuss the calibration of) the cores of solar models with the quantity $d_{n,l}$ (or an average of it) or with an average of the product of $d_{n,l}$ with $\nu_{n,l}$ or some power of ν_0 (e.g., Gough 1983; Christensen-Dalsgaard 1986; Ulrich 1986; Guenther 1989) rather than with A itself. Here, I follow Gough & Novotny (1990), and use an average d_l of $d_{n,l}$ obtained by averaging Equation 7 over a set of frequencies:

$$d_l = 6A\nu_0^2 \langle \nu_{n,l}^{-1} \rangle, \tag{8}$$

where the quantities A and ν_0 have been obtained by regression of the asymptotic formula (Equation 4) with the observed frequencies $\nu_{n,l}$. Roughly, d_l is typically about 10μHz in the present Sun. Also $\nu_0 \simeq 138$ μHz, from which one obtains a range of about $(14, 28)$ of $n + \frac{1}{2}l$ in the frequency range (2 mHz, 4 mHz).

I conclude this section by pointing out that the accuracy of the calibration reported below does not depend directly on the accuracy of this asymptotic discussion. The purpose of this discussion was for designing an appropriate seismic diagnostic which is believed to be sensitive to $t_\odot$, yet is relatively insensitive to other properties of the model. Once the diagnostic has been chosen, the results of this section are used no more. What are compared with observation for the calibration are numerically computed eigenfrequencies of a theoretical model, which are processed in precisely the same fashion as the solar frequencies determined from the observations.

4. The Calibration Diagnostics

The calibration reported in this paper is with the two quantities d_0 and d_1 which can be obtained from whole-disc observations of modes with $0 \leq l \leq 3$. The two signatures measure different averages of $r^{-1}dc/dr$ in the core, since the lower turning points $r_{\rm b}$ are at different locations, and thereby, it is to be hoped, the somewhat different influences of Z_0 and $t_\odot$ can be disentangled. We note, however, that the profile differences in the core, evident in Figures 2a and 2b, are not very dissimilar, so one should anticipate a large degree of cancellation in the disentanglement, and a consequent magnification of the data errors.

One way in which one might think of obviating the error magnification is by using a different diagnostic, such as a measure of the large separation $\langle \Delta_{n,l} \rangle$ (where $\Delta_{n,l} \equiv \nu_{n,l} - \nu_{n-1,l} \simeq \nu_0$), in conjunction with d_l. As I pointed out at the end of §2, models with different Z_0 differ in their outer layers by more than models with different $t_\odot$, and since the value of the large frequency separation is dominated by conditions in the outer layers of the star, it is in principle a better diagnostic of Z_0; indeed, the large separation was the first seismic diagnostic quantity to be used to address the value of the initial chemical composition of the Sun (Christensen-Dalsgaard & Gough 1980). For that reason, Gough & Novotny (1990) discussed using a combination of the large and small separations to determine $t_\odot$. However, the value of the large separation is perhaps too sensitive to the ill-understood very outer layers of the convection zone where the sound speed is lowest, and its use could be subject to unacceptably large systematic errors (although not as large as the raw frequencies themselves). One could, as an alternative, supplement d_l with a measure of the depth of the convection zone, which can be determined from modes of higher degree in a combination that is not excessively influenced by the uncertain outer layers of the convection zone (Christensen-Dalsgaard, Gough, & Thompson 1991). However, there would be uncertainties resulting from this approach too, arising from the influence on the sound speed of shear-induced material redistribution in the tachocline, the rotational shear layer near the base of the convection zone.

I might point out here that aside from in the tachocline the effect of rotation on the distribution of chemical elements is largely considered to be negligible, or at least it is ignored. There have in the past been discussions of material and even heat transport by shear-induced turbulence in the radiative interior, or by gravity waves generated either at the interface between the convective and radiative zones or in the energy-generating core. How important such processes are in influencing the age calibration is an open issue; that they engender additional uncertainty must not be forgotten.

5. A Priori Estimate of the Accuracy Required of the Data

It is evident from Figure 4 that in a time interval of 10^8 yr the integrand $x^{-1}dc/dx$ ($x = r/R$), which determines d_0 and d_1, changes by about -0.1 in the middle of the range of the lower turning points of the low-l modes. Let us estimate by how much that changes d_l. It is a straightforward matter to show that the proportion of the frequency of a mode that is contributed by some region within a star is proportional to the sound travel time through that region (e.g., Gough, 1993). That is roughly the range in acoustic radius $\tau = \int c^{-1}dr$ from top to bottom of the region. One can see in Figure 4 that almost all the evolutionary change in $x^{-1}dc/dx$ occurs below $x \simeq 0.15$ ($\tau \simeq 0.06T$, where $T = \tau(R)$ is the acoustic radius of the Sun); a typical lower turning point is at $x \simeq 0.08$ ($\tau \simeq 0.03T$). Thus the evolution influences approximately a proportion 0.03 of the mode frequency. The change in d_l produced by the change over an interval 10^8 yr is therefore approximately

$$\Delta d_l \simeq \frac{6 \times 0.03 \times 0.1}{4\pi^2 R\bar{k}} \tag{9}$$

in which I have used Equations 6 and 8, approximating $\nu_0\langle\nu_{n,l}^{-1}\rangle$ by $1/\bar{k}$, where $\bar{k}$ is a typical value of $n + \frac{1}{2}l$. In this equation R is measured in Mm, so for $\bar{k} = 21$, $\Delta d_l \simeq 0.03$ μHz. Evaluating d_l to this precision from a set of N observed frequencies $\nu_{n,l}$ requires a mean standard error of $0.03\sqrt{N/2}$ μHz in the frequencies, which is about 0.11 μHz for $N = 28$.

This argument assumes that everything other than the age of the Sun is known, which is not the case. Another uncertainty is the initial hydrogen abundance X_0, or equivalently, the heavy-element abundance Z_0, which is related to X_0 through $t_\odot$ by demanding that the solar model has the observed luminosity $L_\odot$ at $t = t_\odot$. There are spectroscopic measurements of the ratio $\zeta_s \equiv Z_s/X_s$ in the photosphere today, which can be related to the value in the radiative interior by theory. According to Grevesse & Noels (1993), $\zeta_s = 0.0245 \pm 0.005$, the standard error being 20%. Subsequently, Grevesse & Sauval (1998) quoted $\zeta_s = 0.023$, with an uncertainty that "might be of the order of 10%". It is of some concern that even this smaller uncertainty might degrade the age calibration, which is why I try to use two seismic parameters to calibrate for both $t_\odot$ and Z_0 simultaneously. Unfortunately, as will be seen later, this leads to a considerable cancellation and a consequent magnification of the errors.

6. The Age Calibration Procedure

The calibration is accomplished by linearizing the deviations from a standard solar model whose seismic frequencies are close to those of the Sun. Thus if δd_l are the differences between the observed and calculated values of d_l, the logarithmic differences $\delta\ln t_\odot$ and $\delta\ln\zeta_0$ between age $t_\odot$ and initial abundance ratio ζ_0 of the Sun and the model satisfy

$$\delta d_l = (\partial d_l/\partial\ln t_\odot)_{\zeta_0}\delta\ln t_\odot + (\partial d_l/\partial\ln\zeta_0)_{t_\odot}\delta\ln\zeta_0. \tag{10}$$

These are two simultaneous linear algebraic equations, for $l = 0$ and $l = 1$, which can be solved for $\delta\ln t_\odot$ and $\delta\ln\zeta_0$ in terms of δd_l.

The standard model I use is Model S of Christensen-Dalsgaard et al. (1996). It was computed with $\zeta_0 \equiv Z_0/X_0 = 0.0277$ to an age $t_\odot = 4.6$ Gyr, yielding $X_0 = 0.7091$ and a present surface abundance ratio $\zeta_s = 0.0245$.

Because, seismically speaking, Model S is a good representation of the Sun, it is not necessary to compute partial derivatives of the seismic signatures with respect to $t_\odot$ and ζ_0 for precisely that model. It is adequate to use values already computed for a different, yet similar, model. Here I use the model considered by Gough & Novotny (1990). The model was computed with a heavy-element abundance $Z_0 = 0.0201$, also to an age of $t_\odot = 4.60$ Gyr, yielding $X_0 = 0.7036$ and hence $\zeta_0 = 0.0286$, from which, with the help of Model S, one can obtain $\delta\ln\zeta_s \simeq 0.98\delta\ln\zeta_0$. Random errors in the frequency data were propagated through the calibration procedure to provide estimates of the errors in $\delta\ln t_\odot$ and $\delta\ln\zeta_0$.

7. The Age Calibration

The calibration uses the frequencies of the low-l modes obtained recently from whole-disc measurements by the BiSON (Chaplin et al. 1999). In carrying out the calibration it is necessary to select which of the mode frequencies to use. Here I seek a subset which yields a low estimate of the error in $\delta\ln t_\odot$ arising from the data errors. The latter are plotted against $k = n + l/2$ (not to be confused with the magnitude of the total wavenumber in §3) in Figure 5. Not surprisingly, there is some similarity with the widths of the lines in the acoustic power spectrum (cf. Chaplin et al. 1998). There is a characteristic dip in the vicinity of $k = 21$ ($\nu \simeq 3000$ μHz), and then the errors rise with k as the lines broaden due partly to interactions with the turbulence in the upper superadiabatic boundary layer of the convection zone. Also, in common with the line widths the errors decline as k decreases below about 15 ($\nu \simeq 2200$ μHz), although not as steeply as the line widths do, because the signal-to-noise ratio also decreases, rendering the frequencies more difficult to measure and eventually even causing the mode peaks in the spectrum to be obliterated.

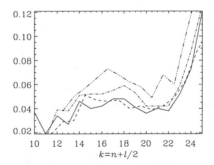

Figure 5.   R.m.s. errors in BiSON frequency data (Chaplin et al. 1999) used for the calibration (μHz): $l = 0$ (continuous), $l = 1$ (dashed), $l = 2$ (dot-dashed), and $l = 3$ (dash-triple-dotted).

I have chosen to calibrate the solar model against data sets containing all the modes with $0 \le l \le 3$ and with values of k satisfying $k_1 \le k \le k_2$.

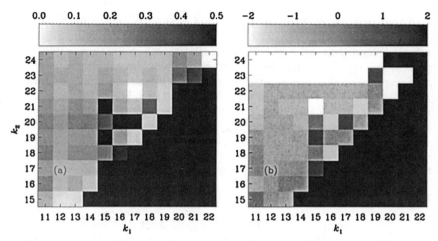

Figure 6. (a) Formal errors in the calibration of $\ln t_\odot$, and (b) associated values of $\delta\ln t_\odot$, for different frequency ranges. There are no calibrations for $k_1 \geq k_2 - 1$, which is blacked out.

The errors in the calibration for $\delta\ln t_\odot$ are depicted in Figure 6a for a range of (k_1, k_2). The general behaviour of the errors in $\delta\ln\zeta_0$ are similar, but the magnitudes are larger. There are four regions of relatively low error, with values below 0.03: they are $(k_1, k_2) = (12, 15)$ and $(13, 15)$, an essentially diagonal strip from $(k_1, k_2) = (17, 19)$ to $(19, 22)$ and extending possibly to $(19, 24)$, the point $(k_1, k_2) = (17, 22)$ and the point $(k_1, k_2) = (22, 24)$. The errors associated with all other mode sets are substantially larger. The corresponding values of $\delta\ln t_\odot$, illustrated in Figure 6b, are 0.076, 0.054 (this is the arithmetic mean over the region), -0.006 and 3.51 respectively. How do we choose between them?

The first solution uses no mode penetrating more deeply than $r_b \simeq 0.10\,R$, and therefore hardly samples the region in which the hydrogen abundance and the sound speed vary substantially with the evolution of the Sun, suggesting that the apparently low error is accidental. One might also have some reservations in accepting the solution in the diagonal strip, because it is right on the boundary of the zone within which it is possible to obtain a solution, and uses the fewest modes possible for computing d_0 and d_1 by the method adopted here. There is of course the strip at constant k_1 extending to large k_2: high-frequency modes (with $k \gtrsim 23$) extend close to the surface of the Sun and sense asphericity due to magnetic activity; modes with different l (and different m) sample the surface differently, and therefore their frequencies are influenced differently, providing a contribution to d_l which the calibration mistakes for a signature of conditions in the core of a spherically symmetrical Sun. There is a suggestion that such contamination is occurring because the formal solutions along the strip become large in magnitude, as is evident in Figure 6b: the solutions for $(k_1, k_2) = (19, 23)$ and $(19, 24)$ for $\delta\ln t_\odot$ are $+0.43$ and -1.80 respectively. A similar remark applies to the solution $+3.51$ at $(k_1, k_2) = (22, 24)$; owing perhaps to the putative systematic error due to asphericity, the formal errors in the (nonlinear) evaluation of d_0 are found to be anomalously low for this range of k, despite the relatively large errors associated with the raw frequencies used in the evaluation.

There remains the solution for $(k_1, k_2) = (17, 22)$, which uses the largest number of modes with $k < 23$ of all the solutions with a standard error in $\delta \ln t_\odot$ less than 0.03, all of which lie in the 'basin' of the error plots in Figure 5. Moreover, it extends to $k = 22$, which provides modes that penetrate most deeply into the core without being subjected too severely (it is hoped) by the asphericity of the surface and the fluctuations due to interactions with the turbulence. I suggest that it is the preferred solution, and therefore expose the calibration in a little more detail. For this range in k, BiSON data yield $d_0 = 9.73 \pm 0.03$ and $d_1 = 10.01 \pm 0.02$; the frequencies of Model S yield $\delta d_0 = -0.15$ and $\delta d_1 = -0.18$. The units are μHz. Moreover, with this mode set the partial derivatives of d_l with respect to $\ln t_\odot$ and $\ln \zeta_0$ are $(\partial d_0/\partial \ln t_\odot)_{\zeta_0} = -4.47, (\partial d_0/\partial \ln \zeta_0)_{t_\odot} = -0.676, (\partial d_1/\partial \ln t_\odot)_{\zeta_0} = -3.58$ and $(\partial d_1/\partial \ln \zeta_0)_{t_\odot} = -0.771$. The solution to Equation 10 is therefore

$$\delta \ln t_\odot = -0.75\delta d_0 + 0.66\delta d_1 = -0.006 \pm 0.026,$$
$$\delta \ln \zeta_0 = 4.0\delta d_0 - 5.0\delta d_1 = 0.30 \pm 0.16. \tag{11}$$

It is evident, as was pointed out by Gough & Novotny (1990), that there is substantial cancellation in Equation 11, which leads to a greater uncertainty than would have been the case had it been acceptable to calibrate the model for $t_\odot$ assuming ζ_0 were known. The solution for $\delta \ln \zeta_0$, and hence for $\delta \ln \zeta_s$, is rather greater than the standard error of 20% quoted by Grevesse & Noels (1993), however, and substantially greater than that suggested by Grevesse & Sauval (1998), although the formal uncertainty is comparable. Although this seismic calibration measures heavy-element chemical composition, principally via the opacity, deep in the radiative interior of the Sun, it is unlikely that error in calculating chemical segregation due to gravitational settling is anywhere near enough to explain the discrepancy with the photospheric abundances. It is likely that the root of the problem lies elsewhere in the solar modeling.

8. Conclusion

The seismic age calibration of the solar models reported here yields

$$t_\odot = 4.57 \pm 0.12 \text{ Gyr}$$

and a surface heavy-element to hydrogen abundance ratio

$$Z_s/X_s = 0.032 \pm 0.004.$$

There are many unresolved uncertainties remaining in the calibration, and the formal uncertainties must certainly be considered as being lower bounds to the real uncertainties. Indeed, even the formal calibration reported here has yielded several possibilities which differ by more than the formal uncertainties, and it could be that one of the others represents the Sun more closely.

It may be noticed that the formal uncertainty in the estimate of $t_\odot$ has decreased from that estimated by Gough & Novotny (1990) for putative data with 0.1 μHz errors by a factor of about 3, which is hardly surprising given both that the current frequency errors are about 0.05 μHz and that I have selected a subset of the modes that yields an estimate with a particularly low formal error.

There is considerable scope for improving the calibration, through both refining the analysis of the seismic data to produce more-appropriate frequencies and through seeking a better seismic diagnostic, perhaps by augmenting the low-degree data with frequencies of modes of intermediate or even high degree. The goal of achieving a precision better than 10^7 yr is not unrealistic.

References

Chaplin, W. J., Elsworth, Y., Isaak, G. R., Miller, B. A., & New, R. 1998, MNRAS, 298, L7

Chaplin, W. J., Elsworth, Y., Isaak, G. R., Miller, B. A., & New, R. 1999, MNRAS, 308, 424

Christensen-Dalsgaard, J. 1986, Proc. IAU Symp. 123, 295

Christensen-Dalsgaard, J., Dilke, F. W. W., & Gough, D. O. 1974, MNRAS, 169, 429

Christensen-Dalsgaard, J., & Gough, D. O. 1980, in Nonradial and Nonlinear Stellar Pulsation, ed. W.A. Dziembowski & H. A. Hill (Heidelberg: Springer-Verlag), 184

Christensen-Dalsgaard, J., Gough, D. O., & Thompson, M. J. 1991, ApJ, 378, 413

Christensen-Dalsgaard, J., et al. 1996, Science, 272, 1286

Dziembowski, W. A. 1983, Sol. Phys., 82, 259

Elsworth, Y., Howe, R., Isaak, G. R., McLeod, C. P., & New, R. 1991, MNRAS, 251, 7P

Gough, D. O. 1983, in Primordial Helium, ed. P. A. Shaver, D. Kunth, & K. Kjär, (European Southern Observatory), 117

Gough, D. O. 1993, in Astrophysical Fluid Dynamics, ed. J-P. Zahn & J. Zinn-Justin (Amsterdam: Elsevier), 399

Gough, D. O. 1995, in GONG '94: Helio- and Astero-seismology From the Earth and Space, ed. R. K. Ulrich, E. J. Rhodes, Jr., & W. Däppen (San Francisco: ASP), 551

Gough, D. O., & Novotny, E. 1990, Sol. Phys., 128, 143

Grevesse, N., & Noels, A. 1993, in Origin and Evolution of the Elements, ed. N. Prantzos, E. Vangioni, & M. Cassé (Cambridge: CUP), 15

Grevesse, N., & Sauval, A. J. 1998, Space Sci. Rev., 85, 161

Guenther, D. B. 1989, ApJ, 339, 1156

Jordinson, C., & Gough, D. O. 2000, in The Impact of Large-Scale Surveys on Pulsating-Star Research, ed. L. Szabados & D. W. Kurtz (San Francisco: ASP), 390

Lamb, H. 1908, Proc. London Math. Soc., 7, 122

Lopes, I. P., & Gough, D. O. 2001, MNRAS, 322, 473

Tassoul, M. 1980, ApJS, 43, 469

Ulrich, R. K. 1986, ApJ, 306, L37

Acknowledgments. I am grateful to D. Bayston, G. Houdek, and C. Simpson for their help in producing the LaTeX version of this paper.

Astrophysical Ages and Time Scales
ASP Conference Series, Vol. 245, 2001
T. von Hippel, C. Simpson, N. Manset

Precision Pulsar Timing

V. M. Kaspi

McGill University, Rutherford Physics Building, 3600 University Street, Montreal, Quebec H3A 2T8, Canada, and Department of Physics and Center for Space Research, Massachusetts Institute of Technology, 70 Vassar Street, Cambridge, MA 02139, USA

Abstract. In this review, after a brief introduction to pulsars and the technique of phase coherent timing of both isolated and binary pulsars, we present a summary of interesting results in the field, biased toward those obtained via long-term monitoring. These include timing of relativistic binary pulsars, pulsar/main sequence star binaries, millisecond pulsars, and anomalous X-ray pulsars.

1. Introduction

Radio pulsar timing is among the most celebrated techniques in precision astronomy. Though perhaps most famous for its use in sensitive tests of general relativity, pulsar timing is in fact useful for studying a very wide variety of astrophysical topics, including astrometry, binary evolution, core collapse supernovae, extrasolar planets, globular cluster structure and dynamics, gravitational waves, interstellar medium physics, the "magnetar" hypothesis, properties of matter at very high densities, pulsar emission mechanisms, measurement of neutron-star masses, stellar winds, and time-keeping metrology.

In this review, I highlight the interesting results that have been found via precision pulsar timing. Clearly "interesting" is subjective and unfortunately space is limited; apologies go to the many diligent pulsar timers whose work I had to omit. One particularly notable omission is any discussion of pulsars in globular clusters; space constraints simply preclude it.

2. Pulsars

Pulsars are rotating, highly magnetized neutron stars. They produce beams of radio emission that can be observed by a fortuitously located astronomer as pulsations, one per rotation period. Note that there exists a distinct class of X-ray pulsars which are powered by accretion from a binary companion. However in this review, I refer only to rotation-powered pulsars, that is, neutron stars whose source of energy is loss of rotational kinetic energy due to magnetic braking (but see §6). This causes the star to slowly spin down.

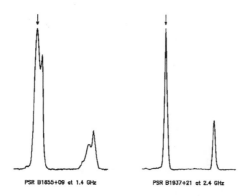

Figure 1. Average pulse profiles for PSRs B1855+09 (*left*) and B1937+21 (*right*). The arrows indicate the reference phases chosen for timing. For both pulsars, one full period is shown (see §5).

There are currently some 1200 pulsars known, all but a handful of which are in the Milky Way, the remainder being in the Magellanic Clouds. Known pulse periods range from a few seconds down to 1.5 ms.

For this review, we emphasize two important properties of pulsars, the ones primarily responsible for making precision timing possible: the stabilities of the average pulse profile and the stellar rotation. By "average pulse profile" we mean the result of the addition of many (typically thousands) of individual pulses, by folding the sampled telescope power output modulo the apparent pulse period. Examples of such pulse profiles are shown in Figure 1. Average profiles are generally observed to be stable, even though individual pulse morphologies vary greatly. Currently there is no theory to explain this observation.

The stability of the stellar rotation is no less relevant. In a reference frame not accelerating with respect to the pulsar, the observed times of pulsations are generally predictable with high precision, in some cases to within a few microseconds over years (see §5) given only the pulse period and spin-down rate. This is perhaps less surprising than the profile stability because of the large stellar moment of inertia and absence of external torques, in strong contrast to accreting neutron stars whose rotation is much noisier (e.g., Bildsten et al. 1997).

3. Pulsar Timing

In pulsar timing, the rotational stability of the pulsar is exploited in order to make precise astrophysical measurements. A standard pulsar timing observation consists of observing a pulsar at a telescope continuously over many cycles. The start time of the observations is recorded with high precision, and the sampled telescope power output is folded at the pulse period. The resulting average pulse profile is cross-correlated with a high signal-to-noise template (e.g., Figure 1) to determine the arrival time of the average pulse. This is then transformed to the Solar System barycentre, a reference frame presumed to be non-accelerating with respect to the pulsar (but see §4.1). This transformation requires precise knowledge of the sky coordinates of the pulsar. In actuality, this is turned around so that if observations are available over more than a year, the known motion of the Earth in its orbit permits a high-precision measurement of the pulsar's position, and sometimes of its proper motion and even parallax. Details of the transformation can be found in various references (e.g., Manchester & Taylor 1977). In a reference frame that does not accelerate with respect to an isolated pulsar, the time evolution of the pulse phase $\phi(t)$ is generally well described in a Taylor expansion:

$$\phi(t) = \phi(t_0) + \nu(t - t_0) + \tfrac{1}{2}\dot{\nu}t^2 + \ldots, \qquad (1)$$

where $\nu \equiv 1/P$ is the spin frequency (P is the period), and $\dot{\nu}$ is its time derivative.

The above procedure for timing a pulsar is repeated typically on a bi-weekly or monthly basis. The time series of pulse times-of-arrival (TOAs) is then compared with a model prediction, where the model consists of ν, $\dot{\nu}$, and astrometric parameters. Specifically, the squares of the residual differences between the initial model-predicted pulse times-of-arrival (TOAs) and the observed TOAs are minimized by varying and hence improving the model parameters in an iterative fashion. Note that by using TOAs, as opposed to measuring the pulse period at each observing epoch, the timing analysis is coherent in the sense that every rotation of the neutron star is accounted for. Thus, the technique is often referred to as "phase coherent timing."

4. Binary Pulsars

If the pulsar is in a binary system, the motion about the binary centre of mass will cause regular delays and advances in observed TOAs just as the Earth's motion around the Sun does. Classically, five additional parameters are required to describe and predict pulse arrival times for binary pulsars, in addition to the spin and astrometric parameters. Conventionally the five Keplerian parameters are the orbital period, P_b, the projected semi-major axis, $a \sin i$, where i is the inclination angle of the orbit, the orbital eccentricity, e, the longitude of periastron, ω measured from the line defined by the intersection of the plane of the orbit and the plane of the sky, and an epoch of periastron, T_0. Only the *projected* semi-major axis is measurable, as pulsar timing is only sensitive to the radial component of the pulsar's motion. Therefore, the component masses cannot be uniquely determined. They can, however, be constrained via the mass function, $f(Mp)$, where

$$f(M_{\rm p}) = \frac{(M_{\rm c} \sin i)^3}{(M_{\rm c} + Mp)^2} = \frac{4\pi^2 (a \sin i)^3}{G M_\odot P_b^2}, \qquad (2)$$

where $M_{\rm p}$ and $M_{\rm c}$ are the pulsar and companion masses, respectively.

In some binary systems, particularly double neutron star binaries, relativistic effects must also be taken into account in order to model the binary orbit and hence observed TOAs properly. The non-classical post-Keplerian (PK) effects to have been measured in a binary pulsar system thus far are: the rate of periastron advance, $\dot{\omega}$, the combined effects of relativistic Doppler shift and time dilation, γ, the rate of orbital decay, $\dot{P}_b$, and r and s, the two parameters describing the Shapiro Delay, or the observed pulse time delay due to the bending of space-time near the pulsar companion, important for highly inclined orbits. The systems for which tests of theories of relativistic gravity are possible are those with N measurable post-Keplerian parameters, where $N > 2$. These systems permit $N - 2$ tests of gravity, as the first two parameters determine the component masses.

4.1. Relativistic Binary Pulsars

The results of long-term timing observations of the relativistic binary pulsar PSR B1913+16 are perhaps the most interesting of all pulsar timing efforts; indeed they have been distinguished with the 1993 Nobel Prize in Physics awarded to the discoverers Joseph Taylor and Russell Hulse. Detailed descriptions and reviews of the results and implications of those timing observations can be found in a variety of references (Taylor & Weisberg 1989; Damour & Taylor 1991, 1992; Taylor et al. 1992; Taylor 1992, 1993).

As reported by Taylor (1993), timing observations PSR B1913+16 made through 1993 (the Arecibo telescope, where the observations were done, became inoperable not long afterward in preparation for a major upgrade) have resulted in the measurement of three post-Keplerian parameters: the rate of periastron advance $\dot{\omega} = 4°.226621 \pm 0°.000011$, the combined time dilation and gravitational redshift $\gamma = 4.295 \pm 0.002$ ms, and the observed orbital period derivative $\dot{P}_b = (-2.4225 \pm 0.0066) \times 10^{-12}$. The first two of these parameters determine the component masses to be $1.4411 \pm 0.0007\ M_\odot$ and $1.3874 \pm 0.0007\ M_\odot$. The third post-Keplerian parameter, $\dot{P}_b$, in principle allows for one test of GR (or other theory of gravity).

However, the observed value of $\dot{P}_b$ must be corrected for the effect of acceleration in the Galactic potential. This correction follows from the simple first-order Doppler effect, where $P_b^{obs}/P_b^{int} = 1 + v_R/c$, where P_b^{obs} and P_b^{int} are the observed and intrinsic values, and v_R is the radial velocity of the pulsar relative to the Solar System barycentre. A changing v_R leads to a Galactic term

$$\left(\frac{\dot{P}_b}{P_b}\right) = \frac{a_R}{c} + \frac{v_T^2}{cd}, \tag{3}$$

where a_R is the radial component of the acceleration, v_T is the transverse velocity, and d is the distance to the pulsar. The second term in this equation is the familiar transverse Doppler or "train-whistle" effect. The best estimate correction factor for PSR B1913+16 is $(-0.0124 \pm 0.0064) \times 10^{-12}$ (Damour & Taylor 1991; Taylor 1992). With this correction applied to $\dot{P}_b^{obs}$, the comparison with the GR prediction can be made; the result (Taylor 1992) is that $\dot{P}_b^{obs}/\dot{P}_b^{GR} = 1.0032 \pm 0.0035$. Note that the uncertainty in this expression is dominated by the uncertainty in the Galactic acceleration term. Since a_R and d are unlikely to be known with much greater precision than is currently available, this particular test of GR will probably not improve much in the near future. Additional tests of GR may still be possible with the PSR B1913+16 system if the parameters r and s can be measured. This may be possible given the recent major upgrade to the Arecibo telescope.

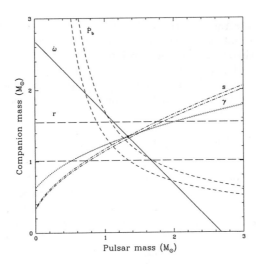

Figure 2.   Companion mass, M_c, versus pulsar mass, M_p, for the relativistic binary pulsar PSR B1534+12, with constraints from the various relativistic parameters indicated (courtesy I. Stairs)

A different relativistic binary pulsar, PSR B1534+12, may provide a better object with which to study general relativity. This 38-ms pulsar is in a 10-hr eccentric orbit with a second neutron star (Wolszczan 1991). PSR B1534+12 offers the hope of additional and more precise tests of GR for a number of reasons. First, its narrow pulse profile permits very high timing precision. Second, the orbital plane of this system is highly inclined, which facilitates the measurements of the relativistic parameters r and s. Thus, in principle, five relativistic parameters are measurable with high precision for PSR B1534+12, which allows two new additional tests of GR that have not been done for PSR B1913+16 (see Damour & Taylor 1992 for details).

Stairs et al. (1998, 1999) report on long-term timing of PSR B1534+12. As expected, they measure the five post-Keplerian relativistic parameters, $\dot{\omega}, \gamma, \dot{P_b}, r$, and s. The results are nicely summarized in Figure 2, where the component masses are plotted on the axes. As each of the five post-Keplerian parameters has a different dependence on the masses, each parameter defines a curve in this plane. If GR holds, then the five curves, as calculated in GR, should meet at a single point. In Figure 2, the curves for $\dot{\omega}, \gamma$ and s agree well (though that for r is not yet precise enough to be constraining). Their intersection implies that the pulsar and companion have masses 1.344 ± 0.002 $M_\odot$ and 1.335 ± 0.002 $M_\odot$, respectively.

As is clear in Figure 2, the curve for $\dot{P_b}$ just misses this intersection point. Stairs et al. (1998, 1999) argue that this discrepancy is a result of an incorrect distance estimate to the pulsar. They then assume GR is correct and determine the distance to the pulsar to be 1.08 ± 0.15 kpc. This demonstrates that the measurement of an improved $\dot{P_b}$ for PSR B1534+12 is unlikely to offer a useful test of GR unless the distance to the source can be determined independently (for example, via a timing or interferometric parallax measurement). However, the expected improved determination of the r parameter should yield a useful test in addition to that from $\dot{\omega}$-γ-s.

4.2. Pulsar/Main Sequence Star Binaries

To date, there are several dozen known binary pulsars, of which but a small minority are relativistic. An even smaller class of binary pulsars are those in which the companion is a massive main sequence star. Though not relativistic, these sources are of astrophysical interest and we briefly discuss results from timing observations of them. Under certain circumstances, even in a clean, classical system, the five Keplerian parameters may be insufficient to fully describe the orbit. In particular, for a companion having a substantial quadrupole moment, deviations from a simple Keplerian orbit can be very significant. In particular, a classical advance of periastron, $\dot{\omega}$, is induced.

The classical $\dot{\omega}$ is quite large in the binary pulsar PSR J0045$-$7319, a 0.9-s pulsar in an eccentric 51-day orbit with a main-sequence B star in the Small Magellanic Cloud (Kaspi et al. 1994a). In addition, in that system, classical spin–orbit coupling induces an apparent variation in the projected semi-major axis, because of a varying orbital inclination angle. This is also a result of the quadrupole moment of the B-star companion, but is only possible if its spin axis is misaligned with the orbital angular momentum (Lai, Bildsten, & Kaspi 1995; Wex 1998). As the angular momenta should have been aligned prior to the supernova that created the pulsar, this provides strong evidence that the pulsar was given a kick at birth (Kaspi et al. 1996).

The same effect may have also been seen in the pulsar main-sequence binary PSR B1259$-$63 (Johnston et al. 1992; Wex et al. 1998). However, this system's 3.4-yr orbital period makes it more difficult to detect, as do the strong eclipses of the radio pulsations near periastron due to the wind of the companion. But one astronomer's noise is another astronomer's signal: these eclipses provide an interesting probe of the wind and magnetic field structure of the Be star companion to PSR B1259$-$63 (e.g., Johnston et al. 1999).

Another interesting pulsar/main-sequence star candidate is PSR 1740$-$3052 (Stairs et al. 2001). This 0.57-s pulsar is in a 231-day binary orbit with a companion that has mass greater than 11 $M_\odot$, as inferred from the mass function (see Equation 2). However, optical/IR observations have not yet revealed a plausible binary companion. At the position of the pulsar is a K supergiant, which should have induced a much larger classical periastron advance than is seen. The pulsar companion, probably a B star, is likely obscured by the K star (Stairs et al. 2001). Continued timing observations, as well as high-resolution IR observations, should help decide.

5. Millisecond Pulsars

The discovery in 1982 of the first millisecond pulsar by Backer et al. heralded a major milestone in pulsar timing. Because pulse arrival times can be measured to within typically ~ 0.01 and sometimes ~ 0.001 of a pulse period, the short pulse periods of millisecond pulsars allow pulse arrival times to be measured typically to within a few microseconds. This permits a wide range of precision astronomy to be done, including, for example, pulsar positions to be measured to microarcsecond precision. Here we highlight some of the main results of millisecond pulsar timing.

5.1. PSRs B1937+21 and 1855+09

The first long-term, high precision millisecond pulsar timing project was carried out at the Arecibo telescope, and was described by Kaspi, Taylor, & Ryba (1994b). In this study, two millisecond pulsars, PSRs B1855+09 and B1937+21, were observed, and results for each were compared with the other. High-precision timing of PSR B1855+09 was begun in 1986 at Arecibo, and yielded daily-averaged TOA uncertainties of $\sim 1\ \mu s$. Data for PSR B1937+21 go back to 1984, with daily-averaged TOA uncertainties of $\sim 0.2\ \mu s$. The published data sets include arrival times obtained through the end of 1992, when the Arecibo telescope shut down for a major upgrade.

PSR B1855+09 is a 5.4-ms pulsar in a circular orbit with a white dwarf. Residuals after removal of the best spin, astrometric and binary parameters are shown in Figure 3. The residuals are dominated by random, Gaussian measurement uncertainties, which indicates that the model describes the data well. For PSR B1855+09, the celestial coordinates and proper motion were determined with uncertainties of 0.12 mas and 0.07 mas yr^{-1} respectively. A significant timing parallax signal was also measured for this pulsar: $\pi = 1.1 \pm 0.3$ mas. The timing model for PSR B1855+09 also includes a relativistic Shapiro delay near superior conjunction. This is measurable because of the plane of the binary is viewed nearly edge-on (see also Ryba & Taylor 1991). The results determine the pulsar and white dwarf masses with high precision. Furthermore, limits on the rate of change of the orbital period of PSR B1855+09 set interesting phenomenological constraints on the rate of change of Newton's constant.

PSR B1937+21 was the first discovered millisecond pulsar, and is still the fastest known rotator, having $P = 1.5$ ms. The residuals after subtraction of the best model including astrometric and spin parameters are shown in Figure 3. For PSR B1937+21, celestial coordinates and proper motion were measured with uncertainties of 0.06 mas and 0.01 mas yr^{-1}. These astrometric parameters can be compared with those measured using VLBI to tie the dynamical and extragalactic reference frames. There is an obvious cubic trend in the residuals that indicates the model does not completely describe the rotation of the neutron star, a trend also observed by Stinebring et al. (1990). Note, however, that the trend is at an extremely low level: the arrival times here have been modeled to better than a few microseconds over nearly a decade. This corresponds to a fractional uncertainty in the determination of the neutron star spin period of a part in 10^{14}. Such precision is unparalleled in astronomy; indeed the rotational stability rivals the stability of most atomic clocks (Taylor 1991).

With long-term timing data for two millisecond pulsars, Kaspi et al. (1994b) could address the issue of the origin of the trend in the timing residuals for PSR B1937+21. They discussed several possible origins of the "noise." Given that a similar signal is largely absent from data from PSR B1855+09, they concludes that its most likely origin is intrinsic to the pulsar itself, as opposed to in the reference terrestrial time standard or the Solar System ephemeris. Indeed similar timing behavior is commonly observed among slow pulsars, and has been shown

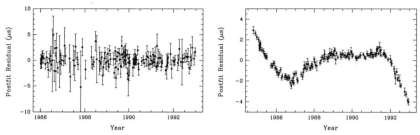

Figure 3.   Post-fit residuals for PSRs B1855+09 (*left*) and B1937+21 (*right*), after Kaspi et al. (1994b).

to be correlated with $\dot{P}$ (e.g., Arzoumanian et al. 1994), in accord with the fact that PSR B1937+21 has a much higher $\dot{P}$ than does PSR B1855+09. This noise is most likely a result of complicated processes within the neutron star itself.

These results therefore suggest that millisecond pulsars having particularly low values of $\dot{P}$, like PSR B1855+09, may be more useful for long-term, high-precision timing. This further suggests that the establishment of a "pulsar timing array," which would consist of numerous millisecond pulsars, timed regularly, distributed isotropically on the sky, could indeed be realized. In addition to yielding plenty of astrophysical results, a timing array could potentially be used to provide a pulsar-based time standard, possibly more stable than today's most stable atomic time standards, as well provide as a check on planetary ephemerides, and detect a stochastic background of gravitational waves (Foster & Backer 1990).

With the goal of a timing array in mind, various groups have been hard at work monitoring several dozen millisecond pulsars. Long-term timing results are just starting to come in for more sources, and are already yielding a variety of interesting results (e.g., Toscano et al. 1999; Wolszczan et al. 2000; Nice, Splaver, & Stairs 2001).

5.2. PSR J0437−4715

PSR J0437−4715 is the brightest millisecond pulsar known, having flux density at 400 MHz of over 600 mJy (Johnston et al. 1993), some 3 and 20 times brighter than PSRs B1937+21 and B1855+09 respectively. PSR J0437−4715 is a 5.75-ms pulsar in a 5.74-day binary orbit with a low-mass white dwarf. The position and proper motion for this pulsar are determined via timing to a precision of ~ 50 mas and ~ 0.07 mas yr^{-1} (Sandhu et al. 1997). The pulsar's large flux density makes it an excellent candidate for a dynamical/extragalactic reference frame tie. Among other interesting results of timing observations of PSR J0437−4715 include a highly significant timing parallax, corresponding to a distance of 178 ± 26 pc, as well as a measurement of the secular change in the orientation of the orbit arising from the system's proper motion (Bell et al. 1997; Sandhu et al. 1997). Intriguingly, although timing residuals over 2.5 yr are only ~ 500 ns, they show systematic effects that depend strongly on subtleties such as how orthogonal polarizations are added to yield average profiles (Britton et al. 2000). With these corrected, residuals of under ~ 100 ns might be possible and would be a record breaker. This pulsar is thus of considerable interest, as it may help determine optimal data analysis strategies for increasing timing precision in all pulsars.

5.3. PSR B1257+12: Pulsar Planets

Perhaps the most famous of the millisecond pulsars is PSR B1257+12, for which timing observations have revealed planet-massed companions (Wolszczan & Frail 1992). The discovery of these planets represented the first detection of extrasolar planets. Thus far, three planets are confirmed, having orbital periods 25 days, 66 ddays, and 98 days. The latter two have minimum masses of 3.4 and 2.8$M_\oplus$, respectively, and the former is Moon-like. The veracity of the latter two has been confirmed via timing, from the signature of post-Keplerian effects due to the planets' mutual gravitational interaction, a relatively strong effect because of the resonant (3:2) orbital periods (Wolszczan 1994). The reality of the Moon-mass planet was questioned as possibly being due to dispersive effects in the solar wind (Scherer et al. 1997). A 25-day periodicity is seen in the Sun, and could perhaps produce a related periodicity in the solar wind. This would affect the timing because of the relative proximity of the pulsar's line-of-sight to the ecliptic. Careful multi-frequency timing observations made over several years clearly confirm that the Doppler shifts are frequency-independent, thus ruling out a dispersive effect (Wolszczan et al. 2000).

6. Anomalous X-ray Pulsars

Recently, the technique of phase-coherent timing that has long been used for radio pulsars has been applied to a very different class of sources, the "anomalous X-ray pulsars" (AXPs). As the name suggests, the nature of these objects is currently uncertain. Indeed, their energy source is not known: though they are spinning down, like rotation-powered pulsars, AXP X-ray luminosities far exceed their spin-down luminosities. For this reason, they were long thought to be accretion powered. However absolutely no evidence for binarity has been seen. Interestingly, of the five known AXPs and one AXP candidate, three are in supernova remnants, which argues that AXPs are young neutron stars. For a detailed review of these objects, see Mereghetti (1999).

There are two competing models for the nature of AXPs. First, it has been proposed that they are isolated young neutron stars with magnetic fields several orders of magnitude larger than those of radio pulsars (Thompson & Duncan 1996). In this "magnetar" model, the X-rays are primarily thermal, produced either via heating from magnetic field decay (Thompson & Duncan 1996) or enhanced initial cooling (Heyl & Hernquist 1997). In a competing model, AXPs are in fact powered by accretion, but from a low-mass, compact disk of material left over from the supernova explosion (Chatterjee, Hernquist, & Narayan 2000).

How can precision timing contribute to our understanding of AXPs? In principle, the timing properties of a magnetar and of a neutron star accreting from a fall-back disk should be quite different. AXPs, if magnetars, should spin down slowly under the influence of the magnetic field, possibly with occasional sudden spin-up events known as "glitches," as are seen in radio pulsars. Some random noise around the steady spin-down might also be present (e.g., Arzoumanian et al. 1994), though at a sufficiently low level that maintaining

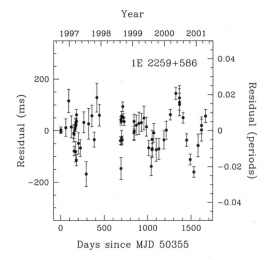

Figure 4.   Timing residuals for AXP 1E 2259+586 over 4.5 yr, after subtraction of a model consisting of P, $\dot{P}$, and $\ddot{P}$. The r.m.s. residual is $\sim 1\%$ of the pulse period.

phase coherence over many years (in the absence of glitches) should be possible. By contrast, accreting sources generally suffer from tremendous torque noise, and more often than not, experience long-term spin-up episodes (Bildsten et al. 1997). Thus the long-term timing properties of AXPs have the potential to distinguish between the two main models for their nature.

For this reason, the five known AXPs have been monitored roughly monthly over the past several years using the Rossi X-ray Timing Explorer (RXTE). Prior to these observations, AXPs had been observed typically once or twice per year, by whatever X-ray telescope happened to be in orbit. The large gaps in the observations precluded phase coherent timing. Indeed the sparse timing data from prior to RXTE were subject to a variety of conflicting interpretations (Heyl & Hernquist 1999; Melatos 1999; Baykal et al. 2000).

For four of the five sources monitored by *RXTE*, the spin downs are extremely regular, indeed comparable to, and in at least one case, more stable than some radio pulsars. For example, AXP 1E 2259+586 has exhibited steady spin down, well described by only three parameters, P, $\dot{P}$ and $\ddot{P}$, over 4.5 yr, with r.m.s. phase residuals of 1% of the pulse period (Figure 4; Kaspi, Chakrabarty, & Steinberger 1999; F. P. Gavriil & V. M. Kaspi, in preparation). Similar results, though over shorter time spans, are obtained for 4U 0142+61 (F. P. Gavriil & V. M. Kaspi, in preparation) and 1E 1841−045 (Gotthelf et al. 2001). 1RXS J1708−4009 also exhibits very steady spin down, although it has experienced a spin-up event consistent with a glitch (Kaspi, Lackey, & Chakrabarty 2000), and post-glitch relaxation (F. P. Gavriil & V. M. Kaspi, in preparation). These observations tend to favor the magnetar model, though they do not prove it conclusively, as the torque noise properties of fall-back disks are not known. One difficulty, however, is that one of the AXPs, 1E 1048−5937, is unlike the others, showing significant deviations from simple spin down that make it impossible to phase connect over more than a few months (Kaspi et al. 2001). The origin of this spin-down behavior is unknown. The timing stability is qualitatively similar to that seen in accreting sources, although the magnetar model can account for this behavior as well. Patient continued monitoring of this source and the others will hopefully clarify the issue.

7. Conclusions

Precision pulsar timing is among the most precise branches of astrophysics, yet, as we hope we have demonstrated in this review, has an amazingly wide variety of astrophysical applications. While demanding patience and dedication, its rewards are ample. Indeed, precision pulsar timing provides strong support for the underlying theme of this meeting, namely that "timing is everything."

References

Arzoumanian, Z., Nice, D., Taylor, J., & Thorsett, S. 1994, ApJ, 422, 671

Backer, D. C., Kulkarni, S. R., Heiles, C., Davis, M. M., & Goss, W. M. 1982, Nature, 300, 615

Baykal, A., Strohmayer, T., Swank, J., Alpar, A., & Stark, M. J. 2000, MNRAS, in press

Bell, J. F., Bailes, M., Manchester, R. N., Lyne, A. G., Camilo, F., & Sandhu, J. S. 1997, MNRAS, 286, 463

Bildsten, L., et al. 1997, ApJS, 113, 367

Britton, M. C., van Straten, W., Bailes, M., Toscano, M., & Manchester, R. N. 2000, in ASP Conf. Ser. 202, Pulsar Astronomy—2000 and Beyond, ed. M. Kramer, N. Wex, & R. Wielebinski (San Francisco: ASP), 73

Chatterjee, P., Hernquist, L., & Narayan, R. 2000, ApJ, 534, 373

Damour, T., & Taylor, J. H. 1991, ApJ, 366, 501

Damour, T., & Taylor, J. H. 1992, Phys. Rev. D, 45, 1840

Foster, R. S., & Backer, D. C. 1990, ApJ, 361, 300

Heyl, J. S., & Hernquist, L. 1997, ApJ, 489, L67

Heyl, J. S., & Hernquist, L. 1999, MNRAS, 304, L37

Johnston, S., Manchester, R. N., Lyne, A. G., Bailes, M., Kaspi, V. M., Qiao, G., & D'Amico, N. 1992, ApJ, 387, L37

Johnston, S., Manchester, R. N., McConnell, D., & Campbell-Wilson, D. 1999, MNRAS, 302, 277

Johnston, S., et al. 1993, Nature, 361, 613

Kaspi, V. M., Bailes, M., Manchester, R. N., Stappers, B. W., & Bell, J. F. 1996, Nature, 381, 584

Kaspi, V. M., Chakrabarty, D., & Steinberger, J. 1999, ApJ, 525, L33

Kaspi, V. M., Gavriil, F. P., Chakrabarty, D., Lackey, J. R., Muno, M. P. 2001, ApJ, in press

Kaspi, V. M., Johnston, S., Bell, J. F., Manchester, R. N., Bailes, M., Bessell, M., Lyne, A. G., & D'Amico, N. 1994a, ApJ, 423, L43

Kaspi, V. M., Lackey, J. R., & Chakrabarty, D. 2000, ApJ, 537, L31

Kaspi, V. M., Taylor, J. H., & Ryba, M. 1994b, ApJ, 428, 713

Lai, D., Bildsten, L., & Kaspi, V. M. 1995, ApJ, 452, 819

Manchester, R. N. & Taylor, J. H. 1977, Pulsars (San Francisco: Freeman)

Melatos, A. 1999, ApJ, 519, L77

Mereghetti, S. 1999, Mem. della Soc. Ast. It., 69, 819

Nice, D. J., Splaver, E. M., & Stairs, I. H. 2001, ApJ, 549, 516

Ryba, M. F., & Taylor, J. H. 1991, ApJ, 371, 739.

Sandhu, J. S., Bailes, M., Manchester, R. N., Navarro, J., Kulkarni, S. R., & Anderson, S. B. 1997, ApJ, 478, L95

Scherer, K., Fichtner, H., Anderson, J. D., & Lau, E. L. 1997, Science, 278, 1919

Stairs, I. H., Arzoumanian, Z., Camilo, F., Lyne, A. G., Nice, D. J., Taylor, J. H., Thorsett, S. E., & Wolszczan, A. 1998, ApJ, 505, 352

Stairs, I. H., Nice, D. J., Thorsett, S. E., & Taylor, J. H. 1999, in Gravitational Waves and Experimental Gravity, ed. J. Tran Thanh Van et al. (Gif-sur-Yvette: Editions Frontières)

Stairs, I. H., et al. 2001, MNRAS, in press (astro-ph/0012414)

Stinebring, D., Ryba, M., Taylor, J., & Romani, R. 1990, Phys. Rev. Lett, 65, 285

Taylor, J. H. 1991, Proc. IEEE, 79, 1054

Taylor, J. H. 1992, Philos. Trans. Roy. Soc. London A, 341, 117

Taylor, J. H. 1993, in Particle Astrophysics, ed. G. Fontaine & J. Trân Thanh Vân, (Gif-sur-Yvette: Editions Frontières), 367

Taylor, J. H., & Weisberg, J. M. 1989, ApJ, 345, 434

Taylor, J., Wolszczan, A., Damour, T., & Weisberg, J. 1992, Nature, 355, 132

Thompson, C., & Duncan, R. C. 1996, ApJ, 473, 322

Toscano, M., Britton, M. C., Manchester, R. N., Bailes, M., Sandhu, J. S., Kulkarni, S. R., & Anderson, S. B. 1999, ApJ, 523, L171

Wex, N. 1998, MNRAS, 298, 997

Wex, N., Johnston, S., Manchester, R. N., Lyne, A. G., Stappers, B. W., & Bailes, M. 1998, MNRAS, 298, 997

Wolszczan, A. 1991, Nature, 350, 688

Wolszczan, A. 1994, Science, 264, 538

Wolszczan, A., & Frail, D. A. 1992, Nature, 355, 325

Wolszczan, A., Hoffman, I. M., Konacki, M., Anderson, S. B., & Xilouris, K. M. 2000, ApJ, 540, L41

Wolszczan, A., et al. 2000, ApJ, 528, 907

Discussion

Anjum Mukadam: Would you be able to detect a planet like Earth at 1 AU from the pulsar?

Vicky Kaspi: It depends from which pulsar. For millisecond pulsars, it is generally very easy given observations spanning at least a full cycle. Now millisecond pulsars are of course a small fraction of the general pulsar population—there are a few dozen millisecond pulsars known while we know 1200 pulsars in total. For the average pulsar it would not be possible though there may be a few exceptions. Note, though, that it would be difficult to detect any planet having orbital period of a year, unless we had a solid measurement of the pulsar position from interferometry. This is because otherwise, the planet's motion would be fit out of the timing residuals as being due to the Earth's motion, i.e., the timing position would be in error.

Marcello Rodonò: A possible reason for having binary pulsars with a non-constant period is maybe due to the fact that the companion star is a magnetically active star. We have been computing some possibility of having a cycle that changes the quadrupole moment and this brings a change in the rotational regime of the star and consequently the orbital motion. So you may have some oscillation due to this magnetic activity on the star.

Vicky Kaspi: I should say that in fact there are some cases of pulsars in orbits around other stars—one I can think of immediately is PSR 1957+20—where such an effect may have actually been seen. In that system, there are definitely deviations from a simple Keplerian orbit that have been proposed as being due to magnetic activity on the companion star. I believe the relevant observations were discussed by Arzoumanian et al. in 1994 in an ApJ Letter.

Astrophysical Ages and Time Scales
ASP Conference Series, Vol. 245, 2001
T. von Hippel, C. Simpson, N. Manset

Dwarf Nova and Accretion Disks

Yoji Osaki

Faculty of Education, Nagasaki University, Nagasaki 852-8521, Japan

Abstract. Dwarf novae are eruptive variable stars exhibiting semi-periodic outbursts that recur on time scales of weeks to months with outburst amplitudes of 2–5 mag. The light curves of dwarf novae are rich in variations in different time scales, ranging from quasi-periodic eruptions with weeks to months, to very strict periodicities such as binary orbital periods of a few to several hours and the so-called superhump periodicity. Some strange phenomena occur in some of dwarf novae, such as superoutbursts and superhumps in SU UMa sub-class. Outburst mechanisms of dwarf novae are discussed. Outbursts and related phenomena can basically be understood by the disk instability model in that intrinsic instabilities in accretion disks cause time-varying light output from the disk. Two different intrinsic instabilities are known in accretion disks of dwarf novae, the thermal instability and the tidal instability. With these two intrinsic instabilities combined, a rich variety in outburst behaviors of dwarf novae can be explained.

1. Introduction

Dwarf novae are eruptive variable stars showing repetitive outbursts typically of amplitude of 2–5 magnitude, outburst duration of a few days to 20 days, and typical recurrence times of 20–300 days. For instance, one of the best observed dwarf novae is a star called SS Cyg, which goes into outburst every 50 days or so from 12th mag at minimum to 8th mag at maximum and its typical outburst lasts for several to ten days. Dwarf novae belong to a more general class of variable stars called "cataclysmic variable stars." Cataclysmic variable stars are semi-detached close binary systems, which consist of a white dwarf primary star and a red dwarf secondary star, with orbital periods ranging from an hour to several hours. The Roche-lobe filling secondary star loses mass from its surface through the inner Lagrangian point and the white dwarf star accretes it via the accretion disk. Although dwarf novae are rather faint stars with a typical absolute magnitude of 3–5 mag in outbursts, more than hundred such stars are known around the solar neighborhood. Besides their own interest, dwarf novae are important in astrophysics because they serve as one of the best laboratories to study accretion disks that play important roles in many different fields in astrophysics ranging from star formation to X-ray binaries, to Active Galactic Nuclei, and to quasars. An extensive monograph on cataclysmic variable stars dealing with both observations and theories by Warner (1995) is now available and reviews on theoretical aspects of dwarf-nova outbursts are found in Cannizzo (1993) and Osaki (1996).

2. Observations of Dwarf Novae

Let us briefly discuss observations of dwarf novae. The most basic observations of dwarf novae are their light curves. The light curves of dwarf novae are very rich in variations in different time scales, ranging from quasi-periodic eruptions with weeks to years to very strict periodicities such as binary orbital periods of order of a few hours and the superhump periodicity. Since the exact times of outbursts in dwarf novae can not be known beforehand, careful watches for these stars are needed, and in this respect amateur astronomer's contributions are very important. The recent progress in applications of cooled CCD cameras to astronomical observations enables amateur astronomers with small telescopes to perform high-precision photometry of rather faint stars like dwarf novae. In fact, besides well known observational groups such as AAVSO (American Association of Variable Star Observers), very successful international networks of amateur and professional observers particularly specializing for observations of cataclysmic variable stars and dwarf novae exist, such as CBA (the Center for Backyard Astrophysics) organized by Joe Patterson in Columbia University and VSNET (Variable Star Network) organized by Taichi Kato in Kyoto University.

As for observations by large telescopes such as Subaru and Gemini, two possible observations suitable for them are mentioned here. First, time-resolved spectroscopy for eclipsing dwarf novae which enables us to make a detailed mapping of accretion disks both in quiescence and in outburst, a technique called Doppler Tomography. Second, search for a brown dwarf secondary in extremely short-period dwarf novae. Some of secondary stars in dwarf novae are suspected to be brown dwarfs, i.e., degenerate dwarfs below the lowest end of the red dwarf main sequence. The secondary star in a cataclysmic variable star is thought to evolve from a lower main sequence to a brown dwarf by transferring its mass to the primary white dwarf. However, no direct observations are not yet available to prove the brown-dwarf nature for any secondary stars. A large telescope is needed to confirm the brown-dwarf nature of a suspected secondary star because it is very faint object. This could be done by IR spectroscopy by finding some molecular features characteristic of a brown dwarf atmosphere. One of possible candidates for such stars will be the secondary star of WZ Sge, one of the best known extreme dwarf novae with an outburst recurrence time of 30 years.

3. Disk Instability Model

Outbursts of dwarf novae are caused by variable accretion onto the white dwarf via accretion disks and the modulation in light output from the accretion disk of dwarf novae is now widely believed to be caused by instabilities in accretion disks, and such a model is called the disk instability model. In the disk instability model, mass is transferred steadily from the secondary star to the accretion disk but accretion onto the white dwarf rather occurs intermittently due to intrinsic instabilities within the accretion disk.

The basic idea of the disk instability model for dwarf nova outburst was first proposed by the present author in 1974 (Osaki 1974). In this model, mass transferred from the secondary star to the accretion disk is not accreted during quiescence but it is simply stored in the outer parts of the disk. When the mass thus accumulated in the disk reaches some critical amount, some kind of instability within the disk sets in and a significant amount of mass is then suddenly accreted onto the central white dwarf and the sudden brightening of accretion disk ensues due to release of the gravitational energy of accreted matter, explaining an outburst of dwarf novae. However, the very physical mechanism for the assumed instability had remained unknown until circa 1980.

An intrinsic instability in accretion disks of cataclysmic variable stars was first discovered around 1980 by Hoshi (1979) and by Meyer & Meyer-Hofmeister (1981). They have found that the local thermal equilibrium solution of accretion disks allows double stable solutions in the outer parts of disks in cataclysmic variables; one solution is a low-viscosity cold disk (in which hydrogen is neutral) and another solution is a high-viscosity hot one (in which hydrogen is ionized) and these two solutions are connected by a thermally unstable solution (where hydrogen is partially ionized). The disk instability model then explains the dwarf nova outbursts as the thermal relaxation oscillations between a hot fully ionized state and a cold unionized state in the bi-stable disk.

The thermal relaxation oscillation is most easily understood locally in terms of the so-called S-shaped equilibrium curve. At each radius in the disk there exist two critical surface densities, Σ_{max}, above which no cold equilibrium state exists and Σ_{min}, below which no hot state exists. At the end of an outburst, all parts of the disk return back to a cold state and the surface density lies between Σ_{max} and Σ_{min}. Addition of matter from the secondary or the viscous evolution within the disk makes the surface density to increase at some part of the disk and as time passes, the surface density will ultimately reach the local critical density Σ_{max} at some point in the disk and it ignites a heating transition to hot state. The heating front propagates both inward and outward, turning all of the disk matter into hot state, causing greatly increased accretion to the central white dwarf, corresponding to an outburst.

Greatly increased accretion now depletes mass in the disk and the disk mass reaches another turning point of the hot branch. Then a downward transition to a low-viscosity cold state ensues, corresponding to a return to quiescence. The cooling transition always occurs at the outer edge of the disk and the cooling front propagates inwards, extinguishing the outburst. The accretion disk thus jumps between these two states discontinuously showing a limit cycle oscillation. This model is now called "thermal instability model" or "thermal limit-cycle model." Several different groups simulated outbursts of dwarf novae based on this model, and they demonstrated this thermal instability model could successfully reproduce outburst light-curves of dwarf novae. It is now well accepted that the basic features of dwarf-nova outbursts are explained by the disk instability (see a review by Cannizzo 1993 for a more detailed account of the thermal instability model).

In late 1980s, another intrinsic instability within the accretion disk, which is now called the "tidal instability" (or tidal eccentric instability), was discovered by Whitehurst (1988). The tidal instability was further studied and confirmed by Hirose & Osaki (1990) and by Lubow (1991). In this instability, the accretion disk is deformed to an eccentric form and the apsidal line of eccentric pattern slowly advances in the prograde direction (i.e., a precession of eccentric pattern) in the inertial frame of reference. It is now well accepted that the "superhump phenomenon" observed during the "superoutburst" of SU UMa-type dwarf novae can be explained by the tidal instability in the accretion disk. The cause of the tidal instability is now well understood as due to a resonance phenomenon between the fluid flow in the accretion disk and the orbiting secondary star via tidal interaction. That is the 3:1 resonance between the flow in the disk and the secondary star and the resonance condition is only realized when the accretion disk is large enough to accommodate the 3:1 resonance radius, which is around 0.47 in units of the binary separation. This condition is met only if the mass of the secondary star is sufficiently low as compared with the mass of the primary star, that is, the mass ratio, $q = M_2/M_1$, of the secondary star to the primary star, is less than about 0.25. The latter condition is in turn satisfied exclusively in cataclysmic variable stars with extremely short orbital periods (i.e., orbital periods less than around 2 hours).

4. Supercycle of SU UMa Stars

The SU UMa-type dwarf novae are one sub-class of dwarf novae, which are characterized by two distinct types of outbursts: more frequent normal outbursts lasting for a few days, and less frequent long and large-amplitude outburst called "superoutbursts" which usually last for about two weeks. Curiously no intermediate types of outbursts are observed. Corresponding to these two types of outbursts, two different outburst cycles are defined: a short normal cycle and a supercycle. A supercycle is defined as a cycle from one superoutburst to the next superoutburst. For instance, in the case of a prototype SU UMa star, VW Hyi, the short normal cycle length is about 30 days while the supercycle length is about 180 days so that several normal outbursts are sandwiched between two consecutive superoutbursts. In general, the supercycle clock is more regular than that of normal cycle, whose timing could differ by a factor two.

The most enigmatic feature of superoutbursts is an occurrence of periodic light humps with amplitude of 0.2–0.3 mag called "superhumps" which repeat with a period very near to the orbital period of the binary but always longer than that by a few percent. Superhumps are observed only during superoutbursts of SU UMa stars and they are not observed either during quiescence or during a normal outburst and thus they are called "superhumps." It is now generally accepted that the superhump phenomenon is caused by the precessing eccentric disk in SU UMa stars. In this picture, when an SU UMa star goes into a superoutburst, the tidal instability sets in, which deforms the accretion disk into an eccentric form. The eccentric disk thus formed will then slowly precesses and the orbiting secondary star gives rise to a periodic tidal stressing to the eccentric disk with a synodic period between the slowly precessing disk and the orbiting secondary star. This produces a periodic light hump, explaining the superhump phenomenon.

The most intriguing question is why a binary system produces two different outbursts with two different time scales. For this, the most promising scenario will be the thermal-tidal instability model (called in short "TTI-model") proposed by the present author (Osaki 1989). This model is essentially based on the disk instability model and it uses two intrinsic instabilities of accretion disks in cataclysmic variable stars, that is, the thermal instability and the tidal instability. Both of the normal outburst and the superoutburst are understood in this model as outbursts caused by the thermal instability, but the normal one is not accompanied by any tidal instability while the superoutburst is that accompanied by a tidal instability.

In the early phase of a supercycle, the accretion disk is supposed to be compact. As the accretion disk undergoes normal dwarf nova outburst, mass accreted onto the central white dwarf during each outburst is less than mass transferred during quiescence so that mass and angular momentum of the disk are gradually built up. The disk's outer radius gradually expands and finally a normal outburst brings the disk to expand beyond the 3:1 resonance radius, triggering the tidal eccentric instability in the accretion disk. Once the disk is deformed into an eccentric form, the tidal stress acting on the disk is greatly increased, producing a periodic light hump (i.e., the superhump). The enhanced tidal stress and tidal torques remove angular momentum from the disk, producing enhanced accretion and longer duration of outburst, i.e., "superoutburst." After the end of the superoutburst, the disk is now compact because of greatly enhanced tidal removal of angular momentum from the disk, thus coming back to the starting point of the supercycle. The same process is then repeated. The supercycle is understood in this model as a tidal relaxation oscillation with a variable disk radius. Figure 1 shows a numerical simulation for a full supercycle of an SU UMa star based on the thermal-tidal instability model.

5. Disk Viscosity and Recurrence Time

The outburst light curves of dwarf novae are simulated by various workers based on the disk instability model, by using the Shakura–Sunyaev α prescription (Shakura & Sunyaev 1973) for the turbulent viscosity. Here the parameter α is non-dimensional quantity describing the strength of turbulent viscosity with the kinematic viscosity, ν, given by $\nu = \alpha c_s H$ where c_s and H denote the sound velocity and the half thickness of the disk, respectively. Although the thermal limit cycle instability works locally under a constant viscosity parameter α, it has turned out that we need to choose different α values for the hot and cold states in order to get a clear-cut outburst-quiescence behavior of dwarf nova outbursts. In the disk instability model, the viscosity parameter in the hot state is constrained to be $\alpha_{hot} \simeq 0.2$ in order to reproduce the decay phase of outburst light curves while that in the cold state is constrained to be $\alpha_{cold} \simeq 0.02 \sim 0.04$ in order to get recurrence times of typical dwarf novae (see e.g., Cannizzo 1993).

For a given binary system, the recurrence time of outburst depends most strongly on the viscosity parameter in the cold state α_{cold} and the mass-supply rate from the secondary to the disk, $\dot{M}$. In general, the theoretical recurrence time increases with a decrease in α_{cold}. The dependence of recurrence time on these two parameters was discussed by Cannizzo, Shafter, & Wheeler (1988). For a given binary system and given viscosity parameters, recurrence time of outbursts is determined by mass-supply rate $\dot{M}$. For systems with shorter re-

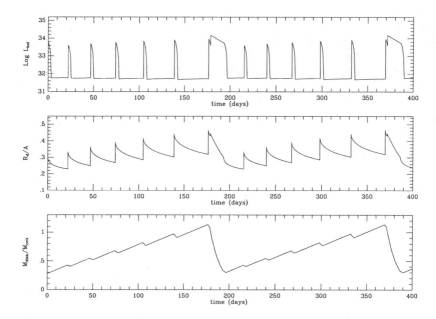

Figure 1.   A numerical simulation of supercycle of an SU UMa-type
star based on the thermal-tidal instability model (Osaki 1989). From
top to *bottom*, the bolometric light curve (*top*), the disk radius, R_d
in units of the binary separation, A (*middle*), and the total disk mass
normalized by the critical mass above which the disk is tidally unstable
(*bottom*).

currence time, the recurrence time is approximately given to be inversely proportional to the square of the mass-supply rate, i.e., $t_{rec} \propto \dot{M}^{-2}$ (see e.g., Osaki
1996). The shortest recurrence time observed in dwarf novae is about 4 days for
a star V1159 Ori. Since a few days are needed for the heating and cooling waves
to traverse in the accretion disks of dwarf novae, a recurrence time of about 4
days may very likely be the shortest possible value.

On the other hand, for long recurrence-time systems, mass-supply rate from
the secondary star must be low. However, in order to explain an extremely long
recurrence time of WZ Sge of 30 years, it is not enough to decrease the mass-
transfer rate but we must choose an extremely low viscosity parameter as low
as $\alpha_{cold} \sim 0.001$ (Smak 1993; Osaki 1995; Howell, Szkody, & Cannizzo 1995).

Let us now discuss on the origin of turbulent viscosity in accretion disks of
dwarf novae. The origin of turbulent viscosity in the accretion disk had remained
unknown for a long time but the situation has recently changed very much as a
very promising mechanism has been found, and it is MHD turbulence produced
by the magneto-rotational instability or the Balbus–Hawley instability (Balbus
& Hawley 1991, 1998). Gammie & Menou (1998) suggested that the MHD turbulence could be decayed in quiescent cold disk because of poor conductivity of
the cold disk matter. Meyer & Meyer-Hofmeister (1999) suggested the following

picture for viscosities in the dwarf-nova disk. The magneto-rotational instability of small-scale magnetic fields in the hot disk produces the high viscosity in outbursts. In quiescence, this is not possible (Gammie & Menou 1998) but the magnetic fields reaching over from the secondary star allows a weaker MHD turbulence and cause the low viscosity in the cold disk. The extreme low viscosity in WZ Sge stars may then be due to the absence of magnetic fields in the case of brown-dwarf secondaries as the brown dwarf nature for the secondary star is suspected in WZ Sge stars. If no magnetic fields are expected in the cold disk of WZ Sge stars, an extremely weak viscosity may be produced by some kind of hydrodynamic instabilities (Papaloizou & Pringle 1985) or tidal spiral shocks (Sawada, Matsuda, & Hachisu 1986).

6. Superhump Periods

Another interesting subject for strict timing-observations in dwarf novae is the superhump periodicity. The observed superhump periods, P_{sh}, in SU UMa stars are in general longer by a few percent that the orbital periods. It has been known from observations that there is a good correlation between the orbital period P_{orb} of the binary and the superhump period excess, i.e., $\epsilon \equiv \Delta P/P_{orb} = (P_{sh} - P_{orb})/P_{orb}$.

The superhump periodicity is explained in the tidal instability model as the synodic period between the slowly and progradely precessing eccentric disk and the orbiting secondary star, that is, the superhump period is given by

$$\frac{1}{P_{sh}} = \frac{1}{P_{orb}} - \frac{1}{P_{prec}},$$

where P_{prec} is the precession period of the eccentric disk. Thus the beat period P_{beat} between the superhump period and the binary orbital period is identified as the precession period of the eccentric disk.

Several physical factors affect the precession rate of the eccentric disk but two of the most important are first, that of the tidal perturbation of the secondary's gravitational fields on the accretion disk, contributing to prograde precession, and second, that of the pressure force in the accretion disk, contributing to retrograde precession. Since observed precession is prograde, the former contribution is larger than the latter. If the former one is dominant, the precession rate of an eccentric disk characterized by ϵ should strongly be correlated with the binary mass ratio $q = M_2/M_1$ as discussed by Patterson (1998), who used the equation

$$\epsilon \simeq 0.23 \frac{q}{\sqrt{1+q}}.$$

If we use this relation, we can estimate the binary mass ratio for cataclysmic variable stars exhibiting the superhump phenomenon. WZ Sge itself is one of such systems with extremely low ϵ and thus the secondary star is suspected to be a brown dwarf with the secondary mass as low as $M_2 \simeq 0.03\ M_\odot$ from the observed superhump period excess to the orbital period.

References

Balbus, S. A., & Hawley, J. F. 1991, ApJ, 376, 214

Balbus, S. A., & Hawley, J. F. 1998, Rev. Mod. Phys., 70, 1

Cannizzo, J. K. 1993, in Accretion Disks in Compact Stellar Systems, ed. J. C. Wheeler (Singapore: World Scientific), 6

Cannizzo, J. K., Shafter, A. W., & Wheeler, J. C. 1988, ApJ, 333, 227

Gammie, C. F., & Menou, K. 1998, ApJ, 492, L75

Hirose, M., & Osaki, Y. 1990, PASJ, 42, 135

Hoshi, R. 1979, Prog. Theor. Phys., 61, 1307

Howell, S. B., Szkody, P., & Cannizzo, J. K. 1995, ApJ 439, 337

Lubow, S. H. 1991, ApJ, 381, 259

Meyer, F., & Meyer-Hofmeister, E. 1981, A&A, 104, L10

Meyer, F., & Meyer-Hofmeister, E. 1999, A&A, 341, L23

Osaki, Y. 1974, PASJ, 26, 429

Osaki, Y. 1989, PASJ, 41, 1005

Osaki, Y. 1995, PASJ, 47, 47

Osaki, Y. 1996, PASP, 108, 39

Papaloizou, J. C. B., & Pringle, J. E. 1985, MNRAS, 217, 387

Patterson, J. 1998, PASP, 110, 1132

Sawada, K., Matsuda, T., & Hachisu, I. 1986, MNRAS, 219, 75

Shakura, N. I., & Sunyaev, R. A. 1973, A&A, 24, 337

Smak, J. 1993, Acta Astronomica, 43, 101

Warner, B. 1995, Cataclysmic Variable Stars (Cambridge: Cambridge University Press)

Whitehurst, R. 1988, MNRAS, 232, 35

Astrophysical Ages and Time Scales
ASP Conference Series, Vol. 245, 2001
T. von Hippel, C. Simpson, N. Manset

The Most Stable Optical Clocks Known

Anjum Mukadam

Department of Astronomy, University of Texas at Austin, Austin, TX 78712-1083, USA

S. O. Kepler

Instituto de Física, Universidade Federal do Rio Grande do Sul, 91501-970 Porto Alegre, RS, Brazil

D. E. Winget

Department of Astronomy, University of Texas at Austin, Austin, TX 78712-1083, USA

Abstract. The hotter ZZ Ceti (HDAV) stars pulsate with a few, extraordinarily stable low-amplitude modes. We expect their pulsation periods to change on the evolutionary, or cooling, time scale. Constraining the rate of period change with time, $\dot{P}$, is equivalent to constraining the drift rate of these clocks. ZZ Ceti stars like G117–B15A and R548 have a measured $\dot{P}$ of the order of 10^{-15} s/s that makes them more stable than atomic clocks and most pulsars.

1. Introduction

Nature has provided white dwarfs (WDs) in three distinct pulsational instability strips along their cooling track. The high temperature instability strip consists of the PNN (Planetary Nebula Nuclei) and the DO (strong He II lines) pre-white dwarf stars at an effective temperature around 80 000 K and hotter. The DB WDs (He I lines; helium atmosphere) are observed to pulsate around 25 000 K. DA WDs (Balmer lines; hydrogen atmosphere) pulsate in an instability strip found between 11 000 K and 12 000 K (Winget 1998). The periods are typically 100–1000 s, consistent with non-radial g-mode pulsations. WDs have high surface gravities ($\log g \sim 8$), so non-radial g-mode pulsations require less energy to reach observable amplitudes because they involve motion mostly along equipotential surfaces. Pulsating WDs are not defective or special in any way; pulsations are an evolutionary feature (McGraw & Robinson 1976; Lacombe & Fontaine 1980).

Cooling of the WD increases the pulsation period, while residual gravitational contraction decreases the period. For high overtone g-modes, we have (Winget, Hansen, & van Horn 1983),

$$\frac{\dot{P}}{P} \approx -\frac{1}{2}\frac{\dot{T}}{T} + \frac{\dot{R}}{R},$$

(1)

where P is the pulsation period, T is the temperature of the driving region and R is the stellar radius. Simple Mestel cooling theory tells us that hot and consequently, more luminous WDs cool faster than the cool WDs (Kepler & Bradley 1995). Hence, we should expect that of all the classes of pulsating WDs, the DAVs (DA Variables) are bound to show the slowest increase in period with time as they are the coolest and therefore have the longest cooling time scale. Furthermore, gravitational contraction is important only for the hot DOV (DO Variable) and PNNV (PNN Variable) stars. Kepler et al. (2000) conclude that $\dot{P}$ is dictated by the rate of cooling alone for the DAV stars.

The DAV WDs, also known as the ZZ Ceti stars, exhibit distinct pulsation trends with temperature. The hot DAV (HDAV) stars present only a few modes with pulsation periods around a few 100 s and low amplitudes. The cooler DAV stars (CDAV) stars show large amplitude and longer period pulsations. They exhibit many pulsation modes, most of which are unstable (Kleinman et al. 1998). We will be concerned only with the stable hot DAV stars in the context of this paper, as they are good time-keepers.

2. Highly Reliable Clocks

Bradley (1996) gives a theoretical 3σ upper limit of $\dot{P} \leq 6.5 \times 10^{-15}$ s/s, corresponding to a time scale $(P/\dot{P})$ larger than 1.2 Gyr. Measuring the change in pulsation period with time, i.e., $\dot{P}$, is thus equivalent to measuring the drift rate of these clocks. HDAV stars like R548 and G117–B15A exhibit extreme frequency stability. G117–B15A is currently the most stable optical clock known with a $\dot{P} \leq (1.9\pm1.4)\times10^{-15}$ s/s (Kepler et al. 2000). R548, also called ZZ Ceti, is the second most stable optical clock with a $\dot{P} \leq (2.4 \pm 2.1) \times 10^{-15}$ s/s (A. Mukadam et al., in preparation). The above constraints have been placed after observing each of these WDs for 3 decades. The uncertainties in $\dot{P}$ are currently of the same order as the changes themselves for both R548 and G117–B15A, so our results are only upper limits and not yet real measurements.

Based on our observations of R548 and G117–B15A, the time of arrival of a pulse maximum can be predicted five years in the future to an accuracy of 2–3 seconds. They are more stable than atomic clocks and most pulsars. The millisecond pulsar PSR B1885+09 is, however, more stable than these amazing WDs. It has a period of 5.36 ms and a measured $\dot{P} = 1.78363 \times 10^{-20}$ s/s (Kaspi, Taylor, & Ryba 1994). The relevant time scale is $P/\dot{P} \sim 9.5$ Gyr. The evolutionary time scales that we computed for G117–B15A and R548 are longer than 3.6 and 2.8 Gyr respectively. PSR B1885+09 is more stable than G117–B15A by a factor of 2.6. We note that these WDs are stable enough to be used in calibrating GPS software.

We assume that the atomic clock is perfect and determine $\dot{P}$ on that basis. Ed Nather pointed out to us that since we measure a stability level of $\sim 10^{-15}$ s/s for pulsating WDs, we can safely conclude that the uncertainty in the atomic clocks is smaller than 10^{-14} s/s, claimed by NIST, by at least a factor of 2.

3. Motion of a Clock

Motion of a clock with uniform velocity towards or away from us produces a Doppler shift in the period, but does not affect the rate of change of period ($\dot{P}$). If the motion were accelerated, however, then there would be a resultant $\dot{P}$ associated with it. There are two possible ways to have such an accelerated motion: an orbital companion or a significant proper motion.

3.1. Orbital Companion

If G117–B15A or R548 had an unseen orbital companion, then its motion about the center of mass of the system would manifest itself as a periodic variation of the arrival time of pulse maxima. Such a variation could, in principle, be distinguishable from the parabolic signature due to cooling of the WD. The period would be the orbital period and the amplitude would allow the estimation of the mass and/or distance of the orbital companion. The variable period resulting from the orbital motion of the clock, would cause a $\dot{P}_{\mathrm{orb}}$ (Kepler et al. 1991), given by

$$\dot{P}_{\mathrm{orb}} = \frac{P}{c} \frac{Gm}{a^2} \sin(i) \tag{2}$$

where P is the pulsation period, m is the mass of the orbital companion, a is the separation between the components and i is the angle of inclination. Any uniform motion along the line of sight would just be interpreted as a correction in pulsation period, ΔP. The amplitude, A, is set by orbital light travel time and can be expressed in terms of the orbital radius $r_\star$ for the DAV

$$(2A)c = 2r_\star \sin(i). \tag{3}$$

If the plane of the orbit is perpendicular to the line of sight, then we cannot detect the companion. Using the relation for center of mass, we can set a limit on the mass m of the orbital companion modulated by a factor of $\sin(i)$.

Detection of an orbital companion around a pulsating WD depends on three criteria, the mass of the companion ($\dot{P}_{\mathrm{orb}} \propto m$), its distance from the WD ($\dot{P}_{\mathrm{orb}} \propto 1/a^2$) and the orbital period T. It is easy to understand the first 2 criteria. If the companion is not massive or far away from the WD, then its gravitational influence may not be detectable. The third criterion is more subtle. When we observe pulsating WDs, we do not directly measure $\dot{P}$. We infer a $\dot{P}$ by comparing our measurements of the phases to what we would expect for a constant period, i.e., using the $O - C$ technique. The phase difference, $O - C$, should increase due to an orbital companion for half an orbital period, after which it must start decreasing. At the end of an orbital period, the $O - C$ must reflect a change from cooling alone. So, the amplitude in the $O - C$ diagram, depends not only on the magnitude of $\dot{P}_{\mathrm{orb}}$, but also on the time for which the phase change was allowed to accumulate, i.e., $T/2$. With this technique, it is easier to detect companions with large orbital periods. However, such detections necessitate long term observations to cover a significant fraction of the orbit and are bound to be cumbersome. The phase changes are cumulative, and so in the limit of slow changes (long orbital periods), our limits improve with the square of time. Nearby planets with shorter orbital periods maybe detected by decreasing the uncertainties on individual phase measurements.

If the WD had an Earth revolving around it at a distance of 1 AU, we would expect a $\dot{P}_{\mathrm{orb}} = 12.5 \times 10^{-15}$ s/s, but the amplitude in the $O - C$ diagram from an Earth-like companion would only be a few ms. Detection of Earth would thus require greater timing accuracy than current observations of R548 and G117–B15A, even though P_{orb} is 4 times larger than that due to cooling. Planets the size of Jupiter are considerably easier to detect than Earth-like planets. For a planet like Jupiter ($M = 318\ M_{\oplus}$) at 5.2 AU, we compute $\dot{P}_{\mathrm{orb}} = 1.5 \times 10^{-13}$ s/s with an amplitude of 3 to 4 s. Using our current limit of $\dot{P}_{\mathrm{orb}} = 3 \times 10^{-15}$s/s, we are able to detect planetary companions of masses $M \geq 56.4 M_{\oplus}$ at distances $a \leq 15.3$ AU

3.2.   Proper Motion

Pulsating WDs also have a potential non-evolutionary secular period change due to proper motion. The size of this effect on $\dot{P}$ was evaluated to be of the order of 10^{-15} s/s by Pajdosz (1995). This effect is insignificant for the DOVs and the PNNVs. However, it is of the same order as the $\dot{P}$ measured for the hot DAVs.

As we noted before, any motion (with a uniform velocity) of a DAV along the line of sight will manifest itself as a correction in the period estimate for the pulsations and will not affect the $\dot{P}$. However, motion perpendicular to the line of sight is equivalent to a centripetal acceleration, with Earth as the reference. This causes a $\dot{P}_{\mathrm{pm}}$. Pajdosz (1995) has shown this to be

$$\dot{P}_{\mathrm{pm}} = 2.43 \times 10^{-18} P[\mathrm{s}] (\mu[''/\mathrm{yr}])^2 (\pi[''])^{-1}, \qquad (4)$$

where μ is the proper motion and π is the parallax. $\dot{P}_{\mathrm{pm}}$ is always positive and should be subtracted out from $\dot{P}_{\mathrm{obs}}$.

For the 213.132 s period in R548, the term $\dot{P}_{\mathrm{pm}}$ is of the order of $(2.2 \pm 0.4) \times 10^{-15}$ s/s (Pajdosz 1995). While for G117–B15A, the magnitude of the correction is estimated to be $(0.92 \pm 0.5) \times 10^{-15}$s/s (Kepler et al. 2000), smaller than in the case of R548, because G117–B15A is farther away.

4.   Constraining the Evolution of a White Dwarf

Constraining the rate of cooling for a WD has important implications in stellar evolution theory. These limits improve our calibration of the cooling curve and provide a way to measure the mean core composition. They may also help in deciphering other phenomena like crystallization and phase separation in massive pulsators, as well as neutrino production for the hottest pulsators (DOVs, PNNVs).

4.1.   Calibration of the Cooling Curve

WDs are very hot initially, just after the ejection and expansion of the planetary nebulae, and cool rapidly. The cooling rate decreases as their temperature drops, allowing even the oldest WDs to remain visible. Detecting WDs with faint absolute luminosities and average masses is then synonymous with detecting the oldest WDs. Also, the exponential decrease in the cooling rate causes a pile up of WDs at lower temperatures. The volume density of WDs per unit absolute bolometric magnitude as a function of their luminosity, i.e., the luminosity function (LF), is expected to show more and more WDs in lower temperature bins. However, the best current observational determinations of the WD LF for the Galactic disk indicate a turn-down in the space density of low luminosity stars (Liebert, Dahn, & Monet 1988; Leggett, Ruiz, & Bergeron 1998; Oswalt et al. 1996). We presume this to be a signature of the finite age of the disk. The luminosity where this turn-down occurs, in conjunction with theoretical cooling calculations, allows us to estimate the age of the Galactic disk (Winget et al. 1987). The determination of the halo LF (Oppenheimer et al. 2001) would enable us to make the same sort of estimate for the halo. This process is referred to as WD Cosmochronology. It involves observational and theoretical uncertainties. Some observational uncertainties come from statistical difficulties in locating the turn-down of the LF accurately. Cool WDs are intrinsically faint and relatively few are currently known. In addition, uncertainties in their bolometric corrections and trigonometric parallaxes prove to be the chief observational errors at present (Méndez & Ruiz 2001). The location of the turn-down is not determined solely by the few WDs detected at low temperatures, but because none are detected at lower temperatures. Most of the theoretical uncertainty in the age estimation comes from uncertainties in the constitutive physics and the basic parameters used in the cooling. These include compositional stratification, crystallization, and associated release of latent heat, as well as phase separation. We can calibrate the cooling curve by measuring the rate of cooling for WDs at different temperatures. We have already established that $\dot{P}$ is related to cooling of the star and hence measuring a $\dot{P}$ is equivalent to measuring the rate of cooling.

The DOV star PG 1159-035 revealed a rate of period change for the 516 s mode (Costa, Kepler, & Winget 1999) to be $(13.0 \pm 2.6) \times 10^{-11}$ s/s. At the cool end, the main periodicity of the DAV G117–B15A shows a $\dot{P} \leq (2.3 \pm 1.4) \times 10^{-15}$ s/s (Kepler et al. 2000). We measured $\dot{P} \leq (2.4 \pm 2.1) \times 10^{-15}$ s/s for R548. These values allow us to begin calibration of the cooling curves; measuring a $\dot{P}$ for a DBV would further improve this calibration.

4.2. Core Composition

The rate of cooling of a WD depends on the core composition and the stellar mass. The heavier the core, the faster the star cools. By estimating the rate of cooling for the variable WDs, and comparing it to theoretical evolutionary models, we effectively measure the mean atomic weight of the core. Results for the DAVs R548 and G117–B15A, two white dwarfs with masses around 0.6 $M_{\odot}$, are consistent with a C/O core, as predicted by evolutionary models, and eliminate models with substantially heavier cores, as they would produce a faster rate of period change than observed.

4.3. Crystallization and Phase Separation

As a WD cools, the thermal energy of the ions becomes much smaller than the energy of the Coulomb interaction between neighboring ions. At this temperature, the interior of the WD begins to crystallize. For a 0.6 $M_{\odot}$ Wood (1992) model, the onset of crystallization is at $T_{\rm eff} = 6000$ K for a C core, and at $T_{\rm eff} = 7200$ K for an O core. These temperatures are much cooler than the DAV instability strip from 11 000–12 000 K. So, ordinarily, one would not be able to study effects such as crystallization and phase separation using asteroseismology. However, massive stars like BPM 37093 ($M \simeq 1\ M_\odot$) should be partly crystallized at 11 000–12 000 K, making them crystallized pulsators (Winget et al. 1997; Montgomery & Winget 1999; Nitta et al. 1999). Crystallization affects $\dot{P}$ in the following ways; it releases latent heat and delays the cooling and, secondly, the outward moving crystallization front causes the periods to increase.

Nonradial g-modes get excluded from the crystallized region, as the crystallization front represents a hard boundary. In a 99% crystallized star, pulsation periods can increase by as much as 30% and more, depending on the mode. Though the fractional change in period is significant, it takes place on an evolutionary time scale as well, and acts to increase $\dot{P}$. Montgomery (1998) calculated $\dot{P}$ to be $\sim 7 \times 10^{-15}$ s/s for periods less than 1000 s and $\sim 5 \times 10^{-15}$ s/s for periods between 500–700 s.

As the star crystallizes, there should be an enhancement of oxygen content in the crystallized region, while the overlying fluid layer will be carbon enhanced, provided the WD interior was initially a mixture of carbon and oxygen (Stevenson 1980; Ichimaru, Iyetomi, & Ogata 1988; Segretain & Chabrier 1993; Montgomery et al. 1999). This phase separation increases the magnitude of the binding energy of the star and along with the release of latent heat, slows cooling. This causes a relative decrease in $\dot{P}$. Thus, $\dot{P}$ may not be a useful diagnostic of the effects of crystallization by itself. If the crystallized mass fraction is independently known, for example, from the period distribution, then $\dot{P}$ will be a very powerful test of the extent and energetics of phase separation.

4.4. Neutrino Flux

The neutrino luminosity of a DOV can be a more efficient means of cooling than the photon luminosity. The ratio of neutrino to photon luminosity can vary from 0.1 to 3, depending on the effective temperature and mass of the star (O'Brien et al. 1998). For the DOV PG 0122+200, O'Brien et al. (1998) quote $\dot{P} < 6 \times 10^{-10}$ s/s. The $\dot{P}$ measurement for PG 1159-035 by Costa, Kepler, & Winget (1999), on the other hand, revealed a value of $(13.0 \pm 2.6) \times 10^{-11}$ s/s. O'Brien et al. (1998) estimate that the ratio of neutrino to photon luminosity for PG 0122+200 is about 2.5 to 2.6, while for PG 1159-035, it is about 0.1. This implies that PG 0122+200 is a good candidate for obtaining the neutrino flux by a measurement of its $\dot{P}$ value. As O'Brien and collaborators point out, such a measurement could prove to be an important test for neutrino rates.

5.   Implications Related to Asteroseismology

5.1.   Trapped Modes

In WDs, compositional stratification occurs due to prior shell burning stages and gravitational settling. Hydrogen, if present, floats on the surface. In such WDs, there is a mechanical resonance effect between the local g-mode oscillation wavelength and the thickness of one of the compositional layers. This mechanical resonance typically serves as a stabilizing mechanism. However, if we observe a transition between trapped modes, $\dot{P}$ may be temporarily quite large (see avoided crossings below). Most often trapped modes are more stable than un-trapped modes. Modes with nodes in the vicinity of the H/He interface tend to be reflected and are consequently trapped in the outer H layer. Such modes are energetically favorable, as the amplitudes of their eigenfunctions near and below the H/He interface are smaller. Mechanical damping is more prominent in the core than in the envelope and therefore modes trapped in the envelope can have kinetic oscillation energies lower by six orders of magnitude, as compared to the adjacent non-trapped modes (Winget, van Horn, & Hansen 1981).

If the 213 s doublet in R548 and the 215 s mode in G117–B15A consist of trapped modes, then indeed we could be measuring the stability of the trapping mechanism and not the cooling. Theoretical calculations indicate that trapped modes would have a $\dot{P}$ smaller than due to cooling by a factor ≤ 2 (Bradley 1993). We do not have true measurements for the $\dot{P}$s, only upper limits, so we cannot tell whether or not these modes are trapped. If they are trapped, then they will prove to be more stable than current constraints by a factor of 2.

5.2. Avoided Crossings

Different pulsation modes sample different regions of the star. Hence, in general, the rates of period change need not be the same. Consider one trapped and one un-trapped mode, so one is changing faster than the other. If the two modes have frequencies very close to each other, then it may be possible for them to interchange their natures. Such an interaction is termed as an avoided crossing (Aizenman, Smeyers, & Weigert 1977; Christensen-Dalsgaard 1981). Trapped modes are more stable than un-trapped modes in general, but they do show an instability in the region of an avoided crossing. In other words, if you were monitoring the $\dot{P}$ for any of these modes, you would observe a rapid change during the crossing, i.e., the $\ddot{P}$ term would be important. The 274 s doublet in R548 could be undergoing such an avoided crossing, and we plan to investigate this possibility more thoroughly.

6. A Summary of the Results

Measuring a rate of change of period with time for hot DAVs like R548 and G117–B15A has led us to conclude that they are the most stable optical clocks known. This has been a crucial aid in understanding white dwarf evolution. We can now measure the core composition of a pulsating WD by comparing the $\dot{P}$ obtained to theoretical evolutionary models. We may learn about crystallization and phase separation by measuring a $\dot{P}$ for massive crystallized pulsators, with a baseline of at least 10 years. We can hope to detect or constrain the presence of orbital companions like planets or brown dwarfs, that DAVs like R548 or G117–B15A might have. We also hope to learn about asteroseismological effects like mode trapping and avoided crossings from $\dot{P}$ values, as this is the only current technique to study them observationally.

References

Aizenman, M., Smeyers, P., & Weigert, A. 1977, A&A, 58, 41
Bradley, P. A. 1993, Ph.D. Thesis, 4
Bradley, P. A. 1996, ApJ, 468, 350
Christensen-Dalsgaard, J. 1981, MNRAS, 194, 229
Costa, J. E. S., Kepler, S. O., & Winget, D. E. 1999, ApJ, 522, 973
Ichimaru, S., Iyetomi, H., & Ogata, S. 1988, ApJ, 334, L17
Kaspi, V. M., Taylor, J. H., & Ryba, M. F. 1994, ApJ, 428, 713
Kepler, S. O., & Bradley, P. A. 1995, Baltic Astronomy, 4, 166
Kepler, S. O., Mukadam, A., Winget, D. E., Nather, R. E., Metcalfe, T. S.,
    Reed, M. D., Kawaler, S. D., & Bradley, P. A. 2000, ApJ, 534, L185
Kepler, S. O., et al. 1991, ApJ, 378, L45
Kleinman, S. J., et al. 1998, ApJ, 495, 424
Lacombe, P., & Fontaine, G. 1980, JRASC, 74, 147
Leggett, S. K., Ruiz, M. T., & Bergeron, P. 1998, ApJ, 497, 294
Liebert, J., Dahn, C. C., & Monet, D. G. 1988, ApJ, 332, 891
Méndez, R., & Ruiz, M. 2001, ApJ, 547, 252
McGraw, J. T., & Robinson, E. L. 1976, ApJ, 205, L155
Montgomery, M. H. 1998, Ph.D. Thesis, 21
Montgomery, M. H., Klumpe, E. W., Winget, D. E., Wood, M. A. 1999, ApJ,
    525, 482
Montgomery, M. H., & Winget, D. E. 1999, ApJ, 526, 976
Nitta, A., et al. 1999, in ASP Conf. Ser. 169, Eleventh European Workshop on
    White Dwarfs, ed. S.-E. Solheim & E. G. Meistas (San Francisco: ASP),
    144
O'Brien, M. S., et al. 1998, ApJ, 495, 458
Oppenheimer, B. R., Hambly, N. C., Digby, A. P., Hodgkin, S. T., & Saumon,
    D. 2001, Science, 292, 698
Oswalt, T. D., Smith, J. A., Wood, M. A., & Hintzen, P. 1996, Nature, 382, 692
Pajdosz, G. 1995, A&A, 295, L17
Segretain, L., & Chabrier, G. 1993, A&A, 271, L13
Stevenson, D. J. 1980, Journal de Physique, 41, 2
Winget, D. E. 1998, Journal of Physics: Condensed Matter, 10, 11247
Winget, D. E., Hansen, C. J., Liebert, J., van Horn, H. M., Fontaine, G., Nather,
    R. E., Kepler, S. O., & Lamb, D. Q. 1987, ApJ, 315, L77
Winget, D. E., Hansen, C. J., & van Horn, H. M. 1983, Nature, 303, 781
Winget, D. E., van Horn, H. M., & Hansen, C. J. 1981, ApJ, 245, L33
Winget, D. E., Kepler, S. O., Kanaan, A., Montgomery, M. H., & Giovannini,
    O. 1997, ApJ, 487, L191
Wood, M. A. 1992, ApJ, 386,539

Acknowledgments. We would like to acknowledge the NSF grant AST-9876730, NASA grant NAG5-9321, and Pronex (CNPq/Brazil) for supporting this work.

Discussion

Tom Andrews: Is it possible to calculate the loss of energy approximately and compare it with your measurements?

Anjum Mukadam: Sure. In 1952, Mestel worked out a simple theoretical estimate for the cooling rate of white dwarfs. There are more state of the art calculations today, (e.g., Bradley, Winget, & Wood 1992, 1992, ApJ, 391, L33) and our constraint on the cooling rate of R548 is consistent with those calculations. Using our best limit for the 213 s doublet, $\dot{P} \leq (2.4 \pm 2.1) \times 10^{-15}$ s/s, we calculate the evolutionary time scale $\mid \frac{P}{\dot{P}} \mid \geq 1.5$ Gyr at the 1σ level. We can place a 3σ limit $\mid \frac{P}{\dot{P}} \mid \geq 0.8$ Gyr. The theoretical 3σ upper limit of $\dot{P} \leq 6.5 \times 10^{-15}$ s/s (Bradley 1996), corresponds to a time scale ≥ 1.0 Gyr. We thus conclude that our limit is consistent with detailed theoretical calculations for cooling of white dwarfs as well as the $\dot{P}$ measurement for G117–B15A.

Astrophysical Ages and Time Scales
ASP Conference Series, Vol. 245, 2001
T. von Hippel, C. Simpson, N. Manset

Time Scale Invariant High–Low Transitions in Neutron Star/Black Hole Systems

H. Inoue

Institute of Space and Astronautical Science, Sagamihara, Japan

Abstract. The Rapid Burster is known to show rapidly repetitive bursts (Type II bursts). An interesting feature of the Type II burst is an approximate proportionality of the burst duration to the time to the next burst. The duty ratio of the burst is almost constant over the large range of the burst fluence. The time sequence from a burst to the following quiescent period can be said to be a time scale invariant high–low (burst phase to quiescent phase) transition. The Galactic superluminal source GRS 1915+105 exhibits a variety of time variation behavior of the X-ray flux in which the flux seems to change between the high-flux and low-flux states. In this high–low transitions, Belloni et al. (1997b) found an approximate proportionality between the duration of the low-flux state and that of the following high-flux state, over a wide range of time scales. This high–low transition can again be said to be time scale invariant, similarly to the case of the Type II bursts. However, an interesting difference between the two time scale invariant high–low transitions is an opposite order of the high- and low-flux states in the time scale invariant sequence. In the case of the Rapid Burster, the high state (burst) is the first. On the other hand, the low state is the first in the case of GRS 1915+105. A limit cycle between an accretion disk in a state of the standard disk and that in a state of the advection dominated accretion flow disk, is discussed to explain the time scale invariant high–low transition, as well as the difference between the neutron star system and the black hole system, qualitatively.

1. The Rapid Burster

The Rapid Burster (RB) is a unique source which exhibits rapidly repetitive X-ray bursts (for a review, see Lewin, van Paradijs, & Taam 1993). The burst activity of this source is known to recur every 6–8 months and continues for three to four weeks while changing its burst pattern (see Guerriero et al. 1999). The rapidly repetitive bursts were considered most probably due to chopped accretion flows caused by instabilities (Lewin et al. 1976), and thus were designated as Type II bursts to discriminate them from Type I bursts, interpreted as thermonuclear flashes on the surface of neutron stars (Hoffman, Marshall, & Lewin 1978). The Rapid Burster shows Type I bursts and hence is believed to be an accreting neutron star.

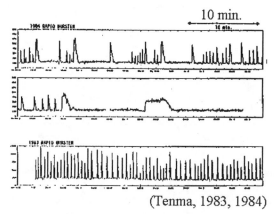

(Tenma, 1983, 1984)

Figure 1.   Various patterns of rapid bursts in the energy range of 1–9 keV. Profiles in the upper two panels and the bottom panel were obtained with Tenma in July 1984, and with Hakucho in August 1983, respectively (Kunieda et al. 1984; Kawai 1985).

Figure 1 shows various patterns of the rapid burst activity. One extreme is a rapid, quasi-periodic repetition of short spiky bursts. The other extreme is a train of long flat-topped bursts lasting as long as 10 minutes. One of the most distinct characteristics of Type II bursts is an approximately linear relation between burst fluence and the time interval to the next burst. Such a linear relation is seen in Figure 2, but the time-averaged luminosity, $E/t_{\rm w}$, seems to be slightly different between the data in 1983 (Kunieda et al. 1984) and those in 1984 (Kawai 1985). Here, E and $t_{\rm w}$ are the burst fluence and the time interval to the next burst (waiting time), respectively. However, if we plot the burst duration, $t_{\rm d}$, defined to be $E/L_{\rm p}$ ($L_{\rm p}$ is the peak luminosity of a burst) against the waiting time, all the data including not only the Tenma data in 1983 and in 1984 but also the EXOSAT data in 1985 (Stella et al. 1988) tend to align on a single linear-like relation over two orders of magnitude of the burst duration. The duty ratio ($t_{\rm d}/t_{\rm w}$) appears to be the fundamental parameter, and the sequence of a burst and the following quiescent period can roughly be said to be time scale invariant. The time scale invariant profile in the decay part of the rapid bursts was also pointed out by Tawara et al. (1985).

2.   Galactic Superluminal Source GRS 1915+105

GRS 1915+105 is a transient X-ray source discovered by Granat (Castro-Tirado, Brandt, & Lund 1992). This source ejected radio components with apparent superluminal motion and is known as the first Galactic superluminal source (Mirabel & Rodriguez 1994).

   The central object in GRS 1915+105 is believed to be a black hole because of similarities to the other Galactic superluminal source, GRO J1655−40 (Zhang

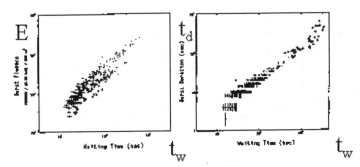

Figure 2.   *Left*: Burst fluence vs. waiting time relation of rapid bursts observed with Hakucho and Tenma in August 1983 (solid diamonds; Kunieda et al. 1984) and with Tenma in July 1984 (crosses; Kawai 1985); *Right*: Burst duration vs. waiting time relation for the same date in the left panel and for the data observed with EXOSAT in August 1985 (solid circles; Stella et al. 1988).

et al. 1994), which has a dynamical mass estimate implying a black hole (Bailyn et al. 1995), and of its X-ray luminosity exceeding the Eddington limit of a neutron star.

GRS 1915+105 is also known to show rapid transitions between high- and low-flux states, which repeat on various time scales. Examples of such high–low transitions are shown in Figure 3. An interesting characteristic of the high–low transitions is a good correlation between the duration of the low state and the following high state (Belloni et al. 1997b). We can see such a trend in the left panel of Figure 3. If we see a train of short bursts as one high state, the similar relation seems to be held even in the right panel.

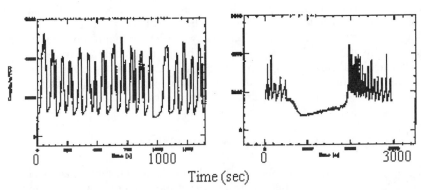

Time (sec)

Figure 3.   Two examples of rapid transitions between the high-flux and low-flux states on various time scales. These are obtained with RXTE (Yamaoka 2001).

3. A Difference Between the Rapid Burster and GRS 1915+105

As shown above, both the Rapid burster and GRS 1915+105 exhibit high–low transitions and each of them has an approximate proportionality between the duration of the high state and that of the low state. Both seem to have a respective clock that controls a sequence of a high–low transition, but its fundamental frequency seems to be reset at the beginning of each high–low transition. However, the sequence is in the opposite sense between the two sources. In the case of the Rapid Burster, the high state (Type II burst) is first. The duration of the high state (Type II burst) has an approximate proportionality to the time to the next burst. On the other hand, in the case of GRS 1915+104, the low state is first. The duration of the low state is proportional to that of the following high state.

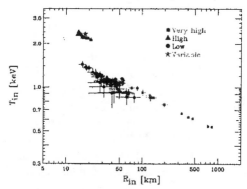

Figure 4. Relation between the highest color temperature, T_{in}, and the innermost radius, R_{in}, of optically thick disks, obtained from spectral fits of a two-component-model to the spectra of GRS 1915+105 observed with RXTE in various occasions (Yamaoka 2001).

4. What Happens in a High–Low Transition?

GRS 1915+105 is characterized by exhibiting various temporal and spectral states. The spectra in various states are commonly approximated by a model comprising a soft component represented by the multi-color disk blackbody (DBB) spectrum and a hard component. The spectrum of the hard component can be determined by analyzing the spectral change on a sub-second time scale on an assumption that the DBB component does not change on such a rapid time scale. Yamaoka (2001) performed such an analysis and found that the hard, rapidly valuable component can be approximated by a broken power-law model. Then, the spectra in the various states observed with RXTE in January 1999–April 2000 were fitted with a model comprising a DBB spectrum and a broken power law spectrum, and the distribution of the best fit R_{in} and T_{in} values are shown in Figure 4. It is seen from this figure that the data points are separated in two distinct groups, the high and the low temperature groups.

This suggests the presence of two states in the accretion disk. The spectral analyses by Yamaoka (2001) further shows that the high and low state of the flux commonly corresponds to the high and low temperature group, respectively. The R_{in} is relatively small in the high state, but it is relatively large in the low state. The similar result was already obtained by Belloni et al. (1997a).

This suggests that a high–low transition of this source could be a transition of a disk between an advection dominated accretion flow (ADAF; Abramowicz et al. 1988; Narayan & Yi 1994) disk and a standard disk (Shakura & Sunyaev 1973). If so, the transition of a disk is well-explained by a so-called limit-cycle in the accretion rate, $\dot{M}$, and the surface density of the disk, Σ (see e.g., Kato, Fukue, & Mineshige 1998).

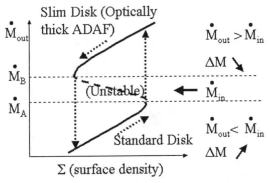

Figure 5.   Schematic diagram of the steady solutions of accretion disks in various disk states on the surface density and the accretion late plane.

When an accretion rate is lower than a critical value, $\dot{M}_A$, there is a steady solution for an accretion disk to be the standard disk. Viscous heating balances with radiative cooling in the disk. The solution on the Σ–$\dot{M}$ plane makes a locus, on which the surface density increases as the accretion rate increases. This locus is known to be stable. However, when the accretion rate reaches a value, $\dot{M}_A$, the radiation pressure becomes dominant over the gas pressure in the disk, and it becomes thermally and secularly unstable.

On the other hand, when an accretion rate is higher than another critical value, $\dot{M}_B$, there is another steady solution for an accretion disk to be the slim disk (optically thick ADAF disk: Abramowicz et al. 1988), and $\dot{M}$ and Σ positively correlates with each other on the locus of the steady solution. In this type of disk, viscous heating balances with advective cooling. The disk on this locus is again known to be stable. However, when the accretion rate is as low as $\dot{M}_B$, the steady solution tends to be a radiation-pressure-dominated standard disk, which is unstable. As a result, when $\dot{M}_A < \dot{M} < \dot{M}_B$, there is no stable solution of the disk.

Let us consider a case in which a matter inflow rate from outside to a place in a disk, $\dot{M}_{in}$, is in the unstable range just mentioned above (see Figure 5). If the surface density is very low, the matter outflow rate to the inside, $\dot{M}_{out}$, should be very low, and hence the surface density, Σ, gradually increases because

$\dot{M}_{in} > \dot{M}_{out}$. However, as Σ reaches the critical value, Σ_A, the disk becomes unstable and would change to another stable solution, a slim disk with the same Σ value. Under the steady solution of the slim disk, $\dot{M}_{out}$ is now larger than $\dot{M}_{in}$, and hence the once accumulated matter now decreases gradually. However, when Σ decreases down to Σ_B, the disk becomes unstable again and will return to the standard disk. Then, the above cycle will be repeated. This is a limit cycle. This can explain not only the high–low transitions, but also the difference between the neutron star system and the black hole system, qualitatively.

The accretion rate through the ADAF disk is expected to be much larger than that through the standard disk in the limit cycle. In the ADAF disk, the gravitational energy released by the matter fall is carried by matter itself and is not efficiently transferred to radiation from the surface of the disk. On the other hand, the gravitational energy-release in a disk is efficiently radiated from the surface of the disk in the standard disk. Hence, the luminosity from the ADAF disk could be smaller than that from the standard disk, even if the accretion rate through the ADAF disk is larger than that through the standard disk. If so, the ADAF-disk and the standard-disk phase should correspond to the low and high state, respectively, in the case of black hole systems. However, in the case of neutron star systems, the gravitational energy carried by matter through the ADAF disk should be finally radiated away from the surface of the neutron star. Thus, the ADAF-disk and the standard-disk phase should correspond to the high and low state, respectively, in this case.

5. Reset of a Clock Controlling a Burst Cycle

As shown earlier, the duration of a Type II burst, t_d has a positive correlation with the time to the next burst, t_w. The phenomenon seems to be such that the duty ratio is kept almost constant, but the fundamental frequency governing one burst-quiescent cycle is reset at the beginning of a burst. This could be explained by introducing a probability process on r, the disk radius, in which the standard disk becomes unstable and changes to the ADAF disk. In the present model, the duration of the burst or the quiescent phase would roughly be determined by r/v, where v is the velocity of matter fall in a disk. Since v is basically proportional to the Kepler velocity and would have a r-dependence as $r^{-1/2}$, the duration would be proportional to $r^{3/2}$. Hence, the longer burst would appear when the instability happens at the larger distance. Then, the matter in the standard disk within r would fall onto the neutron star or the black hole, and it would take a longer time for matter from outside to refill the standard disk.

6. Remaining Questions

The above limit-cycle scenario seems to explain several features of the time scale invariant high–low transitions of both the Rapid Burster and GRS 1915+105. However, there remain some important and interesting questions.

This limit-cycle scenario indicates that the high-flux and the low-flux state should correspond to the standard-disk state and the ADAF-disk state, respectively. However, if we plot relations between R_{in} and L_{DBB} obtained from the

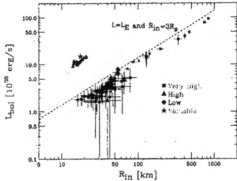

Figure 6. Relation between the bolometric luminosity, L_{DBB}, and the innermost radius, R_{in}, of optically thick disks, obtained from spectral fits of a two-component-model to the spectra of GRS 1915+105 observed with RXTE in various occasions (Yamaoka 2001). A line in which $L_{DBB} = L_{Edd}$ and $R_{in} = 3R_S$ are simultaneously satisfied for any values of M_{BH} is indicated.

spectral fits to the GRS 1915+105 data (Yamaoka 2001), together with a line where $L_{DBB} = L_{Edd}$ and $R_{in} = 3R_S$ are simultaneously satisfied for any values of M_{BH}, the high temperature group is located in a region where $L_{DBB} > L_{Edd}$ and/or $R_{in} < 3R_S$ as seen in Figure 6. This situation in which $L_{DBB} > L_{Edd}$ and/or $R_{in} < 3R_S$ is consistent with a theoretical expectation from the slim disk (Watarai et al. 2000). Hence, from this point of view, the high-flux (high-temperature) state seems to correspond to a case of the slim disk, whereas, R_{in} is sometimes very large in the low-flux (low-temperature) state, as seen in Figure 4. This seems consistent with the above limit-cycle scenario in which an inner part of the disk is considered to change to an ADAF disk through the disk instability at some disk location, in this state. However, the ADAF disk should be optically thin rather than optically thick, otherwise a higher luminosity from the ADAF part should have been observed as in the high state.

The duty ratio between the low and high state in GRS 1915+105 is also inconsistent with the limit-cycle scenario. As seen in Figure 3, the durations of the low and high state in a time scale invariant cycle is about the same. However, the velocity of the matter fall in an ADAF disk is much higher than that in a standard disk, and thus, the duration of the ADAF-disk state should be significantly shorter than that of the standard-disk state. The duty ratio of the rapid bursts is rather consistent with this.

Obviously, the situation is not so simple as the scenario shown in Figure 5, in the case of GRS 1915+105, and it may be necessary to consider a transition between an optically thin ADAF disk to an optically thick ADAF disk in the course of a limit cycle.

Finally, an interesting suggestion from observations of GRS 1915+105 (Mirabel et al. 1998) should be noted. It is that mass ejections seem to take place in association with the high–low transitions. A jet ejection could occur when the disk changes its state. However, it is not clear which the moment of the jet ejection is, a transition from the low state to the high state or the opposite, yet. More detailed observations and considerations of these interesting phenomena are expected in future.

References

Abramowicz, M. A., Czerny, B., Lasota, J. P., & Szuszkiewicz, E. 1988, ApJ, 332, 646

Bailyn, C. D., Orosz, J. A., McClintock, J. E., & Remillard, R. A. 1995, Nature, 378, 157

Belloni, T., Mendez, M., King, A. R., van der Klis, M., & van Paradijs, J. 1997a, ApJ, 479, L145

Belloni, T., Mendez, M., King, A. R., van der Klis, M., & van Paradijs, J. 1997b, ApJ, 488, L109

Castro-Tirado, A. J., Brandt, S., & Lund, S. 1992, IAU Circ. 5590

Guerriero, R., et al. 1999, MNRAS, 307, 179

Hoffman, J. A., Marshall, H. L., & Lewin, W. H. G. 1978, Nature, 271, 630

Kawai, N. 1985, Ph.D. Thesis, University of Tokyo

Kato, S., Fukue, J., & Mineshige, S. 1998, Black-Hole Accretion Disks (Kyoto: Kyoto University Press)

Kunieda, H., et al. 1984, PASJ, 36, 807

Lewin, W. H. G., van Paradijs, J., & Taam, R. 1993, Space Sci. Rev., 62, 223

Lewin, W. H. G., et al. 1976, ApJ, 207, L95

Mirabel, I. F., Dhawan, V., Chaty, S., Rodriguez, L. F., Marti, J., Robinson, C. R., Swank, J., & Geballe, T. 1998, A&A, 330, L9

Mirabel, I. F., & Rodriguez, L. F. 1994, Nature, 371, 46

Narayan, R., & Yi, I. 1994 ApJ, 428, L13

Shakura, N. I., & Sunyaev, R. A. 1973, A&A, 24, 337

Stella, L., Haberl, F., Lewin, W. H. G., Parmar, A. N., van Paradijs, J., & White, N. E. 1988, ApJ, 324, 379

Tawara, Y., Kawai, N., Tanaka, Y., Inoue, H., Kunieda, H., & Ogawara, Y. 1985, Nature, 318, 545

Watarai, K., Fukue, J., Takeuchi, M., & Mineshige, S. 2000, PASJ, 52, 133

Yamaoka, K. 2001, Ph.D. Thesis, University of Tokyo

Zhang, S. N., Wilson, C. A., Harmon, B. A., Fishman, G. J., Wilson, R. B., Paciesas, W. S., Scott, M., & Rubin, B. C. 1994, IAU Circ. 6046

Discussion

Vicky Kaspi: The Rapid Burster is a unique object—very rare—and I just wonder how you can be certain that its properties can be discussed in terms of all neutron stars and why we don't see similar properties in other neutrons accreting X-ray binaries.

Hajime Inoue: I think that recently, I don't remember when, a bursting pulsar was found; that burst has a similar behavior to this source, the Rapid Burster. And also, GRS 1915+105 shows similar quasi-periodic high/low transitions. I think from those phenomena that when the mass accretion rate is very high such kind of things could happen.

Frank Shu: I'd like to follow up on Vicky's question. As I recall when the Rapid Burster was first discovered, the explanation for it was in terms of magnetic field interaction. The magnetosphere seems to play no role in your model; could you comment on that or did I mistake your meaning?

Hajime Inoue: I don't know exactly the theoretical explanation in terms of magnetic field. It will be difficult for such models to explain the time-scale invariance. From observational points of view, at least as I have just said, the bursting pulsar shows a similar phenomena. The bursting pulsar has some stronger magnetic field than the Rapid Burster but the phenomena are similar. We could say it's independent of the strength of the magnetic field. (A late comment: No periodicity was found from the Rapid burster even when it was very dim. The magnetic filed of the Rapid Burster could be very low.)

Astrophysical Ages and Time Scales
ASP Conference Series, Vol. 245, 2001
T. von Hippel, C. Simpson, N. Manset

On the Relationship Between Dynamical Time and Atomic Time

Y. B. Kolesnik

Institute of Astronomy of the RAS, 48 Piatnitskaya Str., 109017 Moscow, Russia

Abstract. The modern conception of a dynamical time scale is presented. The consequences of having a non-traditional space-time metrics on the relationship between dynamical time and atomic time is discussed. The apparent accelerations of the longitudes of Mercury, Venus, and the Earth as found from extensive analysis of optical observations are reported. These can tentatively be interpreted as an indication in favor of an expanding space-time metrics.

1. The Modern Dynamical Time Scales

Initially the dynamical time was introduced in the conception of the Ephemeris Time (ET) and defined as the time-like argument of a gravitational theory. It is assumed to be a uniform ideal time which is not directly measured by clocks but implicitly pervades an inertial dynamical system of reference. The practical determination of the ET was based on the model of macroscopic events in the Solar System. The time scale based on microscopic events is the atomic time (AT). The choice of a frequency standard of the atomic time is conventional. The SI second was chosen so that the seconds of atomic and ephemeris time were of equal duration, and an equivalence of time in quantum physics and time in the Solar System dynamics was implicitly assumed.

In the practical application of the AT to celestial mechanics the effect of relativity and the resulting differences between proper and coordinate time scales must be considered. This is based on a metric accepted by a specific post-Newtonian theory of gravitation and implies both a *secular* and *periodic* rate of the quasi-uniform time scale measured by the atomic clocks on geoid with respect to a coordinate time of the geocentric and barycentric reference systems associated with the respective ephemerides.

The secular divergence between TAI (international atomic time) and a barycentric coordinate time amounts to 49 s in a century and is linear with time. To remove secular divergence the "scaling factors" in units of time was introduced. This approach was revised by the IAU 1992 resolution. Instead of introducing scaling factors, a direct four dimensional coordinate transformation between TCB (coordinated barycentric time) and TCG (coordinated geocentric time) was recommended.

2.   Dynamical Time and the Scale Invariant Metrics of Space-Time

The foundation of the presently accepted physics implicitly imply the article of faith that a *uniform* time really exists, and that this time corresponds to the independent variable of the equations of motion based on the space-time framework of the modern version of general relativity. Therefore we deal with a conventional, theory-dependent concept. Attempts to approximate an ideal time scale are directed toward reconciling dynamical theory with observations which assume that the *uniform* time really exists in nature on the macroscopic level.

The theory of gravitation in general relativity is not scale invariant. However there are arguments in favor of scale invariance from elementary particle theories (Canuto et al. 1977). The generalization of the group of invariances underlying the theory of gravitation was suggested by Dirac (1973) by considering invariance under gauge transformations. When Weil's geometry is applied to the equations of motion an additional term appears, producing the non-Newtonian acceleration proportional to the velocity of an orbiting body in any gravitational system.

If the hypothesis of the scale factor varying linearly with time is applied to the real world, such an acceleration should be observed as the apparent quadratic deviation of the longitudes of orbiting bodies when compared with ephemerides constructed on the basis of presently accepted forms of the equations of motion. Therefore it is in principle possible to check the space-time metric of the real world by observing the bodies in our Solar System.

3.   Using Optical Observations of the Sun and Planets for Checking the Equivalence of the Atomic and Dynamical Time Scales

The very first attempts to check the equivalence of atomic time and dynamical time was made by Oesterwinter & Cohen (1972) who reported a significant linear trend between dynamical and atomic time scales. From early analyses of 15 years of radar ranges, Reasenberg & Shapiro (1978) derived positive accelerations of Mercury and Venus of $\dot{n}/n = (12 \pm 8) \times 10^{-11}$ yr^{-1}, where $\dot{n}$ is the acceleration, and n, the mean motion. Krasinsky et al. (1986) also gave positive accelerations of about $(8.2 \pm 1.6) \times 10^{-11}$ yr^{-1} derived from radar observations taken in 1961–1982. On the other hand, the estimates provided by the JPL team from 20 years of radar ranges of Mercury are in dramatic discordance with the results above $(0.9 \pm 0.7) \times 10^{-11}$ yr^{-1} (Anderson et al. 1991). This discordance can tentatively be explained by the automatic scaling of the time-like arguments applied in the ephemeris construction process at the JPL (Standish 1998) which is virtually equivalent to the respective scaling of the mean motions of all orbiting bodies.

Extending the scope of the investigation, worldwide optical observations of the Sun, Mercury, and Venus made in 60 observatories were incorporated in the present study for comparison with the ephemerides based on post-Newtonian equations of motion. Observations were transformed onto ICRS (Internation Conventional Reference Frame) system and compared with the DE405 ephemeris.

Corrections to the mean longitudes of the Earth, Mercury, and Venus have shown the acceleration factors proportional to the mean motions of planets.

Their numerical estimates for the period 1911–2000 are $(1.51'' \pm 0.10'')$ cy^{-2} for the Sun, $(4.51'' \pm 0.41'')$ cy^{-2} for Mercury and $(2.78'' \pm 0.15'')$ cy^{-2} for Venus. Expressed in terms of the ratio of the acceleration factor $\dot{n}$ to the mean motion n the observed semi-accelerations are equivalent to approximately $\dot{n}/n = 20 \times 10^{-11}$ yr^{-1}.

Since the modern numerical ephemerides have internal errors much below the derived discrepancies these results can tentatively be interpreted as an indication of the validity of the space-time scale invariant metrics not accounted for in the adopted equations of motion used in the ephemeris creation process.

A recently published variant of the steady state theory, the Expanding Spacetime (EST) theory by C. J. Masreliez (2000), is consistent with the reported results. The EST theory predicts a new cosmological effect, the cosmic drag, which is a direct consequence of the geodesic for the expanding spacetime line element. The cosmic drag acts to exponentially decelerate free, slowly moving objects, in proportion to their relative velocities with a time constant equal to the Hubble time. It also diminishes the angular momentum of rotating bodies and systems. For the planets this diminishing angular momentum combined with Newton's law of gravitation induces secular accelerations in the Solar System according to relation $\dot{n}/n = 3/T_{\rm H}$. The semi-accelerations of Mercury, Venus, and the Earth predicted by the EST theory, taking a reasonable Hubble time near 14 billion years, are $5.7''$, $2.3''$, $1.4''$, which are of the same order as the actually detected quadratic terms of the residuals in the secular longitude variations.

References

Anderson, J. D., Slade, M. A., Jurgens, R. F., Lau, E. L., Newhall, X. X., & Standish, E. M. 1991, Proc. ASA, 9, 324

Canuto, V., Adams, P. J., Hseih, S. H., & Tsiang, E. 1977, Phys. Rev. D, 16, 1643

Dirac, P. 1973, Proc. Roy. Soc. London A, 333, 403

Krasinsky, G. A., Aleshkina, E. Y, Pitijeva, E. V., & Sveshnikov, M. L. 1986, in Relativety in Celestial Mechanics and Astronomy, ed. J. Kovalevsky & V. A. Brumberg (Dordrecht: Reidel), 315

Masreliez, C. J. 2000, Ap&SS, 266, 399

Oesterwinter, C., & Cohen, C. J. 1972, Cel. Mech., 5, 317

Reasenberg R. D., & Shapiro I. I. 1978, in On the Measurement of Cosmological Variation of the Gravitational Constant, ed. L. Halpern (Gainesville: University Press of Florida), 71

Standish, E. M. 1998, A&A, 336, 381

Acknowledgments. The travel grant from the LOC of the conference is gratefully acknowledged.

Session 3

Planetary and Debris Disk Systems

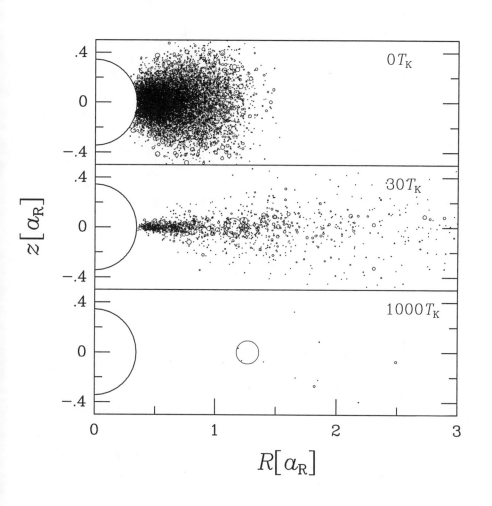

Astrophysical Ages and Time Scales
ASP Conference Series, Vol. 245, 2001
T. von Hippel, C. Simpson, N. Manset

Time Scale of Disk and Planet Formation

Frank H. Shu

*Department of Astronomy, 601 Campbell Hall, University of California,
Berkeley, CA 94720-3411, USA*

Abstract. We review the current theories and observations of the formation of Sun-like stars and protoplanetary disks in distributed and clustered environments. The astrophysical estimates for the temporal durations of various phases of the process are calibrated by theoretical calculations of pre-main sequence contraction tracks for T Tauri stars, a task which is fairly firmly grounded at present. Meteoritical estimates for the ages of the parent bodies of various chondritic components are somewhat compromised by uncertainties concerning whether the ^{26}Al that was alive at the origin of the Solar System was seeded by external nucleosynthetic events or created by internal bombardment via protosolar cosmic rays. In either case, however, planetesimal formation in the solar nebula probably did not extend for longer than three million years. The gaseous component of isolated protoplanetary disks may last approximately ten million years, although systems born in clustered environments close to O stars could lose their disks in a million years or less. Planet formation may be difficult in such circumstances.

Astrophysical Ages and Time Scales
ASP Conference Series, Vol. 245, 2001
T. von Hippel, C. Simpson, N. Manset

The Epoch of Planet Formation

Douglas Lin

Department of Astronomy and Astrophysics, University of California, Santa Cruz, CA 95064, USA

Abstract. Time scales for planets' formation epoch can be constrained by: (1) the evolution of protostellar gaseous and dusty disk cradles out of which they emerge, (2) the dynamical interaction between protoplanets and these disks, and (3) theoretical models for protoplanetary growth. Infrared observation suggests that the building blocks of protoplanets are mostly depleted on the time scale of several million years. Some dynamical properties of recently discovered extrasolar planetary systems may be attained on a similar time scale. However, these time scales are shorter than the theoretically estimated growth time scales for both solid planetesimals and gaseous protoplanetary envelopes. Consistency is sought through several modifications to the present paradigm for protoplanetary formation. Eventually, the incubation duration of protoplanets may be directly deduced from high resolution mappings of protostellar disks.

Astrophysical Ages and Time Scales
ASP Conference Series, Vol. 245, 2001
T. von Hippel, C. Simpson, N. Manset

Solar System Formation Time Scales From Oligarchic Growth

Edward W. Thommes

University of California, Berkeley, CA, USA

Martin J. Duncan

Queen's University, Kingston, Ontario, Canada

Harold F. Levison

Southwest Research Institute, Boulder, CO, USA

Abstract. Runaway growth ends when the largest protoplanets dominate the dynamics of the planetesimal disk; the subsequent self-limiting accretion mode is referred to as "oligarchic growth." We review the oligarchic growth model and derive from it global estimates of the planet formation rate throughout the protoplanetary disk during the time when the nebular gas is still present. For a moderate-mass (5 × minimum-mass) nebula, the Jupiter–Saturn region is predicted to produce giant planet core-sized bodies ($\sim 10~M_\oplus$) in a few million years. However, numerical simulations suggest that accretion actually stalls well before the theoretical maximum mass is reached. Even an implausibly massive nebula is not predicted to produce anything bigger than an Earth mass at the orbit of Uranus by 10 Myr. Subsequent accretion without the random-velocity-damping presence of gas is even slower, yielding estimates of order 10^9 years or more for the formation of the outer giant planets. We review a promising alternative model to in situ formation, which has Uranus and Neptune originating in the more accretion-friendly Jupiter–Saturn region.

1. Introduction

The initial growth mode in a disk of accreting planetesimals is runaway growth (e.g., Wetherill & Stewart 1989; Kokubo & Ida 1996), wherein the mass doubling time for the largest bodies is the shortest. However, when these runaway bodies become sufficiently massive, it is their gravitational stirring which dominates the random velocity evolution of the background planetesimals, rather than the interactions among the planetesimals (Ida & Makino 1993). The subsequent self-limiting accretion mode has been termed "oligarchic growth" (Kokubo & Ida 1998); in this regime, the mass ratio of any two nearby protoplanets approaches unity over time.

In Section 2, we review the condition for the onset of oligarchic growth, and show that it takes place when the largest bodies are still orders of magnitude below an Earth mass. Oligarchic growth is thus much more important than runaway growth in setting the time scale for planet formation. In Section 3, we review the approximate analytic description of oligarchic growth of Kokubo & Ida (1998, 2000), and extend it to obtain a better picture of the relative timing of accretional growth throughout the protoplanetary disk. We find that even for very massive disks, little accretion takes place during the expected gas lifetime ($\lesssim$ 10 Myr) beyond 10 AU. In Section 4, we review a model which makes it possible, despite this time scale problem, to account for the formation of Uranus and Neptune. Section 5 summarizes and discusses the findings.

2. Onset of Oligarchic Growth

Ida & Makino (1993) derive the following condition for the dominance of proto-planet-planetesimal scattering over planetesimal-planetesimal scattering in determining the random velocity evolution of the planetesimal disk:

$$2\sigma_\mathrm{M} M > \sigma_\mathrm{m} m, \tag{1}$$

where M and m are the protoplanet mass and the characteristic planetesimal mass, respectively, σ_m is the surface mass density of the planetesimal disk, and σ_M is the effective surface density due to a protoplanet in the disk. The latter is given by

$$\frac{M}{2\pi a_\mathrm{M} \Delta a_\mathrm{stir}}, \tag{2}$$

where a_M is the semimajor axis of the protoplanet, and Δa_stir is the width of the annulus within the disk gravitationally stirred by the protoplanet. Using the approximate conservation of the Jacobi energy of planetesimals relative to the protoplanet together with some other assumptions, Ida & Makino (1993) show that

$$\Delta a_\mathrm{stir} \sim 5.2 a_\mathrm{M} \langle e_\mathrm{m}^2 \rangle^{1/2}, \tag{3}$$

where $\langle e_\mathrm{m}^2 \rangle^{1/2}$ is the r.m.s. eccentricity of the surrounding planetesimals. They obtain an expression for the equilibrium $\langle e_\mathrm{m}^2 \rangle^{1/2}$ by equating the viscous stirring time scale due to a protoplanet of mass M, with the eccentricity damping time scale due to gas drag for planetesimals of characteristic mass m. The result is

$$\langle e_\mathrm{m}^2 \rangle^{1/2} \sim \frac{1.12 m^{1/6} M^{1/3} \rho_\mathrm{m}^{1/9}}{a^{1/6} C_\mathrm{D}^{1/6} M_\odot^{1/3} \rho_\mathrm{gas}^{1/6} m^{1/9}}, \tag{4}$$

where ρ_m is the volume density of a planetesimal, ρ_gas is the volume density of the nebular gas, and C_D is a dimensionless drag coefficient of order one. Using this result in Equation 1, one obtains an expression for the protoplanet mass at which oligarchic growth commences:

$$M_\mathrm{oli} \sim \frac{5.7 a^{11/10} m^{19/30} \rho_\mathrm{m}^{1/15} \sigma_\mathrm{m}^{3/5}}{C_\mathrm{D}^{1/10} M_\odot^{1/5} \rho_\mathrm{gas}^{1/10}}. \tag{5}$$

To match their numerical simulations, Ida & Makino (1993) use planetesimal masses of 10^{23}–10^{24} g (200–400 km sizes) and thus obtain, for a minimum mass nebula (Hayashi 1981), $M_{\text{oli}} \sim 50\text{--}100$ m, i.e., up to ~ 0.02 $M_\oplus$. Using a perhaps more realistic (e.g., Lissauer 1987) planetesimal mass of 10^{19} g (10 km) gives $M_{\text{oli}} \sim 10^{-6}\text{--}10^{-5}$ $M_\oplus$. In any case, runaway growth is expected to cease long before protoplanets approaching an Earth mass can form. Thus the contribution to a planet's formation time scale from oligarchic growth is much more important than that from runaway growth.

3. Oligarchic Growth Time Scales

When planetesimal random velocities are dispersion-dominated rather than shear-dominated, the mass accretion rate of an embedded protoplanet is well described by the particle-in-a-box approximation (e.g., Safronov 1969):

$$\frac{dM}{dt} \simeq \frac{\sigma}{h} \pi R_{\text{M}}^2 \left(1 + \frac{v_{\text{esc}}^2}{v_{\text{rel}}^2} \right) v_{\text{rel}}, \tag{6}$$

where σ is the planetesimal disk surface density, h the disk scale height, R_{M} the protoplanet radius, v_{esc} the escape velocity from the protoplanet's surface, and v_{rel} the characteristic relative velocity between the protoplanet and the planetesimals. With some further approximations, and the assumption that both dynamical friction and gravitational focusing are effective, this can be rewritten as

$$\frac{dM}{dt} \simeq \frac{C \sigma M^{4/3}}{\langle e_{\text{m}}^2 \rangle a^{1/2}}. \tag{7}$$

$C = 4\pi^{2/3} (3/4\rho_{\text{M}})^{1/3} (G/M_\odot)^{1/2}$, where ρ_{M} is the density of a protoplanet (see e.g., Kokubo & Ida 1996 for details).

Using the oligarchic-regime equilibrium eccentricity of the planetesimal swarm, Equation 4, this becomes

$$\frac{dM}{dt} \sim \frac{6.2 C_{\text{D}}^{1/3} \sqrt{G/M_\odot} \, \rho_{\text{gas}}^{1/3} m^{2/9} \sigma_{\text{m}} M^{2/3}}{\rho_{\text{m}}^{2/9} \rho_{\text{M}}^{1/3} m^{1/3} a^{1/6}}. \tag{8}$$

Kokubo & Ida (2000) obtain estimates of growth time scale, $\frac{M}{dM/dt}$, in this way. To get a better global view of the growth process, we solve the above equation for $M(t)$. First, however, we allow the surface density of planetesimals to vary self-consistently as it is depleted by the growing protoplanets:

$$\sigma_{\text{m}}(M) = \sigma_0 - \frac{M}{2\pi a \Delta a_{\text{M}}}, \tag{9}$$

where σ_0 is the original planetesimal surface density at that location, and Δa_{M} is the spacing between protoplanets. Numerical simulations show that the largest protoplanets in a swarm of planetesimals tend to keep an orbital separation of ~ 10 Hill radii (Kokubo & Ida 1995, 1998). Adopting $\Delta a_{\text{M}} = 10 r_{\text{H}}$ in Equation 9, then substituting that into Equation 8 and solving the resulting

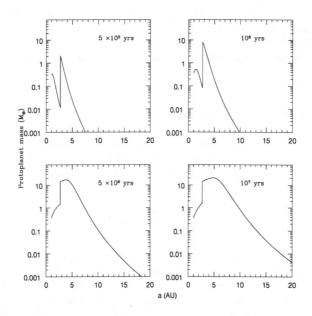

Figure 1.   Protoplanet masses throughout the protoplanetary disk, as given by Equation 10. Gas and solids surface densities are five times those of the Hayashi minimum-mass model ($\propto a^{-3/2}$) beyond the snow line at 2.7 AU, but are more shallow ($\propto a^{-1}$) interior. A 10 km size is used for the planetesimals. Protoplanets and planetesimals are given densities of 3 g cm^{-2} inside the snow line and 1.5 g cm^{-2} outside.

differential equation for $M(t)$, one obtains

$$M \sim \frac{288a^3\sigma_0^{3/2}}{M_\odot^{1/2}} \tanh^3 \left[\frac{0.31C_D^{1/3}G^{1/2}M_\odot^{1/3}\rho_{\text{gas}}^{1/3}\sigma_0^{1/2}}{m^{1/9}a^{7/6}\rho_m^{2/9}\rho_M^{1/3}}t + \tanh^{-1}\left(\frac{0.15M_0^{1/3}M_\odot^{1/6}}{a\sigma_0^{1/2}} \right) \right]. \tag{10}$$

Figure 1 plots this estimate of the protoplanet mass at different times for a protoplanetary disk that has five times the solids and gas densities of the Hayashi minimum-mass nebula beyond the "snow line"—an increase in solids surface density by a factor of slightly more than four at the location in the nebula where water ice is thought to form—at 2.7 AU. We adopt a density profile shallower than that of Hayashi interior to the snow line, to reduce the amount of excess mass in the terrestrial region, where planet formation is very efficient. Such a disk has a surface density of planetesimals between 10 and 15 g cm^{-2} around 5 AU, which is consistent with estimates of the densities needed to accrete Jupiter's solid core (e.g., Lissauer 1987; Pollack et al. 1996; Weidenschilling 1998). One can see protoplanet growth progressing outward in a steep front over time, with a discontinuous jump at 2.7 AU. The limiting mass is that at which a protoplanet has accreted all planetesimals within its 10 r_H annulus. This is analogous to the isolation mass for a single protoplanet (e.g., Lissauer 1987) but larger (about twice) because the overlapping stirred regions of the protoplanets prevent early isolation.

In this model, protoplanets with masses as high as 10 $M_\oplus$ do indeed form after a few million years. These exist within a relatively narrow but expanding annulus, bounded at the inner edge by the snow line, and at the outer edge by the growth front. By 10 Myr, 10 $M_\oplus$ protoplanets have formed out to about 7 AU. This result supports the idea that the solids density enhancement at the snow line serves as the trigger for giant planet core formation (Morfill 1985). The actual location of the snow line is uncertain; models subsequent to that of Hayashi have placed it anywhere from 6 AU (Boss 1995) to 1 AU (Sasselov & Lecar 2000). Indeed, the snow line was probably not stationary as the disk and its temperature profile evolved.

Theoretically, oligarchic growth thus seems well able to account for Jupiter and Saturn, assuming these formed by nucleated gas instability onto a $\sim$ 10 $M_\oplus$ core. However, N-body simulations (E. W. Thommes, M. J. Duncan, & H. E. Levison, in preparation) indicate that the predicted protoplanet masses are upper limits; protoplanet growth stalls once they are not much larger than an Earth mass, due in large part to the rapid orbital decay of planetesimals with large forced eccentricities undergoing gas drag. Using a smaller characteristic planetesimal size speeds accretion but also increases the rate of planetesimal loss to gas drag. Similar results are obtained by Inaba & Wetherill (2001), who perform statistical dynamic simulations which include fragmentation of bodies due to collisions. An earlier simulation (Thommes 2000) suggests that doubling the surface density in the gas giant region, to $\sim$ 30 g cm^{-2} at 5 AU, could be enough to produce 10 $M_\oplus$ bodies, though depending on the density profile, this may imply a disk of questionably large mass.

The ice giants, Uranus and Neptune, fare much worse. Though they have no massive gas envelopes, they must still each capture a few Earth masses of gas to match their present composition; therefore, the protoplanets from which they form ought to reach large masses during the lifetime of the nebular gas. However, even if the disk density is increased by a factor of 15 over the minimum-mass model (yielding a very massive disk with about 0.2 $M_\odot$ interior to 30 AU), the maximum mass of bodies which have formed at 20 AU by 10 Myr is only about an Earth mass. Worse yet, calculations of the much slower oligarchic growth rate after the disappearance of the nebular gas show that even in such a massive disk, the time to grow Neptune approaches a billion years (E. W. Thommes et al., in prepration).

4. Formation of the Ice Giants

The formation of Uranus and Neptune can be more readily accounted for if one does not require them to accrete *in situ*. Thommes, Duncan, & Levison (1999) assume that in addition to the solid cores of Jupiter and Saturn, extra bodies

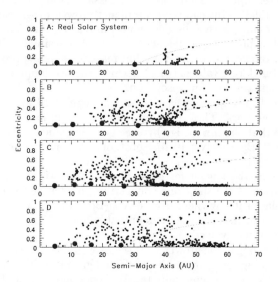

Figure 2.   The present Solar System compared to the outcomes of some of the simulations that produced particularly good analogues to the present Solar System. The innermost large circle is Jupiter, and the other three are bodies of 15 $M_\oplus$. In Panel B, the second-innermost body has Saturn's mass. Small circles are planetesimals (or Kuiper belt objects, in Panel A). The curves show the locus of orbits encountering the outer body; planetesimals remaining in orbits in the region above and to the left of the curves are generally unstable on time scales short compared to the age of the Solar System (from Thommes et al. 1999).

of comparable mass formed in the same region. They perform numerical simulations in which one of the protoplanets abruptly increases its mass to that of Jupiter (and in one case, another protoplanet reaches Saturn's mass), thus simulating the rapid final runaway gas accretion phase. Models of giant planet formation through concurrent gas and solids accretion (Pollack et al. 1996) show a long middle stage of slow gas accretion, lasting for millions of years, during which the body's mass changes very little. This delay gives bodies at slightly larger heliocentric distances time to catch up in their growth, which should lead to a situation similar to the above initial conditions. For a surface density of ~ 10–15 g cm^{-2}, oligarchic growth predicts that between three and five $\sim 10~M_\oplus$ protoplanets will ultimately form in the Jupiter–Saturn region (but see the previous section). For most of the runs, four are used. The simulations show that the mass increase of a giant protoplanet becoming a gas giant destabilizes the orbits of the remaining protoplanets. One or more usually undergo close encounters with "Jupiter," and as a result all of them tend to end up on eccentric, mutually crossing orbits with aphelia in the trans-Saturnian region. Also included in the simulations is a trans-Saturnian planetesimal disk, which serves as a source of dynamical friction for the eccentric protoplanets. As a result, their eccentricities decrease over time, decoupling them from Jupiter and from each other on a time scale of a few Myr. About half of the time, the simulations produce, after 5–10 Myr, a system which is remarkably similar to our own outer Solar System: two would-be giant planet cores have ended up on nearly circular, low-inclination orbits with semimajor axes similar to those of Uranus and Neptune, while the innermost is near the present orbit of Saturn. The timing of the Saturn core's runaway gas accretion phase is therefore not strongly constrained. Figure 2 shows the outcome of some runs, with the present Solar System for comparison.

An interesting additional feature of this model is that early on, the eccentric proto-Uranus and/or -Neptune tend to raise significant eccentricities and inclinations in the region corresponding to the Kuiper belt. Similarly high random velocities are observed there today, and are not readily explained by other mechanisms (e.g., Petit, Morbidelli, & Valsecchi 1999).

5. Summary

Runaway growth allows very short formation times, but only in the early stages of planetesimal accretion; there is a changeover to the self-limiting oligarchic growth mode when the largest bodies are still orders of magnitude below an Earth mass. The time scale of oligarchic growth thus dominates over that of runaway growth, and we use the former alone to obtain a global picture of planet growth throughout the protoplanetary disk. In the terrestrial region, accretion efficiency is high, and oligarchic growth alone is not required to form final bodies of the right masses. Simulations show that a late stage of impacts among protoplanets of up to perhaps a Mars mass is readily able to produce a final system with planet masses and spacings similar to the present-day inner Solar System (e.g., Chambers & Wetherill 1998).

In the giant planet region, however, lower accretion efficiency and the necessity of forming large bodies before the removal of the nebular gas (after $\lesssim 10$ Myr) mean that direct oligarchic growth likely has to be relied on to account for most of the accretion. We find that the analytically predicted outward-sweeping wave of oligarchic growth produces 10 $M_\oplus$ bodies in a protoplanetary disk with five times the mass of the Hayashi (1981) minimum-mass model in a few Myr. These first appear at the snow line, since the surface density enhancement there gives accretion a head start and increases the predicted final masses. Thus, the snow line may indeed play a role in triggering the formation of giant planets, as has been previously proposed (e.g., Morfill 1985). However, simulations in this region (e.g., Inaba & Wetherill 2001; E. W. Thommes et al., in preparation) suggest that the analytic estimates of final protoplanet masses are too high, which is not surprising considering the idealized nature of the calculation.

In 10 Myr, this (optimistic) predicted growth front only produces 10 $M_\oplus$ bodies interior to about 10 AU, even for very massive protoplanetary disks. So, although the solid cores of Jupiter and Saturn can be theoretically accounted for, it seems that Uranus and Neptune are out of luck as far as *in situ* formation is concerned. A promising alternative model of the ice giants' formation is that they shared the same birthplace as Jupiter and Saturn, only to be scattered outward when Jupiter accreted its massive gas envelope (Thommes et al. 1999). The existence of excess large protoplanets is predicted by oligarchic growth, and is supported by the above-mentioned simulations. However, there remains the issue of the smaller-than-predicted final protoplanet masses; displacing the the origin of Uranus and Neptune to the Jupiter–Saturn zone greatly alleviates the time scale problem but by no means completely solves it. More work on accretion in this region is needed.

The results summarized here suggest the following general picture of the planet formation process: oligarchic growth accounts for most of the planetesimal accretion beyond the terrestrial region within the first ten million years, while the gas is still present. Very little accretion takes place in this time beyond about 10 AU, so that the Solar System at this point is quite compact. Bodies large enough to be potential gas giant cores form in a relatively narrow annulus of the disk. The formation of the gas giants then triggers an initially violent evolution of the orbits of the remaining giant protoplanets, which ultimately produces an outer Solar System with the present widely-spaced orbits.

Orbital evolution through disk tides (e.g., Ward 1997) is not considered in this analysis. Fast "Type I" inward migration, thought to affect objects not massive enough to open a gap in the gas disk ($\lesssim 10\ M_\oplus$) constitutes a major problem for any model of giant planet formation, except ones which invoke direct, unnucleated collapse from a gas disk instability (e.g., Boss 1998). Much more work is clearly needed here, but recent results by Papaloizou & Larwood (2000) point to one possible solution. They show analytically that the migration speed should slow with increasing eccentricity, and actually switch direction when an object's radial excursion is similar to the gas disk height ($e \sim 0.07$ at 5 AU). Thus, eccentricity-exciting interactions among growing giant protoplanets may help save them from spiraling into the central star as they form.

References

Boss, A. P. 1995, in Abstracts of the Lunar and Planetary Science Conf. 25, 151

Boss, A. P. 1998, ApJ, 503, 923

Chambers, J. E., & Wetherill, G. W. 1998, Icarus, 136, 304

Hayashi, C. 1981, Prog. Theor. Phys. Suppl., 70, 35

Ida, S., & Makino, J. 1993, Icarus, 106, 210

Inaba, S., & Wetherill, G. W. 2001, 32nd Annual Lunar and Planetary Science Conf., abstract 1384

Kokubo, E., & Ida, S. 1995, Icarus, 114, 247

Kokubo, E., & Ida, S. 1996, Icarus, 123, 180

Kokubo, E., & Ida, S. 1998, Icarus, 131, 171

Kokubo, E., & Ida, S. 2000, Icarus, 143, 15

Lissauer, J. J. 1987, Icarus, 69, 249

Morfill, G. E. 1985, in Birth and Infancy of Stars, ed. R. Lucas, A. Omant, & R. Stora (Amsterdam: Elsevier), 693

Papaloizou, J. C. B., & Larwood, J. D. 2000, MNRAS, 315, 823

Petit, J.-M., Morbidelli, A., & Valsecchi, G. B. 1999, Icarus, 141, 367

Pollack, J. B., Hubickyj, O., Bodenheimer, P., Lissauer, J. J., Podolak, M., & Greenzweig, Y. 1996, Icarus, 124, 62

Safronov, V. S. 1969, Evolution of the Protoplanetary Cloud and Formation of the Earth and Planets (Moscow: Nauka), Engl. transl. NASA TTF-677, 1972

Sasselov, D. D., & Lecar, M. 2000, ApJ, 528, 995

Thommes, E. W. 2000, Ph.D. Thesis, Queen's University, Kingston, Ontario

Thommes, E. W., Duncan, M. J., & Levison, H. F. 1999, Nature, 402, 635

Ward, W. R. 1997, Icarus, 126, 261

Weidenschilling, S. J. 1998, AAS Division for Planetary Sciences Meeting Num. 30, abstract 21.03

Wetherill, G. W., & Stewart, G. R. 1989, Icarus, 77, 330

Discussion

Typhoon Lee: I wonder if you would comment on the recent observation that KBOs seem to be more depleted outside about 70 or 80 AU, even more depleted than the current observed value in the Pluto region?

Martin Duncan: There is a real issue about how far out the Kuiper Belt extends, as you well know. But there are several reasons to think that there are different populations in the Kuiper Belt, and the most recent work by Stern & Levison shows a correlation between the brightness of the objects and their location. This suggests that there may be a size distribution in the sense that as you go further out only smaller and smaller bodies were formed: there was a limit to the accretion. So when people say there's an edge to the Kuiper Belt, they're really talking about objects that are typically 100 km or so across, and there does seem to be a trend that you're not finding things beyond 50 AU in the large sizes. That may suggest either that it was truncated, by a passing star let's say, or that maybe it's just that there weren't any particularly large objects formed out in that region. So I think it's still conceivable that there could be a large amount of mass beyond 50 AU but that it's in much smaller bodies, because the accretion time out there is that much longer.

Bill Ward: If you use the growth of giant planets to initiate the scattering of these cores then isn't there a gas disk present when they are scattered?

Martin Duncan: There may be, but it could well have been diminished by photoevaporation. There are some models in which there's outside-in depletion of these disks. If this is happening in a few Myr time scale while the cores are forming closer in, then by the time Jupiter has accreted its gas, the gas may have been diminished in the outer parts by various evaporative processes. In that case you just have a planetesimal disk out there.

Bill Ward: Well, I'm worried that you'd have inward migration and then you'd bring the cores back in.

Martin Duncan: That's right. So one would have to argue that either the process of migration in a gas disk is different when you have multiple cores (which is of course possible) or that the outer regions of the gas disk have been ablated by the time this process occurred.

Astrophysical Ages and Time Scales
ASP Conference Series, Vol. 245, 2001
T. von Hippel, C. Simpson, N. Manset

Lunar Formation From a Circumterrestrial Disk

E. Kokubo

National Astronomical Observatory, Osawa, Mitaka, Tokyo, Japan

S. Ida

Tokyo Institute of Technology, Ookayama, Meguro, Tokyo, Japan

Abstract. We investigated the evolution of a circumterrestrial disk of debris generated by a giant impact on the Earth and the dynamical characteristics of the moon accreted from the disk by using N-body simulations. We found that in most cases the disk evolution results in the formation of a single large moon on a nearly circular orbit close to the equatorial plane of the initial disk just outside the Roche limit. The efficiency of incorporation of disk material into a moon is 10–55%, which increases with the initial specific angular momentum of the disk. These results hardly depend on the initial condition of the disk as long as the disk mass is 2 to 4 times the present lunar mass and most disk mass exists inside the Roche limit.

1. Introduction

In the "giant impact" scenario, the Moon is accreted from a circumterrestrial disk of debris generated by the impact of a Mars-sized protoplanet with the early Earth (Hartman & Davis 1975; Cameron & Ward 1976). This model has been favored as it can potentially account for major dynamical and geochemical characteristics of the present Earth–Moon system: the system's high angular momentum and depletion of volatiles and iron in the lunar material (e.g., Stevenson 1987). On the other hand, the recent studies of planetary formation revealed that in the late stage of planetary accretion, protoplanets are formed through runaway growth of planetesimals (e.g., Wetherill & Stewart 1989; Kokubo & Ida 1998). In the final stage of the terrestrial planet formation, collisions among protoplanets with sweeping up of residual planetesimals would be likely. Thus, giant impacts may be plausible events.

Giant impacts have been modeled by using a smoothed-particle hydrodynamics (SPH) (e.g., Cameron & Benz 1991; Cameron 1997). It has been found that an impact by a Mars-sized protoplanet usually results in formation of a circumterrestrial debris disk. In most cases, significant amount of the disk material exists within the Roche limit.

The accretion process of the moon from the impact-generated disk was first investigated by Canup & Esposito (1996). They used a gas dynamic (particle-in-a-box) model that can describe accretion of solid particles in the Roche zone where accretion is partially inhibited by the tidal force of the Earth. They showed that, in general, many small moonlets are formed initially rather than a single large moon. However, it is difficult to include global effects, such as radial migration of lunar material and interaction of formed moons, with the disk and collective effects, such as the formation of particle aggregates and spiral arms in the gas model. These effects are potentially very important in the evolution of the disk.

In the present paper, we present high-resolution N-body calculations of the lunar accretion from a circumterrestrial disk (Kokubo, Ida, & Makino 2000). In N-body simulations, global and collective effects are automatically taken into account. We show that the typical outcome of the accretion from the disk is the formation of a single large moon, as long as most disk mass is initially within the Roche limit.

2. Method of Calculation

In N-body simulation, the orbits of particles are calculated by numerically integrating the equation of motion. For numerical integration, we use the predictor-corrector-type Hermite scheme (Kokubo, Yoshinaga, & Makino 1998). For the calculation of mutual gravitational force, which is the most expensive part of N-body simulation, we use the special-purpose computer, GRAPE (Makino et al. 1997).

Collisions between particles play an important role in the evolution of a circumterrestrial disk. It is assumed that two colliding particles rebound with a relative rebound velocity $\vec{v}'$, which is determined by the relative impact velocity $\vec{v}$ and the coefficients of restitution: $\vec{v}'_n = -\epsilon_n \vec{v}_n$ and $\vec{v}'_t = \epsilon_t \vec{v}_t$, where ϵ is the coefficient of restitution ($0 \leq \epsilon \leq 1$) and the subscripts n and t represent normal and tangential components, respectively. The velocity of each particle after the collision is then calculated based on conservation of momentum. We perform simulations with two values for the normal coefficient of restitution, $\epsilon_n = 0.1$ and 0.01; the tangential component is fixed at $\epsilon_t = 1$ for simplicity.

The necessary and sufficient conditions for gravitational binding between two orbiting particles are first, that the Jacobi energy of the two bodies after the collision is negative, and second, that the centers of mass of both colliding bodies are within their mutual Hill sphere (Kokubo et al. 2000). We consider three different accretion (merging) models. For details of the models, see Kokubo et al. (2000). In principle, the results of all the models are essentially the same. The merged spherical body is assigned a total mass equal to that of the colliding bodies, and its position and velocity are set equal to those of the center of mass of the collision.

We start simulations of lunar accretion assuming a solid particle disk. The initial mass distribution of disk particles was modeled by a power-law mass distribution, $n\,dm \propto m^{-\alpha}dm$, where n is the number of particles of mass m. The density of disk particles is $\rho = 3.3$ g cm^{-3} (the bulk lunar density) and the density of the Earth is $\rho_\oplus = 5.5$ g cm^{-3}. Disk particles are assumed to be spheres. The initial disk is axisymmetric, with a power-law surface density distribution given by $\Sigma\,da \propto a^{-\beta}da$, where a is the distance from the Earth, with inner and outer cutoffs, $a_{\rm in}$ and $a_{\rm out}$.

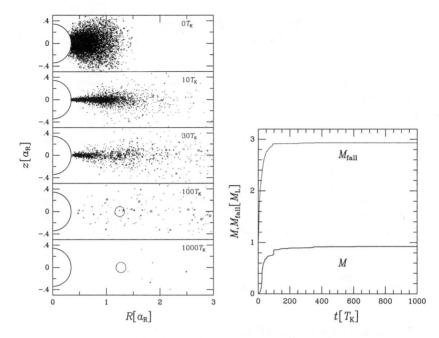

Figure 1.   *Left:* Snapshots of the circumterrestrial disk on the R–z plane at $t = 0, 10, 30, 100$, and $1000\ T_{\rm K}$. The semi-circle centered at the coordinate origin represents the Earth. Disk particles are shown as circles whose size is proportional to the physical size of disk particles. *Right:* The moon mass M (solid curve) and the mass fallen to the earth $M_{\rm fall}$ (dotted curve) as a function of time.

We study the evolution of two initial disk masses, $2M_{\rm L}$ and $4M_{\rm L}$, where $M_{\rm L}$ is the present lunar mass. We also vary the power index of the surface density distribution ($\beta = 1, 3$, and 5) and the outer cutoff of the disk ($a_{\rm out} = 0.5, 1, 1.5$, and $2\ a_{\rm R}$), where $a_{\rm R}$ is the Roche limit radius given by $a_{\rm R} = 2.456(\rho/\rho_\oplus)^{-1/3}R_\oplus$ and $R_\oplus$ is the Earth radius. The effects of the power index of the mass distribution and the initial velocity dispersion on the result are also tested. The power-law exponent of the mass distribution is chosen to be $\alpha = 0.5, 1.5, 2.5$, and ∞, with a dynamic range in mass of $m_{\rm max}/m_{\rm min} = 1000$ for the $\alpha \neq \infty$ cases ($\alpha = \infty$ corresponds to an equal-mass case).

3.   Evolution of a Circumterrestrial Disk

We performed 60 simulations with various disk models and 10^4 initial particles. We followed the disk evolution for $1000\ T_{\rm K}$, where $T_{\rm K}$ is the Kepler period at the distance of the Roche limit and $T_{\rm K} \simeq 7$ hours.

In Figure 1, we show an example of the simulation. The initial disk has a mass $M_{\rm disk} = 4\ M_{\rm L}$ with $\alpha = 1.5$, $\beta = 3$, $a_{\rm in} = R_\oplus$, and $a_{\rm out} = a_{\rm R}$. We adopted $\epsilon_{\rm n} = 0.1$. Figure 1 (left) shows snapshots of the circumterrestrial disk in the

R–z plane for $t = 0, 10, 30, 100$, and 1000 T_K. The circumterrestrial disk first flattens through collisional damping and then expands radially. A single large moon forms around $R \simeq 1.3$ a_R on a nearly non-inclined circular orbit on a time scale of ~ 100 T_K. These are universal characteristics of the accreted moon and appear to be nearly independent of initial disk conditions.

The mass of the largest moon, M, and the mass fallen to the Earth, M_{fall}, are plotted vs. time in Figure 1 (right). The mass of material that escapes from the gravitational field of the Earth is usually smaller than 5% of the initial disk mass. The mass of the moon at $t = 1000$ T_K is 0.9 M_L, while 2.9 M_L of the initial disk mass has fallen to the Earth. The fraction of the disk mass incorporated into the moon varies with the initial disk conditions.

The disk evolution is divided into two stages, namely, the rapid growth and slow growth stages as seen in Figure 1. The duration of the rapid growth stage is ~ 100 T_K, or about a month. In this stage, the redistribution of disk mass through angular momentum transfer, supplies material for accretion outside the Roche limit: most of the disk mass falls to the Earth while some of the mass is transported outward. The formation of the moon is almost completed in this stage. The slow growth stage after ~ 100 T_K is the "cleaning up stage," where the moon sweeps up and scatters away the residual disk mass.

4. Formation of a Single Moon

In order to see why a single large moon is a typical outcome of the disk evolution, we examine the rapid growth stage in detail. Here we use the rubble pile model in which gravitationally bound particles are not merged but form particle aggregates. The evolution of the spatial structure of the disk is most easily seen by the rubble pile model.

Snapshots of the disk in the x–y plane are shown for $t = 0, 1, 5, 10, 20$, and 40 T_K in Figure 2. The initial condition of the disk here is the same as that shown above except here an equal-mass initial distribution was considered. At $t = 40$ T_K, a large bound aggregate with a mass of about one-half the present Moon is formed at $R \simeq 1.3$ a_R.

In a disk, self-gravity tends to produce density contrasts, while the random motion of constituent particles and the tidal force (shear) smooth it. When the effect of the tidal force or the random motion overwhelms that of the self-gravity of the disk, the disk is gravitationally stable and density contrasts do not grow in the disk. In fact, a particulate circumterrestrial disk is marginally stable, but instability still plays an important role in the disk evolution.

The evolution of an initially compact disk in the rapid growth stage is described below:

1. The disk contracts through collisional damping of particles.

2. Particle clumps grow inside the Roche limit as the velocity dispersion of particles decreases.

3. The clumps are elongated by Keplerian shear, which forms spiral arms. The spiral arms are smoothed out as they wind, up and then the formation of spiral arms is repeated.

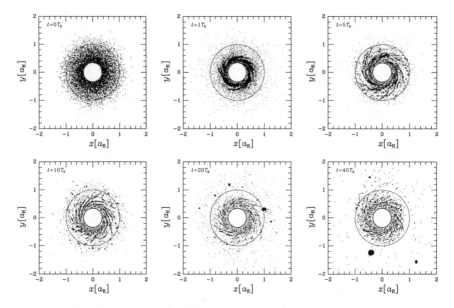

Figure 2.   Snapshots of the circumterrestrial disk on the x–y plane at $t = 0, 1, 5, 10, 20$, and $40\ T_{\rm K}$.

4. Particles are transferred to the outside of the Roche limit through the gravitational torque exerted by the spiral arms.

5. When a tip of the spiral arm goes beyond the Roche limit, it collapses to form a small moonlet (particle aggregate). The rapid accretion of these small moonlets forms a lunar seed.

6. The seed exclusively grows by sweeping up particles transferred beyond the Roche limit.

7. When the moon becomes large enough to gravitationally dominate the disk, it pushes the rest of the inner disk to the Earth.

As a massive and compact particulate disk evolves in the manner described above, a single large moon forms inevitably.

The time scale of the rapid growth stage is of the order of $100\ T_{\rm K}$, relatively independent of the initial conditions. Assuming that the initial disk is contained primarily within the Roche limit, the moon forms from material spreading beyond $a_{\rm R}$, so that the time scale of lunar formation is almost equivalent to the time scale of the mass and angular momentum transfer due to the gravitational torque by the spiral arms. In this case, the time scale of lunar accretion is estimated as

$$T_{\rm g} \sim 10^2 \left(\frac{\Sigma}{0.01 M_{\oplus} a_{\rm R}^{-2}} \right)^{-2} \left(\frac{\Delta R}{0.5 a_{\rm R}} \right)^2 \left(\frac{a}{a_{\rm R}} \right)^{-9/2} T_{\rm K}, \qquad (1)$$

where ΔR is the radial shift of material due to angular momentum transfer (Kokubo et al. 2000). This time scale agrees well with the results of the N-

body simulations. The functional form of T_g shows that the time scale of lunar accretion depends on not the individual mass of disk particles but rather on the surface density of the disk.

In a compact disk, the lunar seed is formed not by gradual pairwise collision of disk particles but collective particle processes: formation of clumps by gravitational instability, angular momentum transfer due to the gravitational torque due to the spiral arm-like structures, and collapse and collision of particle aggregates. The size of the clumps and the spiral arms are in this case determined by the critical wavelength of the disk, which is a function of the surface density. Mass transfer is driven by the gravitational torque by the spiral arms, whose time scale depends on the surface density. Overall, it is the surface density of the disk, rather than the properties of the individual particles, that governs the evolution of the disk.

5.   Dynamical Characteristics of the Moon

We consider the relationship between the dynamical characteristics of the accreted moon and the initial circumterrestrial disk. The orbital elements of the moon are shown in Figure 3 (left). The semi-major axis of the moon in all cases is between a_R and 1.7 a_R, determined mainly by the formation location of the lunar seed and the subsequent interaction with the disk. The lunar seed forms just outside the Roche limit, and it is pushed outward from its birth place somewhat by recoil from the inner disk. The eccentricity and inclination of the moon are small due to dynamical friction and collisional damping; in most cases, they are less than 0.1. These values are almost independent of the detailed initial conditions of the disk. The resultant semi-major axis of the moon is small compared with the present lunar semi-major axis. On a longer time scale, the moon migrates outward by the tidal interaction with the Earth, presumably sweeping up outer residual mass.

In Figure 3 (right), the mass of the accreted moon, M, scaled by the initial disk mass is plotted vs. the initial specific angular momentum of the disk, j_{disk}. The results show that M/M_{disk} increases linearly with j_{disk}. This is because in a small j_{disk} disk, a greater amount of mass must fall to the Earth in order for some mass to spread beyond the Roche limit, yielding a smaller final moon. The fraction of material escaping from the Earth also increases with j_{disk}, although this fraction is usually less than 5% of the disk mass. The overall yield of incorporation of disk material into a moon ranges from 10–55%.

We explain the relationship between the moon mass, M, and the specific angular momentum of the circumterrestrial disk, j_{disk}, by using a conservation of mass and angular momentum argument. From conservation of mass, we have

$$M_{disk} = M + M_{fall} + M_{esc}, \qquad (2)$$

where M_{esc} is the total mass of material that escapes. Conservation of angular momentum gives

$$M_{disk}j_{disk} = Mj + M_{fall}j_{fall} + M_{esc}j_{esc}, \qquad (3)$$

where j, j_{fall}, and j_{esc} are the mean specific angular momenta of the final moon, the mass that impacts the Earth, and the escaping mass, respectively. This con-

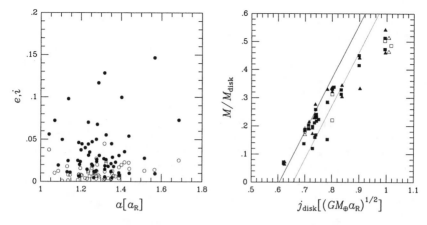

Figure 3.   *Left:* Eccentricity (filled circles) and inclination (open circles) of the moon vs. the semi-major axis of the moon for all the runs. *Right:* Fraction of the initial disk mass incorporated into the moon, M/M_{disk}, plotted against the initial specific angular momentum of the disk, j_{disk}. The triangles correspond to runs with an initial disk mass of $M_{disk} = 2\,M_L$ and the squares correspond to runs with $M_{disk} = 4\,M_L$. The filled triangles and squares are for those runs assuming $\epsilon_n = 0.1$, and the open ones are for those assuming $\epsilon_n = 0.01$. The theoretical estimate is also shown for $M_{esc} = 0$ (solid line) and $M_{esc} = 0.05\,M_{disk}$ (dotted line).

servation argument assumes that the accretion disk is flat and that all material left in Earth's orbit has been accreted into a single moon.

From Equations 2 and 3, we obtain

$$M = \frac{(j_{disk} - j_{fall})M_{disk} + (j_{fall} - j_{esc})M_{esc}}{j - j_{fall}}. \tag{4}$$

Substituting the mean values of specific angular momenta obtained by the simulations into Equation 4, yields

$$\frac{M}{M_{disk}} \simeq 1.9\frac{j_{disk}}{\sqrt{GM_{\oplus}a_R}} - 1.1 - 1.9\frac{M_{esc}}{M_{disk}}. \tag{5}$$

This estimate is also shown in Figure 3 (right).

Since M_{esc} is always much smaller than M_{disk}, we can neglect the M_{esc} terms in Equation 4. In this case M is a function of j, j_{fall}, j_{disk}, and M_{disk}. However, j and j_{fall} are not free parameters but always have almost the same values, since j is determined by the fact that the moon forms just outside the Roche limit and j_{fall} by the fact that remaining particles collide with the Earth. Then, the distribution of the disk mass to the moon and the Earth impactors is determined by the conservation of angular momentum. As the mass of the escapers is small compared with the disk mass, we can predict the mass of the moon from Equation 5 when the mass and the angular momentum of the disk are given.

The results of the N-body simulation deviate a little from the analytical estimate at low (~ 0.6) and high (~ 1.0) j_{disk}. At the low end, the mass of the moon obtained by the simulations is larger than the analytical estimate, because the semi-major axis of these moons is smaller than the mean values used in Equation 4. As the moons in the low j_{disk} cases tend to be smaller in general, they suffer less gravitational recoil the disk and move outward by a smaller distance, yielding a smaller moon semi-major axis than the mean value. For the high j_{disk} cases, the analytical estimate of the lunar mass is larger than that obtained by the N-body simulations. At the end of these simulations, there are still about 1000 particles exterior to the moon, so that accretion is not yet complete. In fact, the sum of the mass of the moon and the mass of the particles bound to the Earth exterior to the moon (which would likely be the final moon mass), is more consistent with the analytical estimate.

6. Summary

We performed the high-resolution N-body simulations ($N = 10\,000$) of lunar accretion from various circumterrestrial disks. We found that as a consequence of the evolution of a particulate circumterrestrial disk, a single large moon on a nearly non-inclined circular orbit is formed just outside the Roche limit. This result hardly depends on the initial condition of the disk, as long as $0.62\sqrt{GM_\oplus a_R} \leq j_{disk} \leq 1.0\sqrt{GM_\oplus a_R}$, $M_{disk} = 2\,M_L$–$4M_L$, and $\epsilon_n = 0.01$–0.1, which may include the plausible conditions for the impact-generated disk. The moon is always formed around $a \simeq 1.3\,a_R$. In this case the mass of the moon is predicted simply by conservation of angular momentum from the initial disk. The accretion yields (the fraction of disk material incorporated into the moon) range from 10% to 55%.

The time scale of lunar accretion is of the order of 100 T_K ($\sim$1 month). This time scale hardly depends on the detailed initial conditions of the disk we simulated, since the time scales of important processes in lunar formation such as mass transfer is regulated mainly by the surface density of the disk.

References

Cameron, A. G. W. 1997, Icarus, 126, 126

Cameron, A. G. W., & Benz, W. 1991, Icarus, 92, 204

Cameron, A. G. W., & Ward, W. R. 1976, in Lunar and Planetary Science VII (Houston: Lunar and Planetary Institute), 120

Canup, R. M., & Esposito, L. W. 1996, Icarus, 119, 427

Hartmann, W. K., & Davis, D. R. 1975, Icarus, 24, 504

Kokubo, E., & Ida, S. 1998, Icarus, 131, 171

Kokubo, E., Ida, S., & Makino, J. 2000, Icarus, 148, 419

Kokubo, E., Yoshinaga, K., & Makino, J. 1998, MNRAS, 297, 1067

Makino, J., Taiji, M., Ebisuzaki, T., & Sugimoto, D. 1997, ApJ, 480, 432

Stevenson, D. J. 1987, AREPS, 15, 271

Wetherill, G. W., & Stewart, G. R. 1989, Icarus, 77, 330

Discussion

Alan Guth: Do you know how much the results depend on the size of the initial particles? I would think that since you're talking about accretion it would make a big difference how small the particles are to start.

Eiichiro Kokubo: Yes, in principle the result depends on the number of particles or the size of particles, but if the number is larger than about 3000, the result doesn't depend on the number or size of particles. Because the basic physics is determined by the surface density of the disk, which is independent of the size or number of particles.

Bill Ward: Just one comment: Al Cameron has done a number of simulations of the impact itself, and my colleague Robin Canup at Southwest Research Institute is doing similar work now but with a different code. She's confirming some of his original claims, which is that most of the material is actually not placed into orbit inside the Roche limit, but that it actually starts with probably half of it outside the Roche limit. Therefore, you don't need to spread material out there to form the Moon, but once the Moon forms then we predict that it actually rams the remainder of the disk into the Earth as it tidally evolves. I'd like to encourage you to look at simulations where the disk is much more extensive in the beginning.

Frank Shu: My impression was that if you evolve the Moon backwards from its present conditions, you find that it is not in the equatorial plane of the Earth at the initial instant. Does this mean that in this hypothesis the Earth had considerable spin at the time of the impact which is misaligned?

Eiichiro Kokubo: It does not necessarily mean as you think. The present inclination of the Lunar orbit against the equatorial plane is about 5°, and this inclination can be raised by the interaction between the debris disk and the Moon. I think Bill has done this work. [To Bill Ward] Do you have any comment?

Bill Ward: We've always thought that one of the main counterarguments against the impact hypothesis was exactly this problematic Lunar inclination, which is a very longstanding puzzle. It seems almost contrary to almost any theory of formation of the Moon except for maybe capturing it. That's the reason I work with Robin Canup on this; we've suggested that it actually is an interaction of the Moon with the remainder of the disk which is taking angular momentum out of it, and the inner 3-to-1 vertical resonance with the Lunar disk is capable of pushing the Moon out of the equatorial plane by about 10° by the time the disk is destroyed, and that seems to fall right on the remaining tidal trajectory that you get when you calculate backwards. We just proposed that last year in an article in Nature.

Astrophysical Ages and Time Scales
ASP Conference Series, Vol. 245, 2001
T. von Hippel, C. Simpson, N. Manset

Radioactivity and Solar System Formation Time Scales

Typhoon Lee

Institute of Earth Science and Institute of Astronomy and Astrophysics, Academia Sinica, Nankang, Taipei, Taiwan

Abstract. Radioactive decays of unstable nuclei are highly reliable clocks because their rates are, with few exceptions, independent of environmental conditions. They date the time elapsed since the last time two samples had the same initial isotopic ratio but different parent/daughter elemental ratios. Such radiometric ages are accurate as long as there were no transport of these elements. For instance, the age of the angrites meteorites has been dated by the ^{235}U decays to be 4557.8 $\pm$ 0.9 Myr. This example shows that the solidification of certain magmas in a differentiated asteroid can be dated to a precision of 1/5000. Radionuclides with shorter half lives ($T_{1/2}$) than that of ^{235}U are extinct by now but are better suited to date events in the early Solar System that evolved on time scales < 1 Myr. The occurrence of extinct nuclides in samples with absolute ages firmly establishes that the formation of our Sun began no earlier than 4.57 Gyr ago. As to the planets, new constraints came from the decay of ^{182}Hf to ^{182}W ($T_{1/2}$ = 9 Myr). This decay has been used to infer that the times of metal segregation in chondrites and differentiated asteroids were within ten million years of each other. In contrast, the estimates for the formation of Earth's core and the Moon were at least 50 Myr later and the Martian core took 30 Myr. Extinct nuclides with $T_{1/2}$ < 6 Myr could have been produced by energetic particles from the young Sun or injected by nearby asymptotic giant branch stars and supernovae. Therefore, their chronological interpretation is currently ambiguous. On the other hand, those with $T_{1/2}$ > 6 Myr could have come from steady state r-process nucleosynthesis in the Galaxy. However, the new ^{182}Hf data imply that two distinct types of r-process sources were required with the one responsible for the lower mass range around 130 amu occurring at least ten times less frequently.

Astrophysical Ages and Time Scales
ASP Conference Series, Vol. 245, 2001
T. von Hippel, C. Simpson, N. Manset

On the Accretion of Distant Planets

William R. Ward

*Department of Space Studies, Southwest Research Institute, 1050
Walnut Street, Boulder, CO 80302, USA*

Craig B. Agnor

*Department of Space Studies, Southwest Research Institute, and
Department of Physics, University of Colorado, Boulder, CO 80302,
USA*

Hidekazu Tanaka

*Department of Space Studies, Southwest Research Institute and
Department of Earth and Planetary Sciences, Tokyo Institute of
Technology, Tokyo 152-8551, Japan*

Abstract. We consider the accretion time scale of planetary objects
at large stellar distances such as Neptune. The shallow stellar potential
well inhibits the formation of large, distant companions in three important ways: (1) orbital periods are longer, and so more remote regions are
dynamically younger; (2) the volume occupied by accreting material is
larger, contributing to low collision frequencies; and (3) the low gravitational binding energy to the star makes it easier for a planetary embryo
to scatter planetesimals out of the system. We suggest that resonant
damping of embryo dispersion velocities by a dissipating remnant of their
precursor gas disk may mitigate this situation and make the existence of
large, outer planets like Neptune easier to explain. Disk torques dynamically cool a system of embryos at the expense of more remote regions of
the gas disk by driving acoustic waves at coorbiting Lindblad resonance
sites. This lowers embryo eccentricities and inclinations; the smaller dispersion velocities increase the embryos' collision cross sections, which in
turn shortens accretion times and results in less material ejected from the
system. If the gas surface density has decreased to the point in which it
is comparable to that of the condensed solids, the embryo's equilibrium
eccentricities become comparable to the normalized scale height of the
gas disk. In such a case, the disk torques causing orbit migration are significantly weakened and the protoplanet could be relatively stable against
type I decay.

1. Introduction

A long standing puzzle regarding the formation of the Solar System concerns
the accretion of the outermost planets, Uranus and Neptune. Despite successes
in modeling the formation of the terrestrial planets, these same concepts seem
to fall short for the most remote planets of our Solar System. For one thing,

111

because orbital periods P in a Keplerian disk are longer at large heliocentric distances ($P \propto a^{3/2}$, where a is the semi-major axis) such regions are dynamically younger. Secondly, the volume of space occupied by planetesimals is so large that collision frequencies are very low—leading to accretion time scales that seem to be unacceptably long, i.e., greater than the age of the Solar System (Lissauer et al. 1995; Levison & Stewart 2001). A third issue relates to the low stellar binding energy of material in distant parts of the disk. This results in large orbital eccentricities $e \sim v/a\Omega$ (where Ω is the mean motion) for a given planetesimal dispersion velocity, v. If a planetary embryo of mass M, radius R, and body density ρ, becomes large enough that its escape velocity, $v_{\rm esc} = \sqrt{2GM/R} = R\sqrt{8\pi G\rho/3}$, exceeds the escape velocity from the star, $\sqrt{2GM_\star/r} = \sqrt{2}r\Omega$, it becomes a very efficient scatterer and can eject much of the material out of the system before it can be accreted. This occurs for embryo masses greater than

$$M/M_\star = \sqrt{3M_\star/4\pi\rho r^3} = 4 \times 10^{-4} \rho_{\rm cgs}^{-1/2}(r/{\rm AU})^{-3/2}\sqrt{M_\star/M_\odot}, \qquad (1)$$

which is only a fraction of an Earth mass in the outer Solar System beyond the orbital distance of Uranus, $r \geq 20$ AU. The masses of Uranus and Neptune exceed this threshold by a significant margin. Indeed, their scattering properties are implicated in the formation of the Oort cloud and the scattered disk component of the Kuiper belt (Duncan & Levison 1997; see also Malhotra, Duncan, & Levison 2000, and references therein).

Here we explore the effect of a small remnant of the gas disk on the accretion process. The H/He content of Neptune is only a few percent that the gas giants. If this is indicative of a reduced, but non-zero gas density in that region during the final assembly of this planet, its presence can have a surprisingly important influence on the dynamics of forming embryos. Resonant interaction of the embryos with the tenuous gas disk will damp embryo dispersion velocities to some fraction of the gas sound speed, largely independent of their mass (Ward 1993). This allows for a significant enhancement in the collision cross section (by a factor $F_{\rm g} \sim 1 + (v_{\rm esc}/v)^2 \sim$ few $\times 10^2$), and shortens the accretion time scale accordingly. We argue here that this behavior, coupled with fresh insight into the phenomenon of protoplanet migration at high eccentricities (Papaloizou & Larwood 2000) provides a new avenue for accretion of the outermost planets.

2. Standard Accretion Theory

The characteristic time scale of the growth of a planetary embryo from a disk of material of surface density $\sigma_{\rm d}$ about a solar mass star can be written

$$\tau_{\rm acc} \sim \frac{\rho R}{\sigma_{\rm d}\Omega}\left(\frac{c_{\rm acc}^{-1}}{F_{\rm g}}\right) = 3\times10^{10}\left(\frac{c_{\rm acc}^{-1}}{F_{\rm g}}\right)\left(\frac{M}{M_\oplus}\right)^{1/3}\left(\frac{r}{30{\rm AU}}\right)^{3/2}\left(\frac{\rho^{2/3}}{\sigma_{\rm d}}\right)_{\rm cgs}{\rm yr}, \quad (2)$$

where $M_\oplus = 6 \times 10^{27}$ g is an Earth mass and $c_{\rm acc}$ is a constant of order unity that depends in part on the I/e ratio of the planetesimals' orbital inclinations, I, and eccentricities, e, (e.g., Lissauer & Stewart 1993). If we set $M \sim 15M_\oplus$, $\rho \sim 1.64$ g cm^{-3} (the density of Neptune), $c_{\rm acc} \sim 1$, $r = 30$ AU, then Equation 2 reads, $\tau_{\text{acc}} \approx 10^{11} \sigma_{\text{d,cgs}}^{-1} F_{\text{g}}^{-1}$ yr. The surface density is of order $\sigma_{\text{d}} \sim M_{\text{d}}/\pi r^2 \sim 10^{-2} (M_{\text{d}}/M_{\oplus})(r/30\text{AU})^{-2} \text{g cm}^{-2}$; note to account for at least the ~ 17 Earth masses contained in Neptune, $\sigma_{\text{d}} \gtrsim 0.2$ g cm^{-2}. The value of the enhancement factor F_{g} describes the increase in the cross section due to gravitational focusing and can affect the time scale estimate considerably. If the dispersion velocity v is due to mutual scattering among field particles of radius R', it will acquire a value on the order of their escape velocities, v'_{esc} (e.g., Safronov 1969) and $F_{\text{g}} \approx (R/R')^2$. This results in $\tau_{\text{acc}} \propto 1/R$ and the well-known accretion runaway behavior. Depending on the size differential, F_{g} could become quite large, although there are limits (e.g., Greenzweig & Lissauer 1990; Ida & Makino 1993).

A principle limiting factor is the influence of the target itself on the field particles' dispersion velocity (e.g., Lissauer 1987). The stirring action of the embryo generates a minimum dispersion velocity of order $v \sim c_{\text{H}} R_{\text{H}} \Omega$, where $R_{\text{H}} \equiv r(M/3M_{\star})^{1/3}$ is the embryo's Hill radius (e.g., Ida & Makino, 1993). This sets the maximum value for F_{g} at $F_{\max} \sim 6 c_{\text{H}}^{-2} (4\pi \rho r^3/3M_{\star})^{1/3} \approx \hat{F} \times 10^3 (r/30$ AU$)$, which is independent of embryo size. Numerical experiments typically yield scaled velocities in the range $c_{\text{H}} \approx 4$–6 (e.g., Tanaka & Ida 1997), so that the value of $\hat{F} \equiv 28/c_{\text{H}}^2$ is likely of order unity. The target-to-field-particle size ratio at which this limit is reached is $R/R' \sim \sqrt{F_{\max}}$. Past that point, growth times are again proportional to R; this is the so-called oligarchic growth. Nevertheless, the rate remains fast enough to form an $M \sim 15$ Earth mass object in $\sim 10^8$ yr if it could continue unabated (e.g., Lissauer et al. 1995, see Figure 1).

However, unabated oligarchic growth is an important caveat because if $\tau_{\text{acc}} \propto R$, there may well be time for other embryos to initiate runaway growth and close the size ratio. As competing embryos develop, they may exhaust the supply of field particles prematurely. Their mutual perturbations then begin to raise the general dispersion velocity, and the enhancement factor diminishes (e.g., Iwasaki et al. 2001). An embryo population undergoes gravitational relaxation at a rate of order (e.g., Stewart & Wetherill 1988)

$$\tau_{\text{scatt}} \sim \frac{c_{\text{scatt}}^{-1}}{\Omega} \left(\frac{M_{\star}}{M} \right) \left(\frac{M_{\star}}{\pi \sigma_{\text{d}} r^2} \right) \left(\frac{v}{r\,\Omega} \right)^4, \tag{3}$$

where $c_{\text{scatt}} \approx 0.4 \ln[(M_{\star}/M)(v/r\Omega)^3]$ is a constant of order a few, and collisional damping at a rate of order Equation 2. This implies that a population in collisional equilibrium tends to relax to a dispersion velocity comparable to the individual escape velocities of the embryos (e.g., Safronov 1969). Again, this results in: (1) an enhancement factor of order unity; (2) frequent scattering out of the system (if $v_{\text{esc}} > r\,\Omega$); and (3) a prohibitively long accretion time for the outermost planets (shown by the ordered growth curve in Figure 1).

3. Disk Damping

If the embryos are embedded in a gas disk of surface density, σ_{d}, and a gas sound speed, c, another damping mechanism operates involving wave interactions at

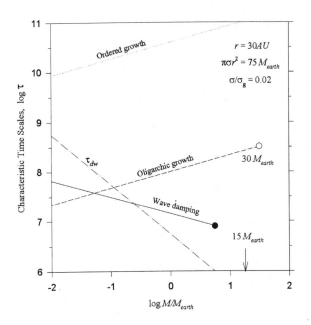

Figure 1.   Comparison of characteristic growth times $\tau \equiv M/\dot{M}$ for three styles of accretion: ordered, oligarchic, and wave assisted growth. Also shown is the characteristic orbital decay time $\tau_{\mathrm{dw}} \equiv r/\dot{r}$ due to disk torques. Solid to gas ratio is taken to be $\sigma_d/\sigma_g = 0.02$.

"coorbiting" Lindblad resonances (Ward 1993, 1988a, 1988b; Artymowicz 1993). The characteristic time scale for this process is

$$\tau_{\mathrm{dw}}^e \sim \frac{c_e^{-1}}{\Omega} \left(\frac{M_\star}{M} \right) \left(\frac{M_\star}{\pi \sigma_g r^2} \right) \left(\frac{c}{r\Omega} \right)^4, \qquad (4)$$

where c_e is a constant of order unity. Equating this to the scattering time scale [Equation 3] provides a new estimate for the dispersion velocity, viz.,

$$v \approx c_{\mathrm{dw}} (\sigma_d/\sigma_g)^{1/4} c, \qquad (5)$$

implying the embryos attain a mildly subsonic velocity in the gas independent of their size (Ward 1988b, 1993); a similar result has also been recently obtained both analytically and numerically by Papaloizou & Larwood (2000). In Equation 5, $c_{\mathrm{dw}} \equiv (c_{\mathrm{scat}}/c_e)^{1/4}$ is also of order unity and relatively insensitive to uncertainties in the other proportionality constants by virtue of its weak power dependence; henceforth we set $c_{\mathrm{dw}} = 1$. For example, in a solar composition gas, the abundance of solid material is only a small fraction of the gas, i.e., $\sigma_d/\sigma_g \approx 0.02$, so that $v \sim 0.4c$. Since the gas disk is thin, with a scale height $h \sim c/\Omega \ll r$, we have $v \lesssim c \ll r\Omega \ll v_{\mathrm{esc}}$, the last inequality being true for embryos larger than Equation 1. The damping furnished by the gas disk prevents the enhancement factor from decreasing to order unity if and when the oligarchic phase of accretion breaks down. The corresponding enhancement factor using Equation 5 reads

$$F_g \sim 1 + (v_{esc}/c)^2 \sqrt{\sigma_g/\sigma_d}. \tag{6}$$

Assuming $v_{esc} \gg c(\sigma_d/\sigma_g)^{1/4}$, the characteristic accretion time scale becomes

$$\tau_{acc} \sim \frac{3c_{acc}^{-1}}{8\Omega} \left(\frac{r}{R}\right) \left(\frac{M_\star}{\pi\sigma_d r^2}\right) \left(\frac{c}{r\Omega}\right)^2 \sqrt{\sigma_d/\sigma_g}. \tag{7}$$

We obtain $\tau_{acc} \approx 1 \times 10^{10} c_{acc}^{-1} \sigma_{d,cgs}^{-1} (M_\oplus/M)^{1/3} (c/r\Omega)^2 \sqrt{\sigma_d/\sigma_g}$ yr for the values employed here, which is shown in Figure 1 for the solar composition gas/solid ratio. Consequently, a 15 $M_\oplus$ object in a gas disk with a normalized scale height of $c/r\Omega \sim O(10^{-1})$ has a characteristic accretion time scale of order $\tau \approx 4 \times 10^7 \sigma_{d,cgs}^{-1} \sqrt{\sigma_d/\sigma_g} \approx 6 \times 10^6$ yr. Note however, that $\tau \propto 1/R$ so that more of the time is expended at the smaller sizes. This enhanced accretion rate due to dispersion velocity damping by a gas disk was first described by Ward (1993), but two provisos must be reiterated: (1) the system should not be over-damped to the degree that individual embryos become isolated on non-crossing orbits[1]; and (2) the system must survive against orbital decay due to disk torques over the accretion time scale.

The first criterion can be satisfied by requiring $v/\Omega \gtrsim M/2\pi\sigma r \equiv \Delta r$, which implies a lower limit on the surface density of solids, viz.,

$$\frac{\pi\sigma_d r^2}{M_\oplus} \gtrsim \frac{1}{2} \left(\frac{M}{M_\oplus}\right) \left(\frac{r\Omega}{c}\right) \left(\frac{\sigma_g}{\sigma_d}\right)^{1/4}. \tag{8}$$

For a 15 $M_\oplus$ planet in a disk with $c/r\Omega \sim O(10^{-1})$, Equation 8 reads $\sim 75(\sigma_g/\sigma_d)^{1/4}$ Earth masses with a corresponding surface density of $\sigma_d \gtrsim 0.7(\sigma_g/\sigma_d)^{1/4}(r/30\text{AU})^2\text{g cm}^{-2}$. For the parameters adopted in Figure 1, over-damping may occur at less than $10M_\oplus$ as indicated by the black circle terminating the time line. A solar composition disk would require $\pi\sigma_d r^2 \sim 200M_\oplus$, $\sigma_d \sim 2$ g cm^{-2} to avoid this.

Concerning the second proviso, the orbital decay time of embryos τ_{dw}^a due to type I decay is

$$\tau_{dw}^a \sim \frac{c_a^{-1}}{\Omega} \left(\frac{M_\star}{M}\right) \left(\frac{M_\star}{\pi\sigma_g r^2}\right) \left(\frac{c}{r\Omega}\right)^2, \tag{9}$$

which exceeds the dispersion velocity damping time [Equation 4] by a factor of order $c_e(r\Omega/c)^2/c_a$, where c_a is the so-called torque asymmetry parameter (Goldreich & Tremaine, 1980; Ward 1986, 1989, 1997; Artymowicz, 1993). For

---

[1] In this case, further encounters must be generated via differential drift rate (Ward 1993; Tanaka & Ida 1999) which results in a different time scale.

our problem, this reads $\tau_{dw}^a \approx 2.7 \times 10^8 c_a^{-1} \sigma_{d,cgs}^{-1} (M_\oplus/M)$ yr. Current best efforts at estimating the torque asymmetry parameter for a body in a circular orbit suggest $c_a = O(1)$ (Tanaka, Takeuchi, & Ward 2001). Equation 9 is also shown in Figure 1 for comparison.

The ratio of accretion time to decay time is

$$\frac{\tau_{acc}}{\tau_{dw}^a} \sim \left(\frac{c_a}{c_{acc}}\right) \frac{\rho R^2 r}{M_\star} \sqrt{\sigma_g/\sigma_d} \approx \frac{1}{3} \left(\frac{c_a}{c_{acc}}\right) \left(\frac{M}{M_\oplus}\right)^{2/3} \sqrt{\sigma_g/\sigma_d}, \qquad (10)$$

which is independent of scale height and surface densities, except for their ratio. The time ratio is thus $\sim 2.3(M/M_\oplus)^{2/3}$ for the conditions of Figure 1, implying that objects a few tenths of an Earth mass are in danger of decaying out of the disk before they can grow much more (e.g., Ward 1996, 1998; Tanaka & Ida 1999). However, this situation could be mitigated by disk dissipation as discussed next.

4.   Remnant Disk

From Equation 7, the e-folding time to grow $10^{-1}M_\oplus$ bodies at $r \sim 30$ AU is few $\times 10^7$ yr. Since the lifetime of gas disks are typically estimated to be of this order, the ratio σ_g/σ_d is likely to be significantly less than its starting value by then. A decreased gas density slows the orbital decay rate proportionally, but has a more modest effect on the accretion time through the dispersion velocity, which approaches the sound speed of the gas. Indeed, this is why the ratio [Equation 10] depends on $\sqrt{\sigma_g/\sigma_d}$. Note that although removing the gas entirely would revert the system to the case where F_g is nearly unity, only a slight residual of the gas disk (i.e., $\sigma_g/\sigma_d \sim O(1)$) is capable of keeping dispersion velocities near Mach one. In fact, the time scale ratio can be made to approach unity for any desired mass, M, by choosing $\sigma_g/\sigma_d \sim 9(c_{acc}/c_a)^2 (M_\oplus/M)^{4/3}$; for $M \sim 17M_\oplus$, this reads $\sim 0.2(c_{acc}/c_a)^2$. Again, the small, but non-negligible, H/He content of Neptune seems consistent with a minor fraction of gas having remained in the vicinity during its formation.

However, there is another key issue that alleviates the situation, that is the recent demonstration by Papaloizou & Larwood (2000) that large eccentricities (i.e., $e \sim v/r\Omega$) also modify the migration rate by elevating the importance of terms in the embryos' disturbing functions that are higher order in e. The added resonant interactions with the disk result in an extra factor of order $\approx [1+(e\,r\,\Omega/c)^5]/[1-(e\,r\,\Omega/c)^4]$ in the migration time scale, τ_{dw}^a. From Equation 5, this results in an additional factor of $[1+(\sigma_d/\sigma_g)^{5/4}]/[1-\sigma_d/\sigma_g]$ in Equation 9. Thus, the migration can slow and even reverse direction as the disk dissipates and the radial excursions of the embryo ($\delta r \sim e\,r \sim v/\Omega$) become comparable to the disk scale height, $h \sim c/\Omega$. Under such conditions, embryos may instead execute a more diffusive migrational behavior, as mutual scattering varies their dispersion velocities around the mean [Equation 5] between subsonic to mildly supersonic values. A strong scattering event tends to increase or decrease the dispersion velocity by amounts comparable to itself so that the PL-correction factor should vary between $\pm O(1)$ between scattering events. An embryo could

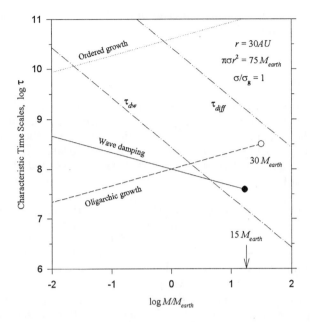

Figure 2.   Same as Figure 1 except that gas to solid ratio is unity. Also shown is the inferred embryo diffusion time scale from incorporating the Papaloizou-Larwood correction factor into the orbit decay rate. For comparison, the orbit decay time without this factor is included as well.

drift a distance $\delta r \sim \pm r(\tau_{\mathrm{scatt}}/\tau_{\mathrm{dw}}^a)$ during this time, where τ_{dw}^a refers to the uncorrected migration rate of Equation 9. This implies a diffusion rate that satisfies $d(\Delta r)^2/dt \approx (\delta r)^2/\tau_{\mathrm{scatt}} \sim r^2\tau_{\mathrm{scatt}}/(\tau_{\mathrm{dw}}^a)^2$. Since equilibrium maintains $\tau_{\mathrm{scatt}} \sim \tau_{\mathrm{dw}}^e$, the time to diffuse a distance r is

$$\tau_{\mathrm{diff}} \sim \tau_{\mathrm{scatt}}(r/\delta r)^2 \sim \tau_{\mathrm{scatt}}(\tau_{\mathrm{dw}}^a/\tau_{\mathrm{scatt}})^2$$
$$\approx \tau_{\mathrm{dw}}^a(\tau_{\mathrm{dw}}^a/\tau_{\mathrm{dw}}^e) \approx \tau_{\mathrm{dw}}^a(r\Omega/c)^2 \gg \tau_{\mathrm{dw}}^a. \qquad (11)$$

With a gas density equal to that of solids, $\tau_{\mathrm{acc}} \sim 1\times10^8 c_a c_{\mathrm{acc}}^{-2}\sigma_{\mathrm{d,cgs}}^{-1}(M_\oplus/M)^{1/3}$ yr and $\tau_{\mathrm{diff}} \approx 2.7\times10^{10}c_a^{-1}(M_\oplus/M)$ yr. Figure 2 shows these time scales as well as τ_{dw}^a without the PL eccentricity correction. The critical mass to prevent over-damping is now $\sim 75\,M_\oplus$ so that the time line is extended to a Neptune mass.

For $M \sim 1\,M_\oplus$, the characteristic accretion time is $\sim 1\times10^8$ yr, which is the most sluggish period of growth. Assuming an exponentially decaying gas density, $\sigma_g \propto e^{-t/\tau_g}$, a decrease by a factor of 0.02 within this time interval requires a disk dissipation e-folding time of $\tau_g \sim 3\times10^7$ yr. Note that recent direct observations of hydrogen suggests an extended the range of possible disk lifetimes around young stars to few $\times10^7$ yr (Thi et al. 2001). Of course, $\tau_g = \sigma_g/\dot\sigma_g$ itself may vary in time, depending on the source(s) of dissipation such as stellar UV and particle fluxes and/or disk viscosity.

5. Summary

We have described a set of circumstances that may help account for the ability of large, (i.e., $15M_\oplus$) planets to form at distances where their orbital velocities are smaller than the planetary escape velocities. We invoke the existence of a small remnant ($\sim 1\%$) of the original gas disk during the late-stage formation of Neptune—an assumption that seems in consonance with its small H/He content. The presence of the gas phase alters the dynamics of accreting embryos in two keys ways:

(1) Disk torques, primarily those generated at coorbiting Lindblad and vertical resonances, damp eccentricities and inclinations among a swarm of mutually interacting embryos. The resulting equilibrium dispersion velocities are comparable to the sound speed of the gas and are independent of embryo mass (Ward 1993). This promotes accretion by enhancing the collisional cross section by a factor $F_g \sim (v_{esc}/c)^2 \approx 3 \times 10^2 (M/M_\oplus)^{2/3}$.

(2) The orbital decay of embryos due to disk torques is largely stalled. This occurs because coupling between the orbit and the disk is weaker for eccentric orbits due to the larger relative velocity between the protoplanet and local disk material (Papaloizou & Larwood 2000). When eccentricities are comparable to the scale height of the gas disk, $e \sim c/r\Omega$, terms in the disturbing function proportional to e^n, $n > 0$, begin to contribute significantly to the migration rate. The resulting net torque is considerably reduced, and can undergo a sign reversal. Instead of secular orbital decay, embryos may execute slow radial diffusion on a much longer time scale. These new insights into protoplanet-disk interactions suggest that an in situ formation of Neptune may be possible in a few $\times 10^8$ yr, and further exploration of accretion in the presence of a remnant gas disk seems warranted. These should include numerical experiments similar to those performed by Papaloizou and Larwood at 1 AU, but conducted for the outer Solar System with a time dependent gas phase. The results of such an investigation may not only illuminate the existence of Uranus and Neptune, but could also place constraints on models of nebula dissipation.

References

Artymowicz, P. 1993, ApJ, 419, 166

Duncan, M. J., & Levison, H. F. 1997, Science, 276, 1670

Goldreich, P., & Tremaine, S. 1980, ApJ, 241, 425

Greenzweig, Y., & Lissauer, J. J. 1990, Icarus, 87, 40

Ida, S., & Makino, J. 1993, Icarus, 106, 210

Iwasaki, K., Tanaka, H., Nakazawa, K., & Emori, H. 2001, PASJ, 53, in press

Levison, H., & Stewart, G. R. 2001, Icarus, in press

Lissauer, J. J. 1987, Icarus, 69, 249

Lissauer, J. J., Pollack, J. B., Wetherill, G. W., & Stevenson, D. J. 1995, in Neptune and Triton, ed. D. P. Cruikshank (Tucson: University of Arizona Press), 37

Lissauer, J. J., & Stewart, G. R. 1993, in Protostars & Planets III, ed. E. H. Levy & J. I. Lunine (Tucson: University of Arizona Press), 1061

Malhotra, R., Duncan, M. J., & Levison, H. F. 2000, in Protostars and Planets IV, ed. V. Mannings, A. P. Boss, & S. S. Russel (Tucson: University of Arizona Press), 1231

Papaloizou, J. C. B., & Larwood, J. D. 2000, MNRAS, 315, 823

Safronov, V. 1969, Evolution of the Protoplanetary Cloud and the Formation of the Earth and Planets (Moscow: Nauka Press); also NASA-TT-F-677, 1072

Stewart, G. R, & Wetherill, G. W. 1988, Icarus, 74, 542

Tanaka, H., & Ida, S. 1997, Icarus, 125, 302

Tanaka, H., & Ida, S. 1999, Icarus, 139, 350

Tanaka, H., Takeuchi, T., & Ward, W. R. 2001, ApJ, submitted

Thi, W. F., et al. 2001, Nature, 409, 60

Ward, W. R. 1986, Icarus, 67, 164

Ward, W. R. 1988a, Icarus, 73, 330

Ward, W. R. 1988b, Abs. Conf. Origin of the Earth, Lunar and Planetary Institute, 96

Ward, W. R. 1989, ApJ, 345, L99

Ward, W. R. 1993, Icarus, 106, 274

Ward, W. R. 1996, in ASP Conf. Ser. 107, Completing the Inventory of the Solar System, ed. T. W. Rettig & J. M. Hahn (San Francisco: ASP), 337

Ward, W. R. 1997, Icarus, 126, 261

Ward, W. R. 1998, in ASP Conf. Ser. 148, Origins, ed. C. E. Woodward, J. M. Shull, & H. A. Thronson (San Francisco: ASP), 338

Acknowledgments. This research was supported by grants from NASA's Origin of Solar Systems Program and Planetary Geology and Geophysics Program. C. B. A. also acknowledges support from NASA's Graduate Student Research Program.

Discussion

Martin Duncan: You talked about growing several of these embryos out to a tenth of an Earth mass or something and then going on. How sensitive is either the tidal migration inward or the damping time to the presence of other comparable-size embryos?

Bill Ward: Well, we really don't know that yet because detailed models are still to be developed, but it's an important consideration. However, you notice that the time scale for accretional growth assisted by wave damping has a downward slope in the figures, indicating that there's actually a runaway going on. The point is that you may well get one object that tends to out pace the others, so fully comparable embryos may not exist. They may be fairly large, but they're probably smaller than the dominant one, which would pretty much control the situation. There's not really an oligarchic growth at this point.

Astrophysical Ages and Time Scales
ASP Conference Series, Vol. 245, 2001
T. von Hippel, C. Simpson, N. Manset

Dust to Dust: Evidence for Planet Formation?

Glenn Schneider, Dean C. Hines, Murray D. Silverstone

Steward Observatory, University of Arizona, Tucson, AZ, USA

Alycia J. Weinberger, Eric E. Becklin

Division of Astronomy and Astrophysics, University of California, Los Angeles, CA, USA

Bradford A. Smith

Institute for Astronomy, University of Hawai'i, Honolulu, HI, USA

Abstract. We discuss the properties of several circumstellar debris disk systems imaged with the Hubble Space Telescope's Near Infrared Camera and Multi-Object Spectrometer in a survey of young stars with known far-IR excesses. These dusty disks around young ($\sim$ 5–8 Myr) unembedded stars exhibit morphological anisotropies and other characteristics which are suggestive of recent or on-going planet formation. We consider evidence for the evolution of populations of collisionally produced disk grains in light of the significant presence of remnant primordial gas in the optically thick disk of the classical T-Tauri star TW Hya; the dust-dominated and Kuiper-belt like circumstellar ring about the young main sequence star HR 4796A; and a possible "intermediate" case, the complex disk around the Herbig AeBe star HD 141569A. Only a small number of debris disks have been imaged thus far in scattered light. Yet all show structures that may be indicative of dust reprocessing, possibly as a result of planet formation, and speak to the contemporary competing scenarios of disk/planet evolution and formation time scales.

1. Introduction

Warm dust around young stars has been inferred from thermal infrared excesses since IRAS, though until recently the expected "cold" dust component had been imaged in circumstellar scattered light only about β Pictoris. To begin to understand both the evolution and dynamics of circumstellar disks, and the nascent planetary systems which they may harbor, a multi-wavelength attack is required. Mid-IR observations can reveal and trace emission from warm dust. Colder disk components can be detected, and on larger spatial scales mapped, in the sub-millimeter. Resolved images of dusty disks, at near-IR and optical wavelengths, provide direct information on the spatial distribution of the disk grains. Until recently, however, such observations have been extremely difficult, given the very high disk-to-star contrast ratios, but when successful provide measurements of radial and azimuthal asymmetries in the dust brightness distributions and scattering properties (colors, phase functions, etc.). The presence of rings, gaps,

121

clumps, warps and central holes seen in scattered light images of circumstellar disks may implicate the existence of embedded or co-orbital perturbers.

Studied extensively for fourteen years since first imaged by Smith & Terrile (1984), the disk around the Vega-like star β Pictoris served as the archetype and sole example of a dusty debris system seen in scattered light. Though now thought to be ~ 20 Myr old, throughout this period its age remained uncertain (and controversial) and was estimated from a few hundred Myr to about 10 Myr. The β Pictoris disk was found to possess at least five asymmetries attributed speculatively to dynamical interactions with unseen planetary bodies (Kalas & Jewitt 1995).

Space-based near-IR and optical coronagraphic imaging with the Hubble Space Telescope second generation instruments NICMOS and STIS has provided a powerful new tool for studying the circumstellar environments of nearby stars. Though only a very small number of spatially resolved images of dusty disk systems currently exist, these systems exhibit a diversity in disk sizes, morphologies, and properties (e.g., Figure 1).

The presence and morphologies of disk asymmetries and the derived properties and spatial distributions of the constituent grains may help to better constrain the ages and evolutionary status of the possible fledging extrasolar planetary systems. The presumed time scales for disk evolution will be tested and refined with an accumulation of observations of disk systems such as these.

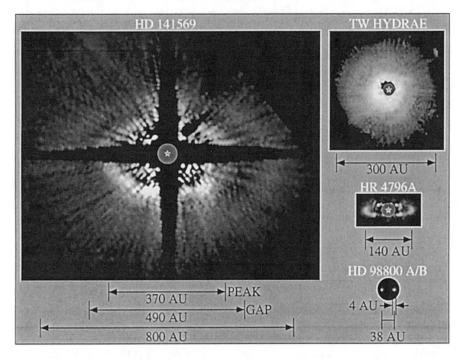

Figure 1.   Comparative sizes and morphologies of dust/debris systems around ~ 5–8 Myr stars imaged by NICMOS.

2. Disk Ages and Evolution

It is commonly conjectured that following the early stages of protostellar collapse, as the rocky cores of giant planets form (in ~ 0.1–1 Myr), primordial dust in the circumstellar environment is both dominated (in the ratio of ~ 100:1) by and locked to gas in the disk. Current theories of circumstellar disk evolution suggest a presumed epoch of planet-building, on the order of 1 Myr following the protostellar collapse, via the formation and agglomerative growth of embryonic bodies. Gaseous atmospheres will then accrete onto the giant planet cores on time scales of a few to about 10 Myr attendant with an expected significant decline in the gas-to-dust ratios in the remnant protostellar environments. During these times primordial dust (i.e., ISM-like grains) would be cleared from "typical" systems on shorter time scales by radiation pressure ($\sim 10^4$ yr), and by Poynting-Robertson drag (~ 1 Myr). In these critical evolutionary phases of newly-formed (or still forming) extrasolar planetary systems, the circumstellar environments become dominated by a second-generation dust population containing larger grains replenished through the collisional erosion of planetesimals, and perhaps, by cometary infall. As the circumstellar regions become optically thin (and the central stars become largely unembedded) the likely-evolving population of dusty debris at these early epochs become more readily observable in scattered light. This scenario suggests a morphological evolutionary sequence which could be modified by the perturbing influence of co-spatial bodies, and which may be explored (and validated) by high contrast imaging.

The presumption of evolution, however, requires a knowledge of the (relative) ages of disk systems, but such knowledge is not necessarily secure. Determining the ages of PMS and ZAMS stars depends strongly upon transforming observable quantities for placement on HR diagrams, and finding their isochronal ages with respect to their birth lines based upon theoretical evolutionary tracks. The fidelity (or lack thereof) of the stellar evolutionary models, and measurement errors in the observables (distances, luminosities, spectral temperatures, etc.) both contribute to uncertainties in derived ages (see Figure 2), which can be significant, particularly for earlier (Vega-like) stars (e.g., β Pic, HR 4796A, HD 141569A). The situation is improved if late spectral type coeval companions can be found, as is the case for HR 4796A (Jura et al. 1993) and HD 141569A (Weinberger et al. 2000), which are discussed in this paper. Three of the four disk systems we discuss (HR 4796A, TW Hya, and HD 98000A/B) are likely members of the TW Hya association (Webb et al. 1999),the nearest site of recent star formation to the Earth. Together with HD 141569A, the small sample of young disk systems we discuss here are of similar ages (~ 5–8 Myr), yet have very different properties.

3. Four ~ 5–8 Myr Disk Systems

Dusty disks were spatially resolved and imaged around three young (< 10 Myr old) stars. These disks show radial and hemispheric brightness anisotropies and complex morphologies, both possibly indicative of dynamical interactions with unseen planetary mass companions. From these and other observations we describe and compare of the properties of these dusty debris systems:

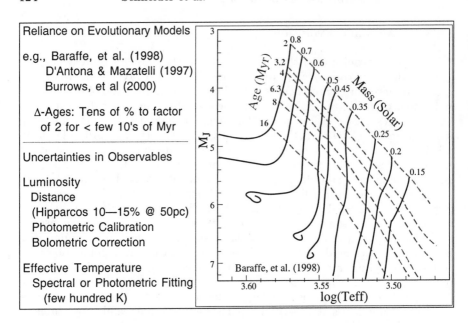

Figure 2.   Stellar (and hence, disk) age determinations are dependent upon uncertainties in both stellar models and measurements.

1. TW Hya (a K7Ve T Tau "old" PMS star) with a pole-on circularly nearly-symmetric disk, a radial break in its surface density of scattering particles, and possibly a radially and azimuthally confined arc-like depression.

2. HD 141569A (a Herbig Ae/Be star, $\sim$ 5 Myr) with a 400 AU radius inclined disk with a 40 AU wide gap at 250 AU.

3. HR 4796A (a A0V star, $\sim$ 8 Myr) with a 70 AU radius ring less than 14 AU wide and unequal ansal flux densities.

   Additionally, our non-detection of scattered light and high precision photometry of a fourth system of similar age, HD 98800 A/B, coupled with mid-IR imaging and longer wavelength flux density measurements, greatly constrain a likely model for the debris about the B component.

3.1.   TW Hydrae (TWA 1)

TW Hya is an isolated classical T Tauri star (Rucinski & Krautter 1983), exhibiting characteristic Hα and UV excesses and an IRAS far-IR excess with $L_{disk}/L_\star \sim 0.3$. Both sub-millimeter continuum (Weintraub, Zuckerman, & Masson 1989) and CO (Zuckerman, Forville, & Kastner 1995) emission have been observed. With a HIPPARCOS-determined distance of 56 ± 7 pc, and estimated age of 6 My, TW Hya is the archetypal member of the young stellar association which bears its name. These characteristics made TW Hya a prime candidate for scattered light disk imaging.

We found TW Hya to harbor an optically thick face-on disk (r ∼ 190 AU) seen in NICMOS F110W and F160W coronagraphic images, also imaged by Krist et al. (2000) with WFPC-2. Additionally, we recently observed TW Hya coronagraphically in the optical with STIS, and find the azimuthally averaged surface brightness profile is globally fit very well at all wavelengths with an $r^{-2.6}$ power law. The disk exhibits essentially gray scattering, implicating a characteristic particle size from the NIR colors of at least $2\mu m$, probably evolved from a primordial ISM-grain population. Mid-IR spectroscopy of this disk, obtained at Keck by Weinberger et al. (2001), shows a broad ∼ $10\mu m$ emission from amorphous silicates. Silicates of a few microns in size can explain the color, thermal extent, and shape of the mid-IR spectrum and further implicate grain growth from the original ISM population.

The brightness profile is explained by an outwardly flared disk with a central hole. Areal scattering profiles in the NIR bands reveal a break in the surface density of scatterers at R ∼ 105 AU, which may be indicative of sculpting of the disk grain distribution. At the same radius, an arc-like depression confined to about 90° in azimuthal extent is seen in both the higher spatial resolution STIS and WFPC-2 images (see Figure 3). This feature might arise from shadowing of the grains due to a discontinuity in the z-height distribution of the flared disk, or to relative deficit of scattering particles at that location. Either might arise from the gravitational effects of an embedded perturber.

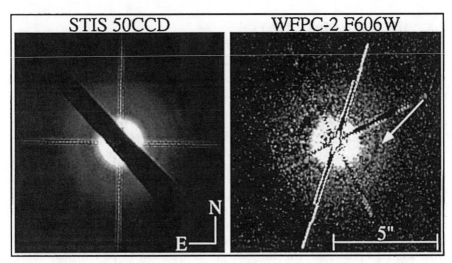

Figure 3.   TW Hya circumstellar disk. *Left:* STIS (0.59 μm, FWHM = 0.45 μm). *Right:* WFPC-2 (0.60 μm, FWHM = 0.20 μm) at the same scale and orientation. Arrow indicates location of the dark arc-like feature (see text) also seen in the STIS image.

3.2.   HD 141569A

The Herbig Ae/Be star HD 141569 (B9V, H = 6.89, d = 99 ± 8 pc, 2.3 $M_{\odot}$) was found to possess extended mid-IR emission to a radius of ∼ 1″.2 at 12.5–20.8μm

Marsh et al. (2001). NICMOS 1.1μm coronagraphic observations revealed a scattered light disk to a radius of at least 400 AU, exhibiting a complex morphology including a 40 AU wide gap in the surface brightness profile at a radius of 250 AU (Weinberger et al. 1999). The disk, with a total 1.1μm flux density of 8 ± 2 mJy beyond 0.''6 (peak surface brightness 0.3 mJy arcsec^{-2} at 185 AU) is inclined to our line-of-site by $51° \pm 3°$. Augereau et al. (1999) saw a similar morphology in lower-resolution 1.6μm observations, also obtained with NICMOS. No significant amount of scattered light was detected closer than r $\sim$ 1.''2, so the regions of warmer dust probed in the mid-IR, and the outer (colder) regions imaged in scattered light are mutually exclusive.

The intrinsic scattering function of the disk results in a brightness anisotropy in the ratio 1.5 ± 0.2, with the brighter side in the direction of forward scattering. The region of the gap may be partially cleared of material by an unseen co-orbital planetary companion. If so, the width/radius ratio of the gap implies a planetary mass of ~ 1.3 Jupiters. This is consistent with a < 3 Jupiter mass point-source detection limit at this radius, where we also estimate the albedo to be 0.35 ± 0.05. HD 141569A is the brightest member of an $\sim$ 6.5:1 hierarchical triple system ($\Delta A(\overline{BC}) = 8.''3$, $\Delta \overline{BC} = 1.''3$), where the presumed coeval M-dwarf dynamical companions were used by Weinberger et al. (2000) to improve the age estimate for the disk system, now ~ 5 Myr. The companions probably influence the dynamics of the circumstellar disk.

3.3.   HR 4796A (TWA 11A)

On 15 March 1998 NICMOS obtained the first scattered light image of a circumstellar debris disk since the discovery of the β Pictoris disk. With an age of 8 ± 2 Myr (Stauffer 1995), a spectral type of A0V, and possessing an M-dwarf companion, HR 4796A is similar to HD 141569A, yet the structure of its disk is very different. The dust in the disk about HR 4796A was found to be confined in a narrow ring 70 AU in radius, as had been suggested from mid-IR imaging by Koerner et al. (1998) and Jayawardhana et al. (1998), and < 14 AU in width (see Figure 1). Earlier, Jura (1991) inferred the presence of large amount of circumstellar dust from IRAS excess and estimated $L_{disk}/L_\star \sim 5 \times 10^{-3}$ (twice that of β Pictoris). Jura et al. (1995) noted their earlier 110K estimate of dust temperature indicated a lack of material at < 40 AU, and inferred grain sizes $> 3\mu$m at $40 < r < 200$ AU given the time scales for disk-clearing. Schneider et al. (1999) reported on the morphology, geometry and photometry of the ring-like disk from well-resolved NICMOS coronagraphic images at 1.1 and 1.6μm from which, unlike TW Hya, the disk grains appeared to be red (J(F110W)-H(F160W) $= 0.6 \pm 0.2$). Augereau et al. (1999) successfully reproduced the observed properties of the disk from all of the then-available observations in a two-component model with a) cold amorphous (Si and H_2O ice) grains $> 10\mu$m in size (cut-off in size by radiation pressure), with porosity ~ 0.6, peaking at 70 AU and b) hot dust at ~ 9 AU of "comet-like" composition (crystalline Si and H_2O), porosity ~ 0.97. They noted that collisionally evolved gains, with bodies as large as a few meters, were required in their model which also gave rise to a minimum mass of a few Earth masses with gas:dust < 1. This is consistent with subsequent sub-millimeter observations by Greaves, Mannings, & Holland (2000) wherein they estimated the total mass in gas as 1–7 Earth masses. This is also consistent with planetesimal accretion calculations by Kenyon et al. (1999) in which they find planet formation at 70 AU is possible in 10 Myr in an initial 10–20 minimum mass solar nebula where dust production is then confined to a ring with Δa = 7–15 AU. Possible evidence for one (or more) unseen planets exists in an $\sim$ 10–15% brightness asymmetry in the NE and SW ansae of the ring seen both in the NICMOS images and by Telesco et al. (2000) in 18.2μm OSCIR images, suggesting a pericentric offset possibly due to a gravitational perturber. While HR 4796B, an M-dwarf companion at a projected distance of 500 AU, may serve to truncate the outer radius of the disk, the narrowness of the ring might implicate co-orbital companions confining the dust through a process akin to the shepherding of ring particles in the Saturnian system.

3.4. HD 98800A/B (TWA4 A/B)

HD 98800, historically classified as a binary comprised of two similar K dwarfs (currently separated by 0″8), was found by IRAS to contain one of the brightest far-IR excesses in the sky. Now a recognized member of the TW Hya Association with a HIPPARCOS-determined distance of 46.7 ± 6 pc, the two PMS components are themselves spectroscopic binaries with periods (Aa+Ab) = 262 d, (Ba+Bb) = 315 d (Torres et al. 1995) and separations of $\sim$ 1 AU. Gehrz et al. (1999) showed the debris is centered on the B component from 4.7 and 9.8μm observations, and 20% of the luminosity of B is emitted in a 164±5 K SED from the mid-IR to the sub-millimeter. The SED is fit very well by a single temperature black-body, indicating that the grains co-exist in a very limited radial zone from the central stars. High-precision NICMOS photometry straddling peak of stellar SEDs by Low, Hines, & Schneider (1999) find T_{eff}(A) = 3831 ± 55 K, T_{eff}(B) = 3459 ± 37 K, and no detectable 0.9–1.9μm excess. From this they suggest the scattered:total light from B is < 6% implying an albedo of < 0.3 for the debris system with an inner radius of 4.5 AU, subtending $\sim$ 20% of the sky seen from the B component. Koerner et al. (2000) confirmed the Ba+Bb circumbinary disk, and that the B components are the source of the large IR excess upon which a silicate feature is imposed. From mid-IR imaging they suggest a disk with properties similar to Low et al. (1999). Unblended optical (0.5–1.0μm) spectra of the A and B components we recently obtained with STIS indicate that the B component closely resembles an M0V star, so it slightly later in spectral type than previously thought (but similar in this regard to TW Hya).

The inferred geometry and properties of the HR 98800B disk bears a resemblance to the zodiacal bands in our own Solar System, and may be similar to the debris system around our Sun as it appeared a few million years after formation. The multiplicity of the system undoubtedly complicates the dynamics, and hence the temporal stability and evolution of the grains. The small size of the B-component circumbinary debris system may be causally related to interactions of the grains with the multiple components in the system.

4. Summary

The four dusty disks considered here exhibit significant variations in sizes, morphologies, and grain properties, despite their similar ages, and differ as well from β Pictoris (of very similar spectral type to HR 4796A and HD 141569A). The desire to construct a "morphological evolutionary sequence" for dusty debris disks is physically complicated by the dispersions in stellar spectral types (and hence masses), compositions and densities of the parent molecular clouds, and disk interactions with stellar and sub-stellar companions. In addition, uncertainties in age determinations by as much as a factor of about two from observable diagnostics and theoretical models, further muddy the waters. The sample of such disks observed to date is very small. Obviously, many more observations are needed to advance our detailed understanding of disk/planet formation mechanisms and time scales.

References

Augereau, J. C., Lagrange, A. M., Mouillet, D., Papaloizou, J. C. B., & Grorod, P. A. 1999, A&A, 348, 557

Gehrz, R. D., Smith, N., Low, F. J., Krautter, J., Nollenberg, J. G., & Jones, T. J. 1999, ApJ, 512, L55

Greaves, J. S., Mannings, V., & Holland, W. S. 2000, Icarus, 143, 155

Jayawardhana, R., Fisher, S., Hartmann, L., Telesco, C., Pina, R., & Fazio, G. 1998, ApJ, 503, L79

Jura, M. 1991, ApJ, 383, L79

Jura, M., Ghez, A. M., White, R. J., McCarthy, D. W., Smith, R. C., & Martin, P. G. 1995, ApJ, 445, 451

Jura, M., Zuckerman, B., Becklin, E. E., & Smith, R. C. 1993, ApJ, 418, L37

Kalas, P., & Jewitt, D. 1995, AJ, 110, 794

Kenyon, S. J., Wood, K., Whitney, B. A., & Wolff, M. J. 1999, ApJ, 524, L119

Koerner, D. W., Jensen, E. L. N., Cruz, K. L., Guild, T. B., & Gultekin, K. 2000, ApJ, 533, 37

Koerner, D. W., Ressler, M. E., Werner, M. W., & Backman, D. E. 1998, ApJ, 503, L83

Krist, J. E., Stapelfeldt, K. R., Ménard, F., Padgett, D. L., & Burrows, C. J. 2000, ApJ, 538, 793

Low, F. J., Hines, D. C., & Schneider, G. 1999, ApJ, 520, L45

Marsh, K. A., et al. 2001, ApJ, submitted

Rucinski, S. M., & Krautter, J. 1983, RMA&A, 7, 200

Schneider, G., et al.
    1999, ApJ, 513, L127

Smith, B. A., & Terrile, R. J. 1984, Science, 226, 1421

Stauffer, J. R., Hartmann, L. W., & Barrado y Navascues, D. 1995, ApJ, 454, 910

Telesco, C. M., et al. 2000, ApJ, 530, 329

Torres, G., Stefanik, R. P., Latham, D. W., & Mazeh, T. 1995, ApJ, 452, 870

Webb, R. A., Zuckerman, B., Platais, I., Patience, J., White, R. J., Schwartz, M. J., & McCarthy, C. 1999, ApJ, 512, L63

Weinberger, A. J., Becklin, E. E., Schneider, G., Smith, B. A., Lowrance, P. J., Silverstone, M. D., Zuckerman, B., & Terrile, R. J. 1999, ApJ, 525, L53

Weinberger, A. J., Rich, R. M., Becklin, E. E., Zuckerman, B., & Matthews, K. 2000, ApJ, 544, 937

Weinberger, A. J., et al. 2001, ApJ, submitted

Weintraub, D. A., Zuckerman, B., & Masson, C. R. 1989, ApJ, 344, 915

Zuckerman, B., Forville, T., & Kastner, J. H. 1995, Nature, 373, 494

Acknowledgments. We thank the other members of the NICMOS IDT for their many contributions to our nearby stars programs, and to Eliot Malumuth, Phil Plait, and Sally Heap for their valuable help with our STIS observations. This work is supported in part by NASA grant NAG 5-3042 and based on observations with the NASA/ESA Hubble Space Telescope.

Discussion

Frank Shu: You mentioned that dust could come from two different sources: the pristine dust that came from the interstellar medium, or dust which is really worn down by planetesimals breaking apart. Have you thought about observational ways to distinguish between the two? I mean, one obvious way that might not work is whether they're amorphous or crystalline to indicate if internally processed...

Glenn Schneider: Yes, and though I did not have time to discuss this there is such work is ongoing for some of these disk systems. Observationally, this is currently best probed by mid-IR spectroscopy of the dust, though we are also attempting optical spectroscopy with STIS. Indeed, in the case of TW Hya, while the disk remained unresolved at these longer wavelengths, Weinberger et al. (2001) detected thermal emission from a broad silicate feature at 9.6 microns which cannot be explained only by amorphous grains. Those Keck observations were of insufficient spectral resolution to identify the likely-attendant crystalline components, though both Fe and Mg species have peaks within the spectral region of the observed broad emission. Augereau et al's. (1999) simultaneous modeling of the HR 4796A disk required both cold amorphous silicates and water ice, and a hot crystalline dust to simultaneously fit both the near-IR scattered, and thermal-IR emitted radiation as previously discussed. With the added optical imaging and, if we are successful, spectroscopy, the additional color information on the resolved dust components may help further delineate the nature of the dust. Clearly, higher spectral resolution observations are required.

Astrophysical Ages and Time Scales
ASP Conference Series, Vol. 245, 2001
T. von Hippel, C. Simpson, N. Manset

Recent Solar Heavy Element Abundance Determinations and Implications for the History of the Solar System

E. Biémont

IPNAS, Bât. B15, University of Liège, Sart Tilman, B-4000 Liège 1, Belgium and Astrophysics and Spectroscopy, University of Mons-Hainaut, B-7000 Mons, Belgium

Abstract. Our knowledge of the chemical composition of the Sun has substantially progressed in recent years, the advances being related more to the progress realized in the determination of the relevant atomic data for a large number of transitions than to the improvement in the resolution reached on the solar spectra or in our knowledge of the model of the outer shells of the Sun. In this context, the laser spectroscopy techniques have played a considerable role for providing transition probabilities accurate within a few percent. A detailed comparison of the solar photospheric abundances and the meteoritic ones (carbonaceous chondrites of type C1) may then shed some light on the history and the formation of the Solar System, particularly for some heavy elements for which accurate radiative parameters were previously missing.

1. The Solar Abundance of Germanium

The photospheric abundance of germanium, quoted by Anders & Grevesse (1989) ($A_{Ge} = 3.41 \pm 0.14$), differs substantially from the meteoritic result, i.e., $A_{Ge} = 3.63 \pm 0.04$. To solve that problem, lifetimes of the 4p5s $^3P^o_{0,1,2}$ and 4p5s $^1P^o_1$ levels of Ge I have been measured by time-resolved laser-induced fluorescence spectroscopy (TR-LIF) at the Lund Laser Center, Sweden and combined with accurate branching fraction (BF) measurements to deduce transition probabilities. In addition, calculations of transition probabilities have been carried out with the relativistic Hartree–Fock (HFR) method (Biémont et al. 1999; Li et al. 1999) in which core-polarization (CP) effects have been introduced.

We have remeasured the equivalent width of the only unblended line (326.9489 nm) on enlargements of the Jungfraujoch spectra (Delbouille, Neven, & Roland 1973). Using the HOLMU model atmosphere (Holweger & Müller 1974) with a microturbulence of 0.85 km s^{-1} and new partition functions for Ge I and Ge II, we have obtained $A_{Ge} = 3.58 \pm 0.05$, a result which is now very close to the meteoritic result.

2. The Solar Content of Lead

Lead has been observed in the solar spectrum and, in the past, 5 lines of Pb I have been used for abundance determination. In the compilation of Anders &

130

Grevesse (1989), the adopted photospheric result is $A_{Pb} = 1.85 \pm 0.05$. The meteoritic result is substantially higher, $A_{Pb} = 2.05 \pm 0.03$.

New accurate radiative lifetimes (2 levels) measured by TR-LIF have been combined with theoretical BF's calculations in order to establish a reliable scale of absolute f values for Pb I (Biémont et al. 2000; Li et al. 1998). Theoretical transition probabilities have also been calculated using the HFR approach with inclusion of extensive configuration interaction (CI) effects. Due to blending problems, we are left, for the solar analysis, with the line at 368.3463 nm, which is the best abundance indicator. The abundance deduced from that line, $A_{Pb} = 2.00 \pm 0.06$, is now in excellent agreement with the meteoritic value.

3. Ce II and the Lanthanides

One hundred and eighty Ce II transitions have been identified in the photospheric spectrum. The cerium abundance is still affected by a large uncertainty, $A_{Ce} = 1.55 \pm 0.20$, while the meteoritic result is very well determined, $A_{Ce} = 1.61 \pm 0.01$ (Anders & Grevesse 1989).

In the recent past however, we have started an extensive investigation of the transition probabilities in lanthanide spectra of astrophysical interest. More details can be found on the website of Mons University where a new database of astrophysical interest (D.R.E.A.M.) has been progressively created at the address: http://www.umh.ac.be/~astro/dream.shtml. The results obtained so far concern Yb II– IV, Lu II– III, La III, Pr III, Er III, Tm III, and Tm II. The relevant references can be found on this site. In the particular case of Ce II, the results are summarized in two papers (Palmeri et al. 2000; Zhang et al. 2001). To provide reliable experimental data for Ce II, new radiative lifetimes of 18 levels have been obtained using the TR-LIF method. The HFR + CP method, combined with a least-squares fitting of the available experimental levels, has also been used. The abundance value derived from the new data is $A_{Ce} = 1.63 \pm 0.04$, which is now very close to the meteoritic result.

4. Conclusions

The new abundance values of the three "heavy" elements (Ge, Pb, Ce), show that, when accurate atomic data are available, the photospheric and meteoritic abundances (in carbonaceous chondrites) do agree.

More particularly, for Ge I, Grevesse & Meyer (1985) had suggested that the discrepancy photosphere could be real and would imply chemical enrichment in CI chondrites. Anders & Grevesse (1989) noticed, however, that germanium is depleted in all other chondrites classes like other elements of similar volatility. This fact apparently reflects volatile loss during chondrule formation. Our present results establish now firmly that germanium has a behavior similar to that of the elements of similar volatility and that abundance in carbonaceous chondrites (CI) and in the photosphere do agree.

An enrichment process for Pb, which is highly volatile in chondrites, although it has been considered as a possibility by some authors, seems to be unlikely. Our present result confirms this point of view. Cerium is a highly refractory element according to the classification adopted for the meteorites and, consequently, should have escaped the chemical fractionation process. Its photospheric abundance is therefore expected to agree with that derived from carbonaceous chondrites. This conclusion is confirmed by the present analysis.

References

Anders, E., & Grevesse, N. 1989, Geochim. Cosmochim. Acta, 53, 197

Biémont, E., Garnir, H. P., Palmeri, P., Li, Z. S., & Svanberg, S. 2000, MNRAS, 312, 116

Biémont, E., Lyngå, C., Li, Z. S., Svanberg, S., Garnir, H. P., & Doidge, P. S. 1999, MNRAS, 303, 721

Delbouille, L., Neven, L., & Roland, G. 1973, Photometric Atlas of the Solar Photospheric Spectrum from $\lambda 3000$ to $\lambda 10\,000$ (Liège: Université de Liège, Institut d'Astrophysique)

Grevesse, N., & Meyer, J.-P. 1985, Goddard Space Flight Center 19th Intern. Cosmic Ray Conf., 3, 5

Holweger, H., & Müller, E. A. 1974, Solar Phys., 39, 19

Li, Z. S., Norin, J., Persson, A., Wahlström, C.-G., Svanberg, S., Doidge, P. S., & Biémont, E. 1999, Phys. Rev. A, 60, 198

Li, Z. S., Svanberg, S., Biémont, E., Palmeri, P., & Zhangui, J. 1998, Phys. Rev. A, 57, 3443

Palmeri, P., Quinet, P., Wyart, J.-F., & Biémont, E. 2000, Phys. Scr., 61, 323

Zhang, Z. G., Svanberg, S., Jiang, Z., Palmeri, P., Quinet, P., & Biémont, E. 2001, Phys. Scr., 63, 122

Acknowledgments. Financial support from the Belgian FNRS is acknowledged. Part of the experimental work reported here has been carried out in collaboration with the Lund Laser Centre, Sweden and Prof. S. Svanberg and his collaborators.

Astrophysical Ages and Time Scales
ASP Conference Series, Vol. 245, 2001
T. von Hippel, C. Simpson, N. Manset

Ages and Abundances Among β Pictoris Stars

Michael L. Edwards, Robert E. Stencel

University of Denver, Department of Physics and Astronomy, 2112 East Wesley Avenue, Denver, CO 80208, USA

Abstract. This work reports on a NASA key project that acquired a magnitude limited sample of 66 stars from the Infrared Space Observatory (ISO) mission, looking for infrared signatures of dusty disks around nearby stars. Age and abundance data have been obtained from the literature to supplement the infrared photometry performed by the ISO satellite in the interest of discovering correlations between infrared excesses, age, and chemical abundance. Preliminary results show a decrease in fractional dust luminosity (τ) with age that fits a power law with an index between -1 and -2.

Stencel & Backman (1994) executed a NASA key project that acquired a magnitude limited sample of 66 stars from the Infrared Space Observatory (ISO) mission, looking for infrared signatures of planet formation around nearby stars (Fajardo-Acosta et al. 1999). Age and abundance data has been obtained from the literature to supplement the infrared photometry performed by the ISO satellite in the interest of discovering correlations between infrared excesses, age and chemical abundance. There are two scenarios where disk stars are identical chemically to non-disk stars. Either the disk will enhance the metal content of the star through accretion or it will lower the metallicity of the parent star by acting as a sink for iron and other heavy elements. The most likely scenario from the evidence we have already collected is that young stars with disks will exhibit a low abundance pattern as the heavy elements in the protostellar cloud form regions of higher density, thus making their own regions of gravitational collapse independent of the main stellar mass. These concentrations could then form into planetesimals of up to a kilometer in size. Then, as these planetesimals collapse into the parent star, the abundances return to more normal levels as time passes, perhaps even increasing the abundances to unusually high levels.

Fractional dust luminosity (τ) values were obtained by fitting fluxes in as many bandpasses as were available from SIMBAD, ISO (Kessler 1996), and IRAS to a photospheric blackbody. After the photospheric component was removed, the remaining fluxes were fitted using a non-linear least squares technique with two parameters, one for normalization and the other representing dust temperature. Due to, in large part, the small number of points being fitted the uncertainties are on the order of 100% for the best cases. Preliminary results show a decrease in fractional dust luminosity (τ) with age that is roughly consistent with a power law with an index between -1 and -2 as shown in Figure 1.

These data lead to the conclusions that after their initial formation, dusty disks are long-lived features of a star system, being gradually replenished by

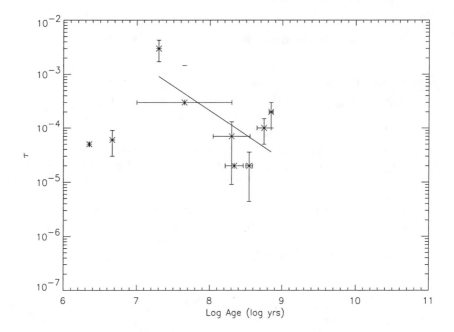

Figure 1.   Derived values of τ to show correlation with age. The line is a power law with an index of roughly -1.0.

some mechanism but not fast enough to prevent slow depletion. The power law nature of the τ to age relation suggests the possibility that the replenishing mechanism is due to collisions between larger planetesimals that would be undetectable by current methods. The lack of a strong correlation between τ and metallicity implies that the dust is not being accreted onto the star but rather dispersing by some other mechanism, perhaps forming the many giant planets that have been recently discovered orbiting other stars. A full reporting of these findings, along with the full thesis text (Edwards 2001), can be found online at http://dim.phys.du.edu/. We are grateful for support of this project under NASA JPL grant 96-1503.

References

Edwards, M. L. 2001, M.Sc. Dissertation, University of Denver

Fajardo-Acosta, S. B., Stencel, R. E., Backman, D. E., & Thakur, N. 1999, ApJ, 520, 215

Kessler, M. F., et al. 1996, A&A, 315, L27

Stencel, R. E., & Backman, D. E. 1994, Ap&SS, 212, 417

Astrophysical Ages and Time Scales
ASP Conference Series, Vol. 245, 2001
T. von Hippel, C. Simpson, N. Manset

Outer Solar System Survey

T. Fuse

Subaru Telescope, National Astronomical Observatory of Japan, Hilo, HI, USA

D. Kinoshita

Graduate University for Advanced Studies, Mitaka, Tokyo, Japan

N. Yamamoto

Science University of Tokyo, Shinjuku, Tokyo, Japan

J. Watanabe

National Astronomical Observatory of Japan, Mitaka, Tokyo, Japan

Abstract. We have carried out survey observations of objects in the outer Solar System, with two telescopes in Japan and one in Hawai'i. These objects should still retain information about the early history of the Solar System. The objective of this survey is to reveal events in the Solar System's evolution and determine the size distribution of the objects in this outer region.

1. Introduction

Since August 1992, more than 300 objects have been discovered in the outer Solar System. The existence of such objects was independently predicted by Edgeworth (1943, 1949) and Kuiper (1951). These objects, called Edgeworth–Kuiper Belt Objects (EKBOs), are believed to be remnants from the early stages of the Solar System formation, and their size should reflect evolutionary events. Therefore, they should contain information about the Solar System's history.

To study those objects, a Solar System Watching Team (SWAT) was formed in Japan. This team has undertaken, since 1998, an outer Solar System survey at two observatories in Japan. The survey looks for bright, and therefore large, EKBOs, because weather and seeing conditions in Japan are unsuitable for the observation of faint objects. The largest objects are important because they give the outer Solar System's accretion time scale. Furthermore, faint EKBOs are surveyed with the Subaru Telescope to determine the size distribution of objects in the outer Solar System.

2.   Observations

2.1.   Kiso Survey

Kiso Observatory is in the western part of Nagano Pref. in Japan. A 1.05-m Schmidt telescope is installed there and can cover 6×6 deg^2 in area at once if a Schmidt plate is available. A 2048×2048 pixels CCD camera is ready for observations and its field of view is 50×50 arcmin^2.

We have been looking for bright EKBOs with the camera since 1998. The total observed area is 34.6 deg^2 and the V-band limiting magnitude is 21.0. No new EKBOs were detected, which means that the surface density of objects brighter than the limiting magnitude is 1.3^{-1} deg^{-2}. This value is consistent with previous estimates.

2.2.   Rikubetsu Survey

Rikubetsu Observatory is in the eastern part of Hokkaido in Japan. It has a 1.15-m telescope with a 1024×1024 pixels CCD camera whose field of view is 10×10 arcmin^2. We have observed 0.36 deg^2 in area with a V-band limiting magnitude of 19.5. We are analyzing the data.

2.3.   Subaru Survey

The 8.2-m Subaru Telescope is on the summit of Mauna Kea in Hawai'i. The telescope has a wide-field camera called Suprime-Cam (Subaru Prime Focus Camera) at the prime focus. The camera has ten 2048×4096 pixels CCDs and its field of view is 28×24 arcmin^2.

Two observational modes are used: (1) a medium survey mode, and (2) a deep survey mode using shift-add methods (Gladman et al. 1998). Three images per field are taken with mode (1), providing us with a wide area survey because of less time spent in this medium depth mode. On the other hand, deep images can be obtained with mode (2) because more than ten individual exposures are added for each field.

We had two nights for EKBO survey observations at the end of February 2001 and have obtained good images. The exposure time was seven minutes. Because of CCD problems with Suprime-Cam, a smaller than usual region, 24 $\times$ 24 arcmin^2, was available per shot. We surveyed five fields with mode (1) and two with mode (2), for a total covered area of 1.12 deg^2. The (1) and (2) R-band limiting magnitudes are 25.4 and 25.9, respectively. The detailed analysis of the data is in progress.

3.   Summary

Our outer Solar System survey for bright EKBOs has been continuing in Japan. A Subaru survey proposal for faint EKBOs will also be submitted to use Open-Use time. The combined data will provide valuable information about the history of the Solar System evolution.

References

Edgeworth, K. E. 1943, Journal of the British Astron. Soc., 53, 181

Edgeworth, K. E. 1949, MNRAS, 109, 600

Gladman, B., Kavelaars, J. J., Nicholson, P. D., Loredo, T. J., & Burns, J. A. 1998, AJ, 116, 2042

Kuiper G. P. 1951, in Astrophysics, ed. J. A. Hynek (New York: McGraw-Hill), 357

Astrophysical Ages and Time Scales
ASP Conference Series, Vol. 245, 2001
T. von Hippel, C. Simpson, N. Manset

The Formation of the Planetary Sequence in a Rotating, Self-Gravitating Disk of Randomly Colliding Planetesimals

E. Griv, M. Gedalin, D. Eichler, E. Liverts

Department of Physics, Ben-Gurion University of the Negev, P.O. Box 653, Beer-Sheva 84105, Israel

Abstract. The kinetic theory is used to study the evolution of the self-gravitating disk of randomly colliding planetesimals. It is shown that as a result of an almost aperiodic Jeans-type instability of small-amplitude gravity disturbances, the disk is subdivided into numerous dense fragments. These can eventually condense into the planetary sequence with mathematical regularity in the spacing formulated empirically by Titius and Bode.

1. Linear Kinetic Theory

Planetary formation is thought to start with dust particle settling to the central plane of a rotating nebula to form a thin dust layer around the central plane. During the early evolution of such a flat disk orbiting the primordial Sun it is believed that on a time scale of $\sim 10^4$ yr the dust particles coagulate into numerous kilometer-sized rocky planetesimals ($\sim 10^{12}$ such bodies) owing to collisional sticking and/or gravitational instability. One can suggest that planets accreted subsequently from a hierarchy of randomly colliding planetesimals with the Titius and Bode (hereafter TB) mathematical regularity in the spacing: $r_n = r_0 A^n$, n = 1, 2, dots, 9, where r_n is the distance of the nth planet and asteroids in astronomical units from the Sun, and $r_0 = 0.2$, $A = 1.73$ are constants. We argue that the classical Jeans instability of small gravity perturbations developing in such a disk leads to the formation of an arrangement of dense fragments—the progenitor of the planetary system.

Let us first rewrite the TB "law" in the form $(2\pi/\ln 1.73)\ln(r_n/r_0) = 2\pi n$. Next, we represent the surface density of the disk σ in the form of the sum of the unperturbed equilibrium density $\sigma_0(r)$ and the small perturbed density

$$\sigma_1(r) = \tilde{\sigma}(r) \cos\left[11.46 \ln(r_n/r_0)\right], \qquad (1)$$

where $\tilde{\sigma} \approx$ constant is the amplitude, r, φ, z are the nebulacentric cylindrical coordinates, and the axis of nebular rotation is taken oriented along the z-axis. The TB relationship and the condition $\tilde{\sigma}(r) > 0$ on the initial phase imply that the maximum values of the perturbed density in Equation 1 coincide with the positions of all the planets (and asteroids).

The collision motion of an ensemble of identical planetesimals in the plane can be described by a self-consistent Boltzmann–Poisson system of equations for the distribution function $f(\vec{r}, \vec{v}, t)$ and the gravitational potential $\Phi(\vec{r}, t)$. The limit, when the collision frequency ν_c is smaller than the rotational frequency Ω, is considered. The functions f and Φ are divided into basic parts $f_0(r, \vec{v})$ and $\Phi_0(r)$, satisfying the equilibrium condition, and small fluctuating parts $f_1(\vec{r}, \vec{v}, t)$ and $\Phi_1(\vec{r}, t)$. If an initial perturbation grows, the system is called unstable. We may seek solutions in the form of normal modes by expanding any perturbation in a Fourier series (the standard WKB approximation)

$$\aleph_1 = \delta\aleph(r) \exp(ik_r r + im\varphi - i\omega t), \qquad (2)$$

where the amplitude $\delta\aleph(r)$ is the slowly varying quantity, k_r and $k_\varphi = m/r$ are the radial and azimuthal components of the wavevector, m is the nonnegative azimuthal mode number (= number of spiral arms), ω is the wavefrequency, and we omit the $\Re$ (it is understood that the real part of all expressions is taken).

Interestingly, and this is the central part of our theory, the TB relationship (1) *satisfies* the conditions of the WKB wave with the effective TB radial wavenumber $k_{\text{eff}} = 11.46/r_n$, $d\ln k_{\text{eff}}/d\ln r = O(1)$, and $k_{\text{eff}}r \gg 1$.

In order to find the perturbed distribution f_1, it is convenient to integrate the Boltzmann kinetic equation along the unperturbed trajectories of particles. In the limit when $\nu_c < \Omega$, in the lowest approximation of the theory we can consider the ordinary epicyclic orbits. This is physically obvious since a typical planetesimal follows at least one circle around a center between collisions: in this limit can one speak of the system's rotation. As a result, by applying the Fourier "ansatz" in Equation 2 from the resulting Boltzmann–Poisson equations, one gets the generalized Lin–Shu–Kalnajs type dispersion relation

$$\frac{k^2 c_r^2}{2\pi G \sigma_0 |k|} = -\kappa \sum_{l=-\infty}^{\infty} l \frac{e^{-x} I_l(x)}{\omega_* - l\kappa + i\nu_c} + 2\Omega \frac{m\rho^2}{rL} \sum_{l=-\infty}^{\infty} \frac{e^{-x} I_l(x)}{\omega_* - l\kappa + i\nu_c}, \qquad (3)$$

where $x = k_*^2 c_r^2/\kappa^2 \simeq k_*^2 \rho^2$, $\rho = c_r/\kappa$ is the mean epicyclic radius, $c_r(r)$ is the radial dispersion of random velocities of planetesimals ("temperature"), $\kappa(r)$ is the epicyclic frequency, $I_l(x)$ is the modified Bessel function, $|L| = |\partial \ln(2\Omega\sigma_0/\kappa c_r)/\partial r|^{-1}$ is the scale of radial inhomogeneity, and $\omega_* = \omega - m\Omega$.

A solution of the dispersion relation, Equation 3, gives the condition for an almost aperiodic ($|\Im\omega_*/\Re\omega_*| \gg 1$ and $\Im\omega_* > 0$), radial Jeans-type instability, namely, $c_r < 3.4G\sigma_0/\kappa$, which is just the familiar Toomre criterion. As the solution also shows, for the azimuthal mode number $m \approx 1$ and the radial wavelength $k_{\text{crit}} \approx \kappa/c_r$ the growth rate of the instability is maximum. This means that of all harmonics of initial perturbation, one perturbation with the maximum of the growth rate, $m \approx 1$, and $k = k_{\text{crit}}$ will be formed asymptotically in time. The numerous condensations with typical dimensions and distances between them $\lambda_{\text{crit}} = 2\pi/k_{\text{crit}} \ll R$, where R is the radius of the system, that arise will remain localized in space because the instability will be an almost aperiodic.

The development of this gravitational Jeans-type instability can result directly in the formation of the system of dense aggregates on a time scale of 10^2–10^3 yr. If the disk is inhomogeneous, the wavelength of a perturbation with a maximum growth rate λ_{crit} will be a function of the radius r. From the formulae above, it follows that the maxima of the perturbed density σ_1 coincide with locations of all the planets as given by the TB relationship (Equation 1), if

$$\frac{2\pi}{\lambda_{\text{crit}}} = \frac{d}{dr} \int^r k_{\text{eff}} dr', \quad \text{that is,} \quad \frac{\kappa(r)}{c_r(r)} = 11.46\frac{r_0}{r}. \tag{4}$$

Equation 4 establishes a law for the dependence of κ and c_r on r in a disk of planetesimals, such that aperiodically growing density perturbations are located at the sites of the planetary orbits in the Solar System.

2. *N*-Body Simulations

One can learn much about the properties of disks of planetesimals experimentally by computer simulation of many-body systems. We analyze the evolution of *N*-body models of self-gravitating disks of planetesimals by direct integration over a time span of Newtonian equations of motion of identical particles. In Figure 1 we show a series of face-on view snapshots from a three-dimensional simulation run of the so-called cold disk, in which the initial dispersion of random velocities of particles was chosen to be less than Toomre's critical dispersion. The time was normalized so that the time $t = 1$ corresponds to a single revolution of the initial disk; the rotation was taken to be counterclockwise. In agreement with the theory, during the first rotation, Jeans-unstable perturbations break the system into several macroscopic fragments. At the final stage at $t \gtrsim 2$ the stable system of a massive sun and a pair of planets has developed.

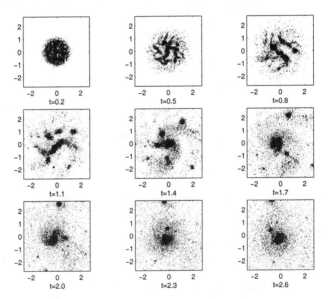

Figure 1.   Time evolution (face-on view) of a Jeans-unstable dynamically cold disk of $N = 30\,000$ identical planetesimals.

Acknowledgments. This work was performed under the auspices of the Israel Science Foundation and the Israel–U.S. Binational Science Foundation.

Astrophysical Ages and Time Scales
ASP Conference Series, Vol. 245, 2001
T. von Hippel, C. Simpson, N. Manset

Star Formation Studies With the Sloan Digital Sky Survey

P. M. McGehee

Los Alamos National Laboratory, MS H820, Los Alamos, NM 87545, USA

Abstract. The determination of time scales associated with planetary formation and circumstellar disk evolution requires large samples of stars having diverse environments and ages. Such a sample can be obtained using the Sloan Digital Sky Survey (York et al. 2000), as it is systematically mapping one-quarter of the entire sky providing photometric data on over 100 million objects in five passbands (Gunn et al. 1998; Fukugita et al. 1996).

Pre-main sequence stars have distinct colors in the SDSS $u'g'r'i'z'$ photometric system as a consequence of their late-type photospheres and strong UV excess driven by the magnetospheric accretion shock. SDSS observations of known Orion population emission line stars cataloged by the Kiso objective prism survey reveal a color-based signature that correlates well with the Hα emission line strength.

As the excess emission is a direct consequence of the presence of a circumstellar disk, we can constrain the duration of the planetary formation process by determining the age of the young star. Follow-on observations of SDSS T Tauri candidates have begun at the Astrophysical Research Consortium's 3.5 meter telescope using medium resolution (R = 5000) spectroscopy. The aim is to place these objects on theoretical evolutionary tracks using spectral indicators for effective temperature and surface gravity and to create a catalog for future studies including a circumstellar disk census.

1. Introduction

The SDSS complements the ROSAT All-Sky Survey (RASS) in its ability to detect actively accreting young stars. While the SDSS T Tauri survey will also include chromospherically active low-mass stars, it can detect Classical T Tauris at 6 magnitudes fainter than the corresponding RASS limit. A recent analysis of RASS observations in the Tucanae association by Stelzer & Neuhaüser (2000) shows the RASS sensitivity limit at 45 parsecs is 2×10^{28} erg s^{-1}. This places the detection limit for analogs of TW Hydrae, the canonical isolated Classical T Tauri ($M_V = 7.3, L_x = 11.5 \times 10^{29}$ erg s^{-1}), at 340 pc, or $V = 15$.

2. Primary Selection

The Wiramihardja et al. (1989) objective prism survey of stars located in the Orion complex provides the estimated Hα emission strength on a numerical scale between 1 (very weak) and 5 (very strong). We examined the correlation of this index with the stellar locus parameters of Fan (1999) and defined $e = 0.33 \times c1 - 0.94 \times c3$.

The coefficients for c1 and c3 were chosen to have a unit basis vector, and to produce a parameter that was reddening independent under $[A_u, A_g, A_r, A_i, A_z] = [1.87, 1.38, 1, 0.76, 0.54] \times A_r$. This latter point was important for target selection since the stellar locus would otherwise move into the search space in regions known to have A_V of 1 to 2.

The strongly emitting stars were closer to the stellar locus in the $r - i$ and $i - z$ colors since these colors are the least affected by the veiling, or excess continuum. The resulting primary selection is $(r - i) \geq 0.3$ and $(i - z) \geq 0.25$ and $e < -0.2$. Orion population stars having weak Hα emission were indistinguishable from the stellar locus in all colors.

3. M Dwarf/White Dwarf Pairs

Examination of candidate SDSS spectra indicated that M dwarf/White dwarf binaries were a major source of contamination. Comparison with the Orion emission line stars and a sample of 39 spectrally identified M dwarf/White dwarf binaries showed that the g-r values for the Orion stars are generally higher (redder), since the T Tauri accretion signature peaks farther into the UV.

The secondary selection aimed at removing the white dwarf binaries is $(g - r) \geq 1.1$ or $(g - r) - 0.45 \times (u - g) \geq 0.7$. This criterion errors on the conservative side, since we see overlap in the region near $(u - g, g - r) = (1.0, 0.6)$.

4. Characterization of Candidates

Spectroscopic examination of candidates are under way using the Astrophysical Research Consortium and Calar Alto 3.5 meter telescopes for characterization of the Balmer line emission generated by the accretion process, computation of the CaH and TiO spectral indices, and possible Lithium detection. These observations will allow verification of the accretion signature and placement of the T Tauris upon theoretical evolutionary tracks thus establishing a catalog of circumstellar disk presence as a function of age and spectral type.

The CaH and TiO spectral indices can be used to estimate the surface gravity of the young star. Using the Grenoble model server of Siess, Dufour, & Forestini (2000) we see an increase of log g by 0.6 to 0.8 dex between the ages of 1 Myr to 10 Myr for low mass stars with the steepest rate of contraction occurring within the first 5 Myr. This provides a good match to the expected time scale for loss of the circumstellar accretion disk.

Early attempts using narrow band photometry (Mould & Wallis 1977) yielded qualitative results while recent approaches used in the star formation community rely upon echelle spectroscopy (Piorno Schiavon, Batalha, & Barbuy 1995). We adopt the method of Kirkpatrick et al. (1991) for determination of the spectral indices.

The Li I absorption signature is a commonly used indicator for youth although the rate of depletion is dependent upon the core temperature, lifetime of the convective zone, and rotation speeds (Duncan & Rebull 1996).

Additional follow-on observations planned include spectroscopic study of fainter targets using 8 to 10 meter class facilities and detailed circumstellar disk surveys using multi-band IR photometry.

5. Summary

Study of SDSS photometric data reveal a color signature for young stars having moderate to very strong Hα emission. While there is some confusion with M dwarf/White dwarf pairs, the objects within this color space appear to be either pre-main sequence stars or chromospherically active low mass stars.

References

Duncan, D. K., & Rebull, L. M. 1996, PASP, 108, 738

Fan, X. 1999, AJ, 117, 2528

Fukugita, M., Ichikawa, T., Gunn, J. E., Doi, M., Shimasaku, K., & Schneider, D. P. 1996, AJ, 111, 1748

Gunn, J. E., et al. 1998, AJ, 116, 3040

Kirkpatrick, J. D., et al. 1991, ApJS, 77, 417

Mould, J. R., & Wallis, R. E. 1977, MNRAS, 181, 625

Piorno Schiavon, R., Batalha, C., & Barbuy, B. 1995, A&A, 301, 840

Siess, L., Dufour, E., & Forestini, M. 2000, A&A, 358, 593

Stelzer, B., & Neuhäuser, R. 2000, A&A, 361, 581

Wiramihardja, S. D., Kogure, T., Yoshida, S., Ogure, K., & Nakano, M. 1989, PASJ, 41, 155

York, D. G., et al. 2000, AJ, 120, 1579

Acknowledgments. The Sloan Digital Sky Survey (SDSS) is a joint project of the University of Chicago, Fermilab, the Institute for Advanced Study, the Japan Participation Group, The Johns Hopkins University, the Max-Planck-Institute for Astronomy, New Mexico State University, Princeton University, the United States Naval Observatory, and the University of Washington. Apache Point Observatory, site of the SDSS, is operated by the Astrophysical Research Consortium. Funding for the project has been provided by the Alfred P. Sloan Foundation, the SDSS member institutions, the National Aeronautics and Space Administration, the National Science Foundation, the U.S. Department of Energy, and Monbusho, Japan. The SDSS website is http://www.sdss.org/.

Astrophysical Ages and Time Scales
ASP Conference Series, Vol. 245, 2001
T. von Hippel, C. Simpson, N. Manset

Observing the Planet Formation Time Scale by Ground-Based Direct Imaging of Planetary Companions to Young Nearby Stars: Gemini/Hōkūpaʻa Image of TWA–5

Ralph Neuhäuser

Institute for Astronomy, University of Hawaiʻi, USA and MPE, D-85740 Garching, Germany

Dan Potter

Institute for Astronomy, University of Hawaiʻi, 2680 Woodlawn Drive, Honolulu, HI 96822, USA

Wolfgang Brandner

Institute for Astronomy, University of Hawaiʻi, USA and ESO, D-85748 Garching, Germany

Abstract. Many extrasolar planets and a few planetary systems have been found indirectly by small periodic radial velocity variations around old nearby stars. The orbital characteristics of most of them are different from the planets in our Solar System. Hence, planet formation theories have to be revised. Therefore, observational constraints regarding young planets would be very valuable. We have started a ground-based direct imaging search for giant planets in orbit around young nearby stars. Here, we will motivate the sample selection and present our direct imaging observation of the very low-mass (15 to 40 Jupiter masses) brown dwarf companion TWA–5 B in orbit around the nearby young star TWA–5 A, recently obtained with the 36-element curvature-sensing AO instrument Hōkūpaʻa of the University of Hawaiʻi at the 8.3m Gemini North telescope on Mauna Kea. We could achieve a FWHM of 64 mas and 25% Strehl. We find significant evidence for orbital motion of B around A.

1. Introduction: Direct Imaging Search for Giant Planets

So far, no direct imaging detection of an extrasolar planet in orbit around a star has been achieved, mainly because of the problem of dynamic range: planets are too faint and too close to their bright primary stars. Neither speckle techniques nor space-based observations have been able to directly detect a planet around another star. From radial velocity observations, it is known, though, that planets and even planetary systems do exist around other stars. One can avoid the problem of dynamic range by observing young nearby stars, where there could be young planets still contracting and accreting, so that they are relatively hot and (self-)luminous, e.g., Burrows et al. (1997).

A well-suited sample for such a program is the TW Hya association (TWA) of a few dozen young ($\sim$ 10 Myr) low-mass pre-main sequence (i.e., T Tauri) stars at a distance of roughly 55 pc (e.g., Webb et al. 1999). Several members of TWA have been observed by the HST NICMOS, where two planet candidates and one brown dwarf companion candidate were detected: a planet candidate near TWA-7 (9.5 mag fainter than the star in H and K at a separation of 2.5″) was also detected by H- and K-band speckle from the ground (Neuhäuser et al. 2000a); an H-band spectrum taken with ISAAC at the VLT has shown that it is a background K-type star (Neuhäuser et al. 2001). The planet candidate near TWA-6 has not yet been confirmed nor rejected, it is 12 mag fainter than the star in H at a separation of 2.5″ (Schneider et al. 2001). The brown dwarf companion candidate 2″ north of TWA-5 (presented first by Lowrance et al. 1999 and Webb et al. 1999, and also observed by Weintraub et al. 2000), has been confirmed by both proper motion and spectroscopy (spectral type M9) by ground-based optical and IR follow-up using FORS2 and ISAAC at the VLT (Neuhäuser et al. 2000b); simultaneously and independently, also Schneider et al. (2001) took a spectrum of this object and confirmed the spectral type first published by Neuhäuser et al. (2000b) using an HST STIS spectrum with smaller wavelength range.

The brown dwarf companion TWA-5 B has a mass of 15 to 40 Jupiter masses (according to different theoretical tracks and isochrones), an age of 12 ± 6 Myr (as TWA-5 A), and it is the 4th brown dwarf companion in orbit around a normal star confirmed by both spectrum and proper motion. It is the one with the lowest mass, possibly only slightly above the deuterium burning mass limit. It is also the first one in orbit around a pre-main sequence star and the first one in orbit around a spectroscopic binary: TWA-5 A is a single- or possibly double-lined SB T Tauri star (Torres, Neuhäuser, & Latham 2001).

2. Gemini Hōkūpaʻa Observation of TWA-5

We observed several TWA members with the University of Hawaiʻi (UH) 36-element curvature-sensing Adaptive Optics (AO) instrument Hōkūpaʻa at the 8.3m Gemini-North on Mauna Kea, Hawaiʻi, in UH pay-back time in February 2001. TWA-5 was observed in the photometric night 23/24 Feb 2001 at sub-arc sec seeing condition. We observed the object with the Wollaston, the dual-imaging polarimeter, in order to detect a possible circumstellar disk around the star (D. Potter et al., in preparation). As usual with Hōkūpaʻa, we used the UH IR camera QUIRC. Individual exposure times were 6 seconds, obtained at different positions and rotations on the chip. After dark and sky subtraction and flat fielding, we shifted and co-added all frames to the final image with 8 min total exposure (Figure 1). We achieved a FWHM of 64 mas and 25% Strehl.

From the similar brightness of the two SB components of TWA-5 A in high-resolution spectra, both stars of the SB are of similar spectral type, hence both are early M-type stars, like the type in (spatially unresolved) spectra of TWA-5 A. The SB orbit has not been solved, yet. If the elongation seen in the saturated part of TWA-5 A in Figure 1 is due to these two stars, then their separation of roughly 50 to 100 mas would correspond to a ~ 10 year orbit at 55 pc, consistent with the radial velocity variation seen since a few years. It may well be possible within a few years to determine the masses of both components dynamically, namely by simultaneous solutions for the astrometric and spectroscopic orbit.

146   *Neuhäuser, Potter, & Brandner*

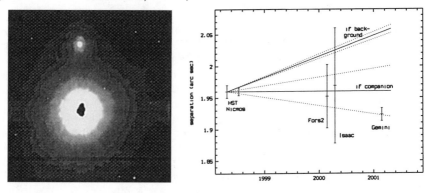

Figure 1.   *Left:* Hōkūpa'a/QUIRC *H*-band image of the faint brown dwarf TWA-5 B (64 mas FWHM) two arcseconds north of the bright TWA-5 A (with saturated center elongated by 40 mas). *Right:* Change in separation between A and B from 1998 to 2001, not consistent with B being a non-moving background object (proper motion of A is known from Tycho) but with B being a companion (dotted lines allow for orbital motion, see Neuhäuser et al. 2000b).

References

Burrows, A., et al. 1997, ApJ, 491, 856

Lowrance, P. J., et al. 1999, ApJ, 512, L69

Neuhäuser, R., Brandner, W., Eckart, A., Guenther, E. W., Alves, J., Ott, T., Huélamo, N., & Fernández, M. 2000a, A&A, 354, L9

Neuhäuser, R., Guenther, E. W., Petr, M. G., Brandner, W., Huélamo, N., & Alves, J. 2000b, A&A, 360, L39

Neuhäuser, R., et al. 2001, in ASP Conf. Ser., From Darkness to Light, ed. T. Montmerle & P. Andre, (San Francisco: ASP), in press (astro-ph/0007305)

Schneider, G., Becklin, E. E., Lowrance, P. J., & Smith, B. A. 2001, in ASP Conf. Ser., Disks, Planets, and Planetesimals (San Francisco: ASP), in press (astro-ph/0007330)

Torres, G., Neuhäuser, R., & Latham, D. W. 2001, in ASP Conf. Ser., Young Stars Near Earth: Progress and Prospects, ed. R. Jayawardhana & T. Greene, (San Francisco: ASP), in press

Webb, R. A., Zuckerman, B., Platais, I., Patience, J., White, R. J., Schwartz, M. J., & McCarthy, C. 1999, ApJ, 512, L63

Weintraub, D. A., Saumon, D., Kastner, J. H., & Forveille, T., 2000, ApJ, 530, 867

Acknowledgments.   We are grateful to the University of Hawai'i Adaptive Optics Group for their support and help with the Hōkūpa'a observations. RN wishes to thank the University of Hawai'i Institute for Astronomy for hospitality during his visit from October 2000 to March 2001.

Session 4

The Galaxy and the Local Group

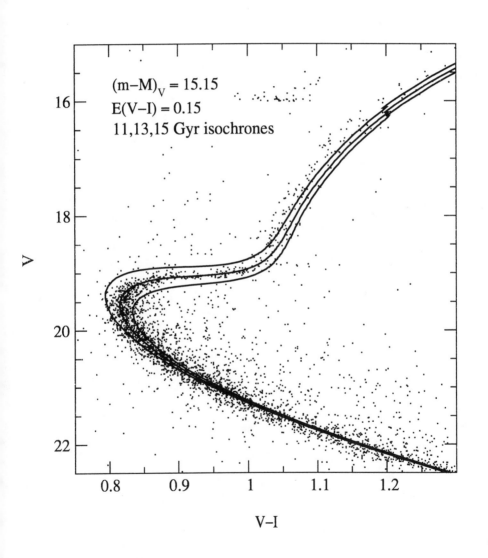

Astrophysical Ages and Time Scales
ASP Conference Series, Vol. 245, 2001
T. von Hippel, C. Simpson, N. Manset

Galaxy Formation One Star at a Time: New Information From the Kinematics of Field Stars in the Galaxy

Timothy C. Beers

Department of Physics and Astronomy, Michigan State University, USA

Masashi Chiba

National Astronomical Observatory of Japan, Japan

Abstract. We review recent studies of the formation and evolution of the Milky Way Galaxy, in particular one based on large samples of non-kinematically selected stars with available proper motions. The Milky Way is a reasonable template for the formation of large spiral galaxies, the only one in which complete kinematical and abundance information can be readily obtained. Ongoing and future projects to obtain proper motions and spectral information for much larger samples of stars that will sharpen our perspective are discussed.

1. Introduction

Observational studies of galaxy formation often conjure visions of deep optical and infrared images of luminous baryonic material caught "in the act" at moderate to high redshift, perhaps inspired by the exquisite views provided in the Hubble Deep Fields. Although such data suggest that at least the basic outlines of a hierarchical assembly model most likely can be applied to many galaxies, the information content of these extremely faint smudges on the night sky is at present limited by the resolution and photon-gathering power of the largest telescopes. By taking the position that galaxy formation is best understood by detailed study of a single galaxy, rather than by obtaining only the sketchiest knowledge of the ancient history of many galaxies, one is led to conclude that the optimal redshift for such studies is at $z = 0$, i.e., the present. We are speaking, of course, of our home galaxy, the Milky Way.

In some respects, we can think of the Milky Way as an analog computer simulation of the stages that many, if not most, large spiral galaxies must have passed through, from the epoch of collapse from the general expansion to the period of the first star formation, and the subsequent evolution leading to the boundary conditions that the Milky Way presents for modern observers. There are numerous clues to the early history of galaxy formation and evolution encoded in the motions and elemental abundances of individual stars in the Galaxy. The challenge to astronomers is to gather, and most importantly, to interpret, this information to provide constraints on the general patterns of (large) galaxy formation throughout the Universe. This approach is not without its limitations, of course. The Analog Milky Way V1.0 is a computer that:

- we live inside of,

- can only directly show us a small portion of the program at once,

- we cannot re-boot,

- runs its CPU at speeds of Gyr, not GHz!

Nevertheless, the Milky Way is the only galaxy for which we can (relatively) easily recover the full six-dimensional phase space information of position and velocity for a substantial number of its luminous constituents while simultaneously obtaining elemental abundance data for the same objects. Moreover, our ability to gather such information is sure to increase rapidly in the near future, with the completion of a number of full (or substantial) sky coverage photometry programs (e.g., 2MASS, SDSS, DENIS), the continuation and extension of several ground-based astrometry programs (e.g., SPM, NPM, USNO), the successful operation of the next generation of astrometric satellites (e.g., DIVA, FAME, SIM, GAIA), as well as the use of wide-field spectroscopic surveys for the inspection of large numbers of spectra of stars in the field of the Milky Way (e.g., 2dF and 6dF).

This conference was convened to discuss the broad issues of time scales in astrophysics. In the context of galaxy formation, this might be interpreted as the estimation of the ages of the oldest (and perhaps more importantly, the youngest) objects that can reasonably be assigned to the various luminous components of the Milky Way that we presently recognize—the halo, the metal-weak thick disk (MWTD), the bulge, the thick disk, and the thin disk. We suggest that the first step toward such a goal is to at least identify the proper *order of formation* of these components. In this brief review, we discuss how the kinematics and abundances of individual stars in a (still small) sample of well-chosen stars can be used to make a tentative inference. Many of the questions we seek to answer will only completely yield when much larger samples of stars with full information become available, but hopefully some of the techniques we explore will provide a good starting point.

2. Kinematic Studies—Past and Present

Kinematic studies of stellar populations in the Galaxy have long been limited by the availability of large samples of stars with measurements of: (1) velocities (radial and tangential), (2) distances (in particular, consistent determinations), and (3) metallicities (accurate and consistent). Such a database is required to constrain plausible scenarios for formation and evolution of the Milky Way. A number of the issues under current discussion include: (1) measures of the local halo velocity ellipsoid and changes with Galactocentric distance, (2) the rotational character of the thick disk and halo, and the existence of gradients in rotation velocity as a function of distance from the Galactic plane, (3) the existence of the MWTD, and the lower limit on its metallicity, (4) correlations (or lack thereof) between orbital eccentricity and the metallicity of halo stars, (5) the global density structure of the halo population, and estimates of the axial ratios of the Galactic halo as a function of distance, (6) searches for kinematic substructure in the halo, and (7) direct comparisons with simulations of galaxy formation.

An ideal sample of stars for exploring these ideas would be unbiased in their selection with respect to both kinematic properties and abundances throughout the Galaxy. Although a number of projects have been initiated that will provide this sort of data (e.g., Morrison et al. 2000; Wyse et al. 2000), even an approximation of the ideal has not yet been achieved. Hence we must make a choice between introduction of: (1) kinematic bias—a tracer sample selected on the basis of stellar motions in the solar neighborhood, e.g., the proper motion selected samples of Ryan & Norris (1991) or Carney et al. (1994), and (2) abundance bias—a tracer sample selected to include stars with metallicities covering the range existing in the Galaxy ($-4.0 \leq [\text{Fe/H}] \leq 0.3$). We have chosen to place our emphasis on the latter, on the grounds that studies of the early formation history of the Galaxy must necessarily include stars of extremely low metallicity, even though they represent a relatively small fraction of the still-shining stars. In this sense an abundance bias is *required* in order to provide sufficient numbers of stars for meaningful investigation, in particular of the halo population.

The study of large *non-kinematically selected* stellar samples began in earnest with the literature assemblage of Norris (1986), who studied the kinematics of ~ 800 stars with available radial velocities, distances, and abundances $[\text{Fe/H}] \leq -0.6$ (the upper cutoff being placed in order to exclude spill-over from the far more numerous thin disk stars). Beers & Sommer-Larsen (1995) supplemented the Norris catalog with the inclusion of a number of stellar samples with measured radial velocities, distances, and abundances published in the literature in the intervening years (in particular from the HK survey stars of Beers, Preston, & Shectman 1992), finishing with a sample of ~ 1900 stars. The catalog of Beers et al. (2000) extended the Beers & Sommer-Larsen (1995) assemblage to include additional stars, in particular RR Lyraes, and presented consistently determined estimates of distance and refined radial velocities and abundances for a total sample in excess of 2000 stars. Of greatest importance, however, was the addition of proper-motion information for over 50% of the cataloged stars, based on several new astrometric programs, including HIPPARCOS (ESA 1997), NPM (Klemola, Hanson, & Jones 1993), SPM (Platais et al. 1998), STARNET (Röser 1996), and ACT (Urban, Corbin, & Wycoff 1998). Additional information is now available from the recently published Tycho II catalog (Høg et al. 2000), but it is not considered in the discussion that follows.

3. Results of the Chiba & Beers Analysis

Based on the large catalog of metal-poor stars with available proper motions, Chiba & Beers (2000) obtain full space motions for some 1200 stars. The local three-dimensional velocity components for this sample, UVW, are shown in Figure 1 as a function of metallicity. This is the first time that such information has become available for an adequate number of low metallicity stars chosen without kinematic bias.

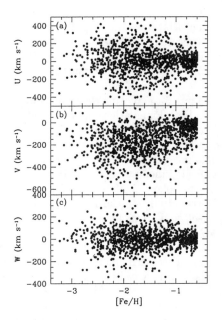

Figure 1.   Distribution of the velocity components (U, V, W) vs. [Fe/H] for the 1203 non-kinematically selected stars from Beers et al. (2000) with available proper motions.

Inspection of panels (a) and (c) of Figure 1 suggests that there appears to be present a core of stars with $-2.0 \leq$ [Fe/H] ≤ -0.6 that is drawn from a low-velocity-dispersion population, in addition to the usual high-dispersion halo population. This population is the MWTD, a component of the Galaxy that was first suggested by Morrison, Flynn, & Freeman (1990), and was the subject of debate in the literature for some time. Its presence now appears indisputable.

There are other pieces of evidence that strongly suggest the presence of a MWTD population. For example, Figure 2 is a comparison of the cumulative distribution of derived eccentricities for stars chosen in two abundance regimes, (a) for [Fe/H] ≤ -2.2, and (b) for $-1.4 <$ [Fe/H] ≤ -1. The different lines correspond to the cases when the range of $|Z|$ is changed. Panel (a) shows that even at quite low abundance, roughly 20% of the stars have $e < 0.4$. The cumulative distribution of e is unchanged when considering subsets of the data with a range of Z, suggesting the absence of any substantial disk-like component below [Fe/H] $= -2.2$. By way of contrast, panel (b) shows that stars with intermediate abundances exhibit (a) a higher fraction of orbits with $e < 0.4$ than for the lower abundance stars, (b) a decrease in the relative fraction of low eccentricity stars as larger heights above the Galactic plane are considered, and (c) convergence at larger heights to a fraction that is close to the 20% obtained for the lower abundance stars. These results imply that the orbital motions of the stars in the intermediate abundance range are, in part, affected by the presence of a MWTD component with a scale height on the order of 1 kpc.

Chiba & Beers (2000) obtained estimates of the local velocity ellipsoids of the halo and thick disk components:

- halo: $(\sigma_U, \sigma_V, \sigma_W) = (141 \pm 11, 106 \pm 9, 94 \pm 8)$ km s^{-1}; [Fe/H] < -2.2

- thick disk: $(\sigma_U, \sigma_V, \sigma_W) = (46 \pm 4, 50 \pm 4, 35 \pm 3)$ km s^{-1}; $-0.7 \leq$ [Fe/H]

Progressing from higher to lower abundances, the velocity dispersions gradually increase as one transitions from disk-like to halo-like behavior. Under the assumption that the MWTD population shares a common velocity ellipsoid with the higher abundance thick-disk stars, a mixture-model analysis suggests that, in the solar neighborhood, the MWTD contributes about 30% of the metal-poor stars in the abundance range $-1.7 <$ [Fe/H] ≤ -1, and perhaps only 10% below [Fe/H] $= -1.7$. It should be recalled, however, that the sample of stars selected by Beers et al. (2000) were contributed primarily from objective-prism surveys that placed a lower cut on Galactic latitude of $|b| > 30°$. Hence, the above fractional contributions of the MWTD at intermediate abundances should be viewed as *lower limits* on the actual fraction. Indeed, a recent analysis of nearby metal-weak giants selected with $|b| < 30°$ suggests that the fraction of MWTD stars at abundances [Fe/H] ≤ -1.7 may be closer to $\sim 40\%$ (T. C. Beers et al., in preparation), much higher than the previously inferred value. Clearly, data obtained from surveys that also target lower Galactic latitudes are needed to resolve this quandary (e.g., Wyse et al. 2000).

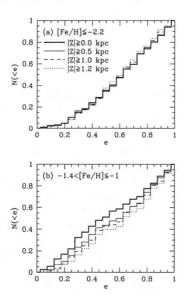

Figure 2.   Cumulative e distributions, $N(< e)$, in the two abundance ranges (a) [Fe/H] ≤ -2.2 and (b) $-1.4 <$ [Fe/H] ≤ -1. The thick solid, thin solid, dashed, and dotted histograms denote the stars at $|Z| \geq 0.0$ kpc (all stars), $|Z| \geq 0.5$ kpc, $|Z| \geq 1.0$ kpc, and $|Z| \geq 1.2$ kpc, respectively.

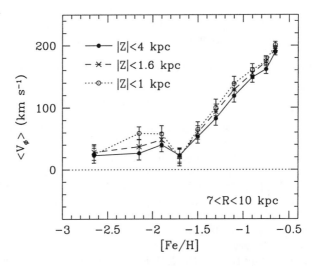

Figure 3.   Distribution of the mean rotational velocity $< V_\phi >$ vs. [Fe/H] for stars closer than 4 kpc from the Sun, assuming an *LSR* rotation velocity of 220 km s^{-1}. Filled circles, crosses, and open circles correspond to the stars at $|Z| < 4$ kpc, $|Z| < 1.6$ kpc, and $|Z| < 1$ kpc, respectively.

Figure 3 is a plot of the mean rotational velocity $< V_\phi >$ as a function of [Fe/H] for stars selected from several cuts in distance above or below the Galactic plane. The general decline in rotation speed with decreasing [Fe/H] marks the transition from a disk population (at the high abundance limit, the canonical thick disk, at the low end, the MWTD) to a halo population. A break at [Fe/H] ~ -1.7 is evident in all three cuts in Z. However, note that below this metallicity, the rotational character of the cuts seem to differ. This feature arises because, close to the Galactic plane, the halo population exhibits a rather strong vertical rotational velocity gradient $\Delta < V_\phi > /\Delta|Z| = -52 \pm 6$ km s^{-1} kpc^{-1}. A smaller, but still significant, gradient appears present in the thick-disk stars as well, $\Delta < V_\phi > /\Delta|Z| = -30 \pm 3$ km s^{-1} kpc^{-1}. The presence of such gradients is a signature of dissipational collapse in both populations.

Before closing this brief summary, it is interesting to view the entire distribution of orbital eccentricities for the stars analysed by Chiba & Beers (2000). Inspection of Figure 4 leaves little doubt that there exist metal-poor stars with eccentricities that populate the entirety of the diagram. This result stands in rather stark contrast to previous claims of an existence of a strong correlation between orbital eccentricity and metallicity (dating back to the classic paper of Eggen, Lynden-Bell, & Sandage 1962) that based their initial selection of stars on high proper-motion surveys.

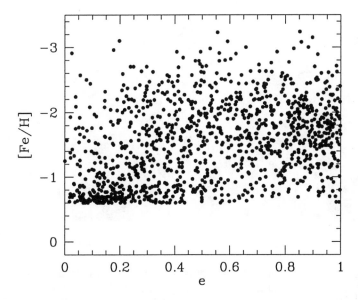

Figure 4.   Relation between [Fe/H] and e for 1203 *non-kinematically selected* stars with [Fe/H] $\leq$ −0.6. Note the diverse range of e even at low metallicities.

4.   Reconstructing the Halo Density Profile From Local Kinematics

A number of authors have followed up the pioneering work of May & Binney (1986), who pointed out that an application of Jeans' theorem might enable the reconstruction of the global structure of the stellar halo from kinematic information of a reasonably large sample of local halo stars. This technique relies on the fact that halo stars that are presently found within a few kpc of the Sun have, during their past motions, explored a substantial fraction of the complete phase space of position and velocity in the Galaxy. Sommer-Larsen & Zhen (1990) developed a maximum-likelihood methodology for the implementation of this idea, and applied the technique to a sample of 118 nearby halo stars with [Fe/H] $\leq$ −1.5. Most recently, Chiba & Beers (2000) have applied the method to a sample of 359 stars with [Fe/H] $\leq$ −1.8 located within 4 kpc of the Sun.

The density distribution for $R > 8$ kpc obtained by Chiba & Beers (2000) is well described by a power-law profile $\rho \propto R^{-3.55\pm0.13}$, similar to that obtained by Sommer-Larsen & Zhen (1990) except in the outermost region, where the more recent analysis does not show a fall-off (likely due to the smaller number of stars considered previously). This result is quite similar to previous estimates of the halo density profile based on counts of globular clusters (Harris 1976; Zinn 1985), and field horizontal-branch and RR Lyrae stars (Preston, Shectman, & Beers 1991, and references therein).

A view of the reconstructed halo is presented in Figure 5a, where contours of constant density are plotted in the (R, Z) plane. The appearance of this figure is quite suggestive. For the outer part of the halo, $R > 15$ kpc, the contours appear roughly spherical, while those in the inner halo become increasingly flatter with

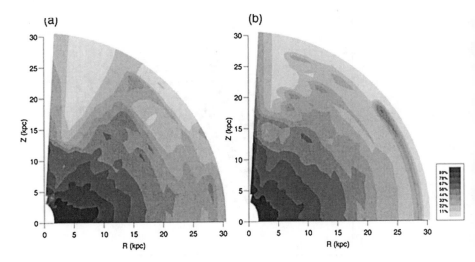

Figure 5.   Equidensity contours for the reconstructed halo in the (R, Z) plane, for stars with [Fe/H] ≤ -1.8. Left and right panels correspond to the density distributions after and before disk formation, respectively.

declining distance. A more quantitative estimate of this behavior, based on the derived axial ratio of the reconstructed halo as a function of distance, reveals that the density distribution in the outer part of the halo, $R \sim 20$ kpc, is quite round. The axial ratio q appears to decrease with decreasing R over $15 < R < 20$ kpc, and the inner part, at $R < 15$ kpc, exhibits $q \sim 0.65$. Similar results, obtained by completely different methods, were found by Preston et al. (1991) and Morrison et al. (2000). Thus, the present stellar halo can be described as nearly spherical in the outer part and highly flattened in the inner part.

Chiba & Beers (2001) extended this type of analysis to investigate the structure of the Galactic halo at a time *before* the disk had a chance to form. Since the bulk of halo stars are found in the inner portion (owing to the large exponent in the density profile), where the gravitational potential arising from a disk dominates over that of the halo, the present-day halo might have had its observed structure affected to a large degree by later disk formation. Chiba & Beers (2001) studied the changes in the derived orbits of their sample of stars when the potential associated with a disk is slowly removed from their model, then reconstructed the density profile of the halo population after it was completely absent. Their result is shown in Figure 5b. The density contours in the (past) inner halo are substantially rounder then inferred for the present halo. The axial ratio obtained for the pre-disk halo is approximately $q = 0.8$ for $R \leq 15$ kpc, transitioning to spherical for $R \sim 20$ kpc. These results suggest that the presently flattened inner halo is a consequence of both an initially slightly flattened inner distribution that has been further flattened by later formation of a disk structure. An initially flattened halo is a signpost of early dissipative formation of this component of the Galaxy.

5. Evidence for Halo Substructure

Several studies (e.g., Helmi et al. 1999; Chiba & Beers 2000) have presented evidence for the existence of kinematic "clumping" of halo stars, a result that is expected if a hierarchical assembly of the Galactic halo applies. Though the numbers of stars in the recognized or suggested structures is presently quite small, we anticipate that future (much) larger data samples will demonstrate further evidence of the phenomenon. Indeed, one of the first exciting results from the SDSS reveals extensive spatially coherent structures of field horizontal-branch and halo blue straggler stars, apparently associated with the trail(s) of debris left by the Sagittarius dwarf (Yanny et al. 2000; Ibata et al. 2001). It should be noted that searches in angular-momentum space (obtainable for stars with full space motions) are inherently much more sensitive than searches based on spatial association (Helmi & White 1999), so efforts to expand the numbers of stars with measured proper motions are of particular value. Studies of the elemental compositions of stars associated with the identified kinematic structures (I. I. Ivans et al., in preparation) will help to establish if a chemical signature (such as low alpha-capture elements) can be assigned to the process.

6. Insights From Numerical Simulations

The hypothesis of the dissipative formation of the inner flattened halo, as well as the later accretion of satellites onto the outer halo, is a natural consequence of the cold dark matter hierarchical clustering model (Bekki & Chiba 2000, 2001). This model postulates that a protogalactic system initially contains numerous subgalactic clumps, comprised of a mixture of primordial abundance gas and dark matter. Once star formation initiates in these clumps, the energy associated with Type II SN explosions quickly drives the gas out into the general ISM of a still-forming galaxy, at the same time touching off formation of second-generation metal-poor stars in the dense shells of these supernovae, as in the models of Tsujimoto, Shigeyama, & Yoshii (1999). The low-mass stars formed in this way are the long-lived fossils we now can observe at extremely low metallicity. Mergers of these small clumps, now composed of dark matter and second-generation stars (but no gas), takes place in a dissipationless manner.

Once the (enriched) gas that was expelled by early supernovae cools, it will fall back into the ever-more massive clumps. In the simulations of Bekki & Chiba (2000, 2001), these larger clumps move gradually toward the central region of the system, due to both dynamical friction and dissipative merging with smaller clumps. Finally, the last merging event occurs between the two most massive clumps, and the metal-poor stars formed inside the clumps are disrupted and spread over the inner part of the halo. The aftermath is characterized by a flattened density distribution. Some fraction of the disrupted gas from the clumps may settle into the central region of the system, and produce a more enriched, more flattened density distribution, providing the raw material for the formation of the Galactic bulge and thick disk. Some of the initially small density fluctuations in the outer region would have gained systematically higher angular momentum from their surroundings, and then slowly fall into the system after most parts of the Galaxy were formed. This may correspond to the process of late satellite accretion, contributing primarily to the outer part of the halo. Thus, the reported initial state of the stellar halo can be explained, at least qualitatively, in the context of a hierarchical clustering scenario.

7. Order of Formation of the Observed Luminous Components of the Galaxy

The picture that emerges from the observations and simulations is one in which the oldest stars that are presently observable in the Milky Way might be found in any of several recognized components. In this interpretation, the inferred order of the formation of the components of the Milky Way is expected to be:

$$\text{inner halo} \rightarrow \text{MWTD} \rightarrow \text{thick disk and Galactic bulge} \rightarrow \text{thin disk.}$$

All the while, the presently observed outer halo is in the process of formation.

We conclude that searches for the most metal-poor (and ancient) stars to be found in the Galaxy should be concentrated on the inner halo. This component is expected to harbor the largest *relative* fraction of second-generation stars. However, it should be noted that the correlation between metal abundance and age is expected to be extremely weak, even for the most metal-deficient stars. The enrichment history of the early Milky Way is driven primarily by the efficiency of star formation during the complex early epochs.

In order to test the ordering suggested above, it is more revealing to consider the *youngest* stars that might be associated with a given observed component of the Galaxy, thus obtaining information on when star formation ceased in that component. Because the second-generation stars share a common origin (in the shells of supernovae exploding in the first clumps of dark matter and primordial gas), and then are re-distributed by subsequent collisions of clumps, representatives of such stars might be found in all but perhaps the thin disk of the Galaxy.

8. The Importance of Future Spectroscopic and Astrometric Surveys

There are several factors that limit our ability to read back the history of the formation of the Milky Way based on present data. The resolution of our vision is severely limited by the (still) relatively small numbers of stars that are presently known with [Fe/H] ≤ -2.0. The ~ 1000 stars below this abundance limit that are currently recognized may *appear* to be a sufficiently large number, but not when one is attempting, as we have, to reconstruct the density structure of the early Galaxy from the phase space of position and velocity that these stars explore. Surveys that *efficiently* identify much larger numbers of extremely metal-poor stars have recently been completed, but are only now starting to be exploited to reveal the riches they contain (e.g., the stellar component of the Hamburg/ESO survey, Christlieb & Beers 2000; the HK-II survey, Rhee, Beers, & Irwin 1999 and Rhee 2000). It should also be recognized that stars of intermediate abundance, $-2.0 \leq$ [Fe/H] ≤ -1.0, are crucial for study of the MWTD–halo interface, and for that reason, are deserving of attention as well. Progress relies on the assignment of telescope time to medium-resolution follow-up surveys, either with rapid single-star measurements with 4m-class telescopes or with multi-fiber, wide-field measurements with instruments such as the 6dF on the UK-Schmidt telescope (Watson et al. 2000).

Even after these stars are recognized, and have had radial velocities and metallicities measured, there still remains the need to determine accurate proper motions and eventually (from astrometric satellites to be launched in the coming decade) accurate distances. Only then will the complete vision be realized.

What can be done in the meantime? In recent years, until funding cuts terminated progress, the most accurate ground-based proper-motion measurements of the stars that occupy the MWTD and halo of the Galaxy have come from the SPM, the Southern Proper Motion survey (e.g., Platais et al. 1998). A proposal to renew the SPM is currently under review at the (US) National Science Foundation. If this proposal is funded, the path is clear for the determination of proper motions, as well as accurately calibrated V magnitudes and $B - V$ colors, for the $\sim 10\,000$ metal-poor (and $\sim 30\,000$ field horizontal-branch and halo blue-straggler) stars in the southern sky identified by the HK-I, HK-II, and Hamburg/ESO surveys, *at least several years before the first data return from future astrometric missions.* With this data in hand, astronomers will be able to refine the issues they might hope to explore with the wealth of data from the astrometric satellites.

We can, and hopefully will, be able to understand the complex process of galaxy formation and evolution for at least *one* galaxy, our home, the Milky Way. Unless the Universe is perverse, what we learn in this process will apply to many, if not most, large spiral galaxies.

References

Beers, T. C., Chiba, M., Yoshii, Y., Platais, I., Hanson, R. B., Fuchs, B., & Rossi, S. 2000, AJ, 119, 2866

Beers, T. C., Preston, G. W., & Shectman, S. A. 1992, AJ, 103, 1987

Beers, T. C., & Sommer-Larsen, J. 1995, ApJS, 96, 175

Bekki, K. & Chiba, M. 2000, ApJ, 534, L89

Bekki, K. & Chiba, M. 2001, ApJ, submitted

Carney, B. W., Latham, D. W., Laird, J. B., & Aguilar, L. A. 1994, AJ, 102, 2240

Chiba, M., & Beers, T. C. 2000, AJ, 119, 2843

Chiba, M., & Beers, T. C. 2001, ApJ, 549, 325

Christlieb, N., & Beers, T. C. 2000, in HDS Workshop, ed. M. Takada-Hidai & H. Ando (Mitaka: NAO), 255

Eggen, O. J., Lynden-Bell, D., & Sandage, A. R. 1962, ApJ, 136, 748

ESA 1997, The Hipparcos and Tycho Catalogues (ESA SP-1200) (Noordwijk: ESA)

Harris, W. E. 1976, AJ, 81, 1095

Helmi, A., & White, S. D. M. 1999, MNRAS, 307, 495

Helmi, A., White, S. D. M., de Zeeuw, P. T., & Zhao, H. S. 1999, Nature, 402, 53

Høg, E., et al. 2000, A&A, 355, L27

Ibata, R., Irwin, M., Lewis, G. F., & Stolte, A. 2001, ApJ, 547, L133

Klemola, A. R., Hanson, R. B., & Jones, B. F. 1993, Lick Northern Proper Motions Program: NPM1 Catalog (NSSDC A1199), (Washington, D.C.: NASA)

May, A., & Binney, J. 1986, MNRAS, 221, 857

Morrison, H. L., Flynn, C., & Freeman, K. C. 1990, AJ, 100, 1191

Morrison, H. L., Mateo, M., Olszewski, E. W., Harding, P., Dohm-Palmer, R. C., Freeman, K. C., Norris, J. E., & Morita, M. 2000, AJ, 119, 2254

Norris, J. E. 1986, ApJS, 61, 667

Platais, I., et al. 1998, AJ, 116, 2556

Preston, G. W., Shectman, S. A., & Beers, T. C. 1991, ApJ, 375, 121

Rhee, J. 2000, Ph.D. Thesis, Michigan State University

Rhee, J., Beers, T. C., & Irwin, M. J. 1999, BAAS, 194, 84.11

Röser, S. 1996, in IAU Symp. 172, ed. S. Ferraz-Mello et al. (Dordrecht: Kluwer), 481

Ryan, S. G., & Norris, J. E. 1991, AJ, 101, 1835

Sommer-Larsen, J., & Zhen, C. 1990, MNRAS, 242, 10

Tsujimoto, T., Shigeyama, Y., & Yoshii, Y. 1999, ApJ, 519, L63

Urban, S. E., Corbin, T. E., & Wycoff, G. L. 1998, ApJ, 115, 2161

Watson, F. G., Parker, Q. A., Bogatu, G., Farrell, T. J., Hingley, B. E., & Miziarski, S. 2000, in Proc. SPIE 4008, Optical and IR Telescope Instrumentation and Detectors, ed. M. Iye & A. F. Moorwood (Bellingham: SPIE), 123

Wyse, R. F. G., Gilmore, G., Norris, J. E., & Freeman, K. C. 2000, BAAS, 197, 04.24

Yanny, B., et al. 2000, ApJ, 540, 825

Zinn, R. 1985, ApJ, 293, 424

Acknowledgments. TCB would like to thank the organizers of this meeting for partial financial support that made his attendance possible, and the National Astronomical Observatory of Japan for granting permission for his absence. MC acknowledges partial support from Grants-in-Aid for Scientific Research (09640328) from the Ministry of Education, Science, Sports and Culture of Japan.

Discussion

Roger Cayrel: I just want to be sure that there is no misinterpretation of your diagram showing a lack of correlation between metallicity and eccentricity. It is certainly true for the sample you have used, but if you do not cut metallicity at -0.6, if you allow the thin disk to be also included, then there is an enormous correlation because 95% of the stars are in almost circular orbits with near solar metallicity.

Tim Beers: Yes, Roger makes a point that is well-taken. When we construct this diagram, keep in mind that we have input a high-metallicity cutoff, so we've suppressed all information on the high metallicity stars. This was really just a historically inherited choice (originally made by Norris 1986), in order to avoid samples such as ours including large numbers of thin-disk abundance stars. In any case, it also helps us maintain a reasonable sized sample, emphasizing the thick-disk and halo populations.

Bengt Gustafsson: There's a recent result which I don't think is published yet by Nissen and a colleague, where they have tried to estimate the ages for stars with metal abundance roughly -1 belonging to the thick disk and the halo, and they find no age difference—they are quite old, both populations. You have two arguments that indicated a real age difference between the thick disk and halo: one is the nice spherical initial condition for the halo if you remove the disk; the other is the fact that there are fewer RR Lyrae stars in the thick disk population. Could you comment further on these points?

Tim Beers: This sort of information—age differences (if any) and membership differences of RR Lyraes in the halo and thick disk—will indeed provide some illumination in the future. But the predictions that come out of the Bekki & Chiba (2000, 2001) simulations, if you actually look at the mean age of the various components, are really quite consistent with their being only small age differences. I just suggested a possible ordering of the components, but the age difference is on the order of hundreds of millions of years, not billions of years. If one had a purely monolithic collapse (a la Eggen, Lynden-Bell, & Sandage, 1962, ApJ, 136, 748) you might expect to see more contrast in the ages of the components.

I think the essential issue, and the one I really hope people will take away from this review, is that the locations of the metal-poor stars at their birth has precious little to do with where they end up. The star formation in the early Galaxy was taking place in those little pockets of primordial gas. This was followed by the dynamical evolution of the system and the re-distribution of those stars into the various components of what we now call the Galaxy. This re-distribution has made the picture difficult to understand from observations alone, but we are making progress.

Astrophysical Ages and Time Scales
ASP Conference Series, Vol. 245, 2001
T. von Hippel, C. Simpson, N. Manset

Globular Cluster Age Dating

Brian Chaboyer

Department of Physics and Astronomy, Dartmouth College, 6127
Wilder Lab, Hanover, NH 03755, USA

Abstract. Astronomers have been determining ages of globular clusters for nearly 50 years. These studies have been driven by the fact that globular clusters are among the oldest objects in the Universe. As such, globular cluster age determinations provide key information on two fundamental questions in astrophysics: How did the Milky Way form? How old is the Universe? The former question requires a study of the relative ages of globular clusters, but the latter question requires the determination of the absolute age of the oldest globular clusters. In this review, the advantages and disadvantages of the various age determination techniques that are in current use are discussed and a review of recent progress in globular cluster age determinations is presented.

1. Introduction

The Milky Way Galaxy contains roughly 150 known globular clusters (Harris 1996). A single globular cluster typically contains several hundred thousand stars in a very compact configuration. Given their compact nature, all of the stars in a given globular cluster lie at the same distance from the Sun. Hence, the relative rankings in apparent brightness are the same as the relative luminosity rankings. Detailed studies of globular clusters have shown that they are each a single-age, single-metallicity system (with one or two exceptions). This makes globular clusters ideal candidates for age determinations.

The globular clusters in the Milky Way belong either to the spherical halo or thick disk population (Zinn 1985; Da Costa & Armandroff 1995). Globular clusters with abundances greater than [Fe/H] = −0.8 form a disk-like subsystem with a scale height of ∼ 1 kpc and a substantial rotation velocity. In contrast, the halo globular cluster system shows no evidence for rotation. Due to their low metallicity and their spatial location, halo globular clusters are thought to have formed very early in the history of the Milky Way. As such, globular clusters are the oldest objects in the Universe whose age we can accurately determine. Studies of globular cluster ages naturally fall into two categories: (1) relative age determinations probe the early formation history of the Milky Way; while (2) absolute age determinations provide a firm lower bound to the age of the Universe. As is well known, the sources of error in relative measurements are often quite different from the sources of error in absolute measurements. For this reason, some globular cluster age determination techniques are best suited for relative age studies, while others are best suited for studies of absolute ages.

In this review, I will provide an overview of different methods which can be used to determine globular cluster ages (§4.). Before discussing these methods, a brief overview of the observed properties of globular cluster color–magnitude diagrams (CMDs) is presented in §2. while §3. contains a brief discussion of theoretical stellar evolution models and isochrones.

2.   Observed Color–Magnitude Diagrams

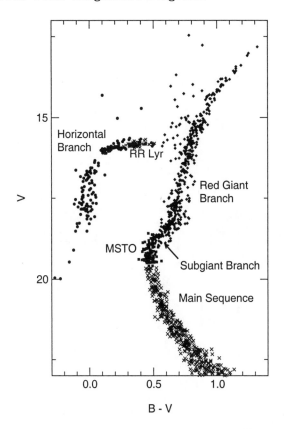

Figure 1.   Color–magnitude diagram of a typical globular cluster, M 15 (data from Durrell & Harris 1993).

Figure 1 shows the color–magnitude for the globular cluster M 15. Since all of the stars in a globular cluster have the same age and chemical composition, their location in the color–magnitude diagram is determined solely by their mass. Higher mass stars have shorter lifetimes and evolve more quickly than low mass stars. The various evolutionary sequence have been labeled on the figure. Most stars are on the main sequence (MS), fusing hydrogen into helium in their cores. For clarity, only about 10% of the stars on the MS have been plotted in Figure 1. Slighter higher mass stars have exhausted their supply of hydrogen in the core, and are in the main sequence turnoff region (MSTO). After the MSTO, the stars are in the subgiant branch (SGB) phase of evolution, where they quickly expand, becoming redder while their luminosity is relatively unchanged. As the stars become brighter they are referred to as red giant branch stars (RGB). These stars are burning hydrogen in a shell outside a helium core. Still higher mass stars have developed a helium core which is so hot and dense that helium fusion is ignited. This evolutionary phase is referred to as the horizontal branch (HB). Some stars on the horizontal branch are unstable to radial pulsations. These radially pulsating variable stars are called RR Lyrae stars, and are important distance indicators.

3.   Theoretical Stellar Models and Isochrones

A basic outline of the procedure which is used to create theoretical stellar models is shown in Figure 2.

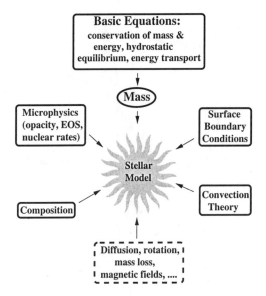

Figure 2.   Overview of the important ingredients needed to create a theoretical stellar model.

A stellar model is constructed by solving the four basic equations of stellar structure for a specified mass and composition. These four, coupled differential equations represent a two point boundary value problem. Two of the boundary conditions are specified at the center of the star (mass and luminosity are zero), and two at the surface. In order to solve these equations, supplementary information is required. The surface boundary conditions are based on stellar atmosphere calculations. The equation of state, opacities and nuclear reaction rates must be known. Finally, as convection can be important in a star, one must have a theory of convection which determines when a region of a star is unstable to convective motions, and if so, the efficiency of the resulting heat transport. Once all of the above information has been determined a stellar model may be constructed. The evolution of a star may be followed by computing a static stellar structure model, updating the composition profile to reflect the changes due to nuclear reactions and/or mixing due to convection, and then re-computing the stellar structure model. This is the standard stellar evolution theory, discussed in numerous textbooks.

In addition to this standard prescription, theoreticians have realized that several other processes, such as atomic diffusion, rotation, and mass loss may also effect the evolution of stars. For the low mass ($M \lesssim 0.8\ M_\odot$) stars in globular clusters, atomic diffusion (whereby heavier elements sink relative to hydrogen) has the largest potential impact on the calculated evolution. Although it is clear from helioseismology that atomic diffusion is occurring in the Sun (e.g., Christensen-Dalsgaard, Proffitt, & Thompson 1993), stellar models which include atomic diffusion are unable to match the observed Li abundance observations in metal-poor stars (e.g., Chaboyer & Demarque 1994).

Probably the least understood aspect of stellar modeling is the treatment of convection. Numerical simulations hold promise for the future (Kim & Chan 1998; Freytag & Salaris 1999), but at present one must view properties of stellar models which depend on the treatment of convection to be uncertain, and subject to possibly large systematic errors. Main sequence and red giant branch globular cluster stars have surface convection zones. Hence, the surface properties of the stellar models (such as its effective temperature or color) are rather uncertain. This is particular true on the RGB, where the predicted color of the RGB depends critically on the exact treatment of convection with the stellar models.

In order to determine the age of a globular cluster, theoretical isochrones are calculated. By interpolating among stellar evolution calculations for stars of different masses, but the same age, one constructs a locus of constant age (isochrone) and compares this to observed CMDs. To make this comparison requires that the theoretical luminosities and effective temperatures be converted to observed magnitudes and colors. This involves a color transformation table which is based upon theoretical model atmosphere calculations (e.g., Lejeune, Cuisinier, & Buser 1997).

Isochrones for old, metal-poor systems have three key properties: (1) the location of the lower main sequence is independent of age; (2) the MSTO region becomes fainter and redder for older systems; and (3) the location of the RGB is very weakly dependent on age. Thus, age determinations for globular clusters focus on the MSTO region.

Given the known uncertainties in the models, the luminosity (absolute magnitude) of the MSTO has the smallest theoretical errors, and is the preferred method for obtaining the absolute ages of globular clusters (e.g., Renzini 1991; Chaboyer 1996). Including atomic diffusion in the theoretical calculations reduces MSTO luminosity based ages by 7% (Chaboyer, Demarque, & Sarajedini 1996b). The absolute magnitude of the MSTO requires that the distance to the globular cluster be known. Furthermore, since the MSTO region is nearly vertical in the observed CMD, it is difficult to determine the magnitude of the MSTO in observations. For these reasons, other age determinations techniques, which utilize the color of the MSTO, may be more suitable for relative age determinations.

4.   Age Determination Techniques

A number of different techniques have been used to estimate the age of globular clusters. These include (1) isochrone fitting; (2) Δ Color (TO − RGB); (3) Δ Magnitude; (4) HB morphology (Lee, Demarque, & Zinn 1994); and (5) luminosity functions (e.g., Jimenez & Padoan 1998). In this review, I will concentrate on the first three techniques, which are the most widely used and accepted.

4.1.   Isochrone Fitting

In isochrone fitting, one simply adjusts the distance modulus and reddening to a given globular cluster until a good match to the globular cluster CMD with a theoretical isochrone is obtained. The age of the globular cluster is simply the age of the isochrone which best matches the data. An example of this is shown in Figure 3 where an age of 13 Gyr is obtained for the globular cluster NGC 6652.

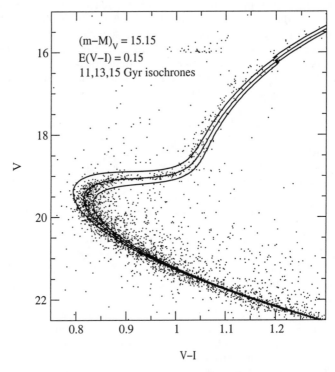

Figure 3.   Sample of isochrone fitting to determine the age of the globular cluster NGC 6652 (data from Chaboyer, Sarajedini, & Armandroff 2000).

Given the large uncertainties in the predicted colors of the RGB, and the sensitivity of the TO region to age, it is best to concentrate on the MS when adjusting the distance modulus and reddening to obtain the best match with the observations. Once a good match is found along the MS, the derived values of reddening and distance modulus may be checked against independent constraints, and the age determined from a comparison of the isochrones to the observations in the TO region.

Even when one follows this procedure, the derived ages are still subject to a large uncertainty. In particular, uncertainties in the metallicity of the cluster usually imply that a range of isochrones with different ages and metallicities provide reasonable matches to the data. Furthermore, this method depends on the colors of the models, which are rather uncertain. As a result, isochrone fits don't, by themselves, provide a robust estimate of the absolute age of a globular cluster. They may provide a useful estimate of the relative age of a cluster, but with a fairly large error bar.

4.2. Δ Color

The principle of this method is straightforward. The color of the MSTO is a strong function of age, while the color of the RGB is relatively insensitive to age. Thus, Δ Color (MSTO − RGB) is sensitive to the age of a globular cluster (Sarajedini & Demarque 1990; VandenBerg, Bolte, & Stetson 1990). This method is sometimes referred to as the "horizontal method," as it utilizes the horizontal axis on a CMD. This method has three advantages: (1) it is independent of the distance, (2) it is independent of reddening, and (3) it is easy to determine the colors of the MSTO and RGB in observations. This latter property implies that, in principal, the Δ Color method yields ages with small errors. The major drawback of this method is that it depends solely on the colors of the models, which have the greatest theoretical uncertainties. For example, inclusion of atomic diffusion or changing the treatment of convection in the stellar models can lead to large changes in the derived ages of a globular cluster. For this reason, the horizontal method is best used to determine *relative* ages of clusters with similar compositions.

A number of studies have utilized the horizontal method to investigate the relative ages of globular clusters. In general, these studies have found that all metal-poor clusters ($[Fe/H] \lesssim -1.7$) are the same age, but that an age spread of a few Gyr appears among the more metal-rich clusters (e.g., Vandenberg et al. 1990; Salaris & Weiss 1998).

4.3. Δ Magnitude

The Δ Magnitude age determination technique uses the difference in magnitude between the MSTO (or the SGB) and the HB as an age diagnostic (e.g., Renzini 1991). This method is sometimes referred to as the "vertical method," as the vertical axis in a CMD is magnitude. This method is usually applied to data in the V passband, and so is also called ΔV. The absolute magnitude of the MSTO (or SGB) depends sensitively on age, while the absolute magnitude of the HB is independent of age for $t \gtrsim 7$ Gyr. Thus, ΔV is sensitive to the age of old stellar systems. The key advantage that ΔV has over other age determination techniques is that its theoretical calibration depends primary upon the luminosity of the stellar models. As the predicted luminosity of these low mass stars are relatively insensitive to the treatment of convection, the theoretical uncertainty associated with ΔV age estimates is minimized. For this reason, ΔV

is the preferred age determination techniques when one is interested in absolute ages, or in comparing the ages of clusters which have different metallicities. Furthermore, ΔV is independent of reddening.

However, ΔV is often difficult to determine in observed data, as the MSTO region is nearly vertical in the CMD. Furthermore, the HB may only contain very red, or very blue stars, in which case some extrapolation to the color of the RR Lyrae stars may be necessary. The difficulty in determining the magnitude of the turnoff in observed data may be bypassed if one uses the magnitude of a point on the SGB as the age indicator (Chaboyer et al. 1996a). As shown by Chaboyer et al. (1996a) the theoretical uncertainty associated with the calibration of the SGB magnitude as a function of age is similar to the uncertainty in the calibration of the MSTO magnitude as a function of age. For this reason, the theoretical errors in the two age determination techniques are similar, but $\Delta V(\text{SGB} - \text{HB})$ is much easier to determine observationally. As such, $\Delta V(\text{SGB} - \text{HB})$ is to be preferred over $\Delta V(\text{TO} - \text{HB})$.

The key disadvantage of the ΔV age determination technique is that it requires knowledge of the absolute magnitude of the HB, $M_V(\text{HB})$. This essentially requires that the distance scale to globular clusters (Population II objects) be known. There are a large variety of methods which may be used to determine $M_V(\text{HB})$, or equivalently the distance scale to globular clusters. As RR Lyrae stars are radially pulsating variable stars on the HB which are standard candles in astronomy, the calibration of the $M_V(\text{HB})$ is often done through a study of RR Lyrae stars and expressed as $M_V(\text{RR})$.

Studies of the Population II distance scale have often assumed that $M_V(\text{RR})$ is a linear function of metallicity, [Fe/H],

$$M_V(\text{RR}) = \alpha\,[\text{Fe/H}] + \beta, \tag{1}$$

where the slope α affects the relative ages at different metallicities and the zeropoint β affects the absolute ages.

Determination of the slope α requires methods which determine precise relative values of $M_V(\text{RR})$ for stars of different metallicities. The semi-empirical Baade-Wesselink method has been applied by Jones et al. (1992) and by Skillen et al. (1993). Reanalysis of these data suggest that $\alpha = 0.22 \pm 0.05$ (Sarajedini, Chaboyer, & Demarque 1997). Observations of HB stars in M 31 by Fusi Pecci et al. (1996) found $\alpha = 0.13 \pm 0.07$.

Recent theoretical efforts have found that the dependence of $M_V(\text{RR})$ on metallicity may not be a simple linear function. Theoretical pulsation models have found that the $M_V(\text{RR})$–[Fe/H] slope increases at higher metallicities (Caputo et al. 2000). Recent stellar evolution calculations find that $M_V(\text{RR})$ depends the HB morphology, in addition to metallicity (Demarque et al. 2000). This is due to the fact that HB stars pass through the RR Lyrae instability strip at different evolutionary stages, depending on their original position on the HB. For example, a globular cluster with a blue HB will have RR Lyrae stars with higher luminosities than a globular cluster with a similar metallicity, but a red HB. This difficulty may be overcome by considering stars near the zero-age HB (ZAHB), for example Salaris, degl'Innocenti, & Weiss (1997) find a slope of 0.21 for the variation of $M_V(\text{ZAHB})$ with metallicity. Demarque et al. (2000) found that a slope of $\alpha = 0.22$ is suitable for globular clusters which are 6 to 20 kpc from the Galactic Center.

Table 1. $M_V(\text{RR})$ calibration at $[\text{Fe/H}] = -1.9$.

Method	$M_V(\text{RR})$	Reference
Statistical parallax	0.70 ± 0.13	Fernley et al. 1998
Trig parallax of HB stars	0.59 ± 0.10	Gratton 1998
Astrometric	0.53 ± 0.11	Rees 1996
White dwarf fitting	0.51 ± 0.14	Renzini et al. 1996
Theoretical HB models	0.42 ± 0.10	Demarque et al. 2000 Salaris et al. 1997
LMC RR Lyr	0.44 ± 0.13	Walker 1992
Main sequence fitting	0.38 ± 0.10	Chaboyer 1999 and references therein

4.4. Absolute Globular Cluster Ages

Absolute globular cluster ages are of particular interest for setting a minimum age for the Universe, and hence constraining cosmological models. For this reason, the absolute ages of the oldest globular clusters is of particular interest. Given that studies of globular cluster ages using Δ Color have shown an age range among globular clusters with $[\text{Fe/H}] \gtrsim -1.7$, studies of absolute globular cluster ages concentrate on the most metal-poor globular clusters.

The ΔV age determination technique is used to obtain robust absolute ages, which have the smallest possible errors. This requires a knowledge of the distance scale to metal-poor globular clusters. There are a number of ways to calibrate the zero-point of this distance scale, and these are summarized in Table 1, which lists the absolute magnitude of the RR Lyrae stars at $[\text{Fe/H}] = -1.9$.

An examination of Table 1 shows a range of 0.32 magnitudes between the brightest and faintest calibrations. By itself, this would suggest that ΔV ages would be uncertain by 32%. However the error in the individual determinations of $M_V(\text{RR})$ is of order 0.12 magnitudes; with 7 independent measurements of $M_V(\text{RR})$ it is not surprising that a range of 0.32 magnitudes has been found. Indeed, only 2 measurements are more than 1σ away from the mean of $M_V(\text{RR}) = 0.51$. Given 7 measurements, we would expect that $7 \times 0.68 = 4.8$ measurements were within 1σ of the mean. Thus, the observations are all consistent with the mean value. The key question is what error one should take in the mean value. A strict statistical analysis would suggest that ±0.04 magnitudes is appropriate; however I prefer to err on the side of caution and take $M_V(\text{RR}) = 0.51 \pm 0.08$ magnitudes. This error implies that the $\pm2\sigma$ range in $M_V(\text{RR})$ encompass nearly the entire range of the individual measurements.

In addition to the uncertainty in the distance scale to the metal-poor globular clusters, it is important to examine possible sources of errors in the theoretical stellar evolution models of main sequence and subgiant branch stars. Such a study was done by Chaboyer et al. (1998) who critically examined the age determination process and adopted a Monte Carlo approach in which all of the important variables used to determine the absolute ages of the oldest globular clusters (such as composition, atomic diffusion, etc.) were varied within their known uncertainties. The results of this study may be combined with the RR Lyrae calibration discussed above, to determine the absolute age of metal-poor globular clusters. Using this approach, the mean age of a sample of 17 globular clusters with [Fe/H] ≤ -1.7 is 13.2 ± 1.5 Gyr.

5.   Summary

Three different techniques to determine the age of a globular cluster have been discussed in detail. All of these techniques exploit, in one way or another, the fact that the MSTO becomes fainter and redder as a stellar system ages. Isochrone fitting (§4.1.) has the advantage that one is using all of the available observational data in order to determine the age of a globular cluster. The disadvantages of isochrone fitting is sensitivity to reddening, distance, and metallicity of the cluster, and to the colors in the theoretical models. As such, isochrone fitting should only be used to derive relative globular cluster ages.

Relative ages may be obtained for globular clusters of the same metallicity using the Δ Color (MSTO $-$ RGB) age determination technique (§4.2.). This technique is independent of reddening and distance and yields much more precise relative ages (for clusters of similar metallicities) than isochrone fitting. Studies using Δ Color (MSTO $-$ RGB) have found that all metal-poor clusters ([Fe/H] $\lesssim$ -1.7) are the same age, but that an age spread of a few Gyr appears among the more metal-rich clusters (e.g., VandenBerg et al. 1990; Salaris & Weiss 1998). As the Δ Color (MSTO $-$ RGB) method depends exclusively on the colors of the theoretical models for its calibration, the Δ Color (MSTO $-$ RGB) technique is subject to large theoretical uncertainties and should not be used for absolute age determinations.

Absolute globular cluster ages are best determined using the ΔV technique, which requires a calibration of the absolute magnitude of RR Lyrae stars. Hence, ages derived using ΔV depend upon the distance scale to globular clusters. There is a fair bit of uncertainty and controversy associated with the distance scale to globular clusters, implying that absolute globular cluster age estimates are subject to similar uncertainties and controversies. A review of the current constraints on the absolute magnitude of RR Lyrae stars in metal poor stars combined with a detailed study of the uncertainties in the theoretical models finds that the oldest globular clusters have an age of 13.2 ± 1.5 Gyr.

References

Caputo, F., Castellani, V., Marconi, M., & Ripepi, V. 2000, MNRAS, 316, 819

Chaboyer, B. 1996, Nucl. Phys. B Proc. Supp., 51B, 10

Chaboyer, B. 1999, in Post-Hipparcos Cosmic Candles, ed. A. Heck & F. Caputo (Dordrecht: Kluwer), 111

Chaboyer, B., & Demarque, P. 1994, ApJ, 433, 510

Chaboyer, B., Demarque, P., Kernan, P. J., & Krauss, L. M. 1998, ApJ, 494, 96

Chaboyer, B., Demarque, P., Kernan, P. J., Krauss, L. M., & Sarajedini, A. 1996a, MNRAS, 283, 683

Chaboyer, B., Demarque, P., & Sarajedini, A. 1996b, ApJ, 459, 55

Chaboyer, B., Sarajedini, A., & Armandroff, T. E. 2000, AJ, 120, 3102

Christensen-Dalsgaard, J., Proffitt, C. R., & Thompson, M. J. 1993, ApJ, 403, L75

Da Costa, G. S., & Armandroff, T. E. 1995, AJ, 109, 2533

Demarque, P., Zinn, R., Lee, Y., & Yi, S. 2000, AJ, 119, 1398

Durrell, P. R., & Harris, W. E. 1993, AJ, 105, 1420

Fernley, J., Barnes, T. G., Skillen, I., Hawley, S. L., Hanley, C. J., Evans, D. W., Solano, E., & Garrido, R. 1998, A&A, 330, 515

Freytag, B., & Salaris, M. 1999, ApJ, 513, L49

Fusi Pecci, F., et al. 1996, AJ, 112, 1461

Gratton, R. G. 1998, MNRAS, 296, 739

Harris, W. E. 1996, AJ, 112, 1487

Jimenez, R., & Padoan, P. 1998, ApJ, 498, 704

Jones, R. V., Carney, B. W., Storm, J., & Latham, D. W. 1992, ApJ, 386, 646

Kim, Y., & Chan, K. L. 1998, ApJ, 496, L121

Lee, Y., Demarque, P., & Zinn, R. 1994, ApJ, 423, 248

Lejeune, T., Cuisinier, F., & Buser, R. 1997, A&AS, 125, 229

Rees, R. F. 1996, in ASP Conf. Ser. 92, Formation of the Galactic Halo .... Inside and Out, ed. H. Morrison & A. Sarajedini (San Fransico: ASP), 289

Renzini, A. 1991, in NATO ASI Ser. C, 348, Observational Tests of Cosmological Inflation, ed. T. Shanks, A. J. Banday, & R. S. Ellis (Dordrecht: Kluwer), 131

Renzini, A., et al. 1996, ApJ, 465, L23

Salaris, M., degl'Innocenti, S., & Weiss, A. 1997, ApJ, 484, 986

Salaris, M., & Weiss, A. 1998, A&A, 335, 943

Sarajedini, A., Chaboyer, B., & Demarque, P. 1997, PASP, 109, 1321

Sarajedini, A., & Demarque, P. 1990, ApJ, 365, 219

Skillen, I., Fernley, J. A., Stobie, R. S., & Jameson, R. F. 1993, MNRAS, 265, 30

VandenBerg, D. A., Bolte, M., & Stetson, P. B. 1990, AJ, 100, 445

Walker, A. R. 1992, ApJ, 390, L81

Zinn, R. 1985, ApJ, 293, 424

Acknowledgments. Research supported in part by NASA grant NAG5-9225.

Discussion

George Isaak: If you drop the very first word, "oldest" globular clusters, what would the mean age be?

Brian Chaboyer: Well, that really depends on what you say about the fraction of young globular clusters. I'd say roughly 30% of globular clusters are 2–3 Gyr younger than the rest. This would probably drop the mean by maybe 1 Gyr.

Roger Cayrel: Does your 13.2 Gyr age includes helium diffusion or not.

Brian Chaboyer: Yes it does, but I include a fairly large error in the rate, so basically I set a rate that's uncertain by ±40%, so when I do this Monte Carlo study I randomly pick a rate for the diffusion within a pretty large error bar. So the average rate of diffusion this corresponds to is lower than the canonical value by about 30%.

Jim Truran: You tend to find that the oldest globular clusters are the least metal rich. They also tend to be clustered toward the bulge, so to some extent that's inconsistent with the more metal-rich bulge stellar population.

Brian Chaboyer: The closest globular clusters I have are 2–3 kpc away from the bulge, so I don't have any metal-rich bulge globular clusters in this plot.

Jim Truran: They should be concentrated toward the center of the Galaxy.

Brian Chaboyer: They're concentrated toward the "inner halo," would be the terminology I'd use. It is true, but you do find older globular clusters with metallicities near −1.2, it's just that you find younger ones with those metallicities as well.

Marcel Arnould: I'm really impressed with your accuracy.

Brian Chaboyer: That's 1σ.

Marcel Arnould: Did you try to take into account, for example, some influence of the convection model, and in particular of models that attempt to get away from mixing length theory?

Brian Chaboyer: I let the mixing length vary anywhere from 1 to 2.5, so I allowed a very large range in mixing length, but I did use mixing length theory for all the models. But I've been using the luminosity at the turnoff as my age indicator. The luminosity is not nearly as sensitive to the mixing length theory as the colors. If I were to use the colors of the turnoff as an age indicator then the error bar would be much larger, like a factor of 4 larger. So by using the luminosities I minimize the effect of the uncertainty on the treatment of convection. If you use a Canuto & Mazzitelli (1991, ApJ, 370, 295) theory of convection, as opposed to mixing length, it tends to give you ages that are different by about 1 Gyr: it is a difference, but it's not a huge difference.

Astrophysical Ages and Time Scales
ASP Conference Series, Vol. 245, 2001
T. von Hippel, C. Simpson, N. Manset

White Dwarf Stars as Cosmochronometers

G. Fontaine

Département de Physique, Université de Montréal, C.P. 6128, Succ. Centre-Ville, Montréal, Québec H3C 3J7, Canada

Abstract. I briefly review the status of white dwarf stars as cosmochronometers in the light of recent significant progress on both the theoretical and observational fronts.

1. Introduction

White dwarfs represent the end products of stellar evolution for the vast majority of stars and, as such, can be used to constrain the ages of various populations of evolved stars. For example, the oldest white dwarfs in the solar neighborhood (the remnants of the very first generation of intermediate-mass stars in the Galactic disk) are still visible, and have been used, in conjunction with cooling theory, to estimate the age of the disk. More recent observations suggest the tantalizing possibility that a population of very old white dwarfs inhabits the Galactic halo. Such a population may contribute significantly to baryonic dark matter in the Milky Way, and may be used to obtain independent estimates of the age of the halo. In addition, white dwarf cosmochronology is likely to play a very significant role in the coming era of giant 8-10 m telescopes when faint white dwarf populations should be routinely discovered and studied in open and globular clusters.

These exciting developments have led to a renewed interest in white dwarf cooling calculations and model atmosphere calculations using upgraded input physics, and extending into the regime of very cool, evolved white dwarfs. A complete review of these developments has been recently presented by Fontaine, Brassard, & Bergeron (2001). Because of limited space, I focus here on a few salient features and, in particular, I leave to Ted von Hippel (these proceedings) the discussion of white dwarfs in clusters.

2. White Dwarf Ages: The Direct Method

The most direct way for estimating the ages of individual stars is to compare theoretical isochrones with observational data points in a mass-effective temperature diagram. For cool and very cool white dwarfs, the mass can be estimated only if a parallax is known, meaning that the method is restricted to *nearby* objects.

Figure 1 shows the M–$T_{\rm eff}$ distribution of cool and very cool white dwarfs in the parallax sample discussed by Bergeron, Leggett, & Ruiz (2001; BLR). This is the very best sample that currently exists for white dwarfs, and is representative

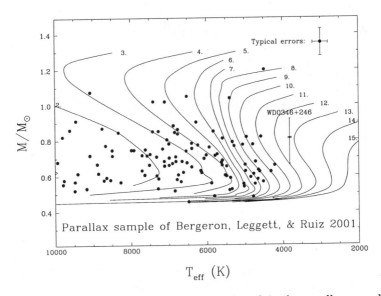

Figure 1.   Distribution of the cool white dwarfs in the parallax sample of BLR in the M–$T_{\rm eff}$ plane (filled circles). The average uncertainties in M and $T_{\rm eff}$ are indicated by the cross in the upper right corner. The solid curves are isochrones based on our new cooling models, expressed in units of Gyr. The individual object WD 0346+246 does not belong to the BLR sample and has halo kinematic characteristics.

of our local region of space, i.e., the thin disk. To compare with theory, I have also plotted isochrones, expressed in units of Gyr. The "S" shape of the isochrones is due, at the top, to the effects of crystallization and Debye cooling (see Fontaine et al. 2001), which are more important in the more massive stars. At the bottom, the finite lifetimes of the progenitors on the main sequence enter into the picture. Of particular interest here are the locations of the two oldest white dwarfs in the BLR sample. The two objects nearly fall on the 11 Gyr isochrone, thus providing a first estimate of the age of the Galactic disk. Note that this estimate is an *upper limit*, as it is based on a set of cooling models that have pure C cores. More realistic models with mixed C/O core compositions would have shorter cooling times.

The case of WD 0346+246 is most interesting as this nearby cool white dwarf has been found to possess halo kinematic characteristics (Hambly, Smartt, & Hodgkin 1997; Hambly et al. 1999; Hodgkin et al. 2000). Although it does not belong to the BLR sample, I included it in Figure 1 because its age can be estimated through the same direct method. In particular, a reliable parallax measurement has been secured (Hambly et al. 1999). Our analysis of the available data suggests that this star may be much older than the disk, with an age of ~ 12.7 Gyr on the basis of our pure C core models. This may be taken as an estimate of the age of the halo, although Fuhrmann (2001, and in these proceedings) argues that WD 0346+246 would rather belong to a thick disk population than to the halo itself. In any case, WD 0346+246 appears to be a likely member of a very old population of white dwarfs that has formed significantly earlier than the (thin) disk. These faint objects, lurking in the halo/thick disk, could contribute in an important manner to the mass budget of baryonic dark matter in our Galaxy.

3. White Dwarf Ages: Luminosity Functions

When individual masses are not available, the age of a white dwarf population can still be estimated through luminosity function studies. This applies to samples for which parallaxes are not available or are unreliable, and to distant objects found in clusters, for example. On the observational side, this means that one must go through detailed stellar counts and statistics, face the problem of completeness, and make sure that the luminosities of the individual stars in the sample are reliable, at least from a statistical viewpoint. In practice, one adopts a "typical" value of the surface gravity, for example $\log g = 8$, and one derives the effective temperatures and absolute magnitudes of the individual objects from the photometric data that must be available. On the theoretical front, this means that beyond white dwarf cooling theory, assumptions must be made on the masses of the progenitors, the relation between the initial and the final mass of the white dwarf, the initial mass function (IMF), and the stellar formation rate. These are all sources of additional uncertainties.

As an example, Figure 2 illustrates how the age of the local disk can be estimated through a comparison of the observational and theoretical luminosity functions of local white dwarfs. On the observational side, Leggett, Ruiz, & Bergeron (1998) and Knox, Hawkins, & Hambly (1999) have recently published their studies of the luminosity function of white dwarfs in the solar neighborhood. The sample of Leggett et al. (1998) is the same one considered originally by Liebert, Dahn, & Monet (1988), except that the former authors have provided much improved estimates of effective temperatures, bolometric corrections, and absolute magnitudes. It contains 43 objects and constitutes a complete proper motion survey. In comparison, the Knox et al. (1999) study is also a complete survey, but it is a colorimetric survey. It contains 58 objects.

I have rebinned the data of Leggett et al. (1998) in order to compare as closely as possible with the published data of Knox et al. (1999). This exercise has allowed me to produce 4 pairs of data points that may be compared directly, as shown in Figure 2. Given the usual uncertainties in this business, it is indeed remarkable that the data points in common between the two studies agree so well. This gives added confidence in the reliability of these results. To provide a basis for comparison, I also plotted, in Figure 2, theoretical luminosity functions assuming various ages for the white dwarf population, from 8 to 14 Gyr, in steps of 1 Gyr. A constant stellar formation rate and a classic Salpeter IMF have been assumed in these illustrative calculations.

The comparison between the observational and theoretical luminosity functions in Figure 2 leads to a determination of the age of the disk. Although the Knox et al. (1999) study has a bright bin that happens to fall quite nicely on the ascending branch of the normalized theoretical luminosity functions, the

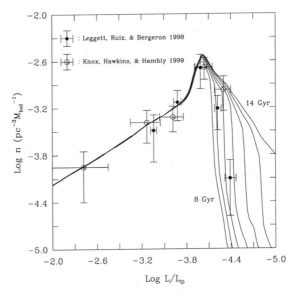

Figure 2.   Comparison of the observational and theoretical luminosity functions of local white dwarfs and the age of the disk. The data points come from two separate studies, one based on a proper motion survey (Leggett et al. 1998) and the other on a colorimetric survey (Knox et al. 1999). The solid curves are theoretical luminosity functions computed on the basis of our recent cooling models with an assumed pure C core composition. These curves are normalized at a point near $L/L_\odot \sim 10^{-3.5}$ on the ascending branch corresponding to the average location of a small cluster of 4 observational points. Various ages for the white dwarf population in the disk, from 8 to 14 Gyr, are considered.

Leggett et al. (1998) study is more useful in the present context as it provides a fainter bin that is critical for the actual comparison with the theoretical curves. Figure 2 indicates that the number density of local white dwarfs generally increases with decreasing luminosities until it reaches a maximum, followed by an important drop-off at still lower luminosities.

The simplest explanation for the observed drop-off of the density of white dwarfs at low luminosities, and the one that has been accepted quite generally, is that the first white dwarfs that were formed in the disk and that are now in our neighborhood, are still bright enough to be visible. Most of them, with representative or average masses have piled up at a luminosity $L/L_\odot \sim 10^{-4}$, while the more massive of them, much less numerous, have trickled down through Debye cooling to lower luminosities during the same time and populate the tail at the faint end of the luminosity function.

A comparison of the curves in Figure 2 with the observational points, particularly the coolest bin, suggests an age slightly less than 11 Gyr for the local disk. This is quite consistent with our above estimate of the age of the oldest white dwarfs in the BLR parallax sample. Again, however, this value is an upper limit to the true age of the disk because I used the same pure C core models.

4. White Dwarfs in the Halo

In the last few years, there has been mounting evidence in favor of the existence of a very old white dwarf population in the Galactic halo, to the point where, today, there can be no doubt that halo white dwarfs (or rather thick disk white dwarfs according to Fuhrmann 2001) *do* exist. In the context of white dwarf cosmochronology, this development is particularly exciting because such a population could be used, in principle, to obtain an independent estimate of the age of the halo. In addition, these old white dwarfs could contribute significantly— perhaps even in a dominant way—to baryonic dark matter in our Galaxy, and, by extension, in other galaxies as well.

The first piece of evidence in favor of an old population of white dwarfs in the Galactic halo is indirect in nature and comes from the MACHO microlensing experiment. I reproduce here a key paragraph taken from one paper published two years ago: "The most straightforward interpretation of the results is that MACHOS make up between 20% and 100% of the dark matter in the halo, and that these objects weigh about 0.5 $M_\odot$. Objects of substellar mass do not comprise much of the dark matter." (Alcock et al. 1999). Cool white dwarfs are, of course, the most likely candidates for subluminous objects that weigh ~ 0.5 $M_\odot$.

The second development is due to Hansen (1998) who published a key paper pointing out that some of the so-called "blue unidentified objects" in the Hubble Deep Field (HDF) could be associated with very old DA white dwarfs in the halo. The presence of such objects there would be in line with the results of the MACHO experiment, but a proof was required to be certain that some of the blue unidentified objects in the HDF are truly halo white dwarfs, namely, they should show small, but measurable proper motions. Ibata et al. (1999) followed up on this idea and, using second-epoch HDF exposures, they reported the probable discovery of detectable proper motions in up to 5 blue unidentified objects in the HDF. Our analysis of the available photometry of these extremely faint objects ($I \sim 28$) indicates that two of the Ibata et al. (1999) high proper motion objects have energy distributions compatible with those of very cool DA white dwarfs. We find that 4–492 has $T_{\text{eff}} \sim 2600$ K and is located at ~ 1.6 kpc, while 4–551 has $T_{\text{eff}} \sim 2300$ K and is located at a distance of ~ 0.8 kpc. The masses of these stars cannot be estimated on the basis of the observations currently available, so their ages cannot be inferred at the moment. Nevertheless, the case for 4–492 and 4–551 as genuine halo white dwarfs appears quite strong.

The third, and more solid, piece of evidence is the documented presence of fast movers in our neighborhood, white dwarfs that have very large space velocities and that are best interpreted as interlopers from the halo. In their proper motion survey, Liebert, Dahn, & Monet (1989) had identified 5 halo candidates, white dwarfs with tangential velocities larger than 250 km s^{-1}. I carried out a critical analysis of that sample in the light of the results of Leggett et al. (1998) which provide much improved estimates of the effective temperatures of these objects, and was left with two possible candidates: WD 1022+009 and

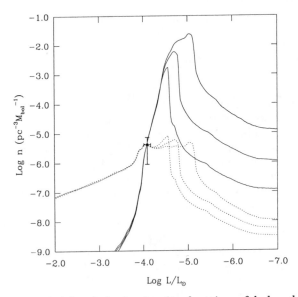

Figure 3.   Models of the luminosity function of halo white dwarfs normalized to the revised data point of Liebert et al. (1989) as discussed in Fontaine et al. (2001). The dotted curves correspond to models based on the classic Salpeter IMF and the solid curves refer to models with a peaked IMF. In both cases, the stellar formation rate is modeled in terms of an initial burst of star formation followed by an exponential decay with a time constant of 0.5 Gyr. The assumed age of the halo is 14, 13, and 12 Gyr, from top to bottom.

WD 2316-064. With $T_{\rm eff} \sim 5000$ K for both stars, those *could* belong to an ancient halo white dwarf population provided they are on the low side of the mass distribution, i.e., with masses around $\sim 0.5\ M_\odot$. This is certainly possible, but it would be nice to have reliable parallaxes for these two objects in order to determine their actual masses. While waiting to secure these additional data, one can still make use of the available observations and construct a single-point luminosity function for halo white dwarfs on the basis of the complete Liebert, Dahn, & Monet (1989) sample. I find, by lumping the two objects in a single luminosity bin, that there should be $n(L) \simeq 10^{-5.39}$ halo white dwarf per pc^3 per unit $M_{\rm bol}$ interval at $L/L_\odot \simeq 10^{-4.09}$. I use this data point in Figure 3.

In a related effort, Ibata et al. (2000) have reported the discovery of two cool high proper motion white dwarfs in the solar neighborhood which they interpret also as interlopers from the halo. Amazingly, one of the Ibata et al. objects is a *rediscovery* of WD 2316–064, one of the original objects in the halo sample of Liebert et al. (1989) and one of the two stars, at $T_{\rm eff} \sim 4740$ K, that I retained as discussed just above. The other star, named F 351–50, appears to be a genuine addition to the putative family of local halo white dwarfs. Our model atmosphere fits to the available photometry published in Ibata et al.

(2000) suggest that it is a very cool white dwarf with $T_{\text{eff}} \sim 3140$ K. A parallax measurement would be necessary to estimate its mass and, ultimately, its age.

In a very important series of papers (Hambly et al. 1997, 1999; Hodgkin et al. 2000), a British group has presented a study of WD 0346+246, another nearby, cool white dwarf with halo kinematic characteristics. Quite interestingly, these authors have measured a parallax for that star, and have published multiband photometry extending into the infrared domain. A preliminary model atmosphere fit to their data suggests that WD 0346+246 has $T_{\text{eff}} \sim 3820$ K and $M \sim 0.8 \, M_\odot$. On the basis of the pure C core models used in this paper, this leads to an age of ~ 12.7 Gyr for WD 0346+246, significantly larger than the estimated age of ~ 11 Gyr for the local disk, as discussed above (see Figure 1).

The idea that an old population of white dwarfs inhabits the halo received a formidable boost recently when Oppenheimer et al. (2001) reported the results of their proper motion survey specifically geared toward uncovering such faint stars. In a most significant discovery, they found a population of 38 cool, nearby, high proper motion white dwarfs that are best interpreted as halo objects that happen to be zooming by us in our region of space. While that population awaits to be fully characterized through follow-up observations, the results of Oppenheimer et al. (2001) already lead to an estimate of the total white dwarf number density in the halo of $\sim 4.3 \times 10^{-4}$ pc^{-3}, which corresponds to about 3% of the dark halo mass (assuming a typical white dwarf mass of 0.6 $M_\odot$). This is a firm *lower limit*, however, as it is very likely that the survey has not probed the entire range of intrinsic luminosities of these objects. Already, as I show in Figure 3, one can infer that the IMF of these halo objects *must* have favored higher mass progenitors as compared to the classic Salpeter law. Indeed, it is not possible to reconcile my single-point luminosity function for halo white dwarfs (the data point shown in Figure 3 and based on the Liebert et al. 1989 sample) with the lower limit obtained by Oppenheimer et al. (2001) on the total number density *if* a Salpeter IMF is used. Indeed, the luminosity integral of the dotted curves gives, respectively, a total halo white dwarf density of 9.7×10^{-6}, 1.1×10^{-5}, and 1.3×10^{-5} pc^{-3} for an assumed halo age of 12, 13, and 14 Gyr. These densities are much lower than the lower limit of Oppenheimer et al. (2001). In contrast, the peaked IMF proposed by Chabrier (1999; see Fontaine et al. 2001) fares a lot better, and the luminosity integral of the solid curves in Figure 3 gives, respectively, a total halo white dwarf density of 6.3×10^{-4}, 3.7×10^{-3}, and 2.3×10^{-2} pc^{-3} for an assumed halo age of 12, 13, and 14 Gyr. These densities correspond to contributions to the dark halo mass of 4.8%, 28%, and 178%(!). The latter value certainly rules out an halo age of 14 Gyr for the particular model shown in Figure 3. Perhaps the most important implication of the Oppenheimer et al. (2001) results at this stage, however, is that the IMF in the halo does not appear to be consistent with the Salpeter law. This may have wide ranging implications on formation scenarios of the Galaxy.

5. Conclusion

White dwarf cosmochronology is still in its infancy, but it already shows its potential as a powerful tool for estimating the ages of the various components of the Galaxy. While models of cooling white dwarfs still need to be improved (see

Fontaine et al. 2001), the method already confines the age of the local disk to less than about 11 Gyr on the basis of pure C core models. Note that the disk could be as young as 8.5 Gyr if real white dwarfs had pure O cores. The largest source of uncertainties indeed comes from the unknown exact proportions of C and O in the cores of field white dwarfs. A very preliminary estimate of the age of the halo based on white dwarf cosmochronology, as discussed here, suggests a value above 12.7 Gyr (from the analysis of the single object WD 0346+246) but below 14 Gyr (from the constraints on the halo white dwarf density). This is significantly larger than the age of ~ 11 Gyr obtained for the disk using the same pure C core models. Finally, in the era of giant 8-10 m ground-based telescopes, the potential of the method for dating open and globular clusters is obvious. There is a lot to learn from a comparison of more standard dating techniques for clusters with white dwarf cosmochronology.

References

Alcock, C., et al. 1999, in ASP Conf. Ser. 165, The Third Stromlo Symposium: The Galactic Halo, ed. B. K. Gibson, T. S. Axelrod, & M. E. Putman (San Francisco: ASP), 362

Bergeron, P., Leggett, S. K., & Ruiz, M. T. 2001, ApJS, 133, 415

Chabrier, G. 1999, ApJ, 513, L103

Fontaine, G., Brassard, P., & Bergeron, P. 2001, PASP, 113, 409

Fuhrmann, K. 2001, A&A, in press

Hambly, N. C., Smartt, S. J., & Hodgkin, S. T. 1997, ApJ, 489, L157

Hambly, N. C., Smartt, S. J., Hodgkin, S. T., Jameson, R. F., Kemp, S. N., Rollerston, W. R. J., & Steele, I. A. 1999, MNRAS, 309, L33

Hansen, B. M. S. 1998, Nature, 394, 860

Hodgkin, S. T., Oppenheimer, B. R., Hambly, N. C., Jameson, R. F., Smartt, S. J., & Steele, I. A. 2000, Nature, 403, 57

Ibata, R. A., Irwin, M., Bienaymé, O., Scholz, R., & Guibert, J. 2000, ApJ, 532, L41

Ibata, R. A., Richer, H. B., Gilliland, R. L., & Scott, D. 1999, ApJ, 524, L95

Knox, R. A., Hawkins, M. R. S., & Hambly, N. C. 1999, MNRAS, 306, 736

Leggett, S. K., Ruiz, M. T., & Bergeron, P. 1998, ApJ, 497, 294

Liebert, J., Dahn, C. C., & Monet, D. G. 1988, ApJ, 332, 891

Liebert, J., Dahn, C. C., & Monet, D. G. 1989, in IAU Coll. 114, White Dwarfs, ed. G. Wegner (Berlin: Springer-Verlag), 15

Oppenheimer, B. R., Hambly, N. C., Digby, A. P., Hodgkin, S. T., & Saumon, D. 2001, Science, in press

Discussion

George Isaak: My question concerns M 67. What is the age of M 67 from main sequence stellar evolution? I'm trying to find out whether there is evidence for carbon or oxygen or some combination of the two.

Gilles Fontaine: I do not know the answer, but someone here ought to know! This question raises a very interesting issue: from the white dwarf point of view, if I repeat the dating exercise with pure oxygen core models, I get a lower limit of 3.5 Gyr. So the estimated age of M 67, according to white dwarf cosmochronology, is between 3.5 (O cores) and 6 Gyr (C cores). If someone were to tell me today that we know the age of M 67 with a relatively good accuracy through the main sequence turnoff method, then I could turn the problem around and adjust the core composition of my models until the white dwarf age would match this accurate value. This would be a wonderful result: the first determination of the mean core composition of white dwarfs in a given homogeneous population!

Ted von Hippel: I wanted to address George Isaak's question. The current estimates for the age of M 67 vary between 4 and 5.2 Gyr, with distance still one of the larger uncertainties. The most recent results give an age of 4.5 $\pm$ 0.5 Gyr, so you probably have a carbon-oxygen white dwarf.

Gilles Fontaine: Thank you Ted, this would indeed imply a mixed C-O core composition as expected from theory. I will certainly do that exercise on the basis of your age estimate.

Tim Beers: Finding the oldest objects has always been a question of great interest. But, at least based on current Galaxy models, the straightforward interpretation is that every component of the Galaxy, except perhaps the thin disk, would in fact have oldest objects of a similar age. So it might be the more interesting question to determine the ages of the youngest members of each Galactic component. That would tell us when the last era of star formation for each component happened, and I think that's an important piece of the puzzle.

Gilles Fontaine: The point is well-taken, yes. Of course, the younger age obtained for the disk white dwarfs (either through comparisons with isochrones as in Figure 1 or using luminosity functions as in Figure 2) compared to my rough estimate of the age of "halo" white dwarfs is related to the fact that these disk white dwarfs most likely belong to the thin disk.

Astrophysical Ages and Time Scales
ASP Conference Series, Vol. 245, 2001
T. von Hippel, C. Simpson, N. Manset

Relative Ages of Galactic Globular Clusters: Clues to the Milky Way Formation Time Scale

A. Aparicio

University of La Laguna and Instituto de Astrofísica de Canarias. Vía Láctea s/n. E38200, La Laguna, Tenerife, Canary Islands, Spain

A. Rosenberg

Instituto de Astrofísica de Canarias. Vía Láctea s/n. E38200, La Laguna, Tenerife, Canary Islands, Spain

I. Saviane

ESO Office Santiago, Alonso de Cordova 3107, Vitacura, Casilla 19001, Santiago 19, Chile

G. Piotto

Dipartimento di Astronomia, Università di Padova. Vicolo dell'Osservatorio 5, I35122, Padova, Italy

M. Zoccali

ESO, Karl Schwarzschild Str. 2, Garching b. München, Germany

Abstract. Based on a new large, homogeneous photometric database of 50 Galactic globular clusters extended out to 25 kpc from the Galactic center, a set of distance and reddening free relative age indicators has been measured: $\delta(V - I)_{@2.5}$ and ΔV_{TO}^{HB}. Using these two independent indicators and two recent updated libraries of isochrones we have found that self-consistent relative ages can be estimated for our GGCs sample. The main results are: (1) most clusters and all with [Fe/h] < −1.2 are old and coeval, (2) there is no trend of age with the Galactocentric distance out to 25 kpc from the Galactic center, (3) there is no age-metallicity trend, and (4) for more metal-rich clusters ([Fe/h] > −1.2) there are indications of a larger age dispersion, of the order of 10–15%. From these results, a tentative interpretation of the Milky Way formation can be given. First, the GC formation process started at the same zero age throughout the halo, at least out to what currently form the ∼ 25 kpc inner halo. The so-called disk globulars were formed at a later time (∼15% lower age). Finally, significantly younger halo GGCs are found at any R_{GC} > 8 kpc. For these, a possible scenario associated with mergers of dwarf galaxies to the Milky Way is suggested.

1. Introduction

Galactic globular clusters (GGC) are the oldest components of the Galactic halo. The determination of their relative ages and of any age correlation with metallicities, abundance patterns, positions and kinematics provides clues on the formation time scale of the halo and gives information on the early efficiency of the enrichment processes in the proto-Galactic material. The importance of these problems and the difficulty in answering these questions is at the basis of the huge efforts dedicated to gather the relative ages of GGCs in the last 30 years or so (VandenBerg, Stetson, & Bolte 1996; Sarajedini, Chaboyer, & Demarque 1997, and references therein).

The methods at use for the age determination of GGCs are based on the position of the turnoff (TO) in the color–magnitude diagram (CMD) of their stellar population. We can measure either the absolute magnitude or the de-reddened color of the TO. However, in order to overcome the uncertainties intrinsic to any method to get GGCs distances and reddening, it is common to measure either the color or the magnitude (or both) of the TO, relative to some other point in the CMD whose position does not depend on age.

Observationally, as pointed out by Sarajedini & Demarque (1990) and VandenBerg, Bolte, & Stetson (1990), the most precise relative age indicator is based on the TO color relative to some fixed point on the red giant branch (RGB). This method is usually called the "horizontal method." Unfortunately, the theoretical RGB temperature is very sensitive to the adopted mixing length parameter, whose dependence on the metallicity is not well established yet. As a consequence, investigations on relative ages based on the horizontal method might be difficult to interpret, and need a careful calibration of the relative TO color as a function of the relative age (Buonanno et al. 1998). The other age indicator, the vertical method, is based on the TO luminosity relative to the horizontal branch (HB). Though this is usually considered a more robust relative age indicator, it is affected both by the uncertainty on the dependence of the HB luminosity on metallicity and the empirical difficulties to get the TO and the HB magnitudes for clusters with only blue HBs.

Given these problems, it is still an open debate whether GGCs are almost coeval (Stetson, VandenBerg, & Bolte 1996) or whether they have continued to form for 5 Gyr or so (i.e., for 30–40% of the Galactic halo lifetime; Sarajedini et al. 1997).

There is a further major limitation to the large scale GGC relative age investigations: the photometric inhomogeneity and the inhomogeneity in the analysis of the databases used in the various studies. Many previous works frequently combine photographic and CCD data, different databases (obtained with different instruments with uncertain calibrations to standard systems and/or based on different sets of standards), or inappropriate color–magnitude diagrams (CMD) were used. This inhomogeneity affects even many recent works (see Stetson et al. 1996 for a discussion).

For this reason, since 1997 (Saviane, Rosenberg, & Piotto 1997) we started a long term project to obtain accurate, homogeneous relative ages by using the horizontal and vertical method in the [Fe/H]–$\delta(V - I)$ plane. Our first observational effort has been aimed at the inner-intermediate halo clusters. It is now complete, and has been published in Rosenberg et al. (1999, 2000a, 2000b).

We provide here a summarized discussion of our database and a few insights about the Milky Way formation time scale that can be inferred from our results. We are currently working on the extension of our data-base to higher metallicity clusters based on HST observations and to GCs in the external halo. Preliminary results of the HST observations are given in Piotto et al. (2000) and are also included here.

2.   The Data

The goal of our observational strategy was to obtain a homogeneous data set which would allow measuring color differences near the TO region with an uncertainty ≤ 0.01 mag, which allows a ≤ 1 Gyr relative age resolution. As a first step, we used 1-m class telescopes to build a large reference sample including all clusters within $(m - M)_V = 16$. The 91cm ESO/Dutch Telescope (for the southern sky GGCs) and the 1m ING/JKT (for the northern sky GGCs) were then used to cover 52 of the scheduled 69 clusters. Finally, only 34 clusters were suitable for the study. The remaining objects were excluded due to several reasons: differential reddening, small number of member stars, large background contamination, or bad definition of the RGB or HB. From 2500 to 20 000 stars per cluster were measured. The typical CMD extends from the RGB tip to ≥ 3 magnitudes below the TO.

A detailed description of the observation and reduction strategies are given in Rosenberg et al. (2000a, 2000b), where the color–magnitude diagrams for the whole photometric sample are also presented. The database is also available on-line at www.iac.es/proyect/poblestelares. Here suffice it to say that the data have been calibrated with the same set of standards, and that the absolute zero-point uncertainties of our calibrations are ≤ 0.02 mag for each of the two bands. Moreover, 3 clusters have been observed both with the southern and northern telescopes, thus providing a consistency check of the calibrations: no systematic differences were found, at the level of accuracy of the zero-points.

As for the HST catalogue, it includes 15 clusters suitable for a relative age determination. More details are given in Piotto et al. (2000).

In our attempt to be as homogeneous as possible, we have adopted the metallicities listed in Rutledge, Hesser, & Stetson (1997) calibrated on the Carretta & Gratton (1997) scale. The main results presented in the following sections would not change adopting the Zinn and West (1984) scale.

3.   Results

Our investigation is based on relative age estimates from two classical reddening and distance independent photometric parameters: ΔV_{TO}^{HB} and $\delta(V - I)_{@2.5}$. These are the magnitude difference between the HB and the TO (vertical method), and the color difference between the TO and the RGB (horizontal method), where the RGB color is measured 2.5 magnitudes above the TO. These quantities are displayed in Figure 1.

In order to interpret the results of our data samples, the theoretical isochrones computed by Straniero, Chieffi, & Limongi (1997) and VandenBerg et al. (2000)

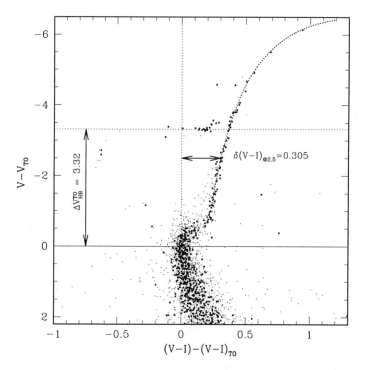

Figure 1. Vertical and horizontal parameters defined on the color-magnitude diagram of NGC 1851.

were used. It is important to notice that these theoretical models are completely independent: indeed, they are obtained with different prescriptions for the mixing-length parameter, the Y vs. Z relation, the temperature–color transformations and bolometric corrections, etc.

The same morphological parameters already defined for the observational CMDs were measured on the isochrones. The trends of the theoretical quantities as a function of both age and metallicity were least-square interpolated by means of third-order polynomials, so that the observed parameters can be easily mapped into age and metallicity variations. Finally, in order to calculate the theoretical values of $\Delta V_{\rm TO}^{\rm HB}$ from the obtained TO magnitudes, we have to assume a relation for the absolute V magnitude of the HB as a function of the metal content. In particular, here we adopted $M_{\rm V}(ZAHB) = (0.18 \pm 0.09) \times ([{\rm Fe/H}] + 1.5) + (0.65 \pm 0.11)$, from the recent investigation of Carretta et al. (2000).

The former procedure provides an absolute age scale for each photometric parameter and each isochrone set. However, we do not need these absolute age calibrations that will have their own zero points, depending on the internal parameters used to compute isochrones. We are rather interested in relative ages, of which we have four independent scales: one for each photometric parameter and isochrone set. The differences among the relative ages resulting from each scale can be taken as an indication of the (internal) uncertainties intrinsic to our present knowledge of stellar structure and evolution.

For each cluster, the four relative age estimates follow very similar trends (see Rosenberg et al. 1999 for details). For simplicity we have averaged them for each cluster and will base the following discussion on the resulting relations. Relative ages are plotted in Figure 2 as a function of metallicity and the Galactocentric distance.

In summary, the principal improvements that we have introduced are: (a) the use of the largest homogeneous CCD database (only two types of instruments have been used and 70% of the objects have been observed with same instrumentation and the same data and photometric reduction has been followed), (b) the use of two independent methods for the age measurement, (c) the use of two recent independent theoretical model libraries, (d) the use of V, I photometry to estimate the horizontal parameter, and (e) an homogeneous metallicity scale.

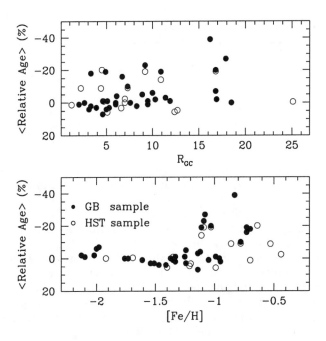

Figure 2.   Mean relative ages for the globular cluster data sample as a function of Galactocentric distance (*top*) and metallicity (*bottom*). Filled and open dots represent ground-based and HST data, respectively.

4. Discussion: Globular Cluster Relative Ages

We arbitrarily divided our GGCs sample into 4 metallicity groups as shown in Figure 2. This figure shows the following important features:

- 34 of 50 clusters are distributed around the mean within an age interval $\Delta_{Age} \leq 10\%$ of the mean.

- Our data do not reveal an age-metallicity relation in the usual sense of age decreasing (or increasing) with metallicity. What is found is an increase of the age dispersion for the metal rich clusters, while the lower metallicity ones ($[Fe/H] \leq -1.2$) seems to be all coeval. This is in agreement with the results of Richer et al. (1996), Salaris & Weiss (1998), Buonanno et al. (1998). On the other hand, Chaboyer, Demarque, & Sarajedini (1996) proposed an age-metallicity relation, of the order $\Delta t_9 / \Delta [Fe/H] \simeq -4$ Gyr dex^{-1}, which is not present in our data set.

- Younger clusters appear at every Galactocentric distance. Those located at larger Galactocentric distances have typical halo kinematics.

5. Discussion: Clues to the Milky Way Formation Time Scale

The age dating progress that has been discussed so far has important consequences on our interpretation of the time scales of the Milky Way formation. In particular, we go from a halo formation lasting for $\sim 40\%$ of the Galactic lifetime (Chaboyer et al. 1996), to the present result of most the halo clusters being coeval.

Besides this basic result, other clues to the Milky Way formation have been obtained from the previous discussion. Going back to Figure 2, a chronological order of structure formation can be inferred. The GC formation process started at the same zero age throughout the halo, at least out to $\sim$ 20 kpc from the center. At later ($\sim 15\%$) times the so-called disk globulars are formed. Finally, significantly younger halo GGCs are found at any $R_{GC} > 8$ kpc (clusters at shorter Galactocentric distances show possible disc kinematics). These clusters could be associated to so-called "streams," i.e., alignments along great circles over the sky, which could arise from these clusters being the relics of ancient Milky Way satellites of the size of a dwarf galaxy (e.g., Lynden-Bell & Lynden-Bell 1995; Fusi-Pecci et al. 1995).

References

Buonanno, R., Corsi, C. E., Pulone, L., Fusi Pecci, F., & Bellazzini, M. 1998, A&A, 333, 505

Carretta, E., & Gratton, R. 1997, A&AS, 121, 95

Carretta, E., Gratton, R., Clementini, G., & Fusi Pecci, F. 2000, ApJ, 533, 215

Chaboyer, B., Demarque, P., & Sarajedini, A. 1996, ApJ, 459, 558

Fusi Pecci, F., Bellazzini, M., Cacciari, C., & Ferraro, F. R. 1995, AJ, 110, 1664

Lynden-Bell, D., & Lynden-Bell, R. M. 1995, MNRAS, 275, 429

Piotto, G., Rosenberg, A., Saviane, I., Zoccali, M., & Aparicio, A. 2000, in The Evolution of the Milky Way, ed. F. Matteucci & F. Giovannelli (Vulcano: Kluwer), 249

Richer, H. B., et al. 1996, ApJ, 463, 602

Rosenberg, A., Aparicio, A., Saviane, I., & Piotto, G. 2000a, A&AS, 145, 451

Rosenberg, A., Piotto, G., Saviane, I., & Aparicio, A. 2000b, A&AS, 144, 5

Rosenberg, A., Saviane, I., Piotto, G., & Aparicio, A. 1999, AJ, 118, 2306

Rutledge, A. G., Hesser, J. E., & Stetson, P. B. 1997, PASP, 109, 907

Salaris, M., & Wiess, A. 1998, A&A, 335, 943

Sarajedini, A., Chaboyer, B., & Demarque, P. 1997, PASP, 109, 1321

Sarajedini, A., & Demarque, P. 1990, ApJ, 365, 219

Saviane, I., Rosenberg, A., & Piotto, G. 1997, in Advances in Stellar Evolution, ed. R. T. Rood & A. Renzini (Cambridge: Cambridge University Press), 65

Stetson, P. B., VandenBerg, D. A., & Bolte, M. 1996, PASP, 108, 560

Straniero, O., Chieffi, A., & Limongi, M. 1997, ApJ, 490, 425

VandenBerg, D. A., Bolte, M., & Stetson, P. B., 1990, AJ, 100, 445

VandenBerg, D. A., Stetson, P. B., & Bolte, M. 1996, ARA&A, 34, 461

VandenBerg, D. A., Swenson, F. J., Rogers, F. J., Iglesias, C. A., & Alexander, D. R. 2000, ApJ, 532, 430

Zinn, R., & West, M. 1984, ApJS, 55, 45

Acknowledgments. This work has received financial support from the Instituto de Astrofísica de Canarias (grant P3-94), the Ministry of Research and Technology of the Kingdom of Spain (grant PB97-1438-C02-01), and the Education and Research Council of the Canary Islands Autonomous Government (grant 1999-008). GP acknowledges partial support from the Italian Space Agency and from the MURST under the program "Stellar Dynamics and Stellar Evolution in Globular Clusters."

Discussion

Oscar Straniero: The vertical method depends on the age and the cluster He abundance. Two clusters with the same age but different He show different Delta-V magnitudes. But the horizontal method ages do not depend on He, so the fact that you get practically the same results with either technique for the bulk of the globular clusters is an implicit illustration that He among globular cluster systems should be almost constant.

Antonio Aparicio: That is right. In fact, I have the hope that, thanks to its extension and internal consistency, our library will be useful to test the results of stellar evolution.

George Isaak: You made the point that the age distribution narrowed with improved quality. If the quality is improved indefinitely, how much better would your results get?

Antonio Aparicio: Well, this is a major question in scientific research. If the quality is improved indefinitely we will get at the end the real age dispersion. By now, we can only say that the point dispersion in our figures is of the order of the error bars. So it seems that the time scale for the Milky Way formation would be even shorter than the ~ 1 Gyr we have found.

Astrophysical Ages and Time Scales
ASP Conference Series, Vol. 245, 2001
T. von Hippel, C. Simpson, N. Manset

White Dwarfs in Open Clusters: New Tests of Stellar Evolution and the Age of the Galaxy

Ted von Hippel

Gemini Observatory, 670 North A'ohōkū Place, Hilo, HI 96720, USA

Abstract. White dwarf cooling theory and very deep observations in star clusters provide a new tool to test stellar evolution theory and time scales. In particular, white dwarf cooling theory is now testing the degree of enhanced core mixing in stars with turnoff ages of 1 to 2 Gyr. More generally, I show the good overall agreement between white dwarf and modern isochrone ages over the range 0.1 to 4 Gyr.

1. Introduction

One of the primary results of stellar evolution theory is a detailed understanding of the ages of star clusters throughout our Galaxy. But what are the limitations of stellar evolution theory and how might an independent technique be applied to determine the ages of star clusters and check stellar evolution theory? In the next section I discuss the difficulties encountered by and the success of stellar evolution theory. In the following section I discuss the viable alternative presented by white dwarf (WD) cooling theory. In the final section I compare the results of stellar evolution and WD cooling ages in open clusters spanning the age range of 0.1 to 4 Gyr.

2. Stellar Ages From Main Sequence Evolution Theory

Stellar structure models assume as basic parameters the hydrogen (X), helium (Y), and heavy element ($1 - X - Y = Z$) fractions, as well as mass (M) and age. Unfortunately, the helium content cannot be measured directly in the vast majority of stars since their stellar atmospheres are too cool to excite helium. Furthermore, the heavy element fraction is an oversimplification; many stars, especially those with significantly non-solar iron abundances, do not have the solar heavy element abundance pattern. The understanding of energy transport in stars is also limited by an incomplete theory of convection (but see recent developments by Canuto & Dubovikov 1998 and Canuto 1999), as well as an incomplete understanding of rotation and opacity, despite extensive and laborious calculations.

The result of an incomplete theory of energy transport, particularly due to an incomplete understanding of convection, is that theorists cannot calculate the radii of stars with surface convection from first principles. The standard approach to this dilemma and the absence of stellar helium abundances is to use the known solar mass, luminosity, radius, temperature, age, and heavy element abundance pattern to constrain the solar helium abundance and determine the solar mixing length. Two unknowns are extracted by fitting models to the Sun since the helium abundance primarily affects luminosity while the mixing length primarily affects radius. Unfortunately, mixing length theory is a gross simplification of convective process and there is no reason why this simple parameterization, fixed for the Sun, should apply to stars with different surface temperatures or abundances. Nevertheless, the solar helium abundance and mixing length form the basis for applying stellar evolution to stars throughout the Galaxy. To be sure, there are helpful constraints on the mixing length from the shapes of the giant branches in open and globular clusters as well as on the helium abundance from the mass–luminosity relation derived from binary stars.

Other problems arise since the mapping between the observational color–magnitude plane and the theoretical temperature–luminosity plane is not precisely known. Even the magnitude and color of the Sun are known to only 3 or 4% ($M_{bol} = 4.72$ to $4.75, B-V = 0.63$ to 0.67).

Given what we are still learning about the Sun, it is remarkable how much we can learn about the ages of star clusters. Particularly encouraging are the advances due to helioseismology measurements and analyses (Gough, these proceedings) which constrain the solar helium abundance, yield a precise solar age, and elucidate key details connected with diffusion and convection (e.g., Elliott & Gough 1999; Baturin et al. 2000).

Assuming theoretical uncertainties are well in hand, due either to direct theoretical advances or careful relative comparisons, one can derive the ages for star clusters depending only on observational parameters. One such useful parameterization (Renzini 1991) is

$$\log T_9 = -0.41 + 0.37 M_V(\text{TO}) - 0.43Y - 0.13[\text{Fe/H}] - 0.12[\text{O/H}],$$

where T_9 is the cluster age in Gyr, and $M_V(\text{TO})$, the absolute magnitude of the main sequence turnoff, depends on the measured apparent magnitude of the turnoff and distance. The above constants in Renzini's parameterization indicate that a given percentage error in distance propagates directly to the same percentage error in age and that the derived age depends critically on the heavy element abundances. The last ten years have seen considerable advances in stellar evolution theory but the above equation still provides a good demonstration of the limitations of isochrone fitting.

When applying stellar evolution isochrones to Galactic clusters the disk is found to contain stars with a wide variety of ages, ranging from newly formed stars to clusters $\gtrsim 8$ Gyr old (e.g., NGC 6791 and Berkeley 17; Phelps, Janes, & Montgomery 1994). Likewise, the thick disk is probably 11 or 12 Gyr old (e.g., 47 Tuc), and the halo is approximately 12 to 14 Gyr old (see contributions by Chaboyer and by Aparicio, these proceedings).

These age results are a remarkable achievement. Given the theoretical and observational difficulties outlined above and the potential for systematic errors, however, it is important to check these results with an independent technique. In the following sections we focus on how white dwarfs provide just this independent technique. The study of radioactive isotopes in stellar atmospheres also holds much promise in this regard, and readers are referred to the many excellent papers on radioactive dating in these proceedings.

3.   Stellar Ages From White Dwarf Cooling Theory

The physics governing white dwarf cooling are largely independent of the physics of main sequence evolution. We thus expect that cooling WDs can provide an independent means of estimating the ages of stellar populations. Additionally, where both WD and main sequence ages are derived for the same star cluster, any conflict between the two results may highlight limitations in WD cooling or main sequence evolutionary theory. Such a test is especially important for checking that the ages derived via the WD luminosity function technique (see contributions by Fontaine, by Leggett et al., and by Harris et al., these proceedings) are on the same age scale as the open and globular cluster ages derived traditionally from stellar evolution theory.

Of course, WD cooling theory has limitations and complications of its own, but rapid progress is being made (see Fontaine, these proceedings). The major difficulties in modeling WD cooling are determining the equation of state in the high density, partially degenerate, partially ionized WD atmospheres, understanding the interplay between convection and gravitational settling, and constraining uncertainties in carbon and oxygen separation during crystallization.

White dwarfs cool through five stages. Stage 1 lasts for 60 to 80 Myr when cooling is dominated by neutrino losses. Stage 2 is usually called fluid cooling and is both well understood and checked by the cooling rate of variable WDs (see Mukadam, these proceedings). In Stage 3 envelope variations as a function of mass and composition become important, so further information and theoretical work is needed on a star-by-star basis. In Stage 4 the WD begins to crystallize. Finally, in Stage 5, after the WD has finished crystallizing, the star enters a state known as Debye cooling in which the latent heat depends on the temperature, and the star cools towards invisibility. Gilles Fontaine (these proceedings) further explains these stages and the present state of their theoretical understanding. For the purposes of comparing WD cooling and stellar evolution in open clusters, the good news is that WD cooling is relatively well-understood for the first 6 to 8 Gyr, depending on the WD mass and atmosphere composition, so that WD age uncertainties due to theoretical difficulties should be less than or approximately equal to main sequence stellar evolution age uncertainties. Readers may wonder why main sequence stellar evolution uncertainties themselves do not play a larger role in WD age uncertainties, since all WDs were once main sequence stars. The answer to this apparent paradox comes from the fact that the oldest WDs in any stellar population came from 5 to 8 solar mass progenitors, whose main sequence lifetimes are $\approx$ 35 to 7 Myr, respectively, so even large uncertainties in the progenitor lifetimes are unimportant for stellar populations more than a few hundred million years old.

The result of WD cooling theory applied to the WD luminosity function indicates that the Galactic disk at the solar annulus is 9 ± 2 Gyr old (Winget et al. 1987; Liebert, Dahn, & Monet 1988; Oswalt et al. 1996; Leggett, Ruiz, & Bergeron 1998; see also Leggett et al., these proceedings).

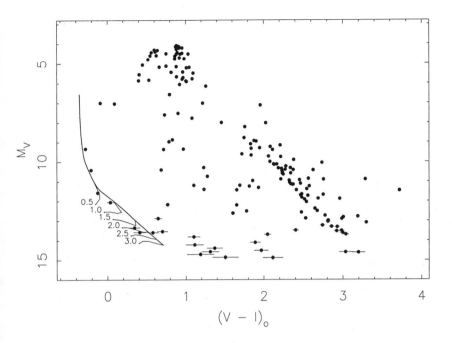

Figure 1. NGC 2420 color–magnitude diagram with the preferred distance modulus, $(m-M)_V = 12.10$, and reddening, $E(B-V) = 0.04$, removed. The 0.5, 1.0, 1.5, 2.0, 2.5, and 3.0 Gyr isochrones (Ahrens 1999) and 0.6 solar mass WD cooling track (Wood 1992) are overplotted.

4. White Dwarf Versus Main Sequence Ages in Open Clusters

To date white dwarfs have been found in a number of open and globular clusters (see von Hippel 1998 for a recent compilation), though the WD cooling limit has not yet been identified in globular clusters due to their great ages and distances. Figure 1 shows a color–magnitude diagram obtained via deep HST imaging of the open cluster NGC 2420 (see von Hippel & Gilmore 2000). Eight objects appear to be excellent WD candidates based on their proximity to the expected WD isochrones and the apparent pile-up near $V = 26$ ($M_V \approx 13.5$). WD isochrone fits to this sequence depend on the assumed distance modulus, which is not precisely known for this cluster. Figure 2 (derived from figure 8 of von Hippel & Gilmore 2000) shows the derived WD age and 1σ uncertainty age range as a function of distance along with the derived ages and distances from recent main sequence stellar evolution studies. The canonical main sequence ages and distance fits are those of Anthony-Twarog et al. (1990; 3.4 Gyr), Castellani, Chieffi, & Straniero (1992; 1.7 Gyr), Castellani, degl'Innocenti, & Marconi (1999; 1.5 Gyr), and Dominguez et al. (1999; 1.6 Gyr). The convective overshoot isochrone age and distance fits are those of Carraro & Chiosi (1994; 2.1 Gyr), Demarque, Sarajedini, & Guo (1994; 2.4 Gyr), Pols et al. (1998; 2.35 Gyr),

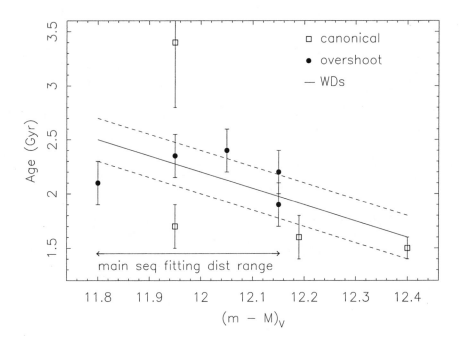

Figure 2.   Comparison between all modern isochrone ages determined for NGC 2420 and the WD cooling age as a function of assumed distance modulus.

Twarog, Anthony-Twarog, & Bricker (1999) employing Bertelli et al. (1994) isochrones (1.9 Gyr), and Twarog et al. (1999) employing Schaller et al. (1992) and Schaerer et al. (1993) isochrones (2.2 Gyr). Models with core convective overshoot are a better match to the WD age and the current distance constraints. As it happens, NGC 2420 is in the key age range of 1 to 2 Gyr where an independent age assessment is the most sensitive to the existence or degree of core convective overshoot, which is a current subject of debate (e.g., Demarque et al. 1994; Dominguez et al. 1999). Other rich open clusters in this important age range are NGC 2477, NGC 752, and possibly NGC 7789.

   Beyond the details of which stellar evolution models provide ages which best match WD ages for a given cluster, we can ask the general question of how the modern but heterogeneous application of stellar evolution models to open clusters compares to the WD ages for these same clusters. Although the WD isochrones have been created and applied by a few groups the range of models applied to clusters in this age range is not as heterogeneous as those within the stellar evolution community. Table 1 presents a list of open clusters for which a WD isochrone (cooling plus precursor) age or limit has been derived, along with recent age determinations from main sequence stellar evolution studies. The first column lists the cluster name, the second column lists the derived WD age, the third and fourth columns list the −1 and +1σ WD ages, respectively, the fifth column lists a recent isochrone age from the literature, and the sixth

Table 1.   White dwarf versus isochrone ages for clusters.

Cluster	WD	WD$_{low}$	WD$_{high}$	Iso	Iso$_{low}$	Iso$_{high}$
NGC 2168	0.094	0.057	0.137	0.160	0.120	0.200
Hyades	0.3	....	....	0.625	0.500	0.675
Praesepe	0.8	0.6	1.0	0.76	0.62	0.82
NGC 2477	1.3	1.1	1.5	1.0	0.8	1.3
NGC 2420	2.0	1.8	2.2	2.2	1.9	2.5
M 67	4.3	3.7	4.5	4.0	3.5	5.0

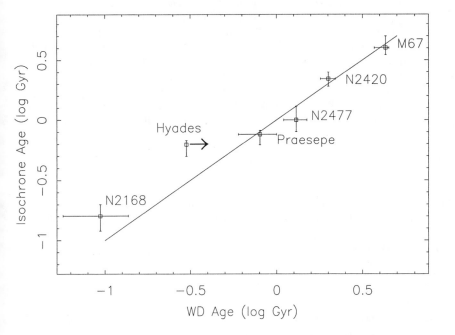

Figure 3.   White dwarf versus isochrone ages from recent studies.

and seventh columns list the -1 and $+1\sigma$ isochrone ages, respectively. The NGC 2168 WD and isochrone ages are from Reimers & Koester (1988a, 1988b). The Hyades WD age is from Weidemann et al. (1992) and the isochrone age is from Perryman et al. (1998). The Praesepe WD and isochrone ages are from Claver (1995). The NGC 2477 WD age is from von Hippel, Gilmore, & Jones (1995) and the isochrone age is from Kassis et al. (1997). The NGC 2420 WD age is from von Hippel & Gilmore (2000) and the isochrone ages are from the references listed above for Figure 2. The M 67 WD age is from Richer et al. (1998) and the isochrone ages are from Demarque, Green, & Gunther (1992) and Dinescu et al. (1995). Figure 3 presents WD versus isochrone ages for these clusters on a logarithmic age scale. The WD age for the Hyades is a lower limit since 50 to 90% of the Hyades has likely evaporated (Weidemann et al. 1992), possibly taking with it some of the oldest WDs. It is clear that even despite the heterogeneous application of stellar evolution models to derive star cluster ages, at the present time there appears to be a good overall agreement between cluster ages derived via the two different techniques over the broad age range of 0.1 to 4 Gyr.

5. Conclusion

White dwarf cooling theory and very deep observations in star clusters provide a new tool to test stellar evolution theory and time scales. In particular, white dwarf cooling is now testing the degree of enhanced core mixing in stars with turnoff ages of 1 to 2 Gyr. More generally, it is also encouraging to see the good overall agreement between WD and modern isochrone ages over the range 0.1 to 4 Gyr. Future application of WD isochrones to open clusters with a variety of ages and metallicities will test the consistency and limitations of WD and stellar evolution theory. Eventually, very deep observations of globular clusters, perhaps with NGST, will yield WD ages for at least a few Galactic globular clusters, comparing the theories of WD cooling with stellar evolution at the greatest possible ages, and thereby improving our estimate in the age of the Galaxy.

References

Ahrens, T. J. 1999, M.S. Thesis, Florida Institute of Technology

Anthony-Twarog, B. J., Kaluzny, J., Shara, M. M., & Twarog, B. A. 1990, AJ, 99, 1504

Baturin, V. A., Däppen, W., Gough, D. O., & Vorontsov, S. V. 2000, MNRAS, 316, 71

Bertelli, G., Bressan, A., Chiosi, C., Fagotto, F., & Nasi, E. 1994, A&AS, 106, 275

Canuto, V. M. 1999, ApJ, 524, 311

Canuto, V. M., & Dubovikov, M. 1998, ApJ, 493, 834

Carraro, G., & Chiosi, C. 1994, A&A, 287, 761

Castellani, V., Chieffi, A., & Straniero, O. 1992, ApJS, 78, 517

Castellani, V., degl'Innocenti, S., & Marconi, M. 1999, MNRAS, 303, 265

Claver, C. F. 1995, Ph.D. Thesis, The University of Texas at Austin

Demarque, P., Green, E. M., & Gunther, D. B. 1992, AJ, 103, 151

Demarque, P., Sarajedini, A., & Guo, X.-J. 1994, ApJ, 426, 165

Dinescu, D. I., Demarque, P., Guenther, D. B., & Pinsonneault, M. H. 1995, AJ, 109, 2090

Dominguez, I., Chieffi, A., Limongi, M., & Straniero, O. 1999, ApJ, 524, 226

Elliott, J. R., & Gough, D. O. 1999, ApJ, 516, 475

Kassis, M., Janes, K. A., Friel, E. D., & Phelps, R. L. 1997, AJ, 113, 1723

Leggett, S. K., Ruiz, M. T., & Bergeron, P. 1998, ApJ, 497, 294

Liebert, J., Dahn, C. C., & Monet, D. G. 1988, ApJ, 332, 891

Oswalt, T. D., Smith, J. A., Wood, M. A., & Hintzen, P. 1996, Nature, 382, 692

Perryman, M. A. C., et al. 1998, A&A, 331, 81

Phelps, R., Janes, K. A., & Montgomery, K. A. 1994, AJ, 107, 1079

Pols, O. R., Schroder, K.-P., Hurley, J. R., Tout, C. A., & Eggleton, P. P. 1998, MNRAS, 298, 525

Reimers, D., & Koester, D. 1988a, ESO Messenger, 54, 47

Reimers, D., & Koester, D. 1988b, A&A, 202, 77

Renzini, A. 1991, in Observational Tests of Cosmological Inflation, NATO Proc. 348, ed. T. Shanks, A. J. Banday, & R. S. Ellis (Dordrecht: Kluwer), 131

Richer, H. B., Fahlman, G. G., Rosvick, J., & Ibata, R. 1998, ApJ, 504, L91

Schaerer, D., Meynet, G., Maeder, A., & Schaller, G. 1993, A&AS, 98, 523

Schaller, G., Schaerer, D., Meynet, G., & Maeder, A. 1992, A&AS, 96, 269

Twarog, B. A., Anthony-Twarog, B. J., & Bricker, A. R. 1999, AJ, 117, 1816

von Hippel, T. 1998, AJ, 115, 1536

von Hippel, T., & Gilmore, G. 2000, AJ, 120, 1384

von Hippel, T., Gilmore, G., & Jones, D. H. P. 1995, MNRAS, 273, L39

Weidemann, V., Jordan, S., Iben, I., & Casertano, S. 1992, AJ, 104, 1876

Winget, D. E., Hansen, C. J., Liebert, J., van Horn, H. M., Fontaine, G., Nather, R. E., Kepler, S. O., & Lamb, D. Q. 1987, ApJ, 315, L77

Wood, M. A. 1992, ApJ, 386, 539

Acknowledgments. This research was supported by the Gemini Observatory, which is operated by AURA, Inc., under a cooperative agreement with the NSF on behalf of the Gemini partnership: the NSF (United States), PPARC (United Kingdom), NRC (Canada), CONICYT (Chile), ARC (Australia), CNPq (Brazil), and CONICET (Argentina).

Discussion

Michael Rich: We (Harvey Richer is the PI) are attempting to do the WD age experiment in the globular cluster M 4. We have 123 orbits on HST: the first data have just been taken, so we'll see if we can reach down to 31st magnitude. From the modeling that we've done with realistic errors, thanks to the blueing of the cooling curves at large ages, we think we can constrain ages between 10 and 13 Gyr.

Ted von Hippel: Good point. Such deep HST observations of M 4 may determine the WD age for this globular cluster. That would be an important result. If it turns out that the globular cluster ages are on the young side of what Brian [Chaboyer] or others say then it may just be possible. If they're on the old side I don't think that HST can go deep enough.

Michael Rich: No, actually, we used Hansen's best tracks and we put in conservative assumptions and we think that there really is the possibility of seeing the color distribution change because so much of the energy emerges, thanks to the molecular hydrogen cooling, around 6000 Å. We'll just have to see.

Oscar Straniero: Ted, let me comment on the comparison of models with and without overshoot and your white dwarf age. What you are really showing is that you need larger mixed cores due to convection or due to other effects. There are a dozen different model ways to increase the size of the mixing zone; convective overshoot is one of them. For example, increasing opacity increases the convectively unstable regions. Rotationally-induced mixing may also produce larger mixed cores. So there are multiple ways to increase the core and get better agreement with the white dwarf age.

Ted von Hippel: I've taken all the main sequence models that have been applied to NGC 2420 to date and I find the one major difference is the use or absence of convective overshooting. Thanks for your clarification. More generally, the white dwarf cooling age for this cluster supports some mechanism that effectively increases the core size.

Astrophysical Ages and Time Scales
ASP Conference Series, Vol. 245, 2001
T. von Hippel, C. Simpson, N. Manset

Stellar Seismology, Stellar Ages, and the Cosmological Constant

G. R. Isaak

School of Physics and Astronomy, University of Birmingham, UK

K. G. Isaak

Cavendish Laboratory, University of Cambridge, UK

Abstract. Solar seismology has allowed precision measurements of both the static and dynamic structure of the Sun. In the near future, seismology of solar-like stars of different ages and masses, necessarily restricted by angular resolution to low l-modes, will allow studies of the internal structure of stars at various stages of evolution. Such studies will test the theory of stellar evolution and allow the determination of ages of stars from the helium content in their cores. Such observations can be made photometrically from space and spectroscopically from the ground. We outline ground-based schemes. By correlating the external properties of nearby stars with their internal properties, it will be possible to extend local studies to distant open and globular clusters, and thereby to obtain an age of the Universe based on many stars. The combination of the age, the density parameter Ω, and Hubble's constant will allow strong limits to be placed on the cosmological constant.

1. Introduction

Recent measurements of distant, type Ia supernovae (Perlmutter et al. 1999; Riess et al. 1998) have been interpreted as implying an accelerating expansion of the Universe and demanding the existence of a cosmological constant and, presumably, a new fundamental interaction (Zlatev, Wang, & Steinhardt 1999). The implication of such interpretations is profound for cosmology and for physics at large. Such extraordinary claims require extraordinarily good evidence, preferably by many, independent means. A possible method of verification is the determination of the ages of Galactic stars, and thereby a lower limit to the age of the Universe. With a Hubble constant $H_0 = 67$ km s^{-1} Mpc^{-1} and a density parameter $\Omega = 1$, the age of the Universe is 9.6 Gyr if the cosmological constant $\Lambda = 0$. Adopting the currently fashionable cosmological vacuum energy contribution of $\Omega_\Lambda = 0.7$, demands an age of 13.8 Gyr.

To measure stellar ages requires calibrating stellar evolution theory. The Sun provides a calibration at one point in mass, composition, and age parameter space. An external calibration, using only the readily measured external parameters has been used in the past. An internal calibration, using parameters of the solar core is also possible. This provides a near-direct measure of the molecular

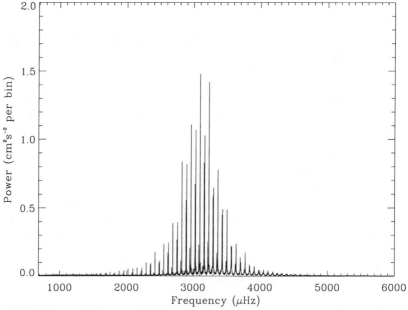

Figure 1.   Power spectrum generated from 7 years of data accumulated from 6 stations of BiSON (Chaplin et al. 1996).

weight in the core and, thereby, a measure of the helium content. If the original amount of helium is presumed known and if gravitational settling is allowed for, the current helium content provides a measure of the amount of thermonuclear conversion of hydrogen into helium and therefore, with a known luminosity, the age of the star. Stellar seismology is capable of providing this information in the near future for stars over a range of ages. A reasonable zero-age calibration, in addition to that given by the Sun, could be provided by a very young star, preferably one with known mass and composition. The absolute error in the theoretically age is then likely to be sufficiently small. This, via a cosmological aeon ladder, ought to allow us to determine the ages of old stars and also of those in old open and globular clusters.

We propose to bootstrap the existing solar seismological calibration to determine the ages of stars, and thus check stellar evolution theory over a wide range of stellar masses, internal composition, and age. Such an undertaking is feasible using current technology and is underpinned by two decades of work that has established the solar acoustic eigenmode spectrum. Here, we outline the relevant key elements of solar seismology, its extension to stars, the sensitivity to age of the relevant eigenfrequency separations, and the feasibility of those measurements from a space platform as well as from the ground.

2.   Solar Seismology

Electromagnetic spectroscopy of the eigenmodes of the atom has provided a wealth of tools with which to probe Nature over the last one and a half centuries. The detailed study of a stellar acoustic spectrum (see Figure 1), and in particular the study of eigenfrequencies, shows equal promise. Unlike many other parameters pertaining to stellar systems, eigenfrequencies are both precise *and* accurate. Frequency measurements with an accuracy of up to a few parts per million have been made with no systematic error. Some statistical errors are present: in particular, those arising from the finite duration of any observation, the finite lifetime of the eigenmode itself, daily aliases, and additional amplitude noise. In spite of these, it was possible already in 1981 (Claverie et al. 1981) to determine the acoustic eigenfrequencies of the individual low l ($l = 0, 1, 2, 3$) modes of the global Sun to better than 1 part in 4000. Currently, the accuracy attained is better than 1 in 10^5 (Chaplin et al. 1998).

Individual eigenfrequencies provide integral measures of the velocity of sound, and thus of $\sqrt{\gamma k T/m}$ over the ray path of the sound wave, with different modes probing different depths. At present the accuracy to which the eigenfrequencies can be measured far outstrips the level to which theoretical models and observational data agree. Thus, little useful science can yet be extracted from direct comparisons between observed eigenfrequencies and theoretical models. A large part of this problem is believed to be the incomplete treatment of the contributions of the surface layers to the travel time of sound waves. By taking *differences* between eigenfrequencies this problem can be substantially eliminated. Differential information gleaned by measuring frequency separations includes: (a) the separation between successive harmonics of the radial and low l non-radial modes—this large separation, Δ, provides a measure of the time taken by sound waves to traverse the diameter of the star—$\Delta \sim (2 \int dr/c)^{-1}$; (b) the small separation, d_0 between radial and quadrupole p modes, closely spaced in frequency; these modes share a very similar path in the outer regions of the Sun, but penetrate to different depths in the core, and thus measure core properties; and (c) an even smaller splitting due to the removal of degeneracy of the non-radial eigenmodes by rotation, the rotational splitting. The number of components, $2l + 1$, can be used to identify the mode, with their spacing providing an integral measure of the internal rotation.

The use of such differencing methods is common in other branches of physics. Perhaps the most familiar example is an atomic analogue: from fine-structure, hyperfine-structure, and isotope shifts of optical atomic transitions it is possible to infer details about atomic and nuclear properties that are too subtle to be inferred theoretically. In a similar way d_0, provides a sensitive measure of the temperature and molecular weight in the stellar core. As the physical conditions vary with mass and age, so too does the separation—in the Sun, its value is close to 9 μHz. As a star evolves, the helium fraction in the nuclear active core increases. The resultant increase in the molecular weight in those regions, sampled more by the radial than the quadrupole modes, results in a decrease in the velocity of sound and thus in the value of d_0. The variation of both d_0 and Δ with evolutionary age and mass has been evaluated by Christensen-Dalsgaard (1984, 1988). For a solar-type star at the midpoint of its life, the sensitivity of the fine-spacing is near 1 μHz Gyr^{-1}. Thus, the accuracies to which Δ and d_0 were measured in the Sun already in 1981, about 0.3 μHz, correspond to a solar age determination error of 0.3 Gyr. Current measurements have errors some twenty times smaller. Thus, it is model-dependent, and *not* observational, error that will dominate.

3. Calibration and Pitfalls

The calibration of any cosmological aeon glass is full of pitfalls. Whilst the value for the age of the meteorite-Earth system, determined using radioactive decays of isotopes of K, Rb, Sm, Lu, Re, Th, and U, has remained within ± 0.01 Gyr of the value of 4.55 Gyr over the last 45 years (Dalrymple 1994), the value for the age of the oldest globular clusters has varied over the range of 5 to 25 Gyr over the same period (Sandage 1962; Gamow 1964). This in itself is an indication that absolute calibration of stellar structure and evolution theory is essential. Such calibration of the basic physics has been provided by solar seismology for the last twenty years, while theoretical solar models have been adjusted repeatedly over that time to approximate the measured eigenfrequencies. The calibration of a model, solar or otherwise, at time t, means that we adjust the proposed model either *quantitatively* by adjusting parameters or *qualitatively* by the addition of a physical mechanism such that model age and radioactive age agree exactly.

Two examples may serve to demonstrate some of the pitfalls in such work. The first comes from the discovery measurements of global solar oscillations (Claverie et al. 1979). One of us, GRI (Isaak 1980), used the extreme high frequency end of the eigenfrequencies of the solar models of Iben & Mahaffy (1976) and Christensen-Dalsgaard, Gough, & Morgan (1979). The theoretical models did not predict the excitation of high n, low l acoustic modes. It was fortuitous that the published values extended into the region in which, unexpectedly (Unno et al. 1979), global solar oscillations were discovered. GRI deduced, by interpolation, that the best fit between Δ and the models implied helium, $Y \leq 0.17$, in agreement with the average solar wind value *but* the observational data, thought to be accurate to the 0.3% level, were indeed correct, however the model was wrong and required substantial quantitative adjustments. Thus, the potentially fundamental cosmological implication of this very low Y inference was false. A second example is based on the much improved measurements of d_0 which BiSON (Birmingham Solar Oscillations Network) has produced since the early 1990's. GRI compared the measured values of d_0 with the much improved models of Christensen-Dalsgaard, concluding that the fit was poor—assigning an age of 5.2 Gyr to the Sun would have provided an excellent match between model and data, however this would have been totally inconsistent with the radioactively-determined solar age, which by then was very secure. In principle, the Sun could be older than the rest of the Solar System, however, the presence of daughter products of short-lived isotopes such as ^{26}Al, ^{107}Pd, and ^{41}Ca in meteorites suggests strongly that the meteorite formation was coeval with the Sun (Guenther 1989; Bahcall, Pinsonneault, & Wasserburg 1995), presumably triggered by a supernova explosion. Models suggest that it takes around 50 Myr for a molecular cloud to collapse onto the zero age main sequence (Schwarzschild 1958; Iben 1965). Thus, the age of the Sun is $4.56 - 0.05 = 4.51$ Gyr, with an uncertainty which would seem to be at most 0.02 Gyr. GRI preferred the radioactive age to the seismological age and speculated, in common with others (Christensen-Dalsgaard, Proffitt, & Thompson 1993), that helium settled gravitationally (Isaak 1993), with the idea shortly afterwards that heavier elements (Proffitt 1994) settled also. These *qualitatively new* additions to the physics of the seismological Sun can nearly, but not perfectly, reproduce the measurements (Elsworth et al. 1995; Chaplin et al. 1997). A cynical reader might wonder as to the relevance of the above: clearly, the measurements are of an integral nature and no unique interpretation is possible. We suggest, however, that the study of solar seismology has substantially enriched the theory of stellar structure and evolution by confronting theory with accurate observation. The future of stellar seismology has even greater potential, if appropriate financial resources were to be made available.

4. Can One Detect Stellar Oscillations?

Soon after the discovery of global solar oscillations (Claverie et al. 1979), it was pointed out (Isaak 1980) that the detection of stellar oscillations of the size seen on the Sun could be achieved in one of two ways: (a) by measuring spectroscopic velocities using large flux collectors, or (b) by photometric measurements using a photometer and very modest optical telescopes situated above the transparency fluctuations and the turbulence of the terrestrial atmosphere. Two years later, GRI also suggested a stellar seismology mission to the European Space Agency; however neither this mission, nor any of its many successors over the last 18 years, have been launched.

Can stellar oscillations really be detected? Here, we assess the feasibility of the detection of solar-like modes in main sequence stars. We assume oscillation amplitudes of a size comparable to those seen on the Sun. There is some evidence to suggest (Houdek et al. 1995) that the more evolved a star, the larger the amplitude of oscillation. We adopt, however, a conservative amplitude of $\sim$ 10 cm s^{-1} in velocity, and 2 parts per million in intensity. We also assume extending spectroscopic measurements from the Sun to the brightest stars is difficult as the fluxes of the Sun:Sirius:3.3 $M_{\odot}$ star are in the ratio $1{:}10^{-10}{:}10^{-12}$.

To determine Δ and d_0 it is necessary to measure the stellar eigenmodes, and to resolve them in order to be able to measure them sufficiently accurately. This requires that each of three distinct, fundamental, and physically-based observational criteria are met. (a) Each mode is measured with a signal-to-noise ratio, S/N > 4. Many claims of the detection of stellar modes have been made over the the last 17 years, largely with low S/N data and complex data analysis. Not one detection to date has, however, been substantiated. (b) The uncertainty principle is satisfied—if we assume that the oscillations have a lifetime greater than the observation time, then to measure with an error of 1 (0.3) μHz, with a commensurate error in stellar age of 1 (0.3) Gyr, requires 2 (6) weeks. (c) The Nyquist sampling theorem is satisfied—at least 2 measurements need to be made in the shortest time period (highest eigenfrequency) of interest. Each of the above need to be met for observations that are shorter than the lifetime (coherence time) of the mode. If the observations are longer than the coherence time, then the power spectra of observations separated by more than the coherence time are independent and the accuracy gain with time is slow. If incoherent, the frequency errors in the power spectrum scale only as the square root of time rather than linearly with time. As an additional constraint, it is clear that the instrumental stability and any resultant systematic errors must not be larger than the statistical noise contributions.

4.1.   Photometry From Space

To achieve the required photometric accuracy of 2 parts per million, roughly equivalent to a velocity amplitude of 10 cm s^{-1} (Isaak 1980), is extremely difficult or even impossible from the ground. Such accuracy can, however, be achieved from above the atmosphere. Whilst expensive, the feasibility of such photometry was amply confirmed when solar oscillations were detected with the Active Cavity Radiometer on the Solar Maximum Mission (Woodard & Hudson 1983). A minimum of 4×10^{12} photons has to be detected, and systematic errors have to be correspondingly small. A 1(0.3) m telescope collects such an integrated flux in just over 1 ($\sim$20) day(s) from an 8th magnitude star.

4.2.   Ground-Based Spectroscopy

Stars subtend angles of less than 10 milliarcseconds. Thus, differential extinction effects across a stellar disk, a source of significant systematic error in global solar oscillation measurements, are negligible. By using differential methods and rapid switching, spectroscopic techniques can reduce systematic errors substantially, as demonstrated by work on the unresolved Sun over more than two decades. By switching between the blue and red wings of a spectral line, common mode effects are reduced substantially. Photon noise is, however, a severe problem as the instantaneous fractional bandwidth of spectroscopic instruments is usually small. Consider three different spectroscopic measurement scenarios: (a) a portion of a spectral line of bandwidth 50 mÅ, (b) all of one spectral line of 0.2 Å width for a slowly rotating star, and (c) the measurement of n spectral lines. Statistically, we gain a factor of two in going from (a) to (b) and an additional factor of $\sqrt{n}$ by going to (c). Whether systematic errors can be kept correspondingly low in (c) is unclear. In an ideal spectrometer, with a throughput of unity, a bandwidth as specified by case (a), fed by a 3 (10) m diameter flux collector, a total of 7×10^4 (7×10^5) photons s^{-1} would be collected from a 3.3 $M_\odot$ star. The latter counting rate is within a factor of two of the BiSON Mk I optical resonance scattering spectrometer viewing the Sun (Brookes, Isaak, & van der Raay 1978). Such a flux collector plus ideal spectrometer of type (a), even of 0.05 Å bandwidth, can readily detect stellar oscillations with typical solar amplitudes.

We discuss another specific spectroscopic system—the magneto-optical-filter (MOF). The MOF was developed by Oehman (1956) and by Cimino, Cacciani, & Sopranzi (1968). An improved version of the MOF (Isaak & Jones 1988) has reached photon noise-limited velocity noise of 35 cm s^{-1} (Bedford et al. 1995). The throughput of that spectrometer is down from the ideal by about 400. Use of two polarisations (gain = 2), a CCD rather than an avalanche photodiode (gain = 4), and Na D1 and Na D2 as well as the potassium resonance line at 7699 Å, restores a factor of 24. The solar spectral line has a slope which is a factor of 4 steeper. The overall expected performance, scaling from the *actual performance*, with a gain of $4 \times \sqrt{24}$ improvement on the Bedford et al. (1995) value, would be a velocity noise level of 2 cm s^{-1} (8 cm s^{-1}) on a 0.3 $M_\odot$ (3.3 $M_\odot$) star with two weeks observing time on a 1.9 m diameter telescope. Clearly, flux collectors of the Hanbury–Brown type with modern active optics and diameters of 6 metres would make the above possible. It should be stressed that the optical resonance spectrometers have an enormous etendue and can, therefore, accept beams from poor quality, i.e., cheaply made flux collectors.

5. Summary

- Global solar oscillations probe a 1 $M_\odot$ Population I star at 4.51 Gyr, and so *calibrate* the aeon glass at one point in time, mass, and composition. Another point, the zero-age point, may be provided by stars which are theoretically reckoned to be extremely young.

- Stellar oscillations of main-sequence stars could calibrate aeon glasses over a range of ages.

- Both photometry from space and spectroscopy from the ground provide sufficient sensitivity to render the detection of stellar oscillations feasible.

- By measuring large and small spacings of the acoustic eigenmode spectrum of old solar-like stars, their ages could be measured to substantially better than 1 Gyr; these internal measurements could then calibrate the external characteristics of the same stars. Transferring this calibration, using external characteristics to more distant Population I and II stars, then allows the ages of old open and globular clusters to be determined.

- Together with the present value of $H_0 = 67$ km s^{-1} Mpc^{-1} and $\Omega = 1$, stellar ages can be used to check on the existence of a cosmological constant, Λ. $\Lambda = 0$ demands an age of the Universe of 9.6 Gyr, whereas the currently fashionable $\Omega_\Lambda = 0.7$ requires an age of the Universe of 13.8 Gyr. Given that any stellar ages provide at best a lower bound to the age of the Universe, if any star were found to have an age in excess of 10 Gyr, then stellar seismology would provide an independent confirmation for a non-zero cosmological constant. Stellar ages below 10 Gyr would, in contrast, provide necessary but not sufficient evidence to exclude a finite Λ if the oldest Galactic stars were to be considerably younger than the Universe.

- We stress that in contrast to the use of SNIa and other techniques, seismological measurements can thus be considered to make possible cosmology studies at zero redshift.

References

Bahcall, J. N., Pinsonneault, M. H., & Wasserburg, G. J. 1995, Rev. Mod. Phys., 67, 781

Bedford, D. K., Chaplin, W. J., Coates, D. W., Davies, A. R., Innis, J. L., Isaak, G. R., Speake, C. C. 1995, MNRAS, 273, 367

Brookes, J. R., Isaak, G. R., & van der Raay, H. B. 1978, MNRAS, 185, 1

Chaplin, W. J., Elsworth, Y., Isaak, G. R., Lines, R., McLeod, C. P., Miller, B. A., New, R. 1998, MNRAS, 300, 1077

Chaplin, W. J., Elsworth, Y., Isaak, G. R., McLeod, C. P., Miller, B. A., New, R. 1997, ApJ, 480, L75

Chaplin, W. J., et al. 1996, Sol. Phys., 168, 1

Christensen-Dalsgaard, J. 1984 in Space Research Prospects in Stellar Activity and Variability, ed. A. Mangeney & F. Praderie (Meudon: Observatoire de Paris), 11

Christensen-Dalsgaard, J. 1988, in IAU Symp. 123 Advances in Helio- and Astroseismology, (Dordrecht; Reidel), 3

Christensen-Dalsgaard, J., Gough, D. O., & Morgan, J. G. 1979, A&A, 73, 121

Christensen-Dalsgaard, J., Proffitt, C. R., & Thompson, M. J. 1993, ApJ, 403, L75

Cimino, M., Cacciani, A., & Sopranzi, N. 1968, Sol. Phys., 3, 618

Claverie, A., Isaak, G. R., McLeod, C. P., van der Raay, H. B., & Cortes, T. R. 1979, Nature, 282, 591

Claverie, A., Isaak, G. R., McLeod, C. P., van der Raay, H. B., & Roca Cortes, T. 1981, Nature, 293, 443

Dalrymple, G. B. 1994, The Age of the Earth (Palo Alto: Stanford University Press)

Elsworth, Y., Howe, R., Isaak, G. R., McLeod, C. P., Miller, B. A., Wheeler, S. J., New, R. 1995, in ASP Conf. Ser. 76, GONG'94, ed. R. K. Ulrich, E. J. Rhodes, & W. Daeppen (San Francisco: ASP), 51

Gamow, G. 1964, in Cosmogony, Encyclopedia Brittanica

Guenther, D. B. 1989, ApJ, 229, 1156

Houdek, G., Rogl, J., Balmforth, N. J., & Christensen-Dalsgaard, J. 1995, in ASP Conf. Ser. 76, GONG'94, ed. R. K. Ulrich, E. J. Rhodes, & W. Daeppen (San Francisco: ASP), 641

Iben, I. Jr. 1965, ApJ, 141, 993

Iben, I. Jr., & Mahaffy, J. 1976, ApJ, 209, L39

Isaak, G. R. 1980, Nature, 283, 644

Isaak, G. R. 1993, in Carnegie Institution of Washington Yearbook 92, 149

Isaak, G. R., & Jones, A. R. 1988, in IAU Symp. 123, Advances in Helio- and Astroseismology, ed. J. Christensen-Dalsgaard & S. Frandsen (Dordrecht: Reidel), 255

Oehman, Y. 1956, Stockholm Observatory Annals, 19, 3

Perlmutter, S., et al. 1999, ApJ, 517, 565

Proffitt, C. R. 1994, ApJ, 425, 849

Riess, A., et al. 1998, ApJ, 116, 1009

Sandage, A. 1962, ApJ, 135, 349

Schwarzschild, M. 1958, Structure and Evolution of the Stars (Princeton: Princeton University Press), 164

Unno, W., Osaki, Y., Ando, H., & Shibahashi, H. 1979, in Non-radial Oscillations of Stars (Tokyo: University of Tokyo Press), 66

Woodard, M. F., & Hudson, H. S. 1983, Sol. Phys., 82, 67

Zlatev, I., Wang, L., & Steinhardt, P. J. 1999, Phys. Rev. Lett., 82, 896

Astrophysical Ages and Time Scales
ASP Conference Series, Vol. 245, 2001
T. von Hippel, C. Simpson, N. Manset

The Dark Side of the Milky Way

Jan Bernkopf, Alex Fiedler

Universitäts-Sternwarte München, Germany

Klaus Fuhrmann

Universitäts-Sternwarte München, Germany and MPE, Garching, Germany

Abstract. The familiar picture of our Galaxy is that of myriad bright stars assembled in a flat disk which we call the Milky Way. But what if, like a shadow, another massive disk population of formerly bright stars inhabited the Galaxy from the beginning, though characterized by somewhat larger scales, but then managed to fade away? Here we present very recent results of such a scenario from the local census of long-lived stars and from the detailed scrutiny of the involved time scales, which has now become possible on the basis of the HIPPARCOS astrometry.

1. Prolog

Field stars of spectral type F and G are intrinsically bright and also long-lived in that they evolve on time scales of billions of years to white dwarfs. Perhaps most F stars may not become as old as the Galaxy, but this should be true for the majority of the G-type stars. This is at least our common wisdom, and has invoked numerous Galactic formation and evolution studies with G stars as tracers, and, among others, G-dwarf problems as their outcome.

The realization that stars arise from various populations and do not exhibit the same metallicity, not even intrinsically, is nowadays by no means challenging, but it was so in the 1940s and 1950s with the landmark papers of Baade (1944) and Chamberlain & Aller (1951). Our generation is instead plagued to internalize that star formation never offers a simple one-to-one metal enrichment, nothing like a well-behaved age-metallicity relation. In fact, while it seems true that the local halo population is almost negligible, the 1980s brought to light that our Galaxy accommodates at a locally small but noticeable level what has since been termed the *thick-disk* population (Gilmore & Reid 1983). Its members have a mean metallicity of [Fe/H] ~ -0.6, but with considerable overlap to the locally dominant (then) *thin disk* with respect to e.g., the α-elements. The thick disk is also evidently a very old population (e.g., Wyse 1996), and it is only within the last few years that a star formation gap between thick disk and thin disk was realized as a possible scenario for the Milky Way evolution (cf. Gratton et al. 1996). This is particularly important, since it potentially provides the ultimate legitimacy for the existence of two *discrete* disk populations, which has been a matter of debate for many years.

Indeed, as we will show in what follows, the early hiatus in star formation must have lasted for no less than three billion years and this again raises the question: *What is a long-lived star if there are two disk populations involved?* Certainly, if we intend to compare the properties of the thin disk to that of the thick disk, we have to resort to only those stars that do not evolve to white dwarfs within 13 or 14 Gyr. But if, as we will see, it happens that, rather unexpected, the *majority* of the ubiquitous thin-disk G-type stars do not withstand this criterion, then what? Most obviously this raises the importance of the thick-disk population considerably, its inconspicuous appearance becomes rather superficial, and we have a vague presentiment that the light of the much younger, bright, and pestering thin-disk stars is much of a shabby trick.

In the end we may find that the Galactic thick disk is as massive as the familiar thin-disk population, but with a principal carrier that consists of stellar remnants, such as that in Figure 1, instead of ordinary stars. If so, we may have to approve the existence of a lot of *baryonic* dark matter to the Milky Way Galaxy, one which may actually cause the Sun to rotate at its ~ 220 km s^{-1} around the Galactic center, and without invoking any non-baryonic components to this.

Figure 1.   Map towards Taurus including the very faint ($V = 19$), but local, white dwarf WD 0346+246, a putative thick-disk object that was serendipitously discovered in 1997 by Hambly, Smartt, & Hodgkin at a distance of 28 pc. Guess where it lurks! *(The interested reader will find the answer, as well as the object's change in position with time, in* ApJ, *489, L159.)*

2.   The Ages and Time Scales of the Local Disk Populations

We will focus our discussion on bright, long-lived FGK stars and on particularly those objects that one can find within 25 pc of the Sun. This is of course only 10% of the scale height of the thin disk and even less than 1% of its scale length. In fact a tiny volume compared to the structure of the Milky Way! But, it is currently the largest sample where our knowledge of the stellar inventory down to the early K stars is thought to be *complete* in terms of accurate astrometric data from the HIPPARCOS satellite. There is also no a priori reason to believe that the solar neighborhood is somehow special compared to other mid-plane regions at this distance from the Galactic center. At the same time, our sample is not too large for an *individual* spectroscopic study at high S/N ($\sim$150–400) and high resolution ($\lambda/\Delta\lambda \sim 60\,000$). All spectra were obtained with the fiber optics Cassegrain échelle spectrograph FOCES (Pfeiffer et al. 1998) on top of the Calar Alto mountain and are thus as well restricted to stars north of $\delta = -15°$. Up to now some 160 stars have been analyzed, which is more than half of the final sample. For details of the applied methods and individual stellar parameters we must refer here to Fuhrmann (1998; 2001), but we will present at least a few explicit results for some of the real *key stars* in what follows.

What is the age of the solar neighborhood? Binney, Dehnen, & Bertelli (2000) have recently tried to answer this question. But, as we think, it is meaningless to ask like this without, first, clarifying the origin of the nearby stars. Instead what one should rather put forward is: *What are the stellar populations that one can eventually identify in the solar neighborhood?* And, if so: *What are the ages of these populations?* As it turns out, and as expected, the Galactic halo and bulge populations are negligible for the nearby stars. But there is a significant thick-disk population that we can separate from that of the thin disk. Figure 2 summarizes these findings for the stars' chemistry with respect to iron and the α-element magnesium. Herein additional stars of the thick disk and halo beyond 25 pc are included as well to better represent the various regimes in somewhat larger numbers. As we can see, there is some degeneracy in the chemical abundances of the disk populations, and the same holds true for the kinematics of the stars (see Figure 8, below). However, what makes the thick disk a discrete entity is its age. Thus, while the thin disk has a mean age that is approximately solar and does not exceed ~ 8 Gyr, the thick disk is instead very old, most likely beyond ~ 12 Gyr, representing eventually the oldest stars in our Galaxy.

To see this, we have searched for nearby subgiants, where detailed evolutionary tracks can nowadays pose meaningful constraints to their ages. This is because, first, the HIPPARCOS data provide accurate absolute visual magnitudes (typically a factor five better than heretofore possible), and, second, the uncertainty in the effective temperature determination is rather insignificant in the subgiant stage of evolution.

But, while we have several thin-disk subgiants within 25 pc at our disposal, the fact that we can identify only a dozen (northern) thick-disk stars within this volume entails that there exists actually no subgiant of this population. Fortunately, the HIPPARCOS data are also fairly accurate out to 50 pc, and in a systematic search we found indeed three bona fide thick-disk subgiants, all observable from Calar Alto, and we took FOCES spectra of these in January

2000. We were also lucky to find a subgiant, HR 7569, among the few objects with intermediate chemistry (designated as "transition stars" in Figure 2) that, as it turns out, provides an upper limit to the *onset* of the thin-disk formation.

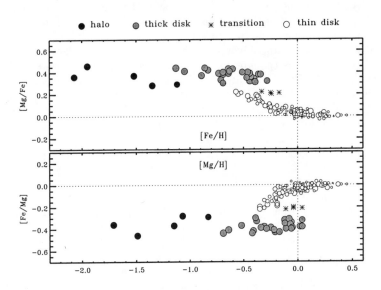

Figure 2.   *Top:* Abundance ratio [Mg/Fe] versus [Fe/H] for the sample stars. *Bottom:* The same data, but with magnesium as the reference element. Except for the few transition stars (the asterisks), depicted circles are in proportion to the stellar ages. Different stellar populations are given with various grayscale symbols as indicated in the legend on top and are based on abundance, kinematics, and age information.

Figure 3 presents a detailed evolutionary track for 70 Vir, the oldest thin-disk subgiant that we have identified so far. For this star we get internal errors of $\Delta\tau(M_{bol}) = 0.5$ Gyr, $\Delta\tau([Fe/H]) = 0.3$ Gyr, and $\Delta\tau(T_{eff}) = 0.1$ Gyr. In Figure 4 we present the results for HR 7569, for which we obtain $\tau = 9.1 \pm 1.0$ Gyr. HD 10519, one of the three thick-disk subgiants, is shown in Figure 5. All relevant data are summarized in Table 1. Two results are striking: first, all three thick-disk subgiants (HR 173, HD 10519, HD 222794) are significantly older than the thin-disk objects, and, second, all are of similar age, as also expected from their common Fe/Mg abundance ratios in Figure 2 (cf. also Mashonkina & Gehren 2000, 2001, for recent intriguing results on Ba and Eu).

Thus there is clear evidence for a significant, if not huge, star formation gap of 3 to 5 Gyr for the two disk populations, and hence *we have every reason to accept their discreteness.* We note here in passing that our stellar interior calculations are done with state-of-the-art input physics and include helium diffusion. We adopt $Y = 0.2723$ for the unknown helium fraction but verified that e.g., $Y = 0.25$ would reduce the nominal ages of our thick-disk subgiants by ~ 0.7 Gyr. Likewise metal diffusion resulted in ~ 1 Gyr younger stars at maximum—except for HR 173, where this mechanism becomes inefficient due to the star's evolved stage near the bottom of the giant branch.

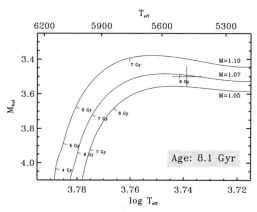

Figure 3. Evolutionary tracks with $M_\star = 1.05$, 1.07, and 1.10 $M_\odot$, at [Fe/H] = -0.11 and [Fe/Mg] = -0.08, as derived for the old thin-disk subgiant 70 Vir. Tick marks are given in steps of 1 Gyr.

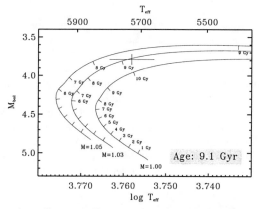

Figure 4. Same as Figure 3 but for the disk transition star HR 7569.

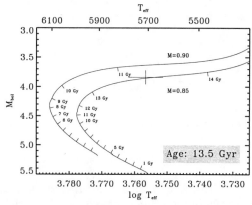

Figure 5. Same as Figure 3 but for the thick-disk subgiant HD 10519.

Table 1.   Basic stellar parameters of the seven nearby subgiants[†].

Object	T_{eff} (K)	$\log g$ (cgs)	[Fe/H]	[Fe/Mg]	M_{bol} (mag)	Mass ($M_\odot$)	Age (Gyr)
109 Psc	5614	3.96	+0.10	+0.00	3.54 ± 0.08	1.16 ± 0.03	6.8 ± 0.6
70 Vir	5481	3.83	−0.11	−0.08	3.50 ± 0.06	1.07 ± 0.03	8.1 ± 0.6
μ Her	5592	3.94	+0.24	+0.00	3.64 ± 0.05	1.17 ± 0.05	7.0 ± 0.7
HR 7569	5729	4.01	−0.17	−0.21	3.79 ± 0.07	1.03 ± 0.03	9.1 ± 1.0
HR 173	5373	3.82	−0.64	−0.39	3.63 ± 0.07	0.84 ± 0.03	13.8 ± 1.3
HD 10519	5710	4.00	−0.64	−0.43	3.84 ± 0.12	0.85 ± 0.04	13.5 ± 1.3
HD 222794	5623	3.94	−0.69	−0.40	3.66 ± 0.08	0.86 ± 0.03	12.5 ± 1.1

† Errors are $\sigma(T_{\mathrm{eff}}) = 70$–$80$ K, $\sigma(\log g) = 0.1$ dex, $\sigma([\mathrm{Fe/H}]) = 0.006$–$0.008$ dex, and $\sigma([\mathrm{Fe/Mg}]) = 0.05$ dex.

3.   Long-Lived or Not Long-Lived

To understand how the Galaxy formed and evolved we must better assess the relative contributions of thick- and thin-disk stars. This not only requires us to identify local thick-disk stars—in fact only a handful of objects—but also to find long-lived thin-disk stars. As mentioned above, it was somewhat surprising that the detailed analyses reveals most G stars to become degenerates *within* 14 Gyr (cf. Figure 6). Indeed, even some K stars, such as 83 Leo A, can be short-lived! This has to do with the notorious shortcomings of spectral classification, which we note only in passing. Its net effect, however, is that the temporally unbiased local normalization of the thick disk is as high as 14–18%. Thus, for a thin disk scale length $h_{\mathrm{r}} = 3.0$ kpc, and a disk scale height ratio ~ 4.7 (Phleps et al. 2000), the mass of the thick disk is *comparable* to that of the thin disk!

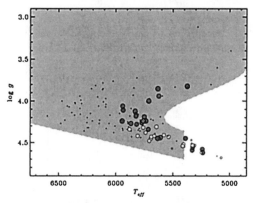

Figure 6.   Kiel diagram of the long-lived disk stars. Grayscaling and symbol sizes have the same meaning as in Figure 2, except for the small black dots, which depict either halo or short-lived disk stars. Shading indicates the selected region of the volume-limited sample, with $T_{\mathrm{eff}} = 5400$ K, taken as the lower limit on the main sequence.

As illustrated in Figure 7, this result is also somewhat dependent on the rather unknown thick disk scale length and it assumes an invariant initial mass function. The latter assumption is presumably not fulfilled, however (Fuhrmann 2001). Instead, there is good evidence (Favata, Micela, & Sciortino 1997) for a thick disk with a low-mass cut at $M \sim 0.7\ M_\odot$ and hence for a top-heavy initial mass function. The resulting high star formation rate at the very early epoch of the Galaxy implies, in turn, a strong Galactic wind and gives thus rise to the hot, massive, and considerably metal-enriched intragroup or intracluster deposits. In this scenario the thick disk is nowadays a massive remnant-dominated population, very much as envisaged by Larson (1986). It may thus not only account for many of the massive compact halo objects [1] (cf. also Gates & Gyuk 2001), but particularly implies that most of the recently detected *blue* white dwarfs (Hansen 1998) in various deep fields (e.g., Knox, Hawkins, & Hambly 1999; Ibata et al. 1999, 2000; Méndez & Minniti 2000; Scholz et al. 2000; Harris et al. 2001) have their origin in this ancient population.

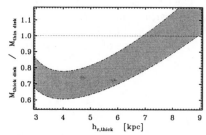

Figure 7.   Disk mass ratio as a function of the scale length of the thick disk. Shading indicates the mass range that results from a local density of 14–18% of thick-disk vs. thin-disk stars.

In the end, we come back to WD 0346+246, the nearby white dwarf of Hambly et al. (1997) in Figure 1 (see also Fontaine, Brassard, & Bergeron, these proceedings), as well as another very cool degenerate LHS 3250, that was recently discussed in Harris et al. (1999). We show simulations for both in Figure 8, the Toomre diagram of our sample stars. With Galactic coordinates $l = 166°$, $b = -23°$, the unknown radial velocity of WD 0346+246 is mostly projected into the U component of the space velocity. The radial velocity of LHS 3250, in turn, varies along its V component. Although by adopting extremely large radial velocities one can always arrive at halo kinematics, the fact that both white dwarfs are very nearby make the assumption of a thick-disk membership at least a factor ten more likely than that of halo objects. We have thus simulated in Figure 8 the velocity ranges $-100 < U < +100$ km s^{-1} and $-160 < V < -30$ km s^{-1} that encompass our current sample ($N \sim 30$) of identified thick-disk stars. From inspection of Figure 8 it is clear that both white dwarfs are indeed bona fide thick-disk members, thereby strongly supporting the notion that our home is embedded in a huge coffin of dead stars.

---

[1] We are then consequently looking for massive compact thick-disk objects or MACTDOs.

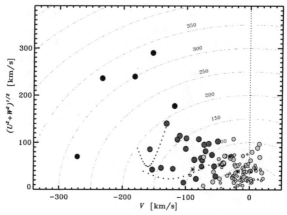

Figure 8.   Toomre diagram for the stars of Figure 2. Circles delineate constant peculiar space velocities $v_{pec} = (U^2 + V^2 + W^2)^{1/2}$ in steps of $\Delta v_{pec} = 50$ km s^{-1}. Likely positions of the *bona fide* thick-disk white dwarfs WD 0346+246 (v-shaped) and LHS 3250 (lower, flat curve) are given by the two dotted traces in steps of $\Delta RV = 10$ km s^{-1}.

References

Baade, W. 1944, ApJ, 100, 137

Binney, J., Dehnen, W., & Bertelli, G. 2000, MNRAS, 318, 658

Chamberlain, J. W., & Aller, L. H. 1951, ApJ, 114, 52

Favata, F., Micela, G., & Sciortino, S. 1997, A&A, 323, 809

Fuhrmann, K. 1998, A&A, 338, 161

Fuhrmann, K. 2001, ApJ, submitted

Gates, E. I., & Gyuk, G. 2001, ApJ, 547, 786

Gilmore, G., & Reid, N. 1983, MNRAS, 202, 1025

Gratton, R., Carretta, E., Matteucci, F., & Sneden, C. 1996, in ASP Conf. Ser. 92, Formation of the Galactic Halo .... Inside and Out, ed. H. Morrison & A. Sarajedini (San Francisco: ASP), 307

Hambly, N. C., Smartt, S. J., & Hodgkin, S. T. 1997, ApJ, 489, L157

Hansen, B. M. S. 1998, Nature, 394, 860

Harris, H. C., Dahn, C. C., Vrba, F. J., Henden, A. A., Liebert, J., Schmidt, G. D., & Reid, I. N. 1999, ApJ, 524, 1000

Harris, H. C., et al. 2001, ApJ, 549, L109

Ibata, R. A., Irwin, M., Bienaymé, O., Scholz, R., & Guibert, J. 2000, ApJ, 532, L41

Ibata, R. A., Richer, H. B., Gilliland, R. L., & Scott, D. 1999, ApJ, 524, L95

Knox, R. A., Hawkins, M. R. S., & Hambly, N. C. 1999, MNRAS, 306, 736

Larson, R. B. 1986, MNRAS, 218, 409

Mashonkina, L., & Gehren, T. 2000, A&A, 364, 249

Mashonkina, L., & Gehren, T. 2001, A&A, submitted

Méndez, R. A., & Minniti, D. 2000, ApJ, 529, 911

Pfeiffer, M. J., Frank, C., Baumueller, D., Fuhrmann, K., & Gehren, T. 1998, A&AS, 130, 381

Phleps, S., Meisenheimer, K., Fuchs, B., Wolf, C. 2000, A&A, 356, 108

Scholz, R.-D., Irwin, M., Ibata, R., Jahreiss, H., Malkov, O. Y. 2000, A&A, 353, 958

Wyse, R. F. G., 1996, in ASP Conf. Ser. 92, Formation of the Galactic Halo .... Inside and Out, ed. H. Morrison & A. Sarajedini (San Francisco: ASP), 403

Discussion

Piotr Popowski: If you have all those white dwarfs then you have a lot of chemical enrichment that we don't see in the Galaxy. Your top-heavy mass function generates many metals and something has to happen to those metals, so what happens to them in your model?

Klaus Fuhrmann: I think it's more the other way round: there is very likely more mass between the galaxies and this is said to be hot and metal-rich (~1/3 solar, or so). If the *stellar* halo consists of only a few billion metal-poor stars, one immediately wonders where the intracluster medium should come from. Here, I think, a more massive thick disk, and in particular one with a top-heavy initial mass function plus the higher metal content, offers a natural explanation.

Piotr Popowski: I think it has been shown that the metals would inflow back even if you blow them away initially.

Klaus Fuhrmann: Yes, part of the material falls back, and then in a second stage—with a delay of a few billion years—the thin disk is formed. But, again, from X-ray data we know as well of the existence of massive, hot intergalactic media, and this X-ray gas is metal enriched.

Astrophysical Ages and Time Scales
ASP Conference Series, Vol. 245, 2001
T. von Hippel, C. Simpson, N. Manset

The Age of the Galactic Bulge [1,2]

R. Michael Rich

Department of Physics and Astronomy, University of California at Los Angeles, Los Angeles, CA, USA

Abstract. The dominant stellar population of the central bulge of the Milky Way is old, with roughly solar metallicity. The age is very similar to that of the old metal rich bulge globular clusters and to 47 Tucanae, which has an age of 13 Gyr. Stellar composition measurements from Keck/HIRES confirm that bulge stars are enhanced in Mg and Ti. New HST/NICMOS data are consistent with an old stellar population dominating the central 100 pc of the Milky Way.

New infrared photometry has been reported for the bulge of M 31. Although bright asymptotic giant branch stars are observed in the infrared, the data are most consistent with an old stellar population.

1. Introduction

Our description of galaxy evolution relies on a balance of data from high redshift, and detailed population studies of the fossil record. Ages, composition, kinematics, and structure of stars preserve evolution in the greatest detail, but are available only for the nearest galaxies. The advent of 8–10m class telescopes and HST marks one of those rare periods where technology conspires to deliver dramatic advances at high redshift. At such times, attention naturally turns away from study of the fossil record, despite the dramatically tighter constraints such work provides. When uncertainties mount in the high-redshift work (as they often do) observers return to the fossil record to resolve the ambiguity. Detailed population studies can give relative ages to accuracies of 1–2 Gyr, and detailed composition measurements that are impossible using integrated light.

1.1. What is the Bulge?

The central bulge of the Milky Way has not easily won recognition as a distinct stellar population. Until the dramatic images obtained by the *COBE* satellite, showing a clear bulge in the infrared, the possibility could not be ruled out that our central population is instead an extension of the thick disk, inner halo, or

---

[1]Based on observations obtained at the W. M. Keck Observatory, which is operated jointly by the California Institute of Technology and the University of California.

[2]Based on observations with the NASA/ESA Hubble Space Telescope, obtained at the Space Telescope Science Institute, which is operated by the Association of Universities for Research in Astronomy, Inc, under NASA contract NAS 5-26555.

perhaps even an intermediate age bar. From the infrared imagery, microlensing studies, and stellar dynamics, we now have good constraints on the total mass of the bulge, of order 2×10^{10} $M_\odot$ and a self-consistent dynamical model (Zhao, Rich, & Spergel 1996). The typical velocity dispersion of the stars is ~ 110 km s^{-1}; the mean abundance of the bulge is ~ -0.3 dex (McWilliam & Rich 1994) and the abundance distribution function is consistent with the closed box chemical evolution model (Rich 1990).

The evolved stars in the bulge are quite different from those found in the typical globular cluster. The well known RR Lyrae stars first identified by Baade are present and represent the metal-poor population. However, the bulge has a range in abundance and the most metal rich giants in the bulge exceed the solar abundance. The horizontal branch in the field color–magnitude diagram (CMD) is dominated by red horizontal branch (HB) (clump) stars, not by the bluer RR Lyrae stars, and in optical colors, the tip of the red giant branch is observed to descend fainter than the red clump luminosity due to blanketing in the M giants (Rich et al. 1998). The asymptotic giant branch (second ascent) stars are late M giants and can reach $M_{\rm bol} = -5$. The presence of such luminous stars in the bulge fueled well-justified speculation that a widespread intermediate age population must be present there. We now know that stars in the 13 Gyr globular cluster age range can reach that luminosity, if they are metal rich (Guarnieri, Renzini, & Ortolani 1997).

Challenged by heavy and variable reddening, extreme image crowding, the spatial depth of the population, and the broad range of metallicity, observers have found it challenging to measure the age. Arp (1965) produced a color–magnitude diagram for the giants in Baade's Window from photographic BV photometry. Van den Bergh (1972) refined further reddening and distance estimates, and argued for a metal-rich giant branch in his photometry of Baade's Window. Blanco's RI photographic plates on the newly commissioned 4m telescope at CTIO led to a breakthrough: for the first time the giant branch was clearly defined, permitting a sample of K giants to be selected for spectroscopy (Whitford & Rich 1983). Van den Bergh & Herbst (1974) developed some age constraints for the Plaut field ($8°S$) of the nucleus. Terndrup (1988) attained an old main sequence turnoff in this field, but could not resolve the metal rich bulge population deep in Baade's Window. The debate concerning the age range in the bulge remains active to this day.

Related to the age is the formation time scale (review in Rich 1999). The time scale for the bulk of star formation can be constrained by modeling the turnoff. One would like to use the asymptotic giant branch (AGB) to constrain residual star formation on Gyr time scales, but the lack of useful correlation between age and tip luminosity for metal rich stars poses a problem. Composition (see Section 3, below) offers some interesting constraint on the formation time scale. To date, the best constraints argue that the bulge is globular cluster age, and formed in < 1 Gyr.

1.2. Historical Perspective

The importance of the bulge, and the description of bulge populations, was well appreciated by the pioneers in the field, especially Walter Baade. Although the actual Galactic center is obscured by tens of magnitudes of extinction, the circumstantial evidence for the nucleus lying toward Sagittarius was strong enough that Baade searched for RR Lyrae stars in that direction. His discovery of a sharply peaked magnitude histogram for the RR Lyrae stars marked the discovery of the bulge as a stellar population (Baade 1951).

By the time of the pivotal 1958 Vatican meeting on stellar populations, the present-day view concerning the age and metallicity of the bulge had been developed and was correct. The presence of RR Lyrae stars argued for an old population, but the presence of late-type M giants (discovered in grism surveys by Victor Blanco) argued for high metallicity. In a final table summarizing the characteristics of stellar populations, the Galactic nucleus was classified as old and metal rich, but some lingering doubt was expressed as to whether the nucleus (bulge) population was as old as the globular clusters. This confluence of thinking was summarized in a prescient and sophisticated analysis of the formation history of the M 31 bulge, the template stellar population for Population II (Baade 1963):

"We must conclude, then, in the central region of the Andromeda Nebula we have a metal-poor Population II, which reaches -3^m for the brightest stars, and that underlying it there is a very much denser sheet of old stars, probably something like those in M 67 or NGC 6752. We can be certain that these are enriched stars, because the cyanogen bands are strong, and so the metal/hydrogen ratio is very much closer to what we observe in the Sun and in the present interstellar medium than to what is observed for Population II. And the process of enrichment probably has taken very little time. After the first generation of stars has formed, we can hardly speak of a 'generation', because the enrichment takes place so soon, and there is probably very little time difference. So the CN giants that contribute most of the light in the nuclear regions of the Nebula must also be called old stars; they are not young."

High resolution spectroscopy was, of course, lacking, and the realization that some globular clusters are very metal poor, and 1/100 solar abundance, also was absent (witness the mention of M 67 and NGC 6752 in the same breath). But the Vatican conference mentions both large age and high metal abundance for the bulge population. For approximately the next 20 years, the bulge would often be described as metal poor, despite the very clear evidence of a wider abundance range, accumulated prior to that time. Ultimately, the observational definition of the globular cluster abundance scale, combined with new work on the bulge, would modify this erroneous picture.

1.3.   The Relationship to High Redshift Studies

Although much can be inferred from the evolution of spiral galaxies as a function of redshift, the well-determined age and age range of one single bulge population (that of the Milky Way) is a powerful constraint.

The classic bulge formation model is that of Eggen, Lynden-Bell, & Sandage (1962) in which bulges form from dissipationless collapse very early on. The widely accepted cold dark matter models have some variations, but basically all concur that elliptical galaxies form from merged disk galaxies (Kauffmann, White, & Guiderdoni 1993; Baugh et al. 1998). At any given redshift, bulges should then be older than ellipticals. Some bulges may be related to bars, which could vertically thicken due to scattering of resonant orbits off of the bar, or due to the dissolution of bars (Combes 2000, and references therein). Bulges with exponential profiles may be more closely related to disks (Carollo et al. 2001).

The most distant normal star forming galaxies are the Lyman-break galaxies (Steidel et al. 1996) which are strongly clustered (Adelberger et al. 1998; Giavalisco & Dickinson 2001). These galaxies have strong outflows and are evidently metal rich; it is therefore reasonable to suppose that they will evolve into spheroids. The connection between the galaxies at $z \sim 3$ and the population of galaxies that begin to fall in well defined Hubble types by $z \sim 1$ requires much effort and lies ahead of us. By $z \sim 1$ we can begin to classify bulges of spirals, and ellipticals and galaxies are well enough resolved to even investigate their star formation histories pixel by pixel. Ellis, Abraham, & Dickinson (2001) studied the evolution of bulges and ellipticals in the Hubble Deep Fields and found that at any given redshift, ellipticals are redder than bulges. Issues such as disk contamination and reddening have rigorously been accounted for. Numerous studies have found evidence for very red ellipticals, and clustering of ellipticals, in the $z \sim 1 - 1.5$ range (see Stockton's review, these proceedings). Locally, Peletier et al. (1999) and numerous other studies argue that spiral bulges are as old as present-day Coma ellipticals. In the section that follows, I argue that the stellar populations of the Galactic bulge are consistent with a large age. How do we reconcile these findings with the high redshift studies? Are most bulges subjected to brief starburst events, due to the availability of gas from the disk or environment, or are elliptical galaxies systematically more metal rich? We have to resolve this tension between the high redshift data and the fossil record.

2.   Age Constraints From the Turnoff

Because of the high reddening, uncertainty in the distance modulus, and presence of foreground stars, one cannot constrain the age of the bulge with confidence by placing isochrones on the CMD. The foreground main sequence disk stars have proven a vexing population, as they overlay the old main sequence turnoff point precisely and appear as an intermediate age population. To increase the contrast between the bulge population and the foreground, one needs to study fields close to the Galactic center, where reddening and crowding become serious. Estimates from modeling and direct starcounts place the foreground contamination at 10–15% (see the CMDs by Feltzing & Gilmore 2000).

The strongest arguments in favor of an intermediate age population in the bulge were based on the presence of luminous AGB stars. An empirical correlation between AGB star luminosity and intermediate populations was established using Magellanic Clouds clusters (Aaronson & Mould 1985). The ~ 1 mag extension at the turnoff point prevented a convincing resolution of the problem.

Ortolani et al. (1995) found a solution to the problem. Both the reddening and distance modulus uncertainties could be eliminated if a differential comparison is made between the bulge luminosity function and that of an old globular cluster of comparable metallicity. The crucial technique is to force fit the bulge field luminosity to the globular cluster luminosity function at the point of the red clump, and to examine the agreement at the turnoff point in detail. Red clump stars are fueled by helium core burning, and the core mass (hence luminosity) has almost no dependence on age and little dependence on metallicity.

The bulge is concentrated spatially and consequently there is little distance dispersion. Therefore, when the red clumps are force fit, it is easily seen that the turnoff rise of bulge and NGC 6553 are identical. Therefore, using the well established ΔV_{TO}^{HB} method of age determination, we are able to show that the bulge and NGC 6553 have identical ages.

The next question is, how old is NGC 6553? The metallicity of the cluster is in some debate right now. Barbuy et al. (1999) find [Fe/H] = −0.5 and alpha elements up by +0.5 dex relative to scaled solar. Cohen et al. (1999) use Keck/HIRES spectroscopy to find [Fe/H] = −0.16 and alpha elements up by +0.3 dex. R. M. Rich, L. Origlia, & S. M. Castro (in preparation) use infrared spectroscopy to find [Fe/H] = −0.3 and alpha elements up by +0.3 dex. Regardless of which value one adopts, NGC 6553 is a good match for the field population of the bulge, and only 0.4–0.5 dex more metal rich than 47 Tuc.

We have determined a white dwarf distance for 47 Tuc (Zoccali et al. 2001) one of the nearest of the metal rich disk globular clusters. The distance modulus translates into a turnoff age of 13 ± 2.5 Gyr. Given the close correspondence between NGC 6553 and 47 Tuc, I adopt an identical age for the Galactic bulge. This makes the bulge of order the age of the halo. There has been recent work that solidifies this large age, and additionally places limits on the time scale of bulge formation. Feltzing & Gilmore (2000) show quantitatively that *all* of the stars lying slightly brighter than the old main sequence turnoff in the bulge fields belong to the foreground. This is also being confirmed in proper motion surveys based on HST data (K. Kuijken & R. M. Rich, in preparation).

2.1. Implications for Red Globular Cluster Systems

As a final side note, in many elliptical galaxies with globular cluster systems, the globular clusters are bimodal in their color distribution. The bluer clusters tend to be spatially extended, while the red clusters follow the spheroid, just as is the case in the Milky Way (the bulge clusters follow the bulge). It is widely accepted (Ashman & Zepf 1992) that the red cluster systems are connected with late mergers. In the Milky Way bulge, we have shown (one example) that the red clusters are old, like those in the halo. However, I would propose that the red clusters in ellipticals should be considered to have the age of the old spheroidal stars, unless proof to the contrary can be developed.

3. Time Scale Constraints From Abundances

Considering that this is a meeting that discusses both ages and time scales, we should address the time scale of formation of the bulge. Strong constraints have been attained from the photometry discussed in the last section. The compositions of stars also have the potential to constrain the time scale for chemical enrichment. Massive star supernovae which explode on 10^6 yr time scales have ejecta rich in alpha-capture elements (O, Mg, Si, etc.), the products of nucleosynthesis in their hydrostatic burning shells. Type Ia SNe experience a deflagration in their explosions, and contribute more iron (but on time scales of $\sim$ 1 Gyr) than the massive star (core detonation) supernovae. The general trend is that the alpha/iron ratio declines with increasing [Fe/H] with the more recently formed stars approaching solar composition. The more rapid the enrichment, the higher

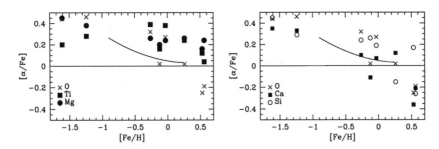

Figure 1.   Abundance trends for Galactic bulge giants, observed with Keck/HIRES (Rich & McWilliam 2000). The solid line shows the mean abundance trend for alpha elements in local disk stars (Edvardsson et al. 1993). Although Mg enhancement is consistent with element production in Type II SNe, SNe models predict lower Ti production. The general enhancement in alpha elements strongly points toward a rapid time scale for bulge formation. The enhancements of Mg and Ti relative to Ca and Si confirm the work of McWilliam & Rich (1994).

the iron abundance of stars that maintain an alpha-enhanced composition. This paradigm, fully described in (cf. McWilliam 1997) permits us to make relative comparisons of enrichment time scales in stellar populations. Figure 1 shows the abundance trends for bulge giants measured from spectroscopy on Keck, using the HIRES echelle (Vogt et al. 1994). A more detailed discussion of recent Keck results on bulge giants is given in Rich & McWilliam (2000).

The bulge shows a unique pattern of enrichment not shared by any other Galactic stellar population. The alpha element Mg remains enhanced even above the solar iron abundance, as does Ti. Oxygen follows a less striking trend, while exceeding the solar abundance at −0.3 dex. Ca and Si are modestly enhanced. This complicated pattern of enhancement is not predicted by supernova models, but is likely consistent with the bulge being rapidly enriched by massive stars. The rapid formation time scale is consistent with the age constraints from the photometry, and with observed large star formation rates at high redshift.

4.   The Age of the Nuclear Population

The stellar population of the Galactic nucleus is one of the most active and dramatic star forming regions in the Galaxy (Morris & Serabyn 1996). Massive star formation dominates the energy input, and includes such dramatic examples as the $10^6 L_\odot$ Pistol Star (Figer et al. 1998) and the Arches cluster, with 10^3 massive O stars and an age of 2 Myr (Figer et al. 1999). Considering the intensity of star formation near the nucleus, one expects to find a stellar population with a wide range of ages, and a present day mass function consistent with a continuous history of star formation. Ground-based imaging of the Galactic Center (Catchpole, Whitelock, & Glass 1990) found evidence for a concentration of the most luminous AGB stars toward the Galactic Center, further strengthening the case that the inner 100 pc of the Galaxy must have a wide age range.

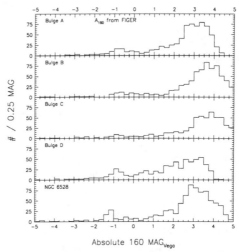

Figure 2.   *H*-band luminosity functions of the bulge using HST/NICMOS (R. M. Rich et al., in preparation). The age-sensitive gap between the red clump ($H \sim -1.25$) and the main sequence rise at $H \sim 3$ is very similar for the Galactic Center fields (10–40 pc from the nucleus) and the old metal rich globular cluster NGC 6528 (*lower panel*). The nucleus is dominated by an old population.

Using the NICMOS infrared imager on board HST, we obtained deep infrared photometry of optically obscured fields near the Galactic center (R. M. Rich et al., in preparation; Figure 2). Much to our surprise, we found a clearly-defined red clump and red giant branch, consistent not with an intermediate age population of ongoing star formation, but rather with an old stellar population. We again use the vertical method of age determination, $\Delta(\mathrm{mag})_{\mathrm{TO}}^{\mathrm{HB}}$, this time comparing our Galactic Center population to NGC 6528 (Figure 2). We conclude that the bulk of the stellar population in these regions is as old as the globular clusters, just as is the case in the outer bulge. In light of the star formation activity in the nuclear region, this is a surprising finding. The luminosity function and color–magnitude diagram of these fields a mere 40 pc from the Galactic Nucleus appears identical to that of bulge globular clusters observed in the same NICMOS bands. The lack of an intermediate age population is surprising; perhaps conditions in the Galactic Center permit only the formation of massive stars, so that longer-lived low mass stars would not survive.

5.   Conclusion

Using the widely accepted vertical method of age determination, the Galactic bulge has approximately the age of the Galactic globular cluster 47 Tuc, 13 ± 2.5 Gyr. The vertical method (using infrared luminosity functions) also shows that the population within 10–40 pc of the nucleus is old. The composition of the bulge K giants is consistent with early, rapid formation of the bulge. These age measurements indicate that the Galactic bulge/bar is among the oldest Galactic stellar populations. If the population formed by the vertical thickening of a massive disk, it must have done so very early in the Galaxy's formation history. Local group bulges are also consistent with this picture. The infrared luminosity functions of M 32 (Davidge et al. 2000) and the bulge of M 31 (Stephens et al. 2001) terminate at $M_{\rm bol} = -5$—confirming early ground-based IR imaging by Rich & Mould 1991. While the old globular cluster NGC 6553 has bright AGB stars, the complicated behavior of metal rich stars on the asymptotic giant branch probably makes it difficult to rule out an intermediate age component in these local group bulges. Better age constraints must come from modeling the color and integrated spectral energy distribution, to constrain trace intermediate-age main sequence stars. Our capability to use the old main sequence turnoff point to constrain the age of the bulge is a golden opportunity not available for even the nearest Local Group bulges. In terms of age, the Galactic bulge much more resembles the halo than it does the disk, and all the data point to it being completely in place very early on in the Galaxy's history.

References

Aaronson, M., & Mould, J. R. 1985, ApJ, 288, 551

Adelberger, K. L., Steidel, C. C., Giavalisco, M., Dickinson, M., Pettini, M., & Kellogg, M. 1998, ApJ, 505, 18.

Arp, H. C. 1965, ApJ, 142, 402

Ashman, K. M., & Zepf, S. E. 1992, ApJ, 384, 50

Baade, W. 1951, Pub. Obs. Univ. Michigan, 10, 7

Baade, W. 1963 in The Evolution of Galaxies and Stellar Populations (Cambridge: Harvard University Press), 256

Barbuy, B., Renzini, A., Ortolani, S., Bica, E., & Guarnieri, M. D. 1999, A&A, 341, 539

Baugh, C. M., Cole, S., Frenk, C. S., & Lacey, C. G. 1998, ApJ, 498, 504

Carollo, C. M., Stiavelli, M., de Zeeuw, P. T., Seigar, M., & Dejhonge, H. 2001, ApJ, 546, 216

Catchpole, R. M., Whitelock, P. A., & Glass, I. S. 1990, MNRAS, 247, 479

Cohen, J. G., Gratton, R. G., Behr, B. B., & Carretta, E. 1999, ApJ, 523, 739

Combes, F. 2000, in ASP Conf. Ser. 197, Dynamics of Galaxies: From the Early Universe to the Present, ed. F. Combes, G. A. Mamon, & V. Charmandaris (San Francisco: ASP), 15

Davidge, T. J., Rigaut, F., Chun, M., Brandner, W., Potter, D., Northcott, M., & Graves, J. E. 2000, ApJ, 545, L89

Edvardsson, B., Andersen, J., Gustafsson, B., Lambert, D., Nissen, P., & Tomkin, J. 1993, A&A, 275, 101

Eggen, O. J., Lynden-Bell, D., & Sandage, A. 1962, ApJ, 136, 748

Ellis, R., Abraham, R. G., & Dickinson, M. 2001, ApJ, 551, 111

Feltzing, S., & Gilmore, R. 2000, A&A, 355, 949

Figer, D. F., Kim, S. S., Morris, M., Serabyn, E., Rich, R. M., & McLean, I. S. 1999, ApJ, 525, 750

Figer, D. F., Najarro, F., Morris, M., McLean, I. S., Geballe, T. R., Ghez, A. M., & Langer, N. 1998, ApJ, 506, 384

Giavalisco, M., & Dickinson, M. 2001, ApJ, 550, 177

Guarnieri, M. D., Renzini, A., & Ortolani, S. 1997, ApJ, 477, L21

Kauffmann, G., White, S. D. M., & Guiderdoni, B. 1993, MNRAS, 297, L23

McWilliam, A. 1997, ARA&A, 35, 503

McWilliam, A., & Rich, R. M. 1994, ApJS, 91, 749

Morris, M., & Serabyn, G. 1996, Nature, 382, 602

Ortolani, S., Renzini, A., Gilmozzi, R. M., Marconi, G., Barbuy, B., Bica, E., & Rich, R. M. 1995, Nature, 377, 701

Peletier, R. F., Balcells, M., Davies, R. L., Andredakis, Y., Vazdekis, A., Burkert, A., & Prada, F. 1999, MNRAS, 310, 703

Rich, R. M. 1990, ApJ, 362, 604

Rich, R. M. 1999 in The Formation of Galactic Bulges, ed. C. M. Carollo, H. C. Ferguson, & R. F. G. Wyse (Cambridge: Cambridge University Press), 54

Rich, R. M., & McWilliam, A. 2000, in Proc. SPIE 4005, Discoveries and Research Prospects from 8- to 10-Meter-Class Telescopes, ed. J. Bergeron (Bellingham: SPIE), 150

Rich, R. M., & Mould, J. 1991, AJ, 101, 1286

Rich, R. M., Ortolani, S., Bica, E., & Barbuy, B. 1998, AJ, 116, 1006

Steidel, C. C., Giavalisco, M., Dickinson, M., & Adelberger, K. L. 1996, ApJ, 462, L17

Stephens, A. W., et al. 2001, AJ, 121, 2597

Terndrup, D. M. 1988 AJ, 96, 884

Van den Bergh, S. 1972, PASP, 84, 306

Van den Bergh, S., & Herbst, E. 1974, AJ, 79, 603

Vogt, S. S., et al. 1994, in Proc. SPIE 2198, Instrumentation in Astronomy VIII, ed. D. L. Crawford & E. R. Craine (Bellingham: SPIE), 362

Whitford, A. E., & Rich, R. M. 1983, ApJ, 274, 723

Zhao, H., Rich, R. M., & Spergel, D. N. 1996, MNRAS, 282, 175

Zoccali, M., et al. 2001, ApJ, in press (astro-ph/0101485)

Acknowledgments. Support for this work was provided by NASA through grants GO-7832, GO-7465, and GO-7826 from the Space Telescope Science Institute, which is operated by AURA, Inc., under NASA contract NAS 5-2655.

Discussion

Scott Trager: In a large sample of elliptical galaxies we have found that Ca tracks Fe much more closely than it tracks Mg, and Mg is clearly enhanced relative to Fe, but Ca is not enhanced in elliptical galaxies. That Ca would track Fe is not predicted by Woosley & Weaver chemical production models from SNe. So what is going on either in the bulge or in elliptical galaxies?

Michael Rich: I answer your comment with yes, I agree. Here's this plot of production factors and the models really don't seem to explain what's going on. It's interesting that you're seeing the same thing in the integrated light.

Bengt Gustafsson: You derive these overabundances of alpha elements relative to Fe for relatively old red giant stars. What chances do we have to derive these abundances for younger objects such as supergiants, for a starbursting region, or even for the interstellar medium?

Michael Rich: The recent infrared echelle spectroscopy of Galactic Center supergiants by Solange Ramirez and the Ohio State group gives us a solar iron abundance as well as [Mg/Fe] for these stars, formed within the last 10^8 yr. With somewhat greater uncertainty, solar compositions are found for the blue supergiants near the Galactic Center by Najarro and Figer. I am not aware of any ISM measurements more recent than those of Lester et al. (1981) which give twice solar Ar abundance. Given the bar morphology of the bulge and the presence of gas and star formation, one may suspect that the gas may be replenished by inflow from the disk, so present day abundances may be lower than those of the oldest stars. Ramirez et al. have greatly contributed to our knowledge of the youngest Galactic Center stellar population, but the time does seem right for new measurements of gas-phase abundances in the region.

Astrophysical Ages and Time Scales
ASP Conference Series, Vol. 245, 2001
T. von Hippel, C. Simpson, N. Manset

Nucleosynthesis Clocks and the Age of the Galaxy

James W. Truran

*Department of Astronomy and Astrophysics, Enrico Fermi Institute,
University of Chicago, Chicago, IL 60637, USA*

Scott Burles

*Department of Astronomy and Astrophysics, University of Chicago,
Chicago, IL 60637, USA*

John J. Cowan

*Department of Physics and Astronomy, University of Oklahoma,
Norman, OK 73019, USA*

Christopher Sneden

Department of Astronomy, University of Texas, Austin, TX 78712, USA

Abstract. Nucleocosmochronology involves the use of the abundances of radioactive nuclear species and their radiogenic decay daughters to establish the finite age of the elements and the time scale for their formation. While there exist radioactive products of several specific nucleosynthesis mechanisms that can reveal the histories of these mechanisms, it is the long-lived actinide isotopes ^{232}Th, ^{235}U, and ^{238}U, formed in the r-process, which currently play the major role in setting the time scale of Galactic nucleosynthesis. Age determinations in the context of Galactic chemical evolution studies are constrained by intrinsic model uncertainties. Recent studies have taken the alternative approach of dating individual stars. Thorium/europium dating of field halo stars and globular cluster stars yields ages on the order of 15±4 Gyr. A solid lower limit on stellar ages is available as well, for stars for which one both knows the thorium abundance and has an upper limit on the uranium abundance. For the cases of the two extremely metal-deficient field halo stars CS 22892–052 and HD 115444, lower limits on the nuclear age of Galactic matter lie in the range 10–11 Gyr. Th/U dating of the star CS 31082–00 gives an age of 12.5 ± 3 Gyr. Observations of thorium and uranium abundances in globular cluster stars should make possible nuclear determinations of the ages of clusters that can be compared with ages from conventional stellar evolution considerations.

1. Introduction

Nuclear chronometers have played an historically important role in the determination of the ages of our Solar System and our Milky Way Galaxy, thus providing important lower limits on the age of the Universe itself. The critical long-lived nuclear radioactivities were long ago identified to be ^{187}Re ($\tau_{1/2} = 4.5 \times 10^{10}$ yr), ^{232}Th ($\tau_{1/2} = 1.4 \times 10^9$), ^{235}U ($\tau_{1/2} = 7.0 \times 10^8$), and ^{238}U ($\tau_{1/2} = 4.5 \times 10^9$), all of which are products of the r-process of neutron capture synthesis (Burbidge et al. 1957; Cameron 1957).

While interesting age constraints can be obtained on the basis of "model independent" chronometric considerations (see e.g., Meyer & Schramm 1986), most determinations of the age of the Galaxy to date have been obtained in the context of models of Galactic chemical evolution (see e.g., the review by Cowan, Thielemann, & Truran 1991a). Such models can be used to explore the sensitivities of the ages thus obtained to the star formation and nucleosynthesis history of Galactic matter. The Re/Os chronometer is of particular concern in this regard, both because the ^{187}Re decay rate is temperature sensitive (a factor when ^{187}Re is recycled through stars) and because both r-process and s-process contributions may be important.

In this paper we will concentrate on recent age determinations for individual stars, built upon the use of thorium and uranium related chronometers. Since the available data is associated with extremely metal-poor stars, formed during the very earliest stages of Galactic evolution, dating of such stars is entirely equivalent to dating the Galaxy as a whole.

2. R-Process Nucleosynthesis Considerations

While the general nature of the r-process of neutron capture synthesis and its contributions to the abundances of heavy elements in the mass range from beyond the iron peak through thorium and uranium is generally understood (Cowan, Thielemann, & Truran 1991b), many of the details remain to be worked out. Of particular concern is the very fact that the astrophysical site in which the r-process occurs remains unidentified. Both meteoritic data (Wasserburg, Busso, & Gallino 1996) and stellar abundance data (e.g., the robust consistency of the heavy ($A \gtrsim 140$) r-process abundance patterns in the most extreme metal-deficient stars with that of Solar System matter) suggest the likelihood of two distinct r-process environments responsible, respectively, for the mass regions $A \lesssim 140$ and $A \gtrsim 140$. The heavy element patterns in low metallicity stars strongly imply that at least the production of r-process nuclei in the heavy element region is associated with stars of short lifetimes (e.g., massive stars and/or supernova Type II).

Excellent reviews of r-process nucleosynthesis have been provided by Hillebrandt (1978), Cowan et al. (1991b), and Meyer (1994). Promising studied sites for the synthesis of the r-process isotopes in the mass range $A \gtrsim 140$ include: (1) the high-entropy, neutrino heated, hot bubble associated with Type II supernovae (Woosley et al. 1994; Takahashi, Witti, & Janka 1994); (2) the ejection and decompression of matter from neutron star–neutron star mergers (Lattimer et al. 1977; Freiburghaus et al. 1999); and (3) the ejection of neutronized matter in magnetic jets from collapsing stellar cores (LeBlanc & Wilson 1970). Note that all three of these mechanisms are tied to environments provided by the evolution of massive stars (and associated Type II supernovae) of short lifetimes ($\tau \lesssim 10^8$ yr), compatible with their presence in the oldest stellar populations of our Galaxy.

3. Dating Individual Stars

The direct determination of the ages of individual stars has recently become possible with the availability of thorium (and most recently uranium) abundances for halo stars. A distinct advantage of this approach is the elimination of uncertainties associated with our imperfect knowledge of the star formation, stellar evolution, and associated nucleosynthesis histories of galaxies, which are a necessary ingredient of chemical-evolution-based models of nucleocosmochronology. There remain, of course, the uncertainties associated with the nuclear properties of heavy actinide nuclei, which dictate the relative levels of production of ^{232}Th, ^{235}U, and ^{238}U in nucleosynthesis sites, and with the character of the astrophysical r-process itself. For example, a critical question associated with the use of the Th/Eu ratio as a chronometer is the robustness of the reproduction of the Solar System r-process pattern over the mass range $130 \lesssim A \lesssim 238$. We should also note that the use of such age determinations for individual stars (particularly the Th/U chronometer) may be *effectively* constrained to low metallicity ([Fe/H] $\lesssim -2.5$), r-process enriched ([r-process/Fe] $\gtrsim 1$) stars like CS 22892–052, CS 31082–001, and HD 115444.

3.1. Th/Eu Dating

The detection of thorium and the determination of its abundance in the extremely metal-deficient halo field star CS 22892–052 first made thorium dating possible (Sneden et al. 1996). In this case, due to the absence of knowledge of the uranium abundance, it is necessary to utilize some stable r-process product. The choice of europium here is a relatively obvious one, since both isotopes of Eu are produced almost exclusively by the r-process.

Data regarding Th and Eu are now available for a number of stars in the halo and in globular clusters. These data are collected in our Table 1. The ages were determined with the assumption that the theoretical r-process production ratio was $(\text{Th/Eu})_{\text{theory}} = 0.51$. Note that the mean thorium/europium age for these stars is 13.6 Gyr, with a spread consistent with the intrinsic observational uncertainty of approximately ± 4 Gyr.

A critical question concerning the use of the Th/Eu chronometer is the robustness of the r-process abundance pattern over a range of mass number $A \approx 140$–238. The abundance patterns in this mass number range for the two extremely metal-deficient stars CS 22892–052 and HD 115444, displayed in Figure 1, both show a remarkable agreement with the Solar System r-process abundance distribution. The quoted abundance level for the star CS 31082–001 (Hill et al., these proceedings), however, is distinctly different from those for the other stars in Table 1 (where we do not quote a Th/Eu age for this star). It is important to examine this behavior for other halo and globular cluster stars to more firmly establish the limits of confidence of the Th/Eu chronometer. We

Table 1.   Thorium/europium ages for individual stars[†].

Object	[Fe/H]	$\log \epsilon_{Th}$	$\log \epsilon_{Eu}$	Δ	(Th/Eu)	τ_*
CS 22892–052	−3.1	−1.57	−0.91	−0.66	0.22	16.8
HD 115444	−3.0	−2.23	−1.63	−0.60	0.25	14.4
HD 115444	−3.0	−2.21	−1.66	−0.55	0.28	12.1
HD 186478	−2.6	−2.25	−1.55	−0.70	0.20	18.9
HD 108577	−2.4	−1.99	−1.48	−0.51	0.31	10.1
M 92 VII–18	−2.3	−1.94	−1.45	−0.49	0.32	9.4
BD +8 2856	−2.1	−1.66	−1.66	−1.16	0.32	9.4
K341 (M 15)	−2.4	−1.47	−0.88	−0.59	0.25	14.4
K462 (M 15)	−2.4	−1.26	−0.61	−0.65	0.22	16.8
CS 31082–001	−3.0	−0.96	−0.70	−0.26	0.55	

† References are (1) Sneden et al. (2000a), (2) Westin et al. (2000), (3) Johnson & Bolte (2001), (4) Sneden et al. (2000b) and Hill et al. (these proceedings).

emphasize that we believe that the Th/Eu chronometer (or perhaps a Th/Pt chronometer) will generally be a more commonly available tool for the dating of halo stars than Th/U—and thus that we should work hard to confirm its reliability.

3.2.   Th/U Dating

When available, it is understood that the Th/U chronometer will give age determinations for which the nuclear uncertainties and astrophysical uncertainties may be smaller than those for the Th/Eu chronometer. Theoretical calculations of r-process nucleosynthesis can be employed to provide estimates of, for example, the ^{232}Th/^{238}U and ^{235}U/^{238}U ratios. Early model predictions for these so-called production ratios typically show rather wide dispersion (Cowan, Thielemann, & Truran 1987). However, more recent theoretical determinations (e.g., Möller, Nix, & Kratz 1997; see also Cowan et al. 1999) are now available.

Recently Cayrel et al. (2001, and in these proceedings) have discovered that the ultra metal poor star CS 31082–001 ([Fe/H] ∼ −3) has even larger n-capture abundance levels than does CS 22892–052. Most interesting is their detection of *many* transitions of Th II and the strongest U II line; the combined abundances of two radioactive elements (with different half-lives) yields an age of the neutron-capture elements in this star of ∼ 13 Gyr. But they also report an *observed* [Th/Eu] ratio that is much larger than in CS 22892–052. This abundance ratio in CS 31082–001 would imply an age of only a few Gyr, obviously inconsistent with this star's metallicity and halo membership (see e.g., the data cited in our Table 1). It is clear that many more Eu and Th abundances need to be determined for halo stars, in order to make some sense of this issue.

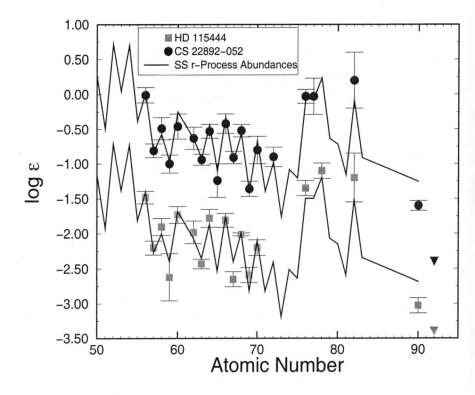

Figure 1. Abundance comparisons between HD 115444 (filled squares) and CS 22892–052 (filled circles). The abundance values for HD 115444 have been vertically displaced by −0.8 for display purposes. The upside-down triangles are upper limits on the uranium abundances. The solid lines are scaled Solar System r-process only abundance distributions.

We also wish to note, however, the *even upper limits on the uranium abundance* allow firm and interesting lower limits on the age of the Galaxy. Figure 2 displays the exponential dependence of the age of a star as a function of the observed Th/U ratio. Here we assume that ^{238}U dominates ^{235}U, and therefore show only ages greater than 5 Gyr. The vertical grey region indicates the range of allowed r-process Th/Eu production ratios (Cowan et al. 1987). Lines of constant $\log(U/Th)$ are displayed. An upper limit on U/Th excludes the lower left half of the plane. We note that the upper limits on the uranium abundances for the two stars CS 22892–052 and HD 115444 allow lower limits on the ages of these stars, respectively, of 10 and 11 Gyr.

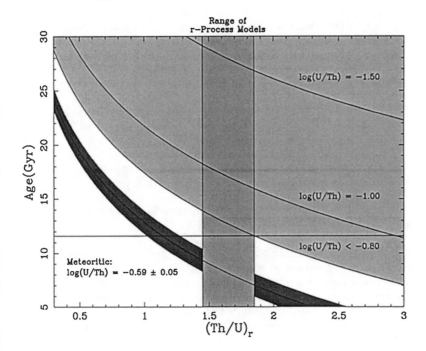

Figure 2.   Stellar ages as a function of the Th/U ratio. Shown is the exponential dependence of the observed Th/U abundance on the age of the system. We assume that ^{238}U dominates ^{235}U and therefore only show ages greater than 5 Gyr. Lines of constant log U/Th ratios are shown in the plane of age vs. ^{232}Th/^{238}U seed ratio. An upper limit on U/Th excludes the lower left half of the plane. A measurement of U/Th is restricted to a narrow band as is the case with meteoritic observations (Anders & Grevesse 1989).

4.   Conclusions

We draw the following conclusions from our overview of considerations of nuclear chronometer dating:

- The important nuclear chronometers ^{232}Th, ^{235}U, and ^{238}U are r-process products.

- The observations of r-process abundance patterns in the oldest ([Fe/H] $\lesssim -2.5$) halo stars serve to confirm the identification of the r-process site with massive star environments and to convince us that we are dating the history of stellar activity in our Galaxy.

- The robustness of the r-process mechanism for the production of nuclei in the region $A \gtrsim 140$ is reflected in the extraordinary agreement (see Figure 1) of the stellar abundance patterns with the Solar System r-process abundances (even at metallicities so low that at best only a few supernovae can have contributed). *The remarkable star CS 22892–052, at [Fe/H] = −3.1, exhibits a pure r-process pattern in the mass range $A \gtrsim 140$, but at an abundance level [r-process/Fe] $\approx$ +1.5.*

- Ages (and/or age constraints) for individual halo field stars or globular cluster stars can be obtained (in principle) from Th/Eu and Th/U, when the abundances of these nuclear species are known.

- → Th/Eu dating of the two well-studied stars CS 22892–052 and HD 115444 give an average chronometric age $\tau_* \sim 15.6 \pm 4.6$ Gyr.

- → Upper limits on the uranium abundances for these two stars provide lower limits on their ages $\tau_* \gtrsim 10$–11 Gyr.

- While both thorium and uranium abundance determinations for individual halo stars will continue to accumulate, it seems likely that uranium abundances will become available for only a small fraction of those for which thorium abundances are known. Continued efforts to place the Th/Eu (or perhaps Th/Pt) chronometer on a firm basis are thus critical to our ability to date the various components of our Galactic halo population.

References

Anders, E., & Grevesse, N. 1989, Geochim. Cosmochim. Acta., 46, 2363

Burbidge, E.M., Burbidge, G.R., Fowler, W.M., & Hoyle, F. 1957, Rev. Mod. Phys., 29, 547

Cameron, A. G. W. 1957, Chalk River Report CRL-41

Cayrel, R., et al. 2001, Nature, 409, 691

Cowan, J. J., Pfeiffer, B., Kratz, K.-L., Thielemann, F.-K., Sneden, C., Burles, S., Tytler, D., & Beers, T. C. 1999, ApJ, 521, 194

Cowan, J. J., Thielemann, F.-K., & Truran, J. W. 1987, ApJ, 323, 543

Cowan, J. J., Thielemann, F.-K., & Truran, J. W. 1991a, ARA&A, 29, 447

Cowan, J. J., Thielemann, F.-K., & Truran, J. W. 1991b, Phys. Rep., i208, 267

Freiburghaus, C., Rembges, J.-F., Rauscher, T., Kolbe, E., Thielemann, F.-K., Kratz, K.-L., Pfeiffer, B., & Cowan, J. J. 1999, ApJ, 516, 381

Hillebrandt, W. 1978, Sp. Sci. Rev., 21, 639

Johnson, J. A., & Bolte, M. S. 2001, ApJ, in press

Lattimer, J. M., Mackie, F., Ravenhall, D. N., & Schramm, D. N. 1977, ApJ, 213, 225

LeBlanc, J. M., & Wilson, J. R. 1970, ApJ, 161, 541

Meyer, B. S. 1994, ARA&A, 32, 153

Meyer, B. S., & Schramm, D. N. 1986, ApJ, 311, 406

Möller, P., Nix, J. R., & Kratz, K.-L. 1997, At. Data Nucl. Data Tables, 66, 131

Sneden, C., Cowan, J. J., Ivans, I. I., Fuller, G. M., Burles, S., Beers, T. C., & Lawler, J. E. 2000a, ApJ, 533, L139

Sneden, C., Johnson, J., Kraft, R. P., Smith, G. H., Cowan, J. J., Bolte, M. S. 2000b, ApJ, 536, L85

Sneden, C., McWilliam, A., Preston, G. W., Cowan, J. J., Burris, D. L., & Armosky, B. J. 1996, ApJ, 467, 819

Takahashi, K., Witti, J., & Janka, H.-T. 1994, A&A, 286, 857

Truran, J. W. 1981, A&A, 97, 391

Wasserburg, G. J., Busso, M., & Gallino, R. 1996, ApJ, 466, L109

Westin, J., Sneden, C., Gustafsson, B., & Cowan, J. J. 2000, ApJ, 530, 783

Woosley, S. E., Wilson, J. R., Mathews, G. J., Hoffman, R. D., & Meyer, B. S. 1994, ApJ, 433, 229

Acknowledgments. Our work on n-capture elements in halo stars has been a collaborative effort over many papers, and we thank our co-authors for their contributions over the years. This research has been supported in part by NSF grant AST 9986974 to the University of Oklahoma (JJC) and by the ASCI/Alliances Center for Astrophysical Thermonuclear Flashes at the University of Chicago under DOE grant B341495 (JWT).

Discussion

Andreas Korn: What do we know about the importance of non-LTE of those elements in the extreme atmospheres of such metal-poor stars?

Jim Truran: I should turn that over to someone like Chris [Sneden].

Chris Sneden: It's unlikely to affect most of those elements because most of them are using transitions of low excitation potential states of majority ions. A few elements you'd worry about but for most of them the atmospheric uncertainties, including worries about non-LTE effects, just disappear.

George Isaak: How large an effect would differential gravitational settling introduce into the Eu/U ratio or Eu/Th ratios?

Chris Sneden: These stars are giants, they're mixed giants.

Jim Truran: You're worried about this because some are these old halo B stars where it could be important; but in these cases it's not.

Typhoon Lee: I worry about studying the r-process in stars with individual or very few progenitors. These stars may not see an averaging effect which you are requiring.

Jim Truran: We're not doing that in what I showed. In those cases we're talking about individual stars. We can expect that there will be variations, but

HD 115444 and CS 22892–052 are very similar; it may be that we're going to find some which are not. We do know that in the halo field stars that there are extraordinary ranges, that there's a tremendous scatter, a factor of 30 or so in the ratio of r-process to iron. That alone gives you some measure of uncertainty.

Douglas Gough: You showed us that there is scatter in the results due to certain uncertainties in the nuclear physics and in the structure of the stars, but the estimate you gave for the age of the Galaxy presumed a very simple law for the rate at which material has been processed during the age of the Galaxy. Star formation which is not uniform can be convolved into Galactic chemical evolution. Does that produce a difference which is significant compared with the uncertainties that you discussed in your talk?

Jim Truran: Studying a stellar population is where placing it into a chemical evolution context become relevant. I prefer to look at individual stars and to say I can understand the Th and U well if I narrow myself down to only nuclear uncertainties in the production of those elements, and that takes away much of the uncertainty that you're referring to. So I agree with you. I was impressed that yesterday we were talking about dating individual white dwarfs. Otherwise it's hard to know with these complicated disk and halo sub-populations what to attribute your measurements to and what the ages are. The U/Th dating I was talking about, as well as the Th/Eu and Th/Pb dating, all involve individual stars, negating most of the chemical effects.

Astrophysical Ages and Time Scales
ASP Conference Series, Vol. 245, 2001
T. von Hippel, C. Simpson, N. Manset

Neutron-Capture Element Abundances and Cosmochronometry

Christopher Sneden

Department of Astronomy, University of Texas, Austin, TX 78712, USA

John J. Cowan

Department of Physics and Astronomy, University of Oklahoma, Norman, OK 73019, USA

Timothy C. Beers

Department of Physics and Astronomy, Michigan State University, East Lansing, MI 48824, USA

James W. Truran

Department of Astronomy and Astrophysics, University of Chicago, Chicago, IL 60637, USA

James E. Lawler

Department of Physics, University of Wisconsin, Madison, WI 53706, USA

George Fuller

Department of Physics, University of California, San Diego, La Jolla, CA 92093-0319, USA

Abstract. Abundance ratios of radioactive to stable neutron-capture elements in very metal-poor stars may be used to estimate the age of our Galaxy. However, extracting accurate ages from these data requires continuing work on many fronts: (1) identifying more low metallicity stars with neutron-capture element excesses, (2) acquiring the best high resolution stellar spectra, (3) improving neutron-capture element transition probabilities, (4) calculating more realistic nuclear models for and interactions among the heaviest elements, and (5) modeling more self-consistent production predictions for these elements in supernovae. This review discusses several of these areas and makes suggestions about how to improve the accuracy of Galactic cosmochronometry.

1. Introduction

The dominant isotopes of elements with atomic numbers $Z > 30$ are synthesized in neutron bombardment reactions during late stellar evolution. Some of the heaviest of these so-called neutron-capture (n-capture) elements are radioactively unstable but long-lived on astrophysically interesting (many Gyr) time scales. In principle, abundance analyses of old, metal-poor stars that compare radioactive to stable n-capture elements can be used to determine the age of the Galactic halo. But there are practical difficulties in applying this idea to most halo stars. In this brief review we focus on studies of n-capture elements relevant to cosmochronometry, discussing in turn general n-capture abundance trends with metallicity, detailed distributions of these elements in a few very well-observed stars, and possible new initiatives to improve the use of n-capture elements in the description of early Galactic nucleosynthesis.

2. Overall N-Capture Abundance Trends With Metallicity

Abundances of n-capture elements vary with respect to those of the Fe-peak by several orders of magnitude from star to star. This scatter is most apparent in the lowest metallicity regimes. The onset metallicity of the n-capture scatter is poorly established, but at least for stars with [Fe/H] < −2, the total range in [<n-capture>/Fe] is ∼ ±1.5 dex (e.g., Gilroy et al. 1988; McWilliam et al. 1995; Ryan, Norris, & Beers 1996; Burris et al. 2000). (Here, [<n-capture>/Fe] stands for abundance ratios [Sr/Fe], [Ba/Fe], [La/Fe], etc.) The abundance scatter is far greater than that ascribable to uncertainties in the observed stellar spectra, atomic data, or model atmosphere parameters. The variation is clearly seen in published spectra of very metal-poor stars (e.g., figure 16, McWilliam et al. 1995; figure 3, Burris et al. 2000; figure 1, Westin et al. 2000). This provides direct evidence that local nucleosynthesis events added to an early Galactic halo ISM that remained poorly mixed on time scales corresponding to the formation of stars with metallicities as high as [Fe/H] ∼ −2.

Abundance ratios among the n-capture elements in very metal-poor stars are distinctly non-solar. Spite & Spite's (1978) high resolution spectroscopic investigation of a few bright metal-poor stars provided the first convincing observational evidence of this phenomenon. In the metallicity domain 0 > [Fe/H] > −2, the surveys cited above have shown that the ratio [Ba/Eu] ∼ 0, but by [Fe/H] ∼ −3 this ratio has declined to ∼ −1. Barium is synthesized most efficiently via slow neutron captures (the *s*-process) while europium is predominantly created via rapid neutron captures (the *r*-process). Thus observed low [Ba/Eu] values in the most metal-poor stars suggest (e.g., Truran 1981) that *r*-process products comprise most of the n-capture abundances at lowest metallicities. This supports the notion that n-capture element production in the early Galaxy should have been from very short-lived, high mass stars that generate large neutron fluxes during their death throes. Products of the *s*-process come from longer-lived (≳ 10^8 yr) low-to-intermediate mass stars. The evident lack of *s*-process contributions at [Fe/H] ∼ −3[1] argues for a very rapid buildup of Galactic metallicity to this level from first onset of star formation.

---

[1] Here we do not consider the so-called CH stars, ones that exhibit extremely large overabundances of C and of n-capture elements evidently created via the *s*-process (Norris et al. 1997a, 1997b; Hill et al. 2000). Most such stars are binaries, undoubtedly victims of mass transfer from higher-mass former AGB companions. These nucleosynthesis events eventually lead to the buildup of *s*-process levels in the Galaxy, but contribute little at lowest metallicities.

Less discussed is the lack of good correlation between abundances of heavier ($Z \geq 56$) and lighter ($Z = 38 - 40$; Sr-Y-Zr) n-capture elements. While the heavier elements appear to be r-process in origin, the lighter elements cannot be fit by the main s-process or r-process or their combination. Instead, it may require a complicated admixture of the *weak* s-process and the r-process to reproduce the Sr-Y-Zr abundance ratios in well-observed very metal-poor stars (e.g., Cowan et al. 1995). Moreover, Wasserburg, Busso, & Gallino (1996) have argued that the Solar System r-process distribution results from two different sources or sites—one for the lighter and one for the heavier n-capture elements. It is also not yet fully understood why the abundances of the Sr-Y-Zr group correlate much more closely with Fe-peak abundances than do the heavier n-capture abundances, although it might suggest a role for the *weak* s-process operating in massive stars early in the history of the Galaxy. The lighter n-capture elements will be considered again in the next section.

3. N-Capture Element Distributions in Well-Observed Halo Stars

To thoroughly examine the details of n-capture abundances in metal-poor stars it is necessary to detect many elements over the entire $Z = 31 - 92$ element range. This in turn requires identification of stars with [<n-capture>/Fe] $\gg$ +0.5, in order to maximize the spectroscopic contrast between the strengths of n-capture and Fe-peak absorption features. The few known extremely n-capture-rich metal-poor stars subjected to detailed abundance scrutiny have been serendipitous discoveries, e.g., HD 115444 (Griffin et al. 1982; Westin et al. 2000), CS 22892–052 (McWilliam et al. 1995; Sneden et al. 2000a, and references therein), and now CS 31082–001 (Cayrel et al. 2001). Much attention has been given to CS 22892–052, and in Figure 1 we summarize the results of Sneden et al. (2000a). This figure also adds revised abundance determinations from the same observed spectra for elements with new transition probability and hyperfine structure data: La (Lawler, Bonvallet, & Sneden 2001a), Ce (Biémont et al. 2000), Tb (Lawler et al. 2001b), and Dy (Wickliffe, Lawler, & Nave 2000). The CS 22892–052 abundances are compared with scaled Solar System r-process, s-process, and total abundance distributions.

Three features stand out in the plots of Figure 1. First, the scaled Solar System r-process distribution almost perfectly fits the observed abundances of the 18 heavier ($Z > 56$) stable n-capture elements; the other two Solar System distributions fail by large margins. While the "solar total" curve appears to hold promise in comparison to the observed abundances, normalizing the curve to Eu or Dy yields a mismatch by $0.5 - 0.9$ dex in the Ba–Ce group. Second, among the long-lived radioactive elements, the observed abundance of Th ($Z = 90$) falls below the "solar r-process" curve, as does the observed upper limit to the abundance of U ($Z = 92$). Thus *if* these elements were originally synthesized in the same proportions with respect to the lighter stable n-capture elements that occurred prior to Solar System formation, then CS 22892–052 is substantially older than the Solar System, and in fact its n-capture elements were synthesized ~ 15 Gyr ago (e.g., Cowan et al. 1999). Third, none of the Solar System abundance distributions provide satisfactory matches to the lighter ($38 \leq Z \leq 48$) n-capture element abundances of this star.

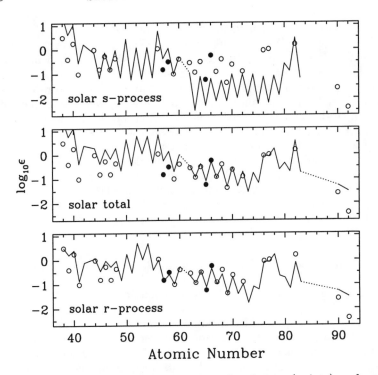

Figure 1.   CS 22892–052 n-capture abundances (points) and scaled
Solar System abundance distributions (solid and dashed lines). Open
circles are from Sneden et al. (2000a), and filled circles are abundances
determined from the same observational data but with new atomic
data (see text). The abundances units are $\log_{10}\epsilon(A) \equiv \log_{10}(N_A/N_H)$
+12.0. The U ($Z = 92$) abundance is an upper limit. Uncertainties
in individual abundances are $\sim \pm 0.05 - 0.10$ dex. The Solar System
distributions are from Burris et al. (2000). The scaling in the middle
and bottom panels force agreement with the Eu abundance; the scaling
in the top panel is approximate, for display purposes only.

The excellent agreement between observed abundances of the heavier stable
elements and the Solar System *r*-process is not confined to CS 22892–052. It is
repeated in each of the few ultra-metal-poor stars ([Fe/H] < −2.5) studied in
comparable detail (Westin et al. 2000; Johnson & Bolte 2001). Note in particular the halo giant BD+17 3248. This star has a nearly pure *r*-process signature
among the n-capture elements, but with [Fe/H] $\sim$ −2.0, it is the highest metallicity example of very *r*-process enriched material. It will be of great interest
to explore further the [Fe/H] > −2 domain to discover at what metallicity the
influence of individual *r*-process synthesis events are lost in the general Galactic
n-capture element buildup.

Recently Cayrel et al. (2001, and in these proceedings) have discovered that
the ultra metal-poor star CS 31082–001 ([Fe/H] $\sim$ −3) may have even larger n-
capture element overabundances than does CS 22892–052. Most interesting is

their detection of *many* transitions of Th II and the strongest U II line; the combined abundances of two radioactive elements (with different half-lives) suggests to them that the age of the n-capture elements in this star is ~ 13 Gyr. Their ongoing study of this star finds an *observed* [Th/Eu] ratio that is much larger than in CS 22892–052 (Sneden et al. 2000a), HD 115444 (Westin et al. 2000), several other halo stars (Johnson & Bolte 2001) and in globular cluster stars (Sneden et al. 2000b). This higher abundance ratio in CS 31082–001 would imply a young age comparable to the Solar System, obviously inconsistent with this star's metallicity and halo membership. This may indicate something unusual about the abundances in CS 31082–001, or it may indicate that the production ratio of [Th/Eu] is not single-valued and may lead to inconsistencies in age determinations (e.g., Goriely & Clerbaux 1999). It is clear that many more Eu and Th abundances need to be determined for halo stars, in order to make some sense of this issue.

Finally, consider the non-conformity of CS 22892–052 abundances in the $38 \leq Z \leq 48$ domain to any of the scaled Solar System curves. Previous studies have indicated that the abundances of Sr–Zr did not fall on the same curve as the heavier elements, but this is the first case where elements between $Z = 40 - 50$ have been detected. The difference between the lighter and heavier n-capture element abundance distributions in CS 22892–052 (as shown in Figure 1) has been explained as the superposition of two distinct r-process events (Wasserburg & Qian 2000), or perhaps resulting from different regions of the same neutron-rich jet of a core-collapse supernova (Cameron 2001) that are responsible for synthesizing the lighter and heavier n-capture elements. The breakpoint between the two r-process signatures is predicted to occur near Ba, as observed in CS 22892–052. However, there has been no success thus far in fitting the individual abundances in this star, although the entire set of n-capture abundances in just CS 22892–052 has been used to constrain nuclear mass models and theoretical r-process predictions (Cowan et al. 1999). Further, since knowledge of the $41 \leq Z \leq 48$ elements is confined so far to just CS 22892–052, what is most needed now are abundances of the lighter n-capture elements in more stars.

4. Some Future Prospects

The link between n-capture elemental abundances and Galactic time scales lies with the long-lived cosmochronometer elements. So far, ratios of Th and U to each other and to other elements do not yield a consistent answer for the age of the Galactic halo when the observed abundances of a (very) few stars are compared to zero-age predictions for them. The [Th/Eu] ratios in CS 22892–052 and HD 115444 suggest ages of ~ 15 Gyr, while that of CS 31082–001 yields ~ 4 Gyr. But the [Th/U] ratio of this star pushes the age back to ~ 13 Gyr. Johnson & Bolte (2001) find that the ages of their five stars with detected Th II features can range from 11–15 Gyr, depending on assumptions about initial Th/Eu production ratios. Clearly both further theoretical (e.g., Cowan, Sneden, & Truran 2001) and observational efforts are needed on this problem. Extensive n-capture abundance distributions should be derived in at least ~ 30 very low metallicity stars, to discover what is the normal abundance level of the radioactive chronometer elements in the Galactic halo.

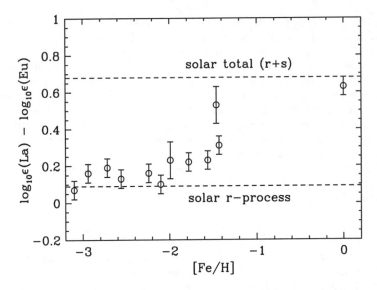

Figure 2.   Ratios of La to Eu abundances in a few representative stars over most of the Galactic halo metallicity range. The La abundances have been derived using new La II laboratory data from Lawler et al. (2001a).

Probably no other *absolute* Galactic time scale information can be deduced from n-capture elements in metal-poor stars. But recent improvements in stellar spectroscopic and laboratory atomic data permit renewed attack on an unsolved *relative* time scale question: what is the Galactic metallicity at which major contributions from s-process nucleosynthesis began to generally influence the halo ISM? Most s-process synthesis is associated with the AGB phases of intermediate-to-low mass stars, $M < 8\ M_\odot$, whose evolutionary time scales are $\geq \sim 10^8$ yr. Metallicity regimes with little or no detectable s-process contributions are those resulting from the first waves of Galactic nucleosynthesis that happened on faster time scales. The approximate [Fe/H] indicating a general rise in the s-process should tell us how much the first nucleosynthesis burst contributed to overall Galactic metallicity.

This question has usually been empirically addressed by trying to find the metallicity domain for substantial movement in [Ba/Eu] ratios from the r-process-dominated value of ~ -0.9 at [Fe/H] ~ -3 toward the Solar System $r + s$ value of ~ 0.0 that appears to be complete by [Fe/H] ~ -1.5. With such data, Gilroy et al. (1988) suggested that the onset of major Galactic s-process occurred at [Fe/H] ~ -2.3, while Burris et al. (2000) argue that this may happen at [Fe/H] ~ -2.8. At present it is impossible to narrow this metallicity down to better than somewhere in the range $-3 \leq$ [Fe/H] ≤ -2.

This situation may never change, because large observed star-to-star scatter exists in [Ba/Eu] ratios at all Galactic halo metallicities. The scatter may be intrinsic to the stars, or it may simply be an artifact of abundance analyses. The problem lies in basic atomic structure. Unlike the first ions of neighboring rare earth elements, Ba II has a structure that in cool stellar atmospheres gives rise to only five very strong transitions from lower energy levels in the visible spectral range. All other Ba II transitions are very weak ones from higher excitation levels. Significant hyperfine and isotopic splitting exists for the strong low excitation lines, and Ba has 5 stable isotopes whose relative abundances are synthesized in different proportions under the r- and s-processes; estimates of these proportions becomes part of the abundance derivation process (e.g., Magain 1995; Sneden et al. 1996). Abundances of Ba also are very sensitive to values of microturbulent velocity assumed in the analyses. All in all, it is difficult to believe [Ba/Eu] ratios in most metal-poor stars to better than an uncertainty of $\sim \pm 0.2$ dex, and this is inadequate to map out the metallicity evolution of the s- and r-processes.

Fortunately other ($Z \geq 56$) elements have very different abundances resulting from the two n-capture events, and in particular [La/Eu] and [Ce/Eu] have r-/s-process sensitivities nearly equal to that of [Ba/Eu] (e.g., see figure 1 of Sneden, Cowan, & Truran 2001). Both La II and Ce II have much more favorable atomic structures than does Ba II, and these elements have one very dominant isotope each. McWilliam (1997) advocated abandoning the use of [La/Eu] in favor of [Ba/Eu] (see his figure 10). But the data depicted in that diagram still had significant scatter. Johnson & Bolte (2001) argue from better data that observed [La/Eu] ratios indicate r-process dominance in stars as metal-rich as [Fe/H] ~ -1.5. Now, armed with higher resolution, higher S/N spectra and employing the new atomic data of Lawler et al. (2001a), we are beginning a large-sample survey of [La/Eu] in metal-poor stars. Excellent line-by-line abundance agreement is seen for La II in the Sun, CS 22892–052, and BD+17 3248 (Sneden et al. 2001) and in Figure 2 we show the run of [La/Eu] with [Fe/H] for a few very metal-poor but n-capture-rich stars. The small star-to-star scatter in [La/Eu] at lowest metallicities is so far very encouraging, but the sample is very small. When this survey is completed we hope to be able to tell with far greater certainty the metallicity at which the s-process makes substantial contributions to most stars' n-capture abundances, and hence be able to tell how fast the Galaxy may have increased its Fe-peak metallicity before the deaths of the first intermediate-mass stars.

References

Biémont, E., Garnir, H. P., Palmeri, P., Li, Z. S., & Svanberg, S. 2000, MNRAS, 312, 116

Burris, D. L., Pilachowski, C. A., Armandroff, T. A., Sneden, C., Cowan, J. J., & Roe, H. 2000, ApJ, 544, 302

Cameron, A. G. W. 2001, Nuc. Phys. A, in press

Cayrel, R., et al. 2001, Nature, 409, 691

Cowan, J. J., Burris, D. L., Sneden, C., McWilliam, A., & Preston, G. W. 1995, ApJ, 439, L51

Cowan, J. J., Pfeiffer, B., Kratz, K.-L., Thielemann, F.-K., Sneden, C., Burles, S., Tytler, D., & Beers, T. C. 1999, ApJ, 521, 194

242   *Sneden et al.*

Cowan, J. J., Sneden, C., & Truran, J. W. 2001, in Cosmic Evolution, ed. E. Vangioni-Flam & M. Cassé (Singapore: World Scientific Publishing), in press

Gilroy, K. K., Sneden, C., Pilachowski, C. A., & Cowan, J. J. 1988, ApJ, 327, 298

Goriely, S., & Clerbaux, B. 1999, A&A, 346, 798

Griffin, R., Gustafsson, B., Viera, T., & Griffin, R. 1982, MNRAS, 198, 637

Hill, V., et al. 2000, A&A, 353, 557

Johnson, J. A., & Bolte, M. 2001, ApJ, in press

Lawler, J. E., Bonvallet, G., & Sneden, C. 2001a, ApJ, in press

Lawler, J. E., Wickliffe, M. E., Cowley, C. R., & Sneden, C. 2001b, submitted

Magain, P. 1995, A&A, 297, 696

McWilliam, A. 1997, ARA&A, 35, 503

McWilliam, A., Preston, G. W., Sneden, C., & Searle, L. 1995, AJ, 109, 2757

Norris, J. E., Ryan, S. G., & Beers, T. C. 1997a, ApJ, 488, 350

Norris, J. E., Ryan, S. G., & Beers, T. C. 1997b, ApJ, 489, L169

Ryan, S. G., Norris, J. E., & Beers, T. C. 1996, ApJ, 471, 254

Sneden, C., Cowan, J. J., Ivans, I. I., Fuller, G. M., Burles, S., Beers, T. C., & Lawler, J. E. 2000a, ApJ, 533, L139

Sneden, C., Cowan, J. J., & Truran, J. W. 2001, in Cosmic Evolution, ed. E. Vangioni-Flam & M. Cassé (Singapore: World Scientific Publishing), in press

Sneden, C., Johnson, J., Kraft, R. P., Smith, G. H., Cowan, J. J., Bolte, M. S. 2000b, ApJ, 536, L85

Sneden, C., McWilliam, A., Preston, G. W., Cowan, J. J., Burris, D. L., & Armosky, B. J. 1996, ApJ, 467, 819

Spite, M., & Spite, F. 1978, A&A, 67, 23

Truran, J. W. 1981, A&A, 97, 391

Wasserburg, G. J., Busso, M., & Gallino, R. 1996, ApJ, 466, 109

Wasserburg, G. J., & Qian, Y.-Z. 2000, ApJ, 529, L21

Westin, J., Sneden, C., Gustafsson, B., & Cowan, J. J. 2000, ApJ, 530, 783

Wickliffe, M. E., Lawler, J. E., & Nave, G. 2000, JQSRT, 66, 363

Acknowledgments. Our work on n-capture elements in halo stars has been a collaborative effort over many papers, and we appreciate our co-authors for their contributions over the years. Emile Biémont, Rica French, Jennifer Johnson, Jennifer Simmerer, and Craig Wheeler are thanked for helpful discussions. We are happy to acknowledge that this research has been supported by various NSF grants to the authors.

Discussion

Typhoon Lee: You said the Ba isotope abundances could be measured?

Chris Sneden: Ba isotopic abundances have been derived in a paper by Magain (1995). They did it without optimal data, but the width of the Ba 4554 Å line will be different in the cases of r-process versus s-process dominance—since Ba has 5 stable isotopes and the isotopic mix is different in the r- and s-process. Their suggestion is that at least in one very metal-poor star the s-process still could account for most of the Ba production. This work needs to be pursued further.

Typhoon Lee: Barium-138 is mostly s-process, so if you detect no Barium-138 that would be very definitive evidence against the s-process. I'm always unhappy about this.

Chris Sneden: What must be understood (actually a couple of us have some new data but it's so far inconclusive) is you really are stuck with doing just the general breadth of that Ba 4554 Å line, and so the results right now are ambiguous.

Andreas Korn: You showed a plot of La versus Eu which had a step-like function proceeding to higher metallicities. Could there be a population transition, or are all these stars genuine halo objects?

Chris Sneden: That's what I'd love to know. You'd like to measure enough stars so that you could actually start in that −1 to −2 metallicity range to separate the thick disk from the true halo stars and see if there is a signature of the two populations. All I can tell you is I would *never* believe it on the basis of Ba-to-Eu ratios, and we can now start to hope the trick can be done better from La-to-Eu. The stumbling block in accuracy right now is not the La, it's the Eu for which the atomic data are now getting somewhat old, and I'm hopeful that the Lawler group or the Biemont group will perform new Eu lab measurements. But your point is *exactly* on the mark.

Astrophysical Ages and Time Scales
ASP Conference Series, Vol. 245, 2001
T. von Hippel, C. Simpson, N. Manset

First Measurement of the Uranium/Thorium Ratio in a Very Old Star: Implications for the Age of the Galaxy

R. Cayrel, M. Spite, F. Spite

Observatoire de Paris, France

V. Hill, F. Primas, P. François

European Southern Observatory, Germany

T. C. Beers

Michigan State University, USA

B. Plez

GRAAL, Université Monpellier-2, France

B. Barbuy

Universidade de São Paulo, Brazil

J. Andersen, B. Nordström

University of Copenhagen, Denmark

P. Molaro, P. Bonifacio

Osservatorio Astronomico di Trieste, Italy

Abstract. During an ESO-VLT large programme devoted to high resolution spectroscopy of extremely metal-poor stars selected from the HK survey of Beers and colleagues, the [Fe/H] = −2.9 giant CS 31082–001 was found to be as enriched in neutron-capture r-process elements as CS 22892–052 but with a much-reduced masking by molecular lines. This allowed the detection and measurement of the uranium line at 3859 Å for the first time in a stellar spectrum. By making use of the relatively short ^{238}U decay (half-life 4.47 Gyr), we obtain a radioactive dating of the formation of the r-process elements in this star, born in the early days of the Galaxy.

1. Introduction

Radioactive decay provides a very direct and accurate measurement of time between the epoch of formation and the present, if one also has a means of estimating the abundance ratio of the radioactive element relative to a stable element (or to another radioactive element having a substantially different half-life). So far only the ratio of the abundance of ^{232}Th to a stable element, usually Eu (also primarily produced by the rapid neutron capture process), has been used as cosmochronometer. We report here the first measurement of the abundance of ^{238}U, which has the great advantage of having a smaller half-life than ^{232}Th (4.47 Gyr instead of 14.05 Gyr). This detection was made in CS 31082–001, an extremely metal-poor star ([Fe/H] = −2.9) selected from the HK survey of Beers and collaborators (see Beers 1999 for a summary). This star is more metal-deficient than the globular clusters, and was likely born in the very early Galaxy. This measurement, reported in a letter to Nature (Cayrel et al. 2001), was made with the ESO/VLT unit 2 (Kuyen) telescope, and the UVES spectrograph. In this presentation we concentrate on a discussion of the abundances of U and Th in CS 31082–001, and the consequences for radioactive dating of the material in this star. In a contribution by Hill et al. (these proceedings) the abundances of the other neutron-capture elements in CS 31082–001 are discussed. In another contribution Toenjes et al. (these proceedings) present computations of the expected production ratios of U/Th, U/Eu, U/Ir, Th/Eu, and others. Christlieb et al. (these proceedings) discuss a strategy for future detection of additional r-process enhanced metal-poor stars that might be used as cosmochronometers (see also the discussion by Sneden et al., these proceedings).

2. The Spectroscopic Observation of CS 31082–001 and Comparison With CS 22892–052

High-resolution VLT/UVES spectra, in the region of the U II line at 3859.57 Å, are shown in Figures 1 and 2, respectively, for the star CS 22892–052, already known to be greatly enriched in r-process neutron-capture elements (Sneden et al. 1996), and the newly discovered star CS 31082–001.

It is quite clear from these spectra that both stars are considerably enriched in r-process elements—the primary difference between the two stars is the much reduced contamination of atomic lines by molecular features (mainly CH and CN) in CS 31082–001. The visibility of the U II line, excellent in CS 31082–001, is spoiled in CS 22892–052 by the presence of the CN line at 3859.67 Å . The already high signal-to-noise ratios of these spectra are shown here in Figures 1 and 2, enhanced by a convolution with a Gaussian kernel having FWHM equal to half the resolution of the spectrograph (only one quarter of the FWHM of the stellar lines—producing a negligible loss in spectral resolution). In the spectrum of CS 22892–052 the U II line is at the limit of the noise for two reasons: it is blended with the stronger CN line and the line is intrinsically weaker than in CS 31082–001. Only upper limits to the uranium abundance have been given so far in CS 22892–052.

Another considerable advantage of the lower blending by molecular lines in CS 31082–001 is that we have identified 14 individual lines of thorium in its spectrum, 10 of which are sufficiently unblended to allow for a precise determination of its abundance. In CS 22892–052 there are thus far only 3 lines that might be used for this abundance measurement (Sneden & Cowan 2000). Figure 3 illustrates this point (to be compared with figure 2 of Sneden & Cowan 2000).

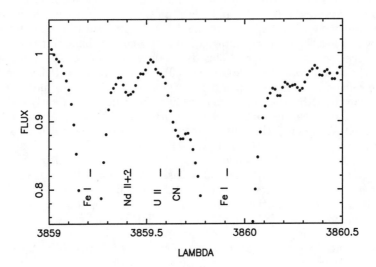

Figure 1.   VLT/UVES spectrum of star CS 22892–052 in the region of the uranium line at 3859 Å.

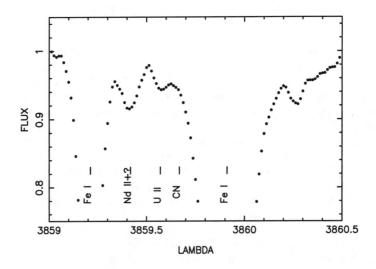

Figure 2.   VLT/UVES spectrum of star CS 31082–001 in the region of the uranium line at 3859 Å.

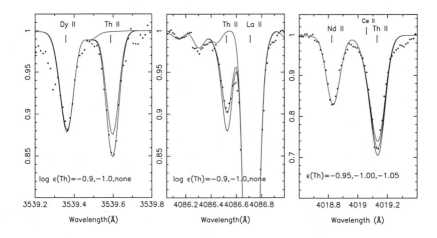

Figure 3.   Synthesis of three thorium lines in CS 31082–0012.

3.   The Abundances of U and Th in CS 31082–001 and the Age Determination

It has been claimed by Goriely & Clairbaux (1999) that the ratio U/Th might be a better cosmochronometer than either Th/Eu or Th/Dy, because of the much smaller mass difference between Th and U than between either one of these two actinides and the lighter lanthanides (see their figures 2-5). It is therefore of particular interest to use the ratio U/Th to determine the age of formation of these elements, presumably in a type-II SN explosion, the products of which were later trapped in the atmosphere of stars such as CS 31082–001, and which have both decayed with a constant rate, ever since. Table 1 summarizes the relationship between the logarithmic concentrations of three chronometer pairs and time.

In Table 1, "r" refers to any stable neutron-capture element produced by the r-process, i.e., with a time between two consecutive neutron captures smaller than the spontaneous decay of the target. Because of the large numerical factor 46.7 in the first expression, resulting from the long half-life of Th, this chronometer is less favourable than the other two. At first view, the second expression is the best choice. However, the warning of Goriely and Clairbaux suggests that it might be quite difficult to reduce the uncertainty in $\log(U/r)_0$, as the first stable r-elements are still far from U and Th in nuclear mass. Therefore, the

Table 1.   Time Δt elapsed during epochs at which two abundance ratios are known[†].

Time elapsed		Expression
Δt	=	$46.7\ [\log(\mathrm{Th/r})_0 - \log(\mathrm{Th/r})_{\mathrm{obs}}]$
Δt	=	$14.8\ [\log(\mathrm{U/r})_0 - \log(\mathrm{U/r})_{\mathrm{obs}}]$
Δt	=	$21.8\ [\log(\mathrm{U/Th})_0 - \log(\mathrm{U/Th})_{\mathrm{obs}}]$

[†] Initial epoch labeled with the subscript "0" and the present epoch labeled with the subscript "obs.'

last expression is probably the safest choice. The increase in the multiplicative coefficient with respect to the second expression is likely to be more than compensated for by the reduction in the uncertainty of the production ratio of $\log(\mathrm{U/Th})_0$.

For the present, the best we can do is to select estimates of the $\log(\mathrm{U/Th})_0$ ratio, and derive Δt from expression 3 of Table 1. In Table 2 we also provide, for comparison, the ages derived from expression 2, using the heaviest observable stable elements Os and Ir. The errors do not include the uncertainties in the theoretical ratios, but do include all other known sources of errors, including uncertainties in the oscillator strengths. Assuming a 0.1 dex uncertainty in $\log(\mathrm{U/Th})_0$, and a 0.15 uncertainty in $\log(\mathrm{U/Th})_{\mathrm{obs}}$, leads to a global uncertainty of 4 Gyr arising from the use of U/Th ratio alone. However, the two independent ratios U/Os and U/Ir lead to a similar value for the age, with smaller net error bars, because of the smaller multiplicative coefficient, 14.8 instead of 21.8. We thus consider 3 Gyr to be a reasonable estimate of the global error, taking into account the results of the three chronometer pairs.

Table 2.   Ages of r-process elements in CS 31082–001 and their associated production ratios.

Pair	Log(Prod. Ratio)	Ref	Log(Observ. Ratio)	Age
U/Th	−0.255	1	−0.74 ± 0.15	10.6 ± 3.3
U/Th	−0.10	2	idem	14.0 ± 3.3
U/Th	−0.16	3	idem	12.6 ± 3.3
U/Os	−1.27	1	−2.19 ± 0.18	13.6 ± 2.7
U/Ir	−1.30	1	−2.07 ± 0.17	11.4 ± 2.5

(1) Cowan et al. (1999), (2) Goriely & Clairbaux (1999), and (3) Toenjes et al. (these proceedings).

4. Relationship Between the Value of $(U/Th)_0$ and the Measured Ratio in CS 310822–001 Versus the Solar (U/Th) Values

The present solar value of $\log(U/Th)$ is -0.59 ± 0.05, according to Grevesse & Sauval (1998). The value in CS 31082–001 is -0.74 ± 0.15 (Table 1). The values at the birth of the Sun are the preceding ones corrected by the same factor: the variation of the ratio during 4.6 Gyr, i.e., 0.211 dex, according to Table 1, or respectively -0.379 and -0.259, still differing by 0.15 dex.

If U and Th are always produced in the proportion $(U/Th)_0$, it is indeed expected that the value of $\log(U/Th)$ is smaller in CS 31082–001 than in the Sun, because it has been decaying over the full age of the Galaxy, whereas in the Sun it has been decaying over any time between 4.6 Gyr and the age of the Galaxy, depending on its epoch of formation and injection into the presolar nebula. If we assume that U and Th are formed together continuously during the time preceding the birth of the Sun, it is possible to derive expressions for the build-up of the concentrations ϵ_U and ϵ_{Th} of U and Th in the interstellar medium as:

$$d\epsilon_U = a_U f(t)\, dt$$

$$d\epsilon_{Th} = a_{Th} f(t)\, dt$$

with:

$$\frac{a_U}{a_{Th}} = (U/Th)_0$$

where $f(t)$ is the common history function of the production of U and Th before the birth of the Sun. After the birth of the Sun, U/Th has evolved identically in the Sun and in CS 31082–001, so the difference between the ratios is only due to the fact that the decay between the origin of the Galaxy and the time of birth of the Sun $t_\odot$ is for CS 31082–001:

$$\frac{\epsilon_U(CS)}{\epsilon_{Th}(CS)} = (U/Th)_0 \exp(-(\alpha_U - \alpha_{Th})t_\odot) \tag{1}$$

where $\alpha_U = 0.1551$ and $\alpha_{Th} = 0.04933$.

For the Sun at the same time $t_\odot$:

$$\epsilon_U(\odot) = \int_0^{t_\odot} a_U f(t) \exp(-\alpha_U(t_\odot - t))\, dt$$

$$\epsilon_{Th}(\odot) = \int_0^{t_\odot} a_{Th} f(t) \exp(-\alpha_{Th}(t_\odot - t))\, dt$$

leading to:

$$\frac{\epsilon_U(\odot)}{\epsilon_{Th}(\odot)} = (U/Th)_0 \frac{\int_0^{t_\odot} f(t) \exp(-\alpha_U(t_\odot - t))\, dt}{\int_0^{t_\odot} f(t) \exp(-\alpha_{Th}(t_\odot - t))\, dt}. \tag{2}$$

If $f(t)$ or $f(t/t_\odot)$ is known, the equations (1) and (2) contain only two unknown quantities, $(U/Th)_0$ and $t_\odot$. It is then possible to derive $(U/Th)_0$ and $t_\odot$, without knowing $(U/Th)_0$ a-priori. Actually, this is not our goal, as we prefer to obtain $(U/Th)_0$ from physicists, and thereby derive as much information on the astrophysical unknowns as possible, rather than derive nuclear properties from astrophysical data. But at least we can see what a particular value of $(U/Th)_0$ implies for the history function of production of the r-process elements in the Solar System. Assuming a uniform production rate, the result is easily derived. Taking the ratio of equation (2) to equation (1) yields:

$$\frac{\epsilon_U(\odot)/\epsilon_{Th}(\odot)}{\epsilon_U(CS)/\epsilon_{Th}(CS)} = \frac{\int_0^{t_\odot} \exp(\alpha_U t)\, dt}{\int_0^{t_\odot} \exp(\alpha_{Th} t)\, dt}. \tag{3}$$

The integration is straightforward, and the value of $t_\odot$ which gives the observed ratio of 1.41 (the antilog of 0.15) is 5.9 Gyr. Adding this to the 4.6 Gyr elapsed since the birth of the Sun gives an age for CS 31082–001 of 10.5 Gyr. Plugging the value of $t_\odot$ into equation (1) gives $(U/Th)_0 = -0.258$, a value almost equal to that listed on the first line of Table 1. It is equally easy to find the result for $f(t) = \exp(-\lambda t)$. The value of the parameter λ leading to the value of $(U/Th)_0 = -0.16$ in the second line of Table 1 is 0.25, giving $t_\odot = 8$ Gyr, or as listed in Table 1, a total age of 12.6 Gyr. This implies a decay of the production rate of about a factor 7 between the epoch of the early Galaxy and the epoch of the birth of the Sun. This simple calculation emphasizes the astrophysical impact of the exact value of $(U/Th)_0$.

5. Conclusion

Uranium has been detected, and had its abundance measured, in a very metal-poor star that was born in the early Galaxy, likely before the formation of the globular clusters. It has been possible, for the first time, to use the ratio U/Th for determining the age of formation of these elements in the early Galaxy. The accuracy of the result, 12.6 ± 3 Gyr, is still rather severely limited by the 0.15 dex uncertainty in the abundance ratio derived from observation, and by the 0.1 uncertainty in the theoretical estimation of the U/Th production ratio. Progress can be expected in the near future in three areas: (i) improvement in the measurement of the oscillator strengths of U and Th, (ii) improvement in the estimation of the production ratio U/Th_0, based on measurement of the abundances of other heavy elements in CS 31082–001 such as Pb, Bi, Os, Pt, and Ir, as well as refinements in the nuclear physics models, and (iii) discovery of other metal-deficient r-process enhanced stars in which U and Th can be measured, to improve the statistics and to provide a valuable check of the stability of the production ratio in several stars.

References

Beers, T. C. 1999, in ASP Conf. Ser. 165, The Third Stromlo Symposium: The Galactic Halo, ed. B. K. Gibson, T. S. Axelrod, & M. E. Putman (San Francisco: ASP), 202

Cayrel, R., et al. 2001, Nature, 409, 691

Goriely, S., & Clairbaux, B. 1999, A&A, 346, 798

Grevesse, N., & Sauval, 1998, SSRv, 85, 161

Sneden, C., & Cowan, J. J. 2000, RMA&A, in press (astro-ph/0008185)

Sneden, C., McWilliam, A., Preston, G. W., Cowan, J. J., Burris, D. L., & Armosky, B. J. 1996, ApJ, 467, 819

Acknowledgments. We are indebted to T. von Hippel for scheduling this talk with great precision, so that it took place just at the end of the embargo period on our letter to Nature. We are also very indebted to ESO for the generous allocation of VLT observing time which allowed this discovery to be made. We thank C. R. Cowley for having called our attention on the fact that uranium had been observed before, in Ap stars.

Discussion

Brian Chaboyer: The error in your age, is that dominated by the production uncertainties or the uncertainties in the measurement of the abundance?

Roger Cayrel: They are just about the same order of magnitude, both the order of 0.1 dex, I would say.

Astrophysical Ages and Time Scales
ASP Conference Series, Vol. 245, 2001
T. von Hippel, C. Simpson, N. Manset

Hope and Inquietudes in Nucleocosmochronology

M. Arnould, S. Goriely

Institut d'Astronomie et d'Astrophysique, Université Libre de Bruxelles, B-1050 Brussels, Belgium

Abstract. Critical views are presented on some nucleocosmochronological questions. Progress has been made recently in the development of the ^{187}Re–^{187}Os cosmochronometry. From this, there is good hope for this clock to become of the highest quality for the nuclear dating of the Universe. The *simultaneous* observation of Th and U in ultra-metal-poor stars would also be a most interesting prospect. In contrast, a serious inquietude is expressed about the reliability of the chronometric attempts based on the classical ^{232}Th–^{238}U and ^{235}U–^{238}U pairs, as well as on the Th (without U) abundance determinations in ultra-metal poor stars.

1. Introduction

The use of radionuclides to estimate astrophysical ages has a long history, a milestone of which is the much celebrated piece of work of Fowler & Hoyle (1960). For long, the field of nucleocosmochronology that has emerged from this paper has been aiming at determining the age T_{nuc} of the nuclides from abundances in the material making up the bulk of the Solar System. If indeed the composition of this material witnesses the long history of the compositional evolution of the Galaxy prior to the isolation of the solar material, a reliable evaluation of T_{nuc} (clearly a lower bound to the age of the Universe) requires (i) the identification of radionuclides with half-lives commensurable with estimated reasonable Galactic ages (i.e., $t_{1/2} \gtrsim 10^9$ y), (ii) the construction of nucleosynthesis models that are able to provide the isotopic or elemental yields for these radionuclides, (iii) high quality data for the meteoritic abundances of the relevant nuclides, and, last but not least, (iv) models for the build up of the abundances of these nuclides in the Galaxy, primarily in the solar neighborhood. All these requirements clearly make the chronometric task especially demanding. While everybody would agree this far, there are different ways to look at the question.

A pessimistic view is that nucleocosmochronology can at best set limits on T_{nuc} through the use of a so-called model-independent approach. Using this formalism, Meyer & Schramm (1986) conclude that $9 \lesssim T_{nuc}(\text{Gyr}) \lesssim 27$. (Only the lower limit is truly model-independent. The upper limit depends on some model assumptions.) This is clearly not a highly constraining range! An optimistic view is that, given the presumed complexity of the chemical evolution of the Galactic disk, it is by far preferable to describe its nucleosynthetic history by a simple function with some adjustable parameters. This has been advocated by Fowler over the years with the use of the so-called exponential model.

A practitioner's view is that it is worth studying nucleocosmochronology in the framework of chemical evolution models which are simple enough not to account for all the dynamical aspects of the formation of the present Galactic disk, but which imperatively satisfy as many observational constraints as possible (e.g., Yokoi, Takahashi, & Arnould 1983; Takahashi 1998).

A new chapter in the story of nucleocosmochronology has been written with the discovery of isotopic anomalies attributed to the *in situ* decay in a minute fraction of the meteoritic material of radionuclides with half-lives in the approximate $10^5 \lesssim t_{1/2} \lesssim 10^8$ yr range. These observations will hopefully provide information on discrete nucleosynthesis events that presumably contaminated the Solar System at times between about 10^5 and 10^8 yr prior to the isolation of the solar material from the general Galactic material. Constraints on the chronology of nebular and planetary events in the early Solar System could be gained concomitantly.

Finally, the observation of Th in some very metal-poor stars and of U in one of them has opened the way to a possible nuclear-based evaluation of the age of individual stars other than the Sun. This clearly broadens the original scope of nucleocosmochronology still further.

In Section 2, we reiterate the inquietude originally expressed by Yokoi et al. (1983) concerning the classical ^{232}Th$-^{238}$U and ^{235}U$-^{238}$U pairs, which may not be as reliable Galactic clocks as it is often stated. In contrast, we express some hope concerning ^{187}Re$-^{187}$Os in Section 3. Our second inquietude relates to the chronological information one may gain from the observation of Th in very low-metallicity stars (Section 4). The situation would be brighter if one could rely on precise measurements of the Th/U ratio in such stars (Section 5). The chronometry using short-lived radionuclides is not discussed here. The interested reader is referred to e.g., Arnould, Meynet, & Mowlavi (2000) for a brief review of this question and references.

2. Inquietude 1: The Trans-Actinide Clocks

The most familiar long-lived ^{232}Th$-^{238}$U and ^{235}U$-^{238}$U clocks based on the present meteoritic content of these nuclides are reviewed by Truran (these proceedings). At several occasions, we have expressed reservations about the use of these pairs as reliable evaluators of $T_{\rm nuc}$ (e.g., Arnould & Takahashi 1990). The first reason for this inquietude relates to the clear necessity of knowing with high precision the production ratios of the involved actinides. Such quality predictions are clearly out of reach at the present time. One is indeed dealing with nuclides that can be produced by the r-process only, which suffers from very many astrophysics and nuclear physics problems, as we have emphasized on many occasions (e.g., Arnould & Takahashi 1999; Goriely & Clerbaux 1999). In particular, the true astrophysical site(s) of the r-process remain(s) so far unknown, preventing any firm (sometimes far-reaching) conclusions to be drawn for nucleosynthesis applications. On the nuclear physics side, the nuclear properties (such as nuclear masses and deformations) of thousands of exotic nuclei located between the valley of β-stability and the neutron drip line have to be known, as well as their transmutation rates through α- or β-decay, various fission channels, as well as through nuclear reactions, such as (n,γ) and (γ,n). Despite much recent experimental effort, none of these quantities are known for the nuclei involved in the r-process, so that they have to be extracted from theory. In addition, the Th and U nuclides are the only naturally-occurring ones beyond ^{209}Bi, so that any extrapolation relying on semi-empirical analyses and fits of the solar r-process abundance curve is in danger of being especially unreliable.

Recent r-process calculations (Goriely & Clerbaux 1999) provide production ratios for $0.8 \lesssim P_{235}/P_{238} \lesssim 1.2$ and $1.4 \lesssim P_{232}/P_{238} \lesssim 2.7$. The extent of these ranges largely forbids the build up of precise nucleocosmochronometries. It has to be noticed that this problem would linger even if a realistic r-process model were given.

A second source of worry relates to the necessity of introducing nucleocosmochronology in Galactic chemical evolution models that satisfy as many astronomical observables as possible, and to carefully check the internal consistency of such extended chemical models. In fact, this step is quite complex. Even if the nucleosynthesis yields of the relevant radionuclides are assumed to be reliably known for all stellar masses and metallicities, which is far from being the case, one still has to worry about such effects as the so-called astration, i.e., the possible more or less substantial destruction during the stellar lifetime of those nuclei which were absorbed from the interstellar medium at the stellar birth. From their constrained one-zone model for the evolution of the composition of the Galactic disk in the solar neighborhood, Yokoi et al. (1983) conclude that the predicted $(^{235}\mathrm{U}/^{238}\mathrm{U})_0$ and $(^{232}\mathrm{Th}/^{238}\mathrm{U})_0$ ratios at the time $T_\odot$ of isolation of the Solar System material from the Galactic material about 4.6 Gyr ago (the subscript 0 refers to this time) is only very weakly dependent on Galactic age, at least in the explored range from about 11 to 15 Gyr. This is analyzed as resulting largely from the expected rather weak time dependence of the stellar birthrate (except possibly at early Galactic epochs, but reliable information on these times is largely erased by the subsequent long period of chemical evolution). Yokoi et al. (1983) also conclude that the predicted abundance ratios at $T_\odot$ are consistent with those derived at the same time from meteoritic analyses if the r-process production ratios lie in the approximate ranges $1 < P_{235}/P_{238} < 1.5$ and $1.7 < P_{232}/P_{238} < 2$. The latter values are consistent with the predicted ones reported above, the former ones being only marginally so. At this point, the estimate of the level of convergence between predicted production ratios and observed abundances is blurred not only by the large uncertainties in the estimated r-process actinide yields, but also by the uncertainties affecting the meteoritic Th and U abundances, which amount to at least 25% and 8%, respectively (Grevesse, Noels, & Sauval 1996).

Finally, most of the huge amount of work devoted to the trans-actinide chronometries (e.g., Truran, these proceedings) relies on the simple exponential model whose substantial mathematical ease is obtained at the expense of a scanty astrophysical content. This situation makes the use of the Th-U chronologies especially unreliable. As demonstrated by Arnould & Takahashi (1990), a given exponential model may predict an increase by a factor of 3 in the T_{nuc} value by just changing the meteoritic $(^{232}\mathrm{Th}/^{238}\mathrm{U})_0$ ratio from 2.32 to 2.67, which is compatible with existing data (Grevesse et al. 1996)!

From the above considerations, one may restate the conclusion drawn by Yokoi et al. (1983) that the relatively large uncertainties in the measured solar

$(^{232}\text{Th}/^{238}\text{U})_0$ and $(^{235}\text{U}/^{238}\text{U})_0$ ratios coupled with the relatively weak dependence with Galactic age of these calculated ratios make it next to impossible to obtain a *reliable* value of T_{nuc} from the trans-actinide chronometries as they stand now. This conclusion holds even in the most favorable situation where the r-process predictions can be made compatible with the abundance measurements in the framework of a given Galactic chemical evolution model. There is no guarantee at this time to reach this necessary compatibility in a natural way, i.e., without having to play around with a rich variety of parameters.

3. Hope 1: The ^{187}Re–^{187}Os Chronometry

First introduced by Clayton (1964), the chronometry using the ^{187}Re–^{187}Os pair is able to avoid the difficulties related to the r-process modeling. True, ^{187}Re is an r-nuclide. However, ^{187}Os is not produced directly by the r-process, but indirectly via the β^--decay of ^{187}Re ($t_{1/2} \sim 42$ Gyr) over the Galactic lifetime. This makes it in principle possible to derive a lower bound for T_{nuc} from the mother-daughter abundance ratio, provided that the cosmogenic ^{187}Os component is deduced from the solar abundance by subtracting its s-process contribution. This chronometry is thus reduced to a question concerning the s-process. This is a good news, as the s-process is, generally speaking, better under nuclear or astrophysics control than the r-process. Other good news comes from the recent progress made in the measurement of the abundances of the concerned nuclides in meteorites, which provide in addition the decay constant of the neutral ^{187}Re atoms. The derived value $\lambda = (1.666 \pm 0.010)^{-11}$ yr^{-1} is substantially more precise than its direct determination (Faestermann 1998). This improved input is essential for the establishment of a reliable chronometry.

Even if the s-process models are in much better shape than the r-process ones, the evaluation of the relative production of ^{186}Os and of ^{187}Os by the s-process is not a trivial matter. One difficulty relates to the fact that the ^{187}Os 9.75 keV excited state can contribute significantly to the stellar neutron-capture rate because of its thermal population in s-process conditions ($T \gtrsim 10^8$ K). The ground-state capture rate measured in the laboratory has thus to be modified. Several estimates of this correction factor F_σ based on preliminary experimental data analyzed in the framework of Hauser-Feshbach models are available (e.g., McEllistrem et al. 1989). The question has been re-examined with the code MOST (Goriely 1998). In particular, the impact on the predictions of uncertainties in various key ingredients of the model (like the Γ_γ-width or the neutron optical potential) has been analyzed (Goriely, unpublished). There is also reasonable hope to reduce the uncertainty on F_σ through further dedicated experiments (Koehler & Mengoni, private communication).

A second difficulty has to do with the possible branchings of the s-process path in the $184 \leq A \leq 188$ region which may affect the evaluated ^{187}Os/^{186}Os s-process production ratio (Arnould, Takahashi, & Yokoi 1984). The modeling of these branchings has been improved substantially with the measurements of radiative neutron capture cross sections in the mass range of relevance (Käppeler et al. 1991), as well as to detailed s-process calculations in realistic model stars. We note in particular that the estimate in the framework of the proton-mixing scenario of thermally pulsing AGB stars (Goriely & Mowlavi 2000) predict branching effects up to 20% on the ^{187}Os/^{186}Os ratio.

The development of the Re–Os chronology also needs a reliable estimate of the ^{187}Os and ^{187}Re yields from stars in a wide range of masses and metallicities. This evaluation requires not only good-quality modeling of the s-process in order to predict the ^{186}Os/^{187}Os ratio, but also of other ^{187}Os and ^{187}Re transmutation channels in stellar interiors which may affect the ^{187}Re/^{187}Os ratio (astration effects). One of these mechanisms is the possibility of enhanced ^{187}Re β-decay in stellar interiors. This is especially the result of the bound-state β-decay phenomenon which has the dramatic effect of reducing the ^{187}Re half-life from about 42 Gyr for the neutral atom to a mere 32.9 ± 2.0 yr for the completely ionized atom. The experiment that measured this half-life (Bosch et al. 1996) is a significant step forward in the establishment of a reliable Re–Os chronometry. It also put on safer grounds the evaluation of the rate of transformation of ^{187}Os into ^{187}Re that can occur in certain stellar layers through continuum electron captures (Arnould 1972; Takahashi & Yokoi 1983). A further complication arising in the evaluation of the astration effects comes from the fact that neutron captures can modify the ^{187}Re/^{187}Os ratio as well (Yokoi et al. 1983).

Clearly, the yield evaluation also requires the modeling of a variety of stars with different masses and metallicities. Model stars adopted in recent studies of the Re–Os chronology are briefly described by Takahashi (1998). The relevant yields can also be found in his work.

Last but not least, it remains to construct a model for the chemical evolution of the matter from which the Solar System formed. This is the hardest part of all, even in the framework of the simple constrained models already referred to above. In their update of the study of Yokoi et al. (1983), Takahashi (1998) concludes that the ^{187}Re–^{187}Os pair leads to T_{nuc} values in the reasonable 15 ± 4 Gyr range. This result may not be as impressive as one may wish. However, there is good hope that the Re–Os chronometry will improve. Since the work of Yokoi et al. (1983), much progress has been made on the nuclear physics and meteoritic sides, and there is likely much more to come. This optimistic note cannot be expressed for the trans-actinide clocks (Section 2). Of course, more sophisticated Galactic evolution models have to be devised. This is certainly difficult, but this does not defy hope.

4.   Inquietude 2: Thorium in Very Metal-Poor Stars

The recent observations of r-nuclides, including Th, in the metal-poor halo stars CS 22892–052 or HD 115444 (see contributions by Sneden and by Truran, these proceedings) have initiated a flurry of nucleocosmochronological excitement, both observationally and theoretically. The search for r-process-rich very low metallicity stars has been given an impressive boost. This is illustrated by the analysis of CS 31082–001. This star is the first one whose U content (not just an upper limit) has been measured (Cayrel et al. 2001). The abundances of other r-nuclides have also been measured in this same star (Hill et al., these proceedings).

On the theoretical side, it is now commonplace to claim that an ideal nucleocosmochronometer has been found at last, the observed Th in the above-mentioned stars providing a clean way of measuring the ages of these stars, and consequently a reliable lower limit for the age of the Galaxy. In this concerted optimism, some individuals (among whom the authors of this contribution) dare expressing some reservation, however.

4.1. Is the R-Process Pattern Universal?

Almost all astrophysicists dealing with the Th data in very low-metallicity stars claim that the observed patterns of r-nuclide abundances are demonstrably solar, implying a universal abundance distribution. This universality is of essential importance for bringing the observed Th to the status of a chronometer. It is clearly as essential not to take it too easily for granted!

Goriely & Arnould (1997) have tackled this question. They reach the conclusion that the CS 22892–052 r-process composition in the best observationally documented $56 \leq Z \leq 76$ range is remarkably close to the *whole* solar distribution. However, they stress that it is very easy to construct artificial r-nuclide distributions that reproduce almost equaly well the CS 22892–052 data in the Z range mentioned above, while they diverge substantially from the solar data outside of this range (including the actinides). Their so-called random distribution even appears to reproduce relatively well (and by far better than the solar distribution) the observed $Z < 56$ abundance distribution.

From these results, Goriely & Arnould (1997) (see also Goriely & Clerbaux 1999) conclude that *the convergence of the solar, CS 22892–052, and HD 115444 abundance patterns in the $56 \leq Z \leq 76$ range in no way demonstrates the universality of the r-process, without excluding it, however*. In fact, they stress that the r-pattern convergence is to a large extent the signature of the nuclear properties of the r-progenitors of the $56 \leq Z \leq 70$ elements, and does not provide any useful information on the stellar conditions under which the r-process may develop. This conclusion fully applies as well to the recent abundance determinations of $50 \leq Z \leq 70$ elements in 22 metal-poor r-process-rich stars (Johnson & Bolte 2001). In fact, only observations of $A = 90 - 130 - 195$ peak elements can shed light on astrophysics issues concerning the r-process.

In addition, there has been some bad news for proponents of r-process universality from a preliminary analysis of the halo star CS 31082–001, which shows large neutron-capture element enhancements comparable to the ones observed in CS 22892–052 (Hill et al., these proceedings). While a large similarity between the $56 < Z < 70$ element patterns in CS 22892–052, HD 15444, and CS 31082–001 is reported, the abundances in the latter star differ significantly from those of the other two stars in the $Z > 70$ range. This includes Th, the Th/Eu ratio in CS 31082–001 being 2.8 times larger than in CS 22892–052. Hence, under the universality assumption, CS 22892–052 (with [Fe/H] $= -3.1$) is older than CS 31082–001 (with [Fe/H] $= -2.9$) by 21 Gyr and would be about 35 Gyr old (see below). In addition to smashing to bits the idea of the r-process universality, the analysis of CS 31082–001 also reinforces our views that the question of the number of r-processes (two being these days in favor in a substantial part of the nuclear physics community) is of no substantial scientific interest. It would be more useful to unravel the basic mysteries of the r-process before trying to establish the number of r-process events!

4.2.   Would a Universal R-Process Help Th Chronometry?

From this point on, we will *assume* that the r-pattern of abundances is universal. To make things clear, this means nothing more nor less than the following: we assume that *all possible blends of r-process events always lead to the same final abundance pattern*, an event being defined here, as in the canonical r-process model, by the ensemble (T, N_n, t_{irr}) of an assumed constant temperature and neutron concentration during the irradiation time t_{irr}. In direct relation with this statement, we remark that calling for a blend of r-process events is not at all equivalent to considering an ensemble of stars able to produce r-nuclides. As an example, a single supernova is most likely the site of some continuum of r-process events. Furthermore, the precise characteristics of the individual events may which contribute to the universal mix are unknown, as well as the relative level of the contribution of each of the events to the mix. Lastly, it cannot be excluded that different combinations of different individual r-events lead to the same final mix.[1]

Once the r-process universality is taken for granted, the procedure for estimating the age of the Th-bearing metal-poor stars is straightforward, at least in its principle. From the best possible fit to the Solar System r-process abundance distribution, the Th r-process production Th_r can be deduced. The confrontation between this value and the observed abundances leads trivially to the stellar ages. Quite clearly, the consideration of very metal-poor stars allows one to make the economy of a Galactic chemical evolution model. This very nice feature is unfortunately compensated by the nightmare of evaluating Th_r with the high accuracy which is indispensable for building up a chronometry.

As mentioned in Section 2, the r-process remains the most complicated nucleosynthetic process to model from the astrophysics as well as nuclear physics point of view and is subject to large uncertainties that transpire directly into the Th_r predictions. The Th predictions are found to be especially sensitive to the nuclear mass evaluations. As shown by Goriely & Clerbaux (1999), different mass models lead to stellar age differences that can amount to more than 20 Gyr for *r-process models fitting equally well the r-process data for CS 22892–052 and HD 115444 in the* $56 \leq Z \leq 82$ *range*. The reader may wonder why the quoted uncertainty is by far larger than the one classically claimed in the literature. The reason may be summarized as follows: it is a common practice to identify the best nuclear mass models from a confrontation between r-process predictions and the solar data (which are identical to the stellar ones through the assumption of universality). This choice is biased, however. It results from implicit assumptions concerning the characteristic r-process events making up the universal mix. Other assumptions may lead to different best nuclear models. In other words, it is meaningless to select best mass models as long as the detailed characteristics of the suite of r-process events contributing to an assumed universal r-process pattern remain unknown. This is indeed clearly the case as even the proper site(s) for the r-process is (are) not identified yet, with all the proposed scenarios facing serious problems. The astrophysics uncertainty

---

[1] To make an analogy, it is well known in statistical mechanics that each macroscopic state of a complex gas system can be obtained through a variety of different superpositions of the states of the individual particles of the gas.

that most critically affect the reliability of the predictions of the Th production is clearly the unknown maximum strength of the neutron irradiation in an r-process event. Age uncertainties amounting to about 16 Gyr can result (Goriely & Clerbaux 1999).

Finally, one does not have to underestimate the uncertainties that still affect the evaluation of the contribution of the r-process to the Solar System Pb–Bi peak, the predicted Th abundance being directly correlated to this contribution. This situation is responsible for an uncertainty of about 20 Gyr in the Th chronometry (Goriely & Clerbaux 1999).

To conclude, it is our opinion that the Th abundances observed in very low-metallicity stars are of no chronometric virtue even if the universality of the r-process is assumed, which is far from being demonstrated. As an example, uncertainties in the age of CS 22892–052 just originating from the errors in the Th abundance determination amounts to about 3.5 Gyr. A comparable precision of theoretical origin would impose a $15 - 20\%$ level of accuracy in the r-process production of Th. This is just impossible to achieve! In fact, Goriely & Clerbaux (1999) report an age for CS 22892–052 lying in the $7 \lesssim T^*[\text{Gyr}] \lesssim 39$ range! The many sources of uncertainty briefly mentioned above indeed blur the picture substantially, at least if no undue resort is made to too many tooth fairies.

5. Hope 2: Th/U in Very Metal-Poor Stars

As stated elsewhere (Arnould & Takahashi 1999; Goriely & Clerbaux 1999), a way to rescue the Th chronometry discussed in Section 4 would be to have accurate measurements of the Th/U ratio in individual very low-metallicity stars. These nuclides are likely to be produced concomitantly, so that one may hope to predict their production ratios more accurately than the ratio of Th to any other r-element, in particular Eu.

A Th/U ratio has been reported recently for the star CS 31082–001 (Cayrel et al. 2001). This is really good news, even if the situation is not free of observational and theoretical difficulties. The former ones are discussed by Cayrel et al. (2001). On the theoretical side, one still faces the severe question of the universality of the r-process (Section 4.1.). In addition, even if reduced, uncertainties remain in the evaluation of the Th/U production ratio. Goriely & Clerbaux (1999) estimate that it is likely to lie in the $1 \lesssim (\text{Th/U})_r \lesssim 1.3$ range. This result, combined with the observed value $\log\epsilon(\text{Th/U}) = 0.74 \pm 0.15$ leads to an age for the considered star of $14 \pm 3 \pm 2$ Gyr (where the errors correspond to observational and theoretical uncertainties, respectively). Clearly, the Th/U chronometry based on data for individual very metal-poor stars has substantial potential. The detection and analysis of other stars similar to CS 31082–001 would represent a major step forward in this direction.

6. Conclusion

Nucleocosmochronology has the virtue of always being able to provide numbers one can interpret as ages. The real challenge is to evaluate the reliability of these predictions. We estimate that the ^{187}Re–^{187}Os will turn out well as a chronometer based on meteoritic data. We are much more pessimistic concerning the classical ^{232}Th–^{238}U and ^{235}U–^{238}U pairs. On the other hand, we consider that it will remain close to impossible for long to date individual stars in a secure way without the help of precise determinations of the Th/U ratio in these stars.

References

Arnould, M. 1972, A&A, 21, 401

Arnould, M., Meynet, G., & Mowlavi, N. 2000, Chem. Geol., 169, 83

Arnould, M., & Takahashi, K. 1990, in New Windows to the Universe, ed. F. Sanchez & M. Vasquez (Cambridge: Cambridge University Press), 355

Arnould, M., & Takahashi, K. 1999, Rep. Prog. Phys., 62, 393

Arnould, M., Takahashi, K., & Yokoi, K. 1984, A&A, 137, 51

Bosch, F., et al. 1996, Phys. Rev. Lett., 77, 5190

Cayrel, R., et al. 2001, Nature, 409, 691

Clayton, D. D. 1964, ApJ, 139, 637

Faestermann, T. 1998, in Proc. of the 9th Workshop on Nuclear Astrophysics, ed. W. Hillebrandt & E. Müller (Garching: MPA), 172

Fowler, W. A., & Hoyle, F. 1960, Ann. Phys., 10, 280

Goriely, S. 1998, in Nuclei in the Cosmos, ed. N. Prantzos & S. Harissopoulos (Gif-sur-Yvette: Editions Frontières), 314

Goriely, S., & Arnould, M. 1997, A&A, 322, L29

Goriely, S., & Clerbaux, B. 1999, A&A, 342, 881

Goriely, S., & Mowlavi, N. 2000, A&A, 362, 599

Grevesse, N., Noels, A., & Sauval, A. J. 1996, in ASP Conf. Ser. 99, Cosmic Abundances, ed. S. S. Holt & G. Sonneborn (San Francisco: ASP), 117

Johnson, J. A., & Bolte, A. 2001, ApJ, in press

Käppeler, F., Jaag, S., Bao, Z. Y., & Reffo, G. 1991, ApJ, 366, 605

McEllistrem, M. T., Winters, R. R., Hershberger, R. L., Cao, Z., Macklin, R. L., & Hill, N. W. 1989, Phys. Rev. C, 40, 591

Meyer, B. S., & Schramm, D. N. 1986, ApJ, 311, 406

Takahashi, K. 1998, in AIP Conf. Proc. 425, Tours Symp. on Nuclear Physics III, ed. M. Arnould et al. (New York: AIP), 616

Takahashi, K., & Yokoi, K. 1983, Nucl. Phys., 404, 578

Yokoi, K., Takahashi, K., & Arnould, M. 1983, A&A, 117, 65

Acknowledgments. SG is a FNRS Research Associate.

Discussion

Michelle Mizuno-Wiedner: Could you say a few words regarding how the Re radioactive lifetime changes under different conditions?

Marcel Arnould: Oh yeah, that's a very interesting question, which I had no time to talk about. In fact, the Re lifetime changes may be dramatic indeed in stellar conditions. For the neutral atom, the half-life is around 42 Gyr. Now, for a fully-stripped ion, the lifetime is close to 33 yr (not Gyr, yr!). And this now has been measured at the storage ring of GSI Darmstadt in Germany (Bosch et al. 1996). I think it's a really great experiment. I have to say that, many years ago, there was a pre-experimental, theoretical evaluation of 14 yr for the stripped Re half-life (Takahashi & Yokoi 1983, Nucl. Phys., 404, 578). This is not bad at all. From the Darmstadt measurement, the situation concerning half-lives is more or less under control. Of course, we have to know what is the ionization stage within each layer of the star, this stage having a large influence on the half-lives. This brings some difficulty. But all in all, when you look at uncertainties coming from this, it's not that big. This is anyway not the biggest source of uncertainty there.

Astrophysical Ages and Time Scales
ASP Conference Series, Vol. 245, 2001
T. von Hippel, C. Simpson, N. Manset

The Oldest Stars: Diffusion and the Spite Plateau

A. Weiss

Max-Planck-Institut für Astrophysik, Karl-Schwarzschild-Strasse 1, 85748 Garching, Germany

M. Salaris

Astrophysics Research Institute, Liverpool John Moores University, Twelve Quays House, Egerton Wharf, Birkenhead CH41 1LD, UK

Abstract. We discuss the effect of diffusion on the surface lithium abundance in low-mass metal-poor stars and show that the constant Li abundance found observationally in the Spite-plateau stars can be reproduced by our models under realistic assumptions concerning evolutionary state and age. We discuss the consequence for the inferred primordial Li abundance and the baryon density of the universe.

1. Introduction

The Sun has taught us that microscopic diffusion is a fundamental physical mechanism, which has to be considered for accurate stellar modeling. In the case of the solar model, only models including diffusion (i.e., sedimentation) are able to agree with the seismically inferred sound speed profile to better than one percent, reproduce the depth of the convective envelope correctly and— starting out from $Y_S(t = 0) = 0.275$—lead to a reduced surface helium content of $Y_S(t = t_\odot) = 0.245$, just as determined by helioseismology. Since diffusion is an intrinsically slow process operating over cosmic times, long-lived low-mass stars, in particular those of Population II, are the evidently best suited candidates to look for further consequences. Atomic diffusion affects the evolution of stars in the deep interior and in the envelope.

In the deep interior, the sinking of helium towards the core leads to an effectively faster nuclear aging of the star, thus reducing its main sequence (MS) lifetime. For age determinations of globular clusters making use of the absolute brightness of the colour–magnitude diagram (CMD) turnoff (TO), diffusion eventually leads to lower ages; the effect accounts for a reduction of the order of 1 Gyr for old globular clusters.

In the stellar envelope diffusion leads to a depletion of heavy elements and helium. This has the direct consequence that the present abundance of metals is lower than the initial one. In Salaris, Groenewegen, & Weiss (2000) and in the poster by M. Salaris (these proceedings) we discuss in detail the consequences for the MS fitting method, the resulting distances, and subdwarf ages.

A general prediction of models including atomic diffusion is that very metal-poor low-mass stars, with their high effective temperatures and thus very shallow convective envelopes, are readily depleted of metals over the MS lifetime (Deliyannis, Demarque, & Kawaler 1990); this prediction can be contrasted with observations. In this contribution, we will report on a new confrontation of theoretical models with the observational result that the ^{7}Li abundance in such stars of $T_{\mathrm{eff}} \gtrsim$ 5900 K forms a well-defined plateau below [Fe/H] ~ -1.5 (Spite & Spite 1982). In the past, this so-called Li- or Spite-plateau has been used repeatedly as the strongest argument against the presence of diffusion, because the degree of depletion by diffusion increases with effective temperature. Stated differently, the necessity for an additional effect counteracting diffusion was claimed (Chaboyer, Demarque, & Pinsonneault 1995; Vauclair 1999). Candidate processes are rotationally induced mixing (Vauclair & Charbonnel 1998) or a rather strong mass loss by a stellar wind (Vauclair & Charbonnel 1995).

The question whether atomic diffusion—essentially gravitational settling— is operating in the oldest stars is one remaining systematic uncertainty in the determination of ages of the oldest Galactic objects.

2. Lithium in Stars

2.1. Destruction and Production

Lithium is the only element which is produced in noticeable amounts during Big Bang Nucleosyntheses (^{7}Li/H $\sim 10^{-10}$–10^{-9}) *and* which is measurable readily in cool stars. However, it is also very fragile and easily destroyed in stellar interiors at $T \geq 2.5 \times 10^6$ K by ^{7}Li$(p, \alpha)^4$He reactions. Such temperatures are reached at the bottom of the convective envelopes of low-mass pre-main sequence stars, below the convective zone in main-sequence stars, and, of course, in the deeper radiative interior of all stars. Later dredge-up events on the Red Giant Branch makes this visible.

On the other hand, ^{7}Li is also produced in the pp-chain, and, if mixed fast enough to cool regions (Fowler-Cameron mechanism) can survive and enrich the envelopes of AGB-stars. In addition, it is created in spallation reactions by cosmic rays. All this implies that a variety of processes lead to a net Galactic chemical evolution of ^{7}Li.

2.2. The Lithium- or Spite-Plateau

It is a well established observational fact that in hot low-metallicity subdwarfs the ^{7}Li/H-ratio is constant, both as a function of metallicity ([Fe/H]) or effective temperature T_{eff}. Figure 1 from a preprint by S. G. Ryan (2000) illustrates the obvious plateau at low metallicities, and Figure 2 (Ryan, Norris, & Beers 1999) the small spread around it.

3. Stellar Models, Diffusion, and Lithium

3.1. Calculations and Models

While stellar models *without* diffusion predict a constant lithium abundance for stars as observed on the Spite-plateau, those including diffusion result in a decline of Li at the highest temperatures, where the convective envelopes get very

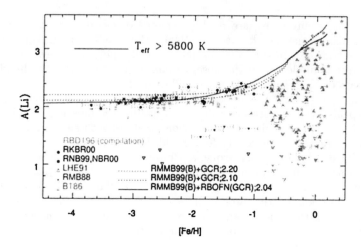

Figure 1.   Abundances of ^{7}Li in low-mass field stars, an illustration of the Spite-plateau visible at the low-metallicity end. The figure represents a collection of various data and some models for the increase in Li due to Galactic Cosmic Ray spallation (lines). In high metallicity stars the various complicating and additional processes lead to a high dispersion. This figure was taken from Ryan (2000).

thin, such that diffusion, operating at the bottom of them, leads to a noticeable depletion even at the surface. At lower temperatures, the convective zones are more extended and the reservoir of lithium is larger. Since no obvious downturn of the Li-abundance at the hot end is observed, this is taken as *the* evidence against diffusion operating in such stars (Vauclair & Charbonnel 1998).

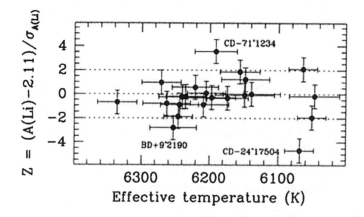

Figure 2.   Spite-plateau as a function of T_{eff} for metal-poor stars, illustrating the small spread around the mean value of [Li] = 2.11 (from Ryan et al. 1999).

We have made new calculations of the evolution of low-mass metal-poor stars. The code and input physics are the same as already used for determining globular cluster ages (see Salaris & Weiss 1998), which resulted in values of 11–12 Gyr for the oldest clusters. For the lithium problem we have taken into account diffusion as in Salaris et al. (2000), where the diffusive speed for lithium was assumed to be the same as that for helium. This assumption was dropped for the present investigation and the diffusive velocity of Li calculated explicitly. Additionally, we have calculated a few pre-main sequence tracks to add the effect of Li-destruction for the lowest masses.

Figure 3 shows how the Li abundance varies along isochrones of cosmological ages: the lower masses have been depleted already on the pre-main sequence, while the downturn at the hot end (the higher masses) is due to diffusion only. However, also the maximum of the isochrone is below the primordial value of [Li] = 2.5 used in these calculations. Nevertheless, the larger parts of the curves are fairly flat. The metals (and thus Li) diffused out of the surface convective layer have accumulated just beneath it. So far, our results are in close agreement with those of previous works on this subject.

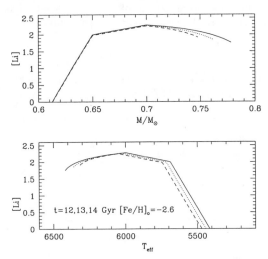

Figure 3.   Lithium abundance along isochrones (12–14 Gyr) as a function of stellar mass (*top*) or T_{eff} (*bottom*). The initial metallicity was set to [Fe/H] = −2.6. Note the declining Li-abundance at the hottest temperatures (highest mass).

3.2.   Reproducing the Spite-Plateau

Setting out from isochrones such as those shown in Figure 3, we simulated observed Li-samples by the following procedure: 1) assume a Salpeter initial mass function, 2) assume an age distribution, 3) draw a sample of 10 000 stars, 4) set an initial [Fe/H]-distribution, such that this sample shows the observed metallicity distribution, 5) from the 10 000-stars sample select randomly an *observed* sample of typical size (120 stars) and [Fe/H] < −2.0; we selected 30 such sam-

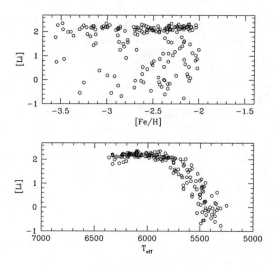

Figure 4.   Simulated observed sample of old low-metallicity stars (see text for the procedure). The panels show [Li] as a function of metallicity and T_{eff}.

ples in total, and 6) for each such star selected, add typical errors in [Fe/H] (0.20 dex), T_{eff} (60 K), and [Li] (0.07 dex).

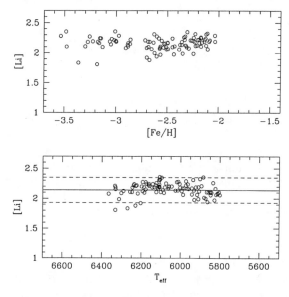

Figure 5.   Simulated Spite-plateau in the presence of diffusion. Note the few stars below the plateau. The lower panel includes both the average lithium abundance of about 30 realizations and its variance ($\langle[\mathrm{Li}]\rangle = 2.14 \pm 0.07$), and the 3σ-range (dashed lines).

The result of such a simulation is shown in Figure 4: in the upper panel, plateau and depleted stars appear for all metallicities. In the lower one, it is apparent that most depleted stars have lower mass and the depletion is that of the pre-MS. About 60 stars are hotter than 5800 K and are used to define the Lithium-plateau, which is repeated in Figure 5. Only a handful of stars is actually below the plateau level—as indeed is the case in the observational samples! An important restriction of our simulation is that the age assumed is between 13 and 14 Gyr (uniformly distributed). If we assume lower ages, the decline in [Li] at the hottest end is much more visible. Altogether, the spread in age, the consideration of observational errors, and most importantly, the higher evolutionary speed close to the turnoff, where the depletion due to diffusion finally shows up, lead to the plateau-like feature. We add that we also find a small gradient in [Li] as function of metallicity: $\triangle[\text{Li}]/\triangle[\text{Fe/H}] = 0.05$ to be compared with 0.12 found by Ryan et al. (1999), and explained by them by Galactochemical evolution.

3.3.   A Consistency Test

Grundahl (1999) has demonstrated that, in Strömgren colours, stars in the metal-poor cluster M 92 overlap with the metal-poor objects ($[\text{Fe/H}] \leq -2.5$) in the sample by Schuster et al. (1996). We have done the same comparison using the simulated sample of field stars of age 13.5–14 Gyr and a simulated cluster of 12 Gyr and $[\text{Fe/H}] = -2.3$ for M 92 (Salaris & Weiss 1998). In this colour–colour diagram (Figure 6), the turnoff (at $c_0 \sim 0.32$ and $(b-v) \sim 0.34$) is most sensitive to age. Evidently, the two—completely independently obtained—age determinations agree very well.

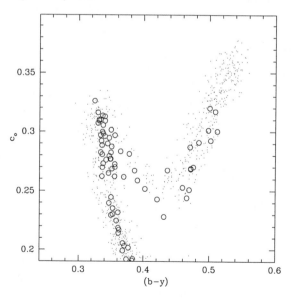

Figure 6.   Comparison between models for metal-poor low-mass field stars of 13.5–14 Gyr (circles) and M 92 of 12 Gyr (dots).

4.   Conclusions

4.1.   Our Result

We have shown that with present-day observational sample sizes and taking into account observational errors, stellar models with diffusion are able to reproduce the Spite-plateau with a large agreement with the observations. The effect of diffusion can therefore not be excluded, and given evidence from the solar model or from HIPPARCOS subdwarfs (see Salaris et al. 2000 and Salaris & Weiss, these proceedings), diffusion seems to be likely to act in old low-mass stars. This would have the consequence of lower ages for globular clusters and field stars!

4.2.   Cosmological Implications

Apart from lowering age determinations for the oldest Galactic objects, our models also allow us to infer the primordial lithium abundance. To reproduce the observed Li-level of the Spite-plateau, we had to set this to [Li] = 2.5 (diffusion leads to a depletion of 0.35 dex). BBN results by Vangioni-Flam, Coc, & Cassé (2000) show that such a high primordial Li-abundance agrees very well with low-deuterium determinations of QSO absorption lines and a high ^{4}He as found by Izotov & Thuan (1998). The corresponding value of η would be 5.1×10^{-10} and (assuming a Hubble constant of $H_0 = 71$ km s^{-1} Mpc^{-1}) a baryon density of $\Omega_B = 0.037$, completely consistent with the value inferred from the BOOMERANG and MAXIMA-1 experiments.

4.3.   A Prediction

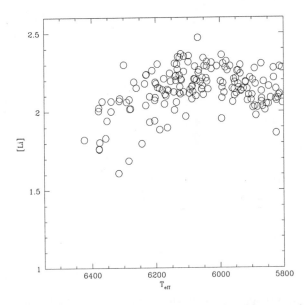

Figure 7.   Simulated sample with 150 stars in the plateau region. The decline of [Li] at the TO, which is due to diffusion, shows up clearly.

   The main reason for the Li-depletion at the TO being hidden in the samples is the fast evolution in this phase. Correspondingly, in larger samples it should be more prominent. Figure 7 displays a simulated sample with 150 stars on the plateau. The effect of diffusion is now visible. Extended observations will thus be able to decide on the question whether diffusion is operating in these stars.

References

Chaboyer, B., Demarque, P., & Pinsonneault, M. H. 1995, ApJ, 441, 865

Deliyannis, C. P., Demarque, P., & Kawaler, S. D. 1990, ApJS, 73, 21

Grundahl, F. 1999, in ASP Conf. Ser. 192, Spectrophotometric Dating of Stars and Galaxies, ed. I. Hubeny, S. Heap, & R. Cornett (San Francisco: ASP), 223

Izotov, Y. I., & Thuan, T. X. 1998, ApJ, 500, 188

Ryan, S. G. 2000, in IAU Symp. 198, The Light Elements and Their Evolution, ed. L. da Silva, M. Spite, & R. de Medeiros (San Francisco: ASP), 249

Ryan, S. G, Norris, J. E., & Beers, T. C. 2000, ApJ, 523, 654

Salaris, M., Groenewegen, M. A. T., & Weiss, A. 2000, A&A, 355, 299

Salaris, M., & Weiss, A. 1998, A&A, 335, 943

Schuster, W. J., Nissen, P. E., Parrao, L., Beers, T. C., & Overgaard L. P. 1996, A&AS, 117, 317

Spite, F., & Spite, M. 1982, A&A, 115, 357

Vangioni-Flam, E., Coc, A., & Cassé, M. 2000, A&A, 360, 15

Vauclair, S. 1999, A&A, 351, 973

Vauclair, S., & Charbonnel, C. 1995, A&A, 295, 715

Vauclair, S., & Charbonnel, C. 1998, ApJ, 502, 372

Discussion

Michael Turner: I like your final conclusion. I'd like to shape it a bit. If you take the deuterium-predicted baryon density, $3 \pm 0.3 \times 10^{-5}$, and ask what that means for the lithium, it means a Li value of about 3.5×10^{-10}, which is exactly what you are getting. Your number for Li is very consistent with Big Bang Nucleosynthesis, though it is a little low for BOOMERANG. The error bars on the microwave background determination of the baryon density are still pretty big, however, and one should walk away with the impression that the glass is half full, that the microwave background is saying we have a low baryon density, nowhere near the matter density. Whether or not they in detail agree we'll find out as more microwave background data come in. Your models predict a depletion of a factor of two?

Achim Weiss: Yes.

Michael Turner: OK, that's very very welcome.

Achim Weiss: Note added after conference: The latest BOOMERANG analysis (Netterfield et al. 2001, astro-ph/0104460) results in $\Omega_B h^2 = 0.022 \pm 0.003$; our Li-based value is 0.0187, which is within 1σ, and now consistent with the former result.

Astrophysical Ages and Time Scales
ASP Conference Series, Vol. 245, 2001
T. von Hippel, C. Simpson, N. Manset

How Good Are New Ages for Old Stars?

Bengt Gustafsson, Michelle Mizuno-Wiedner

Uppsala Astronomical Observatory, Box 515, S-751 20 Uppsala, Sweden

Abstract. The accuracy obtainable with the best dating methods for old field stars is hardly better than 20% for any of the current methods. For a number of problems in the study of the evolution of the Galactic halo and the disk, a reduction of the errors to below 10%, at least in relative ages, would be very rewarding. We discuss the possibilities to achieve this improvement within the foreseeable future.

1. Introduction

Astonishingly little is known about the sequence of events that led to the formation of our Galaxy.

- Is there a real spread in the ages of the halo stars?

- Is the Galactic halo considerably older than the disk?

- Is there an age difference between the metal-rich stars of the halo and those of similar metallicity in the thick disk?

- Is there an age spread for the stars in the thick disk?

- Is the thick disk significantly older than the thin disk?

Although some of these questions are being answered (e.g., as presented in these proceedings in studies of globular clusters by Chaboyer, in kinematic studies of field stars by Beers & Chiba, and in chemical analyses of disk stars by Fuhrmann), most of the answers are still tentative and there remain fundamental uncertainties concerning the early history of our Galaxy. To what extent do we see the remains of one large system that more or less gradually relaxed into a disk while continuously producing stars, and to what extent do we see the results of events like random mergers with smaller or larger galaxies and late-infalling giant gas clouds? (Cf. e.g., the beautiful simulations by Barnes, these proceedings.)

In order to answer such questions, theoretical modeling of the early evolutionary phases of our Galaxy does not suffice. Since the complexity of the Galactic system is high, and probably beyond the realm of contemporary realistic quantitative modeling, empirical data are absolutely crucial. The chemical composition, orbits, and ages of individual stars and star clusters are of particular significance.

The chemical compositions of a vast number of field stars may now be obtained with a high accuracy for many of the elements, thanks to the high-resolution spectrometers at the largest telescopes such as HIRES at the Keck I telescope and UVES at the VLT. Galactic orbits are also known for many stars thanks to the HIPPARCOS parallaxes and proper motions, and to impressive radial-velocity surveys. In work aiming to reconstruct the early structure of the Galaxy, one problem is the evolution of the orbits due to the changes in the gravitational potential. Yet these changes contain important dynamical information and are therefore interesting to study in themselves. The stellar ages, however, are rarely known to an accuracy better than 20%, and this is a major hindrance to the exploration of early Galactic history.

In the present review we shall discuss what accuracies are usually obtained, and may be expected in the near future, for individual ages of old field stars and in particular Population II stars (cf. Chaboyer, these proceedings, on dating globular clusters). One of us recently summarized the different methods for spectroscopic and photometric age determinations for field stars (Gustafsson 1999). An interesting comparison between results from different methods, also when applied to younger stars, was published by Lachaume et al. (1999). The present paper is an update of the former review, and mainly concentrates on the isochrone method. Remarks on the stellar oscillation method will also be made, and we shall begin with some comments on radioactive dating.

2. Radioactive Dating

The recent detection of uranium in the halo star CS 31082–001 (Cayrel et al. 2001, and in these proceedings) has stirred renewed interest in the use of radioactive nuclei, and in particular the ^{232}Th–^{238}U chronometer pair, to estimate stellar ages. The method is attractive because when applied to old stars, it delivers a lower limit for the age of the Galaxy that is independent of models of stellar and galactic chemical evolution (Schramm 1974).

In determining how much of a radioactive element has decayed, the main question is of course how much of that element was present initially. It is possible to estimate this empirically when the stable neutron-capture elements show the same relative abundances as the r-process distribution in the solar photosphere and meteorites; the assumed initial abundance for a chronometer can then be obtained by scaling the element's solar r-process abundance to that of the star and adjusting for radioactive decay over the lifetime of the Solar System. Although this pattern seems to hold very well for the heavy ($Z \geq 56$) neutron-capture elements in a number of metal-poor stars (e.g., Sneden et al. 1996; Westin et al. 2000; Burris et al. 2000; Johnson & Bolte 2001), age estimates that rest entirely on this assumption must be viewed with caution—more so now that Hill et al. (these proceedings) have shown that the pattern fails for several $Z > 70$ elements in CS 31082–001. Even in cases where r-process scaling seems to be valid, uncertainties will arise not only from the line-to-line scatter for each abundance but also from the observed abundance scatter around the Solar System pattern (cf. the different ages obtained by Hannawald et al., these proceedings) and, to a lesser extent, from uncertainties in the observed Solar System distribution itself.

The alternative is to rely instead on Th/U production ratios from nuclear physics models, but the models themselves are often judged by their ability to reproduce Solar System abundances and, more problematically, are still yielding quite disparate predictions. Finally, even if the initial abundance is thought to be known, determinations of the current chronometer abundance will still depend on the quality of available observations, stellar atmosphere models, and atomic transition probabilities.

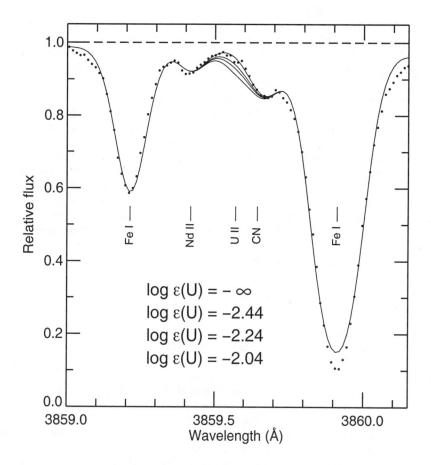

Figure 1.   Observed spectrum of CS 22892–052 (dots) compared to synthetic spectra with different uranium abundances.

Thus the advantage of astronomically model-independent age limits from radioactive dating is somewhat offset by the method's strong sensitivity to both the initial and currently measured chronometer abundances. In light of the thorough discussions already presented in these proceedings (Truran; Sneden & Cowan; Arnould & Goriely), we will confine ourselves to illustrating some of these problems with the example of the metal-poor star CS 22892–052.

Figure 1 shows an observed spectrum of this halo star (V. Hill 2000, private communication) based on public data from the UVES Commissioning observations at the VLT Kueyen Telescope. The comparison between observed and calculated spectra is obviously dependent on the true location of the stellar continuum, which is sensitive to the possible presence of many weak lines, e.g., from molecules like CN in the 3860 Å region. Also, the presence of unknown blending lines in the U II feature is a disturbing possibility. Systematic efforts in molecular and atomic physics to trace and measure such features in the U and Th line regions are of great significance.

Our preliminary analysis (ESO press release 08/00) had estimated a conservative upper limit for the uranium abundance to be $\log \epsilon(U) = -2.5$, assuming a $\log gf$ of -0.105 for the U II transition at 3859.57 Å. Since then, Lundberg et al. (2001) have measured new laboratory lifetimes of U II to get an improved value, $\log gf = -0.167$ (they also review previous gf determinations). The Lund group plans to assign strict error bars to this value soon.

For the production ratio of $\log(\mathrm{Th}/\mathrm{U}) = 0.26$ from Pfeiffer, Kratz, & Thielemann (1997), the thorium abundance $\log \epsilon(\mathrm{Th}) = -1.66$ along with the old limit of $\log \epsilon(U) = -2.5$ gave a minimum age of 12.8 Gyr for the star. The new gf value raises the previous upper limit to $\log \epsilon(U) = -2.44$ for the uranium abundance (seen as the uppermost curve in Figure 1) and reduces the minimum age by almost 1.4 Gyr. Combined with recent production ratios of $\log(\mathrm{Th}/\mathrm{U}) = 0.37$ or 0.16 (Toenjes et al., these proceedings), that abundance limit translates into a minimum age of 9.0 Gyr or 13.5 Gyr, depending on the nuclear mass model.

While work is progressing on atomic data and nuclear models, one should seek a much larger sample of metal-poor, r-process enriched stars (as proposed by Christlieb et al., these proceedings).

3.   Isochrone Ages

Comparison between observed stars and calculated isochrones in the Hertzsprung–Russell (or some related) diagram can be used for age determinations. This method is particularly well suited for evolved stars that are still in the main-sequence band or in the subgiant region. But how accurate are the isochrones? For example, how should one consider the effects of gravitational settling and of mixing processes? What are the consequences of the primitive treatment of convection? And how accurate are the transformations from observed quantities such as colors or the $H\beta$ index, and parallaxes or surface gravity measures with errors in them, to effective temperature and absolute magnitude, respectively? What are the effects of errors in the estimated chemical composition of the star, which would lead to comparisons with improper models?

For old solar-type stars one can give approximate answers to these questions. The effects of uncertainties in gravitational settling of helium and heavier elements toward the stellar center may be estimated to be 1 Gyr or considerably more in age (Salaris, Groenewegen, & Weiss 2000). The phenomenon is, however, complicated, leading both to effects due to the chemical inhomogeneity in the stellar structure (with increased mean molecular weight in the center which decreases the effective temperature of hydrogen-burning stars, and the age) as well as effects due to comparisons with models with systematically too low metal abundances (since those observed in the atmosphere are not representative). These phenomena partly counteract each other, and further work is needed to explore them.

The problematic situation is illustrated by the systematic difference between observed HIPPARCOS subdwarfs and standard model isochrones found by Cayrel et al. (1997) and Nissen, Høg, & Schuster (1997). This was explained as a result of a combination of gravitational settling and departures from LTE in neutral iron, leading to errors in the metallicities ascribed to the stars (Lebreton 2000, and references therein). However, VandenBerg et al. (2000) and Bergbusch & VandenBerg (2001) did not find such large differences between observed stars and isochrones, which could be because they used different estimates of metal abundance. Also, note the results of Gratton et al. (2001) that turnoff stars and stars on the lower red giant branch in the globular cluster NGC 6397 do not show any detectable differences in metallicities. These results, and the observation of Li in turnoff stars, suggest that diffusion is inhibited in the surface layers of Population II dwarfs (D. A. VandenBerg, private communication), although settling of He probably occurs in the stellar interior.

Furthermore, there are uncertainties in convection, as modeled with variations in mixing length in the interval $1 < l/H_p < 3$, and corresponding to a typical spread in age of ± 1 Gyr (Castellani, Degl'Innocenti, & Marconi 1999).

As regards the transformation of observed quantities to the basic model parameters of effective temperature, luminosity and initial metallicity, errors in $T_{\rm eff}$ of 150 K (which may still occur, e.g., for Population II stars), correspond to errors in age of about 0.5–1.5 Gyr, dependent on whether the star is in the main sequence band or on the low subgiant branch. In the latter case the lower age error will apply. Errors in metallicity of 0.2 dex will again correspond to an error of about 1 Gyr in age, as will errors of 0.3 dex in the abundances of the alpha elements (O, Ne, Mg, Si, etc.) relative to iron (VandenBerg & Bell 2001). Basically, increasing alpha-element abundances will increase the opacity, which will decrease both the effective temperature and the luminosity of models. The CNO burning rate will increase, and the estimated age will decrease. Relative errors in parallaxes of 10%, an accuracy which has only been obtained for about 20 Population II stars with HIPPARCOS, correspond to an error in age of about 2 Gyr. If spectroscopic gravities are instead estimated, to an accuracy of 0.1 dex which is not very easily achieved (see e.g., figure 2 in Fuhrmann 1998), and comparison is made with isochrones in the $T_{\rm eff}$–$\log g$ diagram, this again results in errors in age of about 2 Gyr.

The situation is summarized in Table 1. The figures given there are very schematic, e.g., the errors are rather different depending on whether the star is located on the main sequence, at the turnoff point or on the subgiant branch. However, in most cases we have to accept errors in isochrone ages of 3–4 Gyr for old solar-type stars, with considerably greater errors in particular cases.

Can we hope that this situation may be improved in the next few years? The answer is yes, as regards the model problems. The mixing phenomena may be better clarified (for recent attempts, see Weiss & Salaris and Salaris & Weiss, these proceedings). Empirical studies of globular clusters will be helpful (cf. Gratton et al. 2001). Also the uncertainties in convection should be possible to decrease, using more detailed hydrodynamical modeling and comparisons

Table 1.   Sample error budget for ages from isochrones.

Effect	Age Uncertainty (Gyr)
Gravitational settling of heavier elements	$\gtrsim 1$
Inadequate treatment of convection	1
Transformation between observed and model quantities	0.5–1.5
Metallicity, 0.2 dex error	1
[α/Fe], 0.3 dex error	1
Parallax, 10% error	2
log g, 0.1 dex error	2

with cluster data. Uncertainties in effective temperature, metallicity and relative abundances should be possible to diminish to less than half the present values by comparing accurate observations with detailed model atmospheres (e.g., 3D hydrodynamic modeling of subdwarf atmospheres by Asplund & García Pérez 2001; work in preparation by P. S. Barklem on more accurate temperature calibrations of hydrogen line profiles, including hydrogen self-broadening from Barklem, Piskunov, & O'Mara 2000; and the detailed spectroscopy presented in these proceedings by Fuhrmann and by Korn & Gehren). The development is promising in all these areas, and in several of them we already have methods to reduce the corresponding errors in age by a factor of 2. We cannot expect to do much better for Population II stars, however, as long as one major obstacle remains, namely the errors of 10% or more in the parallaxes. The breakthrough may not come until the GAIA satellite's planned launch no later than 2012. However, even before then, the DIVA satellite (planned for 2004) will measure millions of parallaxes, and SIM (planned for 2009) will do so for stars of special interest.

It should finally be noted that a number of the questions listed in the Introduction may be successfully attacked even without a complete understanding of mixing, settling and convection, using strictly differential approaches in comparisons between e.g., halo and thick disk stars of similar effective temperatures and metallicities. If such a program were carried out for stars below the turnoff point, where the isochrones are rather vertical in the HR diagram and the age estimates therefore not very distance dependent, one might hope to achieve relative ages of an accuracy better than 10%.

4.   Stellar Oscillation Ages

Christensen-Dalsgaard (1984), Ulrich (1986), and Gough (1987) were pioneers in exploring the methods to date stars from frequency differences in stellar oscillations, with comparison between modes with low ℓ numbers. These frequency differences measure the mean molecular weight in the stellar center, which is age-dependent due to the gradual conversion of hydrogen to helium. When the method is applied carefully to the Sun it seems that accuracies of about 2% may now be reachable (see the detailed discussion by Gough, these proceedings; also Dziembowski et al. 1999). No oscillation ages have been determined for the two stars with solar-type oscillations discovered until now, the subgiants η Boo (Christensen-Dalsgaard, Bedding, & Kjeldsen 1995) and β Hyi (Bedding et al. 2001), but the frequency patterns at least do not seem inconsistent with their isochronic ages.

Before this method can be applied to a fair sample of Population II dwarfs, however, the means to observe the oscillations for relatively faint stars must be improved greatly. A few Population II stars could possibly be observed with the MONS mission scheduled for 2003; but more extensive exploration will have to await missions like the Eddington space telescope (Giménez & Favata, these proceedings), which could be launched sometime between 2008 and 2013 if implemented by ESA.

However, even with a positive detection of the low-ℓ modes for a Population II star, the interpretation will depend even more strongly than isochrones on the details of mixing and settling, because these phenomena significantly affect the mean molecular weight in the stellar interior. Thus, both phenomena must be explored in some detail, theoretically and empirically, before accurate ages can result. An interesting possibility may be to combine stellar-oscillation data with isochrone comparisons, since mixing and settling effects drive isochrone ages in the opposite direction as they do oscillation ages.

5. Conclusions

In spite of the very interesting discovery of uranium in CS 31082–001 by Cayrel et al. and the possibility of finding more r-element enhanced Population II stars, the use of radioactive dating for Population II stars in general is limited. Also, before nuclear and atomic data (and molecular data for stars with significant blends of molecular bands) have been improved considerably, the errors in radioactive ages will be of the same order of magnitude as the present ones from the isochrone method. The most problematic of the sources of error in isochrone ages for Population II stars is the distance uncertainty. When the GAIA mission has been completed, this method will most probably be the most accurate one. It does require cumbersome efforts on stellar model atmospheres as well as on mixing and settling in stellar interiors, but this work, supported with adequate observations, is in no way impossible today, and will be of wide astrophysical application. The possibility to reach quite accurate results in strictly differential applications already today should be noted.

Isochrone ages will be supported and checked with accurate radioactive dating for a number of r-element rich stars as soon as the nuclear and atomic data have been improved. This method will have the virtue of being relatively model independent as soon as the relative production ratios are well known. The future use of stellar oscillation ages will depend on which modes can be detected and how well mixing and settling in the stellar interior will be understood.

References

Asplund, M., & García Pérez, A. E. 2001, A&A, in press (astro-ph/0104071)

Barklem, P. S., Piskunov, N., & O'Mara, B. J. 2000, A&A, 363, 1091

Bedding, T. R., et al. 2001, ApJ, 549, L105

Bergbusch, P. A., & VandenBerg, D. A. 2001, ApJ, in press

Burris, D. L., Pilachowski, C. A., Armandroff, T. E., Sneden, C., Cowan, J. J., & Roe, H. 2000, ApJ, 544, 302

Castellani, V., Degl'Innocenti, S., & Marconi, M. 1999, MNRAS, 303, 265

Cayrel, R., Lebreton, Y., Perrin, M. -N., & Turon, C. 1997, in Hipparcos, Venice '97, ed. B. Battrick, ESA SP-402, 219

Cayrel, R., et al. 2001, Nature, 409, 691

Christensen-Dalsgaard, J. 1984, in Space Research in Stellar Activity and Variability, ed. A. Mangeney & F. Praderie (Meudon: Obs. de Paris), 11

Christensen-Dalsgaard, J., Bedding, T. R., & Kjeldsen, H. 1995, ApJ, 443, 29

Dziembowski, W. A., Fiorentini, G., Ricci, B., & Sienkiewicz, R. 1999, A&A, 343, 990

Fuhrmann, K. 1998, A&A, 330, 626

Gough, D. 1987, Nature, 326, 257

Gratton, R. G., et al. 2001, A&A, 369, 87

Gustafsson, B. 1999, in ASP Conf. Ser. 192, Spectrophotometric Dating of Stars and Galaxies, ed. I. Hubeny, S. R. Heap, & R. H. Cornett, (San Francisco: ASP), 91

Johnson, J. A., & Bolte, M. 2001, ApJ, in press (astro-ph/0103299)

Lachaume, R., Dominik, C., Lanz, T., & Habing, H. J. 1999, A&A, 348, 897

Lebreton, Y. 2000, ARA&A, 38, 35

Lundberg, H., Johansson, S., Nilsson, H., & Zhang, Z. 2001, A&A, in press

Nissen, P. E., Høg, E., & Schuster, W. J. 1997, in Hipparcos, Venice '97, ed. B. Battrick, ESA SP-402, 225

Pfeiffer, B., Kratz, K.-L., & Thielemann, F.-K. 1997, Z. Phys. A, 357, 235

Salaris, M., Groenewegen, A. A. T., & Weiss, A. 2000, A&A, 355, 299

Schramm, D. N. 1974, ARA&A, 12, 383

Sneden, C., McWilliam, A., Preston, G. W., Cowan, J. J., Burris, D. L., & Armosky, B. J. 1996, ApJ, 467, 819

Ulrich, R. K. 1986, ApJ, 306, L37

VandenBerg, D. A., & Bell, R. A. 2001, New Astron. Rev., in press

VandenBerg, D. A., Swensson, F. J., Rogers, F. J., Iglesias, C. A., & Alexander, D. R. 2000, ApJ, 532, 430

Westin, J., Sneden, C., Gustafsson, B., & Cowan, J. J. 2000, ApJ, 530, 783

Acknowledgments. We would like to thank M. Asplund, P. Barklem, R. A. Bell, G. Bertelli, R. Cayrel, J. Christensen-Dalsgaard, B. Edvardsson, S. Johansson, Y. Lebreton, P. E. Nissen, N. Piskunov, C. Sneden, and D. VandenBerg for helpful comments. MM is grateful to the conference organizers for travel assistance.

Discussion

Jim Truran: In the Edvardsson paper (1993, A&A, 275, 101) there were stars that were referred to as disk stars with ages of maybe 12 Gyr. Are these ages reliable? Would stars in that age range of your sample have much higher velocity dispersions than those at lower ages? Are these thick disk stars?

Bengt Gustafsson: Basically, yes, but one should note that the absolute age calibration in the Edvardsson et al. (1993) paper is still rather shaky. So there's a factor which is unknown. On the other hand, I think some of these stars, the ones at [Fe/H] = −0.5 or so, are quite old in that sample. There is absolutely a very old contribution of such stars in the Galaxy.

George Isaak: I believe you are pessimistic about seismology. Frequencies have no systematic errors, so eigenfrequencies will become extremely accurate with continued observations. If you incorporate the parameters from increasingly accurate observations, and since they pertain to the very interior of the star, to hydrogen-helium aeon glasses, I believe seismology will turn out to be much more immune to theoretical uncertainties than all other methods.

Bengt Gustafsson: The problem I pointed out was simply the dependence on the settling or sedimentation and mixing theory in the interpretation of these possible data in terms of age. That was my worry.

George Isaak: But in the case of the Sun we've accurately calibrated them, i.e., no H mixing and gravitational settling of He and metals.

Bengt Gustafsson: Yes, and that's a great achievement. I owe you a bottle of wine when you get an age for single very metal-poor stars to an accuracy of less than 10% from studying them.

George Isaak: Good, I'll take you up on that.

Astrophysical Ages and Time Scales
ASP Conference Series, Vol. 245, 2001
T. von Hippel, C. Simpson, N. Manset

The Dynamical Relaxation Time Scale of the Milky Way

E. Griv, M. Gedalin, D. Eichler

Department of Physics, Ben-Gurion University of the Negev, P.O. Box 653, Beer-Sheva 84105, Israel

C. Yuan

Academia Sinica Institute of Astronomy and Astrophysics (ASIAA), Taipei 11529, Taiwan

Abstract. We analyse the reaction of collective oscillations excited in the nonresonant interaction between almost aperiodically growing Jeans-type gravity perturbations and stars of a rapidly rotating disk of flat galaxies. An equation is derived which describes the change in the main body of the equilibrium distribution of stars in the framework of the weakly nonlinear theory, that is, a quasi-linearization of a self-consistent Boltzmann–Poisson system of equations. Certain applications of the theory to the problem of collisionless dynamical relaxation of the Milky Way are explored. The theory, as applied to the solar neighborhood, accounts for observed features, such as the Schwarzschild shape for the stellar velocity ellipsoid and the increase in the random stellar velocities with age.

1. Introduction

Dynamical relaxation of a stellar distribution in disk-shaped (i.e., rapidly rotating) galaxies is not completely understood yet. Lynden-Bell (1967) and later Shu (1978) proposed violent relaxation in spherical-like (nonrotating) protogalaxies not in virial equilibrium. The associated relaxation for an individual star to gain or lose energy occurs on an exceedingly short time scale, smaller than one typical radial period and well before the rotating galaxy disk is formed. The relaxation, however, does not stop at this stage. There are numerous observations clearly showing that there exists ongoing relaxation on the time scale of $\lesssim 10$ rotation periods in the collisionless disk of the Milky Way Galaxy (Wielen 1977; Binney & Tremaine 1987, p. 470; Gilmore, King, & van der Kruit 1990, p. 161; Meusinger, Stecklum, & Reimann 1991; Binney 2000). This relaxation of the distribution of stars which were born in the equilibrium disk of the Galaxy results in a randomization of the velocity distribution (Maxwellianization) and a monotonic increase of the stellar random velocity dispersion (heating) from ~ 7 km s^{-1} for freshly made stars to about 50 km s^{-1} for the old disk stars with increasing stellar ages to $\sim 5 \times 10^9$ yr. The latter indicates strongly a significant irregular gravitational field in the Galactic disk (Wielen 1977). The irregular field causes a rapid diffusion of stellar orbits in velocity (and coordinate) space. Various mechanisms for the relaxation have been proposed. See e.g., Grivnev

& Fridman (1990) for a review of the problem. We elaborate upon the idea of the collective relaxation: unstable gravity perturbations in the disk affect the averaged distribution of stars. The instabilities of gravity perturbations and subsequent collective relaxation occur near the equilibrium and the perturbations remain relatively small which makes this process very different from what occurs during violent relaxation.

At the qualitative level, Toomre (1964, p. 1237), Goldreich & Lynden-Bell (1965), Marochnik (1968), and Kulsrud (1972) have suggested gravitational instabilities as a cause of enhanced relaxation in rapidly rotating galaxies. It was stated that because of its long-range Newtonian forces a self-gravitating medium (a stellar gas) would possess collective properties: collective, or cooperative motions in which all the particles of the system participate. These properties would be manifested in the behavior of small gravity perturbations arising against the equilibrium background. Collective processes are analogous to two-body collisions, except that one particle collides with many which are collected together by some coherent process such as a density wave. The collective processes are random, and usually much stronger than the ordinary two-body collisions and lead to a random walk of the particles that rapidly takes the complete system toward thermal quasi-equilibrium.[1]

2. Equilibrium

In the rotating frame of a disk galaxy, the collisionless motion of an ensemble of identical stars in the plane of the system can be described by the Boltzmann kinetic equation for the distribution function $f(\mathbf{r}, \mathbf{v}, t)$ without the integral of collisions (Lin, Yuan, & Shu 1969; Griv & Peter 1996):

$$\frac{\partial f}{\partial t} + v_r \frac{\partial f}{\partial r} + \left(\Omega + \frac{v_\varphi}{r} \right) \frac{\partial f}{\partial \varphi} + \left(2\Omega v_\varphi + \frac{v_\varphi^2}{r} + \Omega^2 r - \frac{\partial \Phi}{\partial r} \right) \frac{\partial f}{\partial v_r}$$

$$- \left(\frac{\kappa^2}{2\Omega} v_r + \frac{v_r v_\varphi}{r} + \frac{1}{r} \frac{\partial \Phi}{\partial \varphi} \right) \frac{\partial f}{\partial v_\varphi} = 0, \qquad (1)$$

where r, φ, z are the Galactocentric cylindrical coordinates, the total azimuthal velocity of the stars was represented as a sum of the random v_φ and the basic rotation velocity $V_{\rm rot} = r\Omega$, v_r is the velocity in the radial direction, and the epicyclic frequency $\kappa(r)$ is defined by $\kappa = 2\Omega[1 + (r/2\Omega)(d\Omega/dr)]^{1/2}$. The quantity $\Omega(r)$ denotes the angular velocity of Galactic nonuniform rotation at the distance r from the center, and κ varies from 2Ω for a rigid rotation to Ω for a Keplerian one. Random velocities are small compared with $r\Omega$. Collisions are neglected here because the collision frequency is much smaller than the cyclic frequency Ω. In Equation (1), $\Phi(\mathbf{r}, t)$ is the total gravitational potential determined self-consistently from the Poisson equation in a suitable form $\nabla^2 \Phi = 4\pi G \int f d\mathbf{v}$. Equation (1) and the Poisson equation give a complete self-consistent description of the problem for disk oscillation modes.

---

[1]It is difficult to explain the observed relaxation by the binary encounters between stars and gaseous clouds (Binney & Tremaine 1987, p. 473; Binney & Lacey 1988; Binney 2000).

The equilibrium state is assumed to be an axisymmetric and weakly in-homogeneous (only along the r-axis) stellar disk. In our simplified model, the gravity perturbation is propagating in the plane of the disk. This approximation of an infinitesimally thin disk is a valid approximation if one considers pertur-bations with a radial wavelength that is greater than the typical disk thickness. We assume that the stars move in the disk plane so that $v_z = 0$. This allows us to use the two-dimensional distribution function $f = f(v_r, v_\varphi, t)\delta(z)$ such that $\int f\, dv_r dv_\varphi dz = \sigma$, where σ is the surface density.

The disk in the equilibrium is described by the following equation:

$$v_r\frac{\partial f_e}{\partial r} + \left(2\Omega v_\varphi + \frac{v_\varphi^2}{r}\right)\frac{\partial f_e}{\partial v_r} - \left(\frac{\kappa^2}{2\Omega}v_r + \frac{v_r v_\varphi}{r}\right)\frac{\partial f_e}{\partial v_\varphi} = 0, \tag{2}$$

where $\partial f_e/\partial t = 0$ and the angular velocity of rotation $\Omega(r)$ is such that the nec-essary centrifugal acceleration is exactly provided by the central gravitational force $r\Omega^2 = \partial \Phi_e/\partial r$. Equation (2) does not determine the equilibrium distri-bution f_e uniquely. For the present analysis we choose f_e in the form of the anisotropic Maxwellian (Schwarzschild) distribution

$$f_e = \frac{\sigma_e(r_0)}{2\pi c_r(r_0)c_\varphi(r_0)}\exp\left[-\frac{v_r^2}{2c_r^2(r_0)} - \frac{v_\varphi^2}{2c_\varphi^2(r_0)}\right] = \frac{2\Omega}{\kappa}\frac{\sigma_e}{2\pi c_r^2}\exp\left(-\frac{v_\perp^2}{2c_r^2}\right). \tag{3}$$

The Schwarzschild distribution function is a function of the two epicyclic con-stants of motion $\mathcal{E} = v_\perp^2/2$ and $r_0^2\Omega(r_0)$, where $r_0 = r + (2\Omega/\kappa^2)v_\varphi$. These constants of motion are related to the unperturbed star orbits:

$$r = \frac{v_\perp}{\kappa}\left[\sin\phi_0 - \sin(\phi_0 - \kappa t)\right]; \quad v_r = v_\perp \cos(\phi_0 - \kappa t);$$

$$\varphi = \frac{2\Omega}{\kappa}\frac{v_\perp}{r_0\kappa}\left[\cos(\phi_0 - \kappa t) - \cos\phi_0\right]; \quad v_\varphi \approx r_0\frac{d\varphi}{dt}$$

$$+ \; r_0\frac{v_\perp}{\kappa}\frac{d\Omega}{dr}\sin(\phi_0 - \kappa t) \approx \frac{\kappa}{2\Omega}v_\perp \sin(\phi_0 - \kappa t), \tag{4}$$

where $v_\perp$, ϕ_0 are constants of integration, $v_\perp/\kappa r_0 \sim \rho/r_0 \ll 1$, ρ is the mean epicycle radius, and we follow Lin et al. (1969), Shu (1970), and Griv & Peter (1996) making use of expressions for the unperturbed epicyclic trajectories of stars in the equilibrium central field $\Phi_e(r)$. In Equations (3) and (4), r_0 is the radius of the circular orbit, which is chosen so that the constant of areas for this circular orbit $r_0^2(d\varphi_0/dt)$ is equal to the angular momentum integral $M_z = r^2(d\varphi/dt)$, and $v_\perp^2 = v_r^2 + (2\Omega/\kappa)^2 v_\varphi^2$. Also, φ_0 is the position angle on the circular orbit, $(d\varphi_0/dt)^2 = (1/r_0)(\partial\Phi_e/\partial r)_0 = \Omega^2$. The quantities Ω, κ, and c_r are evaluated at r_0. In Equation (3) the fact is used that as follows from Equations (4) in a rotating frame the radial velocity dispersion c_r and the azimuthal velocity dispersion c_φ are not independent but connected through $c_r \approx (2\Omega/\kappa)c_\varphi$. In the solar vicinity, $2\Omega/\kappa \approx 1.7$. The distribution function f_e has been normalized according to $\int_{-\infty}^{\infty}\int_{-\infty}^{\infty} f_e dv_r dv_\varphi = 2\pi(\kappa/2\Omega)\int_0^{\infty} v_\perp dv_\perp f_e = \sigma_e$, where σ_e is the equilibrium surface density. Such a distribution function for the unperturbed system is particularly important because it provides a fit to observations (Lin et al. 1969; Shu 1970). It is this equilibrium that is examined for gravitational stability.

3. Collisionless Relaxation

We proceed by applying the procedure of the weakly nonlinear (or quasi-linear) approach (Krall & Trivelpiece 1986, p. 512) and decompose the time dependent distribution function $f = f_0(r, \mathbf{v}, t) + f_1(\mathbf{r}, \mathbf{v}, t)$ and the gravitational potential $\Phi = \Phi_0(r, t) + \Phi_1(\mathbf{r}, t)$ with $|f_1/f_0| \ll 1$ and $|\Phi_1/\Phi_0| \ll 1$ for all $\mathbf{r}$ and t. The functions f_1 and Φ_1 are oscillating rapidly in space and time, while the functions f_0 and Φ_0 describe the slowly developing background against which small perturbations develop; $f_0(t = 0) \equiv f_e$ and $\Phi_0(t = 0) \equiv \Phi_e$. The distribution f_0 continues to distort as long as the distribution is unstable. Linearizing Equation (1) and separating fast and slow varying variables one obtains:

$$\frac{df_1}{dt} = \frac{\partial \Phi_1}{\partial r} \frac{\partial f_0}{\partial v_r} + \frac{1}{r} \frac{\partial \Phi_1}{\partial \varphi} \frac{\partial f_0}{\partial v_\varphi}, \tag{5}$$

$$\frac{\partial f_0}{\partial t} = \left\langle \frac{\partial \Phi_1}{\partial r} \frac{\partial f_1}{\partial v_r} + \frac{1}{r} \frac{\partial \Phi_1}{\partial \varphi} \frac{\partial f_1}{\partial v_\varphi} \right\rangle, \tag{6}$$

where d/dt means the total derivative along the star orbit given by Equations (4) and $\langle \cdots \rangle$ denotes the time average over the fast oscillations. To emphasize it again, unlike Lynden-Bell (1967) and Shu (1978), we are concerned with the growth or decay of small-amplitude perturbations from an equilibrium state.

In the epicyclic approximation, the partial derivatives in Equations (5) and (6) transform as follows (Shu 1970; Griv & Peter 1996):

$$\frac{\partial}{\partial v_r} = v_r \frac{\partial}{\partial \mathcal{E}} - \frac{2\Omega}{\kappa} \frac{v_\varphi}{v_\perp^2} \frac{\partial}{\partial \phi_0}; \quad \frac{\partial}{\partial v_\varphi} \approx \left(\frac{2\Omega}{\kappa} \right)^2 v_\varphi \frac{\partial}{\partial \mathcal{E}} + \frac{2\Omega}{\kappa^2} \frac{\partial}{\partial r} + \frac{2\Omega}{\kappa} \frac{v_r}{v_\perp^2} \frac{\partial}{\partial \phi_0}. \tag{7}$$

To determine oscillation spectra, let us consider the stability problem in the WKB approximation: the perturbation scale is sufficiently small for the disk to be regarded locally as spatially homogeneous. This is accurate for short wave perturbations only, but qualitatively correct even for longer wavelength perturbations, of the order of the disk radius R. In the local WKB approximation in Equations (5) and (6), assuming the weakly inhomogeneous disk, the perturbation is selected in the form of a plane wave (in the rotating frame):

$$\aleph_1 = 0.5\delta\aleph \left(e^{ik_r r + im\varphi - i\omega_* t} + \text{c.c.} \right), \tag{8}$$

where $\delta\aleph$ is an amplitude that is a constant in space and time, m is the non-negative azimuthal mode number (= number of spiral arms), $\omega_* = \omega - m\Omega$ is the Doppler-shifted wavefrequency, and $|k_r|R \gg 1$ (Griv & Peter 1996). The solution in such a form represents a spiral wave with m arms whose shape in the plane is determined by the relation $k_r(r - r_0) = -m(\varphi - \varphi_0)$. With φ increasing in the rotation direction, we have $k_r > 0$ for trailing spiral patters, which are the most frequently observed among spiral galaxies. A change of the sign of k_r corresponds to changing the sense of winding of the spirals, i.e., leading ones. With $m = 0$, we have the density waves in the form of concentric rings.

In Equation (5) using the transformation of the derivatives $\partial/\partial v_r$ and $\partial/\partial v_\varphi$ given by Equations (7), one obtains the solution

$$f_1 = \int_{-\infty}^{t} dt' \left(\mathbf{v}_\perp \cdot \frac{\partial \Phi_1}{\partial \mathbf{r}} \frac{\partial f_0}{\partial \mathcal{E}} + \frac{2\Omega}{\kappa^2} \frac{1}{r} \frac{\partial \Phi_1}{\partial \varphi} \frac{\partial f_0}{\partial r} \right), \tag{9}$$

where $f_1(t' = -\infty) \to 0$. In this equation, using the time dependence of perturbations in the form of Equation (8) and the unperturbed trajectories of stars given by Equations (4) in the exponential factor, it is straightforward to show that (Morozov 1980; Griv & Peter 1996; Griv et al. 2000a, 2000b)

$$
\begin{aligned}
f_1 = \quad - \quad & \Phi_1(r_0)\left[\kappa\frac{\partial f_0}{\partial\mathcal{E}}\sum_{l=-\infty}^{\infty}\sum_{n=-\infty}^{\infty}l\frac{e^{i(n-l)(\phi_0-\zeta)}J_l(\chi)J_n(\chi)}{\omega_*-l\kappa}\right.\\
+ \quad & \left.\frac{2\Omega}{\kappa^2}\frac{m}{r}\frac{\partial f_0}{\partial r}\sum_{l=-\infty}^{\infty}\sum_{n=-\infty}^{\infty}\frac{e^{i(n-l)(\phi_0-\zeta)}J_l(\chi)J_n(\chi)}{\omega_*-l\kappa}\right],
\end{aligned}
\tag{10}
$$

where $J_l(\chi)$ is the Bessel function of the first kind of order l, $\chi = k_* v_\perp/\kappa \sim k_* \rho$, $k_* = k\{1 + [(2\Omega/\kappa)^2 - 1]\sin^2\psi\}^{1/2}$ is the effective wavenumber, ψ is the pitch angle of perturbations, and $\tan\psi = k_\varphi/k_r = m/rk_r$. In the equation above the denominator vanishes when $\omega_* - l\kappa \to 0$. This occurs near corotation ($l = 0$) and other resonances ($l = \pm 1, \pm 2, \cdots$). Clearly, the location of these resonances depends on the rotation curve and the spiral pattern speed (Lin et al. 1969; Binney & Tremaine 1987). We study only the main part of the Galactic disk which lies sufficiently far from the resonances: below in all equations $\omega_* - l\kappa \neq 0$.

Next, we substitute the solution (10) into Equation (6). Taking into account that the terms $l \neq n$ vanish for axially symmetric functions f_0, after averaging over ϕ_0 we obtain the equation for the slow part of the distribution function:

$$
\begin{aligned}
\frac{\partial f_0}{\partial t} = \quad & i\frac{\pi}{2}\sum_{\mathbf{k}}\sum_{l=-\infty}^{\infty}|\Phi_{1,\mathbf{k}}|^2\frac{\partial}{\partial v_\perp}\frac{k_*\kappa}{v_\perp\chi}\frac{l^2 J_l^2(\chi)}{\omega_*-l\kappa}\frac{\partial f_0}{\partial v_\perp}\\
+ \quad & i\pi\sum_{\mathbf{k}}\sum_{l=-\infty}^{\infty}|\Phi_{1,\mathbf{k}}|^2\frac{2\Omega}{\kappa^2}\frac{m}{r}\frac{\partial}{\partial r}\frac{2\Omega}{\kappa^2}\frac{m}{r}\frac{J_l^2(\chi)}{\omega_*-l\kappa}\frac{\partial f_0}{\partial r}.
\end{aligned}
\tag{11}
$$

As is seen, the initial distribution of stars will change upon time only under the action of growing or decaying perturbations ($\Im\omega_* \neq 0$).

As usual in the weakly nonlinear theory, in order to close the system one must engage an equation for $\Phi_{1,\mathbf{k}}$. Averaging over the fast oscillations, we have

$$
(\partial/\partial t)|\Phi_{1,\mathbf{k}}|^2 = 2\Im\omega_*|\Phi_{1,\mathbf{k}}|^2,
\tag{12}
$$

where suffixes $\mathbf{k}$ denote the kth Fourier component. Equations (11) and (12) form the closed system of weakly nonlinear equations for nonresonant Jeans oscillations of the rapidly rotating inhomogeneous disk of stars, and describe a diffusion in configuration space.

The spectrum of oscillations and their growth rate are obtained from Equation (5) and the Poisson equation (Morozov 1980; Griv et al. 2000a, 2000b):

$$
\frac{k^2 c_r^2}{2\pi G\sigma_0|k|} = -\kappa\sum_{l=-\infty}^{\infty}l\frac{e^{-x}I_l(x)}{\omega_*-l\kappa} + 2\Omega\frac{m\rho^2}{rL}\sum_{l=-\infty}^{\infty}\frac{e^{-x}I_l(x)}{\omega_*-l\kappa},
\tag{13}
$$

and $\Im\omega_* \approx \sqrt{2\pi G\sigma_0|k|F(x)}$. In Equation (13), $F(x) \approx 2\kappa^2 e^{-x}I_1(x)/k^2 c_r^2$ is the so-called reduction factor, which takes into account the fact that the wave field only weakly affects the stars with high random velocities, $F(x) \propto \exp(-c_r^2)$ (Lin et al. 1969; Shu 1970; Morozov 1980; Griv & Peter 1996). Also, $I_l(x)$ is a Bessel function of an imaginary argument with its argument $x \approx k_*^2 \rho^2$ and $\rho = c_r/\kappa$ is the mean epicyclic radius. An important feature of the instability under consideration is that it is almost aperiodic (the real part of ω_* almost vanishes in a rotating frame, $|\Re\omega_*/\Im\omega_*| \ll 1$; Griv et al. 2000b, 2000c). A simplification results from restricting the frequency range of the waves examined by taking the low frequency limit ($|\omega_*|$ less than the epicyclic frequency of any disk star). In the opposite case of the high perturbation frequencies, $|\omega_*| > \kappa$, the effect of the disk rotation is negligible and therefore irrelevant (Griv & Peter 1996; Griv et al. 1999a). Then, the series terms of Equations (10)–(13) for which $|l| \geq 2$ may be neglected, and consideration will be limited to the transparency region between the disk turning points (between the inner and outer Lindblad resonances). In this case, in Equations (10)–(13) the function $\Lambda(x) = \exp(-x)I_1(x)$ starts from $\Lambda(0) = 0$, reaches a maximum $\Lambda_{max} < 1$ at $x \approx 0.5$, and then decreases. Hence, the growth rate has a maximum at $x \approx 0.5 < 1$ (Griv, Yuan, & Gedalin 1999b, figure 2).

In general, the growth rate of the Jeans instability is high and comparable to the epicyclic period $|\Im\omega_*| \sim \Omega$; perturbations with wavelength $\lambda_{crit} \approx (2\Omega/\kappa)\lambda_J$ have the fastest growth rate, where $\lambda_J = 2\pi\rho \approx 2\pi c_r/\kappa$ is the familiar Jeans–Toomre wavelength (Morozov 1980; Griv & Peter 1996; Griv et al. 1999a). In the solar vicinity for young, low-dispersion stars (with $c_r \sim 10 \text{ km s}^{-1}$), $\Omega \approx 2 \times 10^{-8}$ yr^{-1} and $\lambda_{crit} \approx 3$ kpc. Interestingly, from observations in the Galaxy, $\lambda = 2 - 4$ kpc. So, the Jeans length is small compared to the disk radius $R \approx 15$ kpc and the radial scale of the most unstable perturbations is small.

4. Astronomical Implications

As an application of the theory we investigate the dynamical relaxation of low frequency and Jeans-unstable ($|\omega_*| < \kappa$ and $\omega_*^2 < 0$, respectively) oscillations in the weakly inhomogeneous Galactic disk. Evidently, the unstable Jeans oscillations must influence the distribution function of the main, nonresonant part of stars in such a way as to hinder the wave excitation, i.e., to increase the velocity dispersion. This is because the Jeans instability, being essentially a gravitational one, tends to be stabilized by random motions (Toomre 1964; Morozov 1980; Griv & Peter 1996; Polyachenko & Polyachenko 1997). Therefore, along with the growth of the oscillation amplitude, random velocities must increase at the expense of circular motion, and finally in the disk there can be established a quasi-stationary distribution so that the Jeans-unstable perturbations are completely vanishing and only undamped Jeans-stable waves remain.[2]

In the following, we restrict ourselves to the most dangerous, in the sense of the loss of gravitational stability, long-wavelength perturbations, χ^2 and $x^2 \ll 1$ (see the explanation after Equation [13] and Griv et al. 1999a). Then in Equations (10)–(13) one can use the expansions $J_1^2(\chi) \approx \chi^2/4$ and $e^{-x}I_1(x) \approx$

---

[2]In turn, the Jeans-stable gravity perturbations are subject to a resonant Landau-type instability developing on a Hubble time scale (Griv et al. 1999b, 2000a, 2000b).

$(1/2)x - (1/2)x^2 + (5/16)x^3$. Equation (11) takes the simple form

$$(\partial/\partial t)f_0 - \mathcal{D}_v(\partial^2/\partial v_\perp^2)f_0 - \mathcal{D}_r(\partial^2/\partial r^2)f_0 = 0, \qquad (14)$$

where $\mathcal{D}_v = (\pi/16\kappa^2)\sum_{\mathbf{k}} k_*^2 \Im\omega_* |\Phi_{1,\mathbf{k}}|^2$, $\mathcal{D}_r = 2\pi \sum_{\mathbf{k}} m^2 |\Phi_{1,\mathbf{k}}|^2/\alpha^2 \Im\omega_*$, $\Im\omega_* > 0$, both $\Im\omega_*$ and $\Phi_{1,\mathbf{k}}$ are functions of t, and we took into account the fact that in actual spiral galaxies with a flat rotation curve $2\Omega/\kappa^2 = r/\alpha$. As is seen, the velocity diffusion coefficient for nonresonant stars $\mathcal{D}_v(t)$ is independent of $\mathbf{v}_\perp$ (to lowest order). This is a qualitative result of the nonresonant character of the stars' interaction with collective aggregates. The term $\propto \mathcal{D}_r(t)$ describes a diffusion of stars' guiding centers in coordinate space.

By introducing the standard definitions $d\tau/dt = \mathcal{D}_v(t)$ and $d/dt = (d\tau/dt)(d/d\tau)$, Equations (12) and (14) are rewritten as follows:

$$(\partial/\partial\tau)f_0 - (\partial^2/\partial v_\perp^2)f_0 - (\mathcal{D}_r/\mathcal{D}_v)(\partial/\partial r^2)f_0 = 0, \quad (\partial/\partial\tau)\mathcal{D}_v = 2\Im\omega_*,$$

which have the solutions

$$f_0(v_\perp, \tau) = \frac{\text{const}}{\sqrt{\tau + c_r^2/2}} \exp\left[-\frac{v_\perp^2}{4(\tau + c_r^2/2)}\right] \qquad (15)$$

and

$$f_0(r, \tau) \approx \frac{\text{const}}{\sqrt{r_0^2 + (\mathcal{D}_r/\mathcal{D}_v)\tau}} \exp\left\{-\frac{r^2}{4\left[r_0^2 + (\mathcal{D}_r/\mathcal{D}_v)\tau\right]}\right\}. \qquad (16)$$

(We have taken into account the observations that most spiral galaxies have an exponential disk with radial surface distribution $\sigma_0(t = 0) \propto \exp[-(r/r_0)^2]$.) Accordingly, during the development of the Jeans instability on the time scale $\sim (\Im\omega_*)^{-1} \sim \Omega^{-1}$ (that is, on the time scale of the epicyclic period of the stars), the Schwarzschild distribution of random velocities (a Gaussian spread along v_r, v_φ coordinates in velocity space) is established. As the perturbation energy increases, the distribution spreads, and the effective temperature grows with time (i.e., Gaussian spread increases): $T = 2\tau \propto \int \mathcal{D}_v(t)dt \propto \sum_{\mathbf{k}} k_*^2 |\Phi_{1,\mathbf{k}}|^2$. That is, the energy of the oscillation field $\propto \sum_{\mathbf{k}} |\Phi_{1,\mathbf{k}}|^2$ plays the role of a temperature T in the nonresonant-particle distribution.

Thus, this mechanism increases (a) the velocity dispersion of stars in Milky Way's disk after they are born (Equation [15]), and (b) the radial spread of the disk (Equation [16]). The collisionless relaxation mimics thermal relaxation in a two-dimensional stellar disk, leading to Maxwell–Boltzmann type velocity distributions with an effective temperature that increases with time. Subsequently, sufficient velocity dispersion prevents the Jeans instability from occurring (Griv & Peter 1996; Griv et al. 1999a). The diffusion of nonresonant stars takes place because they gain kinetic wave energy as the Jeans instability develops.

Concluding, the true time scale for dynamical relaxation in the Milky Way may be much shorter than its standard value $\sim 10^{14}$ yr for the Chandrasekhar–Ogorodnikov (Chandrasekhar 1960; Ogorodnikov 1965) collisional relaxation; it may be of the order $(\Im\omega_*)^{-1} \gtrsim \Omega^{-1} \gtrsim 3 \times 10^8$ yr, i.e., comparable with $2 - 3$ periods of the Milky Way rotation in the solar vicinity. The above relaxation time is in fair agreement with both observations (Binney & Tremaine 1987, p. 470; Gilmore 1990; Grivnev & Fridman 1990; Binney 2000) and N-body simulations (Hohl 1971; Sellwood & Carlberg 1984; Griv et al. 2001).

References

Binney, J. 2000, in Galaxy Disks and Disk Galaxies, ed. F. Bertola & G. Coyne (San Francisco: ASP), in press

Binney, J., & Lacey, C. 1988, MNRAS, 230, 597

Binney, J., & Tremaine, S. 1987, Galactic Dynamics (Princeton: Princeton University Press)

Chandrasekhar, S. 1960, Principles of Stellar Dynamics (New York: Dover)

Gilmore, G., King, I. R., & van der Kruit, P. C. 1990, The Milky Way as a Galaxy, ed. R. Buser & I. R. King (Mill Valley: University Science Books)

Goldreich, P., & Lynden-Bell, D. 1965, MNRAS, 130, 125

Griv, E., Gedalin, M., Eichler, D., & Yuan, C. 2000a, Phys. Rev. Lett, 84, 4280

Griv, E., Gedalin, M., Eichler, D., & Yuan, C. 2000b, Ap&SS, 271, 21

Griv, E., Gedalin, M., Eichler, D., & Yuan, C. 2000c, Plan&SS, 48, 679

Griv, E., Gedalin, M., Liverts, E., Eichler, D., Kimhi, Y., & Yuan, C. 2001, Celest. Mech. Dyn. Ast., submitted

Griv, E., & Peter, W. 1996, ApJ, 469, 84

Griv, E., Rosenstein, B., Gedalin, M., & Eichler, D. 1999a, A&A, 347, 821

Griv, E., Yuan, C., & Gedalin, M. 1999b, MNRAS, 307, 1

Grivnev, E. M., & Fridman, A. M. 1990, Soviet Ast., 34, 10

Hohl, F. 1971, ApJ, 168, 343

Krall, N. A., & Trivelpiece, A. W. 1986, Principles of Plasma Physics (San Francisco: San Francisco Press)

Kulsrud, R. M. 1972, in Gravitational N-body Problem, ed. M. Lecar (Dordrecht: Reidel), 337

Lin, C. C., Yuan, C., & Shu, F. H. 1969, ApJ, 155, 721

Lynden-Bell, D. 1967, MNRAS, 136, 101

Marochnik, L. S. 1968, Soviet Ast., 12, 371

Meusinger, H., Stecklum, B., & Reimann, H.-G. 1991, A&A, 245, 57

Morozov, A. G. 1980, Soviet Ast., 24, 391

Ogorodnikov, K. F. 1965, Dynamics of Stellar Systems (New York: Pergamon)

Polyachenko, V. L., & Polyachenko, E. V. 1997, JETP, 86, 417

Sellwood, J. A., & Carlberg, R. G. 1984, ApJ, 282, 61

Shu, F. H. 1970, ApJ, 160, 99

Shu, F. H. 1978, ApJ, 225, 83

Toomre, A. 1964, ApJ, 139, 1217

Wielen, R. 1977, A&A, 60, 263

Acknowledgments.   We thank Tzi-Hong Chiueh, Alexei Fridman, Muzafar Maksumov, Frank Shu, and Raphael Steinitz for helpful conversations. This research was supported by the Israel Science Foundation, the Israeli Ministry of Immigrant Absorption, and the Academia Sinica in Taiwan. One of us (E. G.) appreciates the hospitality of the ASIAA, where the work was begun.

Discussion

Frank Shu: I don't quite understand how you can get such fast relaxation without very rapidly growing disturbances, and they do grow very rapidly and you would think that fully non-linear effects would have to come into play and that the waves might saturate other effects. Would you care to comment on that?

Evgeny Griv: From our equations we see that the effect is that the typical growth time scale is about one over the growth rate, so it corresponds to about one epicycle rotation period. Therefore, we predict that over one epicycle rotation period the dispersion of random velocities of stars will be increased by a factor of 2 or so at least.

Frank Shu: But your linear wave is growing also on this time scale and in one epicycle it will reach non-linear amplitudes.

Evgeny Griv: Yes, that's true and what can I say? This theory of weak turbulence (quasi-linear theory) describes only the tendency of a disk to be heated by Jeans-unstable density waves. Of course we understand that we still have to develop the theory of strong turbulence, but no one in the world knows how to do that. Nevertheless, our theory accounts for the observed Schwarzschild shape for the velocity ellipsoids of stars and the increase in stellar random velocities with age in the disk of the Galaxy. At least, this theory shows the direction of the disk's evolution.

Harvey Liszt: The classical mechanism for increasing the velocity dispersion of stars with time is to scatter them off molecular clouds, the Spitzer-Schwarzschild mechanism. Is the Spitzer-Schwarzschild mechanism not effective, or are you saying the phenomena you are studying dominate?

Evgeny Griv: You are right. In 1951 Spitzer and Schwarzschild suggested the observed increase of stellar random velocities with age in the Galaxy by interaction of stars with vary massive objects, e.g., massive gaseous clouds. Now we understand that this mechanism is not the dominant one. See the paper by Binney (2000) and textbooks of Binney & Tremaine (1987), Gilmore et al. (1990), and others for for an explanation. Also, such an interaction will lead to equal dispersion in every direction, but what we observe in our Galaxy is the difference between plane and velocity dispersions. So according to Binney (2000), scattering by giant molecular clouds plays only a modest role.

Astrophysical Ages and Time Scales
ASP Conference Series, Vol. 245, 2001
T. von Hippel, C. Simpson, N. Manset

Dating and Reconstructing Star Formation Episodes in the Local Group

David Valls-Gabaud

Laboratoire d'Astrophysique, Observatoire Midi-Pyrénées, 14 Av. E. Belin, 31400 Toulouse, France

Xavier Hernandez

Instituto de Astronomía, UNAM, 04510 Mexico DF, Mexico

Gerard Gilmore

Institute of Astronomy, Madingley Road, Cambridge CB3 0HA, UK

Abstract. The ages and intensities of star formation episodes provide unique constraints on the formation and evolution of galaxies, since the evolution of the star formation rate appears to control their chemical and spectral evolution. Progress in the precise dating of these episodes has been possible through high quality colour–magnitude diagrams of resolved stellar populations in the Local Group. When combined with our variational maximum likelihood technique, we can infer with unprecedented precision a robust, non-parametric reconstruction of the evolution of the star formation rate in a variety of environments. Nearby dwarf spheroidal (dSph) galaxies of the Local Group appear to present a wide variety of star formation histories, from single bursts to extended activity, with no particular epoch being dominant. Star formation in spiral discs can be probed with the HIPPARCOS data for the solar neighbourhood, where over the past 3 Gyr several episodes of star formation have appeared roughly every ∼ 0.5 Gyr, possibly triggered by the interaction with spiral arms.

1. Introduction

Dating star formation episodes has been achieved in the past through a combination of chemical evolution models and assumptions, basically relying on the time scales of type I and II supernovae for heavy element production, plus mixing time scales into the ISM, to derive rough estimates of the duration of star formation. Other attempts use the chemical evolution hypotheses and some age–metallicity relation, which may limit the range of application of such techniques. A far more direct way is the inversion of colour–magnitude diagrams (CMDs) of resolved stellar populations, a technique limited to the nearby, mostly dwarf, galaxies of the Local Group. These dwarf galaxies may have formed before massive galaxies, in the currently popular version of hierarchical structure formation, but may also have formed stars very late if feedback processes suppressed early star formation. Alternatively dwarf galaxies may have formed in the tidal tails of interacting massive galaxies. In the case of late-type dwarfs, studies pioneered by Tosi et al. (1991) via Monte Carlo simulations of CMDs showed intermittent behaviour in their star formation histories, but the duration of the bursts and their amplitudes are still poorly constrained. The small sample of early-type dwarfs of the Local Group may offer a better sample to test theories of dwarf galaxy formation since their metallicities appear to have a small internal dispersion and high quality, deep, high-resolution, narrow-field HST CMDs, reaching well below the main sequence turnoff point, are now available for many of them (see Aparicio 1998; Mateo 1998 for recent reviews).

2. Objective Reconstructions of Star Formation Histories

The unprecedented quality of the current CMDs requires powerful tools to invert them in order to derive the star formation history (SFH) which gave rise to the observed distribution of stars in these diagrams. The methods used so far are based on comparisons between the observed CMD and synthetic CMDs computed assuming a given SFH, and then looking for the best matching SFH. For instance Tolstoy & Saha (1996) use a Bayesian likelihood technique, Mighell (1997) a classical χ^2 optimisation, Ng (1998) a Poisson merit function, while Gallart et al. (1999) and Hurley-Keller, Mateo, & Nemec (1998) minimise a counts in cells statistic for well-defined boxes in the CMD. Alternatively, Dolphin (1997) makes a linear decomposition in terms of fiducial CMDs and solves for the best matching final CMD, in terms of a χ^2 statistic.

One of the problems with most of these techniques is that there is no guarantee that the actual SFH belongs to the set of (usually parametric) SFHs explored within a family of functions defined a priori. For instance, in the case of the Carina dSph (see Figure 2), Hurley-Keller et al. (1998) find the best fitting 3-burst solution, leading to results which are at variance with those from Mighell (1997), who uses a non-parametric approach. In addition, it is unclear to what extent these best matching solutions are actually good fits to the observed CMDs. We have therefore developed an entirely new method based on a combination of Bayesian statistics with variational calculus which does not suffer from the limitations listed above. Full details of the method are given in Hernandez, Valls-Gabaud, & Gilmore (1999, Paper I), and will not be repeated here. Very briefly, the method makes 3 key assumptions: (1) the metallicity of the ensemble of stars is known and has a small dispersion; (2) the initial mass function is given and there are no unresolved binary systems; and (3) both distance and colour excess are known to within some uncertainties. The method then takes as inputs the positions of n stars in a colour–magnitude diagram, each having a colour c_i and luminosity l_i, with (in this example, uncorrelated) associated errors $\sigma(c_i)$ and $\sigma(l_i)$, respectively. Using the likelihood technique, we first construct the probability that the n observed stars resulted from some function $SFR(t)$. This is given by

$$\mathcal{L} = \prod_{i=1}^{n} \left(\int_{t_0}^{t_1} SFR(t)\, G_i(t)\, \mathrm{d}t \right), \tag{1}$$

where

$$G_i(t) = \int_{m_0}^{m_1} \frac{\rho(m;t)}{2\pi\sigma(l_i)\sigma(c_i)} \exp\left(\frac{-D(l_i;t,m)^2}{2\sigma^2(l_i)}\right) \exp\left(\frac{-D(c_i;t,m)^2}{2\sigma^2(c_i)}\right) dm. \quad (2)$$

In this expression ρ is the density of stars of mass m along the isochrone of age t, and only depends on the assumed IMF and the set of stellar tracks (and in particular the durations of the different evolutionary phases). The D factors are the differences in luminosity and colour of the observed star i with respect to the luminosity and colour of a star of mass m at time t. We refer to $G_i(t)$ as the likelihood matrix, since each element represents the probability that a given star i was actually formed at time t with any mass.

Following the discussion of Paper I, we may write the condition that the likelihood has an extremal as the variation $\delta\mathcal{L}(SFR) = 0$, allowing a full variational calculus analysis to be used. We have now transformed what was an optimisation problem, finding the function that maximises the product of integrals defined by Equation 1, into an integro-differential equation with a boundary condition (at either t_o or t_1) which can be solved by iteration to get, non-parametrically, the function $SFR(t)$. Further details of the numerical aspects of the procedure are available in Paper I. It is important to point out that our method has distinctive advantages over other techniques: (1) the variational calculus allows a fully non-parametric reconstruction, free of any astrophysical preconceptions; (2) there is no time-consuming comparisons between CMDs, since the function (not the parameters) $SFR(t)$ that maximises the likelihood is solved for directly; (3) the CPU time required scales linearly with the time resolution required in the reconstruction.

Paper I also presents a detailed study of the influence of the assumptions that were made: the effect of changing the IMF or the presence of unresolved binaries is essentially a normalisation problem, which does not change the overall shape or localisation of a burst of star formation, while an incorrect estimate of the metallicity has drastic effects on the position of a burst, a result of the age-metallicity degeneracy. See also Paper I for the effect of photometric uncertainties in the effective time resolution of the reconstructed SFH. Note that since the IMF and the fraction of binaries (and the distribution function of their mass ratios) are unlikely to be measured, an absolute normalisation of the star formation rate cannot be achieved.

3. Star Formation Histories of Local Group Dwarf Spheroidals

Given the main assumption made so far in order to derive a fully non-parametric variational reconstruction of the SFH, namely a determination of the metallicity of the ensemble of stars, only a few galaxies fulfill this requirement. In addition, in order to minimise any possible systematic effects between different data sets and reduction procedures, as well as crowding/blending effects, we selected from the HST archive an homogeneous sample of five dSph galaxies (Leo I, Leo II, Draco, Carina and Ursa Minor) of the Local Group which were reduced with the same procedures. It is only such an internally consistent data set which allows robust comparisons between different galaxies, with the proviso that the

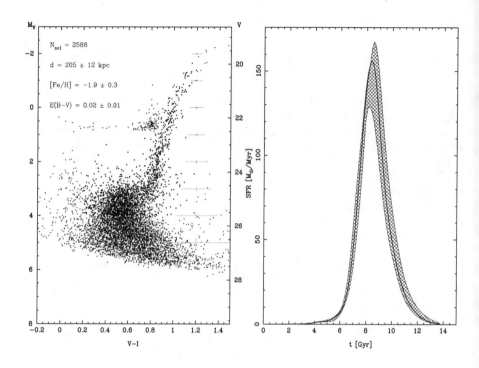

Figure 1.   WFPC2 CMD and reconstructed SFR history of Leo II, a
dwarf galaxy with a single burst of activity 8 Gyr ago.

volumes sampled by the WFPC2 field may not be representative of the entire
galaxy. Full details of the SFH reconstructions for these galaxies are presented
in Hernandez, Gilmore, & Valls-Gabaud (2000a, Paper II).

The dSph in Leo can be used as an example of our procedure. Figure 1
gives the WFPC2 CMD we used as input. Note the increase of the 2σ error bars
with decreasing luminosity and the removal of stars on the horizontal branch
(incompatible with the evolutionary phases included in our isochrones). Since
incompleteness may be important, and isochrones are degenerate at this level,
we also remove stars below $M_V = 4$, so we are left with $N = 2588$ stars in
total. For the fiducial values of distance, metallicity, and colour excess indicated,
the variational method gives the reconstructed SFH indicated with the black
line on the right panel of Figure 1. This function maximises the likelihood, as
defined in Equation 1, for the given data set. To assess the robustness of this
reconstruction, many functions were reconstructed by changing the observational
parameters within their error bounds indicated in Figure 1. The *envelope* of such
solutions is the shaded area on the figure, and represents the uncertainties in
the reconstruction given by the uncertainties in the observations, so that *any*
function that will be contained within these bounds will maximise $\mathcal{L}$, given these
errors. The bulk of the stellar population in Leo II was therefore formed > 6 Gyr
ago, with a peak at 8 Gyr. Given the uncertainties in the photometry, we cannot

resolve any episodes in the star formation rate within the main burst at 8 Gyr, but we can conclude that the burst lasted < 2 Gyr (FWHM).

Having found the best solution (in terms of maximising the likelihood) does not guarantee that it is also a good solution, so we apply again Bayesian techniques to check whether the observed CMD could result from the CMDs produced by the best reconstruction of the SFH. To do this, we do a counts in cells analysis and apply Saha's W statistic (Saha 1998). To compute the distribution function of W, a series of model-model comparisons are made, that is, synthetic CMDs resulting from realisations of the reconstructed SFH are compared pairwise. To check whether the observed CMD could arise from this distribution, the W statistic resulting from comparisons between the observed CMD and a series of CMD realisations are made. If the distribution functions are compatible, at some statistical level, the hypothesis that the observed CMD can be produced by the reconstructed SFH is accepted.

In the case of Leo II, the mean and dispersion of the model-model W distribution is 57 ± 6, while the data-model distribution gives 64 ± 8, so we can deduce that the synthetic CMDs are compatible at better than 1σ with the observed CMD. The best solution is also a good solution.

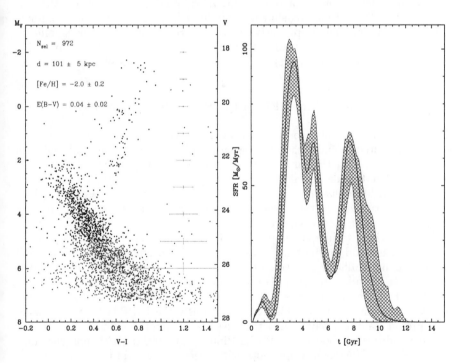

Figure 2.   WFPC2 CMD and reconstructed SFR history of the Carina dwarf, showing extended activity over some 8 Gyr.

The case of the Carina dSph (Figure 2) is more interesting. There are clearly two subgiant branches, identified with two epochs of star formation at

8 and 3 Gyr. Given the 2σ errors indicated in Figure 2 and simulated CMDs with these errors, we can say that the bursting episodes in Carina lasted more than a 1 Gyr, and that there was no epoch in the 2–8 Gyr interval without star formation activity. The W statistic indicates a good solution. It is however puzzling that such an extended star formation history, over 6 Gyr, resulted in such a low metallicity, with a dispersion of 0.2 dex or less. Either these values are only representative of the tip of the giant branch, or the enrichment in metals had a non-standard history.

Leo I presents a similar, extended history of star formation. Some activity was present around 10–13 Gyr ago, compatible with the presence of horizontal branch stars recently detected at the NTT (Held et al. 2000) but not present in our HST data. The bulk of the observed stellar populations was formed continuously, with peaks of activity at 8 and 4 Gyr, in agreement with an independent analysis by Gallart et al. (1999). In this case however, the W statistic indicates that the maximum likelihood solution is not a good solution at the 2σ level. Again, a likely explanation is that the dispersion in metallicity (around 0.3 dex) is too large for the method to work. A full bi-variate non-parametric reconstruction of both $SFR(t)$ and $Z(t)$ is required.

If the Local Group is a representative volume of the Universe, the star formation activity at high redshift was dominated by these dSph that dominate the luminosity function. A properly averaged SFH does not show any significant epoch in the comoving density of SFR (Paper II). There seems to be no link either between the epoch of the bursts and perigalacticon passages of these satellites.

4. The Evolution of the SFR in the Solar Neighbourhood

The evolution of the star formation rate in the Galactic disc is another basic function required to understand the formation of the disc, its chemical evolution and, more generally, the luminosity evolution of spiral galaxies. Previous attempts at determining the SFH in the Galaxy have relied on indirect methods, basically relations between age and some astrophysical property—such as chromospheric activity or metallicity—and then corrected with evolutionary models for the stars which have disappeared from the sample. A good example of complexities of these techniques is provided by Rocha-Pinto et al. (2000b). Here we use the method outlined in § 2. since the solar neighbourhood has a small dispersion in metallicity centred on the solar value. Further details will be found in Hernandez, Valls-Gabaud, & Gilmore (2000b, Paper III).

The first step is to define a volume-limited sample of well-measured stars and the ideal catalogue for this task is obviously the HIPPARCOS catalogue. This provides a direct way to infer the the solar neighbourhood SFH, as opposed to more indirect methods (e.g., Rocha-Pinto et al. 2000a). Although the completeness of HIPPARCOS varies both with spectral type and Galactic latitude, a cut at $V = 7.9$ for the sample with parallax errors smaller than 20% provides a reasonable sub-sample, once binaries and variable stars are removed. A typical cut at $V = 7.25$ produces an absolute-magnitude limited sample complete in volume with a well understood error distribution. The absolute magnitude limit, say at $M_V = 3.15$, implies that only stars younger than about 3 Gyr enter the sample, but the kinematical and geometrical corrections are minimised.

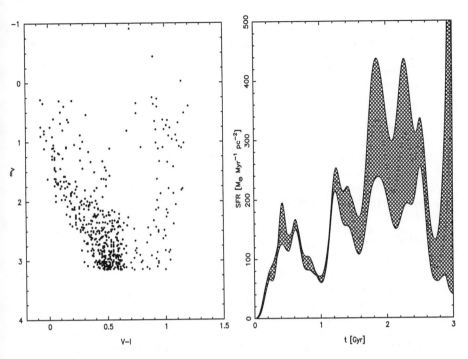

Figure 3.   Colour-magnitude diagram and SFH of the solar neighbourhood subsample observed by HIPPARCOS. Only stars bluer than $V - I = 0.7$ are considered in the reconstruction.

The inversion procedure applied to the actual HIPPARCOS diagram for the solar neighbourhood is shown on Figure 3. The superb quality of the data allows us to reconstruct its SFH with the unprecedented resolution of 50 Myr, and a clear pattern emerges: over a roughly low-level constant SFR, there are several distinct peaks, separated by about 0.5 Gyr. As before, the envelope gives the area where any function will maximise the likelihood, when different M_V cuts are applied. Note that the envelope increases with age, a reflection of larger uncertainty. This best solution is also a good solution (see Paper III).

A possible interpretation of the quasi-periodic episodes in the solar neighbourhood star formation rate may be given in terms of interactions with either spiral arms or the Galactic bar. As the pattern speed and the circular velocity are generally different, the solar neighbourhood periodically crosses an arm region, where the increased local gravitational potential might trigger an episode of star formation. The regularity present in the reconstructed $SFR(t)$ would be consistent with an interaction of the solar neighbourhood with a two-armed spiral pattern that induced the star formation episodes we detect. This is consistent with the current knowledge of the nearby spiral structure and may also be consistent with the creation of the inhomogeneities observed in the HIPPARCOS velocity distribution function, where well-defined branches are associated with moving groups of different ages (Asiain, Figueras, & Torra 1999), although in this case the time scales are much smaller. Of course, other explanations are possible; for example the cloud formation, collision and stellar feedback models predict a phase of oscillatory star formation rate behaviour as a result of a self-regulated star formation régime. Close encounters with the Magellanic Clouds have also been suggested to explain the intermittent nature of the star formation rate, though on longer time scales (Rocha-Pinto et al. 2000b).

5.   Conclusions

Our new maximum likelihood technique, coupled to a variational calculus, allows a robust, non-parametric reconstruction of the evolution of the star formation rate from the information contained in colour–magnitude diagrams. A full Bayesian analysis is also applied to assess whether the best solutions are also good fits to the data. Its current main limitation is the prior knowledge of the metallicity of the stellar ensemble. The typical star formation time scales found in WFPC2 CMDs of Local Group dSph galaxies are at most 2 Gyr (FWHM) and are currently limited by the precision in the photometry. The epochs of star formation episodes are well distributed over a Hubble time, and there is no such thing as a typical star formation history for these morphologically identical galaxies. In the solar neighbourhood observed by HIPPARCOS, we infer—with an unprecedented resolution of 50 Myr—the star formation history over the past 3 Gyr, finding a surprising regularity of star formation episodes separated by some 0.5 Gyr, perhaps triggered by the interaction with the spiral arm pattern.

References

Aparicio, A. 1998, in IAU Symp. 192, The Stellar Content of Local Group Galaxies, ed. P. Whitelock & R. Cannon (San Francisco: ASP), 20

Asiain, R., Figueras, F., & Torra, J. 1999, A&A, 350, 434

Dolphin, A. 1997, New Ast., 2, 397

Gallart, C., Freedman, W. L., Aparicio, A., Bertelli, G., & Chiosi, C. 1999, AJ, 118, 2245

Held, E. V., Saviane, I., Momany, Y., & Carraro, G. 2000, ApJ, 530, L85

Hernandez, X., Gilmore, G., & Valls-Gabaud, D. 2000a, MNRAS, 317, 83

Hernandez, X., Valls-Gabaud, D., & Gilmore, G. 1999, MNRAS, 304, 705

Hernandez, X., Valls-Gabaud, D., & Gilmore, G. 2000b, MNRAS, 316, 605

Hurley-Keller, D., Mateo, M., & Nemec, J. 1998, AJ, 115, 1840

Mateo, M. 1998, ARA&A, 36, 435

Mighell, K. J. 1997, AJ, 114, 1458

Ng, Y. K. 1998, A&AS, 132, 133

Rocha-Pinto, H. J., Maciel, W. J., Scalo, J., & Flynn, C. 2000b, A&A, 358,869

Rocha-Pinto, H. J., Scalo, J., Maciel, W. J., & Flynn, C. 2000a, ApJ, 531, L115

Saha, P. 1998, AJ, 115, 1206

Tolstoy, E., & Saha, A., 1996, ApJ, 462, 672

Tosi, M., Greggio, L., Marconi, G., & Focardi, P. 1991, AJ, 102, 951

Discussion

Michael Rich: It seems to me that for the older systems, just counting stars between the subgiant branches would be a pretty straightforward means of determining the star formation history. There is a great deal of uncertainty in the giant branches; clearly you do not know for sure that the mass function of each burst is identical so trying to disentangle the turnoff is very difficult. Although the results are interesting—for example, in Leo 2 you basically agreed with Mighell & Rich (1996, AJ, 111, 777)—but I see too many uncertainties. There are so many places where the evolutionary tracks overlap and they are otherwise difficult to interpret. Other groups have acknowledged the difficulty of this work. We need to have a comparison of different non-parametric methods on the same stellar populations. I am nervous about the uniqueness of this whole thing, though the results seem quite powerful.

David Valls-Gabaud: There are many things to answer your question. First, the assumption of a constant IMF over time may be an oversimplification, but (1) it is the first order and most conservative approximation one has to make, and (2) even if the IMF is variable, we have shown through our extended simulations that it would only be the amplitude of the bursts that would change from one episode to another, but the relative ages are correct. One could isolate the stars of each burst and try to derive their IMF, provided we knew something about or infer the properties of their binaries. Secondly we are not just fitting the shape of isochrones—we're explicitly taking into account the speed of evolution along the different isochrones. The uniqueness of the solution is guaranteed by the fact that the variational calculus gives a unique solution, and again we have checked that with Monte Carlo simulations. Thirdly we count stars in boxes to check whether the variational maximum likelihood solution is also a good solution. There we count not only stars on the subgiant branches but also all over the colour–magnitude diagram. All this is far more sensitive than the classical isochrone shape-fitting method.

Michael Rich: True, and the theoretical models are not so sensitive either.

David Valls-Gabaud: That's right, but one has to take stellar tracks from theoretical models, unless you want to make comparisons between observed data sets only, in which case you do get relative differences but you can't calibrate them properly. We have checked that using the Geneva or the Padova tracks gives the same solutions, and indeed we do get very similar answers. Perhaps tracks with alpha-enhanced elements will give slightly different solutions, we will check that. The fourth and last point is that indeed I have publicly proposed several times to the other groups involved in these reconstructions to do a double-blind test. An external referee would simulate a color–magnitude diagram with some SFR and chemical evolution histories and would send it to the different groups, without telling them the inputs, save perhaps for the photometric errors, crowding, etc. The referee would then compare the answers that the groups would send back. This seems to be the only way to assess the different biases that exist in different methods, and would convince everyone that the results obtained are trustworthy.

Astrophysical Ages and Time Scales
ASP Conference Series, Vol. 245, 2001
T. von Hippel, C. Simpson, N. Manset

New Searches for R-Process Enhanced Stars

N. Christlieb

Hamburger Sternwarte, University of Hamburg, Germany

T. C. Beers

Department of Physics and Astronomy, Michigan State University, USA

V. Hill, F. Primas

European Southern Observatory, Germany

J. Rhee

Department of Astronomy, University of Virginia, USA

S. G. Ryan

Open University, UK

M. Bessell, J. E. Norris

Mount Stromlo and Siding Spring Observatory, Australia

C. Sneden

Department of Astronomy, University of Texas, USA

B. Edvardsson, B. Gustafsson, T. Karlsson, M. Mizuno-Wiedner

Uppsala Astronomical Observatory, Sweden

Abstract. We discuss strategies for the detection of additional examples of highly r-process-enhanced, ultra-metal-poor stars, such as the two presently known examples of the class, CS 22892–052, and the newly discovered CS 31082–001. We expect that a quick, moderately high resolution spectroscopic survey of a large sample of metal-poor giants may reveal the presence of 10–20 similar stars.

1. Introduction

The extremely metal-poor ([Fe/H] = −3.1), highly r-process element enhanced star CS 22892–052 has been used for nucleochronometric age estimates, through the detection of the radioactive species Thorium, providing a lower limit for the age of the Galaxy. Recently, a second such star has been found. During their VLT/UVES high-resolution study of metal-poor stars selected in the HK survey (Beers 1999), Cayrel et al. (2001, and in these proceedings) reported that CS 31082–001 ([Fe/H] = −2.9) also exhibits r-process elements that are boosted relative to Fe by a factor of 30–50. Even more exciting, this star yielded the first detection of Uranium outside of the Solar System. The addition of U to the arsenal of potential cosmochronometers is a major breakthrough, and in principle provides for improved age estimates relative to previous Th-based chronometers (Toenjes et al., these proceedings). However, as noted by Hill et al. (these proceedings), there are subtle differences in the patterns of heavy elements in CS 31082–001, as compared with CS 22892–052, which serves a warning to astronomers that we must identify the range of behaviors exhibited by a larger sample of these stars before we can fully interpret their utility as chronometers.

The detection of a second star similar to CS 22892–052 indicates that such stars might be found with a frequency as great as 2–3% of the metal-poor giants in the halo. We discuss a strategy for the identification of 10–20 more r-process chronometers based on a quick, moderately high-resolution survey of metal-deficient giants selected from the Hamburg/ESO objective-prism survey (HES; Wisotzki et al. 2000), and from machine scans of the HK survey plates.

2. The HK Survey and the HES

The HK survey and the HES cover an effective area of $\sim 12\,500$ deg^2 in the northern and southern hemisphere. While the faintest objects selected in the HK survey have $B \sim 15.5$, the faintest metal-poor stars from the HES have $B \sim 17.0$. The properties of both objective-prism surveys are summarized in detail by Christlieb & Beers (2000).

In the HES, metal-poor candidates are selected in the database of some 4 million digital spectra by automatic spectral classification (Christlieb 2000). Moderate-resolution spectroscopic follow-up of turnoff stars ($0.3 < B - V < 0.5$) has shown that $\sim 80\%$ of the metal-poor candidates selected from the HES are stars with [Fe/H] < -2.0. For the cooler giants, with a stronger Ca II K line (used as a metallicity indicator in the objective-prism spectra), the yield of metal-poor stars is expected to be even higher.

We have recently completed APM scans of the original HK survey plates, and have developed a new selection algorithm based on Artificial Neural Networks (HK-II, Rhee 2000). The ANN approach has a number of advantages over the previous visual selection, most importantly that it avoids the introduction of a temperature-related bias in the identification of metal-poor stars. Tests conducted to date suggest that the ANN approach used for the HK-II selection should result in a detection efficiency of stars with [Fe/H] < -2.0 roughly three times that of the original visual selection (i.e., 60% as compared to 20–25%).

The high selection efficiency in both surveys would allow one to proceed directly to 8 m-class telescopes in order to find more r-process-enhanced stars, based on moderately high-resolution ($R = 20\,000$) spectra.

3. A Quick Europium Survey Using 8m-Class Telescopes

Our simulations show that stars with strong r-process enhancement can be read-ily recognized with spectra of $R = 20\,000$ and $S/N = 30$ pix^{-1} from their promi-nent Europium lines (see Figure 1). Spectra of this quality can easily be obtained

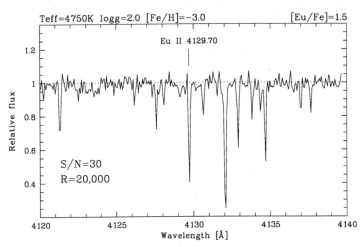

Figure 1.   Simulated spectrum of an r-process enhanced star.

with VLT/UVES or Subaru/HDS in 20 min (including overheads) for a $B = 15$ star, even under unfavourable conditions (full moon, poor seeing). This appar-ent magnitude is about the average brightness of the combined HK survey and HES metal-poor candidates. Therefore, in 10 nights roughly 250 stars could be observed with an estimated yield of 5 new r-process-enhanced stars. In addition, with these circa 1000 stars, one could analyse $N = 10$–12 additional elements for a very large set of metal-poor giants, including α-elements such as Ca and Mg and iron-peak elements such as Co and Ni. This would be *by far* the largest set of stars for which these elements have been analysed, and would be a useful tool to explore the nature of chemical enrichment in the early Galaxy.

References

Beers, T. C. 1999, in ASP Conf. Ser. 165, The Third Stromlo Symposium: The Galactic Halo, ed. B. K. Gibson, T. S. Axelrod, & M. E. Putman (San Francisco: ASP), 202

Cayrel, R., et al. 2001, Nature, 409, 691

Christlieb, N. 2000, Ph.D. Thesis, University of Hamburg

Christlieb, N., & Beers, T. C. 2000, in HDS Workshop, ed. M. Takada-Hidai & H. Ando (Mitaka, Japan: NAO), 255

Rhee, J. 2000, Ph.D. Thesis, Michigan State University

Wisotzki, L., Christlieb, N., Bade, N., Beckmann, V., Köhler, T., Vanelle, C., & Reimers, D. 2000, A&A, 358, 77

Astrophysical Ages and Time Scales
ASP Conference Series, Vol. 245, 2001
T. von Hippel, C. Simpson, N. Manset

Determination of Accurate Stellar Ages Using Detached Eclipsing Binaries

A. Giménez

Centro de Astrobiología (INTA-CSIC), Carretera de Ajalvir km. 4, Torrejón de Ardoz, E-28850 Madrid, Spain

I. Ribas

Villanova University, Villanova, PA 19085, USA and Universitat de Barcelona, Av. Diagonal, 647, E-08028 Barcelona, Spain

E. F. Guinan

Villanova University, Villanova, PA 19085, USA

Abstract. The suitability of using selected detached eclipsing binary systems to determine accurate stellar ages is briefly discussed. The limitations and areas of uncertainty are emphasized, as well as prospects for future work in this field. Special attention is paid to eclipsing binaries in Galactic clusters and Local Group galaxies.

1. Introduction

Double-lined eclipsing binaries are one of the best available sources for accurate masses and radii of stars. Fundamental stellar dimensions with precisions around 1% can be obtained independently of distance, which is otherwise only possible in the case of the Sun. Eclipsing binaries are superior to single stars because they simultaneously yield masses, radii, rotations, chemical compositions, and in some cases, internal structure parameters. Regarding their use for the determination of evolutionary ages, eclipsing binaries in open clusters, associations, or multiple star systems are of special interest, because they can provide additional anchors for the definition of the relevant isochrone for the star system in which they are located. Moreover, these systems offer the opportunity of comparing different methods for the estimation of stellar ages. Now, eclipsing binaries in the Magellanic Clouds can also be studied and attempts are being made to extend these studies even to M 31. Stellar structure and evolution in different stellar formation environments can thus be investigated.

The determination of ages of binary stars is carried out through the comparison of the absolute dimensions of the components with models of stellar evolution computed for a variety of initial chemical compositions. As shown in Figure 1, the currently available precisions in the determinations of absolute dimensions allow for age estimations of eclipsing systems with uncertainties of 2–20%. This accuracy in the age determination is strongly dependent on the evolutionary stage of the system primarily because of the decreasing crowding

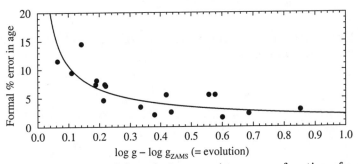

Figure 1.   Variation of the relative error in age as a function of evolution for a sample of eclipsing binaries with accurate dimensions.

of isochrones away from the zero age main sequence. The age uncertainties are based, however, upon the validity of the adopted stellar evolution models used. The most critical problems and limitations of this procedure are discussed below.

2.   The Method and Some Caveats

The stellar parameters that we use to analyze the structure and evolution of the components of a binary star are: age (τ), effective temperature (T), mass (m), radius (R), helium content (Y), and metal content (Z), for each of them, i.e., 12 initially unknown values. Accurate light and radial velocity curves, as well as multicolour photometry, provide excellent determinations of masses, radii, and temperatures of both components. In other words, we can measure 6 parameters and are thus left with only 6 unknowns. But if we assume, as it seems reasonably justified to do so, both components of a system to have the same age and chemical composition, we finally have just 3 parameters to determine, namely, τ, Y and Z. These are obtained by using the corresponding constraints set by the adopted grid of theoretical stellar evolution models since they are functions of the mass, radius, and temperature of each of the components.

   Of course, only binaries with no mass exchange can be used since otherwise the standard models of stellar structure would not be applicable. Unfortunately, many moderate-mass stars in close binaries (short period) tend to fill their corresponding Roche lobe as soon as they leave the main sequence and very few evolved well-detached candidates are available for detailed studies. Very massive systems are even more scarce due to their rapid evolution, and the very low-mass end of the main sequence is difficult to study because of the enhancement of magnetic activity by tidally-induced fast rotation.

   The solution to the multi-parametric functions is sought through the minimization of an ad-hoc χ^2 function written as (Ribas et al. 2000a)

$$\chi^2(Z,Y) = 6 \frac{(\log \tau_1 - \log \tau_2)^2}{\sigma^2_{\log \tau_1} + \sigma^2_{\log \tau_2}} + 2 \frac{(\log T_1^{obs} - \log T_1)^2}{\sigma^2_{\log T_1}} + \frac{(\log T_2^{obs} - \log T_2)^2}{\sigma^2_{\log T_2}}.$$

   Furthermore, we should point out that, even though the determination of the best-fitting age and chemical composition is possible following the method just outlined, a reliable test of the adopted models is only possible if additional information is available, namely a spectroscopic determination of the metal abundance Z. The theoretical models may also have uncertainties that could bias the resulting stellar properties and thus the age determinations. Indeed, there are several physical mechanisms in stellar structure and evolution that are still poorly known or very difficult to account for in practical computer codes. Among them, the most relevant to stellar ages are: (a) rotation, which increases the size of convective cores in massive stars, extending their life within the main sequence and modifying their chemical evolution; (b) convective core overshooting, which also alters the evolution of abundances and increases the duration of the main sequence phase; and (c) settling of helium and heavier elements, which affects the determination of age in low-mass stars.

To illustrate how detached eclipsing binaries are used to provide accurate determinations of stellar ages, some references to individual cases can be given. AG Per (Giménez & Clausen 1994), an unevolved member of a young association (Per OB2), is a system with a relatively low metal content and an eccentric orbit showing apsidal motion. CD Tau (Ribas, Jordi, & Torra 1999) is a member of a wide triple system that allows for a check of the isochrone defined by the close components. AI Phe (Andersen et al. 1988) is a good example of an evolved system beyond the terminal age main sequence. In the LMC, HV 2274 (Ribas et al. 2000b) and HV 982 (Fitzpatrick et al. 2001), both in eccentric orbits, also yield accurate distance determinations to their host galaxy.

3. The Future

The study of eclipsing binaries to derive accurate absolute parameters is proving to be very rewarding and new perspectives are opening for the new millennium. First, the study of systems in other stellar populations and galaxies (members of the halo and the bulge of the Milky Way, the LMC, the SMC, or M 31) will allow tests of the basic assumptions about initial chemical composition in the currently adopted theory of stellar structure. Second, it is extremely important to increase the number of systems with reliable direct (atmospheric) determinations of the value of Z to minimize the grid search as well as to test some of the adopted physics. However, efforts need also to be focused on theoretical aspects to include stellar rotation and other processes in the models in a realistic way. Finally, the combination of better models together with observational improvements should yield stellar ages with accuracies around 2–5%.

References

Andersen, J., Clausen, J. V., Nordstrom, B., Gustafsson, B., & Vandenberg, D. A. 1988, A&A, 196, 128

Giménez, A., & Clausen, J. V. 1994, A&A, 291, 795

Fitzpatrick, E. L., et al. 2001, ApJ, in press

Ribas, I., Jordi, C., & Torra, J. 1999, MNRAS, 309, 199

Ribas, I., Jordi, C., Torra, J., & Giménez, A. 2000a, MNRAS, 313, 99

Ribas, I., et al. 2000b, ApJ, 528, 692

Astrophysical Ages and Time Scales
ASP Conference Series, Vol. 245, 2001
T. von Hippel, C. Simpson, N. Manset

Eddington: A European Space Mission That Will Measure the Age of the Stars

A. Giménez

Centro de Astrobiologia (INTA-CSIC), Carretera de Ajalvir km. 4, Torrejón de Ardoz, E-28850 Madrid, Spain

F. Favata

Space Science Department, ESA, P.O. Box 299, 2200 AG, Noordwijk, The Netherlands

Abstract. Eddington is an F-type mission within the package recently selected by ESA as "reserve" for the period 2006-2012. It has two main scientific goals: the study of stellar interiors by means of asteroseismology of stars all along the HR diagram, and the search for transits of Earth-like planets in front of solar-type stars. The satellite will carry in L2 orbit a 1.2m wide-field 3-reflection telescope with a camera of 20 tiled CCDs. It is expected that the asteroseismology part of the mission will provide ages and chemical compositions within 1% accuracy. Photometry to 1 ppm for $V < 11$ will allow the study of a large number of stars with a broad range of masses, ages, and abundances, both isolated and in stellar clusters.

1. Introduction

Eddington was proposed to the European Space Agency (ESA) at the beginning of 2000 within the F-type class of missions. It was then approved for feasibility studies and selected as a "reserve" in October of the same year. The studies leading to a better definition of the mission and the involved technologies are currently being carried out in order to secure its capability for a full and timely implementation.

Eddington has two main scientific goals:

- to supply an empirical basis to stellar structure and evolution theory, using stellar oscillations, i.e., asteroseismic data,

- to search for habitable extrasolar planets through the measurement of their transit across the stellar disc by means of high-precision photometry.

In the first case, the observation of a large number of stars in different evolutionary stages is necessary. As a main outcome, precise dating (to 1%) of stellar populations and individual stars will be possible complementing and supporting other ESA missions, in particular GAIA. Quantitative astrophysics will be possible with results from Eddington in terms of ages, structure and chemical composition. In fact it will be the dating machine for components of Galactic structure.

Asteroseismology is a known and tested technique for the use of stellar oscillations for the diagnosis of the interior structure and stage of evolution of stars. Oscillation frequencies are determined by the internal structure (pressure, density, core mass, rotation, etc.) because stars are essentially transparent to sound waves. Frequencies are often determined for a broad range of modes, hence detailed information about stellar interiors is available.

Techniques have been tried and tested on the Sun with SoHO, including inversion methods leading to the model that better fits the observed frequencies. But the Sun is just one simple star, with no convective core, slow rotation, relatively unevolved, and comparatively simple material physics. Proper investigations of stellar structure and evolution require the study of a broad range of stars. Eddington will provide this.

2. The Payload

Both science goals of the Eddington mission, the study of stellar structure through asteroseismology and the search for extrasolar planets by means of the transit method, require high-precision differential and long-duration photometry of a large number of stars. The design of the payload is thus made such that both goals are achievable with one single instrument which is as simple and efficient as possible.

Eddington's payload consists of a 1-m class optical telescope providing a field of view of 3 degrees in diameter, and a mosaic CCD camera (called EddiCam). The baseline design of the telescope, not yet studied industrially, is a three-reflection system with no refractive component allowing a compact structure with a large, fully corrected, unvignetted and planar field of view. A single focal plane instrument, EddiCam, consists of an array of 20 tiled CCDs covering the field of view. 16 of these CCDs work in frame transfer mode for asteroseismology and all of them in full-frame mode for planet finding. Large and efficient pixels, 27 μm, allow for full-well capacities above 10^6 e^- and low noise levels. Operation of the detectors at -90 C using passive cooling and the location of the spacecraft in a L2 orbit ensure the required stability, low levels of light contamination and particularly the high duty cycle of 95%.

The observing strategy of Eddington during the part of the mission used for asteroseismology is to devote between 1 and 2 months to a number of selected fields for a total duration of 2 years. In this way as many as around 50 000 stars will be studied allowing a good coverage of the whole H-R diagram, surveying stars with different masses, ages, and chemical abundances. The observation of stellar clusters is very important in this context, limiting the acceptable levels of defocusing and therefore the use of a large number of pixels to measure the photons coming from a given star lead to less stringent pointing and stability requirements.

3. Expected Results

The Eddington mission will study stellar oscillations corresponding to those of the Sun in stars as faint as $V = 11$. A very high duty cycle (95%) leads to simple interpretation of the frequency spectrum and very extended observations during the three years planet-finding phase of the mission will give excellent frequency resolution of slow pulsators using longer integration units.

In terms of possible targets, open clusters with massive pulsating stars, e.g., NGC 6231, will allow the study of pre-supernova evolution. Of course, open clusters with solar-like pulsators, like the Hyades, Pleiades, or Praesepe, will also be studied. Moreover, old, metal-poor stars as samples of the early evolution of the Galaxy will be targeted.

Uncertainties in the derived parameters for a cluster with moderate-mass stars are evaluated to be around 0.1% in age, helium content (Y), and metal abundance (Z) using the measured frequencies. For the same parameters the expected uncertainties are only 1.3%, 0.3%, and 3%, respectively, when only the average frequency separations are used. These values can be compared with currently available precisions, using standard methods, which are larger than 20% in age, and above 10% in Y and Z.

The purpose of Eddington, concerning the study of stellar ages, is to have at the end of the mission a reliable test of the currently adopted models of stellar structure and evolution. Inversion techniques in the measured power spectrum of oscillations will provide the size of the stellar core as well as internal structure parameters, like pressure, density, or rotation. In addition, it will be possible to study the chemical evolution of stars and determine key parameters of stellar interiors, like convective overshoot, mixing, diffusion, etc.

4. Status of the Mission

Eddington was approved as part of ESA's program in the course of the F2/F3 selection round, with "reserve status." The payload was only studied at summary level during assessment phase (feasibility study) and is now the subject of further analysis at Phase-A level which also covers all aspects of the mission.

For the industrial study of the telescope an Invitation To Tender (ITT) has been released in April 2001. This is an open study, not limited to a specific configuration, aimed to maximize performances, for example in field of view. Further an industrial study of CCD performances is being prepared to verify the photometric capabilities of available detectors in the expected conditions of Eddington from all points of view such as orbit or telescope design. Industrial activities at system level will be the subject of another ITT during the coming year.

On the other hand, the camera design and performances are being studied with a PI consortium which is being formed with scientific groups from different ESA member countries. These studies will benefit from the industrial CCD study and is being carried out at the same phase-A level as the telescope. Particular attention is devoted to simulations and analysis of crucial elements of Eddington's scientific performances in light of the different possible detailed camera designs.

Astrophysical Ages and Time Scales
ASP Conference Series, Vol. 245, 2001
T. von Hippel, C. Simpson, N. Manset

The Light Elements in TO and SGB Stars of Two Globular Clusters: Implications for Photometric Age Determinations

VLT Globular Cluster Team: R. G. Gratton (Padova, Italy),
P. Bonifacio (Trieste, Italy), A. Bragaglia (Bologna, Italy), E. Carretta
(Padova, Italy), V. Castellani (Pisa, Italy), M. Centurion (Trieste,
Italy), A. Chieffi (CNR), R. Claudi (Padova, Italy), G. Clementini
(Bologna, Italy), F. D'Antona (Roma, Italy), S. Desidera (Padova,
Italy), P. François (ESO, Germany), F. Grundahl (Aarhus, Denmark),
S. Lucatello (Padova, Italy), P. Molaro (Trieste, Italy), L. Pasquini
(ESO, Germany), C. Sneden (Texas, USA), F. Spite (Meudon, France),
and O. Straniero (Collurania, Italy)

Abstract. We have obtained high-resolution spectra of stars at the
main sequence turnoff (TO) and along the subgiant branch (SGB) in
the globular clusters NGC 6397 and NGC 6752. In both clusters the
[Fe/H] metallicities of the TO and SGB stars agree well, and we obtain
[Fe/H] = −2.03 and −1.42 for NGC 6397 and NGC 6752, respectively.
In NGC 6397, a uniform but low [O/Fe] = +0.21 is obtained, indicating that this small-mass cluster is chemically homogeneous. However,
in NGC 6752 an anticorrelation exists in O vs. Na abundances. This
anticorrelation is often seen in very evolved giants of clusters, but its existence among TO stars of this cluster almost certainly implies that this
abundance signature was the work of previous generation(s) of stars in
the cluster. Oxygen is a key element for cluster photometric age dating
techniques, as it is an important opacity source in stellar interiors. The
low and/or variable derived oxygen abundances of these clusters serve
as cautions against assuming single-valued (and large) O/Fe values in
isochrone computations.

1. Introduction

Globular clusters have generally homogeneous compositions of Fe-peak elements,
with the notable exception of large star-to-star [Fe/H] variations in ω Cen, and
possible (but unconfirmed) much smaller variations in M 22. But there is now
wide-spread evidence for large inter- and intra-cluster scatter in relative abundances [X/Fe] of light elements that are susceptible to high-temperature proton-
capture nucleosynthesis, in reaction chains such as the CN, ON, NeNa, and
MgAl cycles. Abundances of C, N, O, Na, Mg, and Al can vary among stars
of individual clusters by factors of 10 or more, and carbon isotope ratios often
are observed to be ∼4–6, very close to the CN-cycle equilibrium value. The
observational evidence, based mostly on abundance studies of highly evolved
cluster red giants, has been reviewed on several occasions in the past decade,

and for further discussion and references to relevant abundance papers, refer to e.g., Briley et al. (1994), Kraft (1994), and Sneden (2000).

It is clear that surface layers of many globular cluster giants have at some point been subjected to advanced proton-capture fusion, but just where and when has been very unclear. It is possible that the highly convective red giants have dredged up this material from their own interiors, but it is also possible that the stars under observation either were born from (or acquired at a very early stage) material that was proton-capture-processed in a prior star. Arguments can be found both to support these ideas and to call them into question. For example, if the low-mass red giants mixed these processed elements from their own interiors, how did they ever get their hydrogen-burning layers to the required very high temperatures necessary for the MgAl cycle to occur? If these stars instead were passive recipients of material processed by earlier cluster generations of higher mass stars, how did some of the present stars obtain proton-capture material and others avoid it?

A simple observational test of these questions would be to search for abundance variations among the light elements in relatively unevolved stars near cluster main sequences. In a study reported by Gratton et al. (2001), we have determined O, Na, Mg, and Al abundances in low luminosity stars in NGC 6397 ([Fe/H] $= -2.03 \pm 0.05$ in the present study) and NGC 6752 ([Fe/H] $= -1.42 \pm 0.05$). The high resolution, high S/N spectra of these faint stars have been obtained with the new UVES echelle spectrograph of the VLT. This brief report highlights only a few main results of this work.

2.   Abundance Summary and Implications

Five turnoff (TO) and three subgiant branch (SGB) stars were observed in NGC 6397. The metallicity derived from our data is about 0.1–0.2 dex lower than most literature values. In this cluster, we derive [O/Fe] $= +0.21 \pm 0.05$; this is not very large compared to typical halo stars, and (unlike the situation in many other clusters) the O abundance is very uniform from star to star. Mg abundances are also low in this cluster, and [Na/Fe] $= +0.20 \pm 0.05$. The small variations in these abundances suggests that NGC 6397 may be very chemically homogeneous in all elements.

Our results for [O/Fe] in NGC 6397 indicate that care should be taken in assuming oxygen abundances for stellar interiors calculations relevant to isochrone predictions. It appears that different clusters may have different oxygen contents in stars near their main sequence TO positions. This is an additional age determination uncertainty (which can amount to 1–2 Gyr) that cannot be neglected. Unfortunately, another implication of the derived abundances for the light elements in this cluster is that a necessary step in the age determination of an individual cluster must include the acquisition and analysis of high resolution spectra of many cluster members, perhaps at least ten stars to be sure that the range of oxygen abundance in the cluster has been properly sampled.

Nine TO and nine SGB stars were observed in NGC 6752. Our Fe-metallicity value for this cluster is in accord with or larger by 0.1 dex than earlier values. Unlike in NGC 6397, there is a large star-to-star scatter among the light elements in NGC 6752, and a clear O–Na anticorrelation is apparent even among

TO stars. There is also probably a Mg–Al anticorrelation, but Al abundances are more difficult to measure in the TO stars. Light element abundance correlations and anticorrelations conform to proton-capture expectations, but if they have been internally generated via deep mixing they should not appear in TO stars. This suggests that a prior generation or generations of stars was (were) responsible for polluting the presently-observed cluster stars with processed material.

These two clusters illustrate the probable underlying truth that globular cluster environments substantially influence the compositions of their members. The large range in the light elements so evident in globular cluster stars seems not to (ever?) occur in Galactic field stars (e.g., Gratton et al. 2000, and references therein). The relatively invariant abundances of NGC 6397 stars are somewhat similar to those of field stars, whereas the star-to-star variations in especially O and Na of ~ 1 dex seem to be similar to the extremely large variations within M 13 (Kraft et al. 1997). A comprehensive study of TO and SGB abundance patterns in M 13 has yet to be accomplished, and would be of great interest to compare with our results for NGC 6752.

References

Briley, M. M., Bell, R. A., Hesser, J. E., & Smith, G. H. 1994, Can. J. Phys., 72, 772

Gratton, R. G., Sneden, C., Carretta, E., & Bragaglia, A. 2000, A&A, 354, 169

Gratton, R. G., et al. 2001, A&A, 369, 87

Kraft, R. P. 1994, PASP, 106, 553

Kraft, R. P., Sneden, C., Smith, G. H., Shetrone, M. D., Langer, G. E., & Pilachowski, C. A. 1997, AJ, 113, 279

Sneden, C. 2000, in The Galactic Halo: From Globular Cluster to Field Stars, ed. A. Noels, P. Magain, D. Caro, E. Jehin, G. Parmentier, & A. A. Thoul (Li'ege: Inst. d'Ap. Geophys.), 159

Astrophysical Ages and Time Scales
ASP Conference Series, Vol. 245, 2001
T. von Hippel, C. Simpson, N. Manset

Cosmochronometry and New Constraints on R-Process Nucleosynthesis

M. Hannawald, B. Pfeiffer, K.-L. Kratz

Institut für Kernchemie, Universität Mainz, Germany

Abstract. Improved nuclear-physics inputs have resulted in a better reproduction of observed r-process abundances. Based on our new calculations, we have obtained a more precise Galactic age of 14.6 ± 2.4 Gyr from the Th/Ba-Ir ratios of CS 22892–052.

1. Introduction

A correct understanding and astrophysical modeling of r-process nucleosynthesis requires—among several parameters—a reliable knowledge about nuclear properties very far from β-stability. Experiments at CERN/ISOLDE have played a pioneering role in exploring the development of nuclear structure as a function of isospin, e.g., in terms of masses, β-decay properties and level systematics. Initial studies paid attention to individual waiting-point nuclei with magic neutron numbers, related to the formation of the $A \simeq 80$ and 130 Solar System r-process abundance ($N_{r,\odot}$) peaks (Kratz 1993). More recent activities, taking advantage of chemically selective resonance-ionization laser ion-sources are centered on the evolution of shell effects towards the neutron drip-line in the ^{132}Sn region (Kratz 2000). Our new experimental and theoretical nuclear-physics inputs have resulted in a better reproduction of the global isotopic r-process pattern including the Th-U chronometers commonly used to determine the age of the Galaxy (Pfeiffer 1997). Guided by recent observations from low-metallicity halo stars, our calculations support a two-component composition of the r-process separated by the $A \simeq 130$ $N_{r,\odot}$ peak.

2. Progress in Nuclear Mass Models

In the past few years considerable improvement in mass models used in r-process calculations could be achieved. Still up to the mid-1990's, large deficiencies in calculated r-abundances near the magic neutron numbers $N = 82$ and $N = 126$ were observed when applying standard macroscopic-microscopic mass formulae, such as ETFSI-1. The resulting abundance troughs were interpreted to originate from neutron-shell quenching far from β-stability (Kratz 1993; Dobaczewski 1995). Replacing the old mass models with strong shell closures by more recent ones with quenched shell structure, such as ETFSI-Q (Pearson 1996), indeed led to a filling-up of the abundance troughs together with an overall improvement of the r-abundance fit of the heavier elements (Pfeiffer 1997). This is shown in Figure 1 for the $N_{r,\odot}$ isotopic pattern.

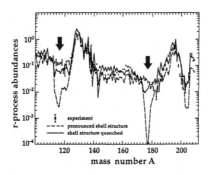

Figure 1.   Comparison of calculated with observed Solar System r-process abundances, using two mass models: ETFSI-1 with pronounced shell gaps and ETFSI-Q with quenched neutron shells far from stability.

3.   Two R-Process Components in the Early Galaxy

Recent results indicate the existence of (at least) two types of r-process. The evidence is based on a variety of observations in different fields. The strongest indication comes from heavy neutron-capture element abundances in very metal-poor halo stars (Westin 2000) as well as in the globular cluster M 15 (Sneden 2000). On the one hand, metallicity-scaled abundances in the Pt peak and down to Ba ($Z = 56$) in all halo stars so far investigated are in remarkable agreement with the solar ($N_{r,\odot}$) pattern, while on the other hand the abundances of low-Z neutron-capture elements ($_{39}$Y to $_{48}$Cd) in CS 22892–052 are lower than solar.

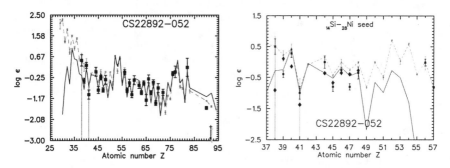

Figure 2.   *Left:* Comparison between observed (filled squares) and calculated (solid line) elemental r-abundances from CS 22892–052. *Right:* Comparison between abundance residuals (filled diamonds) and calculated (full curve) elemental r-abundances.

Taking our waiting-point approach to fit the $N_{r,\odot}$ pattern, we have investigated under which stellar conditions the possible two r-processes have to run. When assuming that the abundances are a living record of the first (few) generation(s) of Galactic nucleosynthesis, the observed pattern beyond $Z \simeq 40$ up to Th-U should most likely be produced by only one (or a few) r-process event(s) in a unique stellar site. This main r-process then produces the low-Z elements $(40 \leq Z \leq 48)$ under-abundant compared to solar, and reaches the full solar values presumably around $_{52}$Te. For CS 22892–052 both the general trend and the detailed structure of the low-z abundance are nicely reproduced in our fit with the ETFSI-Q atomic masses. At the same time, the good overall reproduction of the high-Z elements (beyond $_{56}$Ba) is maintained (see left part of Figure 2). Our approach would imply a roughly constant abundance ratio between the low-Z and high-Z elements. This has recently been confirmed in the case of HD 115444, where our prediction for Ag agrees with observations (Westin 2000).

Consequently, the low Z abundance residuals $(N_{r,\odot} - N_{r,\text{main}} = N_{r,\text{weak}})$ require a separate weak r-process of secondary, yet unknown stellar origin. When assuming seed compositions from $_{14}$Si to $_{24}$Cr or $_{28}$Ni in Solar System fractions, our calculations can reproduce the $N_{r,\text{weak}}$ pattern in CS 22892–052 with neutron densities $n_n \leq 10^{20}$ cm^{-3} and process durations $\tau \sim 0.5 - 1.0$ s (Figure 2, right). These stellar conditions might exist in explosive shell-burning scenarios (Truran 2001). Another outcome of our calculations is that the weak component does not make a significant contribution to the $A \simeq 130$ abundance peak.

4.   New Estimate for the Galactic Age From CS 22892–052

Actinide chronometers have been widely used to determine ages of meteoritic samples and metal-poor stars. Based on our classical r-process calculations (Pfeiffer 1997), the age of the halo-star CS 22892–052 was derived as 15.6 ± 4.6 Gyr from only a single abundance ratio of radioactive ^{232}Th and stable Eu (Cowan 1999), neglecting possible effects from β-delayed fission (βdf). Now, with our improved nuclear-data inputs together with new observations of 15 rare-earth elements and Os and Ir in the third r-abundance peak, we have for the first time obtained an element-consistent picture of the Th-age of CS 22892–052. When using the most recent fission-barrier calculations (Mamdouh 1999), we estimate the βdf-effect on the Th-age to be a maximum of 1.5 Gyr. Taking into account this effect, from the weighted average of Th/Ba–Ir in CS 22892–052, we obtain a lower limit for the Galactic age of 14.6 ± 2.4 Gyr. From this value, we predict a ^{238}U abundance of $\log \epsilon \simeq -2.5^{+0.1}_{-0.2}$ for this halo star.

References

Cowan, J. J. 1999, ApJ, 521, 194

Dobaczewski, J. 1995, Phys. Scr. T., 56, 15

Kratz, K.-L. 1993, ApJ, 403, 216

Kratz, K.-L. 2000, Hyperfine Interaction, 129, 185

Mamdouh, A. 1999, Nucl. Phys. A, 648, 282

Pearson, J. M. 1996, Phys. Lett. B, 387, 455

Pfeiffer, B. 1997, Z. Phys. A, 357, 235

Sneden, C. 2000, ApJ, 536, L85

Truran, J. J. 2001, Nucl. Phys. A, 688, 330

Westin, J., 2000, ApJ, 530, 783

Astrophysical Ages and Time Scales
ASP Conference Series, Vol. 245, 2001
T. von Hippel, C. Simpson, N. Manset

Luminosity Function and Age of White Dwarfs in the Disk

Hugh C. Harris, Conard C. Dahn

U.S. Naval Observatory, P.O. Box 1149, Flagstaff, AZ, USA

S. K. Leggett

Joint Astronomy Centre, 660 North A'ohōkū Place, Hilo, HI, USA

James Liebert

Steward Observatory, University of Arizona, Tucson, AZ, USA

Abstract. An expanded sample of white dwarfs, selected by proper motion, is used to determine an improved luminosity function and to study the makeup of the sample. Most, but not all, are disk stars. The oldest disk star has a cooling time of 9 Gyr.

1. Introduction

The white dwarf luminosity function (WDLF) provides a record of the rate of formation of intermediate-mass stars in the past—the age of the oldest white dwarfs indicates the time when star formation began. Because most white dwarfs in the solar neighborhood are disk stars, the WDLF reflects the star-formation rate and age of the Galactic disk. Improved models of WD cooling times (Fontaine, Brassard, & Bergeron 2001, and these proceedings) have increased our confidence that the age derived from the WDLF is accurate. Observationally, the WDLF from Liebert, Dahn, & Monet (1988; LDM), and the reanalysis of the same sample by Leggett, Ruiz, & Bergeron (1998), provide an age for the disk as accurate as that determined by other methods. Factors currently limiting the accuracy are the small sample size, giving only a few stars with lowest luminosity and oldest age (Wood & Oswalt 1998), and the uncertain changes between an H-dominated and He-dominated atmosphere as a WD cools. This paper addresses both problems by doubling the size of the WD sample used for the WDLF.

2. Expanded Sample

To provide an enlarged sample, we have identified all WDs (91 stars) brighter than the Luyten magnitude limits with $\delta > -20°$ and proper motion $\mu > 0.6''$ yr^{-1}, instead of the limit $\mu > 0.8''$ yr^{-1} used by LDM. Because low-luminosity white dwarfs are distributed nearly uniformly around the Sun, the revised limit was expected to result in a sample roughly $(0.8/0.6)^3 = 2.4$ times larger, and it did. We have also obtained new parallax measurements for 43 stars, thus determining the distances for nearly all stars in the sample, necessary for the LF analysis. The expanded sample includes a new magnetic DQ WD, a high-mass WD, and a cool star with incipient H_2 collision-induced absorption.

This larger sample gives a more precise measure of the WDLF, and it also provides more data with which to address the issue of possible incompleteness. As an example of the improvement in accuracy of the WDLF at the faint end, Figure 1 shows the new sample analyzed in the same way as the LDM sample.

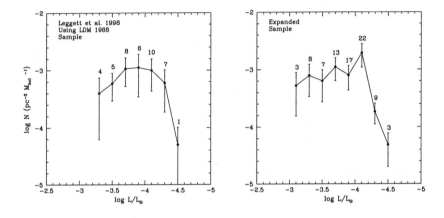

Figure 1.   Increased accuracy of the WDLF at the faint end with the expanded sample (*right*) compared to the earlier sample of LDM (*left*). The number of stars in each luminosity bin is labeled.

3.   Halo Versus Disk Stars

One would like to identify stars belonging to the Galactic halo, in order to determine the exciting and controversial halo WDLF, but also to remove potential halo contamination from the sample used to find an age for the disk. The same is also true for the Galactic thick disk. Halo WDs are well known in local samples (Liebert, Dahn, & Monet 1989). A kinematic identification can be effective, at least for halo stars, based on the tangential velocities (Liebert et al. 1999). Monte Carlo models (like those of Wood & Oswalt 1998) can be used to estimate both completeness and contamination. Figure 2 shows that a 3-component model fits the observed distribution of tangential velocities very well. These models will be explored further in another paper. They give a consistent result that about $15 - 20\%$ of our sample are not disk stars. Note that a tangential velocity limit > 140 km s^{-1}is required in order to isolate a halo population in this sample.

4.   The Disk Age

The abrupt drop in the WDLF occurs at $L/L_\odot = 4.25$. This corresponds to a cooling age (WD cooling time only) of 8.7 Gyr for stars with H atmospheres (e.g.,

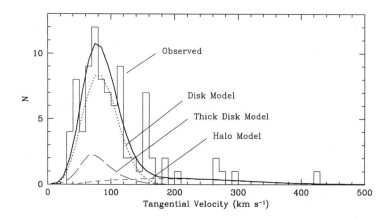

Figure 2. Three-component model that fits the distribution of tangential velocities and proper motions in the sample.

Chabrier et al. 2000). However, the stars in our sample with lowest luminosity appear to have He-dominated atmospheres. They will cool more rapidly, at least during the phases of their evolution that they are He-dominated, so are not indicative of the disk age. The star with lowest luminosity that clearly has an H-dominated atmosphere is LHS 2696 with $L/L_\odot = 4.28$ and a cooling age of 9 Gyr. There are enough stars in the sample to calculate a LF using only stars with H-dominated atmospheres, and we are in the process of doing this.

A potential complication is the evidence that some WDs have their atmospheric composition change between H and He, perhaps switching several times as they cool (Bergeron, Leggett, & Ruiz 2001). If we include in our WDLF only stars that now have H-dominated atmospheres, we still may overestimate the age due to these possible episodes of rapid cooling. This uncertainty will remain until the cooling processes, and associated atmospheric changes, are understood.

References

Bergeron, P., Leggett, S. K., & Ruiz, M. T. 2001, ApJS, 133, 413

Chabrier, G., Brassard, P., Fontaine, G., & Saumon, D. 2000, ApJ, 543, 216

Fontaine, G., Brassard, P., & Bergeron, P. 2001, PASP, 113, in press

Leggett, S. K., Ruiz, M. T., & Bergeron, P. 1998, ApJ, 497, 294

Liebert, J., Dahn, C. C., Harris, H. C., & Leggett, S. K. 1999, in ASP Conf. Ser. 169, Eleventh European Workshop on White Dwarfs, ed. J. E. Solheim & E. G. Meistas (San Francisco: ASP), 51

Liebert, J., Dahn, C. C., & Monet, D. G. 1988, ApJ, 332, 891 (LDM)

Liebert, J., Dahn, C. C., & Monet, D. G. 1989, in IAU Coll. 114, White Dwarfs, ed. G. Wegner (Berlin: Springer-Verlag), 15

Wood, M. A., & Oswalt, T. D. 1998, ApJ, 497, 870

Astrophysical Ages and Time Scales
ASP Conference Series, Vol. 245, 2001
T. von Hippel, C. Simpson, N. Manset

R-Process Pattern in the Very-Metal-Poor Halo Star CS 31082–001

Vanessa Hill

European Southern Observatory, Germany

Bertrand Plez

GRAAL, Université de Montpellier 2, France

Roger Cayrel

DASGAL, Observatoire de Paris, France

Timothy C. Beers

Michigan State University, USA

Abstract. The very-metal-poor halo star CS 31082–001 was discovered to be very strongly r-process-enhanced during the course of a VLT+UVES high-resolution follow-up of metal-poor stars identified in the HK survey of Beers & colleagues. Both the strong n-capture element enhancement and the low C and N content of the star (reducing the CN molecular band contamination) led to the first ^{238}U abundance measurement in a stellar spectrum (Cayrel et al. 2001), and the opportunity to use both radioactive species ^{238}U and ^{232}Th for dating the progenitor to this star. However, age computations all rely on the hypothesis that the r-process pattern is solar, as this was indeed observed in the other famous r-process-enhanced very metal poor stars CS 22892–052 (Sneden et al. 1996, 2000) and HD 115444 (Westin et al. 2000). Here, we investigate whether this hypothesis is verified for CS 31082–001, using a preliminary analysis of over 20 abundances of n-capture elements in the range $Z = 38$–92.

Cayrel et al. (2001, and in these proceedings) discuss the discovery and importance of CS 31082–001, and here we report a summary of a preliminary abundance analysis for this star (Table 1). The n-capture element abundances (relative to iron, [X/Fe]) of CS 31082–001 are compared in Figure 1a to those of CS 22892–052 and HD 115444, showing that the overabundance of the $Z > 56$ n-capture elements in CS 31082–001 is almost identical to that of CS 22892–052, with a mean overabundance of [X/Fe] $\sim +1.7$ dex. These two stars are therefore the most extreme cases of n-capture element enhancement in halo stars, far more extreme than HD 115444. Furthermore, the abundance pattern of the $56 < Z <$ 70 elements in CS 31082–001 are indistinguishable from that of CS 22892–052 or HD 115444. In contrast, the abundance pattern of $Z > 70$ seems to be more abundant in CS 31082–001 than in CS 22892–052 or HD 115444, including Th, which is a factor 4 more abundant in CS 31082–001 than in CS 22892–052.

316

Therefore the $\log \epsilon(\text{Th}/\text{Eu})$ ratio, often used as an age indicator, is a factor 3 larger in CS 31082–001 than in CS 22892–052.

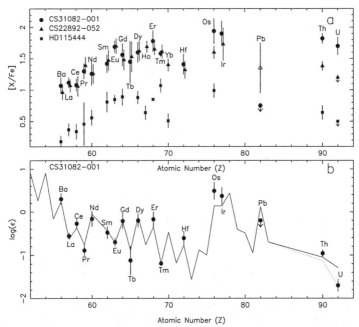

Figure 1. N-capture abundances in CS 31082–001 compared to (*a*) CS 22892–052 and HD 115444, and (*b*) the Solar System r-process (Burris et al. 2000) scaled to match the $56 \geq Z \leq 72$ abundances of CS 31082–001. The abscissa for CS 22892–052 has been shifted by +0.3 for readability. The radioactive (Th and U) Solar System abundances are the values at the formation time of the Solar System. The dotted line shows the abundances observed *today* for these two species.

In Figure 1b, the n-capture element abundances of CS 31082–001 are compared to Solar System r-process pattern (as of Burris et al. 2000) scaled to match the mean $56 \geq Z \leq 72$ n-capture elements abundance of CS 31082–001 $\log \epsilon_{\text{CS 31082--001}} - \log \epsilon_{\text{SS}} = -1.22 \pm 0.03$ ($\sigma = 0.10$ over 13 elements). Here again, whereas the $56 < Z < 70$ elements in CS 31082–001 are very well reproduced by a solar r-process, the $Z > 70$ elements are more erratic. While Os and Ir seem to be more abundant than the scaled solar r-process, Pb is notably underabundant (even the strongest line at 4057 Å was not detected).

This has the interesting consequence that the [Th/Eu] ratio (Eu or any other $56 \leq Z \leq 70$ element) would predict an epoch of formation of the n-capture elements present in CS 31082–001 *later* than the epoch of formation of the n-capture elements which enriched the Solar System! This conflicts with the observed U/Th ratios observed in CS 31082–001 and the Solar System: ^{238}U has a half-life a factor 3 shorter than ^{232}Th, so if the r-process elements of CS 31082–001 were produced after those of the Solar System, the U/Th would be significantly larger in CS 31082–001 than in the Solar-System (dotted line),

318   *Hill et al.*

Table 1.   Neutron-capture elements abundances in CS 31082–001[†].

El	Z	$\log \epsilon$	σ	N_{lines}	El	Z	$\log \epsilon$	σ	N_{lines}
Sr	38	0.68	0.09	4	Tb	65	−1.12	0.33	7
Y	39	−0.16	0.11	9	Dy	66	−0.20	0.16	7
Zr	40	0.47	0.13	5	Er	68	−0.17	0.17	5
Ba	56	0.30	0.13	7	Tm	69	−1.19	0.05	3
La	57	−0.56	0.08	4	Hf	72	−0.61	0.16	2
Ce	58	−0.27	0.10	9	Os	76	0.49	0.20	3
Pr	59	−0.89	0.12	4	Ir	77	0.37	0.2	1
Nd	60	−0.16	0.18	17	Pb	82	< −0.2:		1
Sm	62	−0.48	0.14	9	Th	90	−0.96	0.08	11
Eu	63	−0.70	0.09	9	U	92	−1.70	0.14	1
Gd	64	−0.21	0.18	7					

† for $T_{\text{eff}} = 4825$ K, $\log g = 1.5$, $\xi_{\text{micro}} = 1.8$ km s^{-1}, [Fe/H] $= -2.9$.

which is not observed. In fact, the age of CS 31082–001 predicted from the [Th/Eu] ratio conflicts with those from the [U/Th], [U/Os], or [U/Ir] ratios.

Beyond the issue of the age of this particular star, the fact that the $Z > 70$ elements pattern does not match those of other similar stars (CS 22892–052, HD 115444) nor the Solar-System r-process elements is worrisome concerning the used of Th/Eu (or U/Eu) ratios as age-tracers. The normalization of radioactive elements abundances to elements *in the same mass-range* becomes indispensable.

The reason for the discrepancy of the $Z > 70$ elements could be a direct consequence of chemical inhomogeneities in the early Galaxy: the ISM giving birth to very metal poor stars has probably only been polluted by a very limited number of supernovae, and hence it is possible that we now see the various outcomes of single events. Only *significant samples* of such n-capture enhanced elements will give clues to this issue. Christlieb et al. (these proceedings) suggest one method for quickly achieving this goal.

References

Burris, D. L., Pilachowski, C. A., Armandroff, T. E., Sneden, C., Cowan, J. J., & Roe, H. 2000, ApJ, 544, 302

Cayrel, R., et al. 2001, Nature 409, 691 Sneden, C., Burles, S., Tytler, D., & Beers, T. C. 1999, ApJ, 521, 194

Sneden, C., Cowan, J. J., Ivans, I. I., Fuller, G. M., Burles, S., Beers, T. C., & Lawler, J. E. 2000, ApJ, 533, L139

Sneden, C., McWilliam, A., Preston, G. W., Cowan, J. J., Burris, D. L., & Armosky, B. J. 1996, ApJ, 467, 819

Westin, J., Sneden, C., Gustafsson, B., & Cowan, J. J. 2000, ApJ, 530, 783

Astrophysical Ages and Time Scales
ASP Conference Series, Vol. 245, 2001
T. von Hippel, C. Simpson, N. Manset

Multi-Color Observations of Evolved Stars in ω Centauri

Joanne Hughes

Physics Department, Everett Community College, 2000 Tower Street, Everett, WA 98201, and Visiting Scholar: Astronomy Department, Box 351580, University of Washington, Seattle WA 98195, USA

George Wallerstein

Astronomy Department, Box 351580, University of Washington, Seattle, WA 98195, USA

Floor van Leeuwen

Institute of Astronomy, The Observatories, Madingley Road, Cambridge CB3 0HA, UK

Abstract. We present multi-color observations of the northern population of ω Cen from the main sequence turnoff to high on the red giant branch. We show that the best information about the metallicity and age of the stars can be gained from combining vby, $B - I$, and $V - I$ colors (in the absence of spectroscopy). We confirm our results for the main sequence turnoff stars: there is at least a 3 Gyr age spread, which may be as large as 5–8 Gyr, as suggested by Hughes & Wallerstein (2000) and Hilker & Richtler (2000). The available evidence supports the idea that ω Cen was once the core of a dwarf galaxy, captured and partially stripped by the Milky Way.

1. Observational Evidence

We use proper motion studies to confirm cluster membership (van Leeuwen et al. 2000) at and above the level of the horizontal branch, and we show that the age spread is maintained amongst stars from the subgiant branch through the red giants. We support previous findings that there is another red giant branch, redder (Pancino et al. 2000), and younger than the main giant branch (Hilker & Richtler 2000), but containing relatively few stars.

$B - I$ colors can be affected by excessive CN-content (which can vary for stars with a given value of [Fe/H]), so that the advantage of this longer baseline in temperature is offset by the non-uniform dependence on chemical composition. In Figure 1, we show $(V - I)$ color–magnitude diagrams (CMDs) with isochrones for various metallicities. We derive a distance of 5.2 kpc to give the best fit to the stellar models. Radial velocity and proper motion dispersions indicate that this distance should be 10–15% smaller (van Leeuwen et al. 2000). In all CMDs, the isochrones run from the upper left to lower right as 8, 10, 12, 14, 16 and 18 Gyr. We identify stars as follows: stars with $> 84\%$ probability of cluster membership (small dots), stars with > 95% probability of cluster membership (filled circles), stars at the main sequence turnoff with > 84% probability of cluster membership and colors which fit Malyuto's (1994) formula (larger open circles), and stars confirmed as members by proper motion studies (open stars). We use the giant branch to fix the color/metallicity of the stars and the main sequence turnoff to determine the ages. Conversions of m_0 and $(b-y)_0$ to [Fe/H] are performed using the Hilker (2000) calibration for the giant branches, and the Malyuto (1994) formula (which is only valid for a narrow range in colors) for the main sequence turnoff stars (see Hughes & Wallerstein 2000).

2. Conclusions

We find that it is necessary to use multi-color photometry in the absence of spectroscopy, and that the B-band and the Strömgren v-band are affected by CN-variations. The best temperature index, is thus $(V - I)$. Also, m_1 does not correlate well with [Fe/H] for metal-rich stars in ω Cen.

Though there are suggestions that the spatial distribution of the metal-rich stellar population is different from that of the metal-poor population (Jurcsik 1998), which could be achieved by capturing a smaller metal-rich star cluster, as suggested by Pancino et al. (2000); it is likely that the bulk of ω Cen was the core of a dwarf galaxy (Dinescu, Girard, & van Altena 1999). This model allows for the observed extended periods of star formation using material which had been enriched by asymptotic giant branch stars and Type I and II supernovae at different epochs. Gas and stars could also be added from an external source which has now been tidally stripped from ω Cen by our Galaxy. Chemical enrichment seems non-uniform, indicating that the cluster once had a larger reservoir of gas to draw from. There is good evidence for at least two episodes of star formation in the cluster.

References

Dinescu, D. I., Girard, T. M., & van Altena, W. F. 1999, AJ117, 1792

Hilker, M. 2000, A&A, 355, 994

Hilker, M., & Richtler, T. 2000, A&A, 362, 895

Hughes, J., & Wallerstein, G. 2000, AJ, 119, 1225

Jurcsik, J. 1998, ApJ, 506, L113

Malyuto, V. 1994, A&AS, 108, 441

Pancino, E., Ferraro, F. R., Bellazzini, M., Piotto, G., & Zoccali, M. 2000, ApJ, 534, L83

VandenBerg, D. A., Swenson, F. J., Rogers, F. J., Iglesias, C. A., & Alexander, D. R. 2000, ApJ, 532, 430

van Leeuwen, F., Le Poole, R. S., Reijns, R. A., Freeman, K. C., & de Zeeuw, P. T. 2000, A&A, 360, 472

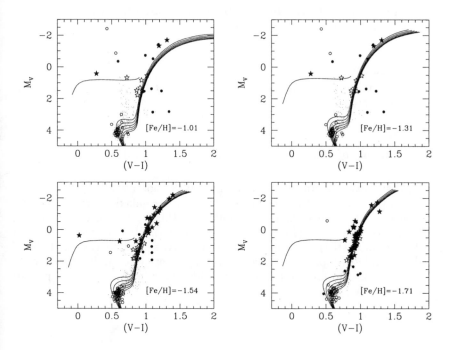

Figure 1.   *Upper left:* Stars with −0.5 > [Fe/H]$_{phot}$ > −1.2 with the
[Fe/H] = −1.01 isochrones of VandenBerg et al. (2000). The red giant
branch is not well-fit by these models, although the main sequence
turnoff suggests a mean age of <8–11 Gyr with a lot of scatter. We
suggest that stars with this range of colors might actually be more
metal-poor than [Fe/H]$_{phot}$ indicates because of the presence of objects
with strong CN-features. *Upper right:* The same group of stars with
the isochrones corresponding to a metallicity of [Fe/H] = −1.31. The
red giant branch is better fit by these models, and the main sequence
turnoff stars are about 8–14 Gyr old, centered at about 12 Gyr. *Lower
left:* Stars with colors corresponding to −1.2 > [Fe/H]$_{phot}$ > −1.6 with
the models with [Fe/H] = −1.54. This is generally a better fit with
the mean age of the main sequence turnoff stars being 14 ± 2.5 Gyr.
The metal-poor group with −1.6 > [Fe/H]$_{phot}$ > −2.2 corresponds to
the first and strongest episode of star formation in the cluster, which
appears to have occurred 14–18 Gyr in the past (*lower right*). The
asymptotic giant branch is shown clearly.

Astrophysical Ages and Time Scales
ASP Conference Series, Vol. 245, 2001
T. von Hippel, C. Simpson, N. Manset

Age Gradients in the Sculptor and Carina Dwarf Spheroidals

Denise Hurley-Keller

Case Western Reserve University, 10900 Euclid Avenue, Cleveland, OH 44106, USA

Mario Mateo

University of Michigan, Ann Arbor, MI 48109, USA

Abstract. We find evidence from our B, R photometry of Sculptor (Scl) and Carina (Car) that the oldest stars in these galaxies form a more extended component than the populations that dominate in the central regions. The inferred age and metallicity gradients explain the horizontal branch (HB) morphology in both galaxies, and suggest that dwarf spheroidals (dSph) are really multi-component systems. In this context, we discuss an interpretation of extratidal stars in these dSph as part of an old halo.

1. Sculptor

The Scl B, R observations were obtained with the Big Throughput Camera at the CTIO 4m telescope. The field covers an area of $\sim 28' \times 28'$. The two left panels of Figure 1 show the observed color–magnitude diagrams (CMDs) of the stars within $\sim 7'$ (Inner) and of the stars beyond $\sim 7'$ (Outer).

We have determined a detailed star formation history for Scl using synthetic color–magnitude diagrams generated from stellar models from the Padua library (Bertelli et al. 1994). Our approach will be described in detail in a later paper. Here we offer an overview of the procedure. We generated a set of simulated CMDs (> 200) with an ancient episode of star formation lasting from 0–4 Gyr and with chemical enrichment consistent with the width of the red giant branch. Approximately 400 000 false stars are used to accurately incorporate observational effects into the model CMDs. We use the distribution of stars across the red giant branch (RGB), the subgiant branch (SGB) and the main sequence turnoff (MSTO) to compare the model CMDs to the observed Scl CMD via the χ^2 statistic.

The right two panels show the model CMDs for the inner and outer regions. For the inner region, the best model has a mean age of 15 Gyr, an age spread of 2–3 Gyr, and an abundance spread ($\Delta[\text{Fe/H}] \sim 0.7$), and provides a good fit to the RGB, SGB, and MSTO. Although we did not use the HB morphology as a constraint, note that this scenario qualitatively reproduces Scl's "second parameter" HB. The outer region models imply a similar age of 15–16 Gyr, but

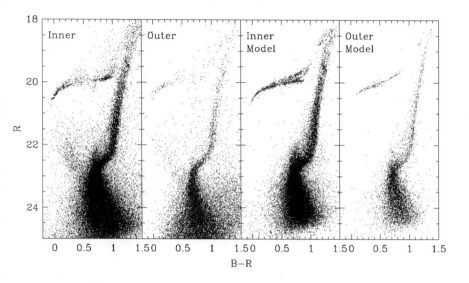

Figure 1.   Observed and model Inner and Outer CMDs for Scl.

an age spread of ≤ 1 Gyr, and a narrower abundance distribution which is more metal poor on average than the inner region.

2.   Carina

Car was observed with MOSAIC II at the CTIO 4m. The images cover an area of $35.4' \times 35.4'$, $\sim 60\%$ of the tidal radius of Car. We separated stars with $r < r_c$ (Inner) from stars with $r > 1.5 r_c$ (Outer) where the core radius, r_c, is $8.8'$ for Car. Differences between the inner and outer CMDs suggest a radial gradient among the stellar populations (Figure 2). The intermediate-age (~ 7 Gyr) subgiant branch (SGB) weakens relative to the old (~ 15 Gyr) SGB from the inner to the outer region, as does the related red clump relative to the (old) HB stars. Differences in star counts in the boxes shown are statistically significant. The intermediate-age population is more centrally concentrated than the old population, and we will further investigate this with detailed modeling as in Scl.

3.   Discussion

The fact that the stellar populations in these dSph are differentiated, and that one is spatially more extended may impact the interpretation of their putative extratidal stars (stars beyond the tidal radius, r_t; Irwin & Hatzidimitriou 1995). These stars have generally been regarded as tidal debris from the dSph, and as such provide an important constraint on the dark matter mass and spatial extent. Their existence implies that the dark matter distribution cannot be

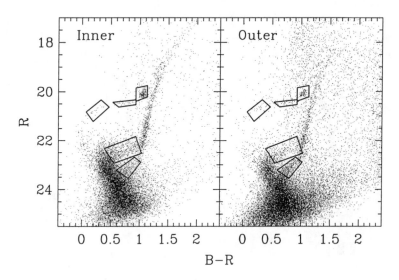

Figure 2.   Inner and Outer Car CMDs. The boxes from faintest to brightest indicate old SGB, intermediate-age SGB, blue and red HBs, and red clump.

much more extended than the stellar component (Moore 1996; Burkert 1997). If the dSph are really multiple (stellar) component systems, however, component King models used to derive r_t are not appropriate and would underestimate the true r_t of the mass distribution. An alternative interpretation is that extratidal stars are part of a stellar halo in the dSph. This scenario is compatible with our observations (at $\leq r_t$) of extended, older components in Scl and Car.

At least one isolated dIrr (DDO 187) has a surface brightness profile with a break similar to that seen in dSph with extratidal stars which is due instead to an older, extended halo (Aparicio, Tikhonov, & Karachentsev 2000). This observational connection between dSph and dIrr is particularly intriguing in light of recent simulations which conclude that rotationally-supported dIrr can transform into dSph dominated by random velocities as they traverse the Milky Way potential over several orbits (Mayer et al. 2001, these proceedings).

References

Aparicio, A., Tikhonov, N., & Karachentsev, I. 2000, AJ, 119, 177

Bertelli, G., Bressan, A., Chiosi, C., Fagotto, F., & Nasi, E. 1994, A&AS, 106, 275

Burkert, A., 1997, ApJ, 474, L99

Irwin, M., & Hatzidimitriou, D. 1995, MNRAS, 277, 1354

Mayer, L., Governato, F., Colpi, M., Moore, B., Quinn, T., Wadsley, J., Stadel, J., & Lake, G. 2001, ApJ, 547, L123

Moore, B. 1996, ApJ, 461, L13

Astrophysical Ages and Time Scales
ASP Conference Series, Vol. 245, 2001
T. von Hippel, C. Simpson, N. Manset

Planetary Nebulae in the Outer Spheroid of M 31

Denise Hurley-Keller, Heather Morrison

Case Western Reserve University, 10900 Euclid Avenue, Cleveland, OH 44106, USA

Paul Harding

University of Arizona, Steward Observatory, Tucson, AZ 85721, USA

George Jacoby

WIYN Observatory, NOAO, 950 North Cherry Avenue, P.O. Box 26732, Tucson, AZ 85726, USA

Abstract. When do galaxy spheroids form? Data from high redshift studies, which could address this question, are hard to interpret because of the possibly severe effects of dust. Local spheroids, however, provide a good stepping-stone to their interpretation. We have surveyed the halo of M 31 for planetary nebulae (PNe) using the Burrell Schmidt in order to study the spheroid of the nearest disk galaxy with a large bulge. PNe can be used as both kinematic tracers of faint stellar components and as an age indicator for very old populations.

Studies of the Milky Way halo and bulge were used for many years to support the idea that galaxy spheroids formed at very early times, because of the large ages of globular clusters. However, recent evidence of a central bar suggests that the bulge is an inner disk phenomenon, bearing no relationship to the halo or to large $r^{1/4}$ bulges in other galaxies (Weiland et al. 1994; Kuijken 1996). The stellar halo itself is difficult to study directly because it is so faint (Morrison 1996). PNe provide an alternative means for studying halos, as the PNe can be much brighter in [O III] λ5007 than the underlying stellar population. They can be used as kinematic tracers to distinguish halo from bulge or thick disk, and potentially as age indicators. The nearest large $r^{1/4}$ bulge is in the galaxy M 31. We have conducted a narrow-band [O III] survey of M 31's outer spheroid, and have identified 149 PNe in a region covering approximately 4 square degrees with the Burrell Schmidt. We will use these objects to study the kinematics and star formation history of M 31's spheroid.

1. Observations

The technique for detecting PNe has been well-established through work on the PN luminosity function as a distance indicator. PNe detection involves imaging in [O III] λ5007 and in a somewhat broader, offset band. Stars are detected through both filters, but PNe only in [O III].

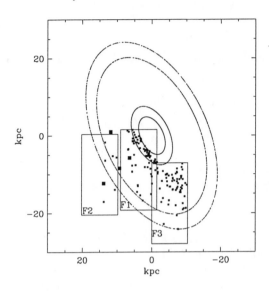

Figure 1.   PNe discovered in our M 31 survey (small squares). Dotted
lines are isophotes of a b/a = 0.6 flattened spheroid.

Figure 1 shows the locations of our Schmidt fields relative to the center
of M 31 (assuming a distance of 725 kpc). Fields 1 and 2 probe mostly halo
regions, with the most distant PN at 17 kpc from the galaxy center along the
minor axis. The larger squares are the locations of the fields from Pritchet &
van den Bergh (1994; PvdB94).The inner regions of field 1 provide a link to
the bulge, while field 3 probes both outer disk and thick disk regions. The
two fields offset from the minor axis will serve to measure the rotation of the
halo/bulge/disk populations. The outermost ellipse is at $\sim$ 15 kpc radius along
the minor axis.

2.   Results

The PN production rate may provide an age indicator for very old populations.
Theory and observation both suggest that populations with low turnoff mass
may underproduce PN because the central stars do not reach the temperature
needed to ionize their ejecta before it disperses into the ISM (Peimbert 1990;
Jacoby et al. 1997). The M 31 bulge is rich in PNe, and this suggests that it
was formed later than the globular clusters.

     The comparison of the minor axis surface brightness profile (PvdB94) with
the number of PNe detected there will constrain possible age differences between
halo and bulge. If PNe are found in much lower numbers than expected, then
the population is as old as the globulars ($\sim$ 10–15 Gyr). Figure 2 shows that
within the counting errors, the PNe follow the same $r^{1/4}$ spatial distribution
as the stars in field 1, suggesting no dramatic change in age within this field.

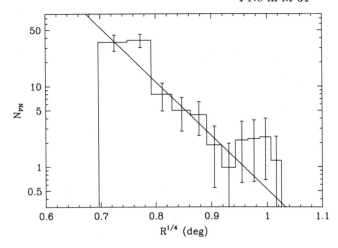

Figure 2.   Number of PNe, corrected for the fraction of the isophote in the field, versus radius for field 1, with arbitrary normalization.

However, M 31's bright bulge may dominate a Milky Way-like halo, if it exists, even at this radius.

Kinematic data from spectroscopy of the PNe, planned for the coming fall, will be a sensitive probe of whether we are looking at a single bulge population which extends to distances of 20 kpc or an inner bulge and a kinematically distinct halo. M 31 bulge kinematics show a low but significant rotation of ~ 50 km s^{-1} and a velocity dispersion of ~ 150 km s^{-1} (McElroy 1983). The halo kinematics, as traced by the globular clusters, show no detectable rotation (Huchra 1993). Evidence that a kinematically distinct halo exists and is significantly older than the bulge would be an important step towards understanding the timing of spheroid formation.

References

Huchra, J. 1993, in ASP Conf. Ser. 48, The Globular Cluster–Galaxy Connection, ed. G. H. Smith & J. P. Brodie (San Francisco: ASP), 420

Jacoby, G. H., Morse, J. A., Fullton, L. K., Kwitter, K. B., & Henry, R. B. C. 1997, AJ, 114, 2611

Kuijken, K. 1996, in IAU Symp. 169, ed. L. Blitz & P. Teuben (Dordrecht: Kluwer), 71

McElroy, D. B. 1983, ApJ, 270, 485

Morrison, H. 1996, in ASP Conf. Ser. 92, The Formation of the Galactic Halo .... Inside and Out, ed. H. Morrison & A. Sarajedini (San Francisco: ASP), 453

Peimbert, M. 1990, RMA&A, 20, 119

Pritchet, C., & van den Bergh, S. 1994, AJ, 107, 1730

Weiland, J., et al. 1994, ApJ, 425, L81

Astrophysical Ages and Time Scales
ASP Conference Series, Vol. 245, 2001
T. von Hippel, C. Simpson, N. Manset

What Are White Dwarfs Telling Us About the Galactic Disk and Halo?

J. Isern

Institute for Space Studies of Catalonia (CSIC/UPC) and Space Science Institute (CSIC), Spain

E. García–Berro

Space Science Institute (CSIC) and Department of Applied Physics (UPC), Spain

M. Salaris

Astrophysics Research Institute, Liverpool John Moores University, UK

Abstract. When white dwarfs are used to determine the age of the Galactic disk, it is implicitly assumed that the star formation rate per unit volume has remained approximately constant over the life of the Galaxy. However, this might not have been the case, and the present disk white dwarf luminosity function can be perfectly fitted with variable star formation rates and ages in the range of 9.5 to 20 Gyr. Concerning the Galactic halo, we find that with its present luminosity function, which is still very uncertain, and using a standard IMF, we can account for only about 1% of its mass.

1. Introduction

White dwarfs have been proposed as chronometers of Galactic evolution because of their large evolutionary time scales and the relative simplicity of their structure (Winget et al. 1987; García-Berro et al. 1988; Hernanz et al. 1994). The tool for doing that is their luminosity function, defined as the number of white dwarfs of a given luminosity per unit magnitude interval. The luminosity function can be computed as:

$$n(l) \propto \int_{M_i}^{M_s} \Phi(M)\,\Psi(T - t_{\text{cool}}(l, M) - t_{\text{MS}}(M))\tau_{\text{cool}}(l, M)\,dM \qquad (1)$$

where l is the logarithm of the luminosity in solar units, M is the mass of the parent star (for convenience all white dwarfs are labeled with the mass of the main sequence progenitor), t_{cool} is the cooling time down to luminosity l, $\tau_{\text{cool}} = dt/dM_{\text{bol}}$ is the characteristic cooling time, M_s and M_i are the maximum and the minimum masses of the main sequence stars able to produce a white dwarf of luminosity l, t_{MS} is the main sequence lifetime of the progenitor of the white dwarf, and T is the age of the population under study. The remaining quantities, the initial mass function, $\Phi(M)$, and the star formation rate, $\Psi(t)$, are not known a priori and depend on the astronomical properties of the stellar population under study. If the properties of white dwarfs are well known and the observational luminosity function is accurate enough, it is then possible to obtain information about the age of the Galaxy (disk, halo or star clusters). The cooling sequences adopted here are those of Salaris et al. (2000).

As it can be seen from Equation 1, it is impossible to separate the age of the Galaxy from the star formation rate because of the extremely long lifetimes of low mass main sequence stars. That is, very old and low mass main sequence stars are able to produce young (in the sense of cooling times) and bright white dwarfs. This implies that the past star formation activity is still influencing the present white dwarf birthrate. Therefore, white dwarfs offer the unique opportunity to obtain the true age of the solar neighborhood instead of just providing a lower limit (Isern et al. 1995).

In principle, the star formation rate can be obtained from Equation 1 by solving the inverse problem. However, since the kernel of the transformation is not symmetric (Isern et al. 1995), it is not possible to obtain a direct solution and uniqueness is not guaranteed. Therefore the procedure consists in minimizing a trial star formation rate.

2. The Disk

There are at present several determinations of the observational white dwarf luminosity function (Fleming, Liebert, & Green 1986; Liebert, Dahn, & Monet 1988, 1989; Evans 1992, Oswalt et al. 1996; Leggett, Ruiz, & Bergeron 1998; Knox, Hawkins, & Hambly 1999). Since all of them are in qualitative agreement, it is meaningful to merge them into a single luminosity function using their error bars as statistical weights (Isern et al. 2001).

As trial functions we have chosen: (1) a constant star formation rate, Ψ, (2) a nearly constant star formation rate plus a relatively extended tail, $\Psi(t) \propto 1/(1 + e^{(t-\tau)/\delta})$, (3) a star formation rate per unit surface that decreases with a time scale τ_s, divided by an scale height above the Galactic plane that decreases with a time scale τ_h: $\Psi(t) \propto e^{-t/\tau_s}/(1 + Ae^{-t/\tau_h})$, and (4) a star formation rate formed by adding bursts of arbitrary intensity and temporal width.

The results can be summarized as follows: (1) The best fit is always obtained by star formation rates that progressively increase with time, then stabilize or even decrease slowly during the last 2–5 Gyr. (2) The age depends on the shape of the trial function and ranges from 9.5 to 20 Gyr. (3) The bright portion of the luminosity function is insensitive to the shape of the star formation rate (Isern et al. 1998) but the dim portion shows clear differences from one star formation rate to another.

3. The Halo

The observed properties of halo white dwarfs are very scarce. They can be summarized as follows: (1) There is a preliminary and incomplete luminosity function (Liebert et al. 1989; Torres, García-Berro, & Isern 1999). (2) Microlens-

330 *Isern, Garcia-Berro, & Salaris*

ing data suggests the existence of a population with a typical mass of $\sim 0.5\ M_\odot$ that could account for 10% of the halo mass. White dwarfs are the most natural candidates to explain such observations. Nevertheless, to avoid problems with nucleosynthesis, energetics, and Type Ia supernovae (Isern et al. 1997; Canal, Isern, & Ruiz-Lapuente 1997), the contribution of white dwarfs to the halo should not exceed 5%. (3) Recently, Oppenheimer et al. (2001) have detected a large number of fast moving white dwarfs that could account for 3% of the halo mass.

The main results can be summarized as follows: (1) For ages in the 12–14 Gyr range, the preliminary luminosity function predicts that $\sim 1\%$ of the halo mass would be made of white dwarfs if a Salpeter-like IMF is adopted. Notice that because of the uncertainties, this figure is not at odds with the results of Oppenheimer et al. (2001). (2) The total number of white dwarfs per steradian ranges from 500 000 to 800 000 depending on whether they are all non-DAs or DAs. This implies that the expected number in a 12 square degree field is of the order of one hundred.

References

Canal, R., Isern, J., & Ruiz-Lapuente, P. 1997, ApJ, 488, L35

Evans, D. W. 1992, MNRAS, 255, 521

Fleming, T. A., Liebert, J., & Green, R. F. 1986, ApJ, 308, 176

García-Berro, E., Hernanz, M., Isern, J., & Mochkovitch, R. 1988, Nature, 333, 642

Hernanz, M., García-Berro, E., Isern, J., Mochkovitch, R., Segretain, L., & Chabrier, G. 1994, ApJ, 302, 173

Isern, J., García-Berro, E., Hernanz, M., Mochkovitch, R., & Burkert, A. 1995, in Lecture Notes in Physics 443, White Dwarfs, ed. D. Koester & K. Werner, (Springer-Verlag), 19

Isern, J., García-Berro, E., Hernanz, M., Mochkovitch, R., & Torres, S. 1998, ApJ, 503, 239

Knox, R. A., Hawkins, M. R. S., & Hambly, N. C. 1999, MNRAS, 306, 736

Leggett, S. K., Ruiz, M. T., & Bergeron, P. 1998, ApJ, 497, 294

Liebert, J., Dahn, C. C., & Monet, D. G. 1988, ApJ, 332, 891

Liebert, J., Dahn, C. C., & Monet, D. G. 1989, in IAU Coll. 114, White Dwarfs, ed. G. Wegner, (Springer-Verlag), 15

Oppenheimer, B. R., Hambly, N. C., Digby, A. P., Hodgkin, S. T., & Saumon, D. 2001, Science Express Research Articles, March 22

Oswalt, T. D., Smith, J. A., Wood, M. A., & Hintzen, P. 1996, Nature, 382, 692

Salaris, M., García-Berro, E., Hernanz, M., Isern, J., & Saumon, D. 2000, ApJ, 554, 1036

Torres, S., García-Berro, E., & Isern, J. 1999, ApJ, 508, L71

Winget, D. E., Hansen, C. J., Liebert, J., van Horn, H. M., Fontaine, G., Nather, R. E., Kepler, S. O., & Lamb, D. Q., 1987, ApJ, 315, L77

Acknowledgments. This work has been supported by the MCYT grants ESP98-1348 and AYA2000-1785 and by the CIRIT.

Astrophysical Ages and Time Scales
ASP Conference Series, Vol. 245, 2001
T. von Hippel, C. Simpson, N. Manset

Nucleosynthesis of Heavy Elements in the First Generation Stars

Nobuyuki Iwamoto

Center for Nuclear Study, University of Tokyo, Wako, Saitama 351-0198 and National Astronomical Observatory, Mitaka, Tokyo 181-8588, Japan

Toshitaka Kajino, Wako Aoki

National Astronomical Observatory, Mitaka, Tokyo 181-8588, Japan

Abstract. We studied the s-process nucleosynthesis in the canonical method by using a large dimension network code. We found that (1) an appropriate neutron exposure exists to best fit the observed abundances in LP 625–44 and LP 706–7 with metallicity [Fe/H] = −2.7, and (2) in this exposure the observational discovery of Pb/Ba ∼ 1 and Ba/Sr ∼ 10 are well explained. High neutron exposure and a small overlap factor are necessary to fit the abundance pattern in these two metal-deficient stars. An equilibrium pattern, except for lead, can be obtained by a single neutron irradiation. A He convective shell developed by a thermal pulse is likely to be a viable site for the s-process in metal-deficient stars.

1. Introduction

It is commonly accepted that $^{13}C(\alpha,n)^{16}O$ is a major neutron source reaction in low mass asymptotic giant branch (AGB) stars. This reaction operates at temperatures greater than 9×10^7 K. Therefore, ^{13}C burns radiatively in a narrow region at the top of the He intershell layer during a long interpulse phase (Straniero et al. 1995).

It is considered that the ^{13}C abundance is independent of metallicity since it is produced by proton capture on ^{12}C newly synthesized via the triple alpha reaction. Because (1) the ^{13}C abundance is not different from solar metallicity stars for metal-deficient stars, and (2) the abundance of seed nuclei is low, it is expected that these stars show an overabundance of heavy s-elements. Observations of metal-deficient stars reveal that heavy s-elements (Ba, La, Nd, and Sm) are significantly enhanced with respect to light s-elements (Sr, Y, and Zr). In particular, theoretical model calculations indicate Pb/Ba > 100 and Ba/Sr ∼ 1 at [Fe/H] = −2.7 (Busso, Gallino, & Wasserburg 1999).

The extremely metal-deficient, carbon-rich stars LP 625–44 and LP 706–7, with metallicity [Fe/H] = −2.7, show a peculiar distribution of s-process elements, Pb/Ba ∼ 1 and Ba/Sr ∼ 10 (Aoki et al. 2000, 2001). These large differences between observations and theoretical predictions may indicate that our interpretation of the evolution of metal-deficient AGB stars is insufficient. Although Ryan et al. (2000) found a good fit to [Pb/Fe] and [Ba/Fe] for LP 625–44, an exceedingly low ^{13}C abundance was necessary to reproduce the observations.

Fujimoto et al. (2000) showed that a He convective shell developed by thermal pulses mixes protons from the H-rich layer to the He intershell layer in extremely metal-deficient AGB stars. This mixing event allows the production of ^{13}C nuclei via the ^{12}C$(p,\gamma)^{13}$N$(\beta^{+}\nu)^{13}$C reaction chain in the He convective shell. The aim of this paper is to investigate the physical environment of the s-process in metal-deficient AGB stars.

2. Results

We calculated s-process nucleosynthesis using a canonical model (Howard et al. 1986) with updated neutron capture reaction rates (Aoki et al. 2001). We assumed the temperature $T_8 = 1$. This assumption is allowed if it is considered that ^{13}C$(\alpha,\text{n})^{16}$O is the dominant reaction for neutron production.

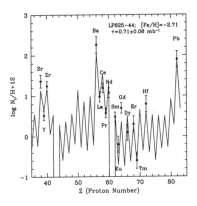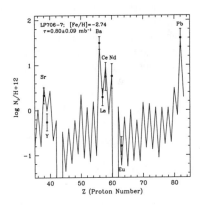

Figure 1.   Best fit to observational results for the extremely metal-deficient stars LP 625–44 *(left)* and LP 706–7 *(right)*.

Figure 1 shows the best fit results for LP 625–44 and LP 706–7. This physical environment for the s-process is a neutron density $N_{\text{n}} = 10^7$ cm^{-3} and an overlap factor $r = 0.1$, the fraction of material having experienced a previous neutron irradiation experiencing another neutron exposure. The resulting neutron exposures $\tau = 0.71\pm0.08$ and 0.80 ± 0.09 mb^{-1} for LP 625–44 and LP 706–7 are relatively high. Neutron exposures are defined as $\tau = N_{\text{n}}v_{\text{T}}\Delta t$, where Δt is the pulse duration and v_{T} is thermal velocity of the neutron, and under constant temperature ($v_{\text{T}} = \text{constant}$) the same value of $N_{\text{n}}\Delta t$ gives the same neutron exposure. Therefore, it is not clear which site is more viable, the thermal pulse or the interpulse phase. It is a significant fact that large neutron exposures and a small overlap factor are needed to reproduce the abundance distribution. It implies that very few thermal pulses could have contributed to make the s-process abundances for these two stars. In fact, almost all elements, except for lead, are made in the first neutron irradiation with the adopted exposure. Although the lead abundance is significantly sensitive to the pulse number, it converges to the equilibrium abundance after a few episode. As a consequence, these results indicate that the abundance distribution of metal-deficient stars may be obtained by a few large neutron exposure (Aoki et al. 2001).

These findings are interesting because this is exactly the behavior recently anticipated by Fujimoto et al. (2000) in what was referred to as Case II'. They proposed that proton mixing into the He intershell layer invoked by upward extension of the He convective shell triggered by a thermal runaway of He shell burning occurs only a few times in metal-deficient stars with [Fe/H] < -2.5. Therefore, only a few occurrences of neutron exposure can be obtained. The metallicity for LP 625–44 and LP 706–7 stars ([Fe/H] $= -2.7$) enters the exact metallicity range. It is possible that s-process elements observed in these two stars are produced by this peculiar mechanism (N. Iwamoto et al., in preparation). We further investigated the time evolution of the He convective shell by calculating the stellar evolution at 2 $M_\odot$ and [Fe/H] $= -2.7$. We found that the penetration of the He convective shell into the H-rich envelope takes place and that the convective shell separates into two parts, where one is sustained by H-burning and the other is sustained by He-burning. In the former convective shell the CN cycle operates, ^{13}C forms, and then significant amounts of neutrons ($N_n > 10^{10}$ cm^{-3} for $X(^{13}$C$) > 0.01$) are produced by ^{13}C$(\alpha,n)^{16}$O. It is expected that the s-process proceeds when seed nuclei capture newly synthesized neutrons (see N. Iwamoto et al., in preparation, for details).

References

Aoki, W., Norris, J. E., Ryan, S. G., Beers, T. C., & Ando, H. 2000, ApJ, 536, L97

Aoki, W., et al. 2001, ApJ, submitted

Busso, M., Gallino, R., & Wasserburg, G. J. 1999, ARA&A, 37, 239

Howard, W. M., Mathews, G. J., Takahashi, K., & Ward, R. A. 1986, ApJ, 309, 633

Ryan, S. G., Aoki, W., Blake, L. A. J., Norris, J. E., Beers, T. C., Gallino, R., Busso, M., & Ando, H. 2000 (astro-ph/0008423)

Straniero, O., Gallino, R., Busso, M., Chieffi, A., Raiteri, C. M., Limongi, M., & Salaris, M. 1995, ApJ, 440, L85

Astrophysical Ages and Time Scales
ASP Conference Series, Vol. 245, 2001
T. von Hippel, C. Simpson, N. Manset

Addressing Ambiguities in the Evolutionary Status of Solar-Mass Stars

Tuba Koktay

Astronomy and Space Science Department, University of Istanbul, Turkey, and Subaru Telescope, 650 North A'ohōkū Place, Hilo, Hawai'i, USA

Abstract. Spectroscopic MK classifications have been obtained for a set of stars having abnormal indices on the Strömgren *uvby* photometric system. Two of these stars, HD 204848 and HD 207687, show extreme carbon anomalies. Moderately high-resolution spectroscopy (0.2 Å) over the wavelength range of 3900 to 8800 Å was compared with synthetic spectra for evolutionary models to give an indication of the true evolutionary status of these stars. The luminosities predicted from the modeling are found to be in agreement with HIPPARCOS observations.

1. Introduction

Approximately 200 stars have been observed spectroscopically at MK dispersion (1–3 Å) from those stars selected by Olsen (1993, 1994) as having very peculiar *uvby* colors, in order to look for spectral anomalies which might account for their abnormal color indices. The most interesting are HD 204848, which is weak-lined, carbon-strong, and nitrogen-weak with a classification of K0III: Fe−2.5, CN−3, CH+1, Ba+1; and HD 207687, which has similar characteristics with a classification of K0III: Fe−2, CN−2, CH+1, Ba+1 (Koktay 1999; Koktay & Garrison 2000). It's very unusual to find stars of this luminosity and temperature class displaying an enhanced carbon abundance, which is generally thought the result of dredge-up occurring as stars pass several times up the asymptotic giant branch. We have constructed a grid of model atmospheres for HD 204848, HD 207687, and the K0III standard star HD 197989, to compare against moderately high dispersion spectra (0.2 Å resolution) and have compared the absolute luminosities calculated from the parallaxes determined by HIPPARCOS (ESA 1997) with the Geneva isochrones to better understand the evolutionary status of these two carbon-peculiar stars.

2. Observations and Spectroscopic Reductions

Observations for the purpose of MK classification were obtained with the CCD spectrograph on the 60-cm Helen Sawyer Hogg Telescope (formerly of the University of Toronto). Observations with a resolution of ~ 0.2 Å were obtained with the 1024×1024 Thompson CCD and Cassegrain spectrograph on the 74-inch telescope of the David Dunlap Observatory (DDO) in Toronto.

The spectra were reduced with the IRAF reduction software provided by the NOAO. The RVSAO package was used to determine the first known radial velocities for HD 204848 and HD 207687 (Table 1).

3. Discussion

Synthetic spectra generated for a range of model parameters and compositions were compared against full-resolution DDO spectra for HD 204848 and HD 207687 (Koktay 1999). The best fit was for $T_{\text{eff}} = 5100 \pm 100$ K, $\log g = 3.0 \pm 0.5$, and $[A/\text{H}] = -1.5 \pm 0.5$ for both stars (Figure 1). The temperature is consistent with a spectral type earlier than K0 and the surface gravity is consistent with a post-main sequence star (i.e., a giant or subgiant). The value for metallicity is not particularly low, and interestingly, no anomalous carbon abundance was needed to reproduce the observed spectra for these stars over a wide wavelength range of 3900–8800 Å.

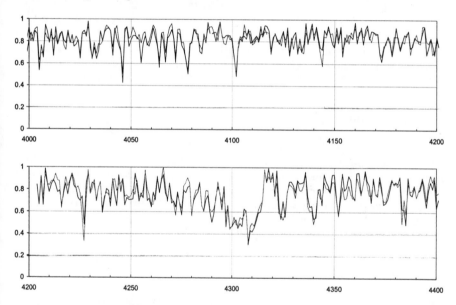

Figure 1.   Portion of the observed spectrum of HD 204848 (light line) compared against the best-fitting synthetic spectrum (heavy line).

Trigonometric parallaxes and apparent magnitudes from the HIPPARCOS catalogue have been used to calculate the absolute visual magnitudes (Table 1) under the assumption that the interstellar extinction for all three stars is zero. The absolute magnitude of HD 197989 ($M_{\text{V}} = 0.88$) is in good agreement with its MK designation. HD 204848 and HD 207687 appear to be about 1.5 and 2.2 magnitudes less luminous, respectively. This would indicate that these two stars are subgiants.

We have calculated for all three stars their space velocities (v_α, v_δ, v) with respect to the Sun using the HIPPARCOS proper motions and our radial veloc-

Table 1.   Radial velocities and other derived parameters.

Star Name	V_r (km s^{-1})	d (pc)	M_V	v_α (km s^{-1})	v_δ (km s^{-1})	v (km s^{-1})
HD 204848	−68	107.0	2.35	−61.1	−61.8	111
HD 207687	−25	80.3	3.07	23.6	3.5	35
HD 197989	−12	22.1	0.88	37.3	34.6	52

ities. All three stars have velocities roughly consistent with being members of the thin or possibly thick disk population.

The position of the K0III standard HD 197989 within the HR diagram overlayed with evolutionary tracks from Schaller et al. (1992) is consistent with it being a 3 $M_\odot$ star near the base of the giant branch. HD 204848 and HD 207687 are also found to reside close to the base of their respective giant branches for stars of 1.85 and 1.5 solar masses. We cannot say with certainty exactly how many times (if at all) these stars have passed up their respective giant and asymptotic giant branches. The estimated minimum age for HD 197989, HD 204848, and HD 207687 are 0.35, 1.5, and 2.6 Gyr, respectively.

4. Conclusions

HD 204848 and HD 207687 appear to be evolved stars of the halo population based on their peculiar CN and CH lines as well as generally weak metal lines. These intriguing stars likely represent an interesting stage of evolution, though we still can't say with certainty exactly where they reside along their evolutionary tracks. They clearly warrant further study, in particular, to determine in detail the abundance ratios of key elements, which might be the product of dredge-up during the asymptotic giant branch phase of evolution.

References

European Space Agency (ESA) 1997, The Hipparcos and Tycho Catalogues, ESA SP-1200

Koktay, T. 1999, Ph.D. Thesis, University of Istanbul

Koktay, T., & Garrison, R. F. 2000, in IAU Symp. 177, The Carbon Star Phenomenon, ed. R. F. Wing (Dordrecht: Kluwer), 141

Olsen, E. H. 1993, A&AS, 102, 89

Olsen, E. H. 1994, A&AS, 104, 429

Schaller, G., Schaerer, D., Meynet, G., & Maeder, A. 1992, A&AS, 96, 269

Astrophysical Ages and Time Scales
ASP Conference Series, Vol. 245, 2001
T. von Hippel, C. Simpson, N. Manset

New Gravities for Old Stars

A. J. Korn, T. Gehren

Universität-Sternwarte München, 81679 Munich, Germany

Abstract. We present the results of calibrating our non-LTE iron model atom with metal-poor HIPPARCOS targets. On the Balmer profile temperature scale, the efficiency of collisions with hydrogen atoms is close to the theoretical prediction, thus not being able to compensate for overionization in Fe I due to strong photoionization. While non-LTE effects are quite small in the Sun (0.04 dex), the iron abundance and gravity of the metal-poor subgiant HD 140283 are affected by 0.11 dex and 0.29 dex, respectively. Among other consequences, isochrone ages of metal-poor stars will require a systematic revision.

1. Non-LTE of Iron in Cool Stars

The concept of local thermodynamic equilibrium (LTE), most often assumed to be valid throughout the atmospheres of solar-like stars, heavily relies on the sufficient strength of collisions to counterbalance radiative processes. While collisions with hydrogen atoms do not scale with metallicity, photoionization will be significantly enhanced in the UV-transparent atmospheres of metal-poor stars making it difficult to avoid departures from LTE for minority species like Fe I. Building on the compilation of energy levels and f values by Nave et al. (1994) and the calculated photoionization cross-sections by Bautista (1997), we put together a comprehensive model atom consisting of 236 terms of Fe I and 267 of Fe II (see Gehren et al. 2001). On the one hand, it was found that Fe II could safely be synthesized under the assumption of LTE, on the other hand, the occupation of Fe I levels turned out to be dictated by the strong (100 times larger than hydrogenic) photoionization rates. As a consequence, the following plasma-diagnostic tools based on Fe I are systematically affected: (1) temperatures derived from the Fe I excitation equilibrium, (2) abundances and microturbulences derived from Fe I, and (3) gravities derived from the ionization equilibrium of Fe I/Fe II.

In the subsequent sections, we will present results for individual calibration stars. It will be shown that our non-LTE calculations are capable of resolving the discrepancies in gravity found by Fuhrmann (1998a, 1998b, 2000).

The analysis of the Sun is based on the KPNO flux atlas (Kurucz et al. 1984), the stellar analyses rest on FOCES spectra with R = 60 000 and S/N > 300 at Hα. Typically 70 Fe I and 15 Fe II lines can be measured in the optical spectra of the stars under investigation ($[m/H] < -2$).

We note that FOCES échelle spectra can be rectified *across* Balmer profiles to an accuracy of 0.5%, a prerequisite for utilizing these lines as a reliable temperature indicator. Hα and the higher Balmer profiles yield identical temperatures, if the efficiency of convection α ($= l/H_p$) is lowered to 0.5 (in the framework of the Böhm-Vitense formulation of convection).

2.   The Sun as a Star

The first and foremost test any cool-star plasma-diagnostic tool has to pass is the reproduction of the solar spectrum. To be able to analyse the most metal-poor stars, we have to synthesize the strongest solar lines. Therefore the correct choice of damping parameters (log C_6, van der Waals) is of central importance. We initially implemented the new C_6 values for Fe I by Anstee & O'Mara (1995, hereafter AOM), but had to realize that they lead to a clear trend of abundance with line strength. To minimize this trend, the microturbulence was increased to 1 km s^{-1}and log C_6 lowered by 0.4 dex resulting in damping constants in between those of AOM and Unsöld (1968).

The choice of damping constants also leaves its imprint on the *absolute* iron abundance one derives for the Sun: While the log C_6(AOM) damping constant leads to an Fe I abundance close to meteoritic, the log C_6(AOM) $-$ 0.4 dex damping constant results in ε(Fe I) $\sim$ 7.6, some 0.05 dex above the value determined from the 35 strongest lines of Fe II using the classical log C_6 damping constant.

There are two lessons to learn from the Sun: (1) different sources of f values are incompatible with one another resulting in substantial scatter of $\sim$0.1 dex (1σ) and (2) even the *internal* consistency within a given source is far from satisfactory (for more details see Gehren et al. 2001; T. Gehren et al., in preparation).

3.   HD 19445, HD 84937, and HD 140283

With a temperature from Balmer profiles and a mass estimate from evolutionary tracks, one can iteratively determine the gravity which best reproduces the parallax measured by HIPPARCOS. Since the volume density of halo stars is low, we have to consider targets out to about 100 pc. In this volume, the programme stars are among the most metal poor and cover stellar evolution from the main sequence to the subgiant branch.

The main free parameter of our non-LTE model, the poorly known efficiency of collisions with hydrogen atoms, has to be calibrated to achieve concordance between the abundances derived from Fe II in LTE and Fe I in non-LTE *at the HIPPARCOS gravity*. As long as H collisions are present at all (parameterized by S_H times the formula of Drawin 1968) the levels are thermalized *relative to one another*, yet heavily underpopulated *relative to* Fe II due to overionization. Thus the abundances scales with S_H enabling us to fine tune the model.

The efficiency of hydrogen collisions needed turns out to scatter around 0.7. Our model is able the resolve the discrepancy between gravities from pressure-broadened wings of Mg Ib lines (in excellent agreement with HIPPARCOS, cf. Fuhrmann 1998b, 2000) and from Fe I/II assuming LTE (see Figure 1). This is particularly noteworthy, as the size of the non-LTE effect (clearly a function of evolutionary stage, metallicity, etc.) is correctly predicted for all three targets.

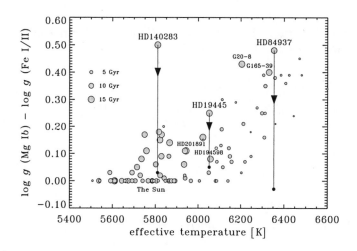

Figure 1.   LTE gravities from Fe I/II vs. Mg Ib in the sample of Fuhrmann (1998b, 2000). Additional stars from Korn & Gehren (2001) are labeled. Iron in LTE yields gravities systematically too low, in particular for hot and metal-poor stars. The black bullets indicate the revised positions of the programme stars using our $S_H = 0.7$ iron non-LTE model. A change of 0.01 dex in gravity corresponds to a 1% deviation with respect to HIPPARCOS.

References

Anstee, S. D., & O'Mara, B. J. 1995, MNRAS, 276, 859

Bautista, M. A. 1997, A&AS, 122, 167

Drawin, H. W. 1969, Z. Physik, 225, 483

Fuhrmann, K. 1998a, A&A, 330, 626

Fuhrmann, K. 1998b, A&A, 338, 161

Fuhrmann, K. 2000, A&A, submitted

Gehren, T., Butler, K., Mashonkina, L., Reetz, J., & Shi, J. 2001, A&A, 366, 981

Korn, A. J., & Gehren, T. 2001, in ASP Conf. Ser. 228, Dynamics of Star Clusters and the Milky Way, ed. Deiters et al. (San Francisco: ASP), 494

Kurucz, R. L., Furenlid, I., Brault, J., & Testerman, L. 1984, Solar Flux Atlas from 296 to 1300 nm, Kitt Peak National Solar Observatory

Nave, G., Johansson, S., Learner, R. C. M., Thorne, A. P., & Brault, J. W. 1994, ApJS, 94, 221

Unsöld, A. 1968, Physik der Sternatmosphären (Berlin: Springer-Verlag)

Acknowledgments.   This work is made possible through a grant from the German National Scholarship Foundation (Studienstiftung des deutschen Volkes). AJK wishes to thank this organization for many years of continuous support.

Astrophysical Ages and Time Scales
ASP Conference Series, Vol. 245, 2001
T. von Hippel, C. Simpson, N. Manset

Did the Inner Halo Globular Clusters Form Earlier Than Those in the Outer Halo?

Jae-Woo Lee, Bruce W. Carney

Department of Physics and Astronomy, The University of North Carolina, Chapel Hill, NC 27599-3255, USA

Laura K. Fullton

Observatoire de Geneve, Switzerland

Peter B. Stetson

Dominion Astrophysical Observatory, Canada

Abstract. We present HST NICMOS and WFPC2 photometry of the metal-poor inner halo globular clusters NGC 6287, NGC 6293, NGC 6541, and an intermediate halo globular cluster NGC 6341 (M 92). Our relative age estimation between NGC 6287 and M 92 using the IR color difference between the main sequence turnoff and the base of the red giant branch shows that they essentially have the same ages within ±2 Gyr. Our HST WFPC2 photometry of NGC 6293 and NGC 6541 also shows that these two clusters have the same ages as M 92 within ±1 Gyr. Within this framework, our results are consistent with the idea that globular cluster formation must have been triggered everywhere at the same time in our Galaxy.

1. Introduction

When did our Galaxy form? In what regions did star formation first begin? How rapidly did the star formation proceed in the regions of different densities? Understanding how the oldest stellar population formed in our Galaxy has always been one of the key quests in modern astrophysics.

The standard inside-out Galaxy formation scenario suggests that the star formation should have started from the high-density regime of the Galaxy due to the smaller free collapse time scale $\tau \propto \rho^{-1/2}$ near the central part of our Galaxy.

During the last decade, a tremendous amount of information has been accumulated regarding the ages, kinematics, and chemical compositions of globular cluster systems. However, the questions associated with the formation and the evolution of our Galaxy still remain unanswered. In particular, due to the high interstellar reddening toward the Galactic center, previous ground-based observations have failed to provide an accurate formation epoch of the inner halo globular cluster system.

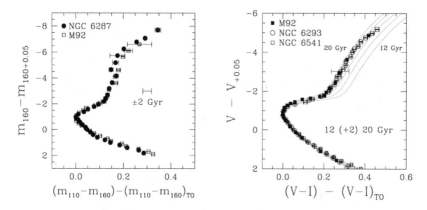

Figure 1. Relative age estimations based on the color difference between the main sequence turnoff and the base of the red giant branch. Our results suggest that inner halo globular clusters NGC 6287, NGC 6293, and NGC 6541 essentially have the same ages as M 92.

We present HST NICMOS/WFPC2 photometry of the three most metal-poor inner halo globular clusters NGC 6287 ([Fe/H] $= -2.12$, $R_{GC} = 1.6$ kpc), NGC 6293 ([Fe/H] $= -2.14$, $R_{GC} = 1.4$ kpc), NGC 6541 ([Fe/H] $= -1.85$, $R_{GC} = 2.2$ kpc), and one of the oldest globular clusters in our Galaxy, NGC 6341 (M 92, $R_{GC} = 9.6$ kpc).

2. Results

The observations for NGC 6293 and NGC 6541 were carried out on 1995 March 23 (UT) and 1994 October 15 (UT) using HST WFPC2 and those for NGC 6287 and M 92 were carried out on 29 January 1998 (UT) and 12 January 1998 (UT) using HST NICMOS with the NIC3 camera. All images were processed using the standard NICMOS/WFPC2 calibration pipeline procedures. PSF photometry for all images have been performed with DAOPHOTII/ALLSTAR and ALL-FRAME (Stetson 1987, 1994), following the standard reduction procedures.

With the final composite color–magnitude diagrams, we derived the fiducial sequences for each cluster and we compared the relative ages using the method proposed by VandenBerg, Bolte, & Stetson (1990). Our results are presented in Figure 1. As can be seen in the figure, NGC 6287 and M 92 have same age within ±2 Gyr, and NGC 6293 and NGC 6541 have essentially the same ages as M 92 within ±1 Gyr.

In Figure 2, we present the relative ages of the globular clusters as functions of metallicity and Galactocentric distance using the combined results of our work, Harris et al. (1997), Rosenberg et al. (1999), and Stetson et al. (1999). In the figure, the metallicity scales for NGC 6287, NGC 6293, and NGC 6541 were taken from the high resolution echelle spectroscopy by J.-W. Lee & B. Carney (in preparation). The figure shows that the ages of the oldest globular clusters in our Galaxy do not vary with metallicity ([Fe/H] ≤ -1.0) or the Galactocentric distance. In particular, our work clearly shows that there is no such gradient in the inner part of our Galaxy. Thus, our results are consistent with the idea that the globular cluster formation must have been triggered everywhere at the same time in our Galaxy. Since Harris et al. (1997) claimed that the very remote halo cluster NGC 2419 ($R_{GC} = 90$ kpc) and M 92 essentially have the same age, our result extends this to the inner halo globular clusters in our Galaxy with $R_{GC} < 2.5$ kpc.

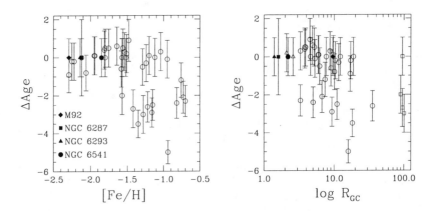

Figure 2.   Comparisons of relative ages of the globular clusters with [Fe/H] and the Galactocentric distance. The ages of the oldest globular clusters do not vary with the Galactocentric distance or the metallicity ([Fe/H] ≤ -1.0).

References

Harris, W. E., et al. 1997, AJ, 114, 1030

Rosenberg, A., Saviane, I., Piotto, G., & Aparicio, A. 1999, AJ, 118, 2306

Stetson, P. B. 1987, PASP, 99, 191

Stetson, P. B. 1994, PASP, 106, 250

Stetson, P. B. et al. 1999, AJ, 117, 247

VandenBerg, D. A., Bolte, M., & Stetson, P. B. 1990, AJ, 100, 445

Acknowledgments.   This is part of the Ph.D. dissertation of J.-W.L. at the University of North Carolina at Chapel Hill. We thank the NSF and the NASA for financial support via NSF grant AST-9988156 and NASA STScI grants GO-06561.02-95A and GO-07318.04-96A.

Astrophysical Ages and Time Scales
ASP Conference Series, Vol. 245, 2001
T. von Hippel, C. Simpson, N. Manset

Horizontal Branch Morphology as an Age Indicator

Young-Wook Lee, Suk-Jin Yoon, Soo-Chang Rey

Center for Space Astrophysics and Department of Astronomy, Yonsei University, Shinchon 134, Seoul 120-749, Korea

Brian Chaboyer

Department of Physics and Astronomy, 6127 Wilder Laboratory, Dartmouth College, Hanover, NH 03755-3528, USA

Abstract. We present our recent revision of model constructions for the horizontal branch (HB) morphology of globular clusters, which suggests the HB morphology is more sensitive to age compared to our earlier models. We also present our high precision CCD photometry for the classic second parameter pair M 3 and M 13. The relative age dating based on this photometry indicates that M 13 is indeed older than M 3 by 1.7 Gyr. This is consistent with the age difference predicted from our new models, which provides further support that the HB morphology is a reliable age indicator in most Population II stellar systems.

1. Introduction

Because the horizontal branch (HB) stars in globular clusters are much brighter than main sequence (MS) stars, the interpretation of HB morphology in terms of relative age differences would be of great value in the study of distant stellar populations where the MS turnoff is fainter than the detection limit. In this paper, we report our progress in the use of HB as a reliable age indicator.

2. New HB Population Models

Some seven years ago, Lee, Demarque, & Zinn (1994, hereafter LDZ) concluded that age is the most natural candidate for the global second parameter, because other candidates can be ruled out from the observational evidence, while supporting evidence do exist for the age hypothesis. Although this conclusion is generally accepted in the community, there is still some debate about this issue (see table 1 of Lee et al. 1999). Among others, critics argue that the relative age differences inferred from the MS turnoffs are often too small compared to the amounts predicted from the HB models for the several second parameter pairs, including the well-known M 3 and M 13 pair.

We found, however, several recent developments can affect the relative age dating technique from HB morphology. First of all, there is now a reason to believe that absolute age of the oldest Galactic globular clusters are reduced to about 12 Gyr, as suggested by the new HIPPARCOS distance calibration and other improvements in stellar models (Reid 1998; Gratton et al. 1997; Chaboyer et al. 1998). As LDZ already demonstrated, this has a strong impact on the relative age estimation from HB morphology. Also, it is now well established that α elements are enhanced in halo populations ($[\alpha/\mathrm{Fe}] = 0.4$). Finally, Reimers' (1975) empirical mass-loss law suggests more mass-loss at larger ages. The result of this effect was also presented in LDZ, but unfortunately the most widely used diagram (their figure 7) is the one based on fixed mass-loss. We found all of the above effects make the HB morphology more sensitive to age (see Y.-W. Lee & S.-J. Yoon, in preparation, for details). Therefore, as illustrated in Figure 1, now the required age difference is much reduced compared to figure 7 of LDZ. Now, only 1.1 Gyr of age difference, rather than 2 Gyr, is enough to explain the systematic shift of the HB morphology between the inner and outer halo clusters. Also, to within the observational uncertainty, age differences of about 1.5–2 Gyr are now enough to explain the observed difference in HB morphology between the remote halo clusters (Pal 3, Pal 4, Pal 14, and Eridanus) and M 3. These values are consistent with the recent relative age datings both from the HST and high-quality ground-based data.

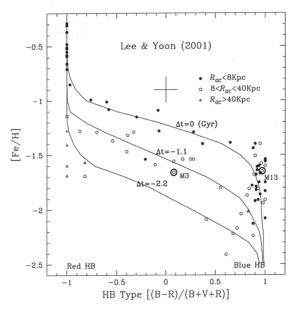

Figure 1.   New HB population models are more sensitive to age than our earlier models. The $\Delta t = 0$ model corresponds to the mean age of the inner halo (R < 8 kpc) clusters. Relative ages are in Gyr.

3.   The Case of M 3 and M 13

We obtained high quality color–magnitude data for the classic second parameter globular clusters M 3 and M 13. The clusters were observed during the same nights with the same instruments (MDM 2.4m), allowing us to determine the accurate relative ages. From the color difference method between the turnoff and the base of the red giant branch (RGB), we now confirm M 13 is indeed 1.7 ± 0.5 Gyr older than M 3 (Figure 2; see Rey et al. 2001 for details). This is consistent with the age difference predicted from our new HB models (see Figure 1), which provides further support that the HB morphology is a reliable age indicator in most Population II stellar systems.

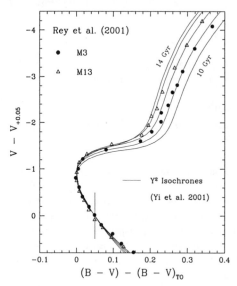

Figure 2.   Fiducial sequence for M 3 compared with that for M 13 following the prescription of VandenBerg, Bolte, & Stetson (1990). Note a separation between the two clusters' RGBs, indicating an age difference.

References

Chaboyer, B., Demarque, P., Kernan, P. J., & Krauss, L. M. 1998, ApJ, 494, 96

Gratton, R. G., Fusi Pecci, F., Carretta, E., Clementini, G., Corsi, C. E., & Lattanzi, M. 1997, ApJ, 491, 749

Lee, Y.-W., Demarque, P., & Zinn, R. 1994, ApJ, 423, 248 (LDZ)

Lee, Y.-W., Yoon, S.-J., Lee, H.-c., & Woo, J.-H. 1999, in ASP Conf. Ser. 192, Spectrophotometric Dating of Stars and Galaxies, ed. I. Hubeny, S. Heap, & R. Cornett (San Francisco: ASP), 185

Reid, I. N. 1998, AJ, 115, 204

Reimers, D. 1975, Mem. Soc. Roy. Sci. Leige, 8, 369

Rey, S.-C., Yoon, S.-J., Lee, Y.-W., Chaboyer, B., & Sarajedini, A. 2001, submitted

VandenBerg, D. A., Bolte, M., & Stetson, P. B. 1990, AJ, 100, 445

Astrophysical Ages and Time Scales
ASP Conference Series, Vol. 245, 2001
T. von Hippel, C. Simpson, N. Manset

Ages of Cool White Dwarfs From Photometric and Spectroscopic Analyses

S. K. Leggett

Joint Astronomy Centre, 660 North A'ohōkū Place, Hilo, HI 96720, USA

P. Bergeron

Département de Physique, Université de Montréal, C.P. 6128, Succ. Centre-Ville, Montréal, Québec H3C 3J7, Canada

María Teresa Ruiz

Departamento de Astronomía, Universidad de Chile, Casilla 36-D, Santiago, Chile

Abstract. Most stars evolve to become white dwarfs, which lose heat slowly; the coolest can constrain the age of the Galaxy. The energy distributions, Hα profiles, and parallaxes of 150 cool white dwarfs are combined and compared against model atmospheres to determine $T_{\rm eff}$ and radius, and constrain atmospheric composition. New evolutionary sequences are used to derive masses and ages. The oldest objects in our local disk sample are found to be 7.9 Gyr or 9.7 Gyr old for thin or thick hydrogen surface layers, respectively.

1. Introduction

The majority of stars evolve to become white dwarfs (WDs) with hydrogen/helium surface layers at high pressure. These objects cool slowly; as the rate of cooling depends on the opacity of the outer envelope, it is important to accurately determine their chemical composition. However the age of a WD may also depends on its chemical evolution *history*, as the composition of the envelope probably changes due to gravitational settling, convective mixing or dredge-up, and accretion from the interstellar medium (e.g., Fontaine & Wesemael 1987). We have been deriving composition, temperature, and where possible radius, mass, and age, to improve our understanding of WD chemical evolution. Descriptions of the model atmospheres, our first analysis, and the full details of the current analysis can be found in Bergeron, Saumon, & Wesemael (1995); Bergeron, Ruiz, & Leggett (1997); and Bergeron, Leggett, & Ruiz (2001).

2. The Sample, Data, and Analysis

The sample consists of all WDs cooler than $\sim 12\,000$ K with trigonometric parallaxes available from van Altena, Lee, & Hoffleit, (1994) or Ruiz, Anguita, & Maza (1989). This produced 150 stars for which optical spectra and optical and infrared photometry were obtained. Spectra around Hα were obtained at KPNO, CTIO, and ESO. *BVRI* photometry was obtained at KPNO and CTIO, and *JHK* photometry at CTIO, the IRTF, and UKIRT. The energy distribution implied by the photometry is fit with a model atmosphere, with T_{eff} and angular diameter as free parameters. Using the parallax, the radius can be obtained as well as the stellar mass using the mass–radius relationship for degenerate stars. We use two mass–radius relationships from Fontaine, Brassard, & Bergeron (2001); both have carbon/oxygen cores, one has a thick envelope of a helium mantle of 1% of the stellar mass and an outermost hydrogen layer of 0.01%; the other thin envelope model has the hydrogen layer reduced to 10^{-10} M_*. The thin envelope model can be used for the non-DA stars. The fit is iterated for a self-consistent value of log g. We find that 80% of the stars can be fit well with pure H or pure He models (although trace amounts of other elements cannot be ruled out). Figure 1 shows our results in terms of mass vs. T_{eff}. Stars with mass < 0.5 $M_{\odot}$ are most likely unresolved binaries or the result of common-envelope evolution, as the Galaxy is too young to produce such objects by normal evolution.

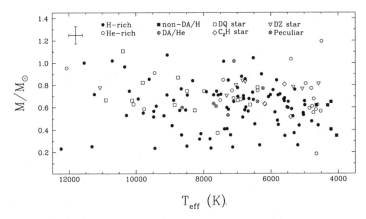

Figure 1. Derived mass vs. T_{eff}. Chemical composition is indicated.

3. Cosmochronology

Figure 1 shows that white dwarf surface chemistry evolves with time; the atmosphere probably goes through periods of convective mixing (a function of temperature and thickness of the surface layer), as well as periods of metal-enrichment from the ISM or via convective dredge-up. Given this, it is not clear which evolutionary models to fit to our data. We present results for both thin and thick hydrogen surface layers; we do not use helium-only envelopes as these show an extreme sensitivity to trace elements leading to a continuum of ages. Figure 2 shows our results compared to both thin and thick hydrogen layer isochrones, with age indicated in Gyr. Less massive (large radii) WDs cool more rapidly, except for the most massive objects for which crystallization sets in. Note that mass must be known to determine age.

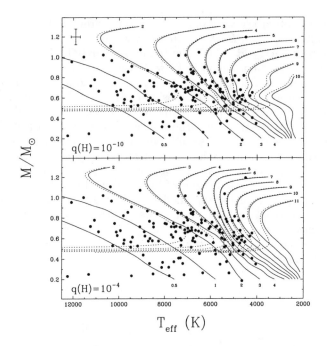

Figure 2.   Mass vs. T_{eff} with isochrones for thin and thick hydrogen layers. Solid line is degenerate cooling time, dashed is MS+WD age.

4.   Conclusion

Neglecting the objects with mass < 0.5 $M_{\odot}$, the coolest star is ER8 which is H-rich and aged 9.7 or 8.5 Gyr by the thick or thin hydrogen layer evolutionary models, respectively. If the two apparently low-mass objects which are cooler than ER8 are in fact unresolved pairs of normal mass WDs, the total age of these H-rich systems, LHS 239 and LP 131–66, are 9–10.3 Gyr. Hence this work gives an age for the local region of the disk of 9.5 ± 1 Gyr.

References

Bergeron, P., Leggett, S. K., & Ruiz, M. T. 2001, ApJS, in press (astro-ph/0011286)
Bergeron, P., Ruiz, M. T., & Leggett, S. K. 1997, ApJS, 108, 339
Bergeron, P., Saumon, D., & Wesemael, F. 1995, ApJ, 443, 764
Fontaine, G., Brassard, P., & Bergeron, P. 2001, PASP, in press
Fontaine, G., & Wesemael, F. 1987, in IAU Coll. 95, Conf. on Faint Blue Stars, (Schenectady: L. Davis Press), 319
Ruiz, M. T., Anguita, C., & Maza, J. 1989, in IAU Coll. 114, White Dwarfs, (Berlin: Springer-Verlag), 122
van Altena, W. F., Lee, J. T., & Hoffleit, E. D. 1994, The General Catalogue of Trigonometric Parallaxes (New Haven: Yale University Observatory)

Astrophysical Ages and Time Scales
ASP Conference Series, Vol. 245, 2001
T. von Hippel, C. Simpson, N. Manset

Revised Ages for the Alpha Persei and Pleiades Clusters

Eduardo L. Martín, Scott Dahm

Institute for Astronomy, University of Hawai'i, USA

Yakiv Pavlenko

Main Astronomical Observatory of Academy of Sciences, Ukraine

Abstract. We derive ages for the Alpha Persei and Pleiades clusters of 77 Myr and 119 Myr, respectively, using lithium abundances and K-band photometry of very low-mass members. These ages are older than the canonical ages of 50 Myr and 70 Myr obtained from isochrone fitting of the upper main sequence.

1. Introduction

The ages of clusters are usually estimated from isochrone fitting to the more massive members. In the case of young open clusters, this method has two problems: (a) there are few massive stars that have evolved away from the main sequence; (b) the ages of the massive stars depend on mass loss, convective overshooting, and rotation. A full model of stellar evolution taking into account all those effects does not exist yet. Magazzù, Martin, & Rebolo (1993) proposed to use observations of lithium in very low-mass stars as a tool to study their age and mass. A derivation of such a lithium test was used by Basri, Marcy, & Graham (1996) to estimate an age for the Pleiades cluster of 115 Myr, based on the detection of lithium in a very low-mass cluster member. Martín et al. (1998, 2000) and Stauffer, Schultz, & Kirkpatrick (1998) provided additional lithium observations of Pleiades very low-mass members. Basri & Martín (1999a) and Stauffer et al. (1999) provided lithium data for very low-mass members of the Alpha Persei open cluster. In this paper we revisit the ages of Alpha Persei and the Pleiades using K-band data to estimate the luminosity of the objects (previous work used I-band data), and synthetic spectra to estimate the lithium abundances. We obtain new age estimates that confirm the known trend for older ages than those estimated from upper main sequence isochrone fitting.

2. Analysis

Unresolved binaries make objects appear brighter than they really are because of the combined light of two components. For equal brightness components, the magnitude decreases by 0.75 mag. Basri & Martín (1999b) found one double-lined spectroscopic binary in the Pleiades with a mass ratio ≥ 0.8. It is possible than many other unresolved binaries exist among the very low-mass population of the Pleiades and other clusters. The effect of unresolved binaries on the age

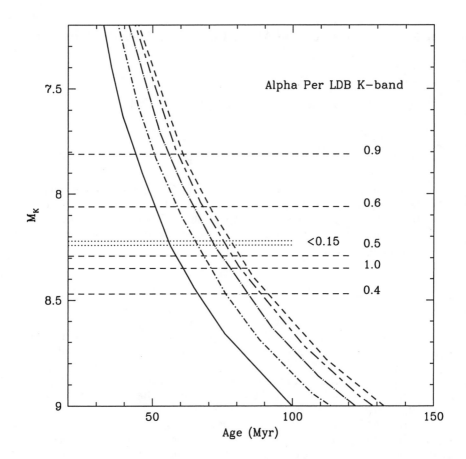

Figure 1.   Lithium isoabundance curves from the Lyon group (Baraffe et al. 1998). Each curve represents a constant amount of lithium left over after depletion from the initial value. Line code (from left to right): solid is 90%, dot and short dash is 50%, dot and long dash is 10%, short dash and long dash is 1%, and dashed is 0.1%. Absolute K-band magnitudes of very low-mass cluster members are marked with horizontal lines. The Li I equivalent width given in the literature is given at the end of each line.

derived from the lithium abundance is to make the cluster appear younger. To minimize this effect we chose to derive the age from the faintest cluster member for which lithium has not been detected. Brighter cluster members for which lithium has been detected are binary candidates.

Previous estimates of the cluster ages from very low-mass members have used I-band photometry. This may be problematic due to variability, such as that reported by Martín & Zapatero Osorio (1997), Terndrup et al. (1999), and

Bailer-Jones & Mundt (2001). In late M-type stars it is likely that the variability is due to magnetic cool spots, and that the amplitude of variability decreases for longer wavelength, although this has never been checked. Additionally, synthetic spectra are not able to reproduce well the *I*-band region because of the presence of TiO and FeH bands, plus the effects of dust condensation for the latest M types. We prefer the *K*-band filter because synthetic models are able to reproduce better the spectrum around 2.1 microns than the spectrum around 0.8 microns. We believe that the *K*-band magnitudes provide a more reliable comparison with theoretical models. The difference in ages obtained using the *I*-band and *K*-band photometry are up to 10 Myr.

Using synthetic spectra, we find that the lithium equivalent width is not very sensitive to effective temperature in the range 3200 K to 2600 K. A cosmic lithium abundance of log N(Li) = 3.1 (in the customary scale of log N(H) = 12) gives an equivalent width of 0.61 Åfor $T_{\rm eff}$ = 3200 K and 0.64 Å for $T_{\rm eff}$ = 2600 K. We also estimate that at log N(Li) = 1.0 (about two orders of magnitude of lithium depletion), the Li I resonance features becomes blended with the surrounding TiO lines and cannot be detected. Stronger lithium lines have been reported by Rebolo et al. (1996) and Stauffer et al. (1998). Such line strengths are not well understood using model atmospheres without dust condensation.

3. Results

Our results for the Alpha Persei open cluster are shown in Figure 1. The lithium depletion boundary age is obtained from the faintest cluster member with no lithium detection. We infer a lithium depletion of two orders of magnitude from synthetic spectra. The age derived for Alpha Per and the Pleiades is 77 Myr and 119 Myr, respectively.

References

Bailer-Jones, C., & Mundt, R. 2001, A&A, 367, 218

Baraffe, I., Chabrier, G., Allard, F., & Hauschildt, P. H. 1998, A&A, 337, 903

Basri, G., Marcy, G., & Graham, J. 1996, ApJ, 458, 600

Basri, G., & Martín, E. L. 1999a, ApJ, 510, 266

Basri, G., & Martín, E. L. 1999b, AJ, 118, 2460

Magazzù, A., Martín, E. L., & Rebolo, R. 1993, ApJ, 404, L17

Martín, E. L., Basri, G., Gallegos, J. E., Rebolo, R., Zapatero Osorio, M. R., & Bejar, V. J. S. 1998, ApJ, 499, L61

Martín, E. L., Brandner, W., Bouvier, J., Luhman, K. L., Stauffer, J., Basri, G., Zapatero Osorio, M. R., & Barrado y Navascués, D. 2000, ApJ, 543, 299

Martín, E. L., & Zapatero Osorio, M. R. 1997, MNRAS, 286, L17

Stauffer, J., Schultz, G., & Kirkpatrick, J. D. 1998, ApJ, 499, L199

Stauffer, J., et al. 1999, ApJ, 527, 219

Terndrup, D. M., Krishnamurthi, A., Pinsonneault, M. H., & Stauffer, J. R. 1999, AJ, 118, 1814

Astrophysical Ages and Time Scales
ASP Conference Series, Vol. 245, 2001
T. von Hippel, C. Simpson, N. Manset

Tidal Streams in the Galactic Halo: Evidence for the Sagittarius Northern Stream or Traces of a New Nearby Dwarf Galaxy

D. Martínez-Delgado, A. Aparicio, R. Carrera

Instituto de Astrofísica de Canarias, La Laguna, Spain

M. A. Gómez-Flechoso

Geneva Observatory, Switzerland

Abstract. We report the detection of a very low density stellar system at 50 ± 10 kpc from the Galactic center. It could form part of the Sagittarius northern stream or, alternatively, could be the trace of a hitherto unknown dwarf galaxy. If it is really associated with Sagittarius, it would confirm predictions of dynamical interaction models indicating that tidal debris from this galaxy could extend along a stream completely enveloping the Milky Way in a polar orbit.

1. Introduction

Standard cosmology predicts that dwarfs were the first galaxies to be formed in the Universe and that many of them merge afterwards to form bigger galaxies such as the Milky Way. This process would have left behind traces such as tidal debris or star streams in the outer halo. Of particular relevance has been the discovery of the Sagittarius (Sgr) dwarf galaxy (Ibata, Gilmore, & Irwin 1994), a Milky Way satellite in an advanced state of tidal disruption. Since its discovery, it was soon clear that its extent was larger than at first assumed, and dynamical models predict that the stream associated with the galaxy should envelop the whole Milky Way in an almost polar orbit.

The first evidence on the Sgr tidal stream was the discovery of Sgr field stars located up to 34°—more than 15 kpc—from the main body of the galaxy (Mateo, Olszewski, & Morrison 1998), forming a narrow extension of its southeast semi-major axis (the Sgr Southern stream). Theoretical models predicted a nearly symmetric tidal extension to the northwest, but the attempts to identified it failed due to the severe differential reddening of these regions close to the Galactic plane. Recently, the Sloan Digitized Sky Survey (SDSS) team discovered the trace of a possibly old stream in the outer Galactic halo (Yanny et al. 2000). Theoretical models of Sgr by Ibata et al. (2001) indicate that the structure found in the SDSS can be explained by the northern stream of the Sgr dwarf galaxy.

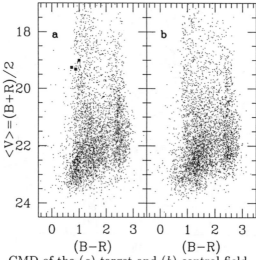

Figure 1. CMD of the (*a*) target and (*b*) control field.

2. Observations and Results

We have studied a $35' \times 35'$ field centered in the region surveyed by the SDSS team in order to obtain a color–magnitude diagram (CMD) reaching down to the turnoff, if any, of the stream stars. Observations were made with Wide Field Camera on the Isaac Newton Telescope, at the Observatorio del Roque de los Muchachos (Canary Islands, Spain).

Figure 1 shows the CMDs of the target (a) and the control (b) fields. Figure 1a shows an denser strip at $(B - R) \simeq 0.8$, $23.5 \leq V \leq 22$ (here using V for $(B - R)/2$), with a color and shape which correspond to what is expected for the upper main sequence of an old stellar population, with the turnoff at $V \simeq 22.6 \pm 0.3$. Two RR Lyrae variable stars have also been found in the field, which are marked with squares in the figure. Their average magnitude is $V = 19.2$, that is, incidentally, the magnitude of an RR Lyr population associated with the main sequence of an old stellar population such as that plotted in Figure 1a.

The distance to the system has been calculated independently from the magnitude of the turnoff and the RR Lyr stars (see Martínez-Delgado et al. 2001). Combining both, the result is $d_0 = 51 \pm 12$ kpc.

3. A Sagittarius Tidal Debris or a New Dwarf Galaxy?

Ibata et al. (2001) identify this SDSS stream as being part of the Sagittarius northern stream. This result suggests that the stellar system reported here could be tidal debris from Sgr, situated some 60° north of its center (see Figure 2).

In order to test this hypothesis, we have run N-body simulations of the Sgr plus Milky Way (MW) systems. The MW and Sgr models correspond to the

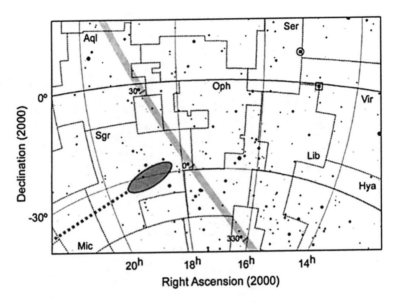

Figure 2.   Position in the sky of our target and control fields, together
with the Sgr dwarf main body and its southern stream (Mateo et al.
1998).

Galaxy and s-B2c dwarf models described by Gómez-Flechoso, Fux, & Martinet
(1999). The result is that our model is consistent with the observations of Sgr
and its tidal tail and is also in good agreement with the unconfirmed detections of
the Sgr tidal tail in the SDSS (see Martínez-Delgado et al. 2001). In addition, our
detection can also be fitted by the Sgr model and is close to our model predictions
for the apocentric distance of the orbit of the Sgr dwarf. This comparison
suggests that we have detected tidal debris from the Sgr northern stream in the
region close to the Sgr apocenter.

However, the lack of kinematic data is the main limitation in our comparison
of the observational data with the Sgr models. Therefore, the possibility that the
stream could be the trace of one or several hitherto unknown, tidally disrupted
dwarf galaxies cannot be rejected.

References

Gómez-Flechoso, M. A., Fux, R., & Martinet, L. 1999, A&A, 347, 77

Ibata, R., Gilmore, G., & Irwin, M. J. 1994, Nature, 370, 194

Ibata, R., Irwin, M., Lewis, G. F., & Stolte, A. 2001, ApJ, 547, L133

Martínez-Delgado, D., Aparicio, A., Gómez-Flechoso, M. A., & Carrera, R. 2001,
    ApJ, 549, L199

Mateo, M., Olszewski, E. W., & Morrison, H. L. 1998, ApJ, 508, L55

Yanny, B., et al. 2000, ApJ, 540, 825

Astrophysical Ages and Time Scales
ASP Conference Series, Vol. 245, 2001
T. von Hippel, C. Simpson, N. Manset

Dwarf Spheroidals With Tidal Tails and Structure Formation

Lucio Mayer

Department of Astronomy, University of Washington, Seattle, WA, USA

Abstract. Recent N-body simulations have shown that dIrrs evolve into dSphs owing to the strong tidal perturbation of the Milky Way. Satellites whose dark matter halos have a core or a Navarro, Frenk, & White (1997) profile with a concentration $c < 5$, undergo severe stripping even on low eccentricity orbits, and their remnants have flat projected stellar profiles at large radii as observed in some of the dSphs. Satellites with higher central densities, as predicted by cold dark matter (CDM) models, are more robust to tides and can reproduce the observations only on nearly radial orbits. Accurate proper motions for Local Group dSphs will allow us to test current models of structure formation.

1. Introduction

Dwarf galaxies in the Local Group (LG) show a striking morphology–density relation; low surface brightness, gas-poor dwarf spheroidals (dSphs) that are supported by the velocity dispersion of their stars are found within 300 kpc from either the Milky Way (MW) or M 31, while gas-rich, rotationally-supported dwarf irregulars (dIrrs) populate the outskirts of the Local Group (Grebel 1999). Recently, high resolution N-body simulations performed on supercomputers with the parallel binary treecode PKDGRAV (J. Stadel & T. Quinn, in preparation) have shown that the strong, time-dependent tidal force exerted by the Milky Way halo induces severe mass loss and non-axisymmetric instabilities in dIrrs, turning them into objects that match all the observed properties of dSphs in < 10 Gyr (Mayer et al. 2001a, 2001b). Tidal stirring leaves unbound stellar tails and streams that keep following the orbit of the satellite for several Gyr. Unbound stars lying or projected close to the spheroidal remnant might be responsible for the outer flattening of the stellar surface density profiles of some dSphs, in particular Carina (Irwin & Hatzidimitriou 1995; Majewski et al. 2000). Here we will show that such features can be used to probe structure formation models.

2. Initial Conditions

We place N-body models of dwarf irregular galaxies on bound orbits in the Milky Way halo, represented by the static external potential of an isothermal sphere with a mass $M_{\mathrm{MW}} \sim 4 \times 10^{12}\ M_\odot$, a virial radius $R_{\mathrm{MW}} \sim 400$ kpc and a core radius $R_c = 4$ kpc (the resulting circular velocity at the solar radius is ~ 220 km s^{-1}). The dwarf galaxy models comprise an exponential stellar disk embedded in a dark matter halo. The sizes and masses of the halos and disks are assigned according to both structure formation models and observations (see Mayer et al. 2001b). The halo of the satellite is either a truncated isothermal sphere or an NFW model with a concentration, $c < 10$ (Navarro, Frenk, & White 1997). The disk scale length, r_h, is determined by the concentration (or the core radius, r_c, in the case of an isothermal halo) and by the the spin parameter, λ, of the halo (see Mo, Mao, & White 1998). We consider different disk/orbit orientations and orbital eccentricities.

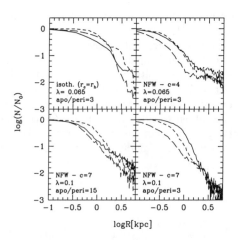

Figure 1.   Projected star counts for the remnants after 8 Gyr for a line of sight along one of the tidal tails (solid line) and along two directions perpendicular to the latter (short and long dashed lines). The orbital eccentricity and the models used are also indicated. The apocenters fall between 120 and 280 kpc; as a reference the present distance of Carina is ~ 110 kpc. The initial mass of the satellites is $M_{\mathrm{sat}} = 10^{-4} M_{\mathrm{MW}}$, and the angle between the spin of the disk and the orbital angular momentum is ~ 63 degrees.

3. Results

The remnants in our N-body simulations exhibit projected profiles with an outer flattening, like some of the dSphs, for a variety of initial conditions; both the internal structure and the orbits of the satellites determine the fraction of the stellar mass that is stripped and that can alter the observed shape of the profile depending on the viewing angle (Figure 1). However, when satellites have a shallow inner density profile, like that of a truncated isothermal sphere or that of an NFW profile with a low concentration ($c = 4$), the observations can be reproduced with orbits having a moderate eccentricity (apo/peri $= 3$), whereas if they have NFW profiles with higher concentrations ($c \geq 7$), as predicted by CDM models on their mass scale, nearly radial orbits are needed for strong mass loss to occur and lead to flattening (Figure 1). This result holds even when the disk scale length is considerably increased owing to a large halo spin parameter; the high concentration moves the tidal radius of the system too far out unless the pericenter of the orbit is very small.

4. Conclusion

The few proper motion measurements available for LG dwarf galaxies seem to indicate low eccentricity orbits (e.g., the Magellanic Clouds in Kroupa & Bastian 1997, and Sculptor in Schweitzer et al. 1995). If this result were to be confirmed by future high-precision astrometric missions, like GAIA or SIM, the outer flattening observed in dSph stellar profiles will be hardly explained within CDM models. Recently, Eke, Navarro, & Steinmetz (2001) have shown that in warm dark matter models small galaxy halos have NFW profiles with $c < 5$; these cosmologies would easily match the observations and, in addition, could yield the correct number of satellites in galaxy halos (Moore et al. 1999) as well as halo profiles soft enough to reproduce the rotation curves of LSB galaxies (de Blok et al. 2001).

References

de Blok, W. J. G., McGaugh, S. S., Bosma, A., & Rubin, V. C. 2001, ApJ, submitted (astro-ph/0103102)

Eke, V. R., Navarro, J. F., & Steinmetz, M. 2001, ApJ, submitted (astro-ph/0012337)

Grebel, E. K. 1999, in IAU Symp. 192, The Stellar Content of Local Group Galaxies, ed. P. Whitelock & R. Cannon (San Francisco: ASP), 1

Irwin, M., & Hatzidimitriou, D. 1995, MNRAS, 277, 1354

Kroupa, P., & Bastian, U. 1997, New Ast., 2, 139

Majewski, S. R., Ostheimer, J. C., Patterson, R. J., Kunkel, W. E., Johnston, K. V., & Geisler, D. 2000, AJ, 119, 760

Mayer, L., Governato, F., Colpi, M., Moore, B., Quinn, T., Wadsley, J., Stadel, J., & Lake, G. 2001a, ApJ, 547, L123

Mayer, L., Governato, F., Colpi, M., Moore, B., Quinn, T., Wadsley, J., Stadel, J., & Lake G. 2001b, ApJ, submitted (astro-ph/0103430)

Mo, H. J., Mao, S., & White, S. D. M. 1998, MNRAS, 296, 847

Moore, B., Ghigna, S., Governato, F., Lake, G., Quinn, T., Stadel, J., & Tozzi, P. 1999, ApJ, 524, L19

Navarro, J. F., Frenk, C. S., & White. S. D. M. 1997, MNRAS, 490, 493 (NFW)

Schweitzer, A. E., Cudworth, K. M., Majewski, S. R., Suntzeff, N. B. 1995, AJ, 110, 2747

Acknowledgments. I thank Tom Quinn and Joachim Stadel for providing me with PKDGRAV and Volker Springel for allowing me to use the code that builds the galaxy models. I thank also Ben Moore, Fabio Governato, and Monica Colpi for many stimulating discussions. Simulations were carried out at CINECA and ARSC supercomputing centers.

Astrophysical Ages and Time Scales
ASP Conference Series, Vol. 245, 2001
T. von Hippel, C. Simpson, N. Manset

MACHO Project Analysis of the Galactic Bulge Microlensing Events With Clump Giants as Sources

P. Popowski[1], T. Vandehei, K. Griest, C. Alcock, R. A. Allsman,
D. R. Alves, T. S. Axelrod, A. C. Becker, D. P. Bennett, K. H. Cook,
A. J. Drake, K. C. Freeman, M. Geha, M. J. Lehner, S. L. Marshall,
D. Minniti, C. A. Nelson, B. A. Peterson, P. J. Quinn, C. W. Stubbs,
W. Sutherland, D. Welch (The MACHO Collaboration)

[1] *Institute of Geophysics and Planetary Physics, Lawrence Livermore National Laboratory, Livermore, CA 94550, USA*

Abstract. We present preliminary results of the analysis of 5 years of MACHO data on the Galactic bulge microlensing events with clump giants as sources. This class of events allows one to obtain robust conclusions, because relatively bright clump stars are not strongly affected by blending. We discuss: (1) the selection of clump giant events, (2) the distribution of event durations, and (3) the anomalous character of event durations and optical depth in the MACHO field 104 centered on $(l, b) = (3°1, -3°0)$. We report the preliminary average optical depth of $\tau = (2.0 \pm 0.4) \times 10^{-6}$ (internal) at $(l, b) = (3°9, -3°8)$, and present an optical depth spatial distribution map. When field 104 is removed from the sample, the optical depth drops to $\tau = (1.4 \pm 0.3) \times 10^{-6}$, in excellent agreement with infrared-based models of the central Galactic region.

1. Introduction

The structure and composition of our Galaxy are outstanding problems in contemporary astrophysics. Microlensing is a powerful tool to learn about massive objects in the Galaxy. The amount of matter between the source and observer is typically described by the microlensing optical depth, defined as the probability that a source flux will be gravitationally magnified by a factor $\geq$ 1.34. Early analyses (Udalski et al. 1994; Alcock et al. 1997) of Galactic center sight lines produced two unexpected results: a very high optical depth of $3 - 4 \times 10^{-6}$ inconsistent with Galactic models and observations and an overabundance of long events. Here we analyze a new set of events to probe these controversial issues.

Blending is a major problem in any analysis of microlensing data involving point spread function photometry. The bulge fields are crowded, so that objects observed at a certain atmospheric seeing are blends of several stars, of which only one is typically lensed. This complicates a determination of an event's parameters and the analysis of the detection efficiency of microlensing events. If the sources are bright one can avoid these problems. Red clump giants are among the brightest and most numerous bulge stars. Therefore, this analysis concentrates on the events where the lensed stars are clump giants.

358

2.   The Data and Selection of Clump Giant Events

The MACHO Project observations were performed with 1.27-meter telescope at Mount Stromlo Observatory, Australia. In total, we collected 7 seasons (1993-1999) of data in the 94 Galactic bulge fields. The data that are currently available for analysis consist of 5 seasons (1993-1997) in 77 fields, and contain the photometry of about 30 million stars, including 2.1 million clump giants.

The events with clump giants as sources have been selected from the sample of all events which contains about $\sim$ 280 candidates. The determination of which of these sources are clump giants is investigated through the analysis of the global properties of the color–magnitude diagram in the Galactic bulge.

Using the accurately measured extinction towards Baade's Window allows one to locate *bulge* clump giants on the reddening-free color–magnitude diagram. This defines the parallelogram-shaped box in the upper left corner of the left panel of Figure 1. With the assumption that the clump populations in the whole bulge have the same properties as the ones in Baade's Window, the parallelogram described above can be shifted by the reddening vector to mark the expected locations of clump giants in different fields. The solid lines are the boundaries of the region where one could find the clump giants in fields with different extinctions. Using this approach, we identified 52 clump events.

3.   Field 104

There is a high concentration of long-duration events in MACHO field 104 at $(l, b) = (3°\!.1, -3°\!.0)$ as 5 out of 10 clump events longer than 50 days are in 104. The analysis of event durations *uncorrected* for efficiencies provides a lower limit on the difference between field 104 and all the other fields. We use the Wilcoxon's test on two samples: events in field 104 and all the remaining ones, and find that the events in 104 differ (are longer) at the level of 2.55σ. In addition, field 104 also has the highest optical depth (see Figure 1). Both of these features can be explained by the concentration of mass along this particular line of sight.

4.   Optical Depth

First, we report the average optical depth of

$$\tau = (2.0 \pm 0.4) \times 10^{-6} \quad \text{at} \quad (l, b) = (3°\!.9, -3°\!.8),$$

which is 1.5–2 times lower than the previously obtained values. We caution that this result is only preliminary, because possible systematic errors may be a fair fraction of the statistical error. We also note that about 40% of the optical depth is in the events longer than 50 days. This is at odds with standard models of the Galactic structure and kinematics.

Second, we plot in the right panel of Figure 1 the spatial distribution of the optical depth. The variation of the optical depth is dominated by the Poisson noise. The gradient of the optical depth is stronger in the b than in the l direction. Again, we note the anomalous character of field 104. It is marked with a black square and has an optical depth of $(1.4 \pm 0.5) \times 10^{-5}$.

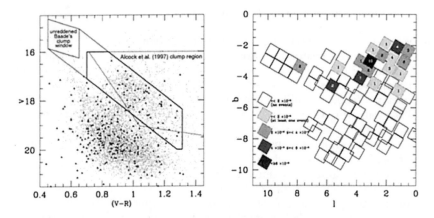

Figure 1.   *Left:* Events marked with filled triangles. The region surrounded by a bold line is our clump region. For comparison, the dotted line indicates the clump region from Alcock et al. (1997). Both selections return very similar events. *Right:* Spatial distribution of optical depth with the numbers of events given in the center of each field. Note anomalous field 104 with 10 events and optical depth of $(1.4 \pm 0.5) \times 10^{-5}$.

5. Conclusions

We presented first results from the analysis of five years of microlensing data toward the Galactic bulge collected by the MACHO Collaboration.

It is possible to select an unbiased sample of clump events based only on an event's position in the color–magnitude diagram. Ten out of 52 clump events have durations > 50 days, which implies that ∼ 40% of the optical depth is in the long events. This is surprising, because long events are the most likely a result of the disk-disk lensing (Kiraga & Paczyński 1994), while clump giants trace the bar rather than the inner disk (Stanek et al. 1994). Field 104 centered on $(l, b) = (3\overset{\circ}{.}1, -3\overset{\circ}{.}0)$ is anomalous: it has longer events and substantially higher optical depth. Both effects can be explained simultaneously by a concentration of mass along this line of sight. The optical depth averaged over the clump giants in 77 fields is $\tau = (2.0 \pm 0.4) \times 10^{-6}$ at $(3\overset{\circ}{.}9, -3\overset{\circ}{.}8)$. When anomalous field 104 is removed, the optical depth drops to $\tau = (1.4 \pm 0.3) \times 10^{-6}$, which is fully consistent with infrared-based models of the Galactic bar.

References

Alcock, C., et al. 1997, ApJ, 479, 119

Kiraga, M., & Paczyński, B. 1994, ApJ, 430, L101

Stanek, K. Z., Mateo, M., Udalski, A., Szymanski, M., Kaluzny, J., & Kubiak, M. 1994, ApJ, 429, L73

Udalski, A., et al. 1994, Acta Astron., 44, 165

Astrophysical Ages and Time Scales
ASP Conference Series, Vol. 245, 2001
T. von Hippel, C. Simpson, N. Manset

Is the Short Distance Scale a Result of a Problem With the LMC Photometric Zero Point?

Piotr Popowski

Institute of Geophysics and Planetary Physics, Lawrence Livermore National Laboratory, Livermore, CA 94550, USA

Abstract. I present a promising route to harmonize distance measurements based on clump giants and RR Lyrae stars. This is achieved by comparing the brightness of these distance indicators in three environments: the solar neighborhood, Galactic bulge, and Large Magellanic Cloud (LMC). As a result of harmonizing the distance scales in the solar neighborhood and Baade's Window, I derive the new absolute magnitude of RR Lyrae stars, $M_V(RR)$ at [Fe/H] $= -1.6$ in the range $(0.59 \pm 0.05, 0.70 \pm 0.05)$. Being somewhat brighter than the statistical parallax solution but fainter than typical results of the main sequence fitting to HIPPARCOS data, these values of $M_V(RR)$ favor intermediate or old ages of globular clusters. Harmonizing the distance scales in the LMC and Baade's Window, I show that the most likely distance modulus to the LMC, μ_{LMC} is in the range 18.24–18.44. The Hubble constant of about 70 km s^{-1} Mpc^{-1} reported by the HST Key Project is based on the assumption that the distance modulus to the LMC equals 18.50. The results presented here indicate that the Hubble constant may be up to 12% higher. This in turn would call for a younger Universe and could result in some tension between the age of the Universe and the ages of globular clusters. I argue that the remaining uncertainty in the distance to the LMC is now a question of one, single photometric reference point rather than discrepancies between different standard candles.

1. Introduction

The determination of the Hubble constant, H_0, is of primary importance in astrophysics. The distance ladder approach implemented by the HST Key Project is one major way to achieve this goal. Most of the extragalactic distances are calibrated with respect to the LMC. Based on the H_0 determination presented by Mould et al. (2000), one has an approximate relation:

$$H_0 = 70 \text{ km s}^{-1} \text{ Mpc}^{-1} \left[1 + 0.46\left(18.5 - \mu_{\mathrm{LMC}}\right)\right].$$

Unfortunately, the distance to the LMC is a subject of strong debate. There are not only disagreements between results from different distance indicators, but even disagreements between different investigations in the case of the same standard candle. Here we check the consistency of two major distance indicators: RR Lyrae stars and clump giants.

2. The Method

The essence of the approach presented here is a comparison between clump giants and RR Lyrae stars in different environments. If answers from two distance indicators agree then either systematics have been reduced to negligible levels in both of them or biases conspire to produce the same answer. This last problem can be tested with an attempt to synchronize distance scales in three different environments, because a conspiracy of systematic errors is not likely to repeat in all environments. Here I show that combining the information on RR Lyrae and clump stars in the solar neighborhood, Galactic bulge, and LMC provides additional constraints on the local distance scale.

3. Assumptions and Procedure

I assume that a universal linear $M_V(RR)$–[Fe/H] relation holds. I adopt a slope of 0.18 ± 0.03 and will calibrate a zero point in the process. I also assume the absolute I-magnitude of clump giants in Baade's Window is approximately known. I adopt $M_{I,BW} = -0.23 \pm 0.04$, which is in agreement with both the HIPPARCOS-based calibration and synthetic population models.

The zero point in the $M_V(RR)$–[Fe/H] relation is calibrated through comparison of clump giants and RR Lyrae stars in the solar neighborhood and Baade's Window. [As discussed by Popowski (2000, 2001), the clump giants in the bulge have unexpectedly redder $(V - I)_0$ colors than their counterparts in the solar neighborhood. The use of either these anomalous, uncorrected $(V - I)_0$ colors or $(V - I)_0$ corrected to the standard value results in two classes of solutions.] A calibrated $M_V(RR)$–[Fe/H] relation is then used to determine $M_V(RR)$ in the LMC. $M_{V,LMC}(RR)$ combined with dereddened apparent V-mags of RR Lyrae stars, $V_{0,LMC}(RR)$, allows one to find μ_{LMC} [$V_{0,LMC}(RR)$ may come either from OGLE or Walker (1992)]. Finally, μ_{LMC} and $I_{0,LMC}(RC)$ can be combined to find $M_{I,LMC}(RC)$. Popowski (2001) gives a more detailed description of the data and methodology (see also Table 1).

4. Results and Sensitivity to Assumptions

The results are presented in Table 1. The first three columns describe assumptions and data input, and the last three columns give the solutions. Most of the significant systematic effects are taken into account through the analysis of four cases presented in Table 1. In particular, μ_{LMC} increases with fainter $V_{0,LMC}(RR)$. Note that, in this method, μ_{LMC} is *insensitive* to the controversial dereddened I-mags of the LMC clump giants, $I_{0,LMC}(RC)$.

5. Overview Conclusions

Requirement of consistency between standard candles in different environments is a powerful tool in calibrating absolute magnitudes and obtaining distances. If the anomalous $(V - I)_0$ color of clump giants in Baade's Window is real, then the distance scale tends to be shorter. In particular, $M_V(RR) = 0.70 \pm 0.05$ at

Table 1.   Various solutions for $M_V(RR)$, $M_{I,\text{LMC}}(RC)$, and μ_{LMC}.

Solution[a]	$I_{0,\text{LMC}}(RC)$	$V_{0,\text{LMC}}(RR)$	$M_V(RR)$[b]	$M_{I,\text{LMC}}(RC)$	μ_{LMC}
anomalous+OGLE	17.91 ± 0.05	18.94 ± 0.04[b]	0.70 ± 0.05	-0.33 ± 0.09	18.24 ± 0.08
anomalous+Walker	17.91 ± 0.05	18.98 ± 0.03[c]	0.70 ± 0.05	-0.42 ± 0.09	18.33 ± 0.07
standard+OGLE	17.91 ± 0.05	18.94 ± 0.04[b]	0.59 ± 0.05	-0.44 ± 0.09	18.35 ± 0.08
standard+Walker	17.91 ± 0.05	18.98 ± 0.03[c]	0.59 ± 0.05	-0.53 ± 0.09	18.44 ± 0.07

[a] The solutions are classified according to $(V - I)_0$ colors in the Galactic bulge (uncorrected = *anomalous* or corrected = *standard*), and the source of LMC RR Lyrae photometry: Udalski et al. (1999) = OGLE or Walker (1992). The errors in $M_V(RR)$ and $M_{I,\text{LMC}}(RC)$ include the uncertainty in the slope of the $M_V(RR)$–[Fe/H] relation as well as 0.1 dex uncertainty of metallicity difference between different stellar systems. Note that $M_{I,\text{LMC}}(RC)$ depends on $I_{0,\text{LMC}}(RC)$, whereas μ_{LMC} does not. As a result, the error in μ_{LMC} is smaller.
[b] at [Fe/H] = -1.6
[c] at [Fe/H] = -1.9

[Fe/H] = -1.6, and the distance modulus to the LMC is in the range $\mu_{\text{LMC}} = 18.24 \pm 0.08$ to $\mu_{\text{LMC}} = 18.33 \pm 0.07$. If the $(V - I)_0$ color of stars in Baade's Window is in error and should be the standard value, then the distance scale tends to be longer. In particular, one can obtain $M_V(RR) = 0.59 \pm 0.05$ at [Fe/H] = -1.6, and the distance modulus from $\mu_{\text{LMC}} = 18.35 \pm 0.08$ to $\mu_{\text{LMC}} = 18.44 \pm 0.07$. The reported μ_{LMC} does not depend on the assumed value of the dereddened I-magnitude of the LMC clump giants. The above results are fully consistent with the tip of the red giant branch (TRGB) method as shown by Bersier (2000) and Udalski (2000).

It is surprising that two reasonable adjustments: one of the $(V - I)_0$ color of clump giants in the Galactic bulge and another of the RR Lyrae photometry in the LMC, could change the estimate of μ_{LMC} by as much as 0.2 magnitudes. The very short distance scale could in principle be a purely photometric phenomenon, namely a problem with calibration of the photometric zero points!

References

Bersier, D. 2000, ApJ, 543, L23

Mould, J. R., et al. 2000, ApJ, 529, 786

Popowski, P. 2000, ApJ, 528, L9

Popowski, P. 2001, MNRAS, 321, 502

Udalski, A. 2000, Acta Astron., 50, 279

Udalski, A., Szymański, M., Kubiak, M., Pietrzyński, G., Soszyński, I., Woźniak, P., & Żebruń, K. 1999, Acta Astron., 49, 201

Walker, A. 1992, ApJ, 390, L81

Astrophysical Ages and Time Scales
ASP Conference Series, Vol. 245, 2001
T. von Hippel, C. Simpson, N. Manset

Chromospherically Young, Kinematically Old Stars

Helio J. Rocha-Pinto, Walter J. Maciel

Instituto Astronômico e Geofísico, Universidade de São Paulo, Av. Miguel Stefano 4200, 04301-904 São Paulo SP, Brazil

Bruno V. Castilho

Laboratório Nacional de Astrofísica, C.P. 21, 37500-000 Itajubá MG, Brazil

Abstract. A group of stars known to have low chromospheric ages but high kinematical ages is investigated. The majority of stars within this group shows a lithium abundance much smaller than that expected for their chromospheric ages, which are interpreted as indicative of their old ages. The results suggest that they can be formed from the coalescence of short-period binaries. Coalescence rates, calculated taking into account several observational data and a maximum theoretical time scale for contact, in a short-period pair, predict a number of coalesced stars similar to that which we have found in the solar neighbourhood.

1. Introduction

The chromospheric activity of a late-type star is frequently interpreted as a sign of youth. Young dwarfs show high rotation rates, and the interaction between rotation and outer envelope convection is expected to drive chromospheric activity. Nevertheless, not only young single stars present high rotation rates. Sometimes, a star suspected of being young can be instead a spectroscopic binary (Soderblom, King, & Henry 1998), not yet investigated by radial velocity surveys.

The term CYKOS (acronym for chromospherically young, kinematically old stars) is used here for all chromospherically active stars which, in a velocity diagram, present velocity components greater than what would be expected for such stars, irrespective of this object being a still undiscovered close binary, a runaway star or another kind of object. CYKOS were originally found by Soderblom (1990) and later confirmed by Rocha-Pinto, Castilho, & Maciel (2000), and we refer the reader to these papers for details.

2. Lithium in CYKOS

Although Li depletion and production in stars are processes not completely well understood, in some cases the Li abundance could be used as a youth indicator. If CYKOS are young objects, as suggested by their chromospheric activity, they

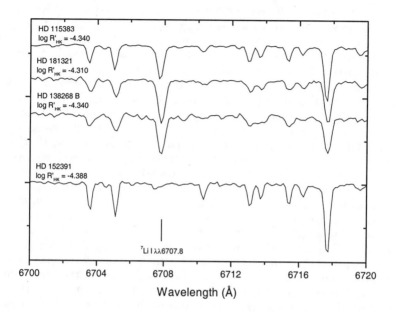

Figure 1.   Comparison between the spectra of a CYKOS and three normal stars having the same activity levels.

must present high Li abundances. On the other hand, if they show very depleted Li, they must be evolved objects, irrespective of their magnetic activity.

We have obtained spectra for 29 stars, amongst CYKOS and normal active stars. In Figure 1 we present a typical comparison between the spectra of a CYKOS and three normal stars having the same activity levels, and supposedly the same age.

In general, the CYKOS show much smaller Li abundances than the abundances of the normal stars with the same chromospheric activity. However, it should be noted that their low abundances agree with their presumed ages coming from their kinematical properties.

3.   Coalescence of Close Binaries

Some of the CYKOS are chromospherically active binaries. Their chromospheric activity results from the synchronization of the orbital with the rotational motion. This is the reason why they have a low chromospheric age, but a higher age from the point of view of its kinematics. Nevertheless, there are CYKOS which are single stars. For them there is no known mechanism that could store angular momentum to be used later by the star. The anomalously low lithium abundance, together with the high velocity components, can only be interpreted as a consequence of old age.

Poveda, Allen, & Herrera (1996) also have found several CYKOS amongst UV Ceti stars, which are known as young low-mass stars. The authors propose that these objects could be low-mass analogues of the well-known blue stragglers. The properties of a supposed low-mass field blue straggler would be very similar to that of some CYKOS, as we will see in what follows.

In short-period binaries we expect the occurrence of synchronization between the orbital and rotational periods. For low-mass stars, the magnetic activity is strong, and increases the angular momentum loss. When both orbital and rotational periods are synchronized, the rotational angular momentum loss occurs at the expenses on the orbital angular momentum. As a result, the period decreases, the components rotates more rapidly, and becomes closer, eventually becoming contact binaries, as those of W UMa-type.

In a coalescence of two low-mass stars (0.5 $M_\odot$ each), the resulting star must present a mass similar to that of the Sun, and a high rotation rate. This rotation rate, together with the convection in the outer envelope, would produce a copious chromospheric activity, in an amount similar to those found in very young stars. In the case of low-mass stars that have not ignited hydrogen in their cores, the just-formed single star would be similar in many respects to a young star, positioning in the zero-age main sequence, just like the blue stragglers. However, this star would inherit the same velocity components of the binary pair from which it was formed. Thus, due to the time before the coalescence, the velocity components are not similar to that of a young star. We would have a star apparently young, but kinematically old. The lithium abundance is also one of the few indicators that can show the real nature of these objects. In spite of not burning hydrogen considerably, stars having around 0.5 $M_\odot$ are highly convective, and Li burning is very efficient in them. Blue Stragglers like these should present small or no Li abundance (Glaspey, Pritchet, & Stetson 1994), since they would be formed by older objects.

Seven stars in our sample seem to fit well within these criteria. Calculations for the coalescence rate between close binaries show that the expected low-mass blue straggler number in the solar neighbourhood agrees within an order of magnitude with this finding. Detailed results will be published elsewhere.

References

Glaspey, J. W., Pritchet, C. J., & Stetson, P. B. 1994, AJ, 108, 271

Poveda, A., Allen, C., & Herrera, M. A. 1996, in RMA&A Conf. Ser. 5, Workshop on Colliding Winds in Binary Systems, 16

Rocha-Pinto, H. J., Castilho, B. V., & Maciel, W. J. 2000, in IAU Symp. 198, The Light Elements and Their Evolution, ed. L. da Silva, M. Spite, & R. de Medeiros (San Francisco: ASP), 512

Soderblom, D. R. 1990, AJ, 100, 204

Soderblom, D. R., King, J. R., & Henry, T. J. 1998, AJ, 116, 396

Acknowledgments.   We have made use of the SIMBAD database, operated at CDS, Strasbourg, France. We acknowledge support by FAPESP and CNPq.

Astrophysical Ages and Time Scales
ASP Conference Series, Vol. 245, 2001
T. von Hippel, C. Simpson, N. Manset

Atomic Diffusion in Stellar Interiors and Field Halo Subdwarfs Ages

M. Salaris

Astrophysics Research Institute, Liverpool John Moores University, Twelve Quays House, Egerton Wharf, Birkenhead CH41 1LD, UK

A. Weiss

Max-Planck-Institut für Astrophysik, Karl-Schwarzschild-Strasse 1, 85748 Garching, Germany

Abstract. We discuss the effect of atomic diffusion on the determination of field halo stars ages using theoretical isochrones. We show that the ages derived from the $T_{\rm eff}$ values of turnoff stars are reduced by 5–6 Gyr if diffusion is taken into account.

1. Introduction

Atomic diffusion is a basic element transport mechanism, which is neglected in the so-called "standard" stellar evolution models. It is driven by pressure (or gravity) gradients, temperature gradients and composition gradients. Gravity and temperature gradients tend to concentrate the heavier elements toward the center of the star, while concentration gradients oppose the above processes. Atomic diffusion acts slowly, with time scales of the order of 10^9 years, therefore it is mainly effective during the main sequence phase and the white dwarf cooling. The occurrence of diffusion in the Sun has been recently demonstrated by helioseismic studies (e.g., Guenther, Kim, & Demarque 1996). Solar models including this process can reproduce much better than standard models the solar pulsation spectrum and the helioseismic values of the helium surface abundance and depth of the convective envelope. Here we discuss the effect of atomic diffusion on the ages of field Population II stars.

2. Evolution of Surface Abundances With Time

Due to atomic diffusion, the abundances of helium (Y) and heavy elements (well parameterized by [Fe/H], since all the relevant metals diffuse approximately with the same speed as Fe) in the envelope of low mass stars decrease steadily during the main sequence phase. Y and [Fe/H] reach a minimum around the turnoff, then they increase along the subgiant branch due to the deepening of the convective envelope, which engulfs almost all the elements previously diffused toward the center. Along the red giant branch the element abundances are restored approximately to their initial values. The variation of [Fe/H] (and Y) along the main sequence is a strong function of the initial metallicity and

367

368 *Salaris & Weiss*

stellar mass. When increasing the initial metallicity and/or decreasing the stellar mass, the maximum [Fe/H] (and Y) depletion at the turnoff decreases, due to the increasing depth of the convective envelope, which implies a larger buffer of metals in the envelope to be diffused downwards.

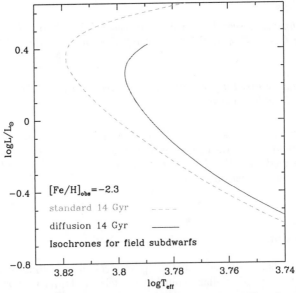

Figure 1. Comparison of two 14 Gyr isochrones showing the same surface [Fe/H] all along the main sequence and turnoff region, computed with and without diffusion.

When comparing theoretical isochrones with the color–magnitude diagram of main sequence stars, it is necessary to take into account the fact that the actual observed stellar chemical composition is different from the initial one when diffusion is accounted for, at odds with the results from standard models, where the surface chemical composition reflects the initial one. In Figure 1 we show isochrones with the same age and the same surface [Fe/H] at each point, computed with and without including diffusion. The main sequence of the isochrone with diffusion is shifted to lower $T_{\rm eff}$, and its turnoff is dimmer and colder than the standard one. This is easily explained by the fact that, in order to show the same [Fe/H] as the standard isochrone, the diffusive one has to be computed with an increased initial [Fe/H] (different increase for different masses) to balance the effect of diffusion (Morel & Baglin 1999; Salaris, Groenewegen, & Weiss 2000).

3. Age of Field Halo Stars

Only very few field halo stars have precise parallax measurements, and in order to estimate the age of this population, one has to use the T_{eff} of the objects at the turnoff.

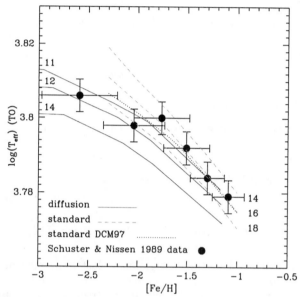

Figure 2. Age determination of the field halo population by comparing the T_{eff} of turnoff stars with the predictions of theoretical isochrones, with and without diffusion. The isochrone ages are in Gyr.

In Figure 2 we compare the T_{eff} of field halo turnoff stars by Schuster & Nissen (1989) with the corresponding temperatures from the standard isochrones by Salaris & Weiss (1998) and D'Antona, Caloi, & Mazzitelli (1997). The standard ages are in the range 18–20 Gyr for the most metal poor objects, and 14–18 Gyr for [Fe/H] $\gtrsim$ −2.0. When including diffusion in the stellar models, the ages are considerably reduced, being in the range 11–14 Gyr at the lowest metallicity end, and at most 12 Gyr for an observed [Fe/H] $\gtrsim$ −2.0.

References

D'Antona, F., Caloi, V., & Mazzitelli, I. 1997, ApJ, 477, 519
Guenther, D. B., Kim, Y.-C., & Demarque, P. 1996, ApJ, 463, 382
Morel, P., & Baglin, A. 1999, A&A, 345, 156
Salaris, M., Groenewegen, M. A. T., & Weiss, A. 2000, A&A, 355, 299
Salaris, M., & Weiss, A. 1998, A&A, 335, 943
Schuster, W. J., & Nissen, P. E. 1989, A&A, 222, 69

Astrophysical Ages and Time Scales
ASP Conference Series, Vol. 245, 2001
T. von Hippel, C. Simpson, N. Manset

Is the Universe Younger Than the Galaxy? The Lesson of the Globular Clusters

Oscar Straniero

Osservatorio Astronomico di Collurania, Teramo, Italy

Alessandro Chieffi

Istituto di Astrofisica Spaziale CNR, Roma, Italy

Inmaculata Dominguez

Universidad de Granada, Granada, Spain

Gianluca Imbriani

Seconda Universita' di Napoli, Caserta, Italy

Marco Limongi

Osservatorio Astronomico di Monte Porzio, Roma, Italy

Abstract. For a long time the globular clusters populating the halo of the Milky Way were considered too old with respect to the age predicted by the widely accepted model of the Universe. Now, a solution of this long-standing problem appears on the horizon.

1. Introduction

In the last couple of decades the most widely used cosmological models have been based on the assumptions that the Universe is flat (i.e., the average density is equal to the critical density or $\Omega = 1$) and the dominant constituent of the Universe was a not well identified kind of cold dark matter with just a few percent of normal baryonic matter. The cosmological constant, Einstein's biggest trouble, was usually ignored. Following this model, the resulting Universe is too young with respect to the oldest stars in the Milky Way, whose ages are derived by using stellar evolution theory. However, recent measurements of the cosmic microwave background coupled with the distance determinations obtained from high redshift supernova experiments seem to solve this longstanding problem. The fluctuations of the microwave background indicate that the Universe is flat (i.e., $\Omega \sim 1$), while the supernova distance scale implies the existence of a large cosmological constant, so that the amount of cold dark matter should be strongly reduced (down to 20–30% of the total density). This implies that the expansion rate was lower in the past and, in turn, the Universe is older than previously believed.

2. Globular Cluster Turnoffs: The Age Indicator

Since the pioneering papers (e.g., Sandage 1962), it was immediately understood that globular clusters (GCs) are among the oldest components of the Milky Way. Their ages may be derived by comparing the observed color–magnitude diagrams with theoretical isochrones. The location in the HR diagram of both the main sequence and the red giant branch do not change with age. On the contrary, the turnoff and the subgiant branch are good age indicators. The turnoff luminosity has been widely used to derive the absolute age of globular clusters. The turnoff color also depends on age, but its use as age indicator is presently restricted due to the large uncertainties affecting the theoretical stellar temperatures and the color–temperature relation. Theoretical isochrones for GC stars firstly appeared at the beginning of the seventies (Iben 1971; Ciardullo & Demarque 1977). Owing to the improvement of our knowledge of the physics governing the behaviour of stellar matter (opacity, equation of state, nuclear reaction rates and the like), the theoretical models of H-burning low mass stars have been recursively updated. Milestones were the paper by Vandenberg & Bell (1985), in which several inputs physics were revised, and the one by Straniero & Chieffi (1991), firstly including electrostatic corrections in the equation of state. This last result has been more recently confirmed by Chaboyer & Kim (1995), by using the more refined OPAL equation of state. As a consequence of these improvements, an approximately 15% reduction of the predicted globular cluster ages has been found. At the end of the eighties and during the nineties it was progressively realized that the α-elements (O, Ne, Mg, Si, Ca, etc.) are overabundant with respect to iron in Population II stars. The effects on the age were deeply studied by Salaris, Chieffi, & Straniero (1993) and substantially confirmed by Vandenberg, Bolte, & Stetson (1996). Another important improvement was the inclusion of microscopic diffusion. The time scale of microscopic diffusion is so long that it is generally neglected in stellar model computations. However, during the long lifetime of an H-burning low mass star, this phenomenon may slowly alter the chemical profile and, in turn, it may lead to a modification of the stellar structure. Globular cluster isochrones including He diffusion were computed by Proffit & Vandenberg (1991) and Chaboyer et al. (1992). More recently Straniero, Chieffi, & Limongi (1997, SCL97) and Castellani et al. (1997) have also included the diffusion of heavier elements. A further reduction of the estimated GC ages (about 10%) was found.

3. The Present State of the Art

Numerous measurements of turnoff luminosities have been listed by Rosenberg et al. (1999). We have obtained the ages of 35 GCs of this database by means of the turnoff luminosity–age relation presented in SCL97 (diffusion isochrones). Distance moduli have been derived by fitting the observed zero age horizontal branch (ZAHB) listed in the same database with the theoretical ZAHB of SCL97. Metallicities and α-elements overabundances have been taken from Carretta & Gratton (1997). The bulk of the GCs is coeval (within a bona fide error of ± 1 Gyr), with interesting exceptions at higher metallicity ([Fe/H] > -1.2). Once the five clusters whose age differ by more than 2 Gyr from that of M 15 are excluded, an average age of 12.95 ± 0.65 Gyr is found. Note that the quoted error is just a standard deviation. Systematic errors in the distance scale (as due to uncertainties in ZAHB models) could largely affect this average value.

4. Further Progress: The Onset of CNO Burning

During most of its life, a low mass star burns H in the center via the proton-proton chain. However, when the central H mass fraction reduces to ~ 0.1, the nuclear energy realized by the H-burning becomes insufficient and the stellar core must contract to extract some energy from its gravitational field. Then, the central temperature (and the density) increases and the H-burning switches from the proton-proton chain to the more efficient CNO-burning. Thus, the escape from the main sequence is powered by the onset of CNO-burning, whose bottleneck is the $^{14}N(p,\gamma)^{15}O$ reaction. A modification of the rate of this reaction alters the turnoff luminosity, but leaves almost unchanged the stellar lifetime, which is mainly determined by the rate of the pp reaction. The minimum energy explored in nuclear physics laboratories is ~ 200 KeV, well above the region of interest for CNO-burning in an astrophysical condition (~ 20–80 keV), so that the values used in stellar model computations are largely extrapolated. How reliable are the extrapolated values of this reaction rate? The complexity of the resonant structure of the ^{15}O compound nucleus makes unpredictable the extrapolation down to typical astrophysical energies. In particular the astrophysical factor for this reaction is strongly dependent on the interference with a subthreshold resonance (7271 keV). Such a big uncertainty demands a measurement of the astrophysical factor at an energy as close as possible the typical Gamow peak in low mass turnoff stars. The Laboratory for Underground Nuclear Astrophysics, already working inside the Gran Sasso mountain (LNGS laboratory of Assergi, Italy), will reach these energies at the end of 2001.

References

Carretta, E., & Gratton, R. 1997, A&AS, 121, 95

Castellani, V., Ciacio, F., degl'Innocenti, S., & Fiorentini, G. 1997, A&A, 322, 801

Chaboyer, B., & Kim, Y. 1995, ApJ, 454, 767

Ciardullo, R. B., & Demarque, P. 1977, Transactions of the Astronomical Observatory of Yale University, (New Haven: Astronomical Observatory)

Iben, I. J. 1971, PASP, 83, 697

Rosenberg, A., Saviane, I., Piotto, G., & Aparicio, A. 1999, AJ, 118, 2036

Salaris, M., Chieffi, A., & Straniero, O. 1993, ApJ, 414, 580

Sandage, A. 1962, ApJ, 135, 349

Straniero, O., & Chieffi, A. 1991, ApJS, 76, 525

Straniero, O., Chieffi, A., & Limongi, M. 1997, ApJ, 490, 425 (SCL97)

Vandenberg, D. A., & Bell, R. A. 1985, ApJS, 58, 561

Vandenberg, D. A., Bolte, M., & Stetson, P. B. 1996, ARA&A, 34, 461

Astrophysical Ages and Time Scales
ASP Conference Series, Vol. 245, 2001
T. von Hippel, C. Simpson, N. Manset

Time Scales and Transitions During Stellar Pulsation and Evolution

R. Szabó, Z. Kolláth, Z. Csubry

Konkoly Observatory, Konkoly-Thege M. u. 13-17., Budapest, Hungary

J. R. Buchler

Physics Department, University of Florida, Gainesville, FL, USA

Abstract. As a consequence of taking into account turbulent convection in numerical hydrodynamical calculations, applying advanced time-frequency analysis, as well as amplitude equation formalism, we can reveal and investigate a number of interesting phenomena. It is now possible to follow consistently a Cepheid or RR Lyrae star throughout its evolutionary path in the whole instability strip and compute corresponding pulsational behavior. We present our theoretical results on mode selection, mode switching, and transitional effects that can occur in these stars on very different time scales. The fundamental question of nonlinear period changes is also discussed.

1. Methods

Using the Florida code, which is a state-of-the-art, nonlinear hydrocode including turbulent convection and is tailored to follow stellar pulsation (Z. Kolláth et al., in preparation) the possible full amplitude and steady pulsational states of a radially pulsating star can be computed. The potential of hydrodynamical modeling of RR Lyrae stars along with applying *amplitude equation (AE) formalism* (Buchler & Goupil 1984) is demonstrated through examples: (1) Starting the hydrocode with different initial conditions we fit time-independent AE coefficients via an analytical signal method (Kolláth & Buchler 2001). It gives a description of the dynamical behavior of a model star with fixed parameters (M, L, T_{eff}). Quasi-evolutionary effects are simulated by interpolating the coefficients in effective temperature. The procedure is performed with constant luminosity given the almost horizontal evolutionary paths (Demarque et al. 2000). If one combines the above T_{eff}-dependent AEs with evolutionary parameter changes, time-dependent AEs can be used that allow the determination of the characteristic time of mode switching that is found to be longer than the thermal time scale (for a detailed discussion see Buchler & Kolláth 2001). (2) For double-mode stars the period ratios are obtained by fitting the AEs. In addition, nonlinear period shifts can be derived. These methods work equally well for Cepheids, and the results are similar.

2.   Evolution Through the Instability Strip

During its evolution a star can undergo different radial pulsational states (fundamental mode, FM, first overtone, O1, and double mode, DM), even hysteresis can occur. In Figure 1 we show the amplitudes of a model RR Lyrae star ($M = 0.77\ M_\odot$, $L = 50\ L_\odot$, $X = 0.75$, $Z = 0.0001$) navigating through the instability strip (left panel). Evolving blueward the star pulsates in FM, and then switches to O1 at point C. Moving to lower temperatures however the star remains in the O1 mode as far as point A, then pulsates in DM (between A and B) before resettling to the FM. Amplitude–amplitude (FM-O1) diagram (Figure 1 right panel) shows the hydrodynamical behavior, i.e., mode-selection at fixed $T_{\rm eff}$. Small, normalized vectors represent the flow-field. We call attention to the stable (0.055, 0) and unstable (0, 0.03), (0.018, 0.028) fixed points along the curved separatrix. Different speeds are coded by different shades (light: rapid, dark: slow motion): thus the time scales of the evolution of the pulsational amplitudes during transients and bifurcations can be seen. The time while the star evolves from the zero fix-point to $(A_0, 0)$ or $(0, A_1)$ is called the *thermal time scale of pulsation*.

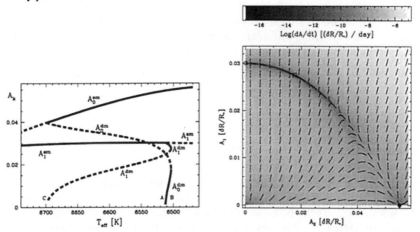

Figure 1.   Mode selection during evolution (*left*). Solid (dashed) curves indicate stable (unstable) single or double mode amplitudes. Amplitude evolution time scales (6518 K) (*right*). See text for details.

3.   Nonlinear Period-Change Rates

Time scales and magnitudes of period changes can significantly differ for single-mode and double-mode pulsators, mainly due to the interaction of involved pulsation modes in latter objects. The sign of the period change can be even different for the two modes (Paparó et al. 1998). On the other hand, period ratios have a great importance in determining the physical parameters of double-mode Cepheids and RR Lyraes. Usually linear periods are used in theoretical considerations and those are generally based on calculations with radiative pulsation models. Thus it is crucial to take into consideration the shifts of period ratios caused by turbulent convection along with nonlinear period shifts. This latter had been impossible to compute until the successful modeling of double-mode pulsation. Figure 2 displays the nonlinear period shifts as a function of the amplitudes. Isoperiod ratio curves remain relatively unchanged. Small temperature differences due to evolution induce small shifts along the separatrix, but sudden changes in the stellar structure cause slow amplitude changes and concomitant period shifts.

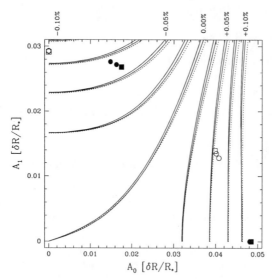

Figure 2. Relative nonlinear period shifts $(P_1/P_0)_{nl} - (P_1/P_0)_{lin}$ normalized by $(P_1/P_0)_{lin}$ for a model RR Lyrae sequence with $M = 0.77\ M_\odot$, $L = 40\ L_\odot$, $T_{eff} = 6400$ (square, dashed lines), 6405, and 6410 K.

References

Buchler, J. R., & Goupil, M. J. 1984, ApJ, 279, 394

Buchler, J. R., & Kolláth, Z., 2001, A&A, submitted

Demarque, P., Zinn, R., Lee, Y.-W., & Yi, S. 2000, AJ, 119, 1398

Kolláth, Z., & Buchler, J. R. 2001, in Astrophysics and Space Science Library Series, ed. M. Takeuti & D. D. Sasselov (Dordrecht: Kluwer), in press

Paparó, M., Saad, S. M., Szeidl, B., Kolláth, Z., Abu Elazm, M. S., & Sharaf, M. A. 1998, A&A, 332, 102

Acknowledgments. This work has been supported by Hungarian OTKA grant T-026031 and NSF grant AST98-19608.

Astrophysical Ages and Time Scales
ASP Conference Series, Vol. 245, 2001
T. von Hippel, C. Simpson, N. Manset

Tuning the Clock: Uranium and Thorium Chronometers Applied to CS 31082–001

R. Toenjes, H. Schatz

National Superconducting Cyclotron Laboratory and Department of Physics and Astronomy, Michigan State University, USA

K.-L. Kratz, B. Pfeiffer

Institut Für Kernchemie, Universität Mainz, Germany

T. C. Beers

Department of Physics and Astronomy, Michigan State University, USA

J. Cowan

Department of Physics and Astronomy, University of Oklahoma, USA

V. Hill

European Southern Observatory, Garching, Germany

Abstract. We employ the classical r-process model to obtain age estimates for the progenitor(s) of the extremely metal-poor ([Fe/H] = −2.9) halo star CS 31082–001, based on the recently reported first observation of uranium in this star. Implications for thorium chronometers are discussed.

1. Introduction

The recent first abundance measurement of uranium (together with thorium) in the extremely metal-poor star CS 31082–001 (Cayrel et al. 2001, and in these proceedings) offers the possibility to determine the age of its progenitor object(s) by comparing the observed abundances with r-process model predictions. Instead of the previously used [Th/X] chronometer (Cowan et al. 1999; Meyer & Truran 2000), there are now three chronometer pairs available, [Th/X], [U/X], and [U/Th] (X represents stable r-process element), that can be combined to yield a more reliable age estimate (Goriely & Clerbaux 1999).

2. Calculations

We calculated the r-process production of ^{238}U and ^{232}Th using the classical site-independent model (Cowan et al. 1999). For comparison, calculations were performed with two nuclear mass models, the ETFSI-Q (Pearson, Nayak, &

Goriely 1996) and the HFBCS-1 (Tondeur, Goriely, & Pearson 2000), which we use here for the first time in a r-process calculation. The β-decay data were the same as in Cowan et al. (1999). New rates for neutron-induced fission, β-delayed fission, and spontaneous fission rates were calculated using the new fission barriers from Mamdouh et al. (1998).

Ages were determined based on the predicted abundance ratios [U/Th], [U/X], and [Th/X]. To compensate for the deficiencies in the abundance predictions of stable r-process elements, and to reduce the impact of observational errors, we also determined ages by replacing X with a fit to the observed abundance distribution assuming a solar pattern. These age estimates are denoted [U*f/U0] and [Th*f/Th0], respectively.

We conducted an extensive investigation of the influence of five types of uncertainty on the age determination. (1) β-decay rates: To obtain a conservative estimate of the possible influence of errors in the β-decay data, we performed a Monte Carlo study varying β-decay half lives by factors between 0.2 and 5. The resulting variance in r-process abundance predictions is Δ_β. (2) Model uncertainties: We determined the range of the four r-process model parameters that still result in a reasonable fit of the solar abundance pattern. This parameter range leads to an uncertainty Δ_{par} in the predicted abundances. While this error is small for [U/Th], it turns out to be unacceptably large for the [U/X] and [Th/X] ages. However, in this work we constrain the simulation parameters by requiring consistency with the [U/Th] age, resulting in the same small $\Delta_{\text{par}} = 0.04$ for all abundance ratios. (3) Observational errors: We denote this uncertainty Δ_{exp}. (4) Mass models: Generally there is good agreement in the abundance predictions based on ETFSI-Q and HFBCS-1, but differences in the prediction of the onset of deformation in the ^{244}Tl region lead to a lower Th abundance prediction and a higher age for ETFSI-Q. (5) Fit to abundance pattern: For the ages based on [U*f/U0] and [Th*f/Th0] the fit to the observed abundances introduces an uncertainty Δ_{logfac}.

3. Results

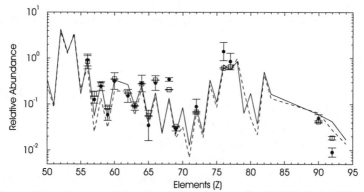

Figure 1.   R-process elemental abundances for ETFSI-Q (solid) and HFBCS-1 (dashed) calculations compared to the scaled solar (open squares) and CS 31082–001 (filled dots) abundances.

Table 1.   Age estimates and errors[†].

Ratio	Model	Log(ϵ_0)	Δ_β	Δ_{exp}	Δ_{logfak}	Δ_{total}	Age (Gyr)
U/Th	ETFSIQ	-0.16	0.07	0.16		0.18	12.6±3.9
U/Th	HFBCS1	-0.37	0.08	0.16		0.18	8.2±4.0
U*f/U0	ETFSIQ	-0.95	0.09	0.14	0.1	0.20	10.5±2.9
U*f/U0	HFBCS1	-1.12	0.09	0.14	0.1	0.20	7.9±2.9
Th*f/Th0	ETFSIQ	-0.79	0.09	0.08	0.1	0.16	5.9±7.5
Th*f/Th0	HFBCS1	-0.76	0.08	0.08	0.1	0.16	7.2±7.3
U/Gd	ETFSIQ	-0.78	0.10	0.23		0.25	10.5±3.7
U/Gd	HFBCS1	-0.90	0.11	0.23		0.26	8.8±3.8
Th/Gd	ETFSIQ	-0.62	0.10	0.20		0.22	6.1±10
Th/Gd	HFBCS1	-0.53	0.11	0.20		0.23	10.2±11

† Log(ϵ_0) is the predicted abundance ratio produced in the r-process and Δ's refer to errors in the predicted abundance ratios. For all ages, $\Delta_{par} = 0.04$.

Figure 1 shows our calculated r-process elemental abundances. In Table 1 we list our resulting age estimates and the various sources of uncertainties. The most robustly predicted abundance ratio is [U/Th], it is therefore the most reliable chronometer. Yet, the mass-model dependence is still substantial, yielding ages of 13±4 Gyr (ETFSI-Q) or 8±4 Gyr (HFBCS-1). With the new Mamdouh et al. (1998) fission barriers, β-delayed fission leads to significant corrections (+0.9 Gyr for the [U/Th] age, −0.8 Gyr for the [U/X] ages and 4 Gyr (!) for the [Th/X] ages). While more theoretical work is needed, this indicates that β-delayed fission should not be neglected. Some of the predicted [U/X] and [Th/X] ages agree well with [U/Th], but others don't. The [U*f/U0] and [Th*f/Th0] ages average over these discrepancies. In particular, [U*f/U0] provides a reasonable age estimate, while [Th/X] suffers from large observational uncertainties. We can now pick the [U/X] and [Th/X] estimates that agree best with the [U/Th] age. These are the ratios based on Gd and Ir (Hill et al., these proceedings) abundances. Our predicted ratios for [Th/Gd] and [Th/Ir] can therefore be used for improved age estimates of stars where no uranium can be detected.

References

Cayrel, R., et al. 2001, Nature, 409, 691

Cowan, J. J., Pfeiffer, B., Kratz, K.-L., Thielemann, F.-K., Sneden, C., Burles, S., Tytler, D., & Beers, T. C. 1999, ApJ, 521, 194

Goriely, S., & Clerbaux, B. 1999, A&A, 346, 798

Mamdouh, A., et al. 1998, Nucl. Phys. A, 644, 389

Meyer, B. S., & Truran, J. W. 2000, Phys. Rep., 333, 1

Pearson, J. M., Nayak, R. C., & Goriely, S. 1996, Phys. Lett. B, 387, 455

Tondeur, F., Goriely, S., & Pearson, J. 2000, Phys. Rev. C, 62, 024308

Acknowledgments.   This work was carried out under NSF contracts PHY 0072636 (Joint Institute for Nuclear Astrophysics) and PHY 9528844.

Session 5

Galaxies

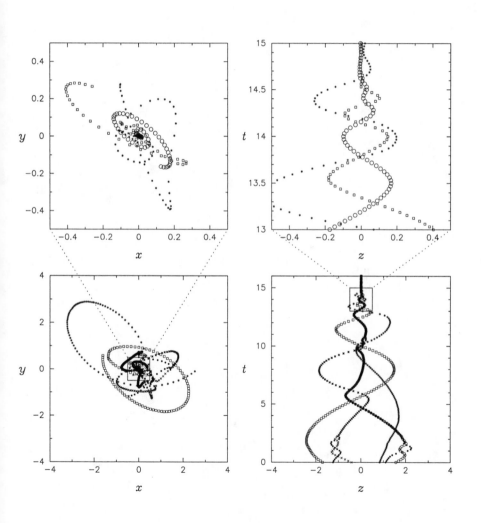

Astrophysical Ages and Time Scales
ASP Conference Series, Vol. 245, 2001
T. von Hippel, C. Simpson, N. Manset

Formation Histories Expected From Cosmological Simulations

Guinevere Kauffmann

Max-Planck-Institut für Astrophysik, Karl-Schwarzschild-Strasse 1,
D-85741 Garching, Germany

Abstract. According to the standard theoretical paradigm, the structures observed in the Universe today were formed by the amplification of small perturbations in an initially Gaussian dark matter density field. Small scale overdensities were the first to collapse, and the resulting objects then merged under the influence of gravity to form larger and larger structures such as clusters and superclusters. Galaxies formed within dense halos of dark matter where gas was able to reach high enough overdensities to cool, condense, and form stars. I discuss our current theoretical understanding of how and when these primordial gas lumps came together to form the Hubble sequence as we know it. A new generation of space- and ground-based telescopes have provided a wealth of information about the evolution of galaxies from an epoch when the Universe was only a tenth of its present age. I discuss how these new observations accord with the standard model. I argue that many of the broad qualitative trends predicted by the theory seem to be supported by the data, but much work remains to be done in understanding how star formation and feedback processes can affect the evolution.

Astrophysical Ages and Time Scales
ASP Conference Series, Vol. 245, 2001
T. von Hippel, C. Simpson, N. Manset

Merger Time Scales

Joshua E. Barnes

Institute for Astronomy, University of Hawai'i, 2680 Woodlawn Drive, Honolulu, HI 96822, USA

Abstract. Two questions are addressed in this review. First, how often do galaxies merge? Second, how long does merging take?

It is meaningless to speak of a merger time scale without specifying the type of galaxies involved. Perhaps the cleanest and best-defined estimate of the local merger time scale was provided by Toomre (1977), who listed $\sim$ 10 strongly interacting pairs of luminous disk galaxies among the $\sim$ 4000 NGC galaxies. Assuming each interaction has a duration of $\sim$ 500 Myr, the merger time scale for luminous disk galaxies is currently $\tau_m \simeq 100$ Gyr.

The fact that τ_m is an order of magnitude greater than the age of the Universe should cause no confusion; this figure implies that the odds of a large disk galaxy merging in the next Gyr are about 1%. If disks have been merging at this rate for the past $\sim$ 12.5 Gyr, and if none of the resulting merger remnants underwent any further morphological transformation, then the NGC catalog should contain $\sim$ 250 merger remnants. This falls a factor of 3 short of the $\sim$ 800 E and S0 galaxies in the NGC catalog. Thus the hypothesis that *most* early-type galaxies were formed by major mergers of disk galaxies requires that mergers be more frequent at earlier epochs.

1. Trends With Redshift

Toomre (1977) advanced a very general argument that the odds of merging were indeed higher in the *recent* past. The morphology and kinematics strongly interacting pairs of disk galaxies indicate that these objects result from close, nearly-parabolic passages of galaxies on long-period, almost radial orbits (Toomre & Toomre 1972). If these pairs have a flat distribution of binding energies, the merger time scale evolves as $\tau_m \propto t^{5/3}$ [$\propto (1+z)^{-5/2}$ in Einstein–de Sitter (EdS) cosmology]. Extrapolating this trend back to $t \to 0$ yields a predicted population of $\sim$ 750 merger remnants in the NGC.

Observations hint at a fairly steep trend in merger time scale with redshift. Galaxies which are classified as "peculiar" are more common at high redshifts (e.g., van den Bergh et al. 1996). In the CFRS (Brinchmann et al. 1998), for example, $\sim$ 10% are peculiar at $z \simeq 0.4$, while $\sim$ 35% are peculiar at $z \simeq 0.8$; these figures are consistent with a peculiar fraction scaling as $(1+z)^5$. It is not clear if this increase is entirely due to a trend in the merger time scale, since major mergers are only one source of peculiar systems, and the gas-rich galaxies found at higher redshifts may be more likely to appear peculiar at a given level of tidal disturbance. But close pairs of galaxies are also more common at higher redshifts (Zepf & Koo 1989; Abraham 1999); if translated into a merger time scale, these observations imply $\tau_\mathrm{m} \propto (1+z)^{-m}$, where $m \simeq 2.5 \pm 1$.

Another way to deduce the past merger rate is to look for evidence of aging merger remnants among nearby galaxies. Early-type galaxies with morphological evidence of merging deviate from the standard relations between color and luminosity and line-strength and luminosity (Schweizer et al. 1990). If such deviations are interpreted as spectral signatures of fading starburst populations, most field E and S0 galaxies have merger ages of $\sim$ 5 to 10 Gyr (Schweizer & Seitzer 1992). Age since last starburst may explain much of the scatter in the fundamental-plane relation (Forbes, Ponman, & Brown 1998). Evidence for a environment-dependent spread in the ages of E and S0 galaxies is discussed by S. Trager (these proceedings).

2. Clustering Theory

In hierarchical theories, structure forms from small-amplitude fluctuations by gravitational clustering. It is convenient to model the power spectrum of these fluctuations by $|\delta_\mathrm{k}|^2 \propto k^n$, where $n \gtrsim -3$. On galactic mass scales, the cold dark matter (CDM) fluctuation spectrum is approximately $n \simeq -2.5$.

When structure forms from a power-law fluctuation spectrum, the characteristic mass scale is $M_*(t) \propto D(t)^{6/(3+n)}$, where $D(t)$ is the *linear growth factor*. In an EdS cosmology, $D(t) \propto (1+z)^{-1} \propto t^{2/3}$, and $M_*(t) \propto t^{4/(3+n)}$. Halos grow by merger and accretion; assuming that the mass spectrum of bound objects remains self-similar, each doubling of the characteristic mass M_* represents a 50% reduction in the number of distinct halos. Thus the merger time scale is roughly the doubling time of M_*, or $M_*(t + \tau_\mathrm{m}) = 2M_*(t)$. In an EdS cosmology, this implies $\tau_\mathrm{m} = (2^{(3+n)/4} - 1)t \propto (1+z)^{-3/2}$.

The merger time scale may be compared to the recovery time scale τ_r on which a newly-formed halo attains equilibrium. The latter is comparable to the dynamical time scale: $\tau_\mathrm{r} \simeq \tau_\mathrm{d} \simeq \varepsilon t$, where $\varepsilon \simeq 0.1$ to 0.2. "Merger mania" ensues if the merger time scale is shorter than the recovery time scale. In an EdS cosmology, this implies $n < 4\ln(1+\varepsilon)/\ln(2) - 3 \simeq -2.5$ for $\varepsilon = 0.1$. Rapid merging at early times may account for the general prevalence of peculiar and interacting morphologies in the Hubble Deep Field (HDF) and similar high-z samples (e.g., van den Bergh et al. 1996).

More sophisticated analytic and numerical models of hierarchical structure formation yield "merger trees," which show how halos assemble from smaller progenitors (Lacey & Cole 1993). The pace of halo assembly is quite sensitive to cosmological parameters. In open- or lambda-CDM models the formation of galactic-mass halos is displaced towards higher redshifts; the merger rate falls dramatically as $z \to 0$. This decline helps explain why mergers are relatively rare at the present epoch. G. Kauffmann (these proceedings) describes semi-analytic models which combine merger trees with rules for galaxy transformation by merging and disk formation; these models reproduce the general distribution of morphological types as well as many of the correlations observed among early-type galaxies.

3. Orbit Decay

Strongly interacting galaxies tend to merge on relatively short time scales. Early numerical experiments showed that equal-mass galaxies that fall together on deeply inter-penetrating, roughly parabolic orbits are strongly decelerated, become bound to one another, and merge at their second passage or soon thereafter (White 1978). High-resolution calculations follow the orbit decay process through many more passages before merging is complete; nonetheless, all of these subsequent passages are packed into an interval much shorter than the time between the first and second pericenters, and so early estimates of orbital decay time scales are still valid.

An example is presented in Figure 1. Here, two spherical Jaffe (1982) models, with density profiles $\rho(r) \propto r^{-2}(r + a)^{-2}$, approach each other on a parabolic orbit[1]. The first pericenter has separation $r_p \simeq 1.2a$ and the following apocenter is at $r_a \simeq 5.3a$. On subsequent passages the pericentric and apocentric separations decay in a roughly self-similar manner; at no point does the orbit become remotely circular!

The self-similar pattern seen here suggests that at any stage of orbit decay, the time left until merger is comparable to the local orbital time. In Jaffe models the orbital time is proportional to the radius, and so the decay time scale is $\tau_{od} \propto d$, where d is the instantaneous distance between the nuclei. Figure 2 tests this prediction; the horizontal axis shows the time until the merger. The vertical axis shows d. The decay time scale is roughly proportional to d over nearly two orders of magnitude, only breaking away from this relationship when d falls below the spatial resolution scale of the simulation.

These results help explain the distribution of inter-nuclear distances in luminous infrared galaxies (Sanders & Miarbel 1996). Once corrections for sample volume and pair fraction have been made, galaxies with $d \simeq 6.4$ kpc are roughly 5 times more common than those with $d \simeq 1.2$ kpc. This is consistent with the expected separation distribution, which is $P(d) \propto \tau_{od} \propto d$.

4. Compact Groups

Compact groups are small, high density systems each containing a few luminous galaxies (Hickson 1982). Group members are often tidally distorted or otherwise peculiar. Such visible signs of dynamical evolution, in combination with theoretical estimates, support the hypothesis that compact groups are prone to rapid merging. Yet as White (1990) notes, "The literature abounds with overly short estimates of the lifetimes of compact groups."

Proper estimates of compact group lifetimes require knowledge of their dynamical time scales. A typical group has a diameter $d \simeq 70$ kpc, and galaxies within it have orbital velocities $v \simeq 300$ km s^{-1}; the time scale for a circular orbit is then $\tau_d \simeq \pi d/v \simeq 0.7$ Gyr. This is nearly an order of magnitude *longer* than the dynamical times sometimes quoted for the most compact groups. How-

---

[1]These results are presented in units in which the Jaffe model scale radius $a = 0.25$ and the time for a circular orbit at this radius is ~ 1.11. The calculation used $N = 2^{18}$ bodies, a force smoothing length of $\epsilon = 0.005$, and a leap-frog time-step of $\Delta t = 2^{-9}$.

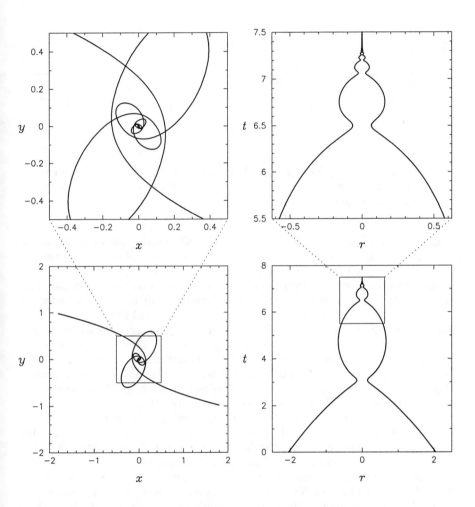

Figure 1.   Orbital decay in an encounter between two Jaffe models.
The lower left-hand plot shows trajectories measured by following the
nuclei of each model projected onto the orbital plane; the panel above
shows enlarged details of later passages. The lower right-hand panel is
a "space-time" diagram; at each t, the separation between the world-
lines shows the distance between the nuclei. Again, details of later
passages are shown in the upper panel.

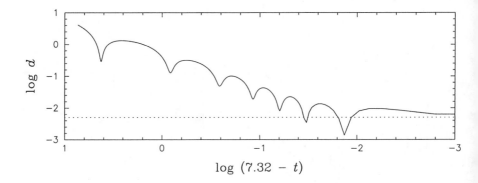

Figure 2.   Separation d plotted against time until merger for an encounter between two Jaffe models. The dotted line indicates the force resolution scale ϵ.

ever, it is quite likely that projection effects make some groups appear smaller than they really are; such systems are not representative of average groups.

Orbital velocities of galaxies in compact groups are comparable to the orbital velocities of stars within these galaxies. Given this fact, a simple cross-section estimate indicates that approximately one pair of galaxies undergo a close encounter—and subsequently merge—per dynamical time, *independent* of the current number of galaxies N in the group. Although this estimate does not take into account the effects of a range of galaxy masses, or the tendency of the most massive galaxies to settle towards the center of the system, it is generally supported by detailed numerical experiments. An example appears in Figure 3. This simulation, previously described by Barnes (1989), begins with a binary hierarchy of $N = 6$ disk galaxies. Galaxies spiral towards the center of the group as their orbits decay, while merging reduces the value of N. As the orbital time scale decreases, the tempo of merging picks up; three of the five mergers in this simulation occur within the final third of the group's lifetime.

If a small group has about one merger per dynamical time, its total lifetime is roughly $\tau_g \simeq N\tau_d$. Using the dynamical time scale estimate given above, a typical compact group of $N = 4$ galaxies has a lifetime $\tau_g \simeq 3$ Gyr. Since most mergers take place towards the *end* of a group's lifespan, the time scale τ_{mg} for a group member to merge is comparable to the group's lifetime; thus $\tau_{mg} \simeq N\tau_d$, or ~ 3 Gyr for a typical group.

An independent estimate of the merger time scale may be made by counting ongoing mergers in compact groups. Zepf (1993) found 19 ongoing mergers among a sample of 328 galaxies in compact groups; applying the same analysis used by Toomre (1977), the estimated merger time scale is $\tau_{mg} \simeq 4$ Gyr. Within the rather large uncertainties, this time scale agrees with the time scale derived from the "one merger per dynamical time" rule. Further observational and theoretical work could improve the accuracy of both estimates, but on the whole the number of mergers observed in compact groups seems consistent with the number predicted from the dynamical models.

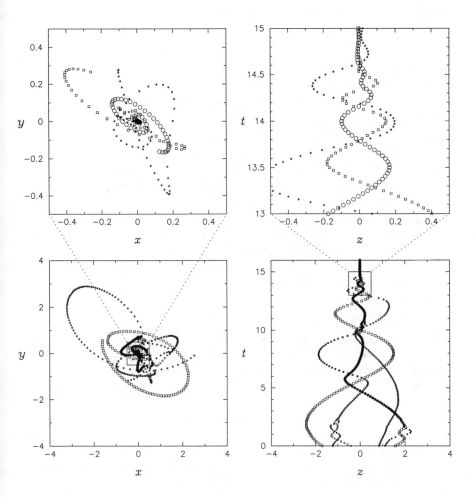

Figure 3. Dynamical evolution of a model compact group. The lower left-hand plot shows trajectories for the entire calculation. The lower right-hand plot shows a space-time diagram. The enlargements above show the "endgame" as the last three galaxies merge.

References

Abraham, R. G. 1999, in Galaxy Interactions at Low and High Redshifts, ed. J.
    E. Barnes & D. B. Sanders (Dordrecht: Kluwer), 11

Barnes, J. E. 1989, Nature, 338, 132

Brinchmann, J., et al. 1998, ApJ, 499, 112

Forbes, D. A., Ponman, T. J., & Brown, R. J. N. 1998, ApJ, 508, L43

Hickson, P. 1982, ApJ, 255, 382

Jaffe, W. 1982, MNRAS, 202, 995

Lacey, C., & Cole, S. 1993, MNRAS, 262, 627

Sanders, D. B., & Miarbel, I. F. 1996, ARA&A, 34, 749

Schweizer, F., & Seitzer, P. 1992, AJ, 104, 1039

Schweizer, F., Seitzer, P., Faber, S. M., Burstein, D., Dalle Ore, C. M., &
    Gonzalez, J. J. 1990, ApJ, 364, L33

Toomre, A. 1977, in The Evolution of Galaxies and of Stellar Populations, ed.
    B. M. Tinsley & R. B. Larson (New Haven: Yale Observatory), 401

Toomre, A., & Toomre, J. 1972, ApJ, 178, 623

van den Bergh, S., Abraham, R. G., Ellis, R. S., Tanvir, N. R., Santiago, B. X.,
    & Glazebrook, K. G. 1996, AJ, 112, 359

White, S. D. M. 1978, MNRAS, 184, 185

White, S. D. M. 1990, in Dynamics and Interactions of Galaxies, ed. R. Weilen
    (Berlin: Springer-Verlag), 380

Zepf, S. E. 1993, ApJ, 407, 448

Zepf, S. E., & Koo, D. C. 1989, ApJ, 337, 34

Acknowledgments. This research was partly supported by Space Telescope Science Institute through grant GO-06430.03-95A. I thank D. B. Sanders for information on the nuclear separations of ULIRGs.

Discussion

Gerhard Hensler: At what time do you start the simulation of the compact group? You must have some evolution of the galaxies, then you approach them and they merge somehow.

Joshua Barnes: I have to admit that what I did here was to build fully-realized galaxies and start them in that initial configuration you saw. To get to this stage [showing the starting frame of the movie] there must have already been a good deal of dynamical evolution. And there's a time scale question—where in the Universe can you make enough of these structures? The answer seems to be that compact groups can be formed by the dynamical decay of loose groups. There are ~ 50 times more loose groups than compact groups in the Universe, and their orbital time scales are roughly an order of magnitude longer, so if a modest fraction of loose groups evolve into compact configurations, then you can get the number of these things that are needed to explain the observations.

Romeel Davé: What role does the dark halo play in governing the time scales of merging?

Joshua Barnes: It's quite important. The heuristic laws that I gave for merger rates assumed that you had an interpenetrating collision where the galaxies reached each other's half-mass radii. Now, if you've just got the visible galaxies then you might think you're dealing with a near-miss situation, but if you put in the dark haloes, then the half-mass radius including the dark halo is a good deal larger, and apparent near-misses become deeply interpenetrating collisions which lead to rapid orbital decay.

Romeel Davé: So does the merger fraction or number of compact groups tell you something about the extent of dark haloes around typical galaxies?

Joshua Barnes: It might if we understood the other aspects of compact groups well enough. But I think that the constraints actually work the other way: studies of dark haloes lead us towards an understanding of compact groups. Trying to use the groups as a tool is tough because they're hard to study; they're relatively rare and catalogs are compromised by projection effects. Once you start worrying about all the details it's really hard to use compact groups to constrain halo structure.

Brad Whitmore: It's always been worrisome that M 31, our nearest galaxy, has a double nucleus. Is it better understood now what's going on there? Is it just a coincidence?

Joshua Barnes: I think Tremaine's model is a brilliant solution to that problem. Basically, what it says is that you're not really looking at two nuclei, you're looking at an equilibrium situation with one black hole and an eccentric disk around it, and it's the regions where the stars pile up in their orbits that seem to be a pair of nuclei. And I think that works very well.

Brad Whitmore: But you've found in general that for other galaxies you don't need any way of kind of hanging up...

Joshua Barnes: That was my point, at least in general. What I'd expect is that your chance of seeing a particular separation should be proportional to the amount until that separation decays, and therefore you expect to see many more larger separations than smaller separations. I think that's actually consistent with the upper end of the IRAS galaxy luminosity distribution.

Astrophysical Ages and Time Scales
ASP Conference Series, Vol. 245, 2001
T. von Hippel, C. Simpson, N. Manset

Time Scales in Starbursts

Claus Leitherer

Space Telescope Science Institute, 3700 San Martin Drive, Baltimore, MD 21218, USA

Abstract. Starbursts are associated with distinct time scales spanning at least three orders of magnitude from 10^6 to 10^9 yr. The triggering of nuclear starbursts occurs over 10^8 to 10^9 yr, when gas is flowing to the galaxy center due to angular momentum loss, thereby raising the nuclear gas density. Star formation then increases as a result of the increased gas pressure. In addition, spontaneous star formation can occur in nuclear and off-nuclear regions if gas clouds are compressed by strong shocks. The star formation time scales in individual starburst regions are short: typically 10^6 to 10^7 yr. The subsequent evolution of a newly formed starburst is determined both by dynamical effects and by the influence of the massive stars on the surrounding interstellar medium over an evolutionary time scale (about 10^7 yr). Eventually starbursts fade since the ambient interstellar gas is removed from the birth sites due to winds and supernova explosion over times scales of order 10^8 to 10^9 yr.

1. Introduction

Star formation in the Universe is not a continuous process. Rather, stars often form in cosmologically brief *bursts* where the formation densities can be several orders of magnitude higher than the Milky Way average (see Leitherer 2000 for a review). Such starbursts account for about a quarter of the high-mass star formation in the local Universe (Heckman 1997), and they are thought to be the dominant star formation mode in the young Universe.

The newly formed massive stars are so luminous (up to 10^6 $L_\odot$) that they can be detected and studied individually at distances of tens of Mpc, and in populations at even the largest redshifts. Moreover, the new-born stars can be inferred indirectly by their effects on the surrounding interstellar medium (ISM), as they energize the ISM by emitting copious ionizing photons and by releasing kinetic energy in stellar winds and supernovae. Some starbursts contribute a small (but still significant) fraction to the total bolometric galaxy luminosity. An example is the giant H II region 30 Doradus in the Large Magellanic Cloud (Walborn 1991). Nuclear starbursts in particular often account for the entire galaxy luminosity. The latter object class is generally referred to as *starburst galaxies* (Terlevich 1997). Prototypical representatives are M 82 (Rieke 1991) or NGC 7714 (Lançon et al. 2001).

This review addresses some of the astrophysically relevant time scales observed in starbursts: (1) the time it takes to provide the gas supply to trigger a starburst; (2) the duration of the star formation event; (3) the propagation of the starburst and its dynamical evolution; (4) changes of the stellar properties over an evolutionary time scale; and (5) the termination of the starburst after a time scale set by the depletion of the gas.

2. Triggering Time Scales

On a global (galaxy-wide) scale, the star formation density Σ is a function of the molecular gas density ρ. This relation is known as the Schmidt Law

$$\Sigma \propto \rho^n, \tag{1}$$

with n between 1 and 2. Kennicutt (1998) found that starbursts are described by this law as well. Figure 1 shows the relation between star formation density and molecular gas density for normal and starburst galaxies. The star formation rates (SFRs) in starbursts are higher because the gas densities are higher. Consequently, mechanisms capable of locally increasing the gas density can potentially trigger a starburst.

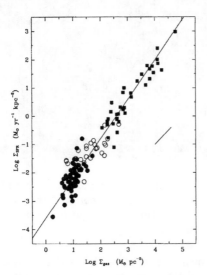

Figure 1. Composite star formation law for normal disk (filled circles) and starburst (squares) galaxies. Open circles indicate centers of normal disk galaxies. The line is a least-squares fit with index $n = 1.4$. From Kennicutt (1998).

Since the seminal work of Tinsley & Larson (1978) it has been known that star formation is enhanced in interacting galaxies. The most extreme cases in terms of luminosity ($L \geq 10^{12} \, L_\odot$) and therefore SFR are ultraluminous infrared galaxies (ULIRGs). Almost all ULIRGs are in interacting or merging systems (Sanders 1997). Numerical simulations (e.g., Hernquist & Mihos 1995; Mihos & Hernquist 1996; Barnes, these proceedings) do indeed support the suggestion of interaction induced starburst activity. Interaction leads to the loss of angular momentum of the disk gas by gravitational torque and dissipation. Angular momentum conservation then requires the gas to flow toward the center of the galaxy, with nuclear gas densities increasing by at least an order of magnitude. Figure 2 is an example of a minor merger where the decay rate of the satellite's orbit is low due to its small mass (Mihos & Hernquist 1996). The nuclear gas density in the simulations increases significantly a few hundred Myr after the onset of the galaxy-galaxy interaction. Applying the Schmidt Law, the models predict nuclear SFRs of tens of $M_\odot$ yr^{-1}, in agreement with observations.

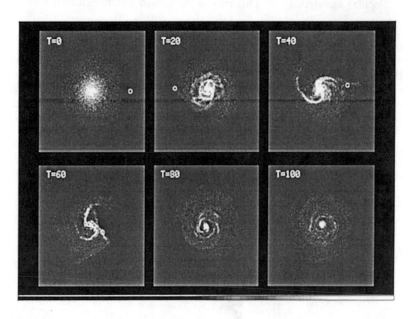

Figure 2.   Numerical simulation of a minor merger (J. C. Mihos 2001, private communication). Shown is gas density. Each time unit T is about 10 Myr. The location of the gas-free low-mass companion is indicated by a circle.

A similar mechanism can work in isolated galaxies. Ho, Filippenko, & Sargent (1996) found SFRs in barred spirals to be higher compared to non-barred spirals. Triaxial deformations, also known as *bars*, can induce angular momentum loss via torque and dissipation (Friedli & Benz 1995). Again, the associated time scales for the gas to flow to the galaxy center is of order 10^8 to 10^9 yr.

The quoted numerical models are not self-consistent in the sense that the applicability of Equation 1 is *assumed*. Mergers, interactions, and bars only maximize the most favorable conditions for elevated star formation: high gas densities. The physical mechanism governing star formation is not understood in detail. High densities do not automatically induce star formation. Depending on the turbulence compression, many dense cloud cores may in fact be stable and never form stars. Generally speaking, high gas densities lead to high gas pressure, and therefore to a high likelihood of induced or spontaneous star formation (e.g., Elmegreen 2000). The relevant time scale is the crossing time scale of a shock wave propagating through the gas clouds. For typical cloud sizes, the observed time scales are of order 10^7 yr, which is short in comparison with the time scale associated with the angular momentum loss discussed earlier.

3. Star Formation Time Scales

What is known about star formation time scales from observations of the *stars* themselves? Star clusters are particularly well suited to address this question. The observational advantage are their secure distances, which allow precise comparisons with theoretical isochrones. The astrophysically relevant issue is that a substantial fraction of the star formation in starbursts occurs in clusters. Meurer et al. (1995) estimate this fraction to be about 20%, but their value is likely to be a lower limit if much of the field star population in starburst galaxies results from the dissolution of previously formed clusters (Tremonti et al. 2001). Star clusters with sizes of a few pc and masses of order 10^5 $M_\odot$ have been found ubiquitously, wherever starburst galaxies were studied at high enough spatial resolution (Whitmore 2001). Therefore one can consider a star cluster as the smallest spatial scale over which star formation occurs.

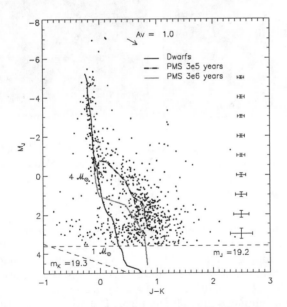

Figure 3.   Color–magnitude diagram $(J-K, J)$ of the stars detected in the central region of NGC 3603. The diagram has been corrected for a global extinction of $A_V = 4.6$ and a distance module of $(m - M) = 14.3$. In addition to the empirical main sequence, the diagram also includes theoretical PMS isochrones for 0.3 Myr and 3 Myr. From Eisenhauer et al. (1998).

The star formation histories of numerous luminous star clusters in the Local
Group of galaxies, including the Milky Way, have been studied with state-of-the-
art color–magnitude analyses. An example is in Figure 3. The Galactic cluster
NGC 3603 is the most massive optically detected star cluster in our Galaxy and
probably the closest counterpart to star clusters observed in starburst galaxies.
Comparison with pre-main sequence (PMS) and main sequence isochrones suggests that all stars down to the detection limit at 1 $M_\odot$ are formed within less
than about 1–2 Myr. This time scale is shorter than the evolutionary time scale
of the most massive stars observed in this cluster ($\sim$ 100 $M_\odot$, with a time scale
of 3 Myr; Schaller et al. 1992). Star formation at both the high- and low-mass
end happens almost instantaneously. Analogous results apply to other clusters
studied (e.g., Massey 1998). Star formation is short for cluster-like systems with
spatial scales of a few pc.

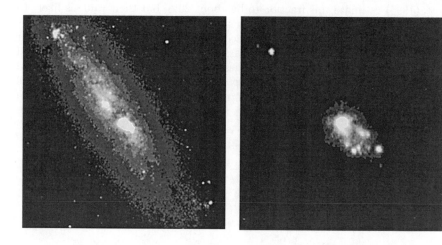

Figure 4.   HST STIS images of NGC 3049. *Left:* $20'' \times 20''$ continuum
+ Hα exposure. *Right:* $5'' \times 5''$ close-up of the nucleus in the far-
ultraviolet ($\sim$1500 Å). $1'' = 100$ pc for a distance of 20 Mpc. From C.
Leitherer et al., in preparation.

What is the relevance of these results to starburst galaxies whose *nuclear*
SFRs are at least an order of magnitude above those of the most luminous
clusters in Local Group galaxies? In Figure 4 HST imagery of the nuclear region
of the starburst galaxy NGC 3049 is shown. The galaxy is a typical metal-rich
($Z \approx Z_\odot$), luminous ($L = 3 \times 10^9 \, L_\odot$) H II galaxy hosting a nuclear starburst 10
to 100 times more luminous than 30 Doradus. The spatial resolution afforded
by HST reveals complex sub-structure in the "nucleus." At least 10 individual
starburst clusters are detected, some of them with properties similar to those of
30 Doradus. Imaging of starburst galaxies with high spatial resolution generally
suggests an absence of one dominating super-massive cluster but rather the
presence of many clusters having masses 10^4 to $10^6 \, M_\odot$ with comparable but
slightly different ages (e.g., Lançon et al. 2001). Most likely, the population of

each individual cluster mirrors the behavior found in local clusters with star formation spreads of less than a few Myr.

We do not know if *all* starburst galaxies behave like NGC 3049, whose luminosity is a factor of 1000 lower than those of the most luminous starburst galaxies. It may very well be that more luminous systems are just bigger, i.e., forming star clusters over a larger area, as suggested by Meurer et al. (1997).

4. Propagation Time Scales

How are the individual clusters seen in starburst regions related? Do they form an age sequence, possibly as a result of positive and negative feedback between star formation and the energy output by newly massive stars? The giant H II region 30 Doradus has been studied with an emphasis on these questions and can provide guidelines for the interpretation of more distant starbursts which lack comparable spatial resolution.

A schematic census of the different stellar generations in 30 Doradus was done by Grebel & Chu (2000). Previous, current, and future star formation activity can be witnessed in this region. Some 20 Myr ago the cluster Hodge 301 may have appeared as the center of 30 Doradus as does today. Its stars may have contributed to the triggering of the current star-formation activity. The current stellar generation in turn seems to be responsible for the onset of star formation seen as protostars. The latter are concentrated on and along the giant shells surrounding the central cluster. The full chronology seen spans over more than 20 Myr, yet we can identify several distinct star formation events, both in space and in time. If observed at larger distance, star formation would appear to occur continuously over a time scale of tens of Myr, yet the underlying physics is much more complex due to the correlation between spatial and temporal properties.

The ultimate fate of the clusters in 30 Doradus depends on their unknown total mass. They are subject to the gravitational field of the Large Magellanic Cloud and will suffer tidal disruption on a time scale of hundreds of Myr (Elson, Fall, & Freeman 1987).

The disruption of clusters by gravitational forces can become the dominant effect governing the morphological evolution of a starbursts located in galaxy centers. Two well studied examples are the Arches and Quintuplet clusters in the center of the Milky Way. These are newly formed (2 to 4 Myr), massive (10^4 to 10^5 $M_\odot$) clusters studied by Figer et al. (1999). The short relaxation times due to the compactness of the clusters and the strong tidal fields near the Galactic center lead to their rapid evaporation. After cluster evaporation, the remaining less massive stars will be part of the surrounding field population (Kim, Morris, & Lee 1999). However, this process alone cannot account for the *entire* high-mass star population as traced by their ionizing radiation. At least a few very massive stars must form outside the most massive clusters.

Observational evidence for dynamical effects on cluster evolution and propagating star formation outside the Local Group of galaxies is sparse and less direct. Puxley, Doyon, & Ward (1997) measured the Brγ recombination line flux and the CO 2-0 band strength around the center of M 83. Over the measured region Brγ and the CO band exhibit the opposite behavior. The data suggest an age gradient in the central region of M 83 if the burst population dominates the emission. The position with deepest CO is significantly older than that having the largest Bγ equivalent width. The age ranges traced are 10 to 20 Myr. This result indicates a difference in the age of the starburst and/or in the star formation history across the region and is reminiscent of the spatial morphology observed in the 30 Doradus region.

5.   Stellar Evolution Time Scales

The evolution of a starburst region, once star formation has ceased, is determined by stellar evolution. Typical evolutionary time scales of massive stars are of order 10^7 yr (Table 1). During the first $\sim$ 3 Myr, the starburst is "photon"-dominated, i.e., the dominant heating of the ISM is by energetic photons from hot main sequence stars which emit about 1/3 of their luminosity shortward of the Lyman break (Leitherer et al. 1999). After evolving off the main sequence, the stars develop strong stellar winds capable of heating the surrounding ISM and creating giant wind-blown shells and bubbles. This phase is particularly pronounced when the Wolf-Rayet stage has its peak around 5 Myr. Subsequently the first Type II supernovae appear, equaling and eventually surpassing the kinetic energy input by stellar winds. After 10 to 15 Myr, the kinetic energy release by winds and supernovae exceeds the heating by stellar ionizing photons, and the starburst becomes "matter"-dominated: supernovae shape the morphology and energetics of the ISM and may regulate subsequent star formation. In the absence of continuous star formation this phase ends after about 50 Myr when intermediate-mass stars reach the main sequence turn-off.

Table 1.   Characteristic evolutionary time scales.

Age (Myr)	Stellar Mass ($M_\odot$)	Observables
2	100	very hot O main sequence stars
5	50	O supergiants and Wolf-Rayet stars
10	20	red supergiants and supernovae
50	8	transition to intermediate-mass stars

Few systematic studies of starburst regions exist which go beyond evolutionary phases dominated by massive stars. Photometric fading of the stellar population is significant and makes any survey of "post-starburst galaxies" prohibitive. (A 300 Myr old population is about 50 times fainter in V than a 3 Myr old population.) The so-called E+A galaxies (Dressler & Gunn 1983) have been suggested to be post-starbursts. Their spectra are dominated by strong Balmer lines in absorption and have negligible emission lines. This spectral morphology is indicative of a galaxy that has no significant current star formation but was forming stars in the past ($<$ 1.5 Gyr). The strong Balmer equivalent widths can only be understood by models seen in a quiescent phase soon after a starburst; for this reason the E+A spectra are often identified with post-starburst galaxies (Poggianti et al. 1999).

Taniguchi, Shioya, & Murayama (2000) proposed an identification of some LINERs as post-starburst galaxies. LINERs are traditionally thought of harboring an active galactic nucleus but alternative interpretations for the origin of the emission-line spectrum have been proposed (see Filippenko 1996 for a review). In the post-starburst model, the ionization sources are planetary nebula nuclei (PNNs) with temperatures of $\sim$100,000 K that appear in the evolution of intermediate-mass stars with mass between 3 and 6 $M_\odot$. The PNN phase lasts until the death of the least-massive stars formed in the starburst, which is about 5×10^8 yr for a stellar IMF truncated at 3 $M_\odot$.

The interpretation of post-starbursts rests on the assumption that the age of the system is large in comparison with the time scale over which star formation occurred. This time scale depends on the spatial scale of the starburst, as discussed in the previous section. Observed values range between $\sim$ 1 Myr for individual star clusters to many tens of Myr for galaxies harboring global starbursts (de Mello, Leitherer, & Heckman 2000; Pettini et al. 2000).

6. Gas Depletion Time Scales

The duration of a starburst (i.e., the period of time over which star formation occurs, as opposed to the age of the starburst) must be short on a cosmological time scale. This can be seen from an illustrative example: infrared-luminous starburst galaxies have SFRs of $\sim$ 100 $M_\odot$ yr^{-1} (Heckman, Armus, & Miley 1990). If these rates are sustained over 10^8 yr, 10^{10} $M_\odot$ of molecular gas are consumed in the star formation process. This is comparable to, or even exceeds the gas reservoir of even the most luminous galaxies. The gas depletion time scale sets a hard upper limit to the duration of a starburst (Weedman 1987).

The gas depletion argument is difficult to apply quantitatively for a variety of reasons. First, the unknown proportion of low-mass stars formed in the starburst makes estimates of the total (integrated over all masses) SFR uncertain by a factor of at least two. The concept of a truncated (i.e., deficient in low-mass stars) initial mass function has sometimes been invoked just to extend the depletion time scale.

The fundamental difficulty in determining the gas depletion time is the break-down of a closed-box model of a starburst, which ignores the various sources and sinks of material. As discussed before, infall of gas to the starburst nucleus results from angular momentum loss due to gravitational torque and dissipation. Another source of gas replenishment is mass return by stellar winds and supernovae. Massive stars have significant mass loss during their evolution and return about 50% of their mass almost instantaneously during the starburst (see Leitherer et al. 1999). Therefore the gas depletion time scale will become almost independent of the rate of star formation, as more material is returned if more stars are formed. (This argument is no longer valid if low-mass stars form in large proportions since their mass return is small.) We should caution, however, that the stellar wind and supernova material are at coronal temperatures due to shock interaction with the ambient ISM. The cooling times are long in comparison with the age of the starburst so that this material may not immediately become part of the cold, star forming gas.

Galactic-scale outflows, or superwinds (Heckman 1997) are a significant sink for the gas reservoir. The combined effects of multiple stellar winds and supernovae are capable of initiating large-scale outflows of interstellar gas. Such outflows have been known from optical and X-ray imagery (Heckman et al. 1990), and they have recently been analyzed by absorption line spectroscopy. Heckman et al. (2001) obtained FUSE far-UV spectra of the dwarf starburst galaxy NGC 1705. These data probe the coronal ($10^5 - 10^6$ K) and the warm (10^4 K) phases of the outflow. The kinematics of the warm gas are compatible with a simple model of the adiabatic expansion of a superbubble driven by the supernovae in the starburst. Radiative losses are negligible so that the outflow may remain pressurized over a characteristic flow time scale of 10^8 to 10^9 yr, as estimated from the size and velocity.

The total mass transported out of the starburst region via galactic superwinds is hard to constrain, given the uncertain ionization corrections and the strength of the observable spectral lines. Attempts were made by Johnson et al. (2000) and Pettini et al. (2000) for a nearby dwarf starburst galaxy and a luminous star-forming galaxy at cosmological distance, respectively. In both cases the mass-loss rate of the ISM is quite similar to the star-formation rate. Taken at face value, this suggests that the available gas reservoir will not only be depleted by the star formation process but, more importantly, by removal of interstellar material. Starbursts may determine their own fate by their prodigious release of kinetic energy into the interstellar medium.

References

de Mello, D. F., Leitherer, C., & Heckman, T. M. 2000, ApJ, 530, 251

Dressler, A., & Gunn, J. E. 1983, ApJ, 270, 7

Eisenhauer, F., Quirrenbach, A., Zinnecker, H., & Genzel, R. 1998, ApJ, 498, 278

Elmegreen, B. G. 2000, ApJ, 530, 277

Elson, R. A. W., Fall, S. M., & Freeman, K. C. 1987, ApJ, 323, 54

Figer, D. F., Kim, S. S., Morris, M., Serabyn, E., Rich, R. M., & McLean, I. S. 1999, ApJ, 525, 750

Filippenko, A. V. 1996, in ASP Conf. Ser. 103, The Physics of LINERS in View of Recent Observations, ed. M. Eracleous, A. Koratkar, C. Leitherer, & L. Ho (San Francisco: ASP), 17

Friedli, D., & Benz, W. 1995, A&A, 301, 649

Grebel, E. K., & Chu, Y.-H. 2000, AJ, 119, 787

Heckman, T. M. 1997, in Star Formation Near and Far, ed. S. S. Holt & L. G. Mundy (Woodbury: AIP), 271

Heckman, T. M. 1997, in RMA&A Conf. Ser. 6, Starburst Activity in Galaxies, ed. J. Franco, R. Terlevich, & A. Serano, 156

Heckman, T. M., Armus, L., & Miley, G. K. 1990, ApJS, 74, 833

Heckman, T. M., Sembach, K. R., Meurer, G. R., Strickland, D. K., Martin, C. L., Calzetti, D., & Leitherer, C. 2001, ApJ, 554, in press

Hernquist, L., & Mihos, J. C. 1995, ApJ, 448, 41

Ho, L. C., Filippenko, A. V., & Sargent, W. L. W. 1996, in ASP Conf. Ser. 91, Barred Galaxies, ed. R. Buta, D. A. Crocker, & B. G. Elmegreen (San Francisco: ASP), 188

Johnson, K. E., Leitherer, C., Vacca, W. D., & Conti, P. S. 2000, AJ, 120, 1273

Kennicutt, R. C. 1998, ApJ, 498, 541

Kim, S. S., Morris, M., & Lee, H. M. 1999, ApJ, 525, 228

Lançon, A., Goldader, J. D., Leitherer, C., & González Delgado, R. M. 2001, ApJ, 552, 150

Leitherer, C. 2000, in Star Formation From the Small to the Large Scale, ed. F. Favata, A. A. Kaas, & A. Wilson (Noordwijk: ESA), 37

Leitherer, C., et al. 1999, ApJS, 123, 3

Massey, P. 1998, in ASP Conf. Ser. 142, The Stellar Initial Mass Function, ed. G. Gilmore & D. Howell (San Francisco: ASP), 17

Meurer, G. R., Heckman, T. M., Leitherer, C., Kinney, A., Robert, C., & Garnett, D. R. 1995, AJ, 110, 2665

Meurer, G. R., Heckman, T. M., Leitherer, C., Lehnert, M., & Lowenthal, J. 1997, AJ, 114, 54

Mihos, J. C., & Hernquist, L. 1996, ApJ, 464, 641

Pettini, M., Steidel, C. C., Adelberger, K. L., Dickinson, M., & Giavalisco, M. 2000, ApJ, 528, 96

Poggianti, B. M., Smail, I., Dressler, A., Couch, W. J., Barger, A. J., Butcher, H., Ellis, R. S., & Oemler, Jr., A. 1999, ApJ, 518, 576

Puxley, P. J., Doyon, R., & Ward, M. J. 1997, ApJ, 476, 120

Rieke, G. H. 1991, in Massive Stars in Starbursts, ed. C. Leitherer, N. R. Walborn, T. M. Heckman, & C. A. Norman (Cambridge: CUP), 285

Sanders, D. B. 1997, in RMA&A Conf. Ser. 6, Starburst Activity in Galaxies, ed. J. Franco, R. Terlevich, & A. Serano, 42

Schaller, G., Schaerer, D., Meynet, G., & Maeder, A. 1992, A&AS, 96, 269

Taniguchi, Y., Shioya, Y., & Murayama, T. 2000, AJ, 120, 1265

Terlevich, R. 1997, in RMA&A Conf. Ser. 6, Starburst Activity in Galaxies, ed. J. Franco, R. Terlevich, & A. Serano, 1

Tinsley, B. M., & Larson, R. B. 1978, ApJ, 221, 554

Tremonti, C. A., Calzetti, D., Heckman, T. M., & Leitherer, C. 2001, ApJ, in press

Walborn, N. R. 1991, in Massive Stars in Starbursts, ed. C. Leitherer, N. R. Walborn, T. M. Heckman, & C. A. Norman (Cambridge: CUP), 145

Weedman, D. W. 1987, in Star Formation in Galaxies, ed. C. J. Lonsdale (Washington: NASA), 351

Whitmore, B. C. 2001, in A Decade of HST Observations, ed. M. Livio, K. S. Noll, & M. Stiavelli (Cambridge: CUP), in press

Acknowledgments. I am grateful to the conference organizers for generous travel support. Chris Mihos kindly provided Figure 2.

Discussion

Gerhard Hensler: When you determined the velocity of the galactic wind, are you sure that this is an external wind, which is already departed from the galaxy, or could it be some clumpy or inhomogeneous structure close to the star clusters themselves?

Claus Leitherer: The material traced by the spectrum I showed is probably still very close to the cluster. When one calculates the densities, one finds that this is certainly not at a distance of 1 kpc or so where one would otherwise observe X-rays, but rather within a few hundred pc. The cold gas is probably the outer wall of a hot bubble. Since this is a very young starburst, there are probably relatively few supernovae. I should be careful when I say that the mass is actually lost, although the velocity is high—400km s^{-1}is relatively high, no matter how much material you put into a massive halo or whatever. I just wanted to give this as an example for a case where the material is seen flowing out. In many cases we cannot be sure of an outflow because the gas could also be infalling; in this case we can really see it is flowing out because we have a blueshift in the spectrum.

Romeel Dave: Another sink I would have thought of would be the fact that the supernovae heat the gas or photodissociate H_2, etc. How important is that relative to the physical removal...?

Claus Leitherer: Yes, of course, when I said that the material is returned by winds and supernovae, I should have added that this is not material which could immediately be used again for star formation. This is hot gas, shocked wind material at temperatures of, say, 10^7–10^8 K, and it takes a long time to cool. The recombination time scales are of the order of hundreds of Myr, which of course is important because otherwise the winds themselves could stall. Therefore this material would certainly not be available for this current starburst; one would have to wait for a few hundred Myr. Assuming it does not leave the galaxy, it could be used for the next starburst. So, you are perfectly right: one would not be able to use the supernova material for the current starburst.

Astrophysical Ages and Time Scales
ASP Conference Series, Vol. 245, 2001
T. von Hippel, C. Simpson, N. Manset

Evolutionary Time Scales of Dwarf Irregular Galaxies

Gerhard Hensler

Institut für Theoretische Physik und Astrophysik, Universität Kiel,
D-24098 Kiel, Germany

Abstract. The evolution of galaxies is perceptible from their stellar populations, the enhancement of elements, gas and stellar kinematics, and vehement episodic phenomena like starbursts, close encounters, etc. Since dwarf irregular galaxies are still actively evolving but not dominated by their gravitation, different evolutionary processes are contributing with various strengths and on variable time scales and characterize their galactic structural properties. Here we study different time scales and compare their relevance for the evolution of dwarf irregulars.

1. Introduction

Dwarf galaxies (DGs) present a variety of morphological types and seem to form at all cosmological epochs and by different processes. In order to understand their different types, the activation of their protogalactic stuff, including collapse and star formation (SF), must have depended on environmental and internal processes and thus involve different characteristic time scales.

Dwarf elliptical (dEs) and dwarf spheroidal galaxies consist of old stellar populations and are suggested to have formed very early from cosmological fluctuations by a short but vehement SF episode. Thereafter, they must have lost their remaining gas in a galactic wind driven by accumulated supernova type II (SNII) explosions (Larson 1974; Dekel & Silk 1986). Another type of DGs are gas-rich and irregularly structured, so-called dwarf irregulars (dIrrs). Their mostly patchy distribution of star-forming regions are caused by large local variations of SF rates from low to extraordinarily high values. Often very massive compact star clusters form within at least one intermediate-age to old stellar population. It is still under debate to what extent morphological transitions are possible where e.g., dEs are rejuvenated by means of gas infall, as in many blue compact DGs, and appear as irregulars.

Empirical studies and theoretical investigations of systems at low gravitational potential have elaborated that their interstellar medium should be balanced by counteracting processes so that the SF is self-regulated under various conditions (see e.g., Franco & Cox 1983; Köppen, Theis, & Hensler 1995, 1998, and references therein). On the other hand, in dEs and in present-day dIrrs with extremely high SF rates, this self-regulation must be overcome.

Because of their low binding energy, DGs are exposed to internal energetic influences as well as to external perturbations. Accumulated SNeII from a massive star cluster lead to hot metal-rich giant gas plumes, so-called superbubbles,

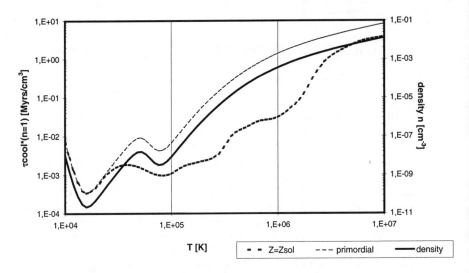

Figure 1.   Cooling time scale τ_{cool} (left-hand scale) in Myr normalized
to $n = 1$ cm^{-3} for primordial (thin dashed line) and solar (thick dashed
line) abundances. The critical density n_{crit} in cm^{-3} (right-hand scale)
for $\tau_{\mathrm{coll}} = \tau_{\mathrm{cool}}$ is drawn as a full thick line.

that expand and drive a galactic wind. The low metal content in DGs is attributed to such gas loss. On the other hand, the hot superbubble gas has to
interact with cool embedded clouds so that gas mixing effects come into play.
Accordingly, the chemical evolution is also influenced, not only on stellar evolution time scales and by the element release, but also on dynamical, mixing
and cooling time scales. This implies large abundance variations due to local
discrepancies between these combined processes.

The evolution of dIrrs is therefore affected by numerous processes on various
time scales which interfere so that their influential strengths and the evolutionary issues must be determined. In the following we wish to describe the most
important time scales and discuss their relevance with respect to the evolution of
dIrrs, in particular, time scales of collapse, cooling, dissipation, SF, large-scale
dynamics, like e.g., gas loss, and gas-phase mixing.

2.   Formation Time Scales

If one considers the formation epoch of DGs on cosmological time scales, the
protogalactic clouds (PGCs) have to decouple from the metagalactic radiation
field. The delayed formation of DGs from cosmological 1σ fluctuations have
e.g., been studied by Kepner, Babul, & Spergel (1997) and Tajiri & Umemura
(1998). From figure 3 of the latter authors one can deduce that a PGC of
10^9 $M_\odot$ baryonic mass is self-shielded against the Lyman continuum radiation
of $I_{21} = 10^{-21}$ ergs s^{-1} cm^{-2} sr^{-1} Hz^{-1} at the H Lyman edge for a radial
extension below 10 kpc. While I_{21} is the standard value at z = 2 (Bechthold

et al. 1987) the resulting particle density $n \sim 10^{-2}$ cm^{-3} is also typical for a cosmological 1σ density fluctuation at 10^4 K.

Whenever a PCG in virial equilibrium and photoionized state cools by means of recombination, it becomes gravitationally unstable. Because of the peak in the cooling function Λ at around 10^4 K (see e.g., Böhringer & Hensler 1989) the gained gravitational energy will not heat the gas and will not retard the collapse if the cooling time scale,

$$\tau_{\text{cool}} = \frac{e}{\Lambda} = \frac{n\,k\,T}{(\gamma - 1)\,\Lambda_0(T)\,n^2} = \frac{3}{2}\,\frac{k\,T}{\Lambda_0(T)\,n},$$

is shorter than the collapse time scale (Hensler et al. 1997),

$$\tau_{\text{coll}} = \sqrt{\frac{3\,\pi}{32\,G\,\rho}} = 1.63 \cdot 10^{15}(\mu\,n)^{-1/2} \text{ s.}$$

τ_{coll} is then called free-fall time τ_{ff}. In Figure 1 τ_{cool} is displayed normalized to unity gas density for primordial and solar abundances, respectively. The curve of $\tau_{\text{cool}} = \tau_{\text{coll}}$ demonstrates that for temperatures between 10^4 K and e.g., 10^5 K, cooling always dominates at reasonable densities. Only for very high temperatures, which are often misleadingly discussed because they represent virial temperatures of massive galaxies without cooling and for low densities, the critical one reaches cosmologically relevant values of 10^{-3}–10^{-2} cm^{-3}.

Density enhancements within the collapsing PGC form gas clouds. Formally the clouds' kinematics correspond to the virial velocity. This random motion leads to inelastic collisions of clouds by which kinetic energy is lost from the system dissipatedly and is transformed into thermal energy of the clouds. If this can be radiated off the system, the individual clouds themselves remain cool and the whole PGC becomes kinetically cooler, and thus it collapses. In this case the collision time scale of a single cloud is responsible for the dissipation so that

$$\tau_{\text{diss}} = (N\,Q\,\bar{v}_{\text{rel}})^{-1},$$

where N denotes the clouds' number density, Q their collisional cross section $\pi\,r_{\text{cl}}^2$ (r_{cl}: cloud radius), and $\bar{v}_{\text{rel}}$ their mean relative velocity that corresponds to the velocity dispersion. The applicability of gas dynamics to collapsing systems and consequences for isotropy and rotation are studied in detail in Hensler et al. (1997). The dissipation time scale, τ_{diss}, is connected with the dynamical time scale, $\tau_{\text{dyn}} \approx \tau_{\text{ff}}$, by (Larson 1969):

$$\tau_{\text{diss}} = 0.23\,\left(\frac{R}{r_{\text{cl}}}\right)^2\,\tau_{\text{ff}}.$$

If one considers typical giant interstellar clouds of 10^6 baryonic $M_\odot$, a $10^9\ M_\odot$ PGC consists of only 10^3 clouds with typical r$_{\text{cl}}$ of 100 pc (Rivolo & Solomon 1987). With a thermal velocity at 10^4 K, τ_{diss} amounts to almost 10^{10} yr. For a system of $10^4\ M_\odot$ clouds and 10 pc radii, τ_{diss} reaches the same value because in this case $N \cdot Q$ remains the same.

In reality, however, τ_{diss} decreases temporally due to the shrinkage of the galaxy extension. For sticky cloud systems, in addition, the cloud number is reduced by inelastic collisions according to $N(t) = N_0(1 + t\ \tau_{\mathrm{diss}}^{-1})^{-1}$ with $\tau_{\mathrm{diss}} = (C_1\ r_0^{-3.5}N_0)^{-1}$ (Theis & Hensler 1993), where C_1 is a function of the cross section Q and of galactic mass M, r_0 is the scale radius e.g., for a Plummer density distribution, and N_0 denotes the initial number of clouds. If the system remains in virial equilibrium, τ_{diss} decreases significantly. Vice versa, an anisotropic velocity dispersion develops for the remaining clouds that prolongs τ_{diss} again (for details see Theis & Hensler 1993). As a further effect, the collisional cross section has to be adapted to the internal density gradient in the clouds and the effect of gravitational focusing, two effects that go into contrary directions for τ_{diss}.

If the evolution of a cloudy PGC is considered in combination with SF (see next section), the velocity dispersion of clouds is increased due to their acceleration by superbubble shells of SNII explosions. Burkert & Hensler (1989) have shown that for $10^6\ M_\odot$ clouds and a SF efficiency of 10%, a critical particle density can be defined by

$$n_{\mathrm{crit}} = 10^6 \left(\frac{\tau_{\mathrm{SF}}}{10^9\mathrm{yr}}\right)^{-1} \left(\frac{\bar{v}_{\mathrm{rel}}}{\mathrm{km\ s}^{-1}}\right)^{-3},$$

(with τ_{SF} as the SF time scale; see next section), above which the collapse can not even be halted by the maintenance of the clouds' velocity dispersion.

3.   Star Formation Time Scales

The inherently formed clouds are able to cool further down and ignite SF. If the clouds are in pressure equilibrium with the intercloud medium in the collapsing PGC their density increases. Because the SF rate Ψ depends on the cool-gas density by a power-law function as $\Psi = \rho^\nu$ the SF rate can be assumed to increase accordingly during the collapse. This behaviour can be perceived in Figure 2 for a dIrr collapse model of $10^9\ M_\odot$ baryonic mass in a $10^{10}\ M_\odot$ dark matter halo (Rieschick & Hensler 2000; Hensler & Rieschick 2001; A. Rieschick & G. Hensler, in preparation). Due to the short lifetimes of the order of 2–20 Myr, massive stars release sufficient energy by means of SNeII and during their life by radiation and stellar winds so that the collapse can be stopped and even be reversed into a reexpansion. In Figure 2 the collapse continues until the peak SF is reached at 200 Myr.

The SF time scale τ_{SF} is typically defined as the gas consumption time:

$$\tau_{\mathrm{SF}} = \frac{M_{\mathrm{H\ I}}}{\Psi}.$$

Self-regulated SF is generally fulfilled in isolated and unperturbed systems and adjusts a quadratic Schmidt law of the SF rate Ψ ($\nu = 2$; Köppen et al. 1995). For this, τ_{SF} goes with ρ^{-1}. While the global τ_{SF} under normal circumstances lasts equal or longer than the Hubble time, the local SF changes on time scales of the gas replenishment after its depletion by SF. After the heating sources by massive stars have ceased due to their deaths, the recovery

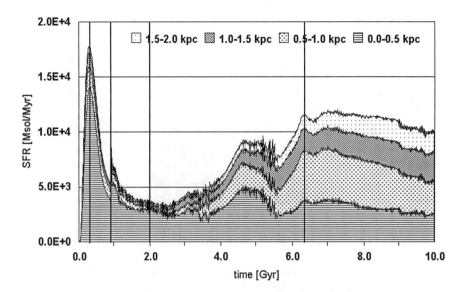

Figure 2.   Cumulative star formation rate (in $M_\odot$ Myr^{-1}) for a 10^9 $M_\odot$ dIrr model for different radial sectors in the equatorial plane (ranges given on top). The absolute values correspond to the differences between two curves. The vertical lines divide the evolution into the different phases (Rieschick & Hensler 2000; Hensler & Rieschick 2001; A. Rieschick & G. Hensler, in preparation).

of SF depends reasonably on the local energetic state and thus on $\tau_{\rm cool}$ and amounts from 100 to several hundred Myr (D'Ercole & Brighenti 1999), but varies significantly in small spatial scales and much shorter than the Hubble time. In the model (Figure 2), oscillations appear between 3–6 Gyr. At stages later than 6 Gyr the SF is self-regulated (Hensler & Rieschick 2001).

If galactic $\tau_{\rm SF}$ is much smaller than the Hubble time, which means that a particular galaxy could not have maintained its present SF over the past because of gas depletion, such an event is denoted as *starburst* (SB) (Leitherer, these proceedings). As another definition for a SB, the ratio of the present SF rate Ψ_0 to $\Psi(t)$ averaged over the past has to reach one order of magnitude and more. From giant ellipticals and dEs, as well as the detected faint blue galaxies at medium redshifts, it is established that the SB phenomenon is a usually short lasting state during the course of galactic evolution and might be repetitive. Presently, SB galaxies of different morphological type are visible in the local Universe, in which SB trigger mechanisms can be studied in detail in nearby objects. In dIrrs the SB sites are mainly centered and compact (blue compact DGs like e.g., NGC 1569, NGC 1705, and many others) and embedded into an old stellar population (e.g., I Zw 18).

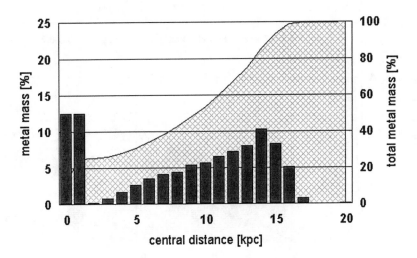

Figure 3.   Local metal deposition from hot supernova type II gas that expands as galactic wind in a 10^9 $M_\odot$ dIrr chemodynamical model (A. Rieschick & G. Hensler, in preparation). The columns of histogram are evaluated on the left-hand scale and reveal the local fraction. The accumulated deposition is displayed by the grey area (right-hand scale). For details see text.

4.   Dynamical Time Scales Gas Infall and Outflow

In some of the dIrrs the luminous galactic body is enveloped by large H I reservoirs with decoupled dynamics (e.g., NGC 4449: Hunter, Elmegreen, & Baker 1998; I Zw 18: van Zee et al. 1998b) or obviously suffering a collision with large intergalactic H I clouds (e.g., He 2–10: Kobulnicky et al. 1995; II Zw 40: van Zee et al. 1998b). If SF is triggered by gas infall and the H I clouds' kinematics replenish the consumed gas on a time scale shorter than the self-regulating energy release by massive stars, the SF continues on a high level. This means that the gas replenishment time scale due to infall

$$\tau_{\rm inf} = \frac{M_{\rm H\,I}}{A} \,,$$

where A denotes the gas infall rate (Köppen & Edmunds 1999), is equal or shorter than the $\tau_{\rm SF}$. Since the infalling clouds have metallicities different to those in a galaxy, probably lower or even pristine, gas infall has significant consequences for abundance ratios in dIrrs (Hensler, Rieschick, & Köppen 1999) and can explain the observed low N/O ratio at simultaneously low oxygen abundance in dIrrs (van Zee, Salzer, & Haynes 1998a). Most frequently, this low metallicity is only attributed to SNII-produced galactic outflows of hot metal-rich gas (Mac Low & Ferrara 1999; D'Ercole & Brighenti 1999). Detailed abundance analyses in hydrodynamical outflow simulations, however, are still missing. As a natural consequence of both observed effects, galactic winds and H I envelopes, respectively, their simultaneous action must be properly considered. In self-consistent chemodynamical models, infall of gas from a surrounding gas reservoir is inherently treated in combination with the outflow of hot metal-enhanced galactic winds and can properly reproduce the observed N/O–O relations (Hensler & Rieschick 2001).

5. Mixing Time Scales for Chemical Evolution

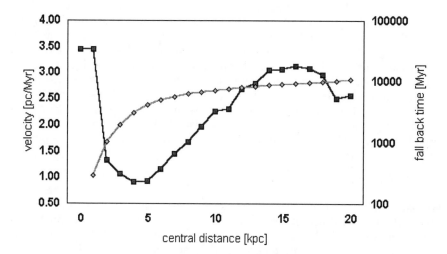

Figure 4.   Mean infall velocity of clouds (thick line, left scale) from an H I reservoir in a 10^9 $M_\odot$ dIrr chemodynamical model (A. Rieschick & G. Hensler, in preparation) and the corresponding fall-back time of metals (grey line, right scale) produced by a former generation of massive stars. The metals are expelled from the dIrr by a galactic wind, condensed on the infalling clouds, and returned smoothly incorporated into the clouds.

Because of the coexistence of cool gas clouds enveloped by hot metal-rich gas and due to the turbulence and fragmentation in superbubble shells, the freshly released SNII-typical elements are mixed into the cool gas. We have analyzed the issues of a 10^9 $M_\odot$ dIrr chemodynamical model (A. Rieschick & G. Hensler, in preparation) with respect to the local gas mixing and metal transport from the hot into the cool gas phase. The results in Figure 3 show that almost 25% of the metals produced in massive stars are deposited close to the SF sites themselves and, by this, lead to a local self-enrichment within 1 kpc on typical time scales in the range of 10 Myr. The remaining 75% of produced SNII metals are carried away from the SF region by the expansion of hot superbubble gas. Nevertheless, this has incorporated the material of evaporated clouds (Hensler et al. 1999). Since the clouds also consist of elements from intermediate-mass stars of an older population, i.e., carbon and nitrogen, the N/O ratio of subsequently arising H II regions is determined by this mixing effect. Because of the (at least adiabatic) cooling of hot gas, condensation onto infalling clouds leads to their polluting the element mix. From Figure 3 one can discern that this amount of metal deposition onto clouds increases reasonably with distance according to the growing cooling and reaches up to a distance of 15 kpc from the dIrr's center.

Although the metals are thus not directly expelled from a dIrr by the galactic wind but incorporated into infalling clouds, the circulation time scale for the metal enrichment via condensation and return in clouds from 8 kpc can last from 1 Gyr at 3 kpc to 10 Gyr from above 10 kpc (Figure 4). This is due to the low velocities of a few km s^{-1}, caused by the permanent compensation of the gravitationally accelerated inflow of the clouds by outflowing gas. If one takes into account that widely distributed gas in dIrrs can be stripped off by the intergalactic medium or by tidal effects because it is only loosely bound, still 50% of the metals from SNeII are transferred to the cool gas within a distance of not more than 8 kpc. Therefore, hydrodynamical models that investigate the expansion of hot SNII gas alone as a tracer of the metal dispersal, overestimate the total metal loss from the galaxy if small-scale mixing effects are neglected.

While local gas phase mixing by means of turbulence and condensation needs only several $\times10^7$ yr (Vieser & Hensler 2000, 2001), the diffusion time scale (Tenorio-Tagle 1996) is much longer except within the hot tenuous plasma.

6. Conclusions

We have illuminated the various time scales (see Table 1) in the evolution of dIrrs. Because of different interacting processes a much more comprehensive analysis and numerical simulations are required in order to understand various details, like e.g., the formation epoch of different types of DGs, the occurrence of SBs, the chemical abundance peculiarities in dIrrs, etc.

Table 1.   Typical time scales that determine the evolution of dIrrs.

Process	Time Scales (Myr)
Cooling of protogalactic clouds	0.1–1
Collapse self-regulation	few 100–1000
Formation of gaseous disks by gas infall	few 1000
Local star formation	100
Gas mixing	10
Element enrichment: local	10–few 10
global	100–1000

References

Bechthold, J., Weymann, R. J., Lin, Z., & Malkan, M. N. 1987, ApJ, 315, 180
Böhringer, H., & Hensler, G. 1989, A&A, 215, 147

Burkert, A., & Hensler, G. 1989, in Evolutionary Phenomena in Galaxies, ed. J. E. Beckmann & B. E. J. Pagel (Cambridge: Cambridge University Press), 230

Dekel, A., & Silk, J. 1986, ApJ, 303, 39

D'Ercole, A., & Brighenti, F. 1999, MNRAS, 309, 941

Franco, J., & Cox, D. P. 1983, ApJ, 273, 243

Hensler, G., & Rieschick, A. 2001, in ASP Conf. Ser., Modes of Star Formation and the Origin of Field Populations, ed. E. Grebel & W. Brandner (San Francisco: ASP), in press

Hensler, G., Rieschick, A., & Köppen, J. 1999, in ASP Conf. Ser. 187, The Evolution of Galaxies on Cosmological Time Scales, ed. J. E. Beckman & T. J. Mahoney (San Francisco: ASP), 214

Hensler, G., Spurzem, R., Burkert, A., & Trassl, E. 1997, A&A, 303, 299

Hunter, D. A., Elmegreen, B. G., & Baker, A. L. 1998, ApJ, 493, 595

Kepner, J. V., Babul, A., & Spergel, D. N. 1997, ApJ, 487, 61

Kobulnicky, H. A., Dickey, J. M., Sargent, A. I., Hogg, D. E., & Conti, P. S. 1995, AJ, 110, 116

Köppen, J., & Edmunds, M. G. 1999, MNRAS, 306, 317

Köppen, J., Theis, C., & Hensler, G. 1995, A&A, 296, 99

Köppen, J., Theis, C., & Hensler, G. 1998, A&A, 328, 121

Larson, R. B. 1969, MNRAS, 145, 405

Larson, R. B. 1974, MNRAS, 169, 229

Mac Low, M.-M., & Ferrara, A. 1999, ApJ, 513, 142

Rieschick, A., & Hensler, G. 2000, in ASP Conf. Ser. 215, Cosmic Evolution and Galaxy Formation, ed. J. Franco et al. (San Francisco: ASP), 130

Rivolo, A. V., & Solomon, P. M. 1987, in Proc. Amherst Symp. Molecular Clouds in the Milky Way and External Galaxies, ed. R. L. Dickman et al. (Berlin: Springer), 42

Tajiri, Y., & Umemura, M. 1998, ApJ, 507, 59

Tenorio-Tagle, G. 1996, AJ, 111, 1641

Theis, C., & Hensler, G. 1993, A&A, 280, 85

van Zee, L., Salzer, J. J., & Haynes, M. P. 1998a, ApJ, 497, L1

van Zee, L., Westphal, D., Haynes, M. P., & Salzer, J. J. 1998b, AJ, 115, 1000

Vieser, W., & Hensler, G. 2000, in Proc. Astrophysical Dynamics, ed. D. Berry et al., Ap&SS, 272, 189

Vieser, W., & Hensler, G. 2001, A&A, submitted

Acknowledgments. The author gratefully acknowledges insight into time scales from discussions with Andreas Rieschick and Daniel Tschoeke and from chemodynamical dIrrs models partly supported by the Deutsche Forschungsgemeinschaft under grant number He 1487/23-1.

Discussion

Alessandra Aloisi: I would like to know what kind of metals do you have inside the outflow that you get? Only the product of supernovae Type II so alpha elements, or what else?

Gerhard Hensler: After the burst or after the star formation locally it's mainly the Type II supernovae elements, the alpha elements, but we also have the other contributions taken into account; I've not shown here those figures but it is the case that we for instance also mix N, O, and also Fe from Type Ia supernovae. This mixture is produced by evaporation of cool clouds nearby the sites of SNII explosions and leads to these very peculiar abundances in comparison to normal galaxies.

Romeel Dave: So what you're saying about the N/O ratio, it looked like the pattern you're seeing is similar to just Type II SNe. Is that right? And so you're reproducing that simply by having the material fall back into the galaxy?

Gerhard Hensler: No, the problem is that normal models for instance only take the galactic wind into account to expel oxygen, but that can only be half of the truth because we have these peculiar abundances. We also have to expel the nitrogen and that happens due to the evaporation of clouds. The clouds consist of nitrogen by intermediate mass stars, and if we have an older population that has already contributed by the nitrogen to the cool phase of the interstellar medium, we can also expel the nitrogen. That is why we get these low N/O ratios.

Astrophysical Ages and Time Scales
ASP Conference Series, Vol. 245, 2001
T. von Hippel, C. Simpson, N. Manset

The Formation and Evolution of Globular Clusters in the Antennae Galaxies

Bradley C. Whitmore

Space Telescope Science Institute, 3700 San Martin Drive, Baltimore, MD 21218, USA

Abstract. Five populations of young, massive, compact clusters have been identified in the "Antennae galaxies," the nearest and youngest example of a prototypical merging galaxy. These clusters have all the attributes expected of young globular clusters. We describe our methods of age-dating the clusters using HST observations ($UBVI$, Hα, GHRS spectra) and then compare our estimates with other studies. The populations range from very red clusters that are still embedded in their dust cocoons ($\lesssim 5$ Myr), to the majority of clusters that have ages in the range 1–20 Myr, to a smaller sample with ages ~ 100 Myr, to a population of clusters that formed during the initial encounter that produced the tidal tails (~ 500 Myr), to the old globular clusters that existed in the galaxies before the merger (~ 15 Gyr). By putting these population subsamples into the correct chronological sequence, we are able to study the formation and evolution of globular clusters in the local universe.

1. Introduction

The discovery of young massive star clusters in merging and starbursting galaxies has revitalized the study of star clusters. The Antennae galaxies are the youngest and closest example in Toomre's (1977) list of prototypical mergers. Hence they represent perhaps our best chance for understanding both the merger process and the formation of globular clusters. Age dating the clusters is important for three basic reasons:

1. The young clusters in merging galaxies provide a timepiece for determining when the interaction occurred. This allows us to put mergers into the correct chronological sequence, and hence test the idea that spiral galaxies can merge to form elliptical galaxies.

2. Most globular clusters are believed to be amongst the oldest artifacts in the universe. However, the massive young clusters being formed in merging galaxies such as the Antennae have all the attributes expected of young globular clusters (e.g., Whitmore et al. 1999). These young clusters provide the unexpected, but very welcome opportunity to study the formation of globular clusters in the local universe.

3. The formation of the clusters also provides insight into the mechanism of star formation, the most fundamental process in astronomy. While we have very detailed models of the structure and evolution of stars (e.g., isochrones in the HR diagram), we have only sketchy ideas of how stars

form to begin with. An obvious approach to solving this problem is to go to where lots of stars are forming, such as merging and starbursting galaxies. When we do this we find that a large fraction of the star formation is in the form of massive, compact star clusters. Understanding how these clusters form should go a long ways toward understanding star formation in general.

This contribution will describe and compare the various methods of age dating young clusters in merging and starbursting galaxies, with a focus on the Antennae galaxies. We will then briefly discuss how we can use these age estimates to study the formation and evolution of young globular clusters. A more comprehensive review, including a listing and more details concerning other galaxies, can be found in Whitmore (2001).

2. Methods for Estimating the Ages of Extragalactic Star Clusters

Measuring the ages of *galactic* star clusters is by now a rather routine process. By plotting the positions of individual stars on a color–magnitude diagram, and then comparing them with theoretical isochrones, accuracies of $\sim 10\%$ are now achievable. The problem is more difficult for *extragalactic* star clusters, where only the integrated properties of the cluster are generally available. However, accuracies of 20–50% are currently being achieved (see Table 1).

2.1. Color–Magnitude Diagrams

As a cluster ages it fades and reddens, as the brightest and bluest stars (i.e., O and B stars) deplete their available fuel and become supernovae. The age of a merger remnant can therefore be crudely estimated based on the brightness and color of its brightest clusters. Figure 15 in Whitmore et al. (1997) shows a simulation based on this simple idea. They assume solar metallicity for the younger population, similar mass functions for the young and old populations, and use the A. G. Bruzual & S. Charlot (1996, private communication) stellar evolution models. The figure shows that at an age of 1–2 Gyr, the colors of the younger, metal-rich clusters are roughly the same as the color of the older metal-poor clusters, making it difficult to distinguish the two populations. However, the brightest young clusters are still 1–2 magnitudes brighter than the brightest old globular clusters.

Figure 1 demonstrates how this works in practice, by using the 10th brightest cluster to compare the various galaxies (from Whitmore et al. 1997). For galaxies with very young clusters, such as the Antennae, distinguishing young clusters from old globular clusters is very easy, since the clusters are $\sim$5 magnitudes brighter and 1 magnitude bluer in $V - I$. It is also fairly easy to age date mergers with ages ~ 500 Myr, such as NGC 3921 and NGC 7252.

For intermediate-age mergers, such as NGC 3610 and NGC 1700, it becomes progressively more difficult (e.g., NGC 3610 yields an age estimate of 4 ± 2.5 Gyr). The case for NGC 1700 is less clear, although a recent study of NGC 1700 by Brown et al. (2000) yielded an estimated age of 3 ± 1 Gyr. Similar results have been determined for NGC 6702, with Georgakakis, Forbes, &

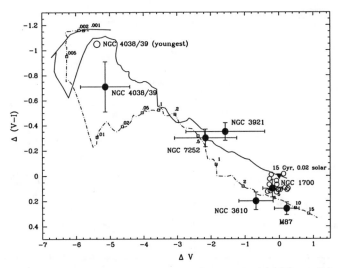

Figure 1.   Plot of the evolution in luminosity (ΔV) and in color ($\Delta(V-I)$) of star clusters, based on the Bruzual–Charlot (1996) tracks for a metal-poor population (solid line) and a solar metallicity population (dashed-dot line). The values are normalized to an old, metal-poor population (filled triangle). Ages in Gyr for the solar metallicity track are marked with squares. See the original article for further details (Whitmore et al. 1997).

Brodie (2001) estimating an age of ~ 2 Gyr. The best example of a so-called "missing-link" galaxy, is currently NGC 1316, as studied by Goudfrooij et al. (2001). Using a combination of photometric and spectroscopic observations they derive an age of 3 ± 0.5 Gyr.

2.2.   Colors–Color Diagrams

More precise age estimates are possible using more colors. This also provides an independent means of solving for the age and the reddening caused by dust. For example, Whitmore et al. (1999) use $UBVI$ photometry and reddening-free Q parameters to determine ages for the clusters in the Antennae (Figure 2). They find evidence for four populations of clusters, ranging in age from < 5 Myr to ~ 500 Myr. They also isolate a population of old globular clusters in this galaxy.

Recently, observations in the UV and in the IR have become available for various galaxies. This helps break the age-metallicity-reddening degeneracy (e.g., Maoz et al. 2001 for two ring galaxies; and Maraston et al. 2001 for NGC 7252).

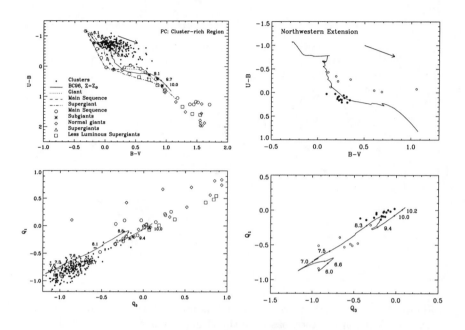

Figure 2.   Color–color diagram and reddening-free Q parameter diagram for clusters in the Antennae. The numbers on the plots are the values of log(age). See Whitmore et al. (1999) for details.

2.3. Hα Emission

Hα emission can be used in two ways to estimate the ages of the younger clusters. The existence of Hα emission itself indicates that a cluster is $\leq$ 10 Myr, since the O and B stars required to ionize the gas only live for this long (e.g., see the Leitherer & Heckman 1995 models). The second, less precise method is to use the size of the Hα ring around a cluster. Whitmore et al. (1999) estimate that the clusters in the western loop of NGC 4038 are 5–10 Myr, since many of them have rings with diameters of $\sim$ 100–500 pc and expansion velocities $\sim$ 25–30 km s^{-1}. The clusters in the overlap region appear to be $\lesssim$ 5 Myr old, since the rings are smaller or non-existent in this region. One of the primary uncertainties from this method is due to the unknown density of gas in the vicinity of the cluster, which may impede the growth of the bubbles more in one region than another.

2.4. Absorption-Line Spectroscopy

The most accurate method of estimating ages is to obtain spectra. Schweizer & Seitzer (1998) obtained UV-to-visual spectra of eight cluster candidates in NGC 7252. They use six absorption lines (Hβ, Ca, Mgb, 3 Fe lines) and determine the ages based on the best simultaneous fit (Figure 3). Solar metallicity provides the best fit. They find that six of the seven clusters they age date have ages in the range 400–600 Myr, roughly consistent with the mean photometric age estimate of 650 Myr from Miller et al. (1997). One cluster turned out to be an emission-line object with an age estimate of < 10 Myr, indicating that cluster formation is still on-going at a low level in the outer parts of the galaxy. Maraston et al. (2001) have reanalyzed the spectra from Schweizer & Seitzer using only the Blamer lines $H\beta$, $H\gamma$, and $H\delta$, and new spectral evolution models that handle AGB stars better. They find good agreement with previous estimates (see Table 1).

Table 1. Age estimates for selected star clusters[†].

Method	Knot B	Knot K	Knot S	WS 80	W3
Q1-Q3[1]	3 Myr	4 Myr	30, 5 Myr	3 Myr	—
GHRS spectra[1]	—	3 ± 1	7 ± 1	—	—
K spectra[2]	~4	—	—	—	—
K spectra[3]	$3.7^{+1.0}_{-0.4}$	8.5–12.8	$8.1^{+2.0}_{-0.2}$	$5.5^{+0.8}_{-0.7}$	—
$V-I$[4]	—	—	—	—	420 ± 110
$B-V$[5]	—	—	—	—	440 ± 60
$V-K$[5]	—	—	—	—	530 ± 30
optical spectra[4]	—	—	—	—	540 ± 30
optical spectra[5]	—	—	—	—	510 ± 10

[†] The first four clusters are in the Antennae. W3 is in NGC 7252. Solar metallicity is assumed. Maraston et al. (2001) prefer 0.5 solar metallicity.
[1] Whitmore et al. (1999)
[2] Gilbert et al. (2000)
[3] Mengel et al. (2000)
[4] Schweizer & Seitzer (1998)
[5] Maraston et al. (2001)

Whitmore et al. (1999) obtained GHRS spectra in the Antennae to yield age estimates of 3 ± 1 Myr and 7 ± 1 Myr for two of the brightest clusters (Figure 4), in fair agreement with the estimates based on the $UBVI$ colors and the $H\alpha$ morphology. Gilbert et al. (2000) and Mengel et al. (2000) have used K-band spectroscopy to estimate ages which are in good agreement with the other age estimates. The K-band is especially important for the very red embedded clusters such as WS80, which is coincident with the strongest ISO source in the entire galaxy.

A comparison of the results in Table 1 shows that the age estimates from the various techniques are in reasonably good agreement, with differences generally in the range 20–50%. We note that the largest outlyer (Q_1-Q_3 vs. spectroscopic estimates for Knot S) is probably due to the fact that the curve folds back upon itself and is almost two-valued for ages near log(age) = 7.5 and 6.7 (see Figure 2). Given the measurement uncertainty, values of 30 and 5 Myr are both possible for these clusters. The lower values would be consistent with the spectroscopic estimates. This highlights one of the primary uncertainties in this method, and demonstrates why a wider variety of colors is needed to break the degeneracy for certain clusters.

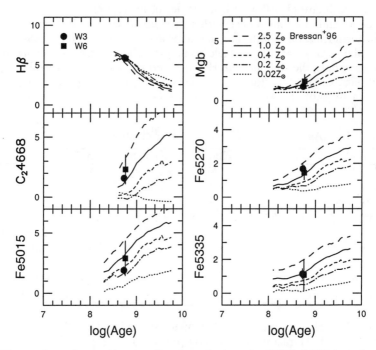

Figure 3.   Age estimates for two clusters in NGC 7252 based on absorption-line strengths for 6 lines (Schweizer & Seitzer 1998).

3.   Formation and Evolution of Globular Clusters in the Antennae

Using the techniques described in Section 2 we are able to identify five populations of star clusters in the Antennae.

The youngest clusters appear to be very red objects (R sample; $\lesssim 5$ Myr), which Whitmore & Schweizer (1995) suggested were only now emerging from their dust cocoons. Several of these have recently been identified as strong IR sources (Vigroux et al. 1996, Wilson et al. 2000, Gilbert et al. 2000, Mengel et al. 2000). In fact, the brightest IR source in the Antennae is one of these very red objects (W80), rather than the nucleus of one of the two galaxies. Wilson et al. (2000) find three separate molecular clouds around W80 within a region of 1 kpc^2, and suggest that cloud-cloud collisions may play an important role in cluster formation. However, the lack of similar morphologies for the other very red objects suggest that this may not be the universal mechanism. Most of the very red objects are in the overlap region between the galaxies, but $\sim 20\%$ can also be found in the Western Loop. Q. Zhang, M. Fall, & B. C. Whitmore (in preparation) found that the red clusters are closely associated with the CO and ISO flux, supporting the idea that they are formed from giant molecular clouds.

The second population of clusters (B1 sample: 1–20 Myr) have been identified using the Q-Q analysis from Whitmore et al. (1999). While many of these clusters are in the overlap region, larger fractions are found in the Western Loop

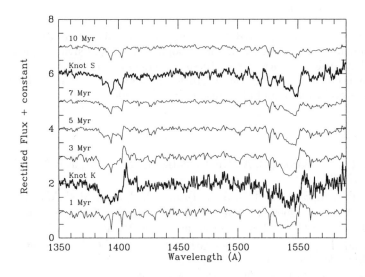

Figure 4.   Age estimates based on GHRS spectra of two clusters in the Antennae (Whitmore et al. 1999).

and Northeastern Star Formation regions. These clusters show very strong correlations with Hα flux, as expected. The fact that the correlation with Hα is much stronger than for the older clusters (Q. Zhang, M. Fall, & B. C. Whitmore, in preparation) demonstrates that the Q-Q analysis can successfully isolate a population of clusters which ages ∼ 10 Myr based on *UBVI* colors alone.

The third population (B2 sample: ∼ 100 Myr) has also been identified using the Q-Q analysis. The majority of these clusters are in the Northeastern Star Formation region. These clusters show much weaker Hα emission than the B1 sample.

The fourth population consists of intermediate-age clusters (I sample: ∼ 500 Myr) which presumably formed during the initial encounter that produced the tidal tails. The clusters are quite widely dispersed, with most of their members in either the Northwestern Extension (see figure 5 in Whitmore et al. 1999) or along the outer edge of the Northeastern Star Formation region as it leads into one of the tidal tails. While the R, B1, and B2 populations appear to arise from a continuous episode of cluster formation, the I population appears to be from a separate episode. Figure 2 shows how clearly this population can be distinguished from the younger populations based on the color–color and Q-Q plots.

The fifth population consists of old globular clusters (O sample; ∼ 15 Gyr) which were preexisting in the two spirals galaxies before they collided. These clusters are most easily found in NGC 4039, due to the smaller numbers of bright young clusters that make it difficult to find these relatively faint clusters. Whitmore et al. (1999) list 11 candidates for old globular clusters in the Antennae.

4. Summary

A variety of methods of age-dating clusters give consistent results and provide the opportunity to isolate five different populations of massive compact clusters in the Antennae. The brightest of these clusters have all the attributes expected of globular clusters (see Whitmore 2001 for details). Hence, we are able to study the entire evolution of globular clusters in this single galaxy.

References

Brown, R. J. N., Forbes, D. A., Kissler-Patig, M., & Brodie, J. P. 2000, MNRAS, 317, 406

Georgakakis, A. E., Forbes, D. A., & Brodie, J. P. 2001, MNRAS, 318, 124

Gilbert, A., et al. 2000, in ASP Conf. Ser., Massive Stellar Clusters, ed. A. Lancon & C. M. Boily (San Francisco: ASP), 101

Goudfrooij, P., Mack, J., Kissler-Patig, M., Meylan, G., & Minniti, D 2001, MNRAS, 322, 643

Leitherer, C., & Heckman, T. M. 1995, ApJS, 96, 9

Maraston, C., Kissler-Patig, M., Brodie, J. P., Barmby, P., & Huchra, J. P. 2001, A&A, submitted (astro-ph/0101556)

Maoz, D., Barth, A. J., Ho, L. C., Sternberg, A., & Filippenko, A. V. 2001 (astro-ph/0103213)

Mengel, S., Lehnert, M. D., Thatte, N., Tacconi-Garman, & Genzel, R. 2000 (astro-ph/0010238)

Miller, B. W., Whitmore, B. C., Schweizer, F., & Fall, S. M. 1997, AJ, 114, 2381

Schweizer, F., & Seitzer, P. 1998, AJ, 116, 2206

Toomre, A. 1977, The Evolution of Galaxies and Stellar Populations, ed. B. M. Tinsley & R. B. Larson (New Haven: Yale University Observatory), 401

Vigroux, L., et al. 1996, A&A, 315, L93

Whitmore, B. C. 2001 (astro-ph/0012546)

Whitmore, B. C., Miller, B. W., Schweizer, F., & Fall, S. M. 1997, AJ, 114, 1797

Whitmore, B. C., & Schweizer, F. 1995, AJ, 109, 960

Whitmore, B. C., Zhang, Q., Leitherer, C., Fall, S. M., Schweizer, F., & Miller, B. W. 1999, AJ, 118, 1551

Wilson, C. D., Scoville, N., Madden, S. C., & Charmandaris, V. 2000, ApJ, 542, 120

Discussion

Masanori Iye: The characteristics of interstellar reddening can be dependent on the size distribution of grains, so to what extent might this affect your analysis? And conversely, can you get from these data some insight about the extinction law of very dense compact regions?

Brad Whitmore: We haven't worked too hard on the characteristics of the dust itself. What we have done is use a variety of different extinction laws to see how it might affect our results, and it didn't affect things very much at all. I believe Wilson et al. (2000, ApJ, 542, 120) find some evidence that the dust properties may be different in one particular area around WS # 80, but in general the dust properties do not appear to be abnormal. The bottom line for our own study is that when we use 3 or 4 different reddening laws, we don't see much of a change in the results. The technique appears to be quite robust.

Josh Barnes: I was wondering how secure is the translation from the observed luminosity function to the inferred mass function of these clusters?

Brad Whitmore: The game of trying to convert from luminosities to mass is fairly difficult. Small observational uncertainties can result in quite a spread in the estimated ages and hence a spread in the mass estimates. In addition, the stellar evolution models tend to have kinks in them, so at certain parts of the curves the procedure becomes quite non-linear. What we try to do for many parts of the analysis is to bundle things into age ranges where it doesn't matter so much. For example, in the figure I showed from the Zhang and Fall paper it looks like there is a gap in the number of clusters between roughly 10 and 15 Myr old. That's not real gap—that's a problem in trying to make the age estimate. In the analysis we therefore break the sample in two at around 10 Myr.

Bob Schommer: Can you make a relative mass estimate of how much mass is involved in the old clusters in the Antennae versus the young systems?

Brad Whitmore: It's been a while since I've looked at those numbers, but I think that in the Antennae, and also in other systems such as NGC 3921 and 7252, it's not like you've actually doubled the number of clusters—it's more like an increase in the number of globular clusters by a factor of ∼ 50%. So you have increased the specific frequency, but you haven't increased it by a factor of 10 or anything. An important thing to mention in the Antennae is that we're still at an early stage in the merger. In the simulation that Josh showed of the Antennae you can see that most of the merging and star formation is still to come. You can tell from the appearance and velocity fields of the two disks that the system has not even undergone violent relaxation yet. Hence, a large number of massive compact clusters are likely to form in the next few 100 million years. [To Schweizer:] Do you remember for NGC 3921 the number...

Francois Schweizer: There was a lower limit of 40% and for NGC 7252 80%.

Astrophysical Ages and Time Scales
ASP Conference Series, Vol. 245, 2001
T. von Hippel, C. Simpson, N. Manset

Galaxy Deconstruction: Clues From Globular Clusters

Michael J. West

*Department of Physics and Astronomy, University of Hawai'i, Hilo, HI
96720, USA*

Abstract. The present-day globular cluster populations of galaxies reflect the cumulative effects of billions of years of galaxy evolution via
such processes as mergers, tidal stripping, accretion, and in some cases
the partial or even complete destruction of other galaxies. If large galaxies
have grown by consuming their smaller neighbors or by accreting material
stripped from other galaxies, then their observed globular cluster systems
are an amalgamation of the globular cluster systems of their progenitors.
Careful analysis of the globular cluster populations of galaxies can thus
allow astronomers to reconstruct their dynamical histories.

1. Introduction

The origin of galaxies is one of the great outstanding problems in modern astrophysics. How and when did galaxies form? How have they evolved over time?
How does environment influence their properties?

One way to unravel the secrets of galaxy formation is by studying their
globular cluster populations. Most galaxies possess globular cluster systems of
various richness, ranging from dwarf galaxies with only a handful of globulars,
to supergiant elliptical galaxies with tens of thousands of globulars surrounding
them (see Harris 1991 or van den Bergh 2000 for reviews). Because globular
clusters are among the oldest stellar ensembles in the universe, they can provide
important clues about the formation of their parent galaxies.

The earliest studies of globular clusters were, by necessity, limited to our
own Galaxy and its nearest neighbors. Over the past few decades, however,
there has been tremendous progress in our understanding of globular cluster
systems of other galaxies. One of the most important recent discoveries in
the study of extragalactic globular cluster systems is that most large galaxies
appear to possess two or more chemically distinct globular cluster populations
(e.g., Gebhardt & Kissler-Patig 1999; Forbes & Forte 2001; Kundu & Whitmore
2001). Some examples are shown in Figure 1, where two peaks are seen in
the distribution of globular cluster metallicities for clusters associated with four
large elliptical galaxies.

A number of different theories have been proposed to explain the origin
of these bimodal globular cluster metallicity distributions. An obvious way to
generate two or more chemically distinct globular cluster populations in galaxies
would be through two or more bursts of globular cluster formation. This might
occur, for example, if mergers of gas-rich galaxies trigger the formation of new

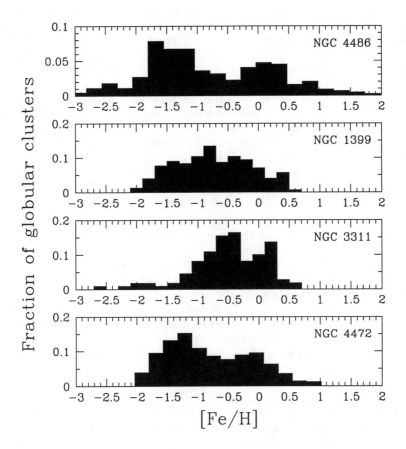

Figure 1.   Observed metallicity distribution of globular cluster systems associated with four giant elliptical galaxies. Note the presence
of two distinct peaks in most cases. The majority of large elliptical
galaxies studied to date exhibit bimodal globular cluster metallicity
distributions.

globulars (Schweizer 1987; Ashman & Zepf 1992), resulting in the birth of multiple generations of globular clusters. Similarly, one might envision a multiphase
galaxy collapse model in which the metal-poor globular clusters formed during
the initial collapse of a protogalactic gas cloud, and the metal-rich globulars
formed some time later (Forbes, Brodie, & Grillmair 1997; Larsen et al. 2001).
In both of these scenarios, the metal-poor globular clusters surrounding galaxies
such as those shown in Figure 1 would be their original population that formed
from low-metallicity gas at early epochs, and the metal-rich globulars would
have formed more recently from gas that was enriched by stellar evolution.

Alternatively, bimodal or multimodal globular cluster metallicity distributions could also arise quite naturally from galaxy mergers and/or accretion of globulars stripped from other galaxies *without needing to invoke the formation of multiple generations of globulars* (Côté, Marzke, & West 1998; Côté et al. 2000). Motivation for this model came from a simple fact: for those elliptical galaxies that exhibit a bimodal globular cluster metallicity distribution, the metallicity of the metal-rich peak shows a clear correlation with parent galaxy luminosity, in the sense that the most luminous galaxies have the most metal-rich globulars (Forbes et al. 1997; Forbes & Forte 2001). However, no such correlation is seen for the metal-poor peak; it appears to be largely independent of parent galaxy luminosity. To my collaborators and me, this suggests that the *metal-rich* globular clusters are innate to large ellipticals, and the metal-poor ones were added later either through mergers or accretion.

2.   Globular Clusters as Diagnostics of Galaxy Mergers

There is no doubt that galaxy mergers have occurred frequently throughout the history of the universe. Figure 2 shows an image of a supergiant elliptical galaxy in which the partially digested remains of several smaller galaxies are still clearly visible. Many large galaxies today may have grown to their present sizes by devouring smaller companions. If so, what becomes of the globular cluster populations of the galaxies that were consumed?

There is also evidence of ongoing galaxy destruction in rich clusters, and countless galaxies may have met their demise over a Hubble time (e.g., Gregg & West 1998; Calcaneo-Rodin et al. 2000). An example is shown in Figure 3. Because they are dense stellar systems, globular clusters are likely to survive the disruption of their parent galaxy, and will accumulate over time in the cores of rich galaxy clusters. The ongoing destruction of the moderate-sized elliptical galaxy shown in Figure 3, for example, will likely strew several hundred globulars into intergalactic space. Some of these may eventually be incorporated into other galaxies, a sort of recycling on cosmic scales (Muzzio 1987). In particular, giant elliptical galaxies at the centers of rich galaxy clusters, which are observed to have enormously rich globular cluster populations, may have inherited myriad intergalactic globular clusters (West et al. 1995).

If large galaxies have grown by consuming smaller neighbors or by accreting material torn from other galaxies, then their present-day globular cluster systems are an amalgamation of the globular cluster systems of their victims. Côté et al. (1998) and Côté et al. (2000) showed that the growth of large galaxies through mergers or accretion will invariably be accompanied by the capture of metal-poor globulars, resulting in bimodal (or even multi-modal) metallicity distributions that are strikingly similar to those see in Figure 1.

Our prescription for building a large elliptical galaxy with a bimodal globular cluster metallicity distribution is remarkably simple:

- We assume that galaxies obey a Schechter-like luminosity function, as is observed. This sets the relative numbers of galaxies of different luminosities that are available for merging.

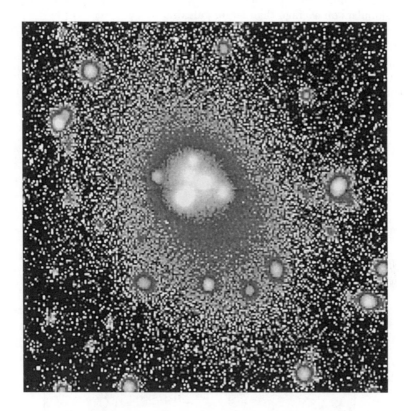

Figure 2. Brightest elliptical galaxy in the cluster Abell 3827. Several smaller cannibalized galaxies are clearly evident in the central regions. Globular clusters belonging to these galaxies are likely to survive the eventual disruption of their parent galaxies and thus will become part of this giant elliptical. If most large elliptical galaxies have grown by cannibalizing smaller neighbors, then their globular cluster populations today are composite systems that can provide information about the progenitor galaxies.

- We assume that each galaxy is born with its own intrinsic globular cluster population, and that the number of globulars per unit galaxy luminosity is constant, which is consistent with observations (Harris 1991). This determines how many globular clusters each galaxy has available to donate during mergers.

- We assume, again from observations, that the mean metallicity of a galaxy's original globular cluster population increases monotonically with parent galaxy luminosity (Côté et al. 1998). Smaller galaxies have metal-poorer globulars on average than larger galaxies.

Beginning with a medium-sized elliptical galaxy as a seed, we allow it to consume its smaller neighbors at random, stopping after enough mergers have occurred to yield a large elliptical. We assume that globular cluster numbers are conserved during mergers, so the larger galaxy gains the globulars from the smaller galaxies that it consumed.

North

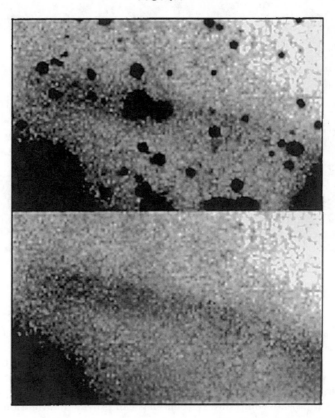

Figure 3.   Tidally disrupted galaxy in the Coma cluster (from Gregg & West 1998). The top panel shows the raw image, and the bottom panel has been cleaned of foreground objects to highlight the ~ 150 kpc long plume of material. The partial, or in some cases complete, disruption of galaxies in dense environments will create a population of intergalactic stars and globular clusters. These freely roaming globulars may be accreted later by other galaxies.

Figure 4 shows some results of Monte Carlo simulations based on the Côté et al. (1998, 2000) model. Our simulations indicate that 80 to 90% of large elliptical galaxies formed in this way exhibit bimodal (or in some cases multimodal) globular cluster distributions. The locations of the metal-rich and metal-poor peaks also agree well with observations (compare Figures 1 and 4). In our model, the metal-rich globular clusters of large ellipticals belonged to the progenitor galaxy seed, and the metal-poor globulars were inherited from the many smaller galaxies that it consumed during its growth, or by accretion of intergalactic globulars that were torn from other galaxies. The globulars gained

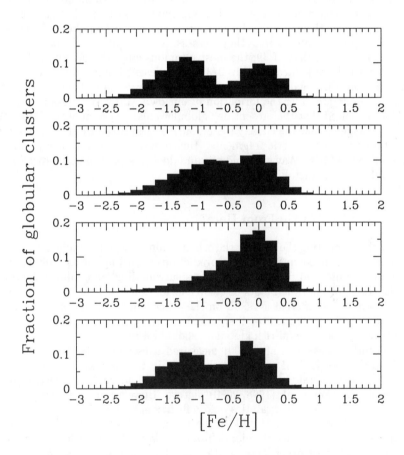

Figure 4.   Results from some simulations based on the dissipationless merger model of Côté et al. (1998). These simulations show that bimodal globular cluster metallicity distributions are easily produced by dissipationless galaxy mergers and accretion, without needing to posit the formation of multiple bursts of globular cluster formation.

from mergers or accretion are predominantly metal-poor because they originate mostly in low-mass galaxies.

It is noteworthy that unimodal globular cluster metallicity distributions also occur from time to time in our model. An example can be seen in Figure 4. This is not surprising, given the stochastic nature of the merger process. For instance, a large elliptical could in principle be built by merging many small dwarf galaxies (resulting in the globular cluster metallicity distribution of the final merger remnant exhibiting a single metal-poor peak), or by merging two or three medium-sized galaxies (which might lead to a single metal-rich peak), or by merging galaxies over a wide range of luminosities (which yields bimodal or multimodal globular cluster metallicity distributions).

If our model is correct, then it offers the exciting possibility of placing some quantitative constraints on the number and types of mergers that galaxies have experienced over their lifetimes by comparing the relative numbers of metal-rich and metal-poor globulars that they possess. Côté et al. (1998) used this reasoning to conclude that M 49, the most luminous elliptical galaxy in the Virgo cluster, must have gained roughly 2/3 of its present luminosity by consuming other smaller Virgo galaxies. More recently, we have applied these same techniques to understanding the formation of our own Milky Way Galaxy. Côté et al. (2000) showed that the present-day globular cluster system of the Milky Way strongly suggests that the Galaxy's spheroid was assembled from a large number of metal-poor protogalactic fragments. Hence even a relatively low luminosity system like the Milky Way spheroid can and does possess a bimodal distribution of globular cluster metallicities.

3.   Where Do We Go From Here?

Clearly the competing theories for the origin of bimodal globular cluster metallicity distributions make quite different predictions regarding the ages of globulars. If the metal-poor and metal-rich globulars surrounding large elliptical galaxies are the result of multiple bursts of cluster formation, then the two populations should have quite different ages. If, on the other hand, bimodal metallicity distributions can be explained by dissipationless merging as described above, then the metal-poor and metal-rich globulars should all be old.

Recently, Beasley et al. (2000) measured ages and metallicities of globular clusters in M 49, and concluded that (within the sizable uncertainties) the metal-poor and metal-rich populations are coeval and old. This clearly seems to support the Côté et al. (1998, 2000) picture. However, more precise data are needed to reduce the uncertainties before firm conclusions can be drawn. With that goal, we have obtained Hubble Space Telescope observations of $\sim 10^3$ globular clusters associated with the giant elliptical galaxy M 87 in order to accurately determine their ages using a powerful narrow-band photometry technique.

The hypothesis that galaxies might accrete substantial numbers of intergalactic globular clusters also needs to be tested with direct observations of these objects to determine if they really exist and in what numbers. Unlike many theories for the origin of globular cluster populations, the intergalactic globulars hypothesis is easily falsifiable; if a significant population of intergalactic globulars is not detected in the cores of galaxy clusters, then this idea will have to be abandoned. My collaborators and I are currently analyzing HST, Keck, Subaru and CFHT images that we obtained to search for intergalactic globulars in the Virgo, Coma and Abell 1185 galaxy clusters.

There is also work to be done on the theoretical front. The simple merger model described here is admittedly somewhat naive. For example, we assumed equal merger probabilities for all galaxies, when in reality there is likely to be some mass dependence of merging. As a step towards more realistic models, Frazer Pearce and I are collaborating in a study of the merger histories of galaxies that form in very high-resolution N-body cosmological simulations. By following the detailed merger histories of galaxies from high redshifts to the present, and inputting simple models of globular cluster formation at different epochs, we will be able to make quantitative predictions regarding the evolution of globular cluster metallicity distributions in galaxies as a function of time.

References

Ashman, K. M., & Zepf, S. E. 1992, ApJ, 384, 50

Beasley, M. A., Sharples, R. M., Bridges, T. J., Hanes, D. A., Zepf, S. E., Ashman, K. M., & Geisler, D. 2000, MNRAS, 318, 1249

Calcáneo-Roldán, C., Moore, B., Bland-Hawthorn, J., Malin, D., & Sadler, E. M. 2000, MNRAS, 314, 324

Côté, P., Marzke, R. O., & West, M. J. 1998, ApJ, 501, 554

Côté, P., Marzke, R. O., West, M. J., & Minniti, D. 2000, ApJ, 533, 869

Forbes, D. A., Brodie, J. P., & Grillmair, C. J. 1997, AJ, 113, 1652

Forbes, D. A., & Forte, J. C. 2001, MNRAS, 322, 257

Gebhardt, K., & Kissler-Patig, M. 1999, AJ, 118, 1526

Gregg, M. D., & West, M. J. 1998, Nature, 396, 549

Harris, W. E. 1991, ARA&A, 29, 543

Kundu, A., & Whitmore, B. 2001, AJ, in press (astro-ph/0103021)

Larsen, S. S., Brodie, J. P., Huchra, J. P., Forbes, D. A., & Grillmair, C. 2001, AJ, in press (astro-ph/0102374)

Muzzio, J. C. 1987, PASP, 99, 245

Schweizer, F. 1987, in Nearly Normal Galaxies, ed. S. Faber (New York: Springer), 18

van den Bergh, S. 2000, PASP, 112, 932

West, M. J., Côté, P., Forman, C., Forman, W., & Marzke, R. O. 1995, ApJ, 453, L77

Acknowledgments. I wish to thank my principal collaborators in this research, Pat Côté, Michael Gregg, Ron Marzke, and Frazer Pearce. This work was supported by NSF grant AST 00-71149.

Discussion

Michael Rich: I support your view; in fact, in the Milky Way there are old metal-rich globular clusters in the bulge and, as I mentioned in my talk, when you age-date them they seem to be rather old—at least of order 13 Gyr. We have a white dwarf distance modulus distance to 47 Tuc which gives a 13 Gyr age, and then we can compare 6528 and 6553 to that, and we think that these red globular clusters were formed along with the Galactic bulge, very early in the history of the Galaxy. So I think that red cluster systems have to do with the formation of the bulge. Now, is the bulge formed in a merger event? Maybe, the simulations show these very early mergers. On the other hand, black hole mass is correlated with bulge velocity dispersion and black holes are very early galaxy formation events. So I don't know, but I would tend to support the view that the red clusters were formed in the metal-rich spheroid.

Michael West: Thanks. Pat Côté, Ron Marke, Dante Minniti, and I published a paper just last year in which we showed that one can account for the properties of the Milky Way spheroid and its globular cluster system if it consumed upwards of 10^3 little dwarf galaxies during its formation.

Hugh Harris: The clusters are usually found more widely distributed spatially than the halo light in their galaxy. How do you interpret that?

Michael West: We think that accretion, not just mergers, has probably played an important role by stripping globular clusters from some galaxies and then adding these to the outer regions of others, especially giant ellipticals in the centers of rich clusters. This would explain why some of large ellipticals, like M 87 for example, have far more globulars per unit galaxy luminosity than expected, and why they have very extended distributions.

Alan Stockton: The problem with making large bulges or large ellipticals from a lot of small things, of course, is the luminosity–color relation, or luminosity–metallicity relation. So you indicated in most cases you favor starting with a fairly large progenitor and then putting a bunch of small things on it. You do still run into any problem with the observed dispersion in that relation?

Michael West: One can imagine making a large elliptical in many different ways, either from the merger of many small dwarfs, in which case the elliptical should be pretty blue as well, or from a number of bigger progenitors which would have had redder colors. The existence of the observed luminosity–color relation for galaxies suggests that in most cases large ellipticals can't have formed just from the merger of many dwarfs—that probably wouldn't work, you're right. It suggests that most large ellipticals must have started from fairly large progenitor seeds. As this seed galaxy consumes smaller neighbors, this will dilute the luminosity–color relation, but won't necessarily obliterate it.

Astrophysical Ages and Time Scales
ASP Conference Series, Vol. 245, 2001
T. von Hippel, C. Simpson, N. Manset

The Formation Time Scales of Giant Spheroids

S. C. Trager

The Observatories of the Carnegie Institution of Washington, 813 Santa Barbara Street, Pasadena, CA 91101, USA

S. M. Faber

UCO/Lick Observatory, Department of Astronomy and Astrophysics, University of California, Santa Cruz, CA 95064, USA

A. Dressler

The Observatories of the Carnegie Institution of Washington, 813 Santa Barbara Street, Pasadena, CA 91101, USA

Abstract. We review current progress in the study of the stellar populations of early-type galaxies, both locally and at intermediate redshifts. In particular, we focus on the ages of these galaxies and their evolution in hopes of determining the star formation epochs of their stars. Due to serious remaining systematic uncertainties, we are unable to constrain these epochs precisely. We discuss our results on the evolution of stellar populations in the context of other observables, in particular the evolution of the Fundamental Plane of early-type galaxies.

1. Introduction

The sequence and time scales of galaxy formation are a major issue in astrophysics today. The formation histories of early-type galaxies, ellipticals and S0's, though seemingly simple systems, remain a puzzle. Long thought to be very old, homogeneous, coeval systems varying only in their metallicities (Baade 1962; Faber 1977; Burstein 1977), accumulating observational evidence on stellar populations of ellipticals and S0's have challenged this view. A number of field elliptical galaxies clearly have had significant recent star formation (see below), and many early-type galaxies have evidence for recent dynamical disturbances (Schweizer & Seitzer 1992). Theoretically, models of the commonly accepted mode of galaxy formation—hierarchical clustering (e.g., Blumenthal et al. 1984; Kauffmann, White, & Guiderdoni 1993)—suggest that merging, accretion, and star formation have continued in the giant early-type galaxy population up until the current epoch, at least in some galaxies in low-density environments. However, these models are unconstrained by detailed observations of the formation epoch (parameterized by the formation redshift z_f of the dominant stellar population) and the time scale of star formation in early-type galaxies.

Detailed studies of the stellar populations of early-type galaxies allow us to determine or at least infer these time scales. In particular, ages derived from stellar populations and their evolution with redshift allow us to determine *when* the stars in giant spheroids form, and their nucleosynthetic properties (their metallicities and relative abundances such as $[\alpha/\text{Fe}]$) allow us to determine *how long* the star formation event(s) lasted.

Unfortunately, such detailed data are difficult to gather due to the *age-metallicity degeneracy* (e.g., O'Connell 1986; Worthey 1994; Trager 1999), which plagues all stellar population studies at some level. As stellar populations age, the populations get cooler and thus redder. Unfortunately, stellar populations also get cooler and redder with increasing metallicity. Colors and metal absorption lines like Mg_2 are therefore degenerate to compensating changes in age and metallicity. Worthey (1994), following earlier work by O'Connell (1980), Rabin (1982), Burstein et al. (1984), and Rose (1985) showed that the Balmer lines of hydrogen can break this age-metallicity degeneracy. These absorption lines originate in the hot main-sequence turnoff (MSTO) stars, and the MSTO temperature is much more sensitive to age than to metallicity in stellar populations (cf. figure 1 of Trager 1999). However, because even the Balmer lines are still somewhat sensitive to metallicity, a combination of Balmer and metal lines are used to determine ages and abundances of early-type galaxies (Figures 1 and 2; e.g., González 1993; Worthey 1994; Jørgensen 1997, 1999, these proceedings; Kuntschner 2000; Trager et al. 2000a, 2000b).

In this contribution, we summarize the current state of our own studies of the stellar populations of both local and distant early-type galaxies. We focus on our attempts to determine the typical formation redshifts of these galaxies in various environments and on the difficulties in measuring—and comparing—the stellar populations of early-type galaxies.

Throughout this contribution, we use the stellar population models of Worthey (1994) for consistency with earlier work, but note that the absolute time scale of these models (that is, MSTO temperature at fixed age) is suspect, with the oldest models being too old by about 25–35% when compared with models based on recent isochrones by the Padova group (e.g., Girardi et al. 2000; cf. Charlot, Worthey, & Bressan 1996). However, the *relative* ages of galaxies are nearly unaffected by the choice of stellar population model (Trager et al. 2000a), which are the data of interest here. Also, it is helpful to keep in mind that the ages (and abundances) presented here are those of single-burst, single-metallicity stellar populations (SSPs); see Trager et al. (2000b) for details of the effects of composite populations on SSP parameters.

2.   The Stellar Populations of Nearby Early-Type Galaxies

At the present, only a limited number of early-type galaxies (less than 100 total ellipticals and S0's) have had absorption line strengths measured accurately enough for detailed stellar population work (that is, $\sigma_{H\beta} < 0.05$ Å, or S/N > 75 Å). Many of these galaxies are shown in Figure 1, in which the galaxies are separated by morphology (elliptical vs. S0) and by environment (cluster vs. field). A full analysis of the stellar populations of the field and Fornax Cluster ellipticals is given by Trager et al. (2000b), but the salient points for a discussion of the formation epochs of early-type galaxies can be gleaned directly from this figure.

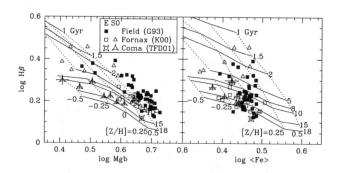

Figure 1.   Line strengths of local early-type galaxies, by morphology and environment. Squares are ellipticals, triangles are S0's, solid points are "field" (isolated, group, and Virgo cluster) galaxies from González (1993), open points are Fornax Cluster galaxies from Kuntschner (2000), and stellated points are Coma Cluster galaxies (S. Trager et al., in preparation). Note the large dispersion in the Hβ strengths—and thus ages—of field ellipticals and Fornax S0's but the rather tight distribution in the Hβ strengths of Fornax ellipticals and Coma galaxies, both ellipticals and S0's.

Field ellipticals. Field ellipticals from González (1993), which include ellipticals in environments ranging from isolated galaxies to Virgo Cluster galaxies, show a wide spread in age. This spread in age should not be interpreted necessarily as indicative of a wide range of formation redshifts for the entire stellar populations of these object. Indeed, Trager et al. (2000b) have interpreted this spread in the context of two-burst models of star formation: Small "frostings" of recent star formation occur on top of massive old stellar populations, with the range in inferred ages corresponding to some combination of the time at which these frostings were formed and the strength of the burst.

Cluster ellipticals. Cluster ellipticals in Virgo (González 1993; Trager et al. 2000b), Fornax (Kuntschner 2000), and in the core of Coma (S. Trager, S. Faber, & A. Dressler, in preparation) appear to be coeval to within a few Gyr, with virtually no recent star formation, except in a few Virgo galaxies. This lack of detectable recent star formation is in agreement with color–magnitude studies of early-type cluster galaxies (e.g., Bower, Lucey, & Ellis 1992).

Cluster S0's. Any distinction between field and cluster S0's is less clearly defined than for ellipticals. The stellar populations of Fornax S0's (Kuntschner 2000) certainly span a large range in age (similar to the stellar populations of field S0's; Fisher, Franx, & Illingworth 1996), but those in the core of Coma (S. Trager et al., in preparation) appear to have a rather tight age distribution, similar to the cluster ellipticals. A detection of any possible environmental effect in the stellar populations of local S0's will

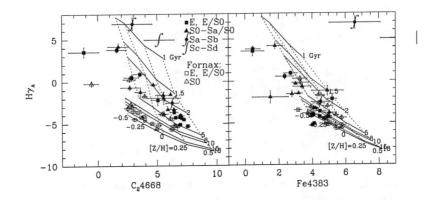

Figure 2.   Evolution of the absorption line strengths of cluster galaxies. Galaxies are coded by redshift (open points are Fornax Cluster galaxies at $z \approx 0$; solid points are galaxies in Abell 851 at $z = 0.41$) and morphology (squares are ellipticals and E/S0 transition cases; triangles are S0's, S0/Sa, and Sa/S0 transition cases; later-type galaxies are integral signs). Evolution is detected at the 5σ level in the Hβ strengths of early-type cluster galaxies, modulo uncertainties in the corrections needed to bring the Fornax galaxies to the same very large physical apertures as the distant galaxies and line strength system calibrations in the distant galaxies (see text).

require a much larger database of high-quality spectra than is currently available.

3.   The Evolution of the Stellar Populations of Cluster Galaxies

The lack of detectable recent star formation in the majority of cluster ellipticals suggests that such galaxies can be used as direct tracers of the evolution of the oldest stellar populations. In this section, we present the first results of our ongoing study of the evolution of the stellar populations of cluster galaxies (Trager 1997; S. Trager, S. Faber, & A. Dressler, in preparation).

Many techniques are available to detect and characterize the evolution of early-type galaxies. Here we concentrate on two methods: direct detection of the evolution of stellar population ages and the evolution of line strength–velocity dispersion relations. (Other methods, such as the evolution of cluster color-magnitude diagrams and the evolution of the morphology-density relation, are described elsewhere; see e.g., Lubin, these proceedings). We compare these two methods with the evolution of the Fundamental Plane of early-type galaxies in the next section.

We begin with the most ambitious method, direct detection of the evolution of stellar populations ages of early-type galaxies. Figure 2 shows our first attempt: a comparison of line strengths of early-type galaxies in Abell 851

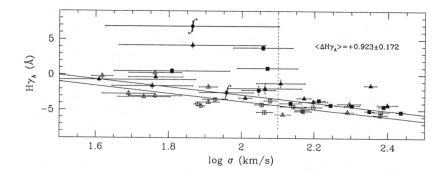

Figure 3.   Evolution of the Balmer line–velocity dispersion relation (cf. Kelson et al. 2001). Assuming all the evolution in Hγ_A is due to age, galaxies at $z \approx 0.4$ are 70% the age of Fornax galaxies of the same velocity dispersion (for typical stellar population models).

(=CL0939+4713), a rich cluster at $z = 0.406$, to those in Fornax. Focusing first on the elliptical galaxies (squares), the ellipticals in Abell 851 are about 40% the age of the Fornax ellipticals ($> 5\sigma$ significance). However, this result is subject to three major uncertainties. First, the distant galaxies are not necessarily on the same line strength system as the local galaxies; in particular, the $C_2$4668 line strengths may be systematically uncertain by up to 0.25 Å, and there may be overall calibration issues with the poorly understood Hγ_A index. Second, the model line strengths of Hγ_A appear to be too weak by about 1 Å on comparison with stellar population parameters derived from Hβ for Fornax galaxies. This offset appears to be due to uncertainties in the empirically-determined fitting functions (Worthey & Ottaviani 1997). While this offset has been corrected in Figure 2, its exact magnitude is still uncertain. Third, because all early-type galaxies have gradients in their line strengths (e.g., Davies, Sadler, & Peletier 1993), galaxies must be measured through the same physical aperture in order to directly compare their line strengths and thus their stellar populations. Unfortunately, the extraction aperture used for the distant galaxies ($1'' \times 2''.4$) projects to a very large aperture on the Fornax galaxies ($\sim 1' \times 2'$), much larger than the apertures through which accurate line strengths for local galaxies have been measured (a result anticipated by Kennicutt 1992). We are therefore forced to apply to the rather uncertain gradients of early-type galaxies to generate aperture corrections for the *local* galaxies. This is currently our largest uncertainty; high S/N, raster-scanned spectroscopy of local early-type galaxies (cf. Kennicutt 1992) will be required to resolve this issue.

Another method is to use the evolution of line strength–velocity dispersion relations. Although the pioneering work of Bender, Ziegler, & Bruzual (1996) used the Mg b–σ relation, using more age-sensitive indices like the high-order Balmer lines (Kelson et al. 2001) provides much better leverage on the evolution. Figure 3 shows the evolution of the Hγ_A–σ relation from $z = 0.41$ (Abell 851) to $z \approx 0$ (Fornax). Assuming that the evolution is *entirely due to age evolution at fixed velocity dispersion*—and not due to metallicity differences between cluster galaxies of the same velocity dispersion in the two clusters, to changes in the *slope* of the relation (if, say, the lower-mass galaxies are younger than the older galaxies; cf. Trager et al. 2000b), or some other manifestation of the age-metallicity degeneracy—then the galaxies in Abell 851 are 70% the age of the Fornax galaxies at fixed velocity dispersion using common stellar population models. This result has the same systematic uncertainties as the previous method—aperture corrections, line strength calibrations, and model uncertainties—and moreover, since only a single absorption line is used, the age-metallicity degeneracy can play a significant role.

4. Discussion and Conclusions

We summarize the results from these methods and the evolution of the Fundamental Plane (van Dokkum & Franx 2001; S. Trager et al., in preparation) in Figure 4; at the moment, these three methods are marginally inconsistent. We have tried to point out the various difficulties involved in measuring these evolutionary indicators: the line strengths required extremely high signal-to-noise and therefore efficient spectrographs on large telescopes, and systematic uncertainties may be large. Systematic uncertainties aside, another effect may be present. Small amounts of recent star formation can strengthen the Balmer lines far out of proportion to the actual mass involved in the burst (Trager et al. 2000b). A recent study of NUV-optical colors of galaxies in Abell 851 (Ferreras & Silk 2000) suggests that such recent star formation may have occurred in many of these objects. This might explain the enhanced $H\gamma_A$ strengths of the distant galaxies without constraints from the evolution of the Fundamental Plane of early-type galaxies (e.g., van Dokkum & Franx 2001), but this explanation appears to be inconsistent with the evolution of the $H\gamma_A$–σ relation.

In conclusion, we have *directly detected evolution* in the stellar populations of early-type cluster galaxies out to a redshift of $z = 0.41$. However, due to systematic uncertainties in measuring and comparing absorption line strengths, the exact amount of the evolution and the significance of this result is still unknown. Formation redshifts of $1 \lesssim z_f \lesssim 5$ are consistent with the current data (Figure 4), but a more precise answer awaits better calibration of both the distant *and* local data.

References

Baade, W. 1962, in Evolution of Stars and Galaxies, ed. C. Payne-Gaposchkin (Cambridge: MIT Press)

Bender, R., Ziegler, B., & Bruzual, G. 1996, ApJ, 463, L51

Blumenthal, G. R., Faber, S. M., Primack, J. R., & Rees, M. J. 1984, Nature, 311, 517

Bower, R. G., Lucey, J. R., & Ellis, R. S. 1992, MNRAS, 254, 601

Burstein, D. 1977, in The Evolution of Galaxies and Stellar Populations, ed. B. M. Tinsley & R. B. Larson (New Haven: Yale University Press), 191

Burstein, D., Faber, S. M., Gaskell, C. M., & Krumm, N. 1984, ApJ, 287, 586

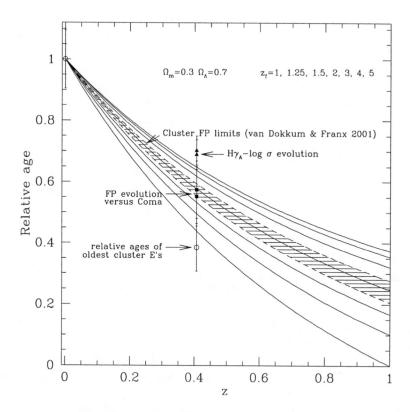

Figure 4.   Constraints on the formation redshift of early-type cluster galaxies from three stellar population evolutionary indicators. The
open circle represents the limits from the relative ages of the oldest
cluster ellipticals in Fornax ($z = 0$) and Abell 851 ($z = 0.41$) from
Figure 2, the solid triangles represent the limits from the evolution of
the $H\gamma_A$–σ relation from Fornax to Abell 851 from Figure 3 (two different stellar population models), and the solid squares represent the
evolution of the Fundamental Plane from Coma to Abell 851 (photometry from Ziegler et al. 1999; authors' own velocity dispersions). The
solid lines represent passively evolving stellar populations with formation redshifts $z_f = 1$, 1.25, 1.5, 2, 3, 4, and 5, from bottom to top.
The hatched region is the best fit from the evolution of the FP from
$z \approx 0$ to $z = 0.83$, including a correction-based model of morphological
transformations onto the FP (van Dokkum & Franx 2001).   Due to
uncertainties in aperture corrections and line strength system calibrations in Figures 2 and 3 and the age-metallicity degeneracy in Figures 3
and 4, the formation redshift of early-type cluster galaxies is still ill-
constrained, although $1 \lesssim z_f \lesssim 5$ is certainly reasonable, with $z_f \approx 2$
preferred by the FP studies.

Charlot, S., Worthey, G., & Bressan, A. 1996, ApJ, 457, 625

Davies, R. L., Sadler, E. M., & Peletier, R. F. 1993, MNRAS, 262, 650

Faber, S. M. 1977, in The Evolution of Galaxies and Stellar Populations, ed. B. M. Tinsley & R. B. Larson (New Haven: Yale University Press), 157

Ferreras, I., & Silk, J. 2000, ApJ, 541, L37

Fisher, D., Franx, M., & Illingworth, G. 1996, ApJ, 459, 110

Girardi, L., Bressan, A., Bertelli, G., & Chiosi, C. 2000, A&AS, 141, 371

González, J. J. 1993, Ph.D. Thesis, University of California, Santa Cruz

Jørgensen, I. 1997, MNRAS, 288, 161

Jørgensen, I. 1999, MNRAS, 306, 607

Kauffmann, G., White, S. D. M., & Guiderdoni, B. 1993, MNRAS, 264, 201

Kelson, D. D., Illingworth, G. D., Franx, M., & van Dokkum, P. G. 2001, ApJ, in press

Kennicutt, R. C. 1992, ApJS, 79, 255

Kuntschner, H. 2000, MNRAS, 315, 184

O'Connell, R. W. 1980, ApJ, 236, 430

O'Connell, R. W. 1986, in Stellar Populations, ed. C. A. Norman, A. Renzini, & M. Tosi (Cambridge: Cambridge University Press), 167

Rabin, D. 1982, ApJ, 261, 85

Rose, J. A. 1985, AJ, 90, 1927

Schweizer, F., & Seitzer, P. 1992, AJ, 104, 1039

Trager, S. C. 1997, Ph.D. Thesis, University of California, Santa Cruz

Trager, S. C. 1999, in ASP Conf. Ser. 191, Photometric Redshifts and High-Redshift Galaxies, ed. R. J. Weymann et al. (San Francisco: ASP), 195

Trager, S. C., Faber, S. M., Worthey, G., & González, J. J. 2000a, AJ, 119, 1645

Trager, S. C., Faber, S. M., Worthey, G., & González, J. J. 2000b, AJ, 120, 165

van Dokkum, P. G., & Franx, M. 2001, ApJ, in press

Worthey, G. 1994, ApJS, 95, 107

Worthey, G., & Ottaviani, D. L. 1997, ApJS, 111, 377

Ziegler, B. L., Saglia, R. P., Bender, R., Belloni, P., Greggio, L., & Seitz, S. 1999, A&A, 346, 13

Acknowledgments. SCT is grateful to D. Kelson for code and welcome criticism, to F. Schweizer for helpful conversations, to the organizers for a truly enjoyable meeting, and to the editors for their patience and understanding in allowing us to present a slightly more pessimistic review than the one delivered at the meeting. Support for this work was provided by NASA through Hubble Fellowship grant HF-01125.01-99A to SCT awarded by the Space Telescope Science Institute, which is operated by the Association of Universities for Research in Astronomy, Inc., for NASA under contract NAS 5-26555; by a Starr Fellowship to SCT; by a Carnegie Fellowship to SCT; by a Flintridge Foundation Fellowship to SCT; by NSF grant AST-9529098 to SMF; and by NASA contract NAS5-1661 to the WF/PC-I IDT.

Discussion

Young-Wook Lee: There is a poster by Hyun-chul Lee and myself where we found that the Hβ dating is seriously affected by the horizontal branch morphology and old populations. The models without the detailed modeling of the horizontal branch and metallicity distribution function, such as the one used in your analysis, would seriously underestimate the ages. What would you say?

Scott Trager: I think that those models don't apply to early-type galaxies in general, and the reason is that you need to move the red clump in the models entirely to the blue horizontal branch to explain the strong Hβ strengths we see, and that's ruled out from two observations. The first (and most compelling) is that M 32 has a very strong red clump and no apparent blue horizontal branch from WFPC2 color–magnitude diagrams. The second is that in the late 1980s and early 1990s, Jim Rose did a survey of 11 nearby early-type galaxies of the 4000 Å region and discovered that there was not enough blue light in the 4000 Å region for there to be more than a few percent by light of hot stars in the A–F range, and that rules out any significant blue horizontal branch which would affect the Hβ strength significantly. Now, granted there could be a small amount of blue horizontal branch stars from a very metal-poor population, but the modeling that we've done suggests that since you're limited to no more than 4 or 5% of the light in a hot population, it turns out that population must be tiny compared to other populations. So it could move galaxies from 13 Gyr to 8 Gyr old if you have a small frosting of blue horizontal branch stars, but it can't make galaxies as young as 2–4 Gyr; there aren't that many horizontal branch stars in M 32 or other young galaxies that we see nearby.

Ofer Lahav: You indicated that the concept of age is ill-defined. And we heard today that one has to be careful when talking about the merger history. The star formation history depends on your weighting scheme towards young or old, so I think this is probably a good forum to bring it up, whether it's useful at all to talk about ages of galaxies, or perhaps one should replace it by talking about a star formation function or merging function, or some better-defined quantity.

Scott Trager: I only the use the concept of age because it's a convenient concept that people understand—after I demonstrate that these ages are weighted to the most recent star formation epochs. But I think that there has to be another word and I don't know what it is—maybe star formation function. Guinevere Kauffmann this morning showed very nice plots of how these things can change as a function of time, and in her papers she has shown very clearly how merging really seriously affects our concept of dating these things. I hope that we've shown that at some level here, because we don't seriously believe the galaxies that we date at ages of 1.5–2 Gyr are really that young; we don't believe that you formed a 10^{11} $M_\odot$ galaxy at $z \sim 0.15$ from whole cloth. Unfortunately, this method does not allow us to probe distinct epochs of star formation yet. When we can learn to do that we can better indicate what the star formation *history* of these galaxies are, rather than what the typical star formation *time scale* of these galaxies are, and there's a distinct difference there that we don't know how to solve yet.

Astrophysical Ages and Time Scales
ASP Conference Series, Vol. 245, 2001
T. von Hippel, C. Simpson, N. Manset

Time Scale of Chemical Enrichment, Transport, and Mixing

Jean-René Roy

Gemini Observatory, 670 North A'ohōkū Place, Hilo, HI 96720, USA

Abstract. We can establish the time scales of various processes or the ages of features such as the bar instabilities in galaxy disks from the observed spatial fluctuations in the chemical abundances of the interstellar medium due to the efficiency of the various transport and mixing mechanisms in gas-rich galaxies. For example, turbulent transport in the shear flow of a differentially rotating disk will efficiently mix the interstellar gas of large disk galaxies so that azimuthal inhomogeneities will persist less than 1 Gyr. On scales of 1 kpc or less, mixing is achieved by gas motions generated by star formation processes that lead to full mixing in a few million years or less.

Astrophysical Ages and Time Scales
ASP Conference Series, Vol. 245, 2001
T. von Hippel, C. Simpson, N. Manset

Dating Stellar Populations in Dwarf Starbursting Galaxies

A. Aloisi, C. Leitherer

Space Telescope Science Institute, 3700 San Martin Drive, Baltimore, MD, USA

Abstract. We present the stellar content and star formation (SF) history of some nearby star-forming dwarf galaxies as derived by HST optical/NIR data. Our major finding is that these stellar systems contain an intermediate-age (> 1 Gyr) stellar component, even if their interstellar medium (ISM) is chemically poorly evolved. We thus conclude that they cannot represent genuine young galaxies in the local Universe.

1. Introduction

In hierarchical clustering scenarios dwarf galaxies represent the building blocks from which larger systems formed by merging. Late-type dwarfs have also been suggested to be the present-day examples of the faint blue galaxy (FBG) population identified at intermediate redshift ($z < 1$). It has been suggested that some dwarf star-forming systems (e.g., I Zw 18 with $Z \sim 1/50\ Z_\odot$) are local counterparts of primeval galaxies in the early Universe due to their relatively low level of chemical evolution. It is thus fundamental to assess if dwarf irregulars are "young" galaxies, forming their dominant stellar population at present (in the last 100 Myr), or "old" systems with a discontinuous and/or low level SF activity over a whole Hubble time.

2. Stellar Content and Star-Formation History

We have inferred the stellar content and SF history of some nearby ($D = 2$–10 Mpc) benchmark dwarf star-forming galaxies. Stars are dated by comparing the stellar distribution in the observed color-magnitude diagrams (CMDs) with the stellar evolution tracks at a certain metallicity. A more quantitative estimate of the SF rates (SFRs) and time scales involved are obtained by interpreting the CMDs with the method of synthetic diagrams.

At $D \sim 10$ Mpc I Zw 18 has been resolved into single stars only with HST. The interpretation of the same WFPC2 optical data is rather controversial. We have detected the presence of stars at least 0.5–1 Gyr old (see Figure 1), in contrast with previous results claiming the absence of a stellar population older than ~ 50 Myr (Aloisi, Tosi, & Greggio 1999). Östlin (2000) has strengthened our finding by dating the age of the oldest NIR stars seen with NICMOS at around 5 Gyr. I Zw 18 has been forming stars at a modest rate

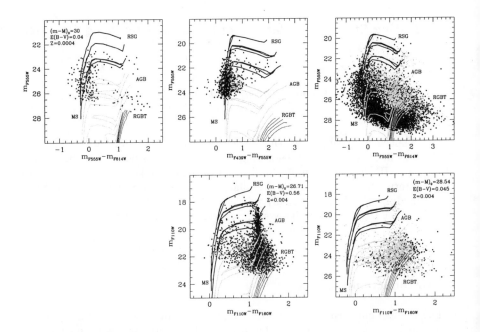

Figure 1.   Optical (*top*) and NIR (*bottom*) diagrams for I Zw 18 (*left*), NGC 1569 (*middle*), and NGC 1705 (*right*). We have considered stars with photometric error $\sigma < 0.2$ for I Zw 18, $\sigma < 0.1$ for NGC 1569 and with no error selection for NGC 1705. In the last case, lighter points are stars with a magnitude in all the four optical/NIR filters. Distance, extinction, and metallicity are indicated for each galaxy. From left to right, black tracks are high-mass stars ($M > 9\ M_\odot$, $\tau < 50$ Myr); grey tracks are intermediate-mass stars ($1.9\ M_\odot < M < 8\ M_\odot$, 50 Myr $< \tau < 1$ Gyr); dark grey tracks are low-mass stars ($M < 1.8\ M_\odot$, $\tau > 1$ Gyr).

(SFR $\sim 10^{-2}\ M_\odot$ yr^{-1} kpc^{-2}) since ~ 1 Gyr ago, with a burst ~ 10 times stronger around 15–20 Myr ago (Aloisi et al. 1999).

NGC 1569 is a gas-rich starburst at 2.2 Mpc resolved into single stars with WFPC2 by our group (Greggio et al. 1998). The optical CMD (Figure 1) indicates that in the last 0.15 Gyr the field has experienced an almost constant SF activity with an exceptionally high SFR ~ 0.5–$3\ M_\odot$ yr^{-1} kpc^{-2}, stopped ~ 5–10 Myr ago. We have observed the field around the two super star clusters (SSCs) with NICMOS (Aloisi et al. 2001). The NIR diagram (Figure 1) shows what we call the red plume at $m_{F110W} - m_{F160W} \sim 1.3$, populated by post-MS red supergiants (RSGs), asymptotic giant branch (AGB) and red giant branch (RGB) stars. The post-MS stars with $M > 9\ M_\odot$ allow us to infer a SFR over the last ~ 30 Myr consistent with the optical data. The presence of many AGB stars with $2 < M < 9\ M_\odot$ confirms a strong continuous SF from ~ 30 Myr to

~ 1 Gyr ago. The resolved objects with $M < 2 < M_\odot$ in their RGB phase indicate ages > 1 Gyr, thus an older SF event at a probably lower rate.

We have resolved NGC 1705 at $D \sim 5$ Mpc with HST-WFPC2 and NIC-MOS (Tosi et al. 2001). Optical/NIR CMDs (Figure 1) show the characteristic blue and red plumes. These features are less strong than in NGC 1569, despite the higher completeness factors. The stellar population of NGC 1705 has a well defined RGB tip, formed by stars with $M < 2 M_\odot$ older than 1 Gyr. The RGB tip allows us to calculate a new distance modulus $(m-M)_0 = 28.54$, corresponding to $D = 5.1$ Mpc. Preliminary results on the simulations of the $\sim 2\,000$ stars with magnitudes measured in all the four optical/NIR filters (Annibali et al. 2001) indicate that this galaxy has experienced a rather continuous SF activity back to several Gyr ago (SFR $\sim 5 \times 10^{-2}$ $M_\odot$ yr^{-1} kpc^{-2}), with a significant enhancement in the field SFR around 10–15 Myr ago.

3. Conclusions

HST observations of local dwarf star-forming galaxies reveal an intermediate-age stellar population ($\tau > 1$ Gyr). We are dealing with stellar systems that cannot be considered primordial. The scenario is a rather continuous SFR over the whole galaxy lifetime. The same conclusions derive from the chemical evolution modeling of I Zw 18 (Legrand 2000): the observed properties are well reproduced by a very low SF activity (SFR $\sim 10^{-4}$ $M_\odot$ yr^{-1}) over a whole Hubble time. Peaks of enhanced SF are not excluded, i.e., the bursts around 15;Myr ago in I Zw 18 and NGC 1705. SFRs in dwarf late-type galaxies are usually lower than required by the Babul & Ferguson (1996) model for the FBG population (SFR $\sim$ 1 $M_\odot$ yr^{-1}). Only NGC 1569 has a SFR compatible with this requirement, at least in the last 0.1 Gyr. Are these galaxies the equivalent at present epochs of the faint blue galaxies, or rather the local counterparts of the low-metallicity absorbing systems in the high-z universe (Legrand 2000)?

References

Aloisi, A., Tosi, M., & Greggio, L. 1999, AJ, 118, 302

Aloisi, A., et al. 2001, AJ, 121, 1425

Annibali, F., Aloisi, A., Greggio, L., Leitherer, C., & Tosi, M. 2001, in Dwarf Galaxies and their Environment, ed. K. de Boer, R. J. Dettmar, & U. Klein (Aachen: Shaker Verlag), in press

Babul, A., & Ferguson, H. C. 1996, ApJ, 458, 100

Greggio, L., Tosi, M., Clampin, M., De Marchi, G., Leitherer, C., Nota, A., & Sirianni, M. 1998, ApJ, 504, 725

Legrand, F. 2000, A&A, 354, 504

Östlin, G. 2000, ApJ, 535, L99

Tosi, M., Sabbi, E., Bellazzini, M., Aloisi, A., Greggio, L., Leitherer, C., & Montegriffo, P. 2001, AJ, submitted

Astrophysical Ages and Time Scales
ASP Conference Series, Vol. 245, 2001
T. von Hippel, C. Simpson, N. Manset

The Ages of Early-Type Galaxies From Infrared Surface Brightness Fluctuations[1]

Joseph B. Jensen

Gemini Observatory, 670 North A'ohōkū Place, Hilo, HI 96720, USA

John L. Tonry, Brian J. Barris, Michael C. Liu

Institute for Astronomy, University of Hawai'i, Honolulu, HI 96822, USA

Abstract. Surface brightness fluctuations (SBFs) are commonly used as a precision distance indicator. When the distance to a galaxy is known, SBFs can also be used to determine the absolute magnitude of the most luminous stars in a galaxy without resolving them individually. We have measured fluctuation brightnesses at 1.6 μm for 62 galaxies using NIC-MOS on the Hubble Space Telescope (HST). Independent distances to these galaxies were taken from I-band SBF and Cepheid variable star measurements. Absolute fluctuation magnitudes were compared to single-age stellar population models. The SBF magnitudes clearly indicate that the early-type galaxies in our sample span a range of ages from 2 to 12 or more Gyr. Because fluctuation measurements are sensitive to the youngest, most-luminous stars, only a fraction of the total stellar mass in a given galaxy may in fact be young. Our results indicate that many early-type galaxies have experienced star formation relatively recently.

1. Measuring Surface Brightness Fluctuations

Because galaxies are composed of discrete stars, Poisson statistics produce a mottling of the galaxy's surface brightness. Nearby galaxies have fewer stars per resolution element and therefore larger amplitude variations in surface brightness. Distant galaxies appear smooth compared to nearby ones. Measurements of the amplitude of the fluctuations can be used to determine the distance to an elliptical or S0 galaxy.

SBFs are measured by first fitting a smooth model to the galaxy light. The spatial power spectrum of the image is fitted with the power spectrum of the point spread function. The variance, in units of flux, is converted to an apparent fluctuation magnitude. The SBF magnitude is approximately equal to the magnitude of a typical red giant star in old stellar populations.

---

[1]Based on observations with the NASA/ESA Hubble Space Telescope, obtained at the Space Telescope Science Institute, which is operated by AURA, Inc., under NASA contract NAS 5-26555.

F160W (1.6 μm) fluctuation magnitudes were measured using images of 62 galaxies taken with NICMOS on the HST. Apparent fluctuation magnitudes were converted to absolute magnitudes using independent distances from I-band SBF data (Tonry et al. 2001) and HST observations of Cepheid variable stars (Ferrarese et al. 2000). The bluer galaxies have intrinsically brighter fluctuations. For galaxies without apparent dust, the data show a tight relationship between SBF magnitude and color (Figure 1).

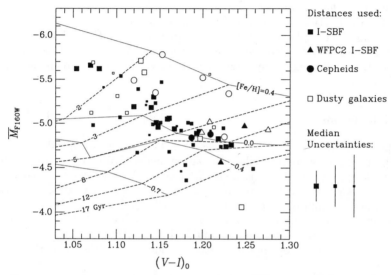

Figure 1.   Absolute fluctuation magnitude $\overline{M}_{F160W}$ plotted as a function of galaxy $(V-I)$ color. The largest symbols indicate galaxies with the smallest uncertainties (< 0.2 mag); open symbols indicate dusty galaxies. The stellar population models shown are those of Liu, Charlot, & Graham (2000).

2.   Stellar Population Models

Single-age, constant-metallicity models show that the degeneracy between age and metallicity seen in broad-band optical colors is broken using near-IR SBFs. The younger models predict brighter SBF magnitudes and bluer colors, because fluctuations are dominated by the brightest stars in a given population. Predicted fluctuation magnitudes are computed by integrating the model luminosity function, and are proportional to the second moment of the luminosity function normalized by the first moment. These predicted SBF magnitudes were derived from the Bruzual & Charlot models by Liu et al. (2000). The SBF measurements follow a trend that tracks roughly along lines of constant metallicity and nearly orthogonally to lines of constant age. The brightest stars in the bluest galaxies appear very young. Exactly the same conclusions can be drawn from comparisons to the Vazdekis models (Blakeslee, Vazdekis, & Ajhar 2001).

3.   Implications for Distance Measurements

Some of the data presented here were collected to better calibrate the F160W SBF distance scale. For galaxies redder than $(V-I) > 1.18$, the slope of absolute fluctuation magnitude with color is essentially flat at -4.82 mag, with an r.m.s. scatter of 0.14 mag. On the blue end, the data are best fit with the line $\overline{M}_{F160W} = -4.81 - 6.88[(V-I) - 1.18]$. The scatter on the blue end is larger (0.24 mag r.m.s.). F160W SBFs can still be used to reliably measure distances provided that the galaxy color is measured.

Absolute fluctuation magnitudes for these data were computed using a calibration based on Cepheid distances. The agreement with the stellar population models implies that the Cepheid distance scale cannot be systematically in error by more than ~ 0.3 mag. Any larger deviations would imply unrealistic ages or metallicities.

4.   Conclusions

1.   The bluer elliptical and S0 galaxies have fluctuation magnitudes that are consistent with very young stellar population models. *None* of the blue half of our sample of 62 galaxies is consistent with the very low metallicities that would be required for these to have old ages comparable to the red giant ellipticals. Only a fraction (approximately 10 to 20%) of the total stellar mass in a galaxy would have to be young to make the SBF magnitudes consistent with young single-age stellar population models (Liu & Graham 2001).

2.   The blue galaxies appear to have higher metallicities than the redder giant ellipticals.

3.   Using H-band SBFs for distance measurements requires knowing the galaxy's color. Galaxies bluer than $(V-I) = 1.18$ have significantly brighter fluctuations than the red giant ellipticals.

References

Blakeslee, J. P., Vazdekis, A., & Ajhar, E. A. 2001, MNRAS, 320, 193
Ferrarese, L., et al. 2000, ApJ, 529, 745
Liu, M. C., Charlot, S., & Graham, J. R. 2000, ApJ, 543, 644
Liu, M. C., & Graham, J. R. 2001, ApJ, submitted
Tonry, J. L., Dressler, A., Blakeslee, J. P., Ajhar, E. A., Fletcher, A. B., Luppino, G. A., Metzger, M. R., & Moore, C. B. 2001, ApJ, 546, 681

Acknowledgments.   We would like to thank the following NICMOS SBF Collaboration team members: Rodger Thompson, Marcia Rieke, Ed Ajhar, Tod Lauer, and Marc Postman. This research was supported in part by NASA grant GO-07453.0196A. J. Jensen acknowledges the support of the Gemini Observatory, which is operated by AURA, Inc. under a cooperative agreement with the NSF on behalf of the Gemini partnership.

Astrophysical Ages and Time Scales
ASP Conference Series, Vol. 245, 2001
T. von Hippel, C. Simpson, N. Manset

Age Estimation of Extragalactic Globular Cluster Systems Using Hβ Index

Hyun-chul Lee, Suk-Jin Yoon, Young-Wook Lee

Center for Space Astrophysics and Department of Astronomy, Yonsei University, 134 Shinchon, Seoul 120-749, Korea

Abstract. After taking into account, for the first time, the detailed systematic variation of horizontal-branch (HB) morphology with age and metallicity, our population synthesis models show that the integrated Hβ index is significantly affected by the presence of blue HB stars. Our models indicate that the strength of the Hβ index increases as much as 0.75 Å due to blue HB stars. According to our models, a systematic difference between the globular cluster system in the Milky Way Galaxy and that in NGC 1399 in the Hβ vs. Mg_2 diagram is understood if globular cluster systems in giant elliptical galaxies are a couple of billion years older, in the mean, than the Galactic counterpart.

1. Introduction

For distant stellar populations, one relies upon the integrated colors or spectra to investigate their ages and metallicities since individual stars are not resolved. Here, we specifically focus on the Hβ index, which is widely used as an age indicator. Most of the previous works (e.g., Worthey 1994), however, have been done on the basis that stars near the main sequence turnoff (MSTO) region are the most dominant sources for the integrated strength of Hβ. Consequently, without meticulous consideration for stars beyond the red giant branch, they claimed that the strength of Hβ depends on the location of the MSTO, which in turn depends on the age at a given metallicity. Several investigators, however, have cast some doubt upon the sensitivity of the Hβ index given the presence of other warm stars, especially blue horizontal-branch (HB) stars (see e.g., Burstein et al. 1984; Jorgensen 1997).

On the observational side, it was barely possible to obtain low signal-to-noise (S/N) spectra of globular clusters in systems outside the Local Group (e.g., Mould et al. 1990). These spectra have been useful only for kinematic information. With the advent of 10 m-class telescopes, however, Kissler-Patig et al. (1998) and Cohen, Blakeslee, & Ryzhov (1998) have successfully obtained relatively high S/N spectra that provide reliable line index calibration for globular clusters in NGC 1399 and M 87, the central giant elliptical galaxies in Fornax and Virgo clusters.

2.   Population Models With Horizontal-Branch Stars

For our population models, the Yale Isochrones (Demarque et al. 1996), rescaled for α-element enhancement (Salaris, Chieffi, & Straniero 1993), and the HB evolutionary tracks by Yi, Demarque, & Kim (1997) have been used. The Salpeter (1955) initial mass function is adopted for the relative number of stars along the isochrones. For the conversion from theoretical quantities to observable quantities, we have taken the most recently compiled stellar library of Lejeune, Cuisinier, & Buser (1998) in order to cover the largest possible ranges in stellar parameters such as metallicity, temperature, and gravity. The detailed calculation method of spectral index is presented in Lee, Yoon, & Lee (2000).

In Figure 1 (left), the variations of Hβ strength as a function of metallicity (Mg$_2$) are plotted at given ages. Here, $\Delta t = 0$ Gyr corresponds the recently favored mean age of Galactic globular clusters (~ 12 Gyr) in the light of new distance scale as suggested by HIPPARCOS (e.g., Chaboyer et al. 1998). Note that, unlike the models without HB stars (*dashed lines*), distinct "wavy" features appear in our models with HB stars (*solid lines*). It is found that the strength of Hβ does not simply decrease with either increasing age or increasing metallicity once HB stars are included in the models. Now it is evident that the blue HB stars around $(B - V)_o \sim 0$ are the key contributors for the strength of Hβ. The differences in Hβ strengths between the models with and without HB stars are as much as 0.75 Å at the peak. In addition, it should be noted that the peak of the Hβ enhancement moves to higher metallicity as age gets older.

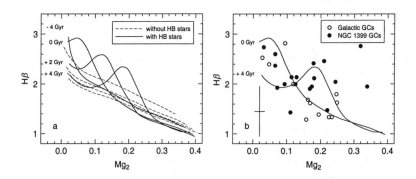

Figure 1.   *Left:* Effect of HB stars on the strength of Hβ as predicted from our models. *Right:* The comparison of Galactic globular clusters with those in NGC 1399. Observational errors are displayed for clusters in NGC 1399. Note that the overall distributions of the globular cluster system in NGC 1399 is different from that in the Milky Way Galaxy, indicating that giant elliptical galaxies may contain globular clusters that are a couple of billion years older, in the mean, than those in the Milky Way.

3. Comparison With Observations

Having confirmed that the detailed modeling of HB is crucial in the use of Hβ index as an age indicator, Figure 1 (right) compares our results with observations of globular cluster systems in the Milky Way Galaxy and in NGC 1399. It is important to note here that all of these observations were carried out at the Keck telescope with the identical instrumental configuration, the Low Resolution Imaging Spectrograph (LRIS; Oke, de Zeeuw, & Nemec 1995). Despite the still large observational uncertainties, it is inferred from Figure 1 (right) that the NGC 1399 globular cluster system is perhaps systematically couple of billion years older, in the mean, than the Galactic counterpart.

It is of considerable interest, in this respect, to find that a similar age difference is inferred from the "metal-poor HB solution" of the UV upturn phenomenon of local giant elliptical galaxies (Park & Lee 1997; Yi et al. 1999). If our age estimation is confirmed to be correct, this would indicate that the star formation in denser environments has proceeded much more rapidly and efficiently, so that the initial epoch of star formation in more massive (and denser) systems occurred several billion years earlier than that of the Milky Way (see also Lee 1992).

References

Burstein, D., Faber, S. M., Gaskell, C. M., & Krumm, N. 1984, ApJ, 287, 586

Chaboyer, B., Demarque, P., Kernan, P. J., & Krauss, L. M. 1998, ApJ, 494, 96

Cohen, J. G., Blakeslee, J. P., & Ryzhov, A. 1998, ApJ, 496, 808

Demarque, P., Chaboyer, B., Guenther, D., Pinsonneault, M., & Yi, S. 1996, Yale Isochrone 1996

Jorgensen, I. 1997, MNRAS, 288, 161

Kissler-Patig, M., Brodie, J. P., Schroder, L. L., Forbes, D. A., Grillmair, C. J., & Huchra, J. P. 1998, AJ, 115, 105

Lee, H.-c., Yoon, S.-J., & Lee, Y.-W. 2000, AJ, 120, 998

Lee, Y.-W. 1992, PASP, 104, 798

Lejeune, T., Cuisinier, F., & Buser, R. 1998, A&AS, 130, 65

Mould, J. R., Oke, J. B., de Zeeuw, P. T., & Nemec, J. M. 1990, AJ, 99, 1823

Oke, J. B., de Zeeuw, P. T., & Nemec, J. M. 1995, PASP, 99, 1823

Park, J.-H., & Lee, Y.-W. 1997, ApJ, 476, 28

Salaris, M., Chieffi, A., & Straniero, O. 1993, ApJ, 414, 580

Salpeter, E. E. 1955, ApJ, 121, 161

Worthey, G. 1994, ApJS, 95, 107

Yi, S., Demarque, P., & Kim, Y.-C. 1997, ApJ, 482, 677

Yi, S., Lee, Y.-W., Woo, J.-H., Park, J.-H., Demarque, P., & Oemler, A., Jr. 1999, ApJ, 513, 128

Acknowledgments. Hyun-chul Lee would like to thank the organizers—especially Ted—of this fantastic conference for their hospitality and for providing financial support.

Astrophysical Ages and Time Scales
ASP Conference Series, Vol. 245, 2001
T. von Hippel, C. Simpson, N. Manset

An Infrared View of Galactic Spheroid Formation

F. J. Masci, C. J. Lonsdale

Infrared Processing and Analysis Center, California Institute of Technology, 770 South Wilson Avenue, Pasadena, CA 91125, USA

R. C. Carlberg

Department of Astronomy, University of Toronto, Toronto, Ontario M5S 3H8, Canada

Abstract. A popular theory for the formation of galactic spheroids involves the infall of gas-rich satellites onto pre-existing dark matter halos. We have explored predictions on counts, colors, and star formation rates in the mid-to-far infrared from this scenario. The upcoming Space Infrared Telescope Facility (SIRTF) mission will provide strong constraints on spheroid evolution and its contribution to the cosmic infrared background and star formation history.

1. Introduction

A number of theories exist for the formation of galactic spheroids and ellipticals: primordial monolithic collapse of individual gas clumps (Eggen, Lynden-Bell, & Sandage 1962), hierarchical merging of pre-formed galaxies (Toomre & Toomre 1972), infall of gas-rich satellites onto pre-existing dark matter disk halos (Cole et al. 1994; Carlberg 1999), and secular evolution where bulges form relatively late by gas inflow from their pre-existing gas-rich outer disk (Norman, Sellwood, & Hasan 1996). A conclusion has yet to be reached, but the observed properties of ellipticals and bulges of spirals strongly support a merger (or satellite accretion) hypothesis as predicted by standard hierarchical models of galaxy formation.

We present here predictions of number counts expected in the SIRTF (MIPS) bandpasses using a scenario where spheroids form by the continuous accretion of gas-rich satellites. Within a CDM framework, a simple model involving wind-regulated accretion of gas-rich satellites was proposed by Carlberg (1999). The model has attained considerable success at explaining some general bulge properties: the Kormendy (density–size) relations, residual angular momentum distributions, and mass–metallicity correlations. Complete details will appear in a forthcoming paper (F. J. Masci et al., in preparation).

2. An Empirical Model for Spheroid Formation

The model (Carlberg 1999) uses a Monte Carlo simulation and takes as input the following: a merger rate (fixed empirically from studies of galaxy clustering),

448

a mass spectrum for the pre-accreted satellites provided by the standard CDM paradigm, and stellar feedback (wind) parameters for determining gas stripping fractions.

Assumptions for deriving far-IR source counts—essentially the merger induced starburst events integrated over time for all forming bulges—are as follows:

1. The final, fully assembled bulges are normalized to the total space density of local bulges (in spirals, S0, E-type, etc.) in the optical. This fixes the comoving volume of the simulation.

2. The star formation rate is given by SFR $= M_{\mathrm{acc}}/\Delta t$ with burst timescale $\Delta t \simeq 10^7$ yr. M_{acc} is the total accreted gas mass.

3. The above SFR is modified by a redshift dependent efficiency parameter

$$\epsilon(z) = 1 - [1 - \epsilon(0)] \exp(-\beta z), \qquad (1)$$

where $\epsilon(0)$ (the local SF efficiency) and β are model dependent parameters.

4. The far-IR luminosity is linked to the above SFR through an empirical relation derived locally by Smith, Lonsdale, & Lonsdale (1998). A Miller & Scalo IMF is assumed with mass range $1\ M_\odot < M < 100\ M_\odot$.

5. An IR SED library for starbursts from the spectrophotometric models of Devriendt, Guiderdoni, & Sadat (1999) is assumed for the k-corrections. A dependence of SED shape on IR luminosity that reflects observed IRAS colors is included.

While assumptions concerning burst timescales and the IMF (2 and 4) are weakly constrained by observations, our model is relatively insensitive to these parameters. Assumption 3 however, concerning the efficiency at which gas is converted into stars as a function of z has a significant affect on the relative number of faint sources predicted.

3. Results

The deep ISO (15 μm) surveys constitute the best available statistics for probing evolution within the range $0 < z \lesssim 1.3$ (Elbaz et al. 1999). In Figure 1 we have assumed a 'minimal' and 'maximal' model defined by two different forms for evolution of the SF efficiency (Equation 1). These correspond to respectively factors of ~ 6.7 and ~ 17 increase in SF efficiency from $z = 0$ to $z = 1$. An open cosmology with $\Omega = 0.3$, $\Lambda = 0$ and $H_0 = 65$ km s^{-1} Mpc^{-1} is assumed.

4. Conclusion

The main conclusion is that evolution in *both* merger rate and star formation efficiency is required, consistent with findings by Roche & Eales (1999) and Blain et al. (1999). We also predict a contribution of $\sim 70\%$ to the IR background at 60 μm and $\sim 10\%$ and $\sim 50\%$ of the global star formation rate density at $z = 0$ and $z = 1.5$, respectively, from bulge-building alone.

450   *Masci, Lonsdale, & Carlberg*

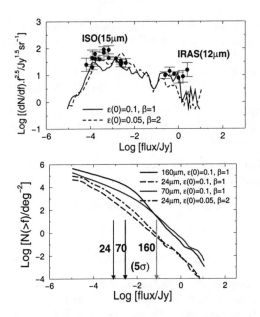

Figure 1.   *Top:* Euclidean normalised 15 μm differential counts with data from Elbaz et al. (1999) (15 μm) and Spinoglio et al. (1995) (12 μm). *Bottom:* Integral counts in SIRTF's MIPS bands. Vertical arrows indicate 5σ sensitivities.

References

Blain, A. W., Jameson, A., Smail, I., Longair, M. S., Kneib, J.-P., Ivison, R. J. 1999, MNRAS, 309, 715

Carlberg, R. C. 1999, in The Formation of Galactic Bulges, ed. C. M. Carollo, H. C. Ferguson, & R. F. G. Wyse (Cambridge: Cambridge University Press), 64

Cole, S., Aragón-Salamanca, A., Frenk, C. S., Navarro, J. F., & Zepf, S. E. 1994, MNRAS, 271, 781

Devriendt, J. E. G., Guiderdoni, B., & Sadat, R. 1999, A&A, 350, 381

Eggen, O., Lynden-Bell, D., & Sandage, A. 1962, ApJ, 136, 748

Elbaz, D., et al. 1999, A&A, 351, L37

Norman, C. A., Sellwood, J. A., & Hasan, H. 1996, ApJ, 462, 114

Roche, N., & Eales, S. A. 1999, MNRAS, 307, 111

Smith, H. E., Lonsdale, C. J., & Lonsdale, C. J. 1998, ApJ, 492, 137

Spinoglio, L., Malkan, M. A., Rush, B., Carrasco, L., & Recillas-Cruz, E. 1995, ApJ, 453, 616

Toomre, A., & Toomre, J. 1972, ApJ, 178, 623

Astrophysical Ages and Time Scales
ASP Conference Series, Vol. 245, 2001
T. von Hippel, C. Simpson, N. Manset

Time Scales of Dynamical Processes in the Intergalactic Medium

M. Rauch

Carnegie Observatories, 813 Santa Barbara Street, Pasadena, CA 91101, USA

W. L. W. Sargent, T. A. Barlow

105-24 California Institute of Technology, Pasadena, CA 91125, USA

Abstract. Measurements of absorption pattern differences produced by high redshift galactic and intergalactic gas on multiple lines of sight to gravitationally-lensed QSOs can be used to investigate the nature and time scales of hydrodynamic disturbances in early galaxies and the intergalactic medium (IGM). The relative differences between the absorption systems as a function of projected separation on the sky constrains the rate of energy input into the IGM and the frequency of recurrent star forming events at high redshift. Both the amplitude of the turbulence and the coherence length of the clouds are consistent with the clouds being produced, stirred, or destroyed by star formation or merger-triggered gasdynamics on a time scale of 10^7–10^8 yr. The results are based on a survey of absorption systems in lensed $z \sim 3$ QSOs with Keck HIRES.

1. Using a Gravitational Microscope

Gravitational lenses are capable of hugely magnifying the angular extent of a background object at cosmological distances. We are using this effect to detect differences between the absorption lines in Keck HIRES spectra of multiple images of lensed background QSOs. It is possible to spatially resolve proper scales down to tens of parsecs at redshift $z \sim 3$. Which differences we see in absorption depends of course strongly on the astrophysical environment the lines of sight are intersecting. Sometimes these are galaxies, most often, however, the absorbers must arise in low (column) density condensations in the intergalactic medium (the so-called Lyman α forest), which may be subject to feedback from nearby galaxies. Observations, interpreted with the help of cosmological hydrosimulations indicate that there is a widely distributed high ionization gas phase with C IV as its most easily detected metal ion.

Already at redshifts 3–4 the high ionization phase is widely polluted with metals, albeit at a low level ($Z \sim 10^{-2.5}\ Z_\odot$; Cowie et al. 1995; Tytler et al. 1995; Womble, Sargent, & Lyons 1996). The origin of the metals is currently not clear. They may be the product of an ancient Population III epoch of nucleosynthesis, raining down on currently forming galaxies, or they may have been expelled more recently, e.g., by winds or SN explosions from nearby galaxies.

451

2. Cosmic Seismometry of the High Ionization Gas Phase

One of the first results of our surveys was that the C IV gas clouds have a finite size and show little structure over regions smaller than a few hundred parsecs. This tells us that the gas, though metal-enriched, is not currently subject to violent hydrodynamic processes as we know them from the interstellar medium.

The finite size of the clouds and the known speed of sound in the IGM allow us to perform "cosmic seismometry," i.e., attempt to measure the frequency of hydrodynamic disturbances in the IGM. Assume that there are recurrent events (SN explosions, stellar winds, mergers) stirring up the C IV-enriched gas with a typical interval τ_s between two such occurrences. Any density gradients in the C IV gas caused by such processes would be damped out by pressure waves propagating with the speed of sound over a spatial distance r, given by the product of the sound crossing time, τ_s, and the sound speed, c_s. Conversely, if there is little structure over a distance r (and we have measured such a minimum size of a few hundred parsecs) then there cannot have been a hydrodynamic disturbance during the past

$$\tau_s \sim \frac{r}{c_s} \approx 1.4 \times 10^7 \left(\frac{r}{300 \text{ pc}}\right) \left(\frac{c_s}{20 \text{ km s}^{-1}}\right)^{-1} \text{yr.} \qquad (1)$$

3. Turbulence and Energy Contents of the C IV Gas

The data allow us to obtain a crude estimate of the turbulent energy in the gas. From the Kolmogorov theory we borrow the notion of a rate of energy transfer, ϵ (in erg g^{-1} s^{-1} or cm^2 s^{-3}). While the Kolmogorov approach may not be applicable (because, e.g., the gas is compressible, may carry magnetic fields, and the energy input may be intermittent), the usual dimensional analysis connecting the energy transfer rate, ϵ, and the r.m.s. velocity v_s between points with spatial separation s, $\epsilon \sim v_s^3/s$, remains valid. The customarily used structure function,

$$B(s) \approx \overline{[v(s') - v(s'')]^2} \approx \overline{v_s^2} \approx (\epsilon s)^{2/3} \qquad (2)$$

(e.g., Kaplan & Pikelner 1970), can be measured from the pairs of C IV column density weighted line of sight velocities, as a function of projected beam separation (the average is taken over all points s' and s'' with separation s).

Taking $s = 300$ pc as a reference point, $B(s) = 100$ km^2 s^{-2}, and the energy transfer rate is found to be

$$\epsilon \sim 10^{-3} \text{ cm}^2 \text{ s}^{-3}. \qquad (3)$$

This is considerably less than values measured, e.g., for the Orion nebula, where $\epsilon \sim 0.1$–1 cm^2 s^{-3} (Kaplan & Pikelner 1970), but it is comparable to the global rate of energy input into the Galactic ISM. Most of the C IV systems appear to consist of rather quiescent gas, unlike any known Galactic environments. C IV absorbing clouds are too smooth for their size to be, e.g., galactic H II regions.

The observed velocity scatter is small but finite. Could it be due to be residual turbulence from an earlier phase of metal ejection? In a steady state, the energy transfer rate ϵ is not only the rate at which energy is fed to the gas, but it also equals the dissipation rate. We can estimate the approximate time scale for dissipation, i.e., the time it takes to transform the mean kinetic energy $1/2\langle v^2 \rangle$ in the gas, at a rate ϵ into heat,

$$\tau_{\text{diss}} \sim \frac{1}{2} \frac{\langle v^2 \rangle}{\epsilon} \sim 9 \times 10^7 \text{ yr.} \tag{4}$$

Apparently, energy is being dissipated at a rate fast enough to destroy turbulence on a time scale of 100 million years. Thus the turbulence we observe near the mean redshift $\langle z \rangle \sim 2.7$ of our sample must have been produced close to the epoch of observation. It is tempting to speculate that this may also be true for a large fraction of the metals themselves. This does not preclude the possibility that the metal enrichment started already at much higher redshifts during a population III phase (at $z > 7$; e.g., Gnedin & Ostriker 1997), but the kinematical signature of such an ancient process would have been lost by now.

The two time scales proposed here agree to within a factor of a few. We tentatively interpret these estimates as measuring the frequency of disturbing events (perhaps starbursts, tidal interactions, or mergers) in the vicinity of the absorbing gas. They indicate that the regions giving rise to C IV absorption, while not *currently* experiencing gasdynamical disturbances, may have done so recently and may continue to do so on a time scale of 10^7–10^8 yr. Interestingly, this is comparable to several time scales that have been identified in the local Universe: e.g., lifetimes of molecular clouds (Shu, Adams, & Lizano 1987), enhancements in the rate of star formation (e.g., Tomita, Tomita, & Saito 1996; Glazebrook et al. 1999; Rocha-Pinto et al. 2000); dynamical ages of superbubbles (Martin 1998); or mergers (Gnedin 1998; Menci & Valdarnini 1994).

References

Cowie, L. L., Songaila, A., Kim, T.-S., & Hu, E. M. 1995, AJ, 109, 1522

Glazebrook, K., Blake, C., Economou, F., Lilly, S., & Colless, M. 1999, MNRAS, 306, 843

Gnedin, N. Y. 1998, MNRAS, 294, 407

Gnedin, N. Y., & Ostriker, J. P. 1997, ApJ, 486, 581

Kaplan, S. A., & Pikelner, S. B. 1970, The Interstellar Medium (Cambridge: Harvard University Press)

Martin, C. L. 1998, ApJ, 506, 222

Menci, N., & Valdarnini, R. 1994, ApJ, 436, 559

Rocha-Pinto, H. J., Scalo, J., Maciel, W. J., & Flynn, C. 2000, ApJ, 531, L115

Shu, F. H., Adams, F. C., & Lizano, S. 1987, ARA&A, 25, 23

Tomita, A., Tomita, Y., & Saito, M. 1996, PASJ, 48, 285

Tytler, D., Fan, X.-M., Burles, S., Cottrell, L., Davis, C., Kirkman, D., & Zuo, L. 1995, in QSO Absorption Lines, ed. G. Meylan (Berlin: Springer-Verlag), 289

Womble, D. S., Sargent, W. L. W., & Lyons, R. S. 1996. in Cold Gas at High Redshift, ed. M. N. Bremer (Dordrecht: Kluwer), 137

Astrophysical Ages and Time Scales
ASP Conference Series, Vol. 245, 2001
T. von Hippel, C. Simpson, N. Manset

Near IR Imaging and Spectroscopy of an Intermediate Mass Black Hole Candidate in M 82

T. Usuda, N. Kobayashi, H. Terada

Subaru Telescope, 650 North A'ohōkū Place, Hilo, HI 96720, USA

S. Matsushita

Harvard–Smithsonian Center for Astrophysics, 60 Keawe Street, Hilo, HI 96720, USA

T. G. Tsuru, T. Harashima

Kyoto University, Sakyo-ku, Kyoto, Japan

T. Matsumoto

Center for Space Research, Massachusetts Institute of Technology, 77 Massachusetts Avenue, Cambridge, MA, USA

R. Kawabe

Nobeyama Radio Observatory, Minamisaku, Nagano, Japan

M. Goto, A. Tokunaga

Institute for Astronomy, University of Hawai'i, 2680 Woodlawn Drive, Honolulu, HI 96822, USA

Abstract. We performed Brγ (2.166 μm) and [Fe II] (1.644 μm) narrow-band ($R \simeq 100$) imaging observations with IRCS on the Subaru Telescope, and showed that Brγ emission exists in a molecular superbubble (MSB; Matsushita et al. 2000) and is accompanied with energetic starburst activity and explosions. The distribution of [Fe II] emission is not associated with the MSB, although the line is a tracer of shock regions. It may be due to high extinction. There exist 2 μm peaks in the central region of M 82, which are massive stellar clusters. One of them agrees with a X-ray point source that is an Intermediate Massive Black Hole (IMBH; Matsumoto et al. 2001) candidate. We acquired IRCS K-band spectra ($R = 300$–400) through three peaks, including the 2 μm counterpart of the IMBH. No broad Brγ or He I emission lines were detected.

1. Introduction

Ongoing star formation can profoundly affect a galaxy's evolution by the enormous radiative and kinematic energy from massive stars and their progeny. It produces an expanding superbubble and large scale outflows, which may trigger the next generation of star formation. Starburst phenomena have a relation with AGNs in terms of the formation of a supermassive black hole and gas fueling. Norman & Scoville (1988) proposed a model for AGNs in which a massive central star cluster builds up and feeds a central black hole as a result of mass loss during post-main-sequence stellar evolution.

The issues of the origin and evolution of galactic activity, including starburst phenomena and AGNs, have not yet been understood clearly. One reason is that we can obtain only spatially integrated informations of external galaxies. Recently, HST and large telescopes like Subaru enable us to obtain spatially-resolved images. Since 0.6 arcsec, which is average seeing size of Subaru, corresponds to ~ 7 pc in the nearby starburst galaxy M 82 ($D = 3$ Mpc), the higher spatial resolution can resolve each massive stellar cluster.

2.   Observations and Data Reduction

Near IR imaging observations of M 82 were carried out on 2000 February 29 using the Subaru Telescope at the summit of Mauna Kea, with the Infrared Camera and Spectrograph (IRCS; Kobayashi et al. 2000) mounted at the $f/12$ Cassegrain focus. We obtained continuum-subtracted Brγ and [Fe II] (1.644 μm) images, using narrow band filters ($R \simeq 100$). The pixel scale was $0''.058$ pixel^{-1} and the field of view was $1'.0 \times 1'.0$. Under a seeing condition of typically $0''.6$ (FWHM), nine exposures of 20 s each were obtained at the two narrow bands. The flat-field frames for the narrow band images were constructed from median-filtered • images of the corresponding sky frames. The K-band spectra were obtained on 2000 December 2 with the Subaru Telescope utilizing the K-grism spectroscopy mode of IRCS, which provides a spectral resolution of 300–400. A $0''.6$ wide slit, aligned east–west, was set at the position of the 2.2 μm secondary peak (Dietz et al. 1986). All the data were reduced by using IRAF.

3.   Results and Discussion

Figure 1 shows continuum subtracted Brγ and [Fe II] maps, on which are superimposed contours of ^{12}CO J=1–0 as imaged with the Nobeyama Millimeter Array (Matsushita et al. 2000). The emission is resolved into some which were not seen in the maps obtained by Satyapal et al. (1995) and Greenhouse et al. (1997). In comparing the distribution of Brγ and [Fe II] with ^{12}CO, the Brγ is strong to the southwest of the galactic center and its peak positions are associated with the expanding molecular superbubble (MSB). The observational evidence suggests that the localized starbursts occurred around the position of 2.2 μm secondary peak ($M \sim 10^7 \ M_\odot$; T. Harashima et al., in preparation) and formed giant H II regions and MSB. On the other hand, the distribution of [Fe II] emission is not associated with MSB, although the line is a tracer of shock regions. It may be due to highly extinction since the visual extinction is estimated to be ~ 25 mag (Puxley 1991).

We obtained K-band spectra of the three 2 μm peaks, namely the secondary peak ("#9" defined in Satyapal et al. 1997), the possible 2 μm counterpart of the IMBH, "#11" (Satyapal et al. 1997), and the 2 μm peak near the ^{12}CO peak ("CO"). We conclude the following two results from the spectra: (1) No broad components (> 1000 km s^{-1}) were detected in Brγ and He I emission, and (2) the age of "CO" cluster is estimated to be 10^{6-7} years and younger than at the other two peaks ("#9" and "#11", estimated to be 10^{7-8} years), from comparing the equivalent width of CO and Brγ with theoretical model (Salpeter IMF, exponential star formation rate, upper mass limit = 30 $M_\odot$, lower mass limit = 1 $M_\odot$). It implies that the superbubble may be producing sequential star formation as it expands.

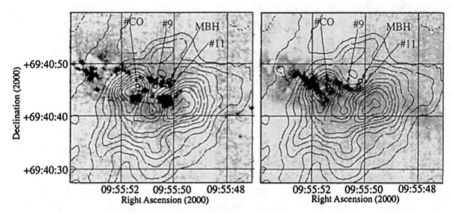

Figure 1.   Brγ (*left*) and [Fe II] (*right*) distributions (grayscale) with ^{12}CO molecular superbubble (contours).

References

Dietz, R. D., Smith, J., Hackwell, J. A., Gehrz, R. D., & Grasdalen, G. L. 1986, AJ, 91, 758

Greenhouse, M. A., et al. 1997, ApJ, 476, 105

Kobayashi, N. et al. 2000, in Proc. SPIE 4008, Optical and IR Telescope Instrumentation and Detectors, ed. M. Iye & A. F. Moorwood (Bellingham: SPIE), 1056

Matsumoto, H., et al. 2001, ApJ, 547, L25

Matsushita, S., Kawabe, R., Matsumoto, H., Tsuru, T. G., Kohno, K., Morita, K.-I., Okumura, S. K., & Vila-Vilaró, B. 2000, ApJ, 545, L107

Norman, C., & Scoville, N. 1988, ApJ, 332, 124

Puxley, P. J. 1991, MNRAS, 249, 11P

Satyapal, S., Watson, D. M., Pipher, J. L., Forrest, W. J., Greenhouse, M. A., Smith, H. A., Fischer, J., & Woodward, C. E. 1997, ApJ, 483, 148

Satyapal, S., et al. 1995, ApJ, 448, 611

Acknowledgments. We would like to thank all members of the Subaru Telescope project team and commissioning teams of Mitsubishi Electric Co. Ltd. and Fujitsu Co. Ltd. for their help in the telescope operation.

Astrophysical Ages and Time Scales
ASP Conference Series, Vol. 245, 2001
T. von Hippel, C. Simpson, N. Manset

Spectro-Chemical Evolution of Low Surface Brightness Galaxies

Gerardo A. Vázquez, J. Jesús González, Leticia Carigi

Instituto de Astronomía, UNAM, Apdo Postal 70-264 CP 04510, México D.F., México

Abstract. The star formation rate (SFR) of low surface brightness galaxies (LSBg) is explored through spectro-chemical evolution models. Models are constrained by present-day SFR, gas fraction μ, and [O/H] abundance. A good agreement with mean broad band colors is obtained with a 14 Gyr-old, exponentially-increasing SFR ($\tau_{\rm SFR} = 4$ Gyr).

1. Introduction

Three different scenarios of star formation (SF) have been suggested for LSBg:

(i) Recent SF. Important contribution in the blue bands produced by young population as suggested by de Block, van der Hulst, & Bothun (1995).

(ii) Different episodes of SF. Quillen & Pickering (1997) discover old stellar population near the nuclear region of LSBg.

(iii) Constant or increasing SF. The presence of old stellar population any way in LSBg has been suggested by different authors (e.g., Jimenez et al. 1998; Gerritsen & de Block 1999; Bell et al. 2000).

This work tests the ability of each scenario to reproduce the mean observed colors of a sample LSBg (van den Hoek et al. 2000) predicted with a spectro-chemical code. Although in this work we show just the results from the these matched model, we suggest how the models change with different values of constraints.

2. Models

Simple closed-box with instantaneous recycling approximation chemical models using yields from Maeder at $Z = 0.001$ are obtained with the code from Vázquez (2001). We use an initial mass function $\propto m^{-2.35}$ in the range $0.1 \leq M/M_\odot \leq 60.0$, and explore three different SFRs: exponentially-increasing, exponentially-decreasing, and constant.

The constrains are taken from van den Hoek et al. (2000):

$$\text{Constraints} \quad \begin{array}{l} \text{Present-day-SFR} \sim 0.02\text{–}0.8 M_\odot \text{ yr}^{-1} \\ M_{\rm gas}/M_{\rm baryonic} = \mu \sim 0.5 - 0.9 \\ \text{[O/H]} \sim -0.6 \end{array} \qquad (1)$$

He have used the mean value of [O/H] and varied the other constraints to generate the models.

The spectral part was covered by the evolutionary synthesis code from G. A. Vázquez, L. Carigi, & J. J. González (in preparation), which considers full integration of population properties following detailed enrichment and SF history from chemical modeling. The code uses Geneva tracks (Charbonnel et al. 1999 and references therein) and the spectral library developed by Lejeune, Cuisinier, & Buser (1997, 1998).

3. Results

With the help of Table 1, we explain in a brief way how the models change with different constrains and parameters (duration of SF τ_{SFR}, end of SF t, and initial mass M_i). All three models may be represented by $e^{-t/\tau_{SFR}}$.

Table 1.   Chemical models.

Model	Ψ_0 $(M_\odot\ yr^{-1})$	M_i $(M_\odot)$	μ	τ (Gyr)	t (Gyr)	$\Psi(t)$ $(M_\odot\ yr^{-1})$	[O/H]
1	8.276(−3)	5.0×10^{09}	0.8	−4.0	14.0	0.2741	−0.2539
2	8.276(−3)	10^{10}	0.9	−4.0	14.0	0.2741	−0.5831
3	0.1132	5.0×10^{09}	0.9	+5.0	14.0	6.884(−3)	−0.5831
4	0.1108	5.0×10^{09}	0.9	+5.0	16.0	4.518(−3)	−0.5831
5	0.1000	5.0×10^{09}	0.7	∞	16.0	0.1000	−0.0464
6	0.1063	5.0×10^{09}	0.7	∞	15.0	0.1063	−0.0464

As we can see from Table 1, the last two columns are the predicted present-day (t) values, all models have high values for μ because models with more consumption produce values for [O/H] higher than the mean value assumed here. Models 3 and 4 with high Ψ_0 do not match the present-day SF and models 5 and 6 do not produce the correct value of [O/H]. We chosen times around age of the Universe. Model 1 does not match the mean value for [O/H] and the best model is 2.

Predictions of best model are compared with mean colors of a LSBg set (van den Hoek et al. 2000) observed by de Block, McGaugh, & van der Hulst (1996) and parameters from the chemical model. Figure 1 compares theoretical and observed colors: a) B vs. $(B-V)$, b) B vs. $(R-I)$, c) B vs. [O/H], and d) B vs. $\log M_{gas}$.

4. Conclusions

A model with an exponentially increasing SFR, low enrichment ([O/H] ~ -0.6), and low gass mass consumption ($\mu = 0.9$) reproduces well the observed B, $(B-V)$, and $(R-I)$ of LSBg. The gas mass is overestimated but the inclusion of winds in chemical models may fit the current values of these galaxies.

These results support the scenario (iii) mentioned in the introduction, that predicts also a low metallicity.

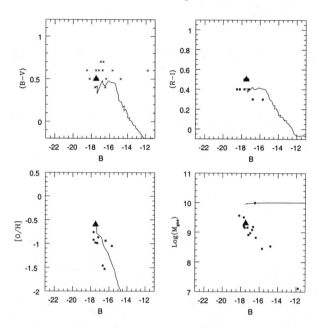

Figure 1.   Predicted evolution up to 14 Gyr (close to the center of each plot). Large triangles represent mean LSBg values. Small symbols indicate individual LSB galaxies (van den Hoek et al. 2000).

References

Bell, E. F., Barnaby, D., Bower, R. G., de Jong, R. S., Harper, D. A., Hereld, M., Loewenstein, R. F., & Rauscher, B. J. 2000, MNRAS, 312, 470

Charbonnel, C., Däpen, W., Schaerer, D., Bernasconi, P. A., Maeder, A., Meynet, G., & Mowlavi, N. 1999, A&AS, 135, 405

de Block, W. J. G., McGaugh, S. S., & van der Hulst, J. M. 1996, MNRAS, 283, 18

de Block, W. J. G., van der Hulst, J. M., & Bothun, G. D. 1995, MNRAS, 274, 235

Guerritsen, J. P. E., & de Block, W. J. G. 1999, A&A, 342, 655

Jimenez, R., Padoan, P., Matteucci, F., & Heavens, A. F. 1998, MNRAS, 299, 123

Lejeune, T., Cuisinier, F., & Buser R. 1997, A&AS, 125, 229

Lejeune, T., Cuisinier, F., & Buser R. 1998, A&AS, 130, 65

Quillen, A. C., & Pickering, T. E. 1997 (astro-ph/9705115)

van den Hoek, L. B., de Block, W. J. G., van der Hulst J. M., & de Jong, T. 2000, A&AS, 357

Vázquez, G. A. 2001, Ph.D. Thesis, Instituto de Astronomía, UNAM

Astrophysical Ages and Time Scales
ASP Conference Series, Vol. 245, 2001
T. von Hippel, C. Simpson, N. Manset

Time Scales and the Formation of Disk Galaxies

P. R. Williams

Astronomical Institute, Tohoku University, Aoba, Sendai 980–8578, Japan and Department of Physics and Astronomy, Cardiff University, P.O. Box 913, Wales, UK

A. H. Nelson

Department of Physics and Astronomy, Cardiff University, P.O. Box 913, Wales, UK

Abstract. Analysis of high resolution numerical simulations of the formation of grand design spiral galaxies has suggested the idea that the general process of disk galaxy formation by gravitational collapse may be usefully divided into phases. In such a picture, each phase of evolution is characterized by the dominant physical processes apparent during that epoch. Such a five phase paradigm may be equivalently considered in terms of the time scales associated with these processes.

The initial conditions used were a uniform density sphere of total mass 5×10^{11} $M_\odot$, initially 10% gas and 90% collisionless dark matter by mass, in solid body rotation with an angular velocity of 0.161 Gyr^{-1}, and with a radial velocity directly proportional to radius. The initial configuration had a radius of 175 kpc, and was expanding at a rate of 560 km s^{-1} Mpc^{-1} measured from the centre of the sphere, resulting in a protogalaxy with spin parameter of $\lambda_i = GJ^{-1}|E|^{1/2}M^{5/2} \simeq 0.06$. An iso–kinetic equation of state with a sound speed of $c_s = 7.5$ km s^{-1} was used, representing the dominant (by mass) cold-cloud component in the ISM of disk galaxies (Spitzer 1978; Stark & Brand 1989), and thus crudely modeling the bulk motions of the ISM. The star formation rate density was determined by a Schmidt law with index 1.5 (Schmidt 1959).

The simulation was carried out using a parallel Treecode–SPH code, and an individual particle timestep integration scheme was used to improve efficiency, see Williams (1998) for further details. In total, 33552 gas and 33401 dark matter particles were used. The conversion of all gas into stars would have resulted in the formation of 64000 star particles. The Cray T3E at the Edinburgh Parallel Computing Centre was utilized as part of the Virgo Consortium's time allocation (Jenkins et al. 1998), the simulation taking roughly 1.3 CPU years.

The evolution of the simulation proceeded as follows. (i) Expansion ($\sim$ 3 Gyr): self-gravity wins over expansion. (ii) Collapse ($\sim$ 3 Gyr): in the central regions, the potential well swiftly deepens, and inhomogeneities begin to grow. Collapse occurs on roughly a free-fall time. The resultant high densities trigger the first significant levels of star formation, which result in the rapid growth of a proto-bulge component. (iii) Fragmentation ($\sim$ 0.5 Gyr): the dark matter reaches an equilibrium state over the region in which the baryonic galaxy will eventually form. The highly inhomogeneous and unstable gas distribution fragments, resulting in clumpy knots of star formation, the largest of these clumps being comparable to dwarf galaxies. Star formation rates peak during this epoch. (iv) Smoothing and merging ($\sim$ 0.5 Gyr): these star-forming clumps, or "sub-galaxies," tidally interact, disrupt and merge with one another, as the galactic disk continues to accrete mass. Such processes act to progressively smooth the evolving galaxy. (v) A quiescent galaxy (after $\sim$ 8 Gyr): the remaining sub-galaxies orbit on near circular paths, and thus are difficult to disrupt. The evolutions of the structural parameters of the numerical galaxy level off to near constant values, changes occurring over timescales much longer than the age of the universe. The final stellar surface density profile, rotation curve, and greyscale images of the gas and stellar distributions are shown in Figure 1.

In recent studies emphasis has been giving to the importance of the detailed dissipational physics at play (Yepes et al. 1997; Hultman & Pharasyn 1999) and the influence of a protogalaxy's surroundings during a galaxy formation event (Navarro, Eke, & Frenk 1996). The idealized and simplistic nature of the dissipational physics modeled, and the simplest of choices for the isolated initial conditions of the protogalaxy mean that this simulation provides a control experiment for these factors. In every measurable respect, the numerical galaxy formed was found to resemble observed disk galaxies of intermediate to late Hubble type. Figure 1 demonstrates only a few of these features, in particular note the clearly delineated stellar bulge and disk components. Thus one must conclude that while the cosmological environment and the details of the ISM undoubtedly have at some level an effect on the evolution of a protogalaxy, they are not prerequisite to the formation of the gross structural components of disk galaxies as we observe them. The physics of the collapse of self-gravitational gas and dissipational dark matter would appear to be sufficient in itself. In the light of the seemingly robust nature of this prescription of physical processes; and the effectiveness of violent relaxation in erasing the detailed phase-space structure of a pre-collapse protogalaxy which is predominantly of dissipationless matter: it may be the case that the timescales given above are surprisingly robust to variations in the minutiae of protogalaxies and galaxy formation physics.

References

Hultman, J., & Pharasyn, A. 1999, A&A, 347, 769

Jenkins, A. J., et al. 1998, ApJ, 499, 20

Navarro, J. F., Eke, V. R., & Frenk, C. S. 1996, MNRAS, 283, L72

Schmidt, M. 1959, ApJ, 129, 243

Spitzer, L. 1978, Physical Processes in the Interstellar Medium (New York: Wiley)

Stark, A. A., & Brand, J. 1989, ApJ, 339, 763

Williams, P. R. 1998, Ph.D. Thesis, Cardiff University

Yepes, G., Kates, R., Khokhlov, A., & Klypin, A. 1997, MNRAS, 284, 235

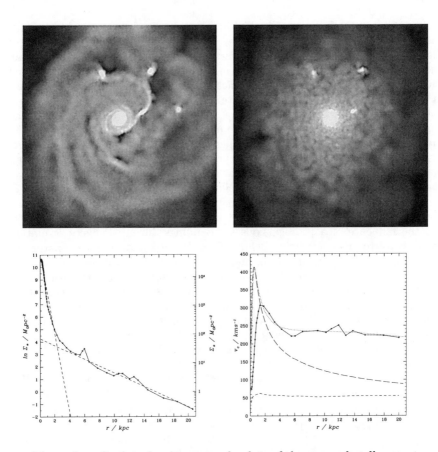

Figure 1.   Surface density greyscale plots of the gas and stellar mass
distributions (*left* and *right*, respectively), the stellar surface density
profile, and rotation curve, taken 8.570 Gyr after the beginning of
the simulation. In the smoothed plot, black and white regions correspond to mass densities of less than 0.015 $M_\odot$ pc^{-2} and greater than
280 $M_\odot$ pc^{-2}, respectively, and the images show a region 42×42 kpc^2.
At this time, 94% of the star mass was found to be older than 10^8 yr.
Thus, the gas image is a better indicator of recent star formation, but
the stellar image is comparable to an infrared observation. The radial
profiles were calculated by summing up the total mass in concentric
annuli centered on the numerical galaxy's centre such that each annulus contained roughly 1000 particles. Least squares fits of exponential
functions to the bulge and disk regions are shown with dashed lines,
The small density peaks observed at $r/$kpc $= 6$, 11, and 13 in the stellar surface density profile are due to the three remaining sub-galaxies,
visible in the above images. In the right graph, the mean gas circular
velocity is shown with a solid line. The circular velocity associated
with the total, gas, and stellar mass within a given radius are shown
with dotted, short dashed, and long dashed lines, respectively.

Astrophysical Ages and Time Scales
ASP Conference Series, Vol. 245, 2001
T. von Hippel, C. Simpson, N. Manset

Protogalaxies and the Hubble Sequence

P. R. Williams

Astronomical Institute, Tohoku University, Aoba, Sendai 980-8578, Japan

Department of Physics and Astronomy, Cardiff University, P.O. Box 913, Wales, UK

A. H. Nelson

Department of Physics and Astronomy, Cardiff University, P.O. Box 913, Wales, UK

Abstract. In a survey of 57 Treecode–SPH starforming simulations, the initial bulk properties of the idealized protogalactic initial conditions used were varied systematically. It was found that there is a well-defined mapping from the initial bulk parameters to the Hubble type of the resultant numerical galaxies in these simulations, although the inverse mapping from Hubble type to bulk initial parameters is degenerate in nature. The implications of these conclusions are briefly explored.

In the related poster paper in these proceedings, "Time Scales and the Formation of Disk Galaxies," based on the analysis of high resolution simulations of similar character to those discussed here, it is suggested that the process of galaxy formation may be profitably analysed into five phases of evolution; and that the relative importance of the physical processes dominant in each of these phases may be controlled principally by the bulk properties of a protogalactic cloud. The natural corollary to this hypothesis is that the Hubble type of a galaxy is therefore fundamentally determined by the bulk properties of a protogalaxy, an assertion probed in the simulations discussed here.

In the parameter survey of initial bulk properties, the total mass of the protogalaxies was fixed at 5×10^{11} $M_\odot$, with initially 10% gas and 90% collisionless dark matter by mass. The initially uniform density spheres were set in solid body rotation with angular velocities chosen to give spin parameters of $\lambda_i \simeq 0.03$, 0.07, and 0.11. A radial velocity field directly proportional to radius was also imposed with a constant of proportionality of 560, 1350, and 2344 km s^{-1} Mpc^{-1}, intended to crudely represent the kinetic energy of matter due to expansion at increasingly higher redshifts. The radii of the protogalaxies were chosen to cover the range of initial densities likely in the early universe. Thus, three bulk properties of the initial protogalactic clouds were varied in this survey. The simulations were carried out using the parallel Treecode–SPH code described in Williams (1998). Roughly 8000 gas and 8000 dark matter particles were used, and the conversion of all gas into stars would have resulted in the formation of 16000 star particles.

The spatial resolution of these simulations was sufficient that observational indicators of morphological type could be reliably measured in the resultant numerical galaxies. Amongst those measured were the scale lengths and central surface densities describing the exponential form of the stellar bulge and disk surface density profiles, bulge-to-disk ratios, rotation velocities, and spiral arm morphology, a subset of which are shown in Figure 1. These parameters were measured when the numerical galaxies had reached a quiescent state. The evolution of these simulations was similar to that described in the related paper mentioned above. Animations of a small sample of these simulations are available at the website http://www.astro.cf.ac.uk/pub/Alistair.Nelson/indexgf.html.

The plots shown in Figure 1 demonstrate the nature of the correlations found between the initial bulk parameters and the final numerical galaxy parameters measured. After consideration of all measured morphological indicators and correlations between them, it was found that these trends produce a consistent mapping from initial bulk properties to morphological type: *for a given radial expansion rate (cosmological epoch), simulations with lower initial density form later type galaxies. For a given cosmological epoch and initial density, simulations with higher rotation form later type galaxies.* These general trends are broadly in keeping with the findings of previous semi-analytic investigations of the origins of the Hubble sequence (Faber 1982a, 1982b; Blumenthal et al. 1984; Evrard 1989); however, in addition, in the simulations discussed here, the ranges of the detailed quantitative morphological indicators measured in the numerical galaxies were found to span the range of those observed in disk galaxies. Only the protogalaxies with extreme initial parameters resulted in morphological indicators not observed.

In conclusion, these simulations tentatively suggest that the origins of the detailed trends and correlations which constitute the modern interpretation of the Hubble classification system may be principally a consequence of variations in the bulk properties of protogalaxies. The fact that such trends and correlations are found observationally at all may be an indication that factors of a statistical nature, factors which would tend to blur underlying trends (such as the power spectrum of density fluctuations within protogalaxies or the frequency of merger events early in a galaxy's evolution), are of secondary importance to the formation of the gross structure of disk galaxies.

References

Blumenthal, G. R., Faber, S. M., Primack, J. R., & Rees, M. J. 1984, Nature, 311, 517

Evrard, A. E. 1989, ApJ, 341, 26

Faber, S. M. 1982a, in Astrophysical Cosmology, ed. H. A. Bruck, G. V. Coyne, & M. S. Longair (Tucson: University of Arizona Press), 191

Faber, S. M. 1982b, in Astrophysical Cosmology, ed. H. A. Bruck, G. V. Coyne, & M. S. Longair (Tucson: University of Arizona Press), 219

Williams, P. R. 1998, Ph.D. Thesis, Cardiff University

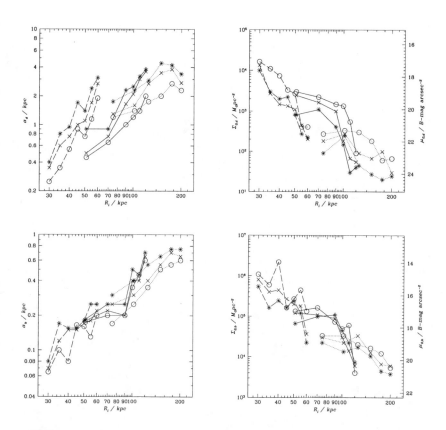

Figure 1. Stellar bulge and disk scale lengths α_b and α_d, and central surface densities $\Sigma_{0,b}$ and $\Sigma_{0,d}$ versus initial radius R_i for the 57 survey simulations. Those with initial spin parameters of $\lambda_i \simeq 0.03$, 0.07, and 0.11 are shown as circles, crosses, and stars, respectively. The dotted, solid, and dashed lines connect simulations with radial expansion rates of 560, 1350, and 2344 km s^{-1} Mpc^{-1}, respectively. In the central surface density plots, the right ordinate shows the stellar central surface densities expressed as surface brightnesses in B magnitudes per square arcsecond.

Session 6

Galaxy Clusters

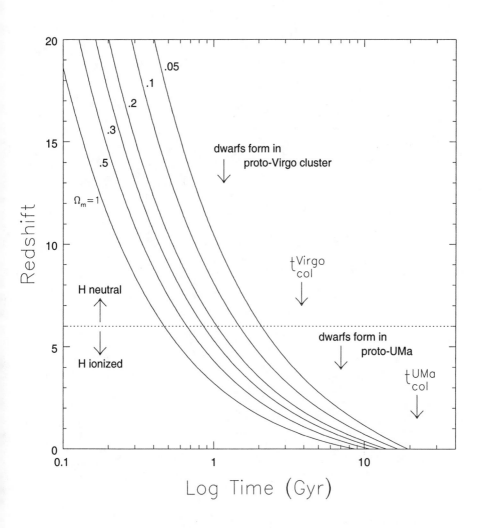

Session 6

Galaxy Clusters

Astrophysical Ages and Time Scales
ASP Conference Series, Vol. 245, 2001
T. von Hippel, C. Simpson, N. Manset

Is Cold Dark Matter Still a Strong Buy? The Lesson From Galaxy Clusters

Fabio Governato

Osservatorio Astronomico di Brera, Milan, Berlusconia, Italy

Sebastiano Ghigna

Astronomy Department, University of Washington, Seattle, WA, USA

Ben Moore

Physics Department, University of Durham, Durham, UK

Abstract. For the last few years the Cold Dark Matter model (ticker: CDM), has been the dominant theory of structure formation. We briefly review the recent advancements and predictions of the model in the field of galaxy clusters. A new set of very high resolution simulations of galaxy clusters show that they have density profiles with central slopes very close to -1.6 and an abundance of subhalos similar to the ones observed in real clusters. These results show a remarkably small cluster to cluster variation and a weak dependence from the particular CDM cosmology chosen. While still a speculative theory with a high prediction/evidence ratio, subject to strong challenges from observational data and competition from other hierarchical theories, we give CDM a rating of "market outperform" and of "long term buy."

1. Introduction

Introduced in the early 1980s (Peebles 1984; Davis et al. 1985), Cold Dark Matter has rapidly become the dominant model within the hierarchical clustering framework. Repeat the mantra with us: "in this theory primordial density fluctuations collapse and merge continuously under the effect of gravitational instability to form more and more massive structures."

One of the most appealing features of CDM is its ability to give a solid framework to provide predictions on the astrophysical properties of cosmological objects, as the number density as a function of mass and redshift and their clustering properties. All on a range of more than 10 orders of magnitude in mass and from redshift ~ 100 to the present.

Not bad.

Being the most massive self bound objects in the Universe, galaxy clusters have received lots of attention, both on the theoretical and the observational side. Statistical properties of the cluster population can be obtained using numerical simulations and/or semi-analytical methods (Governato et al. 1999; Jenkins et

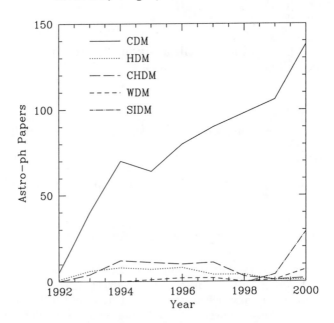

Figure 1.   Number of papers on different hierarchical DM models submitted to the astro-ph database from 1994 to 2000.

al. 2001; Sheth, Mo, & Tormen 2001; but it all started with Press & Schechter 1974). Cool stuff; however this short review focuses on recent results obtained using N-body simulations on the internal structures of clusters within the CDM framework.

2.   Is CDM the Dominant Theory for Cosmic Structure Formation?

As with business companies, there are many, often fuzzy, ways to evaluate the "dominant position" of a theory like influential papers, citations, number of people involved. For CDM a readily available estimate is the number of papers submitted to the arXiv.org e-print archive (Greenspan...I mean...Ginsparg 1996) in the "astro-ph" section. Simple, but fair compared to pro forma earnings, registered users or web page hits often used to evaluate some of Nasdaq's (ex) darlings' performance. Clearly these data show that CDM is the most widely used cosmological theory for structure formation (see Figure 1) at least compared to other dark matter models.

   The number of papers with the word CDM in the abstract has grown at a compound rate of about 15% per year, comparable or higher than the stock market! (the well known Dow Jones and S&P 500 indexes have long term returns of about 10–15% per year). CDM has been able to reinvent itself through the years easily incorporating new experimental evidence that quickly changed our view of cosmology in the last decade. CDM faced its biggest crisis in 1994, due to mounting criticism against its simplest but very successful product SCDM, i.e., a critical Universe, 95% dominated by dark matter. Problems for the model came from lack of power at large scales (Efstathiou et al. 1990), predicted evolution of galaxy cluster numbers stronger than observed (Henry et al. 1992) and the baryon fraction in galaxy clusters too low to be reconciled with observations (White et al. 1993). SCDM had to be recalled from customers and the following year the number of papers containing CDM in their abstract declined almost 10%, while competing models soared, including HDM, a cosmological model already ruled out in the '80s (White, Frenk, & Davis 1983). Indeed alternative hierarchical models enjoyed then a moment of success. CHDM introduced a small component of massive neutrinos (e.g., Ghigna et al. 1997) to increase the amount of large scale power, while other models, like τCDM or Warm Dark Matter (Hogan & Dalcanton 2000) tried to decrease the amount of power at galactic and subgalactic scales. However, these days only Self Interacting Dark Matter (SIDM, see section 4.2) shows a growth rate higher than CDM, but with only a fraction of its market share.

The first robust detection of primordial perturbations in the Cosmic Microwave Background from the COBE satellite suggested that the CDM business model was on the right track, although in need of some major restructuring. After 1998 and observational evidence for an accelerating Universe (Perlmutter et al. 1999) the new "Standard" model became LCDM, a flat Universe with a cosmological constant, $\Omega_0 = 0.3 \sim \Omega_{\mathrm{CDM}}$ and normalization $\sigma_8 \sim 1$. Indeed just a few days ago CDM topped analysts expectations after the findings of optical redshift survey 2dF (Peacock et al. 2001) and the analysis of the full set of BOOMERANG's data (Netterfield et al. 2001), which strongly supported a LCDM universe with baryon abundance close to nucleosynthesis predictions.

Interestingly, the fraction of papers submitted to the astro-ph archive containing the word CDM in the abstract is actually a diminishing fraction of the total number of papers submitted. It was 10% in 1994 and only 2% in the year 2000. Is cosmology going out of fashion? Are we cosmologists losing market share to planet formation, AGNs and, perish the thought, funny variable stars? We offer here the following very speculative (or provocative?) explanation: the total number of papers submitted to the preprint database is growing slower than the total number of world Internet users which doubles every year or so. This is because scientists have likely been faster to adopt the Internet than the average population (no AOL or IOL to fight with); cosmo theorists have been faster than the average astrophysicist population and their number as users of the database got rapidly close to 100%. It is likely that now virtually all of CDM related papers are submitted to astro-ph, while other fields in astrophysics are slower to adopt it as a preferred way to disseminate preprints. The number of generic astrophysics papers submitted (being low at the beginning) has a much larger room to grow compared to that of just CDM papers.

3. Simulations of Galaxy Clusters

With the advent of parallel architectures and dedicated hardware (e.g., GRAPE; Hut & Makino 1999) and cross testing of N-body codes (e.g., Frenk et al. 1999) it has been possible to simulate with accuracy not only the large scale distribution of galaxy clusters, but individual objects at a much larger detail. This is intrinsically a difficult numerical problem, given the large dynamic range across

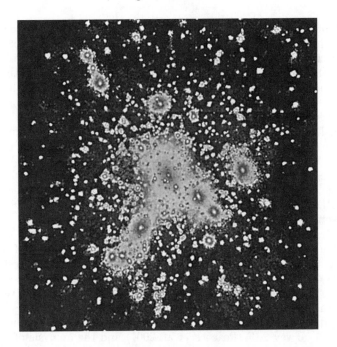

Figure 2.   Phase–density plot of a high SCDM resolution cluster. Box size is twice the virial radius, corresponding to $1.5h^{-1}$ Mpc.

the cluster and the number of dynamical times ($T_{dy} < 0.01/H_0$) at its center. Insufficient dynamical range would cause infalling halos to dissolve in the cluster potential when their central densities became comparable *The effect of increasing the dynamical range in a simulation is to correctly model the evolution of the densest structures (e.g., a subhalo core region), allowing them to survive the tidal forces of the cluster* (Moore, Katz, & Lake 1996).

A new generation of simulations (Figure 2) has allowed us to test CDM under a new, interesting aspect: the internal properties of clusters and galaxies halos, namely the abundance of substructure and the density profile of the parent dark matter halo. A comparison of their results (Ghigna et al. 1998; Brainerd, Goldberg, & Villumsen 1998; Tormen, Diaferio, & Syer 1998; Klypin et al. 1999a; Ghigna et al. 2000; Fukushige & Makino 2001, among many), suggest that (1) a spatial resolution of less than a few percent of the virial radius, (2) half a million particles, (3) several tens of thousands time steps for particles with the largest acceleration, and (4) a surrounding simulated region of several Mpc are required to correctly model the tidal field, the subhalo population and a halo density profile down to a small fraction of its virial radius.

In this review we will briefly discuss previous results from different authors and show findings from a new set of very high resolution simulations (F. Governato et al., in preparation). These simulations explore the cosmic scatter in halo properties (1) within the same (SCDM) cosmology at a fixed mass, (2) with different power spectra (but keeping phases fixed), and (3) at different masses in the same (LCDM) cosmology. The set of simulations presented here satisfies all the requirements of the previous paragraph.

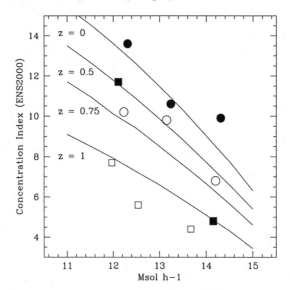

Figure 3.   The concentration parameter (assuming an NFW profile) for a set of halos at different redshifts in a LCDM cosmology. Lines are predictions from ENS. All these halos have at least 30 000 particles within the virial radius. Low z ones have half a million or more.

4.   The Internal Structure of Galaxy Clusters

4.1.   Density Profiles

A significant progress in our understanding of the internal structure of dark matter halos has been the fundamental finding that halos formed in CDM cosmogonies follow a universal profile, with a halo concentration that depends on the amplitude of density fluctuations as well on the ratio of power at small and large scales (Navarro, Frenk, & White 1997, hereafter NFW; Eke, Navarro, & Steinmetz 2000, hereafter ENS).

The proposed density profiles are, among others,

$$\rho/\rho_{\mathrm{crit}} = \frac{\delta_c}{(r/r_{\mathrm{s}})(1 + r/r_{\mathrm{s}})^2}$$

(NFW) or

$$\rho/\rho_{\mathrm{crit}} = \frac{\delta_c}{(r/r_{\mathrm{s}})^{1.5}(1 + (r/r_{\mathrm{s}})^{1.5})}$$

(Moore et al. 1998, 1999a), where δ_c is a function of the so called "concentration parameter" $c = r_{\mathrm{s}}/r_{\mathrm{vir}}$, r_{vir} is the virial radius, and r_{s} is a scale radius.

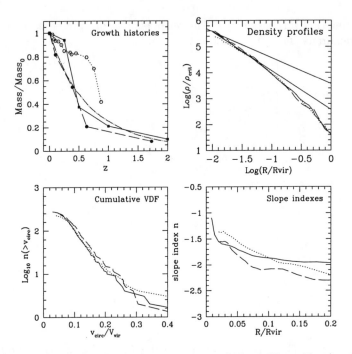

Figure 4.   *Upper left:* Growth histories of three Virgo-like ($\sim 4 \times 10^{14}$ $M_\odot$) SCDM clusters vs. the average predicted with the extended Press & Schechter (1974) formalism. *Upper right:* Density profiles. Straight lines are slopes of -1 and -1.5 respectively. *Lower left:* Substructure abundance as a function of the ratio of circular velocities of subhalos vs. the main one (see also Figure 6). *Lower right:* effective slope of the density profile (the continuous line extends to a very small fraction of the virial radius to show the effect of resolution).

This, after some empirical tuning, allows detailed predictions of the shape of halo profiles. There is general consensus that in all CDM variants halo concentrations decrease at higher z at fixed mass (i.e., at larger $M/M_*(z)$) (see also Bullock et al. 2001). However, parameter space is large and previous works were able to cover a limited part of it at currently state-of-the-art resolution (you need to explore different cosmologies, a large range in redshifts and masses and keep cosmic variance into account before drawing any strong conclusions). Our new simulations are a step in that direction and confirm the ENS predictions for the concentrations of halos (Figure 3).

Note for the profile aficionado: the value of c depends somewhat on the binning method used to measure the density profiles. For results in Figure 3 we used a procedure similar to that used in ENS (V. R. Eke, private communication) namely: (1) ~ 50 logarithmic bins between $0 < r < r_{\rm vir}$, (2) Poisson weighting, and (3) fitting between 0.01 (0.02 for $N_{\rm part} < 10^5$) and $0.75 r_{\rm vir}$.

While interesting, the LCDM model predictions for the concentrations are difficult to test for cluster sized halos. Low concentrations (5–10) imply that the

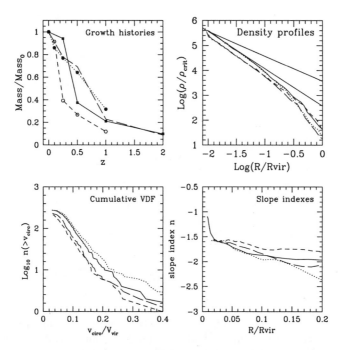

Figure 5. *Upper left:* Growth histories of three Virgo-like clusters in SCDM (continuous line), LCDM (dashed), OCDM (dotted), and TCDM (short dashed) cosmologies. Models have been cluster normalized. *Upper right:* Density profiles. Straight lines are slopes of −1 and −1.5 respectively. *Lower left:* Substructure abundance as a function of the ratio of circular velocities for subhalos vs. the main one. *Lower right:* Effective slope of the density profile.

change in the density profile slope happens at relatively large radii of the order of 100 kpc or larger. Eke, Navarro, & Frenk (1998) and Carlberg, Yee, & Ellingson (1997) reported good agreement between a NFW profile and cluster profiles from galaxy counts, under the assumption that galaxies trace the underlying mass distribution (Carlberg et al. 1996), a somewhat reasonable assumption but difficult to test with simulations, as unwanted numerical effects will tend to underestimate the number of galaxies in the central part of clusters.

Stronger constraints can be placed at galactic scales, both measuring the shape of the rotation curves of individual dwarf and LSB galaxies (Flores & Primack 1994; Moore 1994), or, perhaps more robustly, the mass inside the optical radius (ENS) as low resolution rotation curves have likely been affected by beam smearing (van den Bosch et al. 2000). On the theory side, some claims that LCDM halos were too concentrated (Navarro & Steinmetz 2000) have been retracted (ENS) and careful re-analysis of observational data have been able to set much weaker constraints on the theory (van den Bosch & Swaters 2001).

A tad confusing, isn't it?

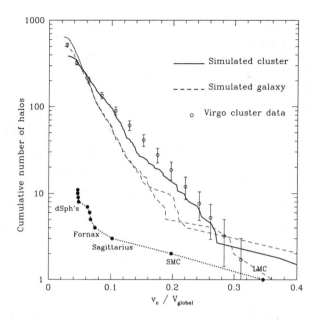

Figure 6. Abundance of cosmic substructure within our Milky Way Galaxy, the Virgo cluster, and our simulated models of comparable masses. We plot the cumulative numbers of halos as a function of their circular velocity ($v_c = \sqrt{(Gm_b/r_b)}$) normalized to the circular velocity, V_{global} of the parent halo. The dotted curve shows the distribution of the satellites within the Milky Way's halo (Mateo 1998), and the open circles with Poisson errors is data for the Virgo galaxy cluster (Binggeli, Sandage, & Tammann 1985), with galaxy luminosities transformed to circular velocities using the Tully–Fisher relation. The second dashed curve shows data for the galaxy at an earlier epoch, 4 billion years ago.

In our opinion the crucial point is the central slope of the density profile in CDM dark matter halos. This is an issue far from being settled. The two proposed profiles have substantially different profile slopes in the inner part of the halo, converging to -1 and -1.5. While current observations seem to be able to accommodate slopes as steep as -1 within a few per cent of the virial radius, halos profiles as steep as -1.5 or more, as shown by Moore et al. (1999a) and recently by Fukushige & Makino (2001) would prove rather difficult to support. Obviously firmer predictions have to be made to use high resolution rotation curves and mass profiles from weak lensing and X-ray observations (e.g., Lombardi et al. 2000) to establish whether there is a strong LCDM crisis.

With the aim of settling the issue of the central slope of the density profile in clusters we have performed a number of high resolution runs of Virgo-sized clusters (a few $\times 10^{14} \, M_\odot$). Three halos were taken from a SCDM cosmology to address the issue of cosmic scatter. Another Virgo-sized halo was run in four different cosmologies (LCDM, SCDM, TCDM, and OCDM), but keeping the same phases.

In all cases the slope of the density profile within $0.01 < r_{\text{vir}} < 0.1$ *is very close to* -1.6, *significantly steeper than the central slope advocated by NFW* (Figures 4 and 5).

4.2. Substructure in Galaxy Clusters

Increasing the dynamical range of numerical simulations showed another major success of the model: as gravitational clustering creates (statistically) small halos first, some of them get gradually subsumed into larger halos. While early works assumed that these subhalos would have been destroyed (White & Rees 1978), high res simulations (e.g., Ghigna et al. 2000 and references therein) showed that they survive within the virialized regions of the parent halos and the abundance of dark substructures predicted by SCDM agrees well with the observed abundance of galaxies inside clusters (Figure 6). *Our set of simulations reveals surprisingly little scatter between different realizations (Figure 4) (contrary to results obtained at much lower resolution) and between different variants of the CDM model (Figure 5).*

Due to the almost power-law shape of the CDM power spectrum $P(k)$ and the long survival times of subhalos, there is also little dependence on the parent halo mass, i.e., once rescaled to the circular velocity of the main halo, the properties of subhalos of galactic and cluster halos look pretty much the same (again Figure 6). While these results are a major success for CDM at cluster scales, galactic dark subhalos are predicted far in excess of the observed population of observed galactic satellites, by almost two orders of magnitude (Moore et al. 1999b; Klypin et al. 1999b).

Several solutions to this puzzle have been suggested within the CDM framework (e.g., the association of dark subhalos with High Velocity Clouds or the effect of an ionizing UV background and SN feedback, e.g., Moore 2001) but the question is still open. Other solutions are being explored. Self Interacting (or Collisional) Dark Matter which IPOed just last year (Spergel & Steinhardt 2000) and WDM (Bode, Ostriker, & Turok 2001) are indeed interesting alternatives (or rather modifications) to LCDM. However, numerical tests of SIDM on the cluster mass scale have given negative (Yoshida et al. 2000) or mixed results (Moore et al. 2000) or require the DM cross section to be a function of velocity, unlikely in the Newtonian regime (Firmani et al. 2000).

4.3. Orbits of Galaxies

Knowing the shape and evolution of orbits of galaxies in clusters is crucial for dynamical estimates of cluster masses (e.g., van der Marel et al. 2000). As tidal stripping is very efficient at decreasing a subhalo mass after the first pericentric passage, subsequent evolution of the subhalo population appears to be very slow, with a time scale likely larger than a Hubble time. Ghigna et al. (1998) showed clearly that orbital properties of subhalos do not differ significantly from those of the underlying DM distribution. Surviving subhalos are on almost radial orbits with a typical pericenter/apocenter ratio of 1:5.

As subhalos orbit inside the dense background (comprising $\sim 85\%$ of the mass of a cluster) they slowly lose orbital energy and sink to the center. However this process is not very efficient. N-body simulations have been combined with semi-analytical models to give insight on the dynamical evolution of the halos identifiable with the hosts of luminous Lyman Break Galaxies (the most massive halos at $z \sim 3$) and the progenitors of present day giant ellipticals (Governato et al. 2001). Orbital shapes of massive halos that fell into the cluster at high z did not show any statistical difference from the global halo population, showing that *orbital decay and evolution of surviving galaxies in clusters is negligible over a Hubble time even for those massive halos that were able to survive as separate entities in the early phases of the cluster formation.*

Colpi, Mayer, & Governato (1999) have proposed a theoretical model for dynamical friction and a fitting formula which keeps orbit eccentricity and the retarding effect of tidal stripping into account:

$$\tau_{\mathrm{DF}} = 1.2 \frac{J_{\mathrm{cir}} r_{\mathrm{cir}}}{[GM_{\mathrm{sat}}/\mathrm{e}] \ln(M_{\mathrm{halo}}/M_{\mathrm{sat}})} \varepsilon^{0.4},$$

where J_{cir} and r_{cir} are, respectively, the initial orbital angular momentum and the radius of the circular orbit with the same energy of the actual orbit and ε is the orbit circularity. The agreement between the semi-analytical approach and N-body simulations is rather remarkable.

5. Discussion

While CDM faces considerable challenges from observational data and competing theories we believe it will still be the reference model for years to come. Recent observational results give support to its business model (but careful investors should perhaps remember the old saying "buy low and sell high"...). The CDM picture gives a coherent frame consistent with large scale structure constraints where galaxies in clusters form in the right numbers and range of masses, almost independently of cosmology. Their sizes and masses are governed by simple and reasonably understood processes like tidal stripping and dynamical friction. Mass attached to individual galaxies is of the order of 15%, a predictions that will be tested by weak lensing measurements in galaxy clusters (e.g., Natarajan et al. 1998). There is mounting evidence that the inner slope of the dark matter profile for clusters in CDM models is close to -1.6, with a small cluster to cluster scatter and weak dependence on the cosmological model.

While CDM seems well positioned, daring colleagues and students in search of market-beating returns should also invest their time and efforts in competing theories which, while riskier and (more) speculative, will offer insight on the physical processes linked to the formation and evolution of galaxy clusters. Space for improvements is getting tight, as constraints from large scale structures improve, and deviations from the currently preferred LCDM model will likely involve galactic and subgalactic scales, with hopefully interesting implications on our understanding of galaxy formation, star formation and feedback on the Intra Cluster/Galaxy Medium, especially at high redshift.

6. Disclosure and Disclaimer (Conforming to SEC Regulations)

FG, SG, and BM have owned shares of CDM since the early 1990s and have started some rather speculative investments in WDM (FG) and Collisional DM

(BM)...so we have some conflicts of interest. Oh well, sue us. This document contains "forward looking statements." These statements are subject to risks and uncertainties and are based on the beliefs and assumptions of the writers based on information currently available. Most important, always remember: past performance is no guarantee of future success!

References

Binggeli, B., Sandage, A., & Tammann, G. A. 1985, AJ, 90, 1681

Bode, P., Ostriker, J. P., & Turok N. 2001 (astro-ph/0010389)

Brainerd, T. G, Goldberg, D. M., & Villumsen, J. V. 1998, ApJ, 502, 505

Bullock, J. S., et al. 2001, MNRAS 321, 559

Carlberg, R. G., Yee H. K. C., & Ellingson, E. 1997, ApJ, 478, 462

Carlberg, R. G., Yee, H. K. C., Ellingson, E., Abraham, R., Gravel, P., Morris, S., & Pritchet, C. J. 1996, ApJ, 462, 32

Colpi, M., Mayer, L., & Governato, F. 1999, ApJ, 525, 720

Davis, M., Efstathiou, G., Frenk, C. S., White, S. D. M. 1985, ApJ, 292, 371

Efstathiou, G., et al. 1990, MNRAS 247, 10P

Eke, V. R., Navarro, J. F., & Frenk, C. S. 1998, ApJ, 503, 569

Eke, V. R., Navarro, J. F., & Steinmetz, M. 2000, ApJ, submitted (astro-ph/0012337)

Firmani, C., D'Onghia, E., Avila-Reese, V., Chincarini, G., & Hernandez, X. 2000, MNRAS 315, 29

Flores, R. A., & Primack, J. R. 1994, ApJ, 427, L1

Frenk, C. S., et al. 1999, ApJ, 525, 554

Fukushige, T., & Makino, J. 2001, ApJ, in press (astro-ph/0008104)

Ghigna, S., Borgani, S., Tucci, M., Bonometto, S., Klypin, A., & Primack, J. R. 1997 ApJ, 479, 580

Ghigna, S., Moore, B., Governato, F., Lake, G., Quinn, T., & Stadel, J. 1998, MNRAS, 300, 146

Ghigna, S., Moore, B., Governato, F., Lake, G., Quinn, T., & Stadel, J. 2000, MNRAS, 554, 616

Ginsparg, P., invited contribution for conference held at UNESCO HQ, Paris, 19–23 Feb 1996 (http://arXiv.org/blurb/pg96unesco.html)

Governato, F., Babul, A., Quinn, T., Tozzi, P., Baugh, C. M., Katz, N., & Lake, G. 1999, MNRAS, 307, 949

Governato, F., Ghigna, S., Moore, B., Quinn, T., Stadel, J., & Lake, G. 2001 ApJ, 547, 555

Henry, J. P., Gioia, I. M., Maccacaro, T., Morris, S. L., Stocke, J. T., & Wolter, A. 1992, ApJ, 386, 408

Hogan, C. J., & Dalcanton, J. J. 2000, Phys. Rev. D, 62, 063511

Hut, P., & Makino, J. 1999, Science, 283, 501

Jenkins, A., et al. 2001, MNRAS, 321, 372

Klypin, A., Gottlöber, S., Kravtsov, A. V., & Khokhlov, A. M. 1999a, ApJ, 516, 530

Klypin, A., Kravtsov, A. V., Valenzuela, O., & Prada, F. 1999b, ApJ, 522, 82

Lombardi, M., Rosati, P., Nonino, M., Girardi, M., Borgani, S., & Squires, G. 2000, A&A, 363, 401

Mateo, M. 1998, ARA&A, 36, 435

Moore, B. 1994, Nature, 370, 629

Moore, B. 2001, in 20th Texas Symposium, ed. J. C. Wheeler & H. Martel (Melville, NY: AIP), in press

Moore, B., Gelato, S., Jenkins, A., Pearce, F. R., & Quilis, V. 2000, ApJ, 535, 21

Moore, B., Ghigna, S., Governato, F., Quinn, T., Stadel, J., & Lake, G. 1999a, ApJ, 524, L19

Moore, B., Governato, F., Quinn, T., Stadel, J., & Lake, G. 1998, ApJ, 499, L5

Moore, B., Katz, N., & Lake G. 1996, ApJ, 457 455

Moore, B., Quinn, T., Governato, F., Stadel, J., & Lake, G., 1999b, MNRAS, 310, 1147

Natarajan, P., Kneib J.-P., Smail, I., & Ellis R. S. 1998, ApJ, 499, 600

Navarro, J. F., Frenk, C. S., & White, S. D. M. 1997, ApJ, 490, 493

Navarro, J. F., & Steinmetz, M. 2000, ApJ, 528, 607

Netterfield, C. B., et al. 2001, ApJ, submitted (astro-ph/0104460)

Peacock, J., et al. 2001, Nature, 410, 169

Peebles, P. J. E. 1984, ApJ, 277, 470

Perlmutter, S., et al. 1999, 517, 565

Press, W. H., & Schechter, P. 1974, ApJ, 187, 425

Sheth, R. K., Mo, H. J., & Tormen, G. 2001, MNRAS, 323, 1

Spergel, D. N., & Steinhardt, P. J. 2000, Phys. Rev. Lett., 84, 3760

Tormen, G., Diaferio, A., & Syer, D. 1998, MNRAS, 299, 728

van den Bosch, F. C., Roberston, B. E., Dalcanton, J. J., & De Blok, W. J. G. 2000, AJ, 119, 1579

van den Bosch, F. C., & Swaters, R. A. 2001, MNRAS, in press (astro-ph/0006048)

van der Marel, R. P., Magorria, J., Carlberg, R. G., Yee, H. K. C., & Ellingson, E. 2000, AJ, 119, 2038

White, S. D. M., Frenk, C. S., & Davis, M. 1983, ApJ, 274, L1

White, S. D. M., Navarro, J. F., Evrard, A. E., & Frenk, C. S. 1993, Nature, 336, 429

Yoshida, N., Springel, V., White, S. D. M., & Tormen, G. 2000, ApJ, 544, L87

Acknowledgments.   The authors thank their colleagues Tom Quinn and Joachim Stadel for allowing them to show results from ongoing projects. Simulations were completed at the ARSC (Fairbanks, AK, USA) and CINECA (Bologna, Italy) supercomputing centers. FG acknowledges generous support from the organizers of this conference. Finally, Hawai'i Volcanoes National Park is a way cool place. Go visit it.

Astrophysical Ages and Time Scales
ASP Conference Series, Vol. 245, 2001
T. von Hippel, C. Simpson, N. Manset

Observational Constraints on Cluster Formation at Low Redshift

Lori M. Lubin

Department of Physics and Astronomy, Johns Hopkins University, Baltimore, MD 21218, USA

Abstract. I present a brief review of some of the current observational data on massive clusters of galaxies at redshifts of $z < 1$. In particular, I describe the evolution over this redshift range of cluster properties, including the global optical and X-ray characteristics, the colors of the cluster galaxies, and the morphological composition of the galaxy population. In addition, I describe the change in the relationship between a galaxy's morphological type and its local environment, more commonly referred to as the morphology–density relation. The implications of the observed evolution on galaxy and cluster formation are discussed.

1. The Optical and X-ray Properties of Clusters of Galaxies

1.1. The Evolution in the Number Density

Clusters of galaxies provide a powerful probe of the nature of galaxy formation and the origin of structure in the Universe. Therefore, quantifying the abundance and dynamical state of clusters is key to understanding the evolution of galaxies and their environment. Because clusters of galaxies are the most massive systems in the Universe, their evolution places strong constraints on cosmology—specifically the mass density of the Universe, Ω, and the normalization of the power spectrum, σ_8. In a low-Ω universe, density fluctuations evolve and freeze out at early times, producing little recent evolution, i.e., redshifts of $z < 1$. In a flat ($\Omega = 1$) universe, fluctuations start growing only recently, thereby producing strong evolution at recent times. Figure 1 shows a plot of the number density of massive clusters ($\gtrsim 8 \times 10^{14} \, h^{-1} \, M_\odot$) as a function of cluster redshift. Existence of massive, relaxed clusters up to $z \sim 1$ implies only a mild negative evolution. At $z \sim 0.8$, the number density of massive clusters is only lower by a factor of ~ 20 (Carlberg et al. 1997; Bahcall & Fan 1998). Compared to numerical simulations, this evolution is completely inconsistent with an $\Omega = 1$ universe. These results, as do all cluster observations, suggest that $\Omega \sim 0.1$–0.4.

1.2. The Evolution in the X-ray Luminosity Function

In addition to optical observations, detailed X-ray studies of distant clusters of galaxies are essential in understanding the effects of the cluster environment on galaxy evolution because the X-ray emitting gas (1) comprises the vast majority of baryons in clusters; (2) is the dominant component in environmental processes like ram pressure stripping; and (3) provides a direct record of the star

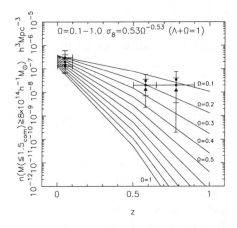

Figure 1.   Evolution of the number density of massive clusters with redshift. Points with error bars indicate the actual data. Solid lines represent the results from numerical simulations with values of Ω ranging from 0.1 to 1. This figure is taken from Bahcall & Fan (1998).

formation history of the cluster galaxies through its metal content. The intracluster medium must, therefore, play a crucial role in cluster evolution. Figure 2 shows the evolution in the X-ray luminosity function for clusters at redshifts of $z > 0.3$. This figure indicates that the cluster X-ray luminosity function does not evolve significantly to $z \lesssim 0.8$ for modest X-ray emitters in the luminosity range of 2×10^{42} to 3×10^{44} h_{50}^{-2} erg s^{-1} (e.g., Ebeling et al. 1997; Nichol et al. 1997; Rosati et al. 1998; Jones et al. 1998); however, there does appear to be a decrease in the number of very X-ray luminous ($L_x \geq 3 \times 10^{44}$ h_{50}^{-2} erg s^{-1}) clusters out to $z \sim 0.9$ (Edge et al. 1990; Henry et al. 1992; Bower et al. 1994; Jones et al. 1998). This evolution implies that there are not as many X-ray luminous clusters at high redshift, suggesting that the X-ray properties of the most massive clusters are evolving with time.

1.3.   The Relation Between the X-ray and Optical Properties

In local clusters, the properties of the galaxies and the intracluster gas are strongly related. In particular, there exists well-defined correlations between the X-ray properties of the gas, such as luminosity (L_x) and temperature (T_x), and the optical properties of the galaxies, such as blue luminosity (L_B) and velocity dispersion (σ). These relations indicate that the galaxies and gas are in thermal equilibrium, i.e., $T_x \propto \sigma^2$ (e.g., Edge & Stewart 1991). Clusters up to $z \sim 0.5$ still exhibit the same X-ray–optical relations (Mushotzky & Scharf 1997). However, at redshifts of $z \gtrsim 0.5$, there are indications that at least some massive clusters do not obey the local relations. Figure 3 shows the relation between L_x and σ for clusters up to $z \sim 1$. All of the optically-selected clusters at $z \geq 0.76$ and at least one X-ray–selected cluster at $z \geq 0.54$ fall well below the local L_x–σ relation (Donahue 1996; Donahue et al. 1998; Gioia et al. 1999; Postman, Lubin, & Oke 2001). Consequently, the X-ray luminosities are low

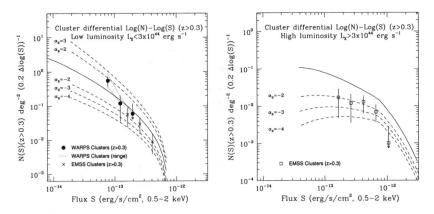

Figure 2.   Comparison between the X-ray luminosity function of local (solid line) and moderate-redshift ($z > 0.3$) clusters (data points) with low luminosities, $L_x < 3 \times 10^{44}\ h_{50}^{-2}$ erg s^{-1} (*left*) and high luminosities, $L_x > 3 \times 10^{44}\ h_{50}^{-2}$ erg s^{-1} (*right*). Dashed lines indicate theoretical X-ray luminosity functions from various evolutionary models. This figure is taken from Jones et al. (1998).

for their velocity dispersions (and therefore their estimated masses). This result indicates that the galaxies and the gas are no longer in thermal equilibrium.

At these early epochs, clusters of galaxies may still be in the process of merging. This process can result in a velocity dispersion which is inflated due to infalling matter or large-scale structure, artificially implying a high mass. However, the X-ray emission can also evolve in a significant way if subclumps in the process of cluster formation are not yet fully virialized. In this case, the gas exists in lower temperature (and lower density) subclumps, substantially reducing its X-ray emission relative to the total mass within the region. Both scenarios suggest that high-redshift clusters are dynamically young. In fact, there are strong indications that a up to 75% of clusters at $z \sim 1$ are experiencing recent formation. The signatures of this formation include strong substructure or a filamentary appearance in the distribution of galaxies, gas, and total mass (e.g., Henry et al. 1997; Gioia et al. 1999; Ebeling et al. 2000). Figure 4 shows indications of this structure in two X-ray–selected clusters at $z \approx 0.8$.

2.   The Galaxy Populations in Clusters of Galaxies

Clusters are ideal laboratories for studying large ensembles of galaxies and the effect of environment on galaxy evolution. In local Abell clusters, early-type galaxies comprise 80% of the cluster population (Dressler 1980a, 1980b). The galaxy content of clusters is part of the general morphology–density relation.

2.1.   Evolution in the Red Sequence

The early-type galaxy population in clusters forms a distinct locus in the color-magnitude (CM) diagram. This locus is referred to as the "red sequence."

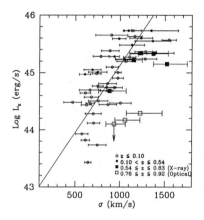

Figure 3. Relation between bolometric X-ray luminosity (h_{50}^{-2} erg s^{-1}) and velocity dispersion (km s^{-1}). Open and closed circles indicate clusters at $z \leq 0.10$ and $0.10 < z \leq 0.54$, respectively (Mushotzky & Scharf 1997). Open and closed squares indicate optically-selected clusters at $0.76 \leq z \leq 0.92$ (Postman, Lubin, & Oke 1998, 2001) and X-ray-selected clusters at $0.54 \leq z \leq 0.83$ (Donahue 1996; Donahue et al. 1998; Gioia et al. 1999), respectively.

This sequence is distinguished by extremely red colors and a tight CM relation (e.g., figure 2 of Gladders et al. 1998). All massive clusters up to redshifts of $z \sim 1$ exhibit a reasonably-strong red sequence in the CM diagram (e.g., Aragón-Salamanca et al. 1993; Lubin 1996; Ellis et al. 1997; Stanford, Eisenhardt, & Dickinson 1995, 1997; Lubin et al. 1998, 2001). By charting the change in the color of the red sequence, it is possible to determine how the stellar populations in early-type galaxies have evolved from $z \sim 1$ to the present day. Figure 5 shows the evolution in the color and color scatter of the red sequence for 15 clusters between $z \sim 0.3$ and $z \sim 0.9$. The broad-band color distribution of the early-type galaxy population show significant bluing; the observed trend is consistent with passive stellar evolution and a relatively well synchronized initial starburst epochs occurring at $z > 2.5$ (e.g., Aragón-Salamanca et al. 1993; Stanford et al. 1995, 1997; Ellis et al. 1997), although continuing star-formation in a fraction of the galaxies is not strongly constrained (van Dokkum & Franx 2001).

2.2. Evolution in the Early-Type Fraction

Even though clusters of galaxies at high redshift still contain a large number of early-type galaxies, the fraction of early-type galaxies is evolving with time. This trend was first observed as the progressive blueing of cluster's galaxy population with redshift (Butcher & Oemler 1984). Butcher & Oemler (1984) found that the fraction of blue galaxies in a cluster is an increasing function of redshift, indicating that clusters at redshifts of $z \sim 0.5$ are significantly bluer than their low-redshift counterparts. At redshifts of $z \sim 0.4$, the fraction of blue galaxies is $\sim 20\%$. High-angular-resolution HST images have revealed that most of these blue galaxies are either "normal" spirals or have peculiar morphologies, resulting

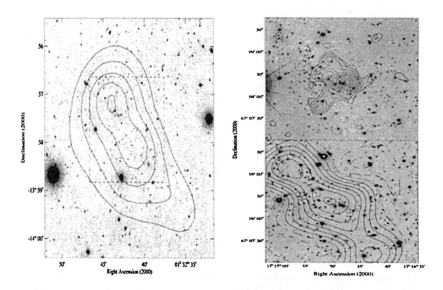

Figure 4.   *Left:* Optical image overlaid with the contours of X-ray flux of Cl J0152.7−1357 at $z = 0.833$ (Ebeling et al. 2000). *Right:* Optical image overlaid with the contours of X-ray flux (*top*) and mass from weak lensing (*bottom*) of RX J1716.6+6708 at $z = 0.809$ (Gioia et al. 1999).

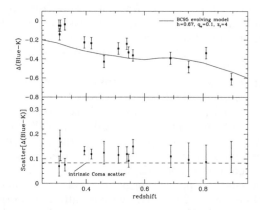

Figure 5.   *Top:* Color of the red sequence (relative to the rest-frame color of the Coma cluster) versus cluster redshift. Solid line indicates the expected color evolution from the 1995 version of the Bruzual & Charlot (1993) evolving model of a single burst of star-formation at $z_f = 4$. *Bottom:* Color scatter in the red sequence versus cluster redshift. Dashed line indicates the intrinsic scatter of the red sequence in the Coma cluster. This figure is taken from Dickinson (1997).

in late-type fractions which are 3 to 5 times higher than the average current epoch cluster (Dressler et al. 1994; Couch et al. 1994; Oemler, Dressler, & Butcher 1997; Dressler et al. 1997).

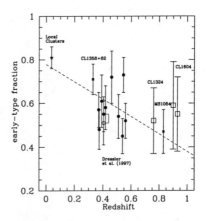

Figure 6.   Early-type fraction versus cluster redshift. Data points are taken from Dressler (1980a), Dressler et al. (1997), Andreon (1998), Lubin et al. (1998, 2001), Fabricant, Franx, & van Dokkum (2000), and van Dokkum et al. (2000). The dashed line indicates best-fit least-squares line. This figure is taken from Lubin et al. (2001).

This trend continues to redshifts of $z \sim 1$ (Lubin et al. 1998, 2001; van Dokkum et al. 2000). Figure 6 shows the evolution of the early-type fraction with cluster redshift. At $z \sim 1$, this fraction is lower than by a factor of 1.5–2.0, compared to local clusters. Consequently, the fraction of late-type (spiral, irregular, and peculiar) galaxies increases with redshift. This evolution implies that early-type galaxies are forming out of the excess of late-type galaxies over this ~ 7 Gyr time scale.

Because there exist strong indications of recent formation activity in high-redshift clusters (see §1.3), significant infall from the the field into the cluster environment is expected (Kauffmann 1995). Since the field galaxy population is comprised mainly of spiral and irregular galaxies, infall is the natural means for providing a reservoir of late-type galaxies. During the infall process and the subsequent interactions in the cluster environment, several processes can transform disk galaxies into spheroids, including galaxy harassment (Moore et al. 1996), ram-pressure stripping (Poggianti et al. 1999), and galaxy–galaxy merging (van Dokkum et al. 2000). Evidence for all three processes is observed in moderate and high-redshift clusters.

3.   The Morphology–Density Relation

The observed evolution in the galaxy populations of clusters also manifests itself as a change in the nature of the morphology–density relation. Locally, there exists a strong correlation between galaxy morphology and environment (Figure 7). As the local density increases, the fraction of E and S0 galaxies increases, while the fraction of spiral galaxies decreases (Hubble 1936; Dressler 1980a, 1980b; Postman & Geller 1984). Studies at intermediate redshift indicate that there may be substantial evolution in the morphology–density (T–Σ) relation (Dressler et al. 1997). Even though this relation exists locally in both open and compact clusters, it is only present in the most centrally condensed, compact clusters at $z \approx 0.5$ (Figure 7). This result suggests that morphological segregation occurs hierarchically over time. Richer, denser clusters, which form at an earlier epoch, are affected first. Smaller, less dense systems are younger dynamically and therefore the segregation has not proceeded as far. This scenario implies that the morphology–density relation is largely tied to galaxy environment, rather than the initial conditions at the time of galaxy formation.

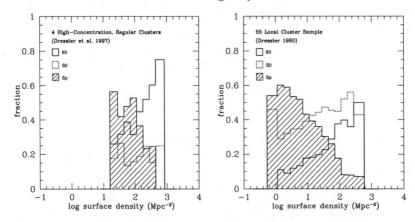

Figure 7. *Left:* Morphology–density relation for the Dressler (1980a) sample of 55 local clusters. *Right:* Morphology–density relation for four high-concentration, intermediate-redshift clusters at $z \sim 0.5$. These figures are adapted from Dressler et al. (1997).

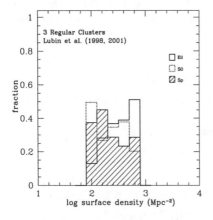

Figure 8. Morphology–density relation for three high-redshift clusters at $z > 0.7$. This figure is taken from Lubin et al. (2001).

Dressler et al. (1997) also find a substantial deficit of S0 galaxies in their clusters at $z \approx 0.5$. They find a ratio of S0 to E galaxies which is ~ 0.5, compared to ~ 2 for local clusters of galaxies (Dressler 1980a, 1980b). Lubin et al. (1998, 2001) and van Dokkum et al. (2000) do not find such a strong deficit, rather they find S0/E ratios of ~ 1–1.5. Despite discrepancies in S0 fractions, Lubin et al. (2001) observe evolution in the T–Σ relation which is qualitatively similar to Dressler et al. (1997). Elliptical galaxies in compact, high-redshift clusters still apparently obey their local T–Σ relation; it is the S0 and spiral populations which have undergone the most significant evolution (Figure 8). At $z > 0.7$, the fraction of spiral galaxies shows no correlation with local density, while the fraction of S0 galaxies shows perhaps a modest anti-correlation. These results suggest that the size and behavior of the elliptical population remains largely the same from redshifts of $z \sim 1$ to the present. However, the decline in S0/E ratio and the evolution in morphology–density relation for S0 galaxies imply that it is the S0 (and spiral) population which undergo the most activity and contribute the most to the observed changes in galaxy properties over this redshift range.

4. Conclusions

Although the number density of massive clusters suggest that they are well-established by $z \sim 1$, substructure, merging signatures, and lack of thermal equilibrium indicate that many clusters are dynamically young. Multi-band observations of the galaxy populations in massive clusters from $z \sim 1$ to $z = 0$ suggest that they all contain significant populations of red, elliptical-like galaxies. The photometric characteristics suggest relatively high formation epochs for these galaxies. However, evolution in the morphological fractions in clusters is observed, indicating that up to 50% of the early-type population has evolved from other galaxy types at $z < 1$. This observed evolution also manifests itself as a significant change in the morphology–density relation with redshift. Only the relationship between ellipticals and their local environment has not changed since a redshift of $z \sim 1$. Rather, it is the S0 and spiral population which is undergoing the most activity within the cluster environment.

References

Andreon, S. 1998, ApJ, 501, 533

Aragón-Salamanca, A., Ellis, R. S., Couch, W. J., & Carter, D. 1993, MNRAS, 262, 764

Bahcall, N. A., & Fan, X. 1998, ApJ, 504, 1

Bower, R. G., Böhringer, H., Briel, U. G., Ellis, R. S., Castander, F. J., & Couch, W. J. 1994, MNRAS, 268, 345

Bruzual, A., & Charlot, S. 1993, ApJ, 405, 538

Butcher, H., & Oemler, A. 1984, ApJ, 285, 426

Carlberg, R. G., et al. 1997, ApJ, 485, 13

Couch, W. J., Ellis, R. S., Sharples, R. M., & Smail, I. 1994, ApJ, 430, 121

Dickinson, M. 1997, in HST and the High Redshift Universe, ed. N. R. Tavir, A. Aragón-Salamanca, & J. V. Wall (Singapore: World Scientific), 207

Donahue, M. 1996, ApJ, 468, 79

Donahue, M., Voit, G. M, Goioa, I., Lupino, G., Hughes, J. P., & Stocke, J. T. 1998, ApJ, 502, 550

Dressler, A. 1980a, ApJS, 42, 565

Dressler, A. 1980b, ApJ, 236, 351

Dressler, A., Oemler, A., Butcher, H. R., & Gunn, J. E. 1994, ApJ, 430, 107

Dressler, A., et al. 1997, ApJ, 490, 577

Ebeling, H., Edge, A. C., Fabian, A. C., Allen, S. W., Crawford, C. S., & Böhringer, H. 1997, ApJ, 479, 101

Ebeling, H., et al. 2000, ApJ, 534, 133

Edge, A. C., & Stewart, G. C. 1991, MNRAS, 252, 414

Edge, A. C., Stewart, G. C., Fabian, A. C., & Arnaud, K. A. 1990, MNRAS, 245, 559

Ellis, R. E., Smail, I., Dressler, A., Couch, W. J., Oemler, Jr., A., Butcher, H., & Sharples, R. M. 1997, ApJ, 483, 582

Fabricant, D., Franx, M., & van Dokkum, P. 2000, ApJ, 539, 577

Gioia, I. M., Henry, J. P., Mullis, C. R., & Ebeling, H. 1999, AJ, 117, 2608

Gladders, M. D., Lopez-Cruz, O., Yee, H. K. C., & Kodama, T. 1998, ApJ, 501, 571

Henry, J. P., Gioia, I. M., Maccacaro, T., Morris, S. L., Stocke, J. T., & Wolter, A. 1992, ApJ, 386, 408

Henry, J. P., et al. 1997, AJ, 114, 1293

Hubble, E. P. 1936, Realm of the Nebulae (New Haven: Yale University Press)

Jones, L. R., Scharf, C., Ebeling, H., Perlman, E., Wegner, G., Maklan, M., & Horner, D. 1998, ApJ, 495, 100

Kauffmann, G. 1995, MNRAS, 274, 153

Lubin, L. M. 1996, AJ, 112, 23

Lubin, L. M., Postman, M., Oke, J. B., Ratnatunga, K. U., Gunn, J. E., Hoessel, J. G., & Schneider, D. P. 1998, AJ, 116, 584

Lubin, L. M., et al. 2001, AJ, submitted

Moore, B., Katz, N., Lake, G., Dressler, A., & Oemler, A. 1996, Nature, 379, 613

Mushotzky, R. F., & Scharf, C. A. 1997, ApJ, 482, 13

Nichol, R. C., Holden, B. P., Romer, A. K., Ulmer, M. P., Burke, D. J., & Collins, C. A. 1997, ApJ, 481, 644

Oemler, A., Dressler, A., & Butcher, H. 1997, AJ, 474, 561

Poggianti, B. M., Smail, I., Dressler, A., Couch, W. J., Barger, A. J., Butcher, H., Ellis, R. S., & Oemler, Jr., A. 1999, ApJ, 518, 576

Postman, M., & Geller, M. J. 1984, ApJ, 281, 95

Postman, M., Lubin, L. M., & Oke, J. B. 1998, AJ, 116, 560

Postman, M., Lubin, L. M., & Oke, J. B. 2001, AJ, submitted

Rosati, P., della Ceca, R., Norman, C., & Giacconi, R. 1998, ApJ, 492, 21

Stanford, S. A., Eisenhardt, P. R. M., & Dickinson, M. 1995, ApJ, 450, 512

Stanford, S. A., Eisenhardt, P. R. M., & Dickinson, M. 1997, ApJ, 492, 461 van Dokkum, P. G., & Franx, M. 2001, ApJ, submitted

van Dokkum, P. G., Franx, M., Fabricant, D., Illingworth, G. D., & Kelson, D. D. 2000, ApJ, 541, 95

Discussion

Scott Trager: It doesn't take long for a galaxy to have an interaction or some star formation event and settle down into looking like an elliptical. Josh explained that this is the case morphologically. From a stellar populations point of view it takes < 1 Gyr for a galaxy to have a major star formation event and then look like an old, dead, red thing. So people who wonder about galaxies in clusters at $z \sim 1$ having formation redshifts of 2 should not be too surprised because it takes very short cosmic epochs for them to look old. We need more precise dating estimates to see the real ages of those objects, and color–magnitude diagrams are a first step in that direction.

Lori Lubin: I think from Josh's simulations within 1 Gyr the indications of the merger are lost, because you can't go deep enough to see the merger signature, it ends up looking like a spheroid. Is that right?

Josh Barnes: You have to look very hard.

Jean-René Roy: Is the dominant factor for this accelerated aging mergers or the tidal field of the cluster?

Lori Lubin: That's unclear. Mergers are unlikely in a cluster environment because the speeds are too high. If there is interaction, probably smaller groups merge into the cluster potential and within the small group mergers are happening and the change in the morphological type occurs. Galaxy harassment may also be contributing to the change in morphological type, but we do not know at this point what the dominant mechanism is. Studying the outskirts of clusters is very interesting as it may be where most of the activity is taking place.

Ofer Lahav: How important is the human component in the visual classification between E and S0 galaxies?

Lori Lubin: Obviously, it's very biased, but the fact is, if you look at automated classifications which use bulge-to-disk ratios, they don't do that much better at high redshift—the scatter is quite large when the signal-to-noise is low. There is the work by Dan Fabricant on a cluster at $z \sim 0.3$. He classified the cluster galaxies but also asked Alan Dressler to classify them independently. Alan consistently found lower numbers of S0s, which is what you see in the classifications of his own clusters. By doing these comparisons, you can quickly tell where the human component is. Again, looking at the total early-type fraction is a very consistent thing to do because you can normally distinguish between spheroids and spiral galaxies; however, there's just a slight distinction between Es and S0s.

Astrophysical Ages and Time Scales
ASP Conference Series, Vol. 245, 2001
T. von Hippel, C. Simpson, N. Manset

Ages and Metal Contents of Early-Type Galaxies

Inger Jørgensen

Gemini Observatory, 670 North A'ohōkū Place, Hilo, HI 96720, USA

Abstract. The mean ages and metallicities of the stellar populations in E and S0 galaxies can be derived using the strengths of spectral absorption lines in the optical, specifically the line indices $H\beta_G$, Mg_2, and $\langle Fe \rangle$, together with stellar population models. This technique has been used for a large sample of E and S0 galaxies in the Coma cluster. The sample covers the core of the cluster as well as galaxies up to 3 degrees from the cluster center. The median value of the mean ages is quite low, about 6 Gyr, indicating that recent star formation has taken place in many of these galaxies. A tight relation between the mean ages, mean metallicities, and the velocity dispersion is found. This relation may be maintained over time by burst of star formation involving a small fraction of the mass. Passive evolution will cause the relation to disappear after a few Gyr. Accurate measurements of line indices for galaxies at redshifts between 0.2 and 0.9 may be used to test the existence of the relation over a large time span.

1. Introduction

The determination of the mean ages and abundances of galaxies may be used to constrain models for galaxy evolution. This is especially the case if determinations can be made for galaxies covering a large range of redshifts, preferably reaching redshifts larger than 0.5. However, the determinations of the ages and abundances is complicated by the so-called age–metallicity degeneracy—the fact that variations in the ages and the metallicities look similar in many of the observables. Worthey (1994) discusses this problem in detail.

For stellar populations where only the integrated light can be studied, the most powerful method for disentangling the ages and the metallicities is to use the strength of the absorption lines. Many authors have used the Lick/IDS system of line indices for this purpose, e.g., Worthey, Faber, & González (1992), Kuntschner & Davies (1998), Trager et al. (2000b). However, the samples used in these studies are in general quite small and not well defined. In this paper we present the preliminary results based on a large magnitude limited sample in the Coma cluster, covering the core of the cluster as well as angular distances from the core as large as 3 degrees.

2.   Sample Properties and Observational Data

The results presented in this paper are based on a large sample of galaxies in the Coma cluster for which we have measurements of the central velocity dispersions and the line indices Mg_2, $\langle Fe \rangle$ and $H\beta_G$ (see Worthey et al. 1994 and Jørgensen 1997 for passband definitions). The sample covers the central $64' \times 70'$ of the cluster (data from Jørgensen 1999) as well as an additional 62 galaxies with angular distances from the cluster center of up to 3 degrees. The data for the outer part of the cluster will be presented in I. Jørgensen (in preparation). The sample contains a total of 177 galaxies and is magnitude limited at $B_{total} = 16$. All 177 galaxies have measurements of the velocity dispersion and Mg_2. $H\beta_G$ and $\langle Fe \rangle$ have been measured for 155 and 133 galaxies, respectively.

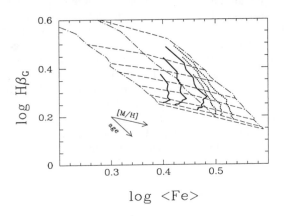

Figure 1.   Single stellar population models from Vazdekis et al. (1996) and Vazdekis (2000). Dashed lines show constant ages (1, 2, 3, 5, 8, 12, 15, and 17 Gyr) and dot-dashed lines indicate constant metallicity ([Fe/H] $= -0.4$, 0.0, 0.4). In addition to the SSP models, the figure shows toy models that are combinations of SSP models. Solid lines represent toy model number 1 and dotted lines represent toy model number 2. See text for details.

3.   Stellar Population Models

We use single stellar population (SSP) models from Vazdekis et al. (1996) and Vazdekis (2000) to interpret the data. The models have a Scalo (1986) initial mass function (IMF) and solar abundance ratios. Figure 1 shows $H\beta_G$ versus $\langle Fe \rangle$ for the models. Because age and metal content are reasonably well separated in $H\beta_G$ and $\langle Fe \rangle$, the models may used to derive luminosity weighted mean ages and metallicities by interpolating between the model predictions.

Real galaxies contain a combination of many stellar populations. Thus, the derived luminosity weighted mean ages and metallicities are dominated by the most luminous (and youngest) stellar populations in the galaxies. To illustrate this, Figure 1 also shows a selection of toy models that combine two stellar populations. Toy model number 1 consists of an old stellar population of age 15 Gyr and [Fe/H] = −0.4, plus a younger stellar population (ages between 1 and 15 Gyr) and [Fe/H] = 0.4. Toy model number 2 is similar, except the old stellar population has [Fe/H] = 0.0. The models are shown for mass fractions in the young stellar population of 0.1, 0.25 and 0.5. The toy-models in which 10 per cent of the mass was involved in a burst of star formation a few Gyr ago has the same ⟨Fe⟩ and HβG as SSP models with ages 5–8 Gyr. Thus, if the SSP models result in low ages, e.g,. 5–8 Gyr, this most likely is caused by some fraction of the mass being involved in star formation within the last 2–5 Gyr, rather than the whole galaxy being young.

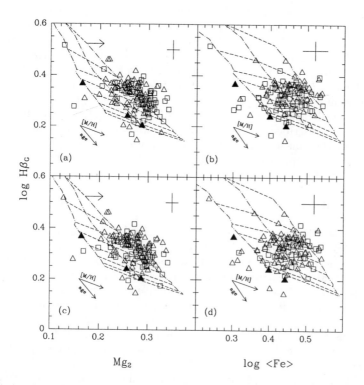

Figure 2.   Line index HβG versus the Mg2 and the ⟨Fe⟩ indices. For (a) and (b) open boxes represent galaxies classified as ellipticals, open triangles represent galaxies classified as S0s, and filled triangles represent galaxies with emission in [O III] 5007 Å. For (c) and (d) open boxes indicate distance from the cluster center, R_{cl} < 0.4 deg; open triangles indicate R_{cl} > 0.4 deg; and filled triangles indicate galaxies with emission in [O III] 5007 Å. Overplotted on all panels are the SSP models from Vazdekis et al. (1996) and Vazdekis (2000).

4.   Line Index Data for the Coma Cluster

Figure 2 shows the line index data for our sample of Coma cluster data together
with the SSP models from Vazdekis et al. (1996) and Vazdekis (2000). In the
$H\beta_G$–Mg_2 diagram the data are systematically offset from the models, reflecting
the above solar abundance ratio in [Mg/Fe] seen for many early-type galaxies
(e.g., Worthey et al. 1992; Jørgensen 1999).

In order to use the SSP models from Vazdekis et al. to derive ages and
metallicities for the galaxies, we need to decide how to handle the non-solar
abundance ratios. We adopt the approach used in Jørgensen (1999). Thus,
we assume that correct mean ages and iron abundances [Fe/H] can be derived
from the $H\beta_G$–$\langle Fe\rangle$ diagram. We then assume that [Mg/Fe] $\neq$ 0 affects the Mg_2
index, but not $\langle Fe\rangle$ or $H\beta_G$. To derive the magnesium abundance, we first apply
an offset to the Mg_2 indices such that there is on average agreement between
ages derived from the $H\beta_G$–$\langle Fe\rangle$ diagram and from the $H\beta_G$–Mg_2 diagram. The
adopted offset is $\Delta Mg_2 = -0.035$ (cf. Jørgensen 1999). The horizontal arrow
on Figure 2 shows the offset as a positive offset to the models. We then derive
ages and metallicities, [M/H], from the $H\beta_G$–Mg_2 diagram. The magnesium
abundances are derived as [Mg/H] = [M/H] − ΔMg_2/0.18, since the Mg_2 index
for a given age depends on the metallicity as $Mg_2 \approx 0.18$[M/H] (Jørgensen 1999).
Obviously, the described method is only approximate and better models taking
into account the non-solar abundance ratios are needed. Trager et al. (2000a)
use an alternative approach for handling non-solar abundance ratios.

5.   Distributions

Figure 3 shows the cumulative frequencies of the derived mean ages and abundances. Panels (a)–(d) show the distributions for the full sample and for subsamples of the E galaxies and the S0 galaxies. Each panel is labeled with the
probability that the distributions of the subsamples are drawn from the same
parent distribution. The probabilities were derived using a Kolmogorov–Smirnov
(K–S) test. The distributions of [Mg/H] and [Mg/Fe] show that the S0 galaxies
on average have slightly lower magnesium abundance that the E galaxies. The
distributions of ages and [Fe/H] are not significantly different for the two subsamples. Because the classification as E or S0 galaxy is a subjective quantity (see
Jørgensen & Franx 1994), the interpretation of this result is not straightforward.
We conclude that the sample of mostly disk-dominated galaxies (the S0 galaxies) has slightly lower [Mg/H] and [Mg/Fe] than the sample of E galaxies which
contain both bulge- and disk-dominated galaxies. Quantitative measurements
of the relative disk luminosities are needed to improve this conclusion.

Panels (e)–(h) compare the distributions for the inner 0.4 degrees of the
cluster with the distributions for the outer part of the cluster. The K–S probabilities that the two subsamples are drawn from the same parent distribution are
given on the panels. None of the distributions show any significant differences
between the inner and the outer part of the cluster.

The sample spans almost a factor 250 in projected surface density of galaxies; many of the outermost galaxies have never crossed the cluster core. Even
so, the derived mean ages and abundances show no significant correlations with

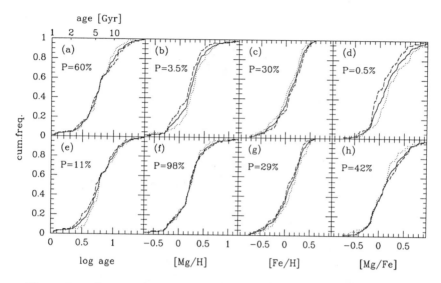

Figure 3.   Distributions in mean ages, mean abundances [Mg/H] and [Fe/H], and the abundance ratios [Mg/Fe]. The distributions are shown as cumulative frequencies. Solid lines represent distributions for the full sample. For (a)–(d) dashed lines indicate distributions for E galaxies and dotted lines show distributions for S0 galaxies. For (e)–(h) dashed lines indicate distributions for galaxies with $R_{cl} < 0.4$ deg and dotted lines represent distributions for galaxies with $R_{cl} > 0.4$ deg. The panels are labeled with the probability that the galaxies in the two subsamples are drawn from the same parent distribution.

the distance from the cluster center. From our analysis of the scaling relations for this sample of galaxies, we find that only the residuals for the $\langle \text{Fe} \rangle$–$\sigma$ relation depends on the cluster center distance; galaxies closer to the center of the cluster have slightly stronger $\langle \text{Fe} \rangle$ than galaxies in the outskirts of the cluster (I. Jørgensen, in preparation).

The median value of the mean ages is quite low, 5.9 Gyr, and 22% of the galaxies have a mean age lower than 4 Gyr. These results indicate that a large number of the galaxies have had some star formation during the last 2–5 Gyr.

6.   The Age–Metal–Velocity Dispersion Relation

The mean abundances derived from the line indices are in Figure 4 shown as a function of the derived mean ages. The abundances [Mg/H] and [Fe/H] are strongly correlated with the ages, while the abundance ratios [Mg/Fe] show a much weaker correlation. Further, [Mg/H] and [Mg/Fe] show a strong correlation with the velocity dispersions. For a given age, galaxies with higher velocity dispersion have higher [Mg/H] and [Mg/Fe]. We have fitted linear relations between the abundances, the ages and the velocity dispersions. We find

$$[Mg/H] = \quad (-0.90 \pm 0.04) \log \text{age} + (1.01 \pm 0.09) \log \sigma - 1.29 \qquad (1)$$

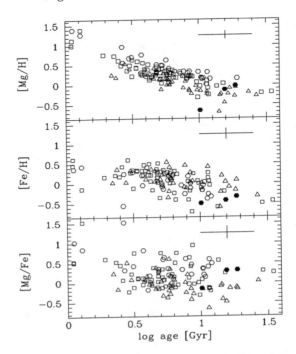

Figure 4.   Mean [Mg/H] and [Fe/H] abundances and the abundance ratios [Mg/Fe] versus the mean ages. Triangles represent $\log \sigma$ in the interval 1.8–2.15, boxes represent $\log \sigma$ in the interval 2.15–2.3, and hexagons show $\log \sigma$ in the interval 2.3–2.65. Solid hexagons indicate galaxies with emission in [O III] 5007 Å.

$$[\text{Fe/H}] = (-0.52 \pm 0.08)\log \text{age} - (0.27 \pm 0.17)\log \sigma - 1.09 \quad (2)$$
$$[\text{Mg/Fe}] = (-0.41 \pm 0.10)\log \text{age} + (1.32 \pm 0.22)\log \sigma - 2.46 \quad (3)$$

The iron abundances, [Fe/H], depend only weakly on the velocity dispersion, but show a strong correlation on the mean ages. The relation for [Mg/H] has significant terms for both the age and the velocity dispersion. This may indicate that the magnesium abundance increases with each subsequent episode of star formation, but that part of the abundance is determined by the velocity dispersion of the galaxies.

The correlation between [Mg/Fe] and the age is quite weak and driven by a few galaxies with very low ages. If we exclude the four galaxies with ages lower than 1.6 Gyr, a Spearman rank order correlation test gives a probability of a correlation of only 4.5 per cent. Further, the relation for [Mg/Fe] has significant intrinsic scatter, while we cannot detect any intrinsic scatter in the relations for [Mg/H] and [Fe/H]. Because [Mg/Fe] depends only weakly on the age we conclude that [Mg/Fe] may be set early in the history of a galaxy and that [Mg/Fe] is changed very little by later episodes of star formation.

Many authors have explored possible explanations for the [Mg/Fe]–σ correlation and for [Mg/Fe] abundance ratios above solar. Magnesium is an α-element and primarily produced in Type II supernovae (SNe), while iron is primarily produced in SNe Ia with some smaller contribution from SNe II. Thus, to explain [Mg/Fe] > 0 and that [Mg/Fe] increases with velocity dispersion, we have to identify processes that either (1) produce more SNe II in high-velocity dispersion galaxies, or (2) retain more of the magnesium produced in the high velocity dispersion galaxies. Two processes are commonly mentioned as possibilities (e.g., Worthey et al. 1992). (1) The IMF is flatter in high velocity dispersion galaxies than in low velocity dispersion galaxies. This would include a larger fraction of the galaxy mass in high mass stars and lead to the desired effect. However, we know of no physical mechanism which will cause this difference in the IMFs. (2) The winds are stronger in low velocity dispersion galaxies than in high velocity dispersion galaxies, and therefore the smaller galaxies lose the enriched gas from the SNe II early in their history. This may also work, except that the resulting mean abundances seem to be too low. It is also unclear how this picture fits with hierarchical clustering models. If the small galaxies lose their SNe II produced magnesium, then we are forced to also assume that hierarchical clustering forms massive E and S0 galaxies by merging only galaxies already massive enough to retain the enriched gas from SNe II.

The age–metallicity–velocity dispersion relation has been established at low redshift. We can roughly outline how it may change with time under different assumptions about galaxy evolution. The relation at low redshift is linear in the logarithm of the age. Thus, if we envision the relation after 5 Gyr of passive evolution (no star formation), then galaxies that currently have low ages will have changed more in the logarithm of the age than galaxies that currently have high ages. Physically this is due to the fact that young stellar populations change line indices faster than old stellar populations. Since the metallicity does not change during passive evolution, the result is that after a few Gyr the lines of constant velocity dispersion are no longer straight in the logarithmic age–metallicity diagram. Thus, if it turns out that the age–metallicity–velocity dispersion relation exist at higher redshifts, e.g., redshift 0.5, and if it is similar but just offset from the relation at redshift zero, then passive evolution is no longer a viable model for the evolution of the early-type galaxies.

If the apparently young galaxies at low redshifts are caused by bursts of star formation involving small mass fractions (10–25%) and such bursts happens fairly frequently, then the age–metallicity–velocity dispersion relation can be maintained if each burst creates stars with a metallicity 0.1–0.6 dex higher than the old stellar population. Further, to match the [Mg/Fe] behavior, the abundance ratios for the burst and the old populations must be similar.

Merging of galaxies with no star formation will result in merger products that deviate from the age–metallicity–velocity dispersion relation. This can be seen by envisioning the merger of two galaxies with the same age, metallicity, and velocity dispersion. The merger product will have the same age and metallicity as the components, but a higher mass and therefore a higher velocity dispersion. Thus, it will not fall on the lines of constant velocity dispersion. Merging accompanied by some star formation may be able to preserve the relation.

7.   Conclusions

We have used line indices in the optical ($H\beta_G$, Mg_2 and $\langle Fe \rangle$) to derive luminosity weighted mean ages and abundances for a large sample of early-type galaxies in the Coma cluster. We find that the median value of the derived mean ages is quite low, 5.9 Gyr, and that 22 per cent of the galaxies have mean ages lower than 4 Gyr. Thus, a larger number of these galaxies have experienced some star formation within the last 2–5 Gyr. To reproduce the low mean ages, 10–25 per cent of the mass may have been involved in the star formation.

Galaxies classified as S0s have on average slightly lower [Mg/H] and [Mg/Fe] than galaxies classified as ellipticals. However, quantitative measurements of the relative disk luminosities are needed to interpret this as an effect of the disk luminosities. We find no significant dependence on the cluster environment (distance from the cluster center) for the derived mean ages and abundances.

The mean ages, the mean abundances and the velocity dispersions form a tight linear relation. This relation is short-lived if the galaxies evolve passively. Burst of star formation involving a small fraction of the mass in the galaxies may maintain the relation over a long period of time. Studies of galaxies with redshifts between 0.2 and 0.9 should resolved whether the relation is maintained over a larger period of the history of the galaxies.

References

Kuntschner, H., & Davies, R. L. 1998, MNRAS, 295, L29

Jørgensen, I. 1997, MNRAS, 288, 188

Jørgensen, I. 1999, MNRAS, 306, 607

Jørgensen, I., & Franx, M. 1994, ApJ, 433, 553

Scalo, J. M. 1986, Fund. Cosmic Phys., 11, 1

Trager, S. C., Faber, S. M., Worthey, G., & González, J. J. 2000a, AJ, 119, 1645

Trager, S. C., Faber, S. M., Worthey, G., & González, J. J. 2000b, ApJ, 120, 165

Vazdekis, A. 2000, http://star-www.dur.ac.uk/~vazdekis/models.html

Vazdekis, A., Casuso, E., Peletier, R. F., & Beckman, J. E., 1996, ApJS, 106, 307

Worthey, G. 1994, ApJS, 95, 107

Worthey, G., Faber, S. M., & González, J. J. 1992, ApJ, 398, 69

Worthey, G., Faber, S. M., González, J. J., & Burstein, D. 1994, ApJS, 94, 687

Acknowledgments.   This research was supported by the Gemini Observatory, which is operated by the Association of Universities for Research in Astronomy, Inc., under a cooperative agreement with the NSF on behalf of the Gemini partnership: NSF (United States), PPARC (United Kingdom), NRC (Canada), CONICYT (Chile), the Australian Research Council (Australia), CNPq (Brazil) and CONICET (Argentina).

Discussion

Guinevere Kauffmann: Are your indexes measured at the centers of the galaxies or do they extend out to large radii? One has to worry a little about extrapolating the conclusions you reach from stellar populations in the centers of ellipticals to the global star formation history.

Inger Jørgensen: They are centrally-measured values and they are corrected to a standard metric aperture size with a diameter of 1.2 kpc. There is no gradient information in our data. I would love to go back and get radial gradients for all these galaxies.

Scott Trager: I can make just a quick comment about the gradients in field galaxies, based on the sample of Gonzales (1993) from his thesis. The gradients in elliptical galaxies tend to indicate in general that they look about 25% older at about half an effective radius than at the center, and that they tend to be about 0.2 dex, or a little more, more metal-poor in the outer parts as well. So if you take a global aperture the galaxies look a little bit older and a little bit more metal-poor. So I am expecting that you'll find the same thing in the Coma galaxies that you have, if you have any longslit data. But it would be nice to get a full-blown set of integral field spectroscopy of local field galaxies so we can measure these quantities at various radii around the galaxy.

Inger Jørgensen: Longslit data for the Coma cluster could effectively be taken with Gemini using GMOS in MOS mode. I plan to apply for the required Gemini time for this project.

Ted von Hippel: I think the interpretation of the Mg–Fe plot you showed earlier is that Mg is produced in the earliest generations by high-mass star formation, while Fe comes from later generations, so this is indicating something about the potential well as a function of time.

Scott Trager: Somewhere between $1/3$ and $1/2$ of the Fe in the solar neighborhood had to come from Type II SNe, so you start out with some fixed value of Mg/Fe from pure Type II SNe, and that will evolve through time. The evolution can be affected by the IMF, by stellar winds, by the newer generations of star formation, etc.

Inger Jørgensen: I think that when we try to settle which of all these possibilities may be the right one, we have to make sure that the results show the dependency on the velocity dispersions of these galaxies. We have heard a lot about mergers over the course of all the talks today. Somehow at redshift zero, basically where the Coma cluster sits, we end up with this overabundance of Mg in the high velocity dispersion galaxies. If the magnesium is formed early by the massive galaxies and it is contained, then you have to have a big potential well already in the start. Maybe the idea of a "big seed" for the merger, which we heard about earlier today, is how the galaxy maintains all this Mg from the earlier stars.

Astrophysical Ages and Time Scales
ASP Conference Series, Vol. 245, 2001
T. von Hippel, C. Simpson, N. Manset

Time Scales in X-ray Clusters of Galaxies

Stefano Ettori

Institute of Astronomy, Madingley Road, Cambridge CB3 0HA, UK

Abstract. I present characteristic time scales related to the physics of X-ray galaxy clusters and focus on the efficiency of the processes of Coulomb collision between ions and electrons and of thermal conduction.

1. Introduction

Clusters of galaxies are the largest gravitationally-bound systems which have approached virial equilibrium on scale of few Mpc and with a total mass of roughly 10^{15} $M_\odot$. The estimate of the cluster mass is fundamental from a cosmological point of view and is obtained from X-ray observations assuming that the distribution of the intracluster medium (ICM) is in hydrostatic equilibrium with the underlying gravitational potential. This assumption relies on the facts that (1) the gas can be treated as a fluid (in general, the time scale of any heating and/or cooling and/or dynamical process is much longer than the elastic collisions time for ions and electrons), and (2) a sound wave in the ICM can cross the cluster in a time shorter than the age of the cluster itself, i.e.,

$$t_{\mathrm{sc}} = 1.2 \times 10^9 \left(\frac{kT_{\mathrm{gas}}}{10 \text{ keV}}\right)^{-0.5} \left(\frac{R}{1 \text{ Mpc}}\right) \text{ yr} < t_{\mathrm{age}} \sim t_{\mathrm{Hubble}}. \tag{1}$$

At radii of few hundred kpc, the ICM has a characteristic density $n_{\mathrm{gas}} \sim 10^{-3}$ cm^{-3}, temperature $T_{\mathrm{gas}} \sim 10^8$ K and a heavy element abundance of about 40 per cent of the solar value. The density drops at larger radii r approximately as r^{-2}. Hence, under these conditions, the gas appears as an optically thin plasma in ionization equilibrium emitting X-rays, where the ionization and emission processes result mainly from collisions of ions with electrons. This emission is mostly due to thermal bremsstrahlung (free–free emission) when $T_{\mathrm{gas}} > 3 \times 10^7$ K, with an emissivity $\propto n_{\mathrm{gas}}^2 T_{\mathrm{gas}}^{1/2}$. In the core, where the density is higher, a larger amount of energy is radiated away and cooling takes place on timescale

$$t_{\mathrm{cool}} = 1.4 \times 10^8 \left(\frac{kT_{\mathrm{gas}}}{2 \text{ keV}}\right)^{0.5} \left(\frac{n_{\mathrm{gas}}}{0.1 \text{ cm}^{-3}}\right)^{-1} \text{ yr} < t_{\mathrm{age}}. \tag{2}$$

This inequality is satisfied in the core of about 70 per cent of nearby clusters (Peres et al. 1998), like A 1795 (Figure 1 from S. Ettori et al., in preparation).

General reviews on the physics of X-ray galaxy clusters are in Sarazin (1988) and Fabian (1994).

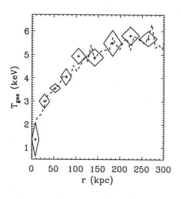 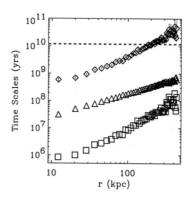

Figure 1.   *Left*: Gas temperature profile in A 1795 from deprojection analyses of Chandra observations that allow spatial resolution on arcsecond scale (S. Ettori et al., in preparation). *Right*: Time scales in the central 500 kpc. The dashed line indicates the age of the Universe for an adopted cosmology of $(H_0, \Omega_{\rm m}, \Omega_\Lambda) = (50 \text{ km s}^{-1} \text{ Mpc}^{-1}, 1, 0)$. The diamonds trace the cooling time, $t_{\rm cool}$, the triangles represent the sound crossing time, $t_{\rm sc}$, and the squares the equipartition time by Coulomb collisions.

2.   Coulomb Collisions

Assuming a polytropic distribution for the gas, the temperature scales with density as $T_{\rm gas} \propto n_{\rm gas}^{\gamma-1}$ where the polytropic index γ ranges between 1 and 5/3, the limits corresponding to the gas being isothermal and adiabatic, respectively. Markevitch (1996) found that $\gamma \approx 1.9$ and 1.7 for the clusters A 2163 and A 665, respectively (> 1.7 and 1.3 and the 90 per cent confidence level). When $\gamma > 5/3$ the gas is convectively unstable, which tends to produce turbulent motions in the ICM, resulting in a mixed, homogeneous gas after several sound crossing times. The detection of a dramatic drop in the temperature profile may be an indication that those clusters are observed in a very unusual, brief stage in their existence, perhaps having experienced a major merger within the previous few billion years.

An alternative possibility to explain the steep temperature decline is that the electron temperature, $T_{\rm e}$, which is the quantity measured by X-ray observations, is not representative of the mean gas temperature, $T_{\rm gas}$. The accretion of infalling material on the cluster potential shock heats the protons to an initial temperature $T_{\rm p}^{\rm i}$, which is representative of their isotropic Maxwellian velocity distribution. The electrons, which are not strongly involved in shock events due to their negligible mass, achieve a low temperature $T_{\rm e}^{\rm i}$ much lower than $T_{\rm p}^{\rm i}$.

Coulomb scattering then equilibrates the temperatures at a rate (Spitzer 1962):

$$\frac{dT_{\rm e}}{dt} = \frac{T_{\rm p} - T_{\rm e}}{t_{\rm eq}}. \tag{3}$$

In this equation, t_{eq}, the equipartition time via Coulomb scattering, is given by

$$t_{eq}(e,p) = \sqrt{\frac{\pi}{2}}\frac{m_p}{m_e}\frac{\theta_e^{3/2}}{n_p c\sigma_T \ln\Lambda} \approx 43\, t_{eq}(p,p) \approx 1870\, t_{eq}(e,e)$$

$$\approx 5.8 \times 10^8 \left(\frac{kT_e}{10\text{ keV}}\right)^{1.5}\left(\frac{n_{gas}}{10^{-3}\text{ cm}^{-3}}\right)^{-1}\text{ yr},\qquad (4)$$

where $\theta_e = kT_e/(m_e c^2)$, n_p is the proton density, σ_T is the Thomson scattering cross section and Λ is the ratio of largest to smallest impact parameters for the collisions ($\ln\Lambda \sim 37.8$, Sarazin 1988; note that a similar term in the corresponding dimensionless proton temperature θ_p is negligible for the conditions of the ICM). The rate is proportional to the gas density and so can be low, and the equilibration time long, at the outer parts of a cluster where the density is least.

Since the total kinetic energy density $U = \frac{3}{2}k(n_p T_p + n_e T_e)$ has to be conserved, one has the following implications: (1) the local mean gas temperature, $T_{gas} = (n_p T_p + n_e T_e)/n_{gas} = (T_p + 1.21 T_e)/2.21$, is constant with time; (2) the initial T_p^i is about 2.2 times the balanced temperature value, T_{gas}^i; and (3) T_e increases (and T_p decreases) with time, with the energy exchange between protons and electrons driven by the relation

$$\frac{dT_e}{dt} = -(n_p/n_e)\frac{dT_p}{dt}.\qquad (5)$$

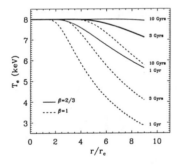 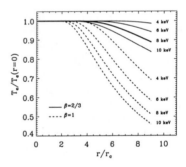

Figure 2.   A β-model (Cavaliere & Fusco-Femiano 1978) with core radius r_c and central density n_0 to describe proton density. (*Left*) The behavior of the electron temperature as a function of R/r_c, when $\beta = 2/3$ (solid line) and $\beta = 1$ (dashed line). The different profiles were calculated with different last merging time, t_m, (upwards: 1, 3, and 10 Gyr), assuming constant n_0 and fixing $T_e(R = 0)$ at 8 keV. The thickest solid line corresponds to $\beta = 2/3$ and $t_m = 3$ Gyr. (*Right*) Dependence of the T_e, normalized to the central value $T_e(R = 0)$, upon different central temperature values (downwards: 4, 6, 8—the thickest one—and 10 keV). All the profiles were calculated at 3 Gyr, assuming $\beta = 2/3$ (solid lines) and $\beta = 1$ (dashed lines).

Rearranging Equation 3 and using the relation between T_{gas} and T_p, I obtain

$$\ln\left(\frac{\sqrt{T_{gas}} + \sqrt{T_e}}{\sqrt{T_{gas}} - \sqrt{T_e}}\right) - \frac{2}{3}\left(\frac{T_e}{T_{gas}}\right)^{3/2} - 2\left(\frac{T_e}{T_{gas}}\right)^{1/2} =$$
$$\sqrt{\frac{2}{\pi}\frac{m_e}{m_p}}\left(\frac{m_e c^2}{kT_{gas}}\right)^{3/2} c\sigma_T \ln\Lambda \, t_m n_{gas}, \qquad (6)$$

where t_m is the time elapsed since the last major heating event. Where $T_e/T_{gas} <$ 0.5, such as in the outer part of the temperature profile given a high central temperature and steep density profile, I can further expand the logarithmic term in Equation 6. Then, for T_{gas} = constant, I obtain that $T_e \propto n_{gas}^{2/5}$.

In Figure 2, I show that electron–proton Coulomb interactions in ICM become inefficient in reaching equipartition and steepen the electron temperature gradient significantly only if (1) the energy per particle is high, (2) the gas density profile is steep, and (3) the time elapsed since the last merger or proton heating event is very short, i.e., $n_{gas} < 3.09 \times 10^{-4} (t_m/1 \text{ Gyr})^{-1} (kT_{gas}/10 \text{ keV})^{3/2}$ cm^{-3}. Local conservation of energy means that the proton temperature drops as the electron temperature rises. Regions of clusters where a large disequilibrium occurs are likely to both be out of hydrostatic equilibrium and to have low X-ray emission.

Applying these considerations to the cluster of galaxies A 2163, we conclude that, more plausibly, the observed gradient is due to a lack of hydrostatic equilibrium following a merger.

Further details on this work are presented in Ettori & Fabian (1998). Other analytic work and simulations on two-temperature ICM are found in Fox & Loeb (1997) and Takizawa (1999).

3. Thermal Conduction

The heat stored in the intracluster plasma is conducted down any temperature gradient present in the gas in a way that can be described through the following equations (Spitzer 1962; Sarazin 1988):

$$q = \kappa \frac{d(kT_e)}{dr}, \qquad (7)$$

where q is the heat flux, T_e is the electron temperature, and κ is the thermal conductivity that can be expressed in term of the density, n_e, the electron mass, m_e, and the electron mean free path, λ_e, as (Cowie & McKee 1977)

$$\kappa = 1.31 \, n_e \, \lambda_e \left(\frac{kT_e}{m_e}\right)^{1/2} = 8.2 \times 10^{20} \left(\frac{kT_e}{10 \text{ keV}}\right)^{5/2} \text{ erg s}^{-1} \text{ cm}^{-1} \text{ keV}^{-1}, \quad (8)$$

where I use $\lambda_e \approx 30.2 \left(\frac{kT_e}{10 \text{ keV}}\right)^2 \left(\frac{n_e}{10^{-3} \text{ cm}^{-3}}\right)^{-1}$ kpc.

If the mean electron free path is comparable to the scale length δr of the temperature gradient, the heat flux tends to saturate to the limiting value which may be carried by the electrons (Cowie & McKee 1977):

$$q_{\mathrm{sat}} = 0.42 \left(\frac{2kT_e}{\pi m_e} \right)^{1/2} n_e kT_e = 0.023 \left(\frac{kT_e}{10 \ \mathrm{keV}} \right)^{3/2} \left(\frac{n_e}{10^{-3} \ \mathrm{cm}^{-3}} \right) \ \mathrm{erg \ cm}^{-2} \ \mathrm{s}^{-1} \tag{9}$$

where the factor 0.42 comes from the reduction effect on the heat conducted by the electrons and that is produced from the secondary electric field that maintain the total electric current along the temperature gradient at zero (Spitzer 1962).

I apply these equations to estimate the efficiency of thermal conductivity in the intracluster medium of A 2142, a cluster of galaxies observed by the X-ray telescope Chandra during its calibration phase in 1999 August. The X-ray image reveals of sharp edges to the surface brightness of the central elliptical-shaped region. The edges are located about 3 arcmin to the northwest and 1 arcmin to the south with respect to the X-ray centre. Markevitch et al. (2000) show that the bright and fainter regions either side of an edge are in pressure equilibrium with each other, but with a dramatic electron temperature decrease on the inside (see Figure 3). Similar breaks in the surface brightness distribution and temperature profile were also detected in A 3667 (Vikhlinin, Markevitch, & Murray 2001) and RX J1720+2638 (Mazzotta et al. 2001).

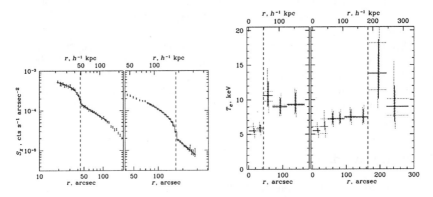

Figure 3. *Left*: Surface brightness profile across the two edges in A 2142. *Right*: Temperature profiles along the same sectors. (From Markevitch et al. 2000).

From the observed plasma properties in A 2142, I can use the equations above to estimate whether thermal conductivity is efficient in erasing the observed temperature gradient. The electron temperature (Figure 3) varies from 5.8 to 10.6 keV on either side of the boundary of the southern edge of the central bright patch in A 2142, and from 7.5 to 13.8 keV at the northern edge. The relative uncertainties on these values are about 20 per cent at the 90 per cent confidence level. The electron density at the edges is $\sim 1.2 \times 10^{-2} \ \mathrm{cm}^{-3}$, $3.0 \times 10^{-3} \ \mathrm{cm}^{-3}$ to the South and North, respectively.

The scale length δr on which this temperature gradient is observed is spatially unresolved in the temperature profile and appears enclosed between 0 and 35 kpc ($H_0 = 70 \ \mathrm{km \ s}^{-1} \ \mathrm{Mpc}^{-1}$) for the Southern edge, 0–70 kpc for the Northern edge. However, the surface brightness profiles show a radially discontinuous derivative at the positions of the sharp edges, on scales of about 10–15 kpc. I adopt hereafter these values as representative of δr, using the larger values only for upper limit purposes. I consider the two extreme cases where (1) $\delta r \gg \lambda_e$ and the heat flux is un-saturated, (2) $\delta r \approx \lambda_e$ and the heat flux is saturated and represented by Equation 9.

The maximum heat flux in a plasma is given by $q = \frac{3}{2} n_e k T_e \bar{v}$, where $\bar{v} = dr/d\tau$ is a characteristic velocity that we are now able to constrain equalizing the latter equation to Equation 7. In particular, given the observed values (and relative errors) of density and temperature across the two edges and δr of (1: non-saturated flux) 10 and 20 kpc (upper limits: 35 and 70 kpc for the edges to South and North, respectively) and (2: saturated flux) $\sim \lambda_e$, the characteristic time, $\delta \tau$, required to erase the electron temperature gradient and due to the action of the thermal conduction alone would be:

$$\delta \tau = \frac{\delta r}{\bar{v}} = \begin{cases} 3.6 \ (< 80) \ \times 10^6 \ \text{yr}, & \delta r \gg 2 \ \text{kpc} \\ 0.3 \ (< 0.4) \ \times 10^6 \ \text{yr}, & \delta r \approx 2 \ \text{kpc} \end{cases} \tag{10}$$

for the Southern edge, and

$$\delta \tau = \frac{\delta r}{\bar{v}} = \begin{cases} 2.4 \ (< 52) \ \times 10^6 \ \text{yr}, & \delta r \gg 12 \ \text{kpc} \\ 1.9 \ (< 2.4) \ \times 10^6 \ \text{yr}, & \delta r \approx 12 \ \text{kpc} \end{cases} \tag{11}$$

for the Northern edge. The upper limits are obtained propagating the uncertainties on the temperature and, for the $\delta r \gg \lambda_e$ condition only, assuming the spatial resolution of the temperature profile as indicative of the length of the gradient. Here I note that the limit on the timescale for saturated flux is the minimum value given the condition of the gas. Any value of $\delta \tau$ estimated on scales considerably larger than the electron mean free path has to be longer than the limit for saturated flux (also significantly, given that the timescale is proportional to δr^2).

When this time interval is compared with the core crossing time of the interacting clumps of about 10^9 yr, I conclude that thermal conduction needs to be suppressed by a factor larger than 10 and with a minimum characteristic value enclosed between 250 and 2500. On the other side, Markevitch et al. (2000) suggest a dynamical model in which the dense cores of the two interacting clumps are moving through the host, less dense, intracluster medium at a subsonic velocity of less than 1000 km s^{-1} and 400 km s^{-1}, for the Northern and Southern edge, respectively, leading to a timescale of about 2×10^7 yr or larger for the cool and hot phases to be in contact. In this scenario, our results implies that the conduction is suppressed *at least* by a factor of 2–200 in the South and 2–32 in the North. However, I note that it is unlikely that the cooler gas can have arrived at its present arrangement and settled in the hotter environment on a timescale much shorter than a core crossing time. The frequency of the occurrence of similar structures in other cluster cores will be important in establishing the timescale for their formation and duration and, hence, improve the constraint on the thermal conduction in the intracluster plasma.

This result is a direct measurement of a physical process in the ICM and implies that thermal conduction is particularly inefficient within $280 h_{70}^{-1}$ kpc of the central core. The gas in the central regions of many clusters has a cooling time lower than the overall age of the system, so that a slow flow of hotter plasma moves here from the outer parts to maintain hydrostatic equilibrium. In such a cooling flow (e.g., Fabian 1994), several phases of the gas (i.e., with different temperatures and densities) are in equilibrium and would thermalize if the conduction time were short. The large suppression of plasma conductivity in the cluster core allows an inhomogeneous, multi-phase cooling flow to form and be maintained, as is found from spatial and spectral X-ray analyses of many clusters (e.g., Allen et al. 2001).

How the conduction is reduced by a so large factor is still unclear. Binney & Cowie (1981) explain the reservoir for heat observed in the region of M 87 as requiring an rms field strength considerably larger than the component of the field parallel to the direction along which conduction occurs. (The transport processes are reduced in the direction perpendicular to magnetic field lines.) This would imply either highly tangled magnetic fields or large-scale fields perpendicular to the lines connecting the hotter to the cooler zones. Such fields could become dynamically important. Chandran & Cowley (1998) use asymptotic analysis and Monte Carlo particle simulations to show that tangled field lines and, with larger uncertainties, magnetic mirrors reduce the Spitzer conductivity by a large factor. Via a phenomenological approach Tribble (1989) argued that a multiphase intracluster medium is an inevitable consequence of the effect of a tangled magnetic field on the flow of the heat through the cluster plasma. Electromagnetic instabilities driven by the temperature gradient (e.g., Pistinner, Levinson, & Eichler 1996) can represent another possible explanation for the suppression of thermal conductivity. Finally, note that cooler gas dumped in the cluster core by a merger (see e.g., Fabian & Daines 1991) would be part of a different magnetic structure to the hotter gas and so thermally isolated.

Further details on this work are presented in Ettori & Fabian (2000).

4.   Last Merging Time in X-ray Cluster Cores

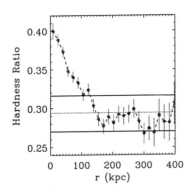
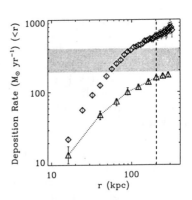

Figure 4.   *Left*: Ratio between counts with energy < 1.5 keV and energy > 1.5 keV. *Right*: Cumulative distribution of the gas deposited after cooling. In both the plots a break in the profiles is observed at about 120 kpc.

Radiative cooling of the inner, denser parts of the cluster plasma causes a flow of gas to move inward under the influence of the gravity to maintain the hydrostatic equilibrium. This *cooling flow* (e.g., Fabian 1994) is the natural state for relaxed systems and is shown by simulations to be interrupted when mergers between clumps with comparable size are taking place (e.g., McGlynn & Fabian 1984). Allen et al. (2001) discuss how an age, or last merging time, of a cooling flow can be determined looking at the extension of the cool emission in cluster cores. In Figure 4, edges at about 120 kpc are shown for A 1795 both in the X-ray colour brightness profile and in the mass deposition rate distribution. This extension is converted in an age of about 5×10^9 yr from the cooling time profile in Figure 1. Other methods to estimate a cooling flow age, and related implications on the formation and evolution of cooling flows in cluster cores, are presented in Allen et al. (2001).

References

Allen, S. W., Fabian, A. C., Johnstone, R. M., Arnaud, K. A., & Nulsen, P. E. J. 2001, MNRAS, 322, 589

Binney, J., & Cowie, L. L. 1981, ApJ, 247, 464

Cavaliere, A., & Fusco-Femiano, R. 1978, A&A, 70, 677

Chandran, B. D. G., & Cowley, S. C. 1998, Phys. Rev. Lett., 80, 3077

Cowie, L. L., & McKee, C. F. 1977, ApJ, 211, 135

Ettori, S., & Fabian, A. C. 1998, MNRAS, 293, L33

Ettori, S., & Fabian, A. C. 2000, MNRAS, 317, L57

Fabian, A. C. 1994, ARA&A, 32, 277

Fabian, A. C., & Daines, S. J. 1991, MNRAS, 252, 17P

Fox, D. C., & Loeb, A. 1997, ApJ, 491, 459

Markevitch, M. 1996, ApJ, 465, L1

Markevitch, M., et al. 2000, ApJ, 541, 542

Mazzotta, P., et al. 2001, ApJ, in press (astro-ph/0102291)

McGlynn, T. A., & Fabian, A. C. 1984, MNRAS, 208, 709

Peres, C. B., Fabian, A. C., Edge, A. C., Allen, S. W., Johnstone, R. M., & White, D. A. 1998, MNRAS, 298, 416

Pistinner, S., Levinson, A., & Eichler, D. 1996, ApJ, 467, 162

Sarazin, C. L. 1988, X-ray Emission from Clusters of Galaxies (Cambridge: Cambridge University Press)

Spitzer, L. 1962, Physics of Fully Ionized Gases (New York: Wiley)

Takizawa, M. 1999, ApJ, 520, 514

Tribble, P. C. 1989, MNRAS, 238, 1247

Vikhlinin, A., Markevitch, M., & Murray, S. S. 2001, ApJ, 549, L47

Acknowledgments. Andy Fabian and the members of the X-ray Group in Cambridge are thanked for discussions and cooperation.

Discussion

Masanori Iye: To what extent does a departure from the isothermal picture affect the mass estimate of the cluster?

Stefano Ettori: The mass is proportional to the temperature. With respect to the isothermal assumption, the plasma temperature tends to be lower in the central part just because it cools, whereas it is almost flat moving outwards. Thus the mass estimates are lower in the centre.

Astrophysical Ages and Time Scales
ASP Conference Series, Vol. 245, 2001
T. von Hippel, C. Simpson, N. Manset

Faint End of the Galaxy Luminosity Function: A Chronometer for Structure Formation?

R. Brent Tully

Institute for Astronomy, University of Hawai'i, 2680 Woodlawn Drive, Honolulu, HI 96822, USA

Abstract. There is accumulating evidence that the faint end of the galaxy luminosity function might be very different in different locations. The luminosity function might be sharply rising in rich clusters and flat or declining in regions of low density. If galaxies form according to the model of hierarchical clustering, then there should be many small halos compared to the number of big halos. If this theory is valid, then there must be a mechanism that eliminates at least the visible component of galaxies in low density regions. A plausible mechanism is photoionization of the intergalactic medium at a time before the epoch of galaxy formation in low density regions but after the epoch of formation for systems that ultimately end up in rich clusters. The dynamical time scales accommodate this hypothesis in a flat Universe with $\Omega_m < 0.4$. If this idea has validity, then upon surveying a variety of environments, it is expected that a dichotomy will emerge. There should be a transition between high density/high frequency of dwarfs to lower density/low frequency of dwarfs. This transition should ultimately be understood by the matching of three timing considerations: (1) the collapse time scale of the transition density, (2) the time scale of reionization, and (3) the linkage given by the cosmic expansion time scale as controlled by the dark matter and dark energy content of the Universe.

1. Introduction

This discussion summarizes ideas developed by Tully et al. (2001). According to the popular cold dark matter (CDM) hierarchical clustering model of galaxy formation there should be numerous low mass dark halos still around today. The approximation by Press & Schechter (1974) that initial density fluctuations would grow according to linear theory to a critical density and then collapse and virialize leads, with a CDM-like power spectrum, to a prediction of sharply increasing numbers of halos at smaller mass intervals. Cosmological simulations are now being realized with sufficient mass resolution to distinguish dwarf galaxies and this modeling basically confirms expectations of the existence of low mass halos (Klypin et al. 1999; Moore et al. 1999).

Indeed, dwarf galaxies are found in abundance in some environments. In the past, most observational effort has gone into studies in rich clusters because the statistical contrast is highest against the background (Smith, Driver, & Phillipps 1997; Trentham 1998; Phillipps et al. 1998) and the small, but dense

Fornax Cluster (Kambas et al. 2000). The general conclusion from these studies has been that, yes, there are substantial numbers of dwarfs of the spheroidal type. There would seem to be reasonable agreement with expectations of CDM hierarchical clustering theory.

However, there has been a suspicion that there might not be the expected abundance of dwarfs in environments less extreme in density than the rich clusters. Klypin et al. (1999) and Moore et al. (1999) have pointed out the apparent absence of large numbers of dwarfs in the Local Group. It is to be appreciated that the task of identifying extreme dwarfs is not trivial. They are tiny and faint. At substantial distances their surface brightnesses are faint against the sky foreground and close up they resolve into swarms of very faint stars. So dwarfs were not being found in the expected numbers but is this because of observational limitations?

2. The Ursa Major Cluster

Motivated by the speculation that the occurrence of dwarfs might be correlated with local density, we made extensive observations in the nearest environment where the density is low (dynamical time is long) yet where there are enough galaxies for a meaningful statistical discussion. We studied the Ursa Major Cluster, a structure fortuitously at about the same distance as the Virgo Cluster and which subtends a comparable amount of sky. The total light in bright galaxies in Ursa Major is about 1/4 that in Virgo but dynamical evidence suggests that the mass in Ursa Major is down by a factor 20 from that associated with Virgo (Tully & Shaya 1999). Roughly 12 square degrees of the Ursa Major Cluster were surveyed with deep CCD imaging with wide field cameras on the Canada–France–Hawaii Telescope and in the 21 cm H I line with the Very Large Array. The footprint of our observations is shown in Figure 1. Results of the two aspects of the survey are reported respectively by Trentham, Tully, & Verheijen (2001) and Verheijen et al. (2000). The important conclusion is that the luminosity function is flat at the faint end in the Ursa Major Cluster, as seen in Figure 2. Whereas Phillipps et al. (1998) found ~ 700 galaxies per square degree with $-16 < M_R < -11$ in Virgo, we find ~ 3 galaxies per square degree in the same magnitude interval in Ursa Major. At the bright end, at $M_R < -17$, the number density of galaxies in Virgo is only 2.5 times higher than in Ursa Major so there is a relative difference of two orders of magnitude in counts at the faint end of the luminosity function between the two locations. The VLA survey confirms that there is no significant population of faint but H I rich systems in Ursa Major.

The Ursa Major luminosity function resembles the luminosity function of the Local Group and, indeed, of other nearby groups. Normalizing to the occurrence of luminous galaxies, there is a shortfall of one to two orders of magnitude from the numbers of dwarfs seen in the Virgo Cluster. The Ursa Major Cluster has a lot of galaxies but in other respects it resembles the nearby groups. It is a loose irregular cluster filled with H I-rich spirals with a crossing time comparable to a Hubble time. From the evidence at hand, such environments host relatively few faint dwarfs. Yet other environments with short crossing times, like Virgo, Fornax, and rich Abell clusters, seem to have large numbers of dwarfs.

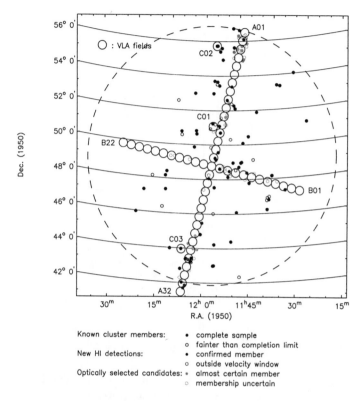

Figure 1.   Ursa Major Cluster survey. The area covered by the VLA survey and with comparable fields by the CFHT wide field CCD survey is indicated by the pattern of circles. Before the survey began, 79 galaxies were known to be associated with the Ursa Major Cluster. The VLA H I survey detected only 10 more members within the survey footprint. The CCD survey has revealed another 3 dozen probable or possible members.

3.   Squelched Galaxies

Hierarchical clustering theory anticipates that there should be numerous dwarf galaxies relative to giant galaxies and this situation is found in rich clusters. This theory predicts that the relative number of dwarfs is even higher in low density regions (Sigad et al. 2000) yet far fewer are found. Apparently we need to explain the *absence* of small galaxies in low density environments. At first thought, it would seem that the rich clusters are more hostile, the low density regions more benign for the survival of small galaxies. In very low density groups dynamical collapse times can be of order the age of the Universe and many galaxies should not have had time to interact with any other galaxy. Hence probably the answer to our problem does not lie with tidal interactions between systems. We need to call upon a mechanism that *allows* small galaxies to form in rich clusters but *thwarts* small galaxy formation in places of low density.

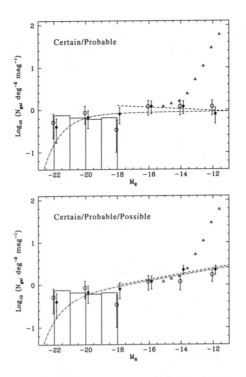

Figure 2.   Ursa Major Cluster luminosity function. The histogram
is the luminosity function for the complete bright sample and the
points with error bars pertain to the area of the cluster covered by the
VLA/CCD survey. The top panel only includes certain and probable
cluster members but the bottom panel also includes possible cluster
members. Small triangles illustrate the flare-up at faint magnitudes
found in the Virgo Cluster by Phillipps et al. (1998).

A plausible squelching mechanism is photoionization of the intergalactic
medium before the epoch of galaxy formation. Efstathiou (1992) discussed the
inhibiting effect on the formation of dwarfs due to the suppression of cooling of a
primordial plasma of hydrogen and helium. Thoul & Weinberg (1996) took the
discussion further with recourse to high resolution hydrodynamic simulations.
These authors argue that gas heating before collapse is more important than in-
hibition of line cooling. The suppression of galaxy formation occurs below a mass
threshold. The UV background heats the precollapse gas to roughly 25 000 K.
This temperature is much less than that associated with the virial energy of a
large galaxy, hence has negligible effect on the collapse of baryons into a massive
potential well. However, for a sufficiently small galaxy this heating is comparable with, or can dominate, the gravitational energy. Thoul & Weinberg find
there is essentially total suppression of baryon collapse for systems with circular
velocities $V_{circ} < 30$ km s^{-1} and, by contrast, little effect on galaxy formation
for systems with $V_{circ} > 75$ km s^{-1}. It follows that luminosity functions would

be unaffected above $M_R^{b,i} \sim -18.6 + 5 \log h_{75}$ $(M_B^{b,i} \sim -17.8 + 5 \log h_{75})$ but truncated below $M_R^{b,i} \sim -16$ $(M_B^{b,i} \sim -15)$. Here, $h_{75} = H_0/75$ km s^{-1} Mpc^{-1} and superscripts b, i indicate corrections are made for Galactic and internal obscuration.

The Thoul & Weinberg model assumes galaxy collapse occurs after reheating of the intergalactic medium. The collapse time scale (Gunn & Gott 1972) is

$$t_{\rm col} = 1.4 \times 10^{10} (R_{\rm vir}^3/M_{14})^{1/2} h_{75}^{-1} \text{ yr}$$

where $R_{\rm vir}$ is the virial radius in Mpc and M_{14} is the virial mass in units of 10^{14} $M_\odot$. Values for $R_{\rm vir}$ and M_{14} can be extracted from Tully (1987) for the Virgo and Ursa Major clusters ($R_{\rm vir} = 0.79$ and 0.98, respectively; $M_{14} = 8.9$ and 0.5 respectively). Hence, rough dynamical collapse times for these clusters are $t_{\rm col}^{\rm Virgo} \sim 3.3$ Gyr and $t_{\rm col}^{\rm Uma} \sim 19$ Gyr. The dense, elliptical dominated Virgo Cluster formed a *core* long ago and the loose, spiral dominated Ursa Major Cluster is still in the process of collapsing. Of course, galaxies continue to fall in and enlarge the Virgo Cluster to this day and, on the other hand, substructure in Ursa Major would have shorter dynamical collapse times than the entire entity.

Smaller mass scales collapse before larger mass scales. Dwarfs must form before their host cluster forms. For the present discussion, the rough approximation is assumed that structure on the mass scale of dwarfs formed at $\sim 1/3$ the time of the collapse of the cluster core. The progression of hierarchical collapse and merging can be followed in semi-analytic models (e.g., Somerville & Primack 1999; Springel et al. 2000). Elaborations on these points will be provided in the discussion by Tully et al. (2001).

These formation time scales in hand, we now ask whether the dwarfs should have formed before or after reionization of the intergalactic medium by the UV radiation of AGNs or hot stars. Observations constrain the epoch of reionization to $z > 6$ (Fan et al. 2000), which can be understood on theoretical grounds (Gnedin & Ostriker 1997). In Figure 3 we see the relationship between redshift and the age of the Universe for a wide range of topologically flat cosmological models. If baryon collapse into small galaxies can only occur before reionization then Figure 3 tells us that if the epoch of reionization is as late as $z_{\rm ion} \sim 6$ then dwarfs with $t_{\rm col} \sim 1$ Gyr could form in a universe with matter density $\Omega_{\rm m} \sim 0.2$ and vacuum energy density $\Omega_\Lambda \sim 0.8$.

We conclude that it is very plausible that small mass halos in the proto-Virgo region collapsed before reionization but almost certainly small mass halos in the proto-Ursa Major region collapsed after the Universe was reionized. Hence this single mechanism could explain why there are many visible dwarf galaxies in dense environments and few in low density regions. Interestingly, this mechanism only works in a universe with relatively low matter density, say $\Omega_{\rm m} < 0.4$, $\Omega_\Lambda > 0.6$. In a universe with $\Omega_{\rm m} = 1$, structure forms at low redshift: $t_{\rm col} \sim 1$ Gyr corresponds to $z \sim 3$.

It would follow that if a range of cluster environments is explored then there should be a break: denser clusters with short dynamical times will have many dwarfs and less dense clusters with long dynamical times will have few dwarfs. The collapse time scale associated with the break point density would reflect the time of reionization of the Universe.

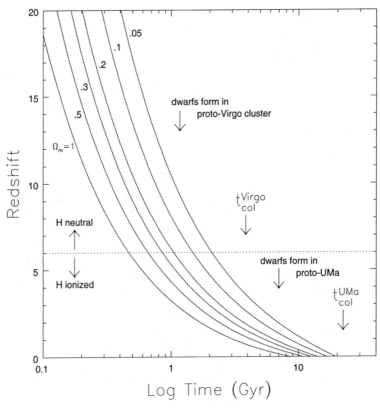

Figure 3.   Redshift vs. age of the Universe for a range of flat world models, from $\Omega_m = 1$, $\Omega_\Lambda = 0$ on the bottom to $\Omega_m = 0.05$, $\Omega_\Lambda = 0.95$ on top. The arrows indicate the rough epochs of galaxy formation in the Virgo and Ursa Major clusters and the collapse time scales of the clusters. Intergalactic reionization must have occurred at $z_{ion} > 6$, that is, above the horizontal dotted line.

4.   Summary

The faint end of the luminosity function of galaxies might be steeply rising in the dense environment of rich clusters but flat or falling in the low density regions of groups. Galaxy formation models anticipate the mass function is sharply rising at the low mass end. It seems something is suppressing the visible manifestations of small galaxies in low density environments.

Reionization of the Universe at $z_{ion} > 6$ could inhibit the collapse of gas in low mass potential wells for late forming galaxies. Dynamical collapse times inferred from the observed densities of clusters are consistent with the picture that dwarf halos formed *before* reionization in high density regions and *after* reionization in low density regions, but only if structure is forming at high redshift; i.e., $\Omega_m < 0.4$ in a flat Universe.

References

Efstathiou, G. 1992, MNRAS, 256, 43p

Fan, X., et al. 2000, AJ, 120, 1167

Gnedin, N. Y., & Ostriker, J. P. 1997, ApJ, 486, 581

Gunn, J. E., & Gott, J. R. 1972, ApJ, 176, 1

Kambas, A., Davies, J. I., Smith, R. M., Bianchi, S., & Haynes, J. A. 2000, AJ, 120, 1316

Klypin, A., Kratsov, A. V., Valenzuela, O., & Prada, F. 1999, ApJ, 522, 82

Moore, B., Ghigna, S., Governato, F., Lake, G., Quinn, T., Stadel, J., & Tozzi, P. 1999, ApJ, 524, L19

Phillipps, S., Parker, Q. A., Schwartzenberg, J. M., & Jones, J. B. 1998, ApJ, 493, L59

Press, W. H., & Schechter, P. 1974, ApJ, 187, 425

Sigad, Y., Kolatt, T. S., Bullock, J. S., Kravtsov, A. V., Klypin, A. A., Primack, J. R., & Dekel, A. 2000, MNRAS, submitted (astro-ph/0005323)

Smith, R. M., Driver, S. P., & Phillipps, S. 1997, MNRAS, 287, 415

Somerville, R. S., & Primack, J. R. 1999, MNRAS, 310, 1087

Springel, V., White, S. D. M., Tormen, G., & Kauffmann, G. 2000, MNRAS, submitted (astro-ph/0012055)

Thoul, A. A., & Weinberg, D. H. 1996, ApJ, 465, 608

Trentham, N. 1998, MNRAS, 294, 193

Trentham, N., Tully, R. B., & Verheijen, M. 2001, MNRAS, in press (astro-ph/0103039)

Tully, R. B. 1987, ApJ, 321, 280

Tully, B., & Shaya, E. 1999, in Evolution of Large Scale Structure, ed. A. J. Banday, R. K. Seth, & L. N. Da Costa (Heidelberg: Springer-Verlag), 296

Tully, R. B., Somerville, R. S., Trentham, N., & Verheijen, M. A. W. 2001, ApJ, submitted

Verheijen, M., Trentham, N., Tully, R. B., & Zwaan, M. 2000, in Mapping the Hidden Universe, ed. R. C. Kraan-Korteweg, P. A. Henning, & H. Andernach (San Francisco: ASP), 263

Acknowledgments. My collaborators in this research are Rachel Somerville, Neil Trentham, and Marc Verheijen. Financial support has been provided by a NATO travel grant.

Discussion

Scott Trager: The CfA luminosity function by Marzke et al. shows a very steep upturn at the very faint end, and I'm not sure I can reconcile this with the flat luminosity for Ursa Major. Are they just sampling denser regions than you are?

Brent Tully: Well, the CfA survey sampled a wide swath of sky, averaging over denser and less dense regions. Maybe that's what you get when you average over a contribution that's flaring up and a contribution that's flat.

Romeel Davé: One of the concerns I have about this model and a similar one proposed by Bullock et al. is that reionization is not independent of galaxy formation. Reionization occurs because galaxies form, so presumably in the denser regions where you're forming these dwarfs, there's already a strong photoionizing flux and it seems like a strange model to assign a single-value time for reionization. Can you comment on that?

Brent Tully: Clearly what I've suggested is an oversimplification and, following your reasoning, we can anticipate that the timing of reionization is different from place to place. But if the general idea is correct, the neat thing is this—we'll find in the fullness of time that there are some places where the ratio dwarf/giant is high and other places where the ratio dwarf/giant is low and the distinction between these places is discrete and characterized by the local density or dynamical collapse time.

Astrophysical Ages and Time Scales
ASP Conference Series, Vol. 245, 2001
T. von Hippel, C. Simpson, N. Manset

The Oldest Stellar Populations at $z \sim 1.5$

Alan Stockton

Institute for Astronomy, University of Hawai'i, 2680 Woodlawn Drive, Honolulu, HI 96822, USA

Abstract. There are at least three reasons for being interested in galaxies at high redshifts that formed most of their stars quite quickly early in the history of the Universe: (1) the ages of their stellar populations can potentially place interesting constraints on cosmological parameters and on the epoch of the earliest major episodes of star formation, (2) their morphologies may provide important clues to the history and mechanisms of spheroid formation, and (3) they are likely to identify the regions of highest overdensity at a given redshift. We describe a systematic search for galaxies at $z \sim 1.5$ having essentially pure old stellar populations, with little or no recent star formation. Our approach is to apply a "photometric sieve" to the fields of quasars near this redshift, looking for companion objects with the expected spectral energy distributions. Follow-up observations on two of the fields having candidates discovered by this technique are described.

1. Introduction

One of the uncertainties in our picture of galaxy formation in the early Universe is in understanding how elliptical galaxies and the massive bulges of early-type spirals developed. The standard cold dark matter scenario implies bottom-up formation—low mass systems form first, possibly as objects something like the ubiquitous star forming dwarf galaxies found at lower redshifts; then these objects merge to form larger entities including, in the denser regions, large spheroidal systems. However, there are at least two nagging worries regarding this scenario. Firstly, there is a correlation between color (i.e., metallicity) and luminosity in elliptical galaxies (Bower, Lucey, & Ellis 1992; Ellis et al. 1996). As Peacock (1999) has observed, "It seems as if the stars in ellipticals were formed at a time when the depth of the potential well that they would eventually inhabit was already determined." Secondly, the very tight correlation found between stellar velocity dispersion in bulges and black hole mass (Ferrarese & Merritt 2000; Gebhardt et al. 2000) seems to demonstrate an intimate connection between the formation of spheroids and the formation of supermassive black holes at their centers. It is certainly not obvious that this correlation could be produced from successive mergers of small building blocks, since supermassive black holes do not seem to be associated with pure disks or irregulars. In fact, both of these observations would appear to fit more comfortably with mono-lithic or quasi-monolithic collapse pictures of spheroidal formation; however, this statement is more a reflection of our current uncertainty than an endorsement of such models. What is clear is that direct observational constraints on formation mechanisms, environments, and formation epoch for spheroidals are necessary in attempting to sort out these difficulties.

2. The Earliest Major Episodes of Star Formation and Constraints on Cosmological Parameters

Figure 1 (*left*) shows the K' magnitude of L^* elliptical galaxies as a function of redshift, assuming only passive evolution of a solar metallicity stellar population formed essentially instantaneously at cosmic epochs of either 0.5 or 1.0 Gyr. At $z = 1.5$, such galaxies would have $K' \sim 19.5$, so they are quite easily detectable. Both we and others (e.g., Dunlop 2000 and references therein) are finding galaxies about 1 mag brighter than this, presumably on the high-luminosity tail of the luminosity function and indicating that the stellar content of some early-type galaxies is essentially fully in place at very high redshifts.

If we could determine precise ages for old populations at high redshifts, we could potentially place interesting constraints on cosmological parameters as well as on formation epochs. This possibility is shown in Figure 1 (*right*): if one could demonstrate an age of ≥ 4 Gyr at $z = 1.5$, with currently reasonable values of h_0 $(= H_0/100$ km s^{-1} Mpc$^{-1})$ and Ω_m all open models would be eliminated, and even Λ-dominated models with $z_f \sim 10$ are only barely consistent.

3. Identifying Old Galaxies at High Redshifts

There have been three main approaches to identifying old galaxies at high redshifts: (1) looking for very red objects among weak radio sources (Dunlop et al. 1996; Dunlop 2000; Spinrad et al. 1997), (2) wide-field multicolor photometric surveys (Thompson et al. 1999; Daddi et al. 2000), and (3) identifying red objects in radio source fields (this paper; see also Cimatti et al. 1997). We have been examining fields of radio-loud QSOs with $1.4 < z < 1.78$. We use a "photometric sieve" approach, which gives us high observing efficiency and clearly distinguishes objects with old stellar populations and little reddening from heavily reddened objects. Using the NASA Infrared Telescope Facility, we first image the fields of interest in the K' band, looking for objects with $18 \leq K' \leq 19.5$ within a $30''$ radius of the quasar. For fields with such objects, we then obtain J-band imaging, looking for objects with $J-K' \sim 2$. At this point we have eliminated typically 80% of our original fields; for the remainder, we must now obtain CCD imaging on the short side of the 4000 Å break, which occurs between the I and J bands for this redshift range. We usually try to obtain at least R and I photometry, although we have used a variety of standard and non-standard bands. Recently, a similar approach has been proposed by Pozzetti & Mannucci (2000). Of the 208 fields in our sample, we have at least some observations for about 60%; we have eliminated 74 fields as having no further interest, and we have 7 fields with quite firm old-galaxy candidates. It is significant that 5 of these 7 have more than one good candidate, including one field with 3. Spectral energy distributions for some of these objects are shown in Figure 2.

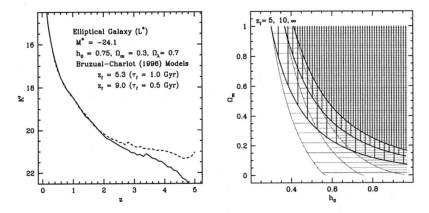

Figure 1.   *Left:* K' magnitude for a passively evolving L^* elliptical galaxy, assuming an instantaneous burst with formation redshifts of 5.3 (dotted line) or 9.0 (solid line). *Right:* Constraints on the matter density parameter, Ω_m, and the Hubble parameter, h_0 ($= H_0/100$ km s^{-1} Mpc^{-1}), assuming the confirmation of a galaxy at $z = 1.5$ with a 4.0-Gyr-old stellar population. The hatched area to the right of each curve shows excluded regions, assuming star formation redshift z_f of 5, 10, and ∞, as indicated at the top. Gray curves are for open models ($\Omega_{total} = \Omega_m$); black curves are for flat models ($\Omega_m + \Omega_\Lambda = 1$). Even the latter require $z_f > 10$ to be consistent with $\Omega_m \sim 0.3$ and $h_0 \sim 0.7$.

4.   An Example: The Field of TXS 0145+386

Figure 3 shows the field of the $z = 1.446$ quasar TXS 0145+386, in bands ranging from R to J. Two EROs, G1 and G2, are marked, and their spectral-energy distributions are shown at the top of Figure 2. We have obtained a spectrum of G1, unfortunately through cirrus, so the S/N is not sufficient to measure age-diagnostic spectral features. However, it does confirm a redshift of 1.4533 from a weak but broad [O II] $\lambda 3727$ line, which possibly indicates the presence of a hidden active nucleus.

Figure 4 shows adaptive optics (AO) imaging of G1. With images having a FWHM of 0″.16, the galaxy is seen to have a generally symmetric elliptical profile, with some faint irregular structure extending to the east. The other candidate (G2) is too far from the guide star for AO imaging, but its morphology on other images suggests that its structure is less regular (see Figure 3), in spite of evidence that its stellar population is as old as that of G1 (Figure 2). Candidates in other fields also seem to show a variety of morphologies. The implication seems to be that, even with apparent stellar ages of 3–5 Gyr, these galaxies may not all be completely relaxed systems in their outer parts, and we may be able actually to observe some aspects of the final stages of bulge formation.

There is an apparent concentration of very faint galaxies in the region around the quasar, G1, and G2, as seen in Figure 3. Figure 5 gives another

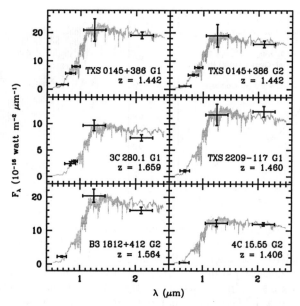

Figure 2.   Examples of old galaxies found in fields of quasars at $z \sim$ 1.5. Vertical bars show 1σ photometric errors; horizontal bars show filter FWHM. Gray traces are 4-Gyr Bruzual & Charlot (1996) models.

view of this clustering. It is often suggested that powerful radio sources can be used as markers to locate regions of high density at high redshifts. However, while strong radio sources are undoubtedly statistically in regions of higher density than is the average galaxy, there appears to be a large dispersion in the densities of radio source environments. Some seem to be in fairly rich clusters (Dickinson, Dey, & Spinrad 1995; Chapman, McCarthy, & Persson 2000), but some seem not to be (e.g., Stockton & Ridgway 1998). It may well be that the presence of nearly fully formed galaxies comprising old stellar populations is a more reliable indicator of a rich cluster: under most plausible formation scenarios, processes of galaxy evolution will proceed more rapidly in strongly overdense regions. The identification of such galaxies in radio source fields may be one of the best ways of finding rich clusters at redshifts beyond the practical range of current wide-area X-ray surveys.

5.   Determining Ages of Stellar Populations

The main difficulty in attempting to use old galaxies at moderately high redshifts to constrain cosmological parameters is in establishing a robust age for the stellar population. Both we (Stockton, Kellogg, & Ridgway 1995) and Dunlop et al. (1996; see also Spinrad et al. 1997) have attempted to use spectroscopic age diagnostics. The spectral features of interest fall in the rest-frame near-UV (2600–3200 Å) and potentially give good age discriminants over the age range $\sim$ 1–5 Gyr. Figure 6 shows our preliminary reduction of a spectrum of one of two EROs in the field of the $z = 1.406$ quasar 4C 15.55, which fairly closely matches a 3-Gyr old model.

The major uncertainty in the results so far is in the reliability of the spectral synthesis models (e.g., Bruzual & Magris 1999; Dunlop et al. 1996; Dunlop 2000; Heap et al. 1998; Yi et al. 2000; Nolan et al. 2001). However, as Dunlop (2000) has emphasized, much of this disagreement is due to the inclusion of broad-band colors in determining the ages; from the spectroscopic age diagnostics alone, the age dispersion is much smaller. In addition to being sensitive to reddening, ages from colors are especially dependent on getting rather uncertain late stages of stellar evolution right. On the other hand, while ages derived from restframe near-UV absorption features require more observing time, they have important advantages: (1) they are potentially capable of higher precision and depend essentially only on the turnoff age of main sequence F stars, where the models are on much firmer ground; (2) they are almost totally insensitive to the IMF of

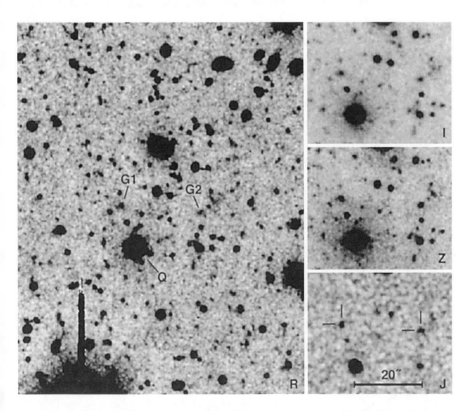

Figure 3.   Images of the field of TXS 0145+386 in R, I, Z, and J bands. The R and Z images were obtained with LRIS on Keck II, and the I and J images were obtained with the UH 88-inch telescope. The quasar, Q, and the two red galaxies, G1 and G2, are marked. Note the evidence for a cluster of faint objects in the vicinity of the quasar and red galaxies.

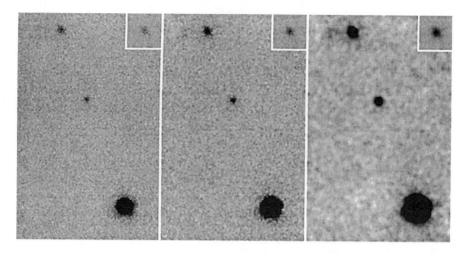

Figure 4.   Adaptive optics imaging of the field of TXS 0145+386 at K', obtained with PUEO on the Canada–France–Hawaii telescope. The left panel shows the unsmoothed image, with FWHM = 0″16. The candidate old galaxy G1 is at the top, the nearly stellar object near the center is a compact galaxy with $z = 0.7883$, and the quasar itself is the bright object at the bottom. The center and right panels show the same image, smoothed with a Gaussian with $\sigma = 1$ and $\sigma = 3$ pixels, respectively. Insets show G1 at lower contrast. The structure seen to the east of G1 in the right panel appears in both nights' data. Panels are $10'' \times 15''$.

the stellar population; and (3) they are also insensitive to reddening. The most serious worry has been the age–metallicity degeneracy (e.g., Worthey 1994), which can affect both colors and spectral features. However, recent work by Nolan et al. (2001) indicates that, with sufficiently good data, it is possible to break this degeneracy from the rest-frame near-UV spectroscopy alone. Thus, while it is still quite reasonable to *select* candidates on the basis of colors, it is essential to obtain spectroscopy in the rest-frame near-UV in order to determine robust lower limits to the age of the stellar population. In the meantime, models of increasing sophistication are being developed by several groups, and such refinements as incorporating α-enhanced stellar atmospheres and evolutionary tracks are likely to be available soon.

6.   Higher Redshifts

In parallel with completing our survey for old galaxies in the fields of quasars with $z \sim 1.5$, we are extending the search to higher redshifts. Now that it seems clearly established that there are galaxies at $z \sim 1.5$ with ages of $\gtrsim 3$ Gyr, we are seeking to identify precursors to these objects at $z \sim 2.5$ with CISCO on the 8.2-m Subaru telescope. At this redshift, the oldest galaxies should be ~ 1.5 Gyr younger than at $z \sim 1.5$.

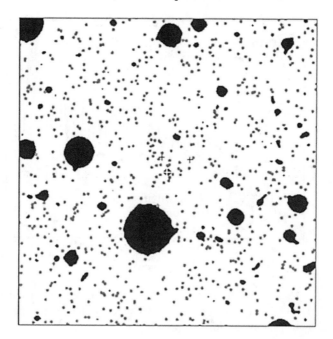

Figure 5.   Clustering of galaxies with $24.5 < Z_{AB} < 26$ (gray dots) in the field of the quasar TXS 0145+386 (white cross on black object at center). The two red galaxies, G1 and G2, are marked with gray crosses. Regions in black are obscured by bright stars or galaxies. The region shown is 4.'3 on a side.

References

Bower, R. G., Lucey, J. R., & Ellis, R. S. 1992, MNRAS, 254, 601

Bruzual A., G., & Charlot, S. 1996, ftp://gemini.tuc.noao.edu/pub/charlot/bc96

Bruzual A., G., & Magris C., G. 1999, in Cosmological Parameters and the Evolution of the Universe, ed. K. Sato, (Dordrecht: Kluwer), 152

Chapman, S. C., McCarthy, P., & Persson, S. E. 2000, AJ, 120, 1612

Cimatti, A., Villani, D., Pozzetti, L., & di Serego Alighieri, S. 2000, MNRAS, 318, 453

Daddi, E., Cimatti, A., Pozzetti, L., Hoekstra, H., Röttgering, H. J. A., Renzini, A., Zamorani, G., & Mannucci, F. 2000, A&A, 361, 535

Dickinson, M., Dey, A., & Spinrad, H. 1995, in Galaxies in the Young Universe, ed. H. Hippelein (New York: Springer-Verlag), 164

Dunlop, J. S. 2000, in The Hy-Redshift Universe, ed. A. J. Bunker & W. J. M. van Breugel (San Francisco: ASP), 133

Dunlop, J. S., Peacock, J. A., Spinrad, H., Dey, A., Jimenez, R., Stern, D., & Windhorst, R. 1996, Nature, 381, 581

524   *Stockton*

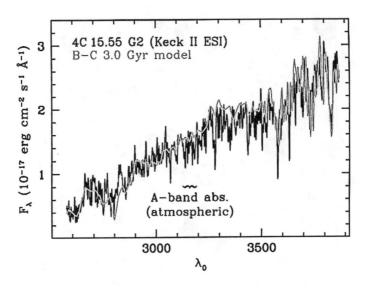

Figure 6.   Spectrum of 4C 15.55 G2 ($R = 26.6$) obtained with ESI
on Keck II. The gray trace shows a Bruzual & Charlot (1996) 3-Gyr
spectral synthesis model for comparison.

Ellis, R. S., Colless, M., Broadhurst, T., Heyl, J., & Glazebrook, K. 1996, MN-
     RAS, 280, 235

Ferrarese, L., & Merritt, D. 2000, ApJ, 539, L9

Gebhardt, K., et al. 2000, ApJ, 539, L13

Heap, S., et al. 1998, ApJ, 492, L131

Nolan, L. A., Dunlop, J. S., Jimenez, R., & Heavens, A. F. 2001, MNRAS,
     submitted (astro-ph/0103450)

Peacock, J. 1999, Cosmological Physics (Cambridge: Cambridge University
     Press), 413

Pozzetti, L., & Mannucci, F. 2000, MNRAS, 317, L17

Spinrad, H., Dey, A., Stern, D., Dunlop, J, Peacock, J., Jimenez, R., & Wind-
     horst, R. 1997, ApJ, 484, 581

Stockton, A., Kellogg, M., & Ridgway, S. E. 1995, ApJ, 443, L69

Stockton, A., & Ridgway, S. E. 1998, AJ, 115, 1340

Thompson, D., et al. 1999, ApJ, 523, 100

Worthey, G. 1994, ApJS, 95, 107

Yi, S., Brown, T. M., Heap, S., Hubeny, I., Landsman, W., Lanz, T., & Sweigart,
     A. 2000, ApJ, 533, 670

Acknowledgments.   I wish to acknowledge and thank my collaborators in
these projects, Gabriela Canalizo, Susan Ridgway, Toshinori Maihara, and Bill
Vacca. This work has been partially supported by NSF Grant AST95-29078. It
has also made use of the NASA/IPAC Extragalactic Database (NED), which is
operated by the Jet Propulsion Laboratory, California Institute of Technology,

under contract with the National Aeronautics and Space Administration. The observations described in this paper were obtained with the NASA Infrared Telescope Facility, the University of Hawai'i 88-inch Telescope, the Canada–France–Hawaii Telescope, and the Keck Observatories, and I wish to express my gratitude to those who have provided these facilities and those who maintain and operate them.

Discussion

Hyun-chul Lee: This may not be a direct question to you, but since you are talking about the oldest stellar populations, I wondered whether there may be *older* stellar populations than the oldest Milky Way globular clusters. The Milky Way is just a typical galaxy, so other galaxies may have older globular cluster systems. In addition, my poster suggests that NGC 1399, the Fornax giant elliptical, could be a couple of Gyr older than our Galaxy. Could you comment?

Alan Stockton: Well, you're right, that's probably not a question for me. The oldest globular clusters of course have some chemical enrichment already so there must have been an earlier generation of stars. I'm sure there are people in the audience who could deal with this issue better than I can.

Michael Rich: I was just going to mention that, as you've noticed from some of the talks, it gets very difficult to accurately age-date stellar populations older than about 12 Gyr. One point that has not been mentioned is that the oldest and most metal-poor globular clusters in the Local Group—NGC 2419, Hodge 11, and M 92 as the template oldest, most metal-poor stellar population in a globular cluster—all seem to be identical in their color–magnitude diagrams. Using both the vertical and horizontal methods of age determination, these systems are coeval and it suggests a universal time of ignition in the Local Group. Whether there are older stars somewhere else would be interesting and maybe as we get more uranium measurements we will improve our constraints on field stars in the halo. But presently, while there is a range in the accepted age of the oldest stars, there is also some evidence for a population of "most ancient" stars.

Wendy Freedman: I realize this is a hard question to answer because there are uncertainties related to the IMF and age and metallicity, but, if you were to allow for some range of uncertainty in various parameters and you tried to estimate an age plus a 1σ uncertainty, what sort of accuracy would you obtain?

Alan Stockton: Yes, it's very difficult, and it seems to have gotten more difficult as I've sat through the first few days of this conference. I think there's at least a hope that we may not be too badly off with respect to the age–metallicity degeneracy because spectroscopy in the near-UV is dominated by the lower metallicity stars. Another reason to be somewhat optimistic is that the spectrum one sees in the near-UV for stellar populations with ages $\lesssim 5$ Gyr is essentially just that of a turnoff star—one can actually use an appropriate F-star stellar spectrum as a template, and it will look just about as good as the spectral synthesis models do. And so it mostly comes down to the errors involved in determining the age of a mid-to-late F star of a given chemical composition (plus, of course, the uncertainties due to the S/N one can obtain for the spectrum of a galaxy with $R \sim 25.5!$). This is somewhat oversimplified— there are problems with, say, the effects of α-enhancement, which, as I indicated, are just now being addressed. But to take a stab at answering your question, my guess is that the uncertainty in the ages from the stellar evolutionary models alone is around 10%. Considering everything, I'm hopeful that in maybe 3 or 4 years we'll be able to say with some conviction that we can date a 4 Gyr-old stellar population at $z \sim 1.5$ to ± 1 Gyr, but we'll just have to wait and see.

Session 7

The Universe

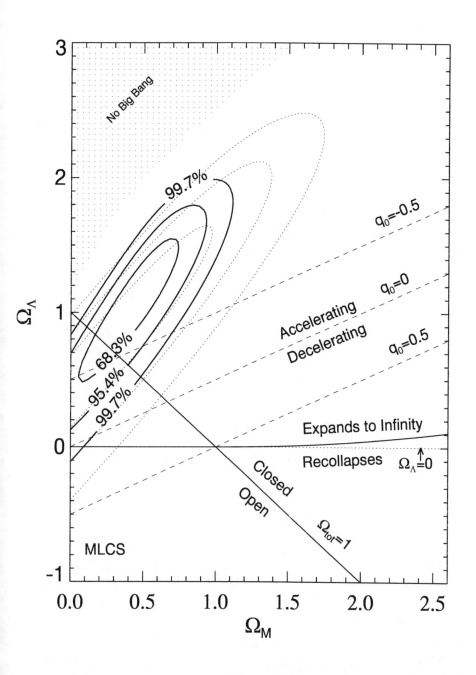

Astrophysical Ages and Time Scales
ASP Conference Series, Vol. 245, 2001
T. von Hippel, C. Simpson, N. Manset

Time at the Beginning

Michael S. Turner

Departments of Astronomy and Astrophysics and of Physics, Enrico Fermi Institute, The University of Chicago, Chicago, IL 60637-1433, and NASA/Fermilab Astrophysics Center, Fermi National Accelerator Laboratory, Batavia, IL 60510-0500, USA

Abstract. Age consistency for the Universe today has been an important cosmological test. Even more powerful consistency tests at times as early as 10^{-32} s lie ahead in the precision era of cosmology. I outline tests based upon cosmic microwave background (CMB) anisotropy, Big Bang Nucleosynthesis (BBN), particle dark matter, phase transitions, and inflation. The ultimate cosmic time scale—the fate of the Universe—will be in doubt until the mystery of the dark energy is unraveled.

1. Introduction

The cosmic clock ticks logarithmically. The Universe's past can be divided into three well-defined epochs: the quantum era (10^{-42} s to 10^{-22} s); the quark era (10^{-22} s to 10^{-2} s); and the Hot Big Bang Era (10^{-2} s to 10^{17} s). Earlier than the quantum era is the Planck era ($t < 10^{-42}$ s); it holds the key to understanding the birth of the Universe, but requires a quantum theory of gravity to proceed. Just recently the Universe has entered a new era, of undetermined duration, where dark energy and its repulsive gravity are dominating the dynamics of the Universe (more later).

The third era derives its name from the very successful cosmological model that describes it; the important events include the synthesis of the light elements, the last scattering of radiation, the formation of large-scale structure, and the onset of accelerated expansion. The earlier two eras are still largely terra incognita, though we have some tantalizing ideas: the transition from quark/gluon plasma to hadronic matter, the electroweak phase transition, and the birth of particle dark matter during the quark era; and inflation and the origin of the matter–antimatter asymmetry during the quantum era (see Figures 1 and 2).

There is no doubt that cosmology is in the midst of the most exciting period ever. Progress in understanding the origin and evolution of the Universe is proceeding at a stunning pace. I mention a few of the highlights of the past decade: the COBE discovery of the small density inhomogeneities ($\delta\rho/\rho \sim 10^{-5}$) that seeded large-scale structure, the mapping of CMB anisotropy on scales down to 0.1 degree and the determination of the shape of the Universe (flat!), identification of the epoch of galaxy formation, determination of the Hubble parameter to a precision of 10%, and the discovery of cosmic accelerated expansion.

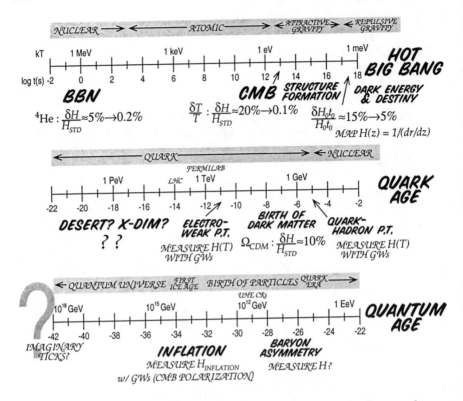

Figure 1.   The Cosmic Clock ticks logarithmically. Major epochs
are listed in the upper two timelines. The lower timeline indicates
opportunities to measure ages.

There is much more to come. These recent discoveries and those still to
be made will test the framework of the standard cosmology and begin to open
the quark and quantum eras. We are already on the way to establishing a new
standard cosmology: a flat Universe comprised of two-thirds dark energy and
one-third dark matter that is accelerating today and whose seeds for structure
came from quantum fluctuations stretched to astrophysical size during inflation.
Much remains to be done to put the new cosmology on the same firm footing as
the Hot Big Bang Model; much of cosmology over the next two decades will be
devoted to this (see e.g., Turner 2001a).

The richness and redundancy of the measurements to be made will also
allow a number of consistency tests of the Big Bang framework and its theoretical
foundation, general relativity. Many of these tests involve cosmic time scales,
and that is the subject of my talk.

2. $H_0 t_0$: The Sandage Consistency Test

The product of the present age and Hubble constant is a powerful consistency test. The standard Hot Big Bang has not always cleared this hurdle, cf. the late 1940s when the Hubble age of about 2 billion years fell short of the age of the Solar System by a factor of two (see e.g., Kragh 1996) or the more recent scares when some measurements of H_0 and t_0 drifted upward.

Recognizing the importance of age consistency, Sandage has devoted much of his career to measuring the Hubble constant H_0 and independent indicators of the age of the Universe (and pioneered many of the methods); hence the title of this section.

The nature of the test has changed over Sandage's career; the importance has not. Until recently one would have written $H_0 t_0$ in terms of one parameter, Ω_M, the fraction of critical density in matter, also assumed to be the total fraction of critical density contributed by all forms of matter and energy ($\equiv \Omega_0$). The discovery of accelerated expansion increased the number of parameters to two, Ω_M and Ω_Λ (fraction of critical density in a cosmological constant; $\Omega_0 = \Omega_M + \Omega_\Lambda$). The realization that accelerated expansion is here to stay, while the cosmological constant may not, added another parameter, $w = p_X/\rho_X$ (the equation of state of the dark energy, see below). The determination from CMB anisotropy measurements that the Universe is flat, effectively reduced the number of parameters to two, Ω_M and w:

$$
\begin{aligned}
H_0 t_0 &= f(\Omega_0) & &\to 1998 \\
&= f(\Omega_M, \Omega_\Lambda) & 1998 &\to 1999 \\
&= f(\Omega_M, \Omega_X, w) & 1999 &\to 2000 \\
&= f(\Omega_M, w) & 2000 &\to \quad ? \\
&= f(w) & ? &\to \quad ?.
\end{aligned}
$$

For the first two cases, there are well known analytic formulae for f; for the others there are not.

There is hope, that in the next ten years Ω_M will be pinned down by a combination of CMB, cluster, and large-scale structure measurements, reducing the number of parameters to one again. At the moment, my analysis of these data gives $\Omega_M = 0.33 \pm 0.035$ (Turner 2001b), and I believe there will be significant improvement over the next decade.

The function f that accounts for the effect of slowing/speeding up on the age of the Universe is given by (for a coasting universe $f = 1$)

$$
f \equiv H_0 t_0 = \int_0^\infty \frac{dz}{(1+z) H(z)/H_0}
$$

with

$$
H(z)/H_0 = \left[\Omega_M (1+z)^3 + \Omega_X (1+z)^{3(1+w)} + (1 - \Omega_0)(1+z)^2 \right]^{1/2}, \quad (1)
$$

where the small contribution from the CMB and relativistic neutrinos ($\Omega_R \sim 10^{-4}$) has been neglected and w has been assumed to be constant.

To illustrate the status of this test, for H_0 and t_0 I will use the values reported at this meeting, $H_0 = 72 \pm 7$ km s^{-1} Mpc^{-1} (Freedman, these proceedings) and $t_0 = 13.5 \pm 1.5$ Gyr (Chaboyer, these proceedings). This leads to

$$H_0 t_0 = 0.94 \pm 0.14.$$

Taking $\Omega_0 = 1$ and $\Omega_M \simeq 0.35$, this is consistent with $w < -\frac{1}{3}$ (at 1σ)—the new cosmology passes the Sandage consistency test with flying colors! While not a tight constraint on w, it does provides more evidence for dark energy (i.e., a smooth component with large, negative pressure): if the dark energy were pressureless ($w = 0$), then $H_0 t_0 = 2/3$, which is clearly inconsistent with current data.

At present, the Sandage test has a precision of about 15%. What kind of improvement could one hope for over the next decade? Together, physics-based measures of H_0 (especially the CMB) and distance scale measures might well pin down the Hubble constant to a few percent. However, it is difficult to imagine independent measures of the age achieving similar accuracy. One irreducible uncertainty for all methods, is the time to formation (of stars, globular clusters, white dwarfs, etc.). Reducing this uncertainty to 0.5 Gyr would still leave a 5% uncertainty in $H_0 t_0$. As I will discuss, other tests of age consistency, at much earlier times, are likely to be significantly more precise.

2.1. Mapping the Expansion Rate

The expansion rate at any epoch is related to the age of the Universe. For example: during a matter-dominated epoch, the scale factor $R(t) \propto t^{2/3}$ and $t = 2/3H(t)$; and during a radiation-dominated epoch $R(t) \propto t^{1/2}$ and $t = 1/2H(t)$. The expansion rate is more accessible than the age at early times.

The expansion rate determines a key observable: the comoving distance to an object at redshift z,

$$H_0 r(z) = \int_0^z \frac{\mathrm{d}z}{H(z)/H_0}, \qquad (2)$$

where a flat Universe has been assumed. The cosmological volume element, luminosity distance, and angular-diameter distance are all directly related to $r(z)$:

$$
\begin{aligned}
\mathrm{d}V/\mathrm{d}z\mathrm{d}\Omega &= r(z)^2/H(z), \\
d_\mathrm{L} &= (1+z)r(z), \text{ and} \\
d_\mathrm{A} &= r(z)/(1+z).
\end{aligned}
$$

A number of probes of the expansion history from $z = 0$ to $z \sim 2$ have been discussed recently. They include gathering a large sample of type Ia supernovae (SNAP) and counting halos or clusters of galaxies. From these experiments, individually or taken together, one could imagine mapping out $H(z)$ back to redshift $z \sim 2$ (see e.g., Huterer & Turner 2000; Tegmark 2001). Note that the volume element and luminosity distance taken together, can in principle directly yield $H(z)$.

Loeb (1998) has gone one step further, suggesting ultra-precise redshift measurements of the Lyman-alpha forest over a decade or more could be used to determine $H(z)$ directly(!). The idea is to measure the tiny time variation of the redshifts of thousands of Lyman-alpha clouds,

$$\delta z = [-H(z_1) + (1 + z_1)H_0]\delta t$$
$$\sim 0.2\,(\delta t/10\,\text{yr})\,\text{m s}^{-1} \quad \text{for } z = 3, \tag{3}$$

where δt is the time interval of the observations. Whether or not this bold idea can be carried out remains to be seen, but it certainly is exciting to think about.

3. CMB Anisotropy and Acoustic Peaks

As is now very familiar, the power spectrum (in multipole space) of CMB anisotropy arises due to acoustic oscillations in the baryon-photon fluid around the time of last scattering (see e.g., Hu, Sugiyama, & Silk 1997). At this time, the baryons (and electrons) are falling into the dark-matter potential wells; still coupled to photons (through Thomson scattering off electrons), the infall is resisted by photon pressure and oscillations ensue. At maximum compression the photons are heated and at maximum rarefaction they are cooled.

The CMB is a snapshot of the Universe at 400 000 yr; regions caught at maximum compression (rarefaction) lead to hot (cold) spots on the microwave sky. While the physics is most easily explained in real space, the signal is best seen in multipole space, as a series of peaks and valleys. The power at multipole l is largely due to k-modes satisfying $k \sim lH_0$.

The condition for maxima in the power spectrum is $\omega_n t_{\text{LS}}/\sqrt{3} = n\pi$, where $n = 1, 2, 3 \cdots$; the odd peaks are compression peaks and the even peaks are rarefaction peaks, $\omega = (1 + z_{\text{LS}})k$ is the physical oscillation frequency at the time of last scattering, and $t_{\text{LS}} \simeq 400\,000\,\text{yr}$ is the age of the Universe then. Thus, in a flat Universe, the harmonic peaks occur at multipoles

$$l_n \sim \frac{n\pi}{H_0 t_{\text{LS}}(1 + z_{\text{LS}})} \sim 200n.$$

While the above formulae are approximate, they capture the physics. In particular, that the positions of the peaks depends upon the age of the Universe at last scattering. There will be sufficient redundancy in the information encoded in the 3000 or so multipole amplitudes that will be measured to not only determine cosmological parameters, but also to check the consistency of the standard relationship between age and energy density.

In particular, an analysis by Lopez et al. (1999) has projected that the Planck mission will be able to peg the neutrino contribution to the energy density of the Universe at last scattering to about 1%. The expansion rate is related to the energy density, $H^2 = 8\pi G\rho/3$, and neutrinos contribute about one-fifth of the total energy density at the time of last scattering. A quick estimate from Lopez et al. (1999) indicates that ultimately, CMB anisotropy will provide an age consistency test at about 0.1% precision, 400 000 yr after the beginning.

The limitations of this test should be noted, however. Actually determining the age of the Universe at last scattering, whose redshift is readily determined by thermodynamic considerations $(1 + z_{\text{LS}}) \simeq 1100)$, is pegged to the present Hubble constant, $t_{\text{LS}} \simeq H_0^{-1}/[\Omega_{\text{M}}(1 + z_{\text{LS}})^{3/2}]$, and thus can be no more accurate than the age of the Universe itself. The CMB consistency above actually probes the relationship between expansion rate and the energy density of the Universe, a test of the Big Bang framework and general relativity.

4. Big Bang Nucleosynthesis

Big Bang Nucleosynthesis (BBN) is a cosmological experiment of great importance. In essence, it is a quenched nuclear reactor whose by-products are the primeval mix from which the first generation stars were born.

The expansion rate of the Universe controls the quench rate: $|\dot{T}/T| = H$. The yields of D, ^{3}He, ^{4}He, and ^{7}Li, the primary products of BBN, are sensitive to the quench rate, and thus to the expansion rate,

$$\frac{\delta X}{X} \sim \frac{\delta H}{H_{\mathrm{STD}}}, \tag{4}$$

where the coefficient of proportionality varies from about 1.5 for ^{7}Li to about 0.6 for ^{4}He.

Measurements of the primeval deuterium abundance (see e.g., O'Meara et al. 2001) together with predictions for its BBN yield pin down the baryon density, $\Omega_{\mathrm{B}}h^2 = 0.02 \pm 0.001$ (Burles, Nollett, & Turner 2001a). Using this value, the other light-element abundances can be predicted, e.g., $Y_{\mathrm{P}} = 0.2472 \pm 0.0005$. Of the four light elements, the primeval ^{4}He abundance is known most accurately, $Y_{\mathrm{P}} = 0.244 \pm 0.002$ (though there is still much debate about systematic error; see e.g., Burles, Nollett, & Turner 2001b; Olive, Steigman, & Walker 2000).

The sensitivity of the abundance of ^{4}He to the expansion rate is $\delta Y_{\mathrm{P}} = 0.15(\delta H/H_{\mathrm{STD}})$. The current agreement of prediction with observation translates to a 2% test of the consistency of the Big Bang prediction for the expansion rate around 1 s, or about 5 times better than the Sandage test! With improved measurements of the baryon density, both from BBN and CMB anisotropy (see below), one might hope to see an improvement of a factor of five or so in the next decade.

Finally, as S. Carroll & M. Kaplinghat (in preparation) have recently emphasized, BBN can be discussed without reference to the standard cosmological model. The yields can be analyzed completely in terms of the quench rate, $|\dot{T}/T|$, and nuclear input data with only the assumption of the Robertson–Walker line element (and not the Friedmann equations which relate the quench rate to the temperature). BBN not only offers a window on the very early Universe, but an almost model-independent time scale test.

5. CMB + BBN and the Baryon Density

Together, the CMB and BBN offer a remarkable test of the consistency of the standard cosmology and general relativity. As mentioned in the previous section, measurements of the primeval deuterium abundance together with precise theoretical predictions can be used to infer the baryon density:

$$\Omega_{\mathrm{B}}(\mathrm{BBN})h^2 = 0.02 \pm 0.001. \tag{5}$$

Since the first measurements of the primeval deuterium abundance in 1998, BBN has provided the best determination of the baryon density (see e.g., Schramm & Turner 1998).

The CMB provides an independent measure of the baryon density based upon the physics of gravity-driven acoustic oscillations 400 000 yr after BBN. Specifically, it is the ratio of the heights of the odd to even acoustic peaks that is sensitive to the baryon density (and insensitive to other cosmic parameters; see e.g., Hu et al. 1997). The ratio of the heights of the first and second peaks is

$$\frac{\text{peak}_1}{\text{peak}_2} \simeq 2 \left(\Omega_\text{B} h^2 / 0.02 \right)^{2/3}, \tag{6}$$

where the peak heights are measured in units of μK^2.

The BOOMERANG and MAXIMA experiments were the first CMB experiments to probe the first and second peaks (and to show that the Universe is flat; de Bernardis et al. 2000; Hanany et al. 2000); based upon their results a value for the baryon density was inferred (Jaffe et al. 2001):

$$\Omega_\text{B} h^2 = 0.032^{+0.005}_{-0.004}.$$

While this baryon density is about 2σ higher than the BBN value, it confirmed a key prediction of BBN—a baryon density far below that of the best estimates of the total matter density ($\Omega_\text{M} \sim 0.3$ vs. $\Omega_\text{B} \sim 0.05$).

Only a couple of months after this meeting, Carlstrom's DASI experiment at the South Pole announced even more accurate and independent measurements of the first three acoustic peaks (Pryke et al. 2001) and arrived at a slightly lower baryon density:

$$\Omega_\text{B}(\text{CMB}) h^2 = 0.022^{+0.004}_{-0.003}, \tag{7}$$

which is bang on the BBN value! BOOMERANG analyzed more data and refined their beam map and pointing model and arrived at an identical value (Netterfield et al. 2001).

The agreement between these two numbers is stunning. Using the fact that $(\text{D/H})_\text{P} \simeq 3 \times 10^{-5} (\Omega_\text{B} h^2 / 0.02)^{-1.6}$ and the sensitivity of D/H to the expansion rate,

$$\frac{\delta(\text{D/H})}{(\text{D/H})} \simeq 1.4 \frac{\delta H}{H_\text{STD}},$$

it follows that

$$\frac{\delta(\Omega_\text{B} h^2)}{\Omega_\text{B} h^2} \sim -0.9 \frac{\delta H}{H_\text{STD}}.$$

Thus, the agreement of the BBN and CMB baryon densities checks the consistency of the standard expansion rate (at about 1 s) to a precision of about 15%. Eventually, both determinations of the baryon density should achieve about 1% or so accuracy, and a factor of ten in the precision of this test might be expected.

6. Particle Relics

The evidence for particle dark matter has only gotten stronger: firmer evidence for $\Omega_\text{B} \ll \Omega_\text{M}$ (discussed above), the many successes of the CDM scenario of structure formation (and no viable model for structure formation without particle dark matter), the detection of acoustic peaks as predicted by inflation and CDM (Netterfield et al. 2001; Pryke et al. 2001), and growing circumstantial evidence for supersymmetry.

The most promising particle candidate is the lightest supersymmetric particle, which in most models is a neutralino of mass 100 to 300 GeV (see e.g., Jungman, Kamionkowski, & Griest 1996). Relic neutralinos remain numerous today because of the incompleteness of neutralino annihilations in the early Universe.

At temperatures when $kT \gg m_\chi c^2$, neutralinos and anti-neutralinos are present in numbers comparable to that of photons; as the temperature drops neutralinos must annihilate to maintain thermal abundance (a factor $e^{-m_\chi c^2/kT}$ less than that of photons). Eventually, annihilations cannot keep pace with the quench rate (the annihilation rate per neutralino, $\Gamma = n_\chi < \sigma v >_{\text{ann}}$, falls rapidly as the neutralino abundance decreases exponentially) and the neutralino abundance freezes out. Freeze out occurs at a temperature, $kT_f \sim m_\chi c^2/30$, corresponding to a time of around 10^{-7} s. The mass density contributed by relic neutralinos is given by

$$\Omega_\chi h^2 \quad \propto \quad \frac{H(kT \sim m_\chi c^2/30)}{< \sigma v >_{\text{ANN}}}. \tag{8}$$

While an approximation, this formula captures the essence of the neutralino production process (see e.g., Kolb & Turner 1990).

Here is the future time scale test: $\Omega_\chi h^2$ will be measured by CMB anisotropy experiments to percent level precision. The properties of the neutralino can be measured at an accelerator lab (next linear collider?) to 10% precision (Brhlik, Chung, & Kane 2001). From this, the expansion rate at $kT = m_\chi c^2/30$ can be inferred and compared to the standard formula. (To be more precise, the measured properties of the neutralino and the Boltzmann equation in the expanding Universe can be used to predict the relic mass density and compared with the value inferred from CMB measurements.) If the neutralino is indeed the dark-matter particle (or another particle that can be produced in the lab), we can look forward to a 10% consistency test of the expansion rate at a time of less one microsecond!

7. Phase Transitions

If current ideas about particle physics are correct, then, during its earliest moments, the Universe should have gone through a series of phase transitions associated with symmetry breaking (e.g., QCD, electroweak, grand unification, compactification?). During a first-order phase transition, the Universe gets shaken up as bubbles of the new phase expand and collide. This can lead to the production of prodigious amounts of gravitational radiation, resulting in $\Omega_{\text{GW}}h^2 \sim 10^{-10}$ today. Laboratory-based knowledge of particle physics (e.g., for the electroweak phase transition, the mass of the Higgs boson) can be used to predict with precision the spectrum and amount of gravitational radiation (Kosowsky, Turner, & Watkins 1992); both depend directly upon the expansion rate at the temperature at which the phase transition takes place. If the stochastic background of gravitational waves from a phase transition can be detected and the particle physics independently probed in the laboratory a cosmological time scale test can be carried out at very early times (e.g., for the electroweak phase transition, $t \sim 10^{-11}$ s).

8. Inflation

Inflation also produces gravitational waves, by a different mechanism—de Sitter-space produced quantum fluctuations in the space-time metric. The stochastic background of gravity waves produced by inflation have wavelengths from 1 km to the beyond the size of the present horizon. Gravity waves are one of the three key predictions of inflation (together with a flat Universe and a nearly scale-invariant spectrum of adiabatic, Gaussian density perturbations; see e.g., Turner 1997a).

Detection of gravitational waves, either by their imprint on CMB anisotropy and/or polarization or directly by a gravity-wave detector (LIGO?, LISA?), would not only confirm a key prediction of inflation, but would also reveal the time scale for inflation: The level of gravitational radiation produced by inflation is directly related to the expansion rate during inflation:

$$h_{\mathrm{GW}} \sim H_{\mathrm{I}}/m_{\mathrm{Pl}}.$$

Measuring the dimensionless metric strain h_{GW} gives the Hubble parameter during inflation in units of the Planck energy ($= 1.22 \times 10^{19}$ GeV). Further, if the spectrum of gravitational radiation can be probed, then there is a consistency test of inflation (and cosmology): $T/S = -5n_T$, where T (S) is the contribution of gravity waves (density perturbations) to the quadrupole CMB anisotropy, and n_T is the spectral index of the inflation-produced gravitational waves (Turner 1997a, 1997b).

As a practical matter, gravity waves from inflation can probably only be detected if $H_{\mathrm{I}} > 3 \times 10^{12}$ GeV, corresponding to a time scale of $H_{\mathrm{I}}^{-1} < 10^{-30}$ s. Said another way, *if* gravity waves from inflation are detected we will be probing the Universe at a time at least as early as 10^{-30} s.

9. Dark Energy and Destiny

For decades cosmologists have believed that geometry (or equivalently Ω_0) and the fate of the Universe were linked. The shape of the Universe has been determined through measurements of CMB anisotropy, $\Omega_0 = 1.0 \pm 0.04$ (Pryke et al. 2001; Netterfield et al. 2001; Hanany et al. 2001), but we are further away than ever from determining the destiny of the Universe. This is because two-thirds of the critical density is in dark energy rather than matter, and the connection between geometry and destiny only applies when the Universe is comprised of matter (or more precisely, stress-energy with $p > -\rho/3$; see Krauss & Turner 1999).

The discovery of accelerated expansion through type Ia supernovae distance measurements (see Tonry, these proceedings) was surprising, but can easily be accommodated within the framework of general relativity and the Hot Big Bang Cosmology: In general relativity, the source of gravity is $\rho + 3p$, so that a fluid that is very elastic (negative pressure comparable in magnitude to energy density) has repulsive gravity. The simplest example is the quantum vacuum (mathematically equivalent to a cosmological constant), for which $p = -\rho$.

Unfortunately, all estimates of the energy of the quantum vacuum exceed by at least 55 orders-of-magnitude what is required to explain the acceleration of the Universe, suggesting to many that "nothing weighs nothing" and that something else with large negative pressure is causing the accelerated expansion (e.g., a rolling scalar field, a.k.a. as mini inflation or quintessence, or a frustrated network of cosmic defects; see Turner 2000 or Carroll 2001). Borrowing from Zwicky, I have coined the term "dark energy" to describe this stuff, which is clearly nonluminous and more energy-like than matter-like (since $|p|/\rho \sim 1$). It seems to be very smoothly distributed and its primary effect is on the expansion of the Universe. The first handle we will have in determining its nature is measuring its equation of state, $w = p/\rho$, through cosmological observations (see e.g., Huterer & Turner 2000).

Until we figure out the nature of the dark energy, the ultimate time scale question is on hold. Accepting that the Universe is flat (or at least that our bubble is; see Guth, these proceedings), the possibilities for destiny are wide open: continued accelerated expansion (and the almost complete "red out" of the extragalactic sky in 150 billion years) if the dark energy is vacuum energy; eternal slowing if the dark energy dissipates and matter takes over the dynamics; or even recollapse if the dark energy dissipates revealing a small, negative cosmological constant (Krauss & Turner 1999).

10.   Concluding Remarks

Cosmology is in the midst of a Golden Age (see e.g., Turner 2001a). As this meeting has illustrated, other areas of astronomy have not been left behind either. Astronomy in general is in the midst of the most exciting period of discovery ever.

The enormous variety and range of time scales in astrophysics makes the subject rich. Cosmology provides an especially good illustration, because the cosmological clock ticks logarithmically. Consistency checks on the cosmological clock have and will continue to play an important role in validating the standard cosmological model. While the consistency of the present age and expansion rate (Sandage test) is important and will improve in accuracy, from its present 15% to perhaps 5%, there are many other age consistency tests in cosmology, whose precision may well approach a few tenths of a percent (e.g., BBN and CMB), and extend to times as early as 10^{-30} s.

References

de Bernardis, P., et al. 2000, Nature, 404, 955

Brhlik, M., Chung, D., & Kane, G. 2001, Intl. J. Mod. Phys., D10, 367

Burles, S., Nollett, K., & Turner, M. S. 2001a, Phys. Rev. D, 63, 063512

Burles, S., Nollett, K., & Turner, M. S. 2001b, ApJ, 552, L1

Carroll, S. 2001, http://www.livingreviews.org/Articles/Volume4/2001-1carroll

Hanany, S., et al. 2000, ApJ, 545, L5

Hu, W., Sugiyama, N., & Silk, J. 1997, Nature, 386, 37

Huterer, D., & Turner, M. S. 2000, Phys. Rev. D, in press (astro-ph/0012510)

Jaffe, A., et al. 2001, Phys. Rev. Lett., 86, 3475

Jungman, G., Kamionkowski, M., & Griest, K. 1996, Phys. Rep., 267, 195

Kolb, E. W., & Turner, M. S. 1990, The Early Universe (Redwood City: Addison-Wesley)

Kosowsky, A., Turner, M. S., & Watkins, R. 1992, Phys. Rev. Lett., 69, 2026

Kragh, H. 1996, Cosmology and Controversy (Princeton: Princeton University Press)

Krauss, L., & Turner, M. S. 1999, Gen. Rel. Grav., 31, 1453

Loeb, A. 1998, ApJ, 499, L111

Lopez, R., et al. 1999, Phys. Rev. Lett., 82, 3952

Netterfield, C. B., et al. 2001 (astro-ph/0104460)

Olive, K., Steigman, G., & Walker, T. P. 2000, Phys. Rep., 389, 333

O'Meara, J. M., Tytler, D., Kirkman, D., Suzuki, N., Prochaska, J. X., Lubin, D., & Wolfe, A. M. 2001, ApJ, 552, 718

Pryke, C., et al. 2001, ApJ, submitted (astro-ph/0104490)

Schramm, D. N., & Turner, M. S. 1998, Rev. Mod. Phys., 70, 303

Tegmark, M. 2001, Phys. Rev. D, submitted (astro-ph/0101354)

Turner, M. S. 1997a, in Generation of Cosmological Large-Scale Structure, ed. D. N. Schramm & P. Galeotti (Dordrecht: Kluwer), 153

Turner, M. S. 1997b, Phys. Rev. D, 55, R435

Turner, M. S. 2000, Phys. Rep., 334, 619

Turner, M. S. 2001a, PASP, 113, 653

Turner, M. S. 2001b, ApJ, submitted (astro-ph/0106035)

Discussion

Taka Kajino: I'm a bit surprised to hear you refer to a concordance between CMB and Big Bang Nucleosynthesis concerning the number of baryons. You referred to the upper limit of ^{4}He and also higher depletion of lithium, but you also concluded that Ω_B from these two calculations is a factor of 2 different, so what do you think of the precision of these data, especially from the CMB?

Michael Turner: $\Omega_B h^2 = 0.02 \pm 0.001$ (68% cl) is what's predicted from Big Bang Nucleosynthesis. For the microwave background, it's still very early, but I think the best analysis is the analysis by the BOOMERANG and MAXIMA teams themselves (Jaffe et al. 2001), and they get $\Omega_B h^2 = 0.032^{+0.005}_{-0.004}$ so you can compare these two numbers and it's about a 2σ discrepancy. We can argue if it's 2.5 or 2.4σ; it's about a 2σ discrepancy. A number of groups have done the BBN analysis; I happen to believe our group has the best determination (Burles et al. 2001). There's not much argument about the central value, just the error bar, and all agree that the BBN prediction is for $\Omega_B h^2$. If you put in the Hubble constant you get 0.04 ± 0.01 and 0.06 ± 0.015, where the uncertainty in the Hubble constant blows up the error bars to 20%. There are still important issues for Big Bang Nucleosynthesis: for lithium, the amount of diffusion and potential depletion in Population II, and that's why I have the big error bar for lithium. There's a raging debate about the third significant figure for ^{4}He, but since the helium depends upon the baryon density logarithmically, it doesn't have much leverage at getting at the baryon density. Deuterium—Dave Schramm taught us this 20 years ago—is the baryometer. *Note added after the conference: Just 3 months after our meeting, DASI and BOOMERANG announced new, more precise results for the baryon density determined from CMB anisotropy: $\Omega_B h^2 = 0.022\pm0.003$. The BBN and CMB baryon densities are now in beautiful agreement.*

Wendy Freedman: Can you comment on how we can improve the measurement precision of the various cosmological parameters in the future?

Michael Turner: All will improve dramatically with measurements of large-scale structure and CMB anisotropy. Let me focus on dark energy. If I thought that the dark energy were Λ, then I think I would answer by saying that the microwave background and large-scale structure will give us everything we need to know. The key thing is that we don't know that it's Λ, and so we need another method to get the dark energy, and one way is a new assault on high-redshift supernovae. In the absence of knowing what the dark energy is, it is (at least) a two parameter problem: Ω_X and w, the equation of state of the dark energy. The projected error ellipse for Planck is quite big and will only pin down w to ±0.25. Planck is not going to tell us very much about what the dark energy is. A dedicated satellite experiment like SNAP with high-quality observations of 3000 supernovae could pin down w to ±0.05. That could teach us a lot about dark energy. For example, if $w = -2/3 \pm 0.05$ we would know that *Lambda* is out as a candidate for dark energy. There are other ways of getting at dark energy which involve counting objects. Number count tests have gotten more complicated since we now know that the Universe has evolved a lot between now and redshift 2, so understanding the evolution of the objects—be they halos or clusters—is crucial. Marc Davis and Jeff Newman think that you can match the projected precision of SNAP by counting halos in the DEEP survey (assuming only Poisson errors). Others have discussed using clusters. The dark energy problem is important and here to stay. To solve it, we need to explore a variety of approaches.

Alfred Vidal-Madjar: I wanted to come back to BBN and the measurement of deuterium because I think you should be more cautious about this number. As you said, deuterium is a strong constraint but Tytler & Burles' (e.g., 1998, ApJ, 507, 732) evaluation is the result of an extremely selective process from high-z quasars with a lot of hydrogen interlopers so they select the lowest possible deuterium abundance. If it's due only to interlopers the work is fine but if it's another reason then the low abundance could be wrong. So it's very important to be more cautious. With spacecraft we are starting to observe deuterium within the Galaxy and we see that it may be varying from region to region and maybe selecting only the lower deuterium abundance value from quasars could lead to the wrong answer. Suppose the real value from the light elements is on the other side of the region where you can find an agreement with helium and deuterium, what would be the answer to reconcile that with the BOOMERANG results?

Michael Turner: Let me give two answers. Number one: I think there's a very good case for the deuterium at 3×10^{-5}. Tytler and collaborators now have four systems and five other upper limit systems. As you say, this is a very important measurement and so we should keep looking at deuterium critically. As you correctly say, if you find something at the deuterium position at a higher value, the first thing that you should think about is an interloper. If you read Adams (1976), he said make these measurements, find putative deuterium systems, and look for a lower shelf, which tells you the primeval values. Values above that are possible caused by interlopers. Number two: I think the leverage at looking for the primeval deuterium value within our Galaxy is poor because the material in our Galaxy that we can see has been processed through stars where the deuterium is burned. But it also supports this value of 3×10^{-5} because our interstellar medium has a value of about 1.5×10^{-5} and that says that if the primeval value is 3×10^{-5}, the amount of astration is about a factor of 2. If you were to ask Jim Truran or another expert on Galactic chemical evolution what fraction of the material in the local ISM has been through stars, they would say about 50%, so that's very consistent. On the other hand, if the primeval deuterium were 10 times higher, that would mean that more than 90% of the material in the local interstellar medium had already been through stars. We can't rule that out but it would not fit into the standard models of Galactic chemical evolution.

Astrophysical Ages and Time Scales
ASP Conference Series, Vol. 245, 2001
T. von Hippel, C. Simpson, N. Manset

The Hubble Constant and the Expansion Age of the Universe

Wendy L. Freedman

Observatories of the Carnegie Institution of Washington, 813 Santa Barbara Street, Pasadena, CA 91101, USA

Abstract. The Hubble constant, together with the total energy density of the Universe, sets the size of the observable Universe and its age. Excellent progress has been made recently toward the measurement of the Hubble constant: a number of different methods for measuring distances have been developed and refined, and a primary project of the Hubble Space Telescope has been the accurate calibration of this parameter. The recent progress in these measurements is summarized. Currently, for a wide range of possible cosmological models, the Universe appears to have a kinematic age less than approximately 14 billion years. Combined with current estimates of stellar ages, the results favor a low-matter density universe. They are consistent with either an open universe or a flat universe with a non-zero value of the cosmological constant.

1. Determining the Expansion Age of the Universe

In standard Big Bang cosmology, the Universe expands uniformly: in the nearby Universe, $v = H_0 d$, where v is the recession velocity of a galaxy at a distance d, and H_0 is the Hubble constant, the expansion rate at the current epoch. The inverse Hubble constant H_0^{-1} sets the age of the Universe, t_0, and the size of the observable Universe, $R_{obs} = ct_0$, given a knowledge of the total energy density of the Universe. In Big Bang cosmology, the Friedmann equation relates the density, geometry and evolution of the universe:

$$H^2 = \frac{8\pi G \rho_m}{3} - \frac{k}{a^2} + \frac{\Lambda}{3},$$

where the average mass density is specified by ρ_m. The curvature term is specified by $\Omega_k = -k/a_0^2 H_0^2$, and for the case of a flat universe ($k = 0$), $\Omega_m + \Omega_\Lambda = 1$. Given an independent knowledge of the other cosmological parameters (H_0, Ω_m, Ω_Λ, and Ω_k), a dynamical age of the Universe can be determined by integrating the Friedmann equation.

Consider three different cosmological models. In the case of a flat, matter–dominated $\Omega_m = 1$ universe, the age is given simply by

$$t_0 = \frac{2}{3} H_0^{-1}.$$

For an open, low–density $\Omega_m < 1$ universe,

$$t_0 = H_0^{-1} \frac{\Omega_m}{2(\Omega_m - 1)^{\frac{3}{2}}} \left[\cos^{-1}(2\Omega_m^{-1} - 1) - \frac{2}{\Omega_m}(\Omega_m - 1) \right]^{\frac{1}{2}}$$

(Kolb & Turner 1990). For $\Omega_m = 0.3$,

$$t_0 \approx \frac{4}{5} H_0^{-1}.$$

Finally, for the case of a flat universe with $\Omega_\Lambda > 0$,

$$t_0 = \frac{2}{3} H_0^{-1} \Omega_\Lambda^{-\frac{1}{2}} \ln \left[\frac{1 + \Omega_\Lambda^{\frac{1}{2}}}{(1 - \Omega_\Lambda)^{\frac{1}{2}}} \right],$$

where, for $\Omega_m = 0.3$, $\Omega_\Lambda = 0.7$,

$$t_0 \approx H_0^{-1}.$$

In principle, with an accurate measurement of H_0 and the age of the universe measured independently of the expansion, t_0, (as discussed in many other articles in these proceedings), the product of $H_0^{-1} t_0$ can provide a powerful constraint on cosmology.

2. Measuring the Hubble Constant

More than seven decades have now passed since Hubble (1929) initially published the correlation between the distances to galaxies and their recession velocities, thereby providing evidence for the expansion of the Universe. But establishing an accurate cosmological distance scale and value for the Hubble constant have proved extremely challenging. Primarily as a result of new instrumentation at ground-based telescopes, and most recently with the availability of the Hubble Space Telescope (HST), the extragalactic distance scale field has been evolving at a rapid pace. Still, until very recently, a factor of two uncertainty in the value of H_0 has persisted for a variety of reasons (e.g., Freedman 1997). Since the 1980s, linear detectors, replacing photographic plates, have enabled much higher accuracy measurements, corrections for the effects of dust, and measurements to much greater distances, all combining to increase the precision in the relative distances to galaxies. Prior to HST, however, very few galaxies were close enough to allow the discovery of Cepheid variables, upon which the absolute calibration of the extragalactic distance scale largely rests.

Determination of the Hubble constant is extremely simple in principle: measure the recession velocities and the distances to galaxies at sufficiently large distances where deviations from the smooth Hubble expansion are small, and the Hubble constant follows immediately from the slope of the correlation between velocity and distance. However, progress in measuring H_0 has been limited by the fact that there exist few methods for measuring distances that satisfy many basic criteria. Ideally, a distance indicator should be based upon well-understood physics, operate well out into the smooth Hubble flow (velocity-distances greater than $\sim 10\,000$ km s^{-1}), be applied to a statistically significant sample of objects, be empirically established to have high internal accuracy, and most importantly, be demonstrated empirically to be free of systematic errors. The above list of criteria applies equally well to classical distance indicators as to other physical methods (in the latter case, for example, the Sunyaev–Zeldovich effect or gravitational lenses). At the present time, an ideal distance indicator or other method meeting all of the above criteria does not exist, and measurement of H_0 as high as 1% accuracy is clearly a goal for the future. However, as described below, an accuracy of H_0 to 10% has now likely been reached.

3.   Recession Velocities

Since the velocity of recession of a galaxy is proportional to its distance (Hubble's law), the farther that distance measurements can be made, the smaller the proportional impact of peculiar motions on the expansion velocities. For a galaxy or cluster at a recession velocity of $10\,000$ km s^{-1}, the impact of a peculiar motion of 300 km s^{-1} is 3% on H_0 for that object. This uncertainty is reduced by observing a number of objects, well-distributed over the sky, so that such motions can be averaged out. Moreover, given the overall mass distribution locally, a correction for peculiar motions can be applied to the velocities (over and above corrections for the Earth's, Sun and our Milky Way's motion in the Local Group). For type Ia supernovae, the distant indicator which currently extends the farthest ($v \sim 30\,000$ km s^{-1}), the effects of peculiar motions are a small fraction of the overall error budget.

3.1.   Distances to Galaxies

The most accurate of the primary distance indicators are the Cepheid variables, stars whose outer atmospheres pulsate regularly with periods ranging from 2 to about 100 days. Cepheids are bright, young stars, abundant in nearby spiral and irregular galaxies. The underlying physics of the pulsation mechanism is simple and has been studied extensively. Empirically it has been established that the period of pulsation (a quantity independent of distance) is very well correlated with the intrinsic luminosity of the star. The dispersion in the Cepheid period-luminosity relation in the I band (~ 8000 Å) amounts to about 20% in luminosity. From the inverse square law, this corresponds to an uncertainty of about 10% in the distance for a single Cepheid. With a sample of 25 Cepheids in a galaxy, a statistical uncertainty of about 2% in distance can be achieved. Hence, Cepheids provide an excellent means of estimating distances to resolved spiral galaxies.

The reach of Cepheid variables as distance indicators is limited. With available instrumentation, for distances beyond 20 Mpc or so, brighter objects than ordinary stars are required; for example, measurements of luminous supernovae or the luminosities of entire galaxies. Secondary methods (for example, type Ia supernovae, the Tully–Fisher relation, the fundamental plane, type II supernovae, and surface brightness fluctuations) provide several means of measuring *relative* distances to galaxies. The absolute calibration for all of these methods is presently established using the Cepheid distance scale. Although references are occasionally made to the "Cepheid distance scale" and the "supernova distance scale," the supernova distance scale is not independent of, but is built upon, the Cepheid distance scale. The same holds true for all of the other methods listed above.

3.2. Systematic Effects in Distance Measurements

Many distance indicators have sufficiently small scatter that with the current numbers of Cepheid calibrators, the statistical precision in their distance scales is 5% or better. The total uncertainty associated with the measurement of distances is higher, however, because of complications due to other astrophysical effects. Many of these systematic effects are common to all of these measurements, although their cumulative impact may vary from method to method. For any given method, there may also be systematic effects that are as yet unknown. However, by comparing several independent methods, a limit to the total systematic error in H_0 can be quantified.

Dust grains in the regions between stars, both within our own Galaxy and in external galaxies, scatter blue light more than red light, with a roughly $1/\lambda$ dependence. The consequences of this interstellar dust are two-fold: (1) objects become redder (a phenomenon referred to as reddening), and (2) objects become fainter (commonly called extinction). If no correction is made for dust, objects appear fainter (and therefore apparently farther) than they actually are.

A second potential systematic effect is that due to chemical composition or metallicity. Stars have a range of metallicities, depending on the amount of processing by previous generations of stars that the gas (from which they formed) has undergone. In general, older stars have lower metallicities, although there is considerable dispersion at any given age. Metals in the atmospheres of stars act as an opacity source to the radiation emerging from the nuclear burning. These metals absorb primarily in the blue part of the spectrum, and the radiation is thermally redistributed and primarily re-emitted at longer (redder) wavelengths.

4. The H_0 Hubble Space Telescope Key Project

The H_0 Key Project was designed to use Cepheid variables to determine primary distances to a representative sample of nearby galaxies in the field, groups, and clusters. Details of the Key Project and a final summary of the results are given by Freedman et al. (2001). The target galaxies were chosen so that each of the secondary distance indicators with measured high internal precisions could be calibrated in zero point, and then intercompared on an absolute basis. The Cepheid distances were then used for secondary calibrations out to cosmologically significant distances, with a goal of measuring H_0 to an accuracy of $\pm 10\%$, including systematic errors. The excellent image quality of HST extends the limit out to which Cepheids can be discovered by a factor of ten from ground-based searches, and the effective search volume by a factor of a thousand. Furthermore, HST offers a unique capability in that it can be scheduled optimally and independently of the phase of the Moon, the time of day, or weather, and there are no seeing variations. Before the launch of HST, most Cepheid searches were confined to our own Local Group of galaxies, and the very nearest surrounding groups, and the numbers of Cepheid calibrators for

various methods was dismally small (5 for the Tully–Fisher relation, one for
for the surface-brightness fluctuation method, and *no* Cepheid calibrators were
available for Type Ia supernovae).

Calibrating five secondary methods with Cepheid distances, Freedman et
al. (2001) find $H_0 = 72 \pm 3$ (random) ± 7 (systematic) km s^{-1} Mpc^{-1}. Type Ia
supernovae are the secondary method which currently extends out to the greatest distances, ~ 400 Mpc. All of the methods: Types Ia and II supernovae, the
Tully–Fisher relation, surface brightness fluctuations, and the fundamental plane
(Gibson et al. 2000; Sakai et al. 2000; Ferrarese et al. 2000; Kelson et al. 2000;
Mould et al. 2000) are in extremely good agreement: four of the methods yield
a value of H_0 between 70–72 km s^{-1} Mpc^{-1}, and the fundamental plane gives
$H_0 = 82$ km s^{-1} Mpc^{-1}. As described in detail in Freedman et al. (2001), the
largest remaining sources of error result from (a) uncertainties in the distance to
the Large Magellanic Cloud, which provides the fiducial period–luminosity relation; (b) photometric calibration of the HST Wide Field and Planetary Camera
2; (c) metallicity calibration of the Cepheid period–luminosity relation; and (d)
cosmic scatter in the density (and therefore, velocity) field that could lead to
observed variations in H_0 on very large scales. These systematic uncertainties
affect the determination of H_0 for all of the relative distance indicators, and
they cannot be reduced by simply combining the results from different methods:
they dominate the overall error budget in the determination of H_0.

In the top panel of Figure 1, a Hubble diagram of distance versus velocity is
plotted. The slope of this diagram yields the Hubble constant in km s^{-1} Mpc^{-1}.
The Hubble line plotted has a slope of 72. Two features are immediately apparent from Figure 1. First, all five secondary indicators plotted show excellent
agreement. Now that Cepheid calibrations are available for all of the methods
shown here, there is not a wide dispersion in H_0 evident in this plot. Second,
although the overall agreement is very encouraging, and each method exhibits
a small, internal or random scatter, there are measurable systematic differences
among the different indicators at a level of several percent.

5. H_0 From Methods Independent of Cepheids

At present, to within the uncertainties, there is broad agreement in H_0 values for completely independent techniques. Published values of H_0 based on
the Sunyaev–Zeldovich (SZ) method have ranged from ~ 40–80 km s^{-1} Mpc^{-1}
(e.g., Birkinshaw 1999). The most recent two-dimensional interferometry SZ
data for well-observed clusters yield $H_0 = 60 \pm 10$ km s^{-1} Mpc^{-1}. The systematic uncertainties are still large, but the near-term prospects for this method
are improving rapidly as additional clusters are being observed, and higher-
resolution X-ray and SZ data are becoming available (e.g., Reese et al. 2000).
A second method for measuring H_0 at very large distances, also independent of
the need for any local calibration, comes from the measurement of time delays in
gravitational lenses. H_0 values based on this technique appear to be converging
to the mid-60s km s^{-1} Mpc^{-1} range (Williams & Saha 2000). As more lenses
with time delays are discovered and monitored, this method also is likely to
improve substantially in the near future. Recent results from measurements of

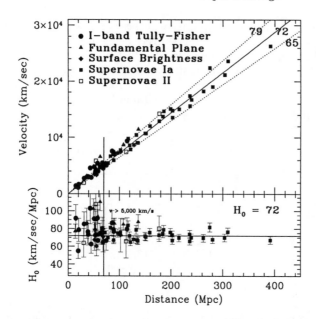

Figure 1.   *Top*: Hubble diagram of distance versus velocity for secondary distance indicators calibrated by Cepheids. The symbols are
as follows: Type Ia supernovae are solid squares, Tully–Fisher clusters
(*I*-band observations) are solid circles, Fundamental Plane clusters are
triangles, surface brightness fluctuation galaxies are diamonds, Type
II supernovae are open squares. A slope of $H_0 = 72$ km s^{-1} Mpc^{-1} is
shown, flanked by $\pm 10\%$ lines, and a distance to the LMC of 50 kpc has
been adopted. The Cepheid distances have been corrected for metallicity. Beyond 5000 km s^{-1} (indicated by the vertical line), both numerical simulations and observations suggest that the effects of peculiar
motions are small. *Bottom*: Residuals in H_0 as a function of velocity.

cosmic microwave background anisotropies are also consistent (de Bernardis et
al. 2000) with a value of $H_0 \approx 70$ km s^{-1} Mpc^{-1}.

6.   The Expansion Age

Up until very recently, the strong motivation from inflationary theory for a flat
Universe, coupled with a strong theoretical preference for $\Omega_\Lambda = 0$, favored a
matter-dominated $\Omega_m = 1$ Universe. In addition, the ages of globular cluster
stars were estimated for many years to be ~ 15 Gyr (Chaboyer et al. 1996).
However, for a value of $H_0 = 72$ km s^{-1} Mpc^{-1}, the $\Omega_m = 1$ model yields a very
young expansion age of only 9 ± 1 Gyr, significantly younger than the earlier
globular cluster age estimates. For $H_0 = 72$ km s^{-1} Mpc^{-1}, $\Omega_m = 0.3$, the
age of the Universe increases from 9 to 11 Gyr. Recently, new data on type Ia
supernovae from two independent groups have provided evidence for a non-zero
vacuum energy density corresponding to $\Omega_\Lambda = 0.7$ (Riess et al. 1998; Perlmutter

et al. 1999). If confirmed, the implication of these results is that the deceleration of the Universe due to gravity is progressively being overcome by a cosmological constant term, and that the Universe is in fact accelerating in its expansion. Allowing for $\Omega_\Lambda = 0.7$, under the assumption of a flat $(\Omega_m + \Omega_\Lambda = 1)$ Universe, increases the expansion age yet further to $t_0 = 13.5$ Gyr. The effect of different Ω values on the expansion age is shown in Table 1. The errors in the age reflect a 10% uncertainty in H_0 alone.

Table 1.   Ages for different values of cosmological parameters.

H_0	Ω_m	Ω_Λ	t_0 (Gyr)
70	0.2	0	12 ± 1
70	0.3	0	11 ± 1
70	0.2	0.8	15 ± 1.5
70	0.3	0.7	13.5 ± 1.5
70	1.0	0	9 ± 1

A non-zero value of the cosmological constant helps to avoid a discrepancy between the expansion age and other age estimates. An expansion age of 13.5 ± 1.5 Gyr is consistent to within the uncertainties with recent globular cluster ages, which have been revised downward to 12–13 Gyr based on a new calibration from the HIPPARCOS satellite (Chaboyer et al 1998; see also these proceedings).

7.   Future Improvements: An Accurate Astrometric Zero Point for the Distance Scale

A critical issue affecting the local determinations of H_0 remains the zero-point calibration of the extragalactic distance scale (more specifically, the Cepheid zero point). The most promising way to resolve this outstanding uncertainty is through accurate geometric parallax measurements. New satellite interferometers are currently being planned by NASA (the Space Interferometry Mission, SIM) and the European Space Agency (a mission known as GAIA) for the end of the next decade. These interferometers will be capable of delivering 2–3 orders of magnitude more accurate parallaxes than HIPPARCOS (i.e., a few microarcsec astrometry), reaching $\sim 1000\times$ fainter limits. Critical for both H_0 and globular cluster age determinations, accurate parallaxes for large numbers of Cepheids, RR Lyrae variables and subdwarfs will be obtained. Moreover, in addition to improving the calibration for the distance to the LMC, it may be possible to measure rotational parallaxes for several nearby spiral galaxies, with distances accurate to a few percent.

8.   Summary

Recent results on the determination of H_0 are encouraging. A ten-year HST Key Project to measure the Hubble constant has just been completed. HST was used to measure Cepheid distances to 18 nearby spiral galaxies. Calibrating 5 secondary methods with these revised Cepheid distances yields $H_0 = 72 \pm 3$ (random) ± 7 (systematic) km s^{-1} Mpc^{-1}, or $H_0 = 72 \pm 8$ km s^{-1} Mpc^{-1}, combining the total errors in quadrature. Other independent means of measuring H_0 (the Sunyaev–Zeldovich method, gravitational lens time delays, and cosmic microwave background anisotropies) are consistent with these results, to within current uncertainties. However, the need to improve the accuracy in the determination of H_0 is certainly not over. For an r.m.s. uncertainty of 10%, the 95% confidence range restricts the value of H_0 only to $58 < H_0 < 86$ km s^{-1} Mpc^{-1}, underscoring the importance of reducing remaining errors in the distance scale (e.g., zero point, metallicity).

A value of $H_0 = 72$ km s^{-1} Mpc^{-1} yields an expansion age of ~ 13 Gyr for a flat Universe (consistent with the recent cosmic microwave background anisotropy results) if $\Omega_m = 0.3$, $\Omega_\Lambda = 0.7$. Combined with the current best estimates of the ages of globular clusters (~ 12.5 Gyr), these results favor a Λ-dominated universe.

References

Birkinshaw, M. 1999, Phys. Rep., 310, 97

Chaboyer, B., Demarque, P., Kernan, P. J., & Krauss, L. M. 1996, Science, 271, 957

Chaboyer, B., Demarque, P., Kernan, P. J., & Krauss, L. M. 1998, ApJ, 494, 96

de Bernardis, P., et al. 2000, Nature, 404, 955

Ferrarese, L., et al. 2000, ApJ, 529, 745

Freedman, W. L. 1997, in Critical Dialogs in Cosmology, ed. N. Turok (Singapore: World Scientific), 92

Freedman, W. L., et al. 2001, ApJ, in press (astro-ph/0012376)

Gibson, B. K., et al. 2000, ApJ, 529, 723

Hubble, E. 1929, Publ. Nat. Acad. Sci., 15, 168

Kelson, D. D., et al. 2000, ApJ, 529, 768

Kolb, E. W., & Turner, M. S. 1990, The Early Universe (New York: Addison-Wesley)

Mould, J. R., et al. 2000, ApJ, 529, 786

Perlmutter, S., et al. 1999, ApJ, 517, 565

Reese, E. D., et al. 2000, ApJ, 533, 38

Riess, A. G., et al. 1998, AJ, 116, 1009

Sakai, S., et al. 2000, ApJ, 529, 698

Williams, L. L. R., & Saha, P. 2000, AJ, 119, 439

Acknowledgments. I thank the organizers of this meeting for a very stimulating conference covering a very wide range of topics. I also would like to thank my many collaborators on the extragalactic distance scale over the years, most recently those involved in the HST Key Project.

Discussion

Piotr Popowski: Do you find it disturbing that most of your error in H_0 comes from uncertainty in the distance to the Large Magellanic Cloud?

Wendy Freedman: Yes, I find it distressing in the sense that having spent enormous time getting decreased uncertainties at larger distances, it then turns out that the overall uncertainty is dominated by errors in our own backyard. And that's certainly the case now—I didn't go into this in detail, but the dominant sources of uncertainty in the current value of H_0 are the distance to the LMC, the distance in our photometric calibration for the Wide Field and Planetary Camera detector, and uncertainties due to the metallicity in the Cepheid distance scale—the LMC is a major component of that. I would add that we (S. E. Persson et al., in preparation) recently obtained JHK period–luminosity relations for a sample of about 90 Cepheids in the in the Large Magellanic Cloud. We calibrate these P–L relations with Cepheids in the Galaxy, and we're finding a distance modulus very much in agreement with the distance modulus to the Large Magellanic Cloud that we've adopted, a distance of 50 kpc or 18.5 magnitudes in distance modulus. But I think it's going to require measurements from SIM, some parallax satellite, before we're going to get to the precisions below 5% which will allow us to do better.

Jean-René Roy: Is the discrepancy between your team and Sandage's dominated by the way you treat extinction?

Wendy Freedman: No, not any more. That was true about 10 years ago when Sandage didn't make any corrections for extinction and he'd done his measurements in the B-band, which are affected most by extinction. He and his group have now adopted the same method that we have for correcting for extinction; that is they observe at V and at I and use a standard extinction correction, so it doesn't come down to that difference at all. Unfortunately, it's not just one reason for the discrepancy, there are several reasons. But the effect is systematic (see Gibson et al. 2000, ApJ, 529, 723; Freedman et al. 2001, ApJ, 552, May 20).

Young-Wook Lee: The Galactic globular clusters certainly contain the oldest stars in our Galaxy and maybe the oldest in the Local Group. But there is no guarantee that they are the oldest in our Universe; Hyun-chul Lee and I found some evidence that giant elliptical galaxies may contain the oldest populations, a few Gyr older than our Galactic globular clusters. This is not a crazy idea because there is every reason to believe that star formation will occur earlier in dense regions, where we find giant elliptical galaxies. I would like to caution you and others to keep this in mind until we have better data.

Wendy Freedman: No, I agree with your comment entirely. I think it has to be looked at as a lower limit and that's all that you have direct empirical evidence for. I was concerned about this point and a few years ago I asked Pierre Demarque whether anybody had run models at significantly larger ages.

Particularly, star formation is likely to occur earlier in deeper potential wells, and you can imagine that in elliptical galaxies it could have occurred earlier. The problem is there's no direct empirical evidence. But it's certainly something to keep in mind.

Michael Rich: I want to comment on this idea that the ellipticals may have stars billions of years older than the oldest globular clusters. There is going to be a test done in the next couple of years which should settle that issue. The ultraviolet rising flux which is observed in ellipticals at present day is due to hot horizontal branch stars, and if you watch that flux disappear as a function of redshift you can constrain the ages of the oldest stars in elliptical galaxies relative to the globular cluster age scale. And I think that would be a robust way—certainly we would be able to constrain ages of order 2 or 3 Gyr that way, probably better.

Wendy Freedman: I think this would shed some light on the issue of whether there are older clusters but modulo dependence on metallicity and various other factors might be difficult to calibrate. But you'd certainly see if there were a problem there with older clusters.

Bob Schommer: On the issue of the zero point and the LMC calibration, there is another potential calibrator, the maser galaxy NGC 4258. Would you comment on its consistency with the LMC distance or zero point?

Wendy Freedman: Yes, NGC 4258, the maser galaxy. There's been an independent determination of the distance to this galaxy using both radial velocity and proper motion measurements for the rotation of masers in the center of this galaxy, which gives a geometric measure of the distance to that system. A few of us (Newman et al., ApJ, in press) got a Cepheid distance to that galaxy a couple of years ago to provide an independent test of the distance using that method. Initially the agreement was reasonable, that is within 1σ, but there was a systematic difference in the sense that would predict that the LMC were closer if you used the calibration based on the masers. There's a new calibration that we've adopted for the Cepheids which is based on a sample of 660 OGLE stars in the Large Magellanic Cloud—our earlier calibration was based on 32 stars—also there were refinements to the WFPC2 zero point. The upshot of all of this was that the agreement between the maser and the Cepheid distances turned out to be better. So they're within a fraction of sigma, which is a nice independent test of the Cepheid scale. But I would emphasize that there are uncertainties at a level of $\pm10\%$ (1σ). At the moment, the 2σ limits for distances published for the LMC are quite large and we want to do better, unquestionably.

Astrophysical Ages and Time Scales
ASP Conference Series, Vol. 245, 2001
T. von Hippel, C. Simpson, N. Manset

Tunable Filter Surveys for Star Formation Across Different Lookback Times

Heath Jones

European Southern Observatory Chile, Alonso de Cordova 3107, Vitacura, Casilla 19001, Santiago, Chile

Joss Bland-Hawthorn

Anglo-Australian Observatory, P.O. Box 296, Epping, NSW 1710, Australia

Abstract. Tunable filters such as the Taurus Tunable Filter (TTF; Bland-Hawthorn & Jones 1998) offer a direct way of searching for red-shifted emission-lines that are the signatures of star-formation in distant galaxies. The use of a tunable narrow-band to scan in wavelength produces samples with well-defined volumes and redshift intervals. Such star-forming galaxy samples differ significantly from those of traditional redshift surveys, in which galaxies are first broadband-selected. Here I summarise results from tunable filter surveys of star-forming galaxies at Hα redshifts of $z = 0.08$, 0.24, and 0.4, corresponding to look-back times of $\sim$ 1, 4, and 5 Gyr. The motivation for the surveys are also discussed, as well as their implications for current models of cosmic star-formation history.

1. Introduction

The union of both high and low redshift measurements of star-formation activity in normal galaxies has drawn attention over the past few years towards a determination of the cosmic star-formation history of the universe (Figure 1). Through a variety of indicators, the volume-averaged star formation rate (SFR) is seen to decline over the past 10 Gyr to the levels we see today (Madau, Pozzetti, & Dickinson 1998). Such a scenario is encouraging as it unifies many separate results into a consistent picture for the first time.

However, it remains a composite picture, with points spanning a diverse range of star-formation indicators and galaxy selection criteria. Here I discuss how this inhomogeneity is problematic for our understanding of cosmic star-formation history. I will then summarise recent results from the *Taurus Tunable Filter Galaxy Survey* (Jones & Bland-Hawthorn 2001), which unifies the *selection* and star-formation *measurement* of the sample, through the same quantity: Hα emission-line luminosity.

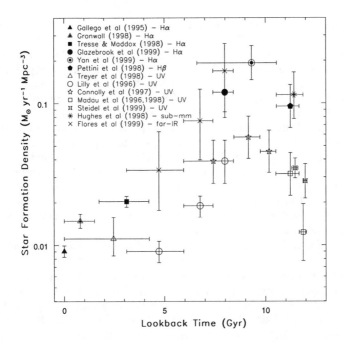

Figure 1. Composite cosmic star-formation history as obtained through a variety of both selection and measurement techniques. Most of the measurements (but not all) make use of the UV continuum (open points) or Balmer lines (solid points) to determine star-formation rate.

2. Selection Versus Measurement Issues

There are two steps involved in the derivation of the star-formation density at a given redshift: (1) the *selection* of a suitable galaxy sample, and, (2) *measurement* of the star-formation rates in each galaxy, via various calibrators at different wavelengths. These calibrators range from flux measurements of nebular emission-lines (such as Hα, Hβ and [O II]), to continua in the ultraviolet, far-infrared and radio. At the basis of all such star-formation calibrators is a conversion between the amount of flux measured at a chosen wavelength and the total stellar mass.

While the relative merits of different star-formation calibrators have been discussed at length by many authors (Kennicutt 1998; Schaerer 1999; Glazebrook et al. 1999; Tresse & Maddox 1998; Kennicutt 1992), the first issue—namely, *selection of a suitable galaxy sample*—remains largely untested. This is unfortunate, since the selection techniques used for the samples in Figure 1 vary across the range of lookback times. In part, the mixture of selection criteria is due to the availability of various selection features at different redshifts. However, it is also because many of the samples were not specifically selected with the measurement of star-formation history in mind. Rather, volume-averaged star-formation rates have usually been derived later as an afterthought.

As Figure 1 shows, the techniques used to *measure* star-formation rates also vary across the plot. The primary concern here is the extent to which flux measures are influenced by factors unrelated to the instantaneous SFR. For example, measurements in the ultraviolet suffer much uncertainty due to the high levels of extinction present at these wavelengths. Furthermore, the light from longer-lived OB stars can taint ultraviolet flux measures of the newly-born stars in unknown amounts. The far-infrared SFR indicator of Flores et al. (1999) is based on the fact that the absorption cross-section of dust peaks at the same UV wavelengths at which young stellar populations emit, therefore providing a measure of what the dust absorbs and re-emits in the infrared. In practice, the relation becomes unreliable in the presence of different dust types (of varying optical depth), and the contribution of active galactic nuclei to the far-infrared flux. Both the far-infrared and sub-millimetre samples may be sensitive to galaxies not present in the optical, because they measure UV light reprocessed into thermal radiation.

The use of nebular emission lines in the optical as a direct measure of SFR is well known (see the reviews by Kennicutt 1998 and Schaerer 1999). These lines re-emit the ionising UV flux (shortward of the Lyman limit, $\lambda < 912$ Å) produced by newly-formed stars. The strength of the hydrogen recombination lines is directly proportional to this and many calibrations for $H\alpha$ exist in the literature (Kennicutt 1998; Madau et al. 1998; Gallego et al. 1995; and references therein). $H\beta$ is less useful, having a third the strength and therefore being more prone to the underlying stellar absorption. [O II] has been widely employed to delineate star-forming populations in samples for which $H\alpha$ was inaccessible due to redshift, and calibrations in terms of either the line flux or a combination of [O II] equivalent width and continuum luminosity exist (Kennicutt 1992). Although a useful SFR measure, [O II] is inferior to $H\alpha$ due to its dependence on both the energy of the UV ionising flux and also the metal content of the intervening gas (Kennicutt 1992). It is also subject to greater extinction corrections, (~ 2 times those for $H\alpha$ for a galaxy with $A_V = 1$ mag of visual extinction). Furthermore, SFR measures based on [O II] equivalent widths are subject to large measurement uncertainties introduced by the faint galaxy continua on this quantity. In short, $H\alpha$ is the best indicator of SFR among the optical emission-lines, as has been noted by several authors (Kennicutt 1998; Schaerer 1999; Glazebrook et al. 1999; Tresse & Maddox 1998; Kennicutt 1992).

Therefore, surveys that use the $H\alpha$ to *both* select *and* measure star-forming galaxies in a given volume, offer the best chance at homogenizing our current picture of star-formation history in the optical. A tunable filter is well-suited to doing both, provided some additional means exist to resolve the redshift/line ambiguity in the lines detected.

3.   Tunable Filter Imaging

Tunable filters are Fabry-Perot interferometers that differ from conventional devices in two important ways. First, the plates are operated at much smaller plate spacings than the Fabry-Perot instruments usually used for astronomy. The effect of this is to widen the central interference region of the chosen wavelength. A conventional Fabry-Perot, with a plate spacing of many tens or even

(a) Conventional Fabry-Perot   ## (b) Tunable Filter

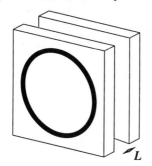   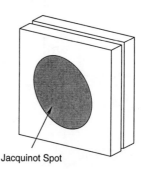

Figure 2.   Comparison between the use of (*a*) a conventional Fabry-Perot interferometer (with plate spacing *L*) and (*b*) a tunable filter. The narrower plate spacing of the tunable filter broadens the central interference region (Jacquinot spot) to cover a substantial portion of the full beam.

hundreds of microns, presents an interference region as a narrow ring on the sky, with very small area (Figure 2a). A tunable filter, with a plate spacing of no more than a few microns, aims to provide a broadened central interference region, known as the Jacquinot spot (Figure 2b). The latter is more useful for survey work, where one seeks a common wavelength transmitted across the full field and where lower spectral resolutions are desired.

Another way in which a tunable filter differs from a conventional Fabry-Perot is in its ability to access a much wider range of plate spacings. Conventional devices are most commonly used to scan through a relatively small range of wavelengths around a single spectral feature. However, a tunable filter, aiming to access as broad a tunable range as is possible, needs to access a much wider range of plate settings. This is made possible by having a stack of piezo-electric transducers (PZTs) to control plate spacing, instead of the usual single-layer.

The red Taurus Tunable Filter (TTF) is a tunable Fabry-Perot Interferometer covering 6500–9600 Å (Bland-Hawthorn & Jones 1998). In effect, the TTF affords monochromatic imaging with an adjustable passband of between 6 and 60 Å. TTF has an important advantage in the detection of faint line emission over conventional redshift surveys. Deep pencil-beam surveys typically pre-select objects down to $B \sim 24$ (Glazebrook et al. 1995) or $I \sim 22$ (Lilly et al. 1995), since these represent the practical spectroscopic limits on the 4-meter telescopes used. However, these limits are set assuming the objects are dominated by continuum. Objects with continuum flux beyond the spectroscopic limit but with emission-lines above the limit could be detected by these same instruments, albeit through the lines alone. However, because the initial selection is made from the broadband flux, such objects are excluded *at the outset* on the basis of having too faint a continuum level.

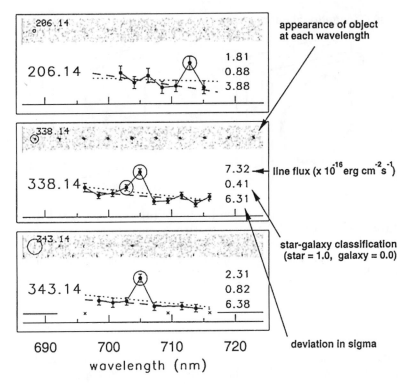

Figure 3.   Example emission-line detections with TTF obtained by scanning around 710 nm. Individual images are 9″ on a side with north up, east left.

4. The Taurus Tunable Filter Galaxy Survey

4.1. Survey Description

The TTF Field Galaxy Survey is a survey for redshifted emission-line galaxies in the field (Jones & Bland-Hawthorn 2001). All fields were taken with TTF in TAURUS-2 at the Anglo-Australian Telescope (AAT). The survey comprises 15 scans of 10 slices, at random high-galactic latitude fields scattered around the sky. By *slice*, we mean an exposure at a particular wavelength, or the image obtained by co-adding many such exposures. The total sky coverage is 0.27 square degrees, with scans distributed between the 707/26 (R_1), 814/33 (R_5) and 909/40 (R_8) TTF order sorting filters. Mean passbands with FWHM ($\delta\lambda$) of 12.9, 16.4 and 22.3 AA were used to cover 707/26, 814/33 and 909/40 respectively. Passbands were stepped in increments of $1.3\,\delta\lambda$ as a compromise between velocity coverage and sampling continuity. Table 1 gives the observational characteristics for scans in each of the three spectral regions, in terms of Hα and [O II]. Figure 3 shows examples of three emission-line galaxies detected during the survey.

Table 1.   Narrow-band scan coverage.[a]

Filter $\lambda/\Delta\lambda$ (nm)	Redshift Range ($z_1 \leq z \leq z_2$)	Fields	Total Volume (Mpc^3)	$\log(L_{line})$[b] (ergs s^{-1})
Hα line:				
707/26	$0.062 \leq z \leq 0.093$	7	1170	39.43
814/33	$0.221 \leq z \leq 0.260$	5	6660	40.44
909/40	$0.359 \leq z \leq 0.411$	3	9850	40.87
[O II] line:[c]				
707/26	$0.870 \leq z \leq 0.924$	7	49500	41.67
814/33	$1.149 \leq z \leq 1.219$	5	50300	41.93
909/40	$1.393 \leq z \leq 1.484$	3	42500	42.12

[a] Assuming $H_0 = 50$ km s^{-1} Mpc^{-1} and $q_0 = 0.5$. Circular TTF field of 9' diameter.
[b] Hα or [O II] log line luminosity at a detected flux of 1×10^{-16} erg cm^{-2} s^{-1}.
[c] Not as prevalent as Hα at the flux limits of the survey.

Only scans taken under photometric conditions were included, reaching $\sim 5 \times 10^{-17}$ erg cm^{-2} s^{-1} as a 3σ-detection in a 2″ aperture. For a 20 Å bandpass, this flux is equivalent to 2.5×10^{-18} erg cm^{-2} s^{-1} Å or $I = 21.8$ as broadband continuum. At the opposite extreme, if the $\sim 5 \times 10^{-17}$ erg cm^{-2} s^{-1} Å is detected as *line only*, (with no continuum), it is equivalent to a broadband limit of $I = 26.8$. This is sufficient for the detection and measurement of star-formation rates amounting to no more than a few tenths solar masses per year, (comparable with SMC-type levels at the nearest redshifts and the LMC at $z_{H\alpha} \sim 0.4$).

In such a blind search at these wavelengths, the emission-lines most commonly detected in galaxies are Hα, [O II], Hβ and [O III]. The detection of Lyα would be rare at such flux levels while the lines of [S II] are too far into the red to be found in an appreciable volume. The TTF passband is too wide to separate Hα and [N II].

Detailed estimates suggest that around the flux limits of the survey, Hα dominates over [O II], representing more than 90% of the full emission-line sample in the redder 814/33 and 909/40 bands, (see figure 3 of Jones & Bland-Hawthorn 2001). In the 707/26 band, the Hα and [O II] galaxies comprise approximately equal numbers ($\sim 50\%$) with comparatively fewer numbers of galaxies seen in [O III] and Hβ. The dominance of Hα at stronger fluxes in all bands suggests that samples restricted to these values will essentially constitute Hα, with little contamination from background [O II]. In the case of 707/26, this means truncating the emission-line sample artificially higher (~ 1.5 to 2×10^{-16} erg cm^{-2} s^{-1}) than the natural flux limits of the survey. The trade-off is that there are far fewer galaxies with higher line fluxes. In short, we expect to Hα galaxies to dominate samples at fluxes exceeding $\sim 2 \times 10^{-16}$ erg cm^{-2} s^{-1} for the 707 interval, and $\sim 0.5 \times 10^{-16}$ erg cm^{-2} s^{-1} for the 814 and 909 intervals.

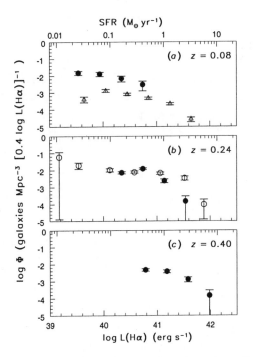

Figure 4.   Preliminary Hα luminosity functions from the TTF Field Galaxy Survey (filled circles) at mean redshifts of (*a*) $\bar{z} = 0.08$, (*b*) $\bar{z} = 0.24$, and (*c*) $\bar{z} = 0.4$. No extinction corrections have been applied to the luminosities. The other points are the Hα luminosity functions from Gallego et al. (1995; open triangles) for $z \leq 0.045$ and Tresse & Maddox (1998; open circles) at $z \sim 0.2$.

4.2.   Preliminary Hα Luminosity Functions

The most direct approach to finding evolutionary trends is to compare Hα luminosity functions directly (Figure 4). Hα fluxes are corrected here for the effect of [N II] using the mean flux ratio Hα/(Hα+[N II]) = 0.69, obtained by Tresse & Maddox (1998). Figure 4a demonstrates the excess in the $z = 0.08$ counts over the faint-end of the Gallego et al. (1995; open triangles) distribution by almost an order of magnitude. Much closer agreement exists between the TTF values and those of a preliminary Hα luminosity function from the KPNO International Spectroscopic Survey (KISS; C. Gronwall 2000, private communication). Taken together, these results suggest that the Gallego et al. (1995) survey underestimates the faint-end of the local Hα luminosity function, and hence, the star-formation content of the local universe. For the $z = 0.24$ interval (Figure 4b) we find broad agreement with the Hα luminosity function derived by Tresse & Maddox (1998; open circles). Due to the volume-limited nature of the survey there are insufficient numbers to define the bright-end "knee" of the distribution. The measurements at $z = 0.40$ (Figure 4c) are the first to use Hα in the ∼ 5 Gyr of lookback time separating the samples of Tresse & Maddox (1998) at $z \sim 0.2$ and Glazebrook et al (1999) at $z \sim 0.9$.

The preliminary Hα luminosity functions argue for an upwards revision of the star-formation content for the local universe. However, follow-up spectroscopy is needed to confirm the nature of the Hα detections and to correct individual Hα fluxes for the effects of extinction and [N II]. This spectroscopic follow-up is currently underway, and should provide estimates for the chemical evolution and AGN number-density of the sample, as well as an improved derivation of the Hα star-formation densities.

References

Bland-Hawthorn, J., & Jones, D. H. 1998, in Proc. SPIE 3355, Optical Astronomical Instrumentation, ed. S. D'Odorico (Bellingham: SPIE), 855

Flores, H., et al. 1999, ApJ, 517, 148

Gallego, J., Zamorano, J., Aragón-Salamanca, A., & Rego, M. 1995, ApJ, 455, L1

Glazebrook, K., Blake, C., Economou, F., Lilly, S., & Colless, M. 1999, MNRAS, 306, 843

Glazebrook, K., Ellis, R., Colless, M., Broadhurst, T., Allington-Smith, J., & Tanvir, N. 1995, MNRAS, 273, 157

Jones, D. H., & Bland-Hawthorn, J. 2001, ApJ, 550, 593

Kennicutt, R. C., Jr. 1992, ApJ, 388, 310

Kennicutt, R. C., Jr. 1998, ARA&A 36, 189

Lilly, S., Le Fèvre, O., Crampton, D., Hammer, F., & Tresse, L. 1995, ApJ, 455, 50

Madau, P., Pozzetti, L., & Dickinson, M. 1998, ApJ, 498, 106

Schaerer, D. 1999, in Building Galaxies: From the Primordial Universe to the Present, ed. F. Hammer (Gif-sur-Yvette: Editions Frontières), 389

Tresse, L., & Maddox, S. J. 1998, ApJ, 495, 691

Acknowledgments. It is a pleasure to thank the many colleagues who have made many helpful suggestions and comments about this work over the years. In particular, I am grateful to Joss Bland-Hawthorn, Matthew Colless, Caryl Gronwall, and Karl Glazebrook.

Discussion

Gavin Dalton: How can you tell the difference between Hα at low redshift and [O II] at high redshift? This could substantially affect your faint-end counts.

Heath Jones: The bottom line is that you can't do that with the survey information from the tunable filter alone; you need to go to a multi-object spectrograph. You could try to tune the filter to a very narrow passband and try to separate the lines, but this then compromises you in terms of survey coverage—you'd lose much more time stepping the narrow passband. So I think the best approach is to obtain spectroscopic follow-up to a set of TTF candidates, and we are currently in the process of obtaining optical spectra using VLT.

Recent developments at the Anglo-Australian Observatory now make it possible to undertake multi-object spectroscopy using the same TAURUS-2 focal reducer that carries TTF. This means spectroscopic follow-up can be done during the same observing run by simply exchanging the tunable filter for a grating and mask. However, without any follow-up spectroscopic information, one is forced to make estimates about the amount of contamination in the way that we've done here. [O II] generally comes in at the fainter flux limits which, and because we're scanning over a relatively narrow volume, the flux directly scales to luminosity. So that means that at the faint luminosity end that you speak of, we are getting more and more into the realm of [O II] and the other major emission lines at higher redshift.

Alan Stockton: I actually think the problem is not [O II] but [O III] at slightly lower redshift. I've done some random fields and a good fraction of the strong emission lines you pick up are actually [O III]. One thing that isn't appreciated is how many of these very low-metallicity, high-ionization galaxies there are out there, once you get out to redshifts of 0.5 or so.

Heath Jones: Yes my estimates would agree with you. Our calculations (shown in figure 3 of Jones & Bland-Hawthorn 2001) are an estimate of what the cumulative contributions from each of the major emission lines would be if one takes a line-selected narrow-band sample down to some flux limit. So at the brighter fluxes it is indeed Hα, but as soon as you get down to some cross-over point, (which is at a different place for the different wavelength regimes), you suddenly get all of these emitters in different lines at higher redshift. So your comment was that [O III] would be coming through in equal numbers: certainly we would expect to encounter them in roughly equal proportions to Hα and [O II] at the fainter limits of the 814 and 909 nm wavelength intervals; in the 707 nm interval they dominate over [O II] at brighter flux limits even still. However, Hα dominates over all of them at the brightest flux limits.

One of the main drivers of this survey was to use the scanning narrow band to survey these redshifts in an unprecedented way. In doing so, we expected to get many faint emission-line galaxies that are missed by broad-band selection, simply because they're swamped by the background. We believe we have detected significant numbers of such galaxies, but our programme of spectroscopic follow-up needs to be followed through to ascertain the nature of these faint sources.

Astrophysical Ages and Time Scales
ASP Conference Series, Vol. 245, 2001
T. von Hippel, C. Simpson, N. Manset

The Evolution of the Lyman Alpha Forest From $z \sim 3 \to 0$

Romeel Davé

Steward Observatory, University of Arizona, 933 North Cherry Avenue, Tucson, AZ, USA

Abstract. I review results obtained from studies of the high-redshift Lyα forest and present new results from HST/STIS spectra of low-redshift quasars in comparison with cosmological hydrodynamic simulations. The evolution of the Lyα forest from $z \sim 3 \to 0$ is well-described by current structure formation models, in which Lyα forest absorbing gas at all redshifts traces moderate-overdensity large-scale structures. I describe some of the insights provided by hydrodynamic simulations into the observed statistical trends of Lyα absorbers and the physical state of the absorbing gas.

1. Introduction

Quasar absorption lines have been used as probes of the high-redshift universe for some time (e.g., Sargent et al. 1980), since the Lyα (1216 Å) transition is redshifted into the optical at $z \gtrsim 2$. HIRES (Vogt et al. 1994) on the Keck 10-m telescope has been at the forefront in providing high S/N, high resolution quasar spectra that has enabled dramatic advances in our understanding of the high-redshift intergalactic medium (Rauch 1998). Conversely, studies of low-redshift Lyα absorbers require ultraviolet spectroscopy, which has only recently become routinely possible with the Hubble Space Telescope. The HST Quasar Absorption Line Key Project (Bahcall et al. 1993) observed over 80 quasars using the Faint Object Spectrograph (FOS), and found a surprisingly large number of absorbers as compared to an extrapolation from high-redshift (Bahcall et al. 1996; Weymann et al. 1998). Recently the deployment of the Space Telescope Imaging Spectrograph (STIS) has provided a dramatic increase in resolving power (resolution 7 kms as opposed to FOS's 230 km s^{-1}), thereby fully resolving all Lyα absorbers seen in UV quasar spectra. While only the brightest quasars and AGN may be easily observed with STIS, this has still resulted in significant gains in our understanding of low-redshift Lyα absorbers and their relation to high-redshift absorbers. In these proceedings I report some recent results from HST/STIS.

In conjunction with these observations, hydrodynamic simulations of structure formation in currently favored cosmological scenarios have elucidated a new paradigm for the nature of Lyα forest absorbers, particularly at high redshift. These simulations indicate that Lyα absorbers arise in highly photoionized diffuse intergalactic gas tracing the dark matter in non-equilibrium large-scale structures. Their temperature is set by a balance between photoionization heating from a metagalactic UV background (presumably from quasars; Haardt & Madau 1996) and adiabatic cooling due to Hubble expansion, resulting in a tight relation $T \propto \rho^{0.6}$. These simulations are able to reproduce various statistical properties of high-redshift Lyα absorbers in detail, spanning the range of damped Lyα systems to the weakest detected with HIRES, all within a model that generically arises in CDM cosmologies (Hernquist et al. 1996; Davé et al. 1997). Figure 1 shows a comparison of a portion of the HIRES spectrum of Q 1422+231 (Songaila & Cowie 1996) with a simulated spectrum covering the same redshift interval, having resolution and noise properties that emulate the data. Can you tell which is which? (the answer is in the Acknowledgments).

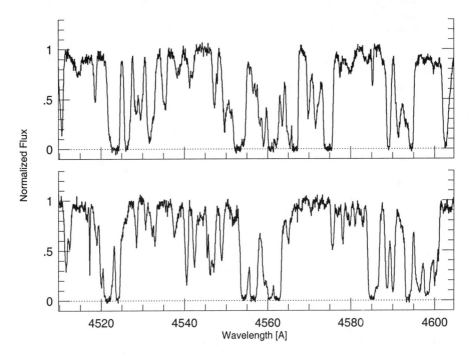

Figure 1.   A 100 Å section from the spectrum of Q 1422+231, and an artificial spectrum drawn from a hydrodynamic simulation of a ΛCDM universe, having resolution and noise characteristics constructed to be similar to Q 1422. The fluctuating absorption pattern seen in the simulated spectrum (which one is it?) arises naturally from large-scale structures generated by hierarchical collapse, and is statistically indistinguishable from that seen in Q 1422.

Due to the relatively simple relationship between H I and the underlying dark matter, a formula known as the "Fluctuating Gunn-Peterson Approximation" (FGPA; see Croft et al. 1998 for full formula) provides a remarkably accurate description of Lyα absorbing gas:

$$\tau_{\rm HI} \propto \rho^{1.6} \Gamma_{\rm HI}^{-1}, \tag{1}$$

where $\tau_{\rm HI}$ is the H I optical depth, ρ is the density of dark matter (or baryons), and $\Gamma_{\rm HI}$ is the metagalactic H I photoionization rate incident on the absorber.

2. Results From High-Redshift Lyα Forest Studies

The FGPA has been combined with HIRES quasar spectra to provide stringent constraints on many aspects of the physics of high-redshift intergalactic gas. For instance, a measurement of the mean optical depth together with an independent estimate of $\Gamma_{\rm HI}$ result in a robust estimate of $\Omega_{\rm b} \approx 0.02h^{-2}$ (Rauch et al. 1997), in good agreement with D/H measurements (Burles & Tytler 1998) and recent CMB analyses (e.g., Pryke et al. 2001).

The FGPA also implies that a given Lyα optical depth (or flux, in the optically-thin regime) corresponds to a particular density of gas at a particular temperature. Hence ionization corrections can be accurately obtained to constrain the metallicity of the diffuse high-redshift intergalactic medium (IGM) from metal-line observations of C IV (Songaila & Cowie 1996; Rauch, Haehnelt, & Steinmetz 1997; Davé et al. 1998). These studies found [C/H] ≈ -2.5 for a Haardt & Madau (1996) ionizing background shape, and significantly lower for a spectrum with fewer high-energy photons (as would be expected if Helium had not yet reionized; see Heap et al. 2000). Furthermore, recent VLT/UVES observations of O VI absorption suggests the presence of some metals even in voids (Schaye et al. 2000), presenting a significant theoretical challenge to models of metal ejection and transport in the IGM (e.g., Aguirre et al. 2001).

The most ambitious use of the FGPA, and the one for which it was originally developed, is a reconstruction of the matter power spectrum at $z \sim 3$ from Keck/HIRES quasar spectra. Since the optical depth in the Lyα forest traces the matter density, fluctuations in the optical depth reflect a 1-D probe of the amplitude and shape of the matter power spectrum, at scales of ~ 1–$10h^{-1}$ Mpc (comoving). For a flat Universe, this measurement yields a constraint on the matter density of $\Omega_{\rm m} \approx 0.5 + 0.29(\Gamma - 0.15)$, where Γ is the power spectrum shape parameter (Croft et al. 2000). This value is slightly higher than in the "concordance model" ($\Omega_{\rm m} \approx 1/3$; Bahcall et al. 1999), but the latest CMB measurements combined with the H_0 Key Project value (Mould et al. 2000) suggest a higher $\Omega_{\rm m}$ as well (Pryke et al. 2001). Thus the Lyα forest provides an independent avenue for doing "precision cosmology," alongside CMB, Type Ia supernovae, etc.

3. Evolution of the IGM to Low Redshift

Recently, hydrodynamic simulations have been evolved to redshift zero to investigate the nature of low-redshift Lyα absorbers and the evolution of the Lyα forest. They indicate that while many of the baryons have moved into galaxies, clusters, and a diffuse shocked "warm-hot intergalactic medium" (Cen & Ostriker 1999; Davé et al. 2001), roughly one-third of all baryons continue to reside in photoionized gas tracing large-scale structure.

These simulations are able to explain a wide range of Key Project observations quite naturally within the context of hierarchical structure formation scenarios. For instance, the Key Project data showed that high-redshift ($z \gtrsim 2$)

Lyα absorbers are evolving away quite rapidly, whereas for $z \lesssim 1.5$ the absorber number density evolution slows abruptly and dramatically (Bahcall et al. 1996). In the simulations of Davé et al. (1999) and Theuns, Leonard, & Efstathiou (1998), this occurs due to the diminution of the quasar population providing the metagalactic photoionizing flux. As a result, the rapid evolution of absorbers due to Hubble expansion in the high-z Universe is countered by the increase in neutral fraction of Lyα absorbing gas at $z \lesssim 2$. These simulations disfavor a scenario in which a different population of absorbers dominates at low redshift as compared to high redshift, and instead suggest that the nature of absorbing gas is quite similar.

Weymann et al. (1998) found that stronger absorbers disappear more rapidly than weaker absorbers, and in fact the weakest absorbers seen in the FOS sample (rest equivalent width $W_r \approx 0.1$ Å) show an *increase* in number between $z \sim 1.5 \rightarrow 0$. Davé et al. (1999) suggested that this arose due to differential Hubble expansion, in particular that low-density regions expand faster than high-density ones and therefore provide an increasing fractional cross-section of absorption to lower redshifts.

Motivated by the good agreement between simulations and FOS observations, Davé et al. (1999) investigated the physical state of the gas giving rise to Lyα absorption at low redshift. They found that the gas follows similar physical relationships as deduced at high-redshift, and is thus of the same basic character, but that a given column density absorber corresponds to different physical densities at different redshifts.

The effect is shown in Figure 2. The top left panel shows the density-column density relation at $z = 3$ from a ΛCDM simulation. The tight relation reflects the accuracy of the FGPA; much of the scatter about the best-fit relation (solid line) comes from errors in Voigt profile fitting (noise with S/N $= 30$ was added to the artificial spectra to roughly match observations). At low redshift, the relationship persists, though there is more scatter due to the presence of absorbers arising in shock-heated intergalactic gas (Davé et al. 1999; Cen & Ostriker 1999). The best-fit slope of the ρ–$N_{\rm HI}$ relation is similar, but the amplitude has shifted considerably from $z = 3 \rightarrow 0$. This shift arises because of the interplay between Hubble expansion and the evolution of the ionizing background.

The bottom panel of Figure 2 focuses on the evolution of this trend with redshift. Here, we show the typical column density of an absorber arising in gas with three different densities, as a function of redshift. For instance, gas at the mean density (which typically remains around that density over a Hubble time) produces a strong absorber of $N_{\rm HI} \approx 10^{14}$ cm^{-2} at $z = 3$, but by $z = 0$ it produces only a very weak absorber of $N_{\rm HI} \approx 10^{12.5}$ cm^{-2}. In general, a given density corresponds to a column density $\sim 30\times$ lower at $z = 0$ as compared to $z = 3$. Thus studying physically and dynamically equivalent absorbers at high and low redshift requires comparing across different column densities.

Using this trend, the knowledge and intuition gained from studies of the Lyα forest at high redshift can be translated to expectations at low redshift. For instance, it is known that most absorbers with $N_{\rm HI} \gtrsim 10^{14.5}$ cm^{-2} at $z \sim 3$ are enriched (Songaila & Cowie 1996); thus one expects that absorbers with

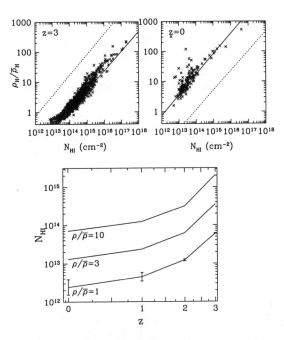

Figure 2.   *Top panels*: Relation between column density of absorber and maximum physical density (in units of mean cosmic density) at the absorber peak optical depth, $z = 3$ and $z = 0$. Best-fit relation is shown as the solid line and is reproduced in the other panel as the dotted line. *Bottom panel*: For three selected densities, the evolution of the corresponding absorber column density as a function of redshift. The $z \sim 0$ forest is qualitatively quite similar to the $z \sim 3$ forest, except column densities are shifted down by ~ 1.5 dex.

$N_{HI} \gtrsim 10^{13}$ cm^{-2} should be enriched to at least a similar level (though detecting these metal lines will be challenging).

4.   Low-Redshift Lyα Forest Observed With HST/STIS

A new window on the low-redshift Lyα forest has opened with the deployment of STIS aboard HST. STIS's E140M Echelle grating provides ~ 7 km s^{-1} resolution, a level of detail comparable to HIRES albeit with somewhat more noise. STIS can thus probe weak forest absorbers that are predicted to be physically similar to those at high-redshift, and allows a detailed comparison to be made with hydrodynamic simulations.

Davé & Tripp (2001) compared the quasar spectra of PG 0953+415 and H 1821+643 to carefully-constructed artificial spectra drawn from a ΛCDM simulation. The statistical properties of absorbers in the simulations were in remarkable agreement with observations, as shown in Figure 3. The column density distributions agree down to the smallest observable absorbers, in both slope and amplitude. Interestingly, the slope of the column density distribution (from $\frac{d^2N}{dz\,dN_{HI}} \propto N_{HI}^{-\beta}$) is measured to be $\beta = 2.0 \pm 0.2$, which is considerably steeper than at high redshift ($\beta \sim 1.5$; Kim et al. 1997). This indicates, in agreement with Weymann et al. (1998), that stronger absorbers have evolved away faster than weaker ones. Our result is also in broad agreement ($\sim 1\sigma$ higher) with Penton, Shull, & Stocke (2000), who found $\beta \sim 1.8$ from GHRS data.

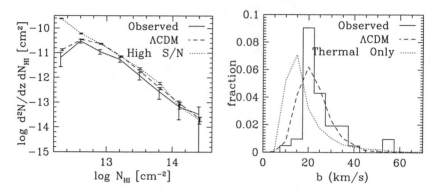

Figure 3.   Column density (*left*) and b-parameter (*right*) distributions from HST/STIS observations of PG 0953+415 and H 1821+643 (solid lines), compared with artificial spectra (dashed lines). Dotted line in left panel shows the result of increasing S/N by a factor of 5 in the artificial spectra, indicating that the absorber sample is complete down to $N_{HI} \lesssim 10^{13}$ cm^{-2}. Dotted line in right panel shows the simulated line widths purely from thermal broadening, indicating that thermal broadening provides a substantial component of line widths at low redshift.

The distribution of line widths (b-parameters) is also in excellent agreement with observations. The median b-parameter is around 21 km s^{-1}, but this value is sensitive to the column density detection threshold, because the column density and b-parameter are correlated. This can be understood from the fact that higher column density systems arise in higher density gas that has higher temperature. Since thermal broadening contributes significantly to line widths (compare the dotted and dashed lines in the right panel of Figure 3), higher column density absorbers will tend to have larger widths. Thus Shull et al. (2000), who find a median b-parameter of ≈ 30 km s^{-1} from FUSE and GHRS data, are not in disagreement with our result because their sample has a higher column density detection limit.

The large thermal component of line widths facilitates a measurement of the typical temperature of Lyα absorbing gas at low redshift. By comparing the observed median b-parameter to simulations, Davé & Tripp (2001) found that purely photoionized Lyα absorbers (those that have not been shock-heated) at the mean density have a temperature of $T \approx 5000$ K, in agreement with Ricotti, Gnedin, & Shull (2000).

The agreement in amplitude of the column density distribution can be translated into a constraint on the H I photoionization rate Γ_{HI}, since varying Γ_{HI} would shift the simulated column density distribution proportionally in the horizontal direction (assuming optically thin lines, which is valid for most STIS absorbers). Davé & Tripp (2001) obtain $\Gamma_{HI} \sim 10^{-13.3\pm0.7}$ s^{-1}, with the errors dominated by systematic uncertainties in modeling, providing among the most precise constraints on the metagalactic Γ_{HI} to date.

5. Conclusion

The Lyα forest has proved to be a powerful tool for constraining a wide variety of physical and cosmological parameters, when hydrodynamic simulations and the FGPA are used in conjunction with high-resolution quasar spectra. Much progress has already been made at high redshift, where the metallicity, ionization state, and mass fluctuation spectrum have all been derived from Lyα forest data.

Hydrodynamic simulations of structure formation also provide an excellent description of low redshift Lyα forest absorbers, particularly for the weak absorbers that reside in moderate-overdensity regions analogous to high-redshift absorbers. HST/STIS has opened up a new era in the study of the low-redshift IGM, which promises to progress considerably further with the deployment of the Cosmic Origins Spectrograph aboard HST (in 2003). Initial results indicate that the low-redshift Lyα forest can yield important constraints on physical parameters in the IGM, as well as provide a more complete picture of the evolution of baryons in the Universe.

References

Aguirre, A., Schaye, J., Hernquist, L., Weinberg, D., Katz, N., & Gardner, J. 2000, ApJ, in press (astro-ph/0006345)

Bahcall, J. N., et al. 1993, ApJS, 87, 1

Bahcall, J. N., et al. 1996, ApJ, 457, 19

Bahcall, N., Ostriker, J. P., Perlmutter, S., & Steinhardt, P. J. 1999, Science, 284, 1481

Burles, S., & Tytler, D. 1998, ApJ, 507, 732

Cen, R., & Ostriker, J. P. 1999, ApJ, 514, 1

Croft, R. A. C., Weinberg, D. H., Katz, N., & Hernquist, L. 1998, ApJ, 495, 44

Croft, R. A. C., et al. 2000, ApJ, in press (astro-ph/0012324)

Davé, R., Hellsten, U., Hernquist, L., Katz, N., & Weinberg, D. H. 1998, ApJ, 509, 661

Davé, R., Hernquist, L., Katz, N., & Weinberg, D. H. 1999, ApJ, 511, 521

Davé, R., Hernquist, L., Weinberg, D. H., & Katz, N. 1997, ApJ, 477, 21

Davé, R., & Tripp, T. M. 2001, ApJ, in press

Davé, R., et al. 2001, ApJ, in press (astro-ph/0007217)

Haardt, F., & Madau, P. 1996, ApJ, 461, 20

Heap, S. R., et al. 2000, ApJ, 534, 69

Hernquist, L., Katz, N., Weinberg, D. H., & Miralda-Escudé, J. 1996, ApJ, 457, L51

Kim, T. S., Hu, E. M., Cowie, L. L., & Songaila, A. 1997, AJ, 114, 1

Mould, J. R., et al. 2000, ApJ, 529, 786

Penton, S. V., Shull, J. M., & Stocke, J. T. 2000, ApJ, 544, 150

Pryke, C., Halverson, N. W., Leitch, E. M., Kovac, J., Carlstrom, J. E., Holzap-
    fel, W. L., & Dragovan, M. 2001, ApJ, submitted (astro-ph/0104490)

Rauch, M. 1998, ARA&A, 36, 267

Rauch, M., Haehnelt, M. G., & Steinmetz, M. 1997, ApJ, 481, 601

Rauch, M., et al. 1997, ApJ, 489, 7

Ricotti, M., Gnedin, N. Y., & Shull, J. M. 2000, ApJ, 534, 41

Sargent, W. L. W., Young, P. J., Boksenberg, A., & Tytler, D. 1980, ApJS, 42,
    41

Schaye, J., Rauch, M., Sargent, W. L. W., & Kim, T.-S. 2000, ApJ, 541, L1

Shull, J. M., et al. 2000, ApJ, 538, L13

Songaila, A. & Cowie, L. L. 1996, AJ, 112, 335

Theuns, T., Leonard, A., & Efstathiou, G. 1998, MNRAS, 297, L49

Vogt, S. S., et al. 1994, in Proc. SPIE 2198, Instrumentation in Astronomy VIII,
    ed. D. L. Crawford & E. R. Craine (Bellingham: SPIE), 362

Weymann, R., et al. 1998, ApJ, 506, 1

Acknowledgments. I thank Todd Tripp, Lars Hernquist, Neal Katz, David Weinberg, and Ray Weymann for collaborative efforts on these projects and helpful discussions. In Figure 1, the Q 1422 spectrum is in the top panel.

Discussion

Brian Chaboyer: How well do you know the ionizing flux as a function of redshift, and how does this affect your results?

Romeel Davé: The ionizing flux at high-redshift is reasonably well constrained because it is likely that quasars are the dominant sources; it's also constrained from other measurements such as the proximity effect. At low redshift, things get much hairier of course, the uncertainties are much larger, but we can actually measure it from the Lyman alpha forest. At high redshift, given a J_ν you measure tau and you get Ω_b, but what we did at low redshift was: given an Ω_b, we measure τ and we get the photoionization rate. The value that you get is in fact in excellent agreement with what's predicted by, for instance, Haardt & Madau, purely from the quasar population. So that would suggest that certainly quasars are at least an important contributor to the low redshift ionizing medium.

Astrophysical Ages and Time Scales
ASP Conference Series, Vol. 245, 2001
T. von Hippel, C. Simpson, N. Manset

Old Galaxies at High Redshift

G. B. Dalton

Astrophysics, Keble Road, Oxford OX1 3RH, UK

Abstract. We review the observational data available on the class of galaxies known as Extremely Red Objects (EROs), and present new data on their apparent surface density, which now appears to be much higher than has previously been claimed. We discuss the division of these objects into two distinct groups: young galaxies with dramatic ongoing star formation and old galaxies with highly evolved stellar populations. Age-determination for the latter group using population synthesis models is complicated by an intrinsic degeneracy between the age and initial metallicity of the stellar populations.

1. Introduction

The typical mix of stellar populations in present day galaxies gives rise to a distribution of optical–infra-red colours with mean $(R - K) \approx 3$ and an upper extremum of $(R - K) \sim 4.5$. A observations extend to fainter magnitudes, and consequently to higher redshifts, the mean galactic colour shifts slightly to the red, and the extreme colour, as represented by the population of elliptical galaxies in rich cluster environments, is correspondingly reddened to $(R-K) \sim 5$ for $z \sim 1$. The deepest complete observations that are currently available are those of the Hubble Deep Field (HDF). From the K-magnitude–redshift relation for the HDF, one would infer that at $K = 19.0$ there are no galaxies more distant than $z \sim 1.25$, or, conversely, that any galaxy above $z = 1.5$ should have $K \gtrsim 20$.

The first indication of galaxies with excessively red optical–NIR colour came from the observations of Elston, Rieke, & Rieke (1988) who found two galaxies at $K \sim 16.5$ with $(R - K) > 5$ in a single field of area 10 arcmin^2, which they initially flagged as candidates for primordial galaxies. However, both were subsequently identified as elliptical galaxies at $z \sim 0.8$ (Elston, Rieke, & Rieke 1989), and so proved to be neither primordial galaxies, or EROs, in the sense that no great effort is required to reproduce these objects with spectrum synthesis codes.

The observation by Hu & Ridgeway (1994) of two galaxies with $(I - K) \sim 6.5$, presents a somewhat different story. Initially flagged as likely elliptical galaxies above $z = 2$, these were subsequntly shown to be extremely young, starbursting galaxies at $z \sim 1.5$, with the extremely red optical–NIR colours produced by the high dust content of the galaxies. These galaxies thus defined the primary classification of EROs, and provide a plausible interpretation of the colours of other galaxies which are observed to have similarly red colours.

569

The second class of EROs reverts to the elliptical galaxies observed by Elston et al. (1989), but is really defined by the observations of LBDS 53W091 (Spinrad et al. 1997). This galaxy has $(R - K) = 5.8$, and was shown to have an absorption line spectrum with a reshift of $z = 1.55$, illustrating that ellipticals do indeed exist with colours that are redder than those predicted by traditional spectrum synthesis models.

Subsequent to these defining observations, a number of authors have presented infra-red observations which identify further examples of EROs, and there is a wide range of estimates for the surface densities of the ERO population available in the literature which we summarize in Table 1. In this paper we present preliminary results from a new survey of more than 4000 arcmin² of moderately deep K-band imaging data, with accompanying deep multicolour optical photometry. We adopt the most common definition of EROs as being galaxies with $(R - K) > 6$, since this colour is difficult to obtain from conventional sprectrum synthesis models, given our K-band magnitude limit. Spectroscopic and high-resolution imaging follow-up of a small subset of these data serves to illustrate further the dichotomy of EROs discussed above. We present robust estimates of the ERO surface densities from this survey, with the indication that there is indeed a substantial variation in the surface density of these objects along different sight lines, indicating that at least one of the two populations exhibits significant clustering.

Table 1.   Estimates of ERO surface densities in the current literature.

Authors	Area ($\Box'$)	Depth	Density ($\Box'$)$^{-1}$	Definition
Hu & Ridgway (1994)	10	$K < 19$	0.01	$(I - K) > 5$
Cowie et al. (1994)	5.9	$K < 21$	0.7	$(I - K) > 5$
Thompson et al. (1999)	154	$K < 19$	0.039 ± 0.016	$(R - K) > 6$
Liu et al. (2000)	6.2	$K < 19$	0.58	$(R - K) > 6$
Hall et al. (2001)	31 flds	var	0.117 ± 0.03	$(R - K) > 6$
McCracken et al. (2000)	94	$K < 20$	0.5	$(I - K) > 4$
Yan et al. (2000)	16	$H < 21.5$	0.94 ± 0.24	$(R - H) > 5$
Daddi et al. (2000)	447	$K < 19.2$	0.10 ± 0.01	$(R - K) > 6$

2.   EROs in the Oxford–Dartmouth Thirty-Degree Survey

The objective of the Oxford–Dartmouth Thirty-degree (ODT) survey is to obtain multicolour optical and NIR imaging of three distinct regions around the sky, each covering roughly 10 square degrees, with magnitude limits given in Table 2. The primary goal of the survey is to study the evolution of galaxy clustering.

The INT data have been obtained as one of the INT Wide Field Survey programmes with the INT Wide Field Camera. This gives a 36' field of view which is sampled by four EEV 2048 × 4096 CCDs with a pixel scale of 0.33″.

Table 2.   Magnitude limits for the ODT survey.

Band	Telescope	5-σ limit	Current Area ($\Box°$)
U	INT	26	15
B	INT	26	15
V	INT	25.5	15
R	INT	25.25	15
i'	INT	24.5	15
Z	INT	22	15
K	MDM	19	2
1.4Ghz	VLA	70μJy	2.5

The K-band data has been obtained with the ONIS instrument at the MDM 1.3-m telescope which currently gives a $5' \times 10'$ field of view with $0.55''$ pixel sampling on an ALADDIN InSb detector. The ONIS detector has a large number of bad pixels in one quadrant, which gives rise to variations in the detection completeness within a target frame as well as the usual weather-dependent variations which give rise to detection completeness variations between target frames within a single field. The level of the incompleteness due to the presence of these bad pixels has been modeled extensively using the `artdata` utility within `IRAF`,

Figure 1.   Illustrative K-band completeness map for the ODT Andromeda field. Light areas are highly complete. The $5' \times 10'$ field size is clearly visible, as is the region at the NE of each frame, which is affected by large numbers of bad detector pixels.

and gives rise to a completeness map for each of the three fields, and example of which is shown in Figure 1. Full details can be found in E. J. Olding et al. (in preparation).

We have used the INT WFC data to perform star–galaxy separation to give a matched B, R, K galaxy catalogue for each survey field (V and i' data has recently been added). In Figure 2 we show the cumulative surface densities for EROs and very red galaxies, after removal of stellar objects and correction for the completeness variations. We use the three independent measurements of ERO surface densities presented here to determine an overall estimate of 0.25 ± 0.05 arcmin^{-1}. A comparison between this number and the spread shown in Table 1 shows that we detect a significantly higher density than would be inferred from the published data. Given the extent of the variation that we see between our three survey fields, we conclude that this apparent discrepancy may be due to true cosmic variance, implying that we are not yet able to declare that a *fair sample* of the high redshift Universe has been observed.

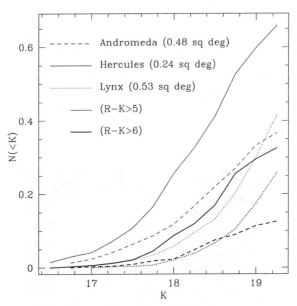

Figure 2.   Cumulative surface densities for red $(R - K > 5)$ and extremely red $(R - K > 6)$ galaxies for each of the three ODT fields.

3.   Follow-up Observations

We visually inspected all the EROs in the Andromeda field, and arbitrarily selected a pair of EROs from an apparent ERO cluster for spectroscopic follow-up with Subaru's OHS instrument. The resultant spectra are shown in Figure 3. The brighter of the two objects shows a strong absorption feature which can be identified as the Mg(5175) feature, yielding a redshift of $z = 1.35$. No features were identifiable in the second spectrum, but it was noted that the integrated spectral light from this object fell short of the H-band magnitude inferred from the acquisition image. A subsequent deep UFTI image of this field reveals that this object does indeed have an extended spiral morphology, whilst the other 7 EROs which fall in the $90 \times 90''$ UFTI field are extremely compact. We therefore conclude that the fainter object is a member of the HR 10 class of EROs, whereas the other objects are most likely representative of a cluster of elliptical galaxies at the measured redshift.

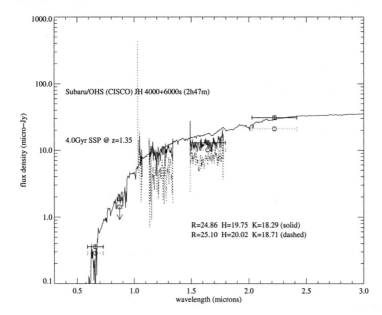

Figure 3.   *J*- and *H*-band spectra of two EROs from the Andromeda field obtained with OHS on the Subaru telescope. The solid line shows a best-fit SED to the brighter of the two objects.

4.   The Time Scale for ERO Formation

We now consider the inferred ages of these extremely red ellipticals. In a passive/pure luminosity evolution scenario, a uniform stellar population of a particular IMF is assumed to form over a very brief evolutionary time scale. The colors of a model galaxy formed in this way provide a useful boundary on the expectations of colors of observed galaxies. In Figure 4, the $(R - K)$ color of passively-evolving galaxies as a function of redshift viewed from are shown (for a Kennicutt IMF), for formation redshifts of $z_{\mathrm{form}} = 3$ and 5 (solid and dashed lines, respectively). For both sets, the colors are computed for a range of five metallicities from $Z = \frac{1}{20}$–$2 \times Z_{\odot}$, the corresponding to solar metallicity. These caution against a straightforward interpretation of narrow ranges of $(R - K)$ in color-selected galaxies corresponding to well-defined redshift ranges. At the same time, these models demark an extreme envelope of colors that unreddened

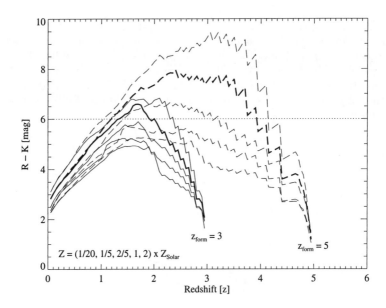

Figure 4.   Illustration of the degeneracy between age and metallicity (see text).

and ancient galaxies may take, and highlight the need to obtain spectra of sufficient signal-to-noise ratio to allow metallicity estimates to be obtained.

References

Cowie, L. L., Gardner, J. P., Hu, E. M., Songaila, A., Hodapp, K. W., & Wainscoat, R. J. 1994, ApJ, 434, 114

Daddi, E., Cimatti, A., Pozzetti, L., Hoekstra, H., Röttgering, H. J. A., Renzini, A., Zamorani, G., & Mannucci, F. 2000, A&A, 361, 535

Elston, R., Rieke, G. H., & Rieke, M. J., 1988, ApJ, 331, 77

Elston, R., Rieke, G. H., & Rieke, M. J., 1989, ApJ, 341, 80

Hall, P. B., et al. 2001, AJ, 121, 1840

Hu, E. M., & Ridgway, S. E., 1994, AJ, 107, 1303

Liu, M. C., Dey, A., Graham, J. R., Bundy, K. A., Steidel, C. C., Adelberger, K., & Dickinson, M. E. 2000, AJ, 119, 2556

McCracken, H. J., Metcalfe, N., Shanks, T., Campos, A., Gardner, J. P., & Fong, R. 2000, MNRAS, 311, 707

Spinrad, H., Dey, A., Stern, D., Dunlop, J. S., Peacock, J. A., Jiminez, R., & Windhorst, R., 1997, ApJ, 484, 581

Thompson, D., et al. 1999, ApJ, 523, 100

Yan, L., McCarthy, P. J., Weymann, R. J., Malkan, M. A., Teplitz, H. I., Storrie-Lombardi, L. J., Smith, M., & Dressler, A. 2000, AJ, 120, 575

Astrophysical Ages and Time Scales
ASP Conference Series, Vol. 245, 2001
T. von Hippel, C. Simpson, N. Manset

High Redshift Quasars and Star Formation History

M. Dietrich, F. Hamann

Department of Astronomy, University of Florida, 211 Bryant Space Science Center, Gainesville, FL 32611-2055, USA

Abstract. Quasars are among the most luminous objects in the Universe, and they can be studied in detail up to the highest known redshift. Assuming that the gas associated with quasars is closely related to the interstellar medium of the host galaxy, quasars can be used as tracers of the star formation history in the early Universe. We have observed a small sample of quasars at redshifts $3 \lesssim z \lesssim 5$ and present results using N V/C IV and N V/He II as well as Mg II/Fe II to estimate the date of the first major star formation epoch. These line ratios indicate solar and supersolar metallicities of the gas close to the quasars. Assuming times of $\tau_{\rm evol} \simeq 1$ Gyr, the first star formation epoch can be dated to $z_{\rm f} \simeq 10$, corresponding to an age of the Universe of less than 5×10^8 yr ($H_0 = 65$ km s^{-1} Mpc^{-1}, $\Omega_{\rm M} = 0.3$, $\Omega_\Lambda = 0.7$).

1. Introduction

In the context of cosmic evolution, the epoch of first star formation in the early Universe is of fundamental importance. During the last few years, several galaxies (cf. Dey et al. 1998; Weymann et al. 1998; Spinrad et al. 1998; Chen, Lanzetta, & Pascarelle 1999; van Breugel et al. 1999; Hu, McMahon, & Cowie 1999) and quasars (Fan et al. 1999, 2000a, 2000b; Zheng et al. 2000; Stern et al. 2000) at redshifts of $z \geq 5$ have been detected. Because quasars are among the most luminous objects in the Universe, they are valuable probes of conditions at early cosmic times. One particularly important diagnostic is their gas metallicity. If the gas near high redshift quasars is related to the interstellar matter of the young host galaxies, quasars can be used to probe the star formation and chemical enrichment history of those galactic environments. Recent studies of quasars at moderately high redshifts ($z \gtrsim 3$) show solar and enhanced metallicities in the line emitting gas (cf. Hamann & Ferland 1993; Osmer, Porter, & Green 1994; Ferland et al. 1996; Hamann & Ferland 1999; Dietrich & Wilhelm-Erkens 2000). These results require a rapid and efficient phase of star formation in the early Universe, e.g., in the dense galactic or protogalactic nuclei where quasars reside.

In the following, we present results of an ongoing study of quasars at redshifts $3 \lesssim z \lesssim 5$. The emission line ratios of N V 1240 to C IV 1549 and He II 1640 are used as well as Mg II 2798 vs. Fe II UV. The relative strength of these ratios indicates that the first epoch of star formation started at redshifts $z \geq 10$. In current cosmological models, the age of the Universe at these redshifts is less than 5×10^8 yr ($H_0 = 65$ km s^{-1} Mpc^{-1}, $\Omega_M = 0.3$, $\Omega_\Lambda = 0.7$; cf. Carroll, Press, & Turner 1992).

2. Observations

The observations of high redshift quasars were carried out at several observatories during 1993 and 2000. We used telescopes at Calar Alto Observatory (Spain), McDonald Observatory (Texas, USA), ESO's La Silla and Paranal Observatories (Chile), Keck Observatory (Hawai'i, USA), and CTIO (Chile) (Table 1).

Table 1. Observing log of the studied high redshift quasars.

Quasar	z	Observatory	$\lambda\lambda$-range (Å)	Date
UM 196	2.81	Calar Alto, 3.5m	3800–8200	Aug 1993
BRI 0019−1522	4.52	CTIO, 4m	11500–23600	Sep 2000
Q 0044−273	3.16	Paranal, 8.2m	3800–9400	Jul 1999
UM 667	3.13	Calar Alto, 3.5m	3800–8200	Aug 1993
Q 0046−282	3.83	Paranal, 8.2m	3800–9400	Jul 1999
Q 0103+0032	4.44	CTIO, 4m	11500–23600	Sep 2000
Q 0103−294	3.12	Paranal, 8.2m	3800–9400	Jul 1999
Q 0103−260	3.36	Paranal, 8.2m	3800–9400	Jul 1999
		La Silla, 3.5m	9500–24800	Oct 1999
Q 0105−2634	3.48	La Silla, 3.5m	9500–24800	Oct 1999
4C 29.05	2.36	Calar Alto, 3.5m	3800–8200	Aug 1993
Q 0216+0803	2.99	McDonald, 2.7m	3800–7800	Jul 1995
PSS J0248+1802	4.44	CTIO, 4m	11500–23600	Sep 2000
Q 0256−0000	3.37	La Silla, 3.5m	9500–24800	Oct 1999
Q 0302−0019	3.29	La Silla, 3.5m	9500–24800	Oct 1999
PC 1158+4635	4.73	Keck, 10m	12700–24700	May 2000
HS 1425+60	3.19	Calar Alto, 3.5m	3800–8200	Aug 1993
Q 1548+0917	2.75	McDonald, 2.7m	3800–7800	Jul 1995
PC 1640+4711	2.77	McDonald, 2.7m	3800–7800	Jul 1995
HS1700+64	2.74	Calar Alto, 3.5m	3800–8200	Aug 1993
PKS 2126−15	3.28	Calar Alto, 3.5m	3800–8200	Aug 1993
PC 2132+0216	3.19	Calar Alto, 3.5m	3800–8200	Aug 1993
Q 2227−3928	3.44	La Silla, 3.5m	9500–24800	Oct 1999
Q 2231−0015	3.02	Calar Alto, 3.5m	3800–8200	Aug 1993
		McDonald, 2.7m	3800–7800	Jul 1995
BRI 2237−0607	4.57	CTIO, 4m	11500–23600	Sep 2000
UM 659	3.04	Calar Alto, 3.5m	3800–8200	Aug 1993
Q 2348−4025	3.31	La Silla, 3.5m	9500–24800	Oct 1999

The redshift range of $2.7 \lesssim z \lesssim 3.3$ was chosen to ensure that most of the diagnostic ultraviolet lines are shifted into the optical regime, in particular the N V 1240, C IV 1549, and He II 1640 emission lines. The quasars which we observed in the near infrared domain ($\sim$ 1–2.5 μm) were selected for their brightness and for a suitable redshift ($3.3 \lesssim z \lesssim 4.7$) that the Mg II 2798 and the broad Fe II emission features in the ultraviolet were shifted to the J or H-bands.

2.1. The Method

Quasars show a prominent emission line spectrum which provides information on the physical conditions of the gas, i.e., temperature, density, ionization state, and the chemical composition. Although the ratios of strong emission lines like Lyα 1215 to C IV 1549 are quite insensitive to the metallicity, other ratios can provide indirect constraints.

The key to using emission line ratios to estimate the metallicity is the different production rates of primary elements like carbon and secondary elements, like nitrogen. N is selectively enhanced by secondary processing at moderate to high metallicities, leading to N increasing as roughly Z^2 (cf. Hamann & Ferland 1993; Vila-Costas & Edmunds 1993). Recent model calculations provide evidence for a strong metallicity dependence of emission line ratios involving such elements. Hence, N V 1240 vs. C IV 1549 and N V 1240 vs. He II 1640 are of particular interest for determining the chemical composition of the gas (cf. Hamann & Ferland 1999 for a review).

The different time scales of the enrichment of gas with α-elements (e.g., O and Mg) and iron are another important aspect using emission line ratios to probe the star formation history. α-elements are produced predominantly in massive stars on short time scales. These elements are released from massive-star supernovae (Types II,Ib,Ic). The dominant source of iron is ascribed to intermediate mass stars in binary systems ending in supernova type Ia explosions (cf. Wheeler, Sneden, & Truran 1989). The amount of iron returned to the interstellar medium in SN II ejecta is rather low (e.g., Yoshii, Tsujimoto, & Kawara 1998). The significantly different time scales of the release of α-elements and iron to the interstellar medium results in a time delay of the order of $\sim$ 1 Gyr. Detecting strong Fe II emission at high redshift can be taken as an indication that the star formation of the stars which had released the iron had occurred $\sim$ 1 Gyr earlier. The viability of the Fe II/Mg II emission line ratio as an abundance indicator was discussed by Hamann & Ferland (1999).

3. Results

3.1. N V 1240 versus C IV 1549 and He II 1640 Line Ratios

The quasars observed in the optical wavelength range were used to determine the N V 1240/C IV 1549 and N V 1240/He II 1640 emission line ratios. To measure the N V 1240 line strength we had to deblend the Lyα 1215, N V 1240 emission line complex. We also deblended the C IV 1549, He II 1640, O III] 1663 emission line complex to measure He II 1640 (cf. Dietrich & Wilhelm-Erkens 2000 for more details of the deblending). The measured line ratios of N V 1240/C IV 1549 and N V 1240/He II 1640 are compared to theoretical predictions (Figure 1). Both line ratios are in good agreement with results obtained by Hamann & Ferland (1992, 1993) for quasars at similar redshift. The measured line ratios were used

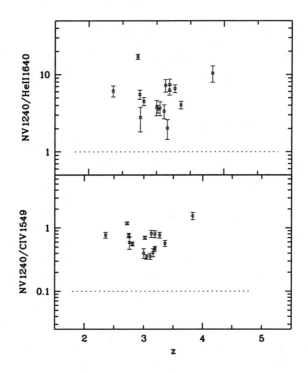

Figure 1.   N v 1240/C iv 1549 and N v 1240/He ii 1640 as a function
of redshift. The dotted lines indicate the line ratios, which are expected
for typical conditions of the BELR gas assuming solar metallicity.

to calculate an average line ratio yielding N v 1240/C iv 1549 = 0.7 ± 0.3 and
N v 1240/He ii 1640 = 5.9 ± 3.6. The dotted lines in Figure 1 indicate the line
ratios expected for *typical* conditions of the broad emission line region (BELR)
assuming solar metallicities (Hamann & Ferland 1999). The observed line ratios are obviously larger than those for solar metallicities indicating supersolar
abundances.

The conversion of observed emission line ratios to relative abundances is affected by several uncertainties. One has to consider only lines, such as N v 1240,
C iv 1549, and He ii 1240, that originate in the same region of the BELR under
comparable conditions of the gas. A detailed discussion of the current limitations of the method can be found in Baldwin et al. (1996), Ferland et al.
(1996), or Hamann & Ferland (1999). Our abundance estimates are based on
the model calculations presented by Hamann & Ferland (1992, 1993). They
computed abundances for a large range of evolutionary scenarios and input the
results into numerical models of the BELR. They varied the slope of the IMF,
the evolutionary time scale for the star formation, as well as the low mass cutoff
of the IMF. They concluded that the high metallicities observed in high redshift
quasars can be achieved only in models with rapid star formation (RSF) and
a shallow IMF (slightly favoring massive stars compared to the solar neighborhood), comparable to models of giant elliptical galaxies. It is reassuring that the

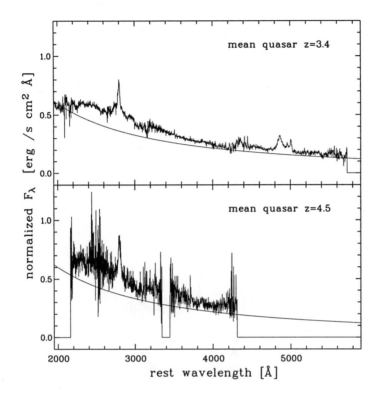

Figure 2.   Mean quasar spectra at $\bar{z} = 3.4$ and $\bar{z} = 4.5$ together with a power law continuum fit with $\alpha = -0.5$ ($F_\nu \propto \nu^\alpha$).

rapid star formation scenario indicates the same range of metallicities based on both the N V 1240/C IV 1549 and N V 1240/He II 1640 line ratios. We estimated an abundance of $Z \simeq 8 \pm 4Z_\odot$ given by our observed N V/C IV and N V/He II within the framework of the RSF model (cf. Dietrich & Wilhelm-Erkens 2000).

3.2.   Mg II 2798 vs. Fe II UV Line Ratio

The line ratio of α-element vs. iron emission can be used as a cosmological clock because the time scales for the release of α-elements and iron to the ISM are significantly different. The enrichment delay is of the order of ~ 1 Gyr (cf. Wheeler et al. 1989; Yoshii et al. 1998). The best indicator of α-elements vs. iron in quasars is the strength of Mg II 2798 emission compared to broad blends of Fe II multiplets spanning several hundred Ångströms (rest-frame) on either side of the Mg II line (cf. Wills, Netzer, & Wills 1980, 1985; Zheng & O'Brien 1990; Boroson & Green 1992; Laor et al. 1995; Vestergaard & Wilkes 2001).

Very few quasars at redshifts larger than $z = 3$ were observed for the wavelength region covering Mg II 2798 to Hβ, [O III] 4959,5007 (Hill, Thompson, & Elston 1993; Elston, Thompson, & Hill 1994; Kawara et al. 1996; Taniguchi et al. 1997; Yoshii et al. 1998; Murayama et al. 1998). Recently, Thompson, Hill, & Elston (1999) studied a few quasars at average redshifts of $\bar{z} = 3.4$ and

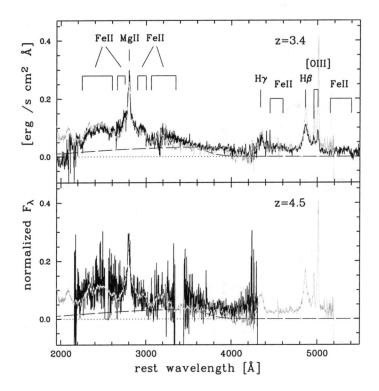

Figure 3.   Comparison of the continuum subtracted mean quasar
spectra at $\bar{z} = 3.4$ (*top*) and $\bar{z} = 4.5$ (*bottom*). The grey curve shows
the local mean quasar spectrum. The long dashed line indicates the
strength of the Balmer continuum emission.

$\bar{z} = 4.5$, respectively. They found no significant difference in the strength of the
ultraviolet Fe II emission relative to Mg II 2798, which suggests an age of the
Universe of more than 1;Gyr at $z \simeq 4.5$.

In contrast to earlier studies, our data cover a much wider range of rest
frame wavelengths, $\lambda\lambda 2100$–5600 Å and $\lambda\lambda 2100$–4300 Å (Figure 2). These wide
and continuous wavelength range enabled us to investigate the strong ultraviolet
Fe II emission based on a reliable continuum fit which was hard to achieve in
earlier studies with smaller and non-continuous wavelength coverage.

Due to the huge number of individual Fe II emission lines ($\sim 10^5$) it is not
practical to treat them individually. As suggested and demonstrated by Wills
et al. (1985), the reconstruction of a quasar spectrum by several well defined
components, i.e., (1) a power law continuum, (2) a Balmer continuum emission
spectrum, (3) a template for the Fe II emission, and (4) a template spectrum for
the broad emission lines, is the best approach to measuring the strength of the
Fe II emission. We are presently involved in a collaboration (Verner et al. 1999)
to use state-of-the-art computer models as well as empirical Fe II,Fe III emission

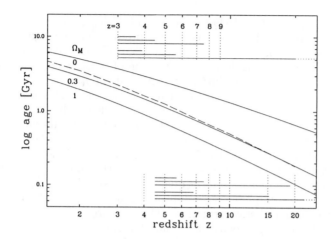

Figure 4.   Estimate of z_f for several combinations of Ω_M and τ_{evol} ($H_0 = 65$ km s^{-1} Mpc^{-1}). The three solid lines show the age of the Universe as a function of redshift ($\Omega_M = 0,0.3,1$). The long dashed line shows the effect of $\Omega_\Lambda = 0.7$ on the age ($\Omega_M = 0.3$). The horizontal lines indicate z_f for $\tau_{evol} = 0.5,1,2$ Gyr ($\Omega_M = 0$ and 0.3) for $z = 3$ and $z = 4.5$.

templates (Vestergaard & Wilkes 2001) to quantify the abundance sensitivities of the Fe II line emission.

To obtain a first estimate of the iron emission strength in comparison to quasars in the local Universe, we compared the rest-frame quasar spectra in our samples ($\bar{z} = 3.4$ and $\bar{z} = 4.5$) to a mean quasar spectrum. The mean quasar spectrum was calculated from a subset of a large quasar sample (> 700 quasars) which we compiled from ground-based observations and from archive spectra measured with IUE and HST (M. Dietrich & F. Hamann, in preparation). The mean spectrum used for this comparison is based on 101 quasars with redshift $z \leq 2$ and luminosities in the same range as our $\bar{z} = 3.4$ and $\bar{z} = 4.5$ quasars.

The mean quasar spectra of our samples at $\bar{z} = 3.4$ and $\bar{z} = 4.5$ are shown in Figure 2 together with power law continuum fits. The continuum fits were subtracted and the pure emission line flux was compared. For wavelengths $\lambda \geq 3200$ Å much of the emission can be attributed to Balmer continuum emission, but for $\lambda \leq 3200$ Å most of the emission is due to broad Fe II emission features (Figure 3). The relative emission strength of the Fe II emission of the mean high redshift quasars are nearly identical ($\lesssim 15\%$) compared to the mean $z \leq 2$ quasar (Figure 3). The mean quasar spectra at $\bar{z} = 3.4$ and $\bar{z} = 4.5$ themselves differ by less than $\sim 20\%$. This can be taken as an indication for no significant evolution in α-element vs. iron in quasars from the local Universe to $z \sim 4.5$.

4.   Summary and Discussion

The presented study of quasars at redshifts $z \simeq 3$ provides evidence for higher solar abundances of the line emitting gas. This result is based on the emission line ratios of N v 1240 vs. C iv 1549 and He ii 1640. Using assumptions about stellar evolution time scales which are necessary to produce solar or higher metallicities, the beginning of the first star formation epoch can be estimated. In Figure 4 the age of the Universe is displayed as a function of redshift z for several settings of H_0 and Ω_M. With an evolutionary time scale of $\tau_{evol} \simeq 1$ Gyr ($\Omega_M = 0.3$) or $\tau_{evol} \simeq 2$ Gyr ($\Omega_M = 0$) based on normal chemical evolution models, the beginning of the first violent star formation episode can be dated to a redshift of $z_f \simeq 6\text{-}8$ based on N v 1240/C iv 1549 and N v 1240/He ii 1640 (Figure 4).

Assuming an evolutionary time scale of $\tau_{evol} \simeq 1$ Gyr for the progenitor stars of type SN Ia, we used the Mg ii 2798/Fe ii UV emission ratio as a tracer of star formation history. The similar Mg ii/Fe ii UV emission ratios in our high redshift quasars compared to local quasars suggests an age of the Universe of ~ 1 Gyr at $z \simeq 4.5$, implying a redshift of $z_f \simeq 8\text{-}15$ ($\Omega_M = 0\text{-}0.3$) for the epoch of the first substantial star formation. The measured Mg ii/Fe ii UV emission ratio probably also suggests at least solar abundances.

We concluded, therefore, that high redshift quasars indicate a redshift of $z_f \simeq 10$ for the first major star formation epoch, corresponding to an age of the Universe of $\lesssim 5 \times 10^8$ yr ($H_0 = 65$ km s^{-1} Mpc^{-1}, $\Omega_M = 0.3$, $\Omega_\Lambda = 0.7$).

References

Baldwin, J. A., et al. 1996, ApJ, 461, 664

Boroson, T. A., & Green, R. F. 1992, ApJS, 80, 109

Carroll, S. M., Press, W. H., & Turner, E. L. 1992, ARA&A, 30, 499

Chen, H.-W., Lanzetta, K. M., & Pascarelle, S. 1999, Nature, 398, 586

Dey, A., Spinrad, H., Stern, D., Graham, J. R., & Chaffee, F. H. 1998, ApJ, 498, L93

Dietrich, M., & Wilhelm-Erkens, U. 2000, A&A, 354, 17

Elston, R., Thompson, K. L., & Hill, G. J. 1994, Nature, 367, 250

Fan X., et al. 1999, AJ, 118, 1

Fan X., et al. 2000a, AJ, 119, 1

Fan X., et al. 2000b, AJ, 120, 1167

Ferland, G. J., Baldwin, J. A., Korista, K. T., Hamann, F., Carswell, R. F., Phillips, M., Wilkes, B., & Williams, R. E. 1996, ApJ, 461, 683

Hamann, F., & Ferland, G. J. 1992, ApJ, 381, L53

Hamann, F., & Ferland, G. J. 1993, ApJ, 418, 11

Hamann, F., & Ferland, G. J. 1999, ARA&A, 37, 487

Hill, G. J., Thompson, K. L., & Elston, R. 1993, ApJ, 414, L1

Hu, E. M., McMahon, R. G., & Cowie, L. L. 1999, ApJ, 522, L9

Kawara, K., Murayama, T., Taniguchi, Y., & Arimoto, N., 1996, ApJ, 470, L85

Laor, A., Bahcall, J. N., Jannuzi, B. T., Schneider, D. P., & Green, R. F. 1995, ApJS, 99, 1

Murayama, T., Taniguchi, Y., Evans, A. S., Sanders, D. B., Ohyama, Y., Kawara, K., & Arimoto, N. 1998, AJ, 115, 2237

Osmer, P. S., Porter, A. C., & Green, R. F. 1994, ApJ, 436, 678

Spinrad, H., Stern, D., Bunker, A., Dey, A., Lanzetta, K., Yahil, A., Pascarelle, S., & Fernández-Soto, A. 1998, AJ, 116, 2617

Stern, D., Spinrad, H., Eisenhardt, P., Bunker, A. J., Dawson, S., Stanford, S. A., & Elston, R. 2000, ApJ, 533, L75

Taniguchi, Y., Murayama, T., Kawara, K., & Arimoto, N. 1997, PASJ, 49, 419

Thompson, K. L., Hill, G. J., & Elston, R. 1999, ApJ, 515, 487

van Breugel, W., De Breuck, D., Stanford, S.A., Stern, D., Röttgering, H., & Miley, G. 1999, ApJ, 518, L61

Verner, E. M., Verner, D. A., Korista, K. T., Ferguson, J. W., Hamann, F., & Ferland, G. J. 1999, ApJS, 120, 101

Vestergaard, M., & Wilkes, B. J. 2001, ApJS, in press

Vila-Costas, M. B., & Edmunds, M. G. 1993, MNRAS, 265, 199

Weymann, R. J., Stern, D., Bunker, A., Spinrad, H., Chaffee, F. H., Thompson, R. I., & Storrie-Lombardi, L. 1998, ApJ, 505, L95

Wheeler, J. C., Sneden, C., & Truran, J. W. 1989, ARA&A, 27, 279

Wills, B. J., Netzer, H., & Wills, D. 1980, ApJ, 242, L1

Wills, B. J., Netzer, H., & Wills, D. 1985, ApJ, 288, 94

Yoshii, Y., Tsujimoto, T., & Kawara, K. 1998, ApJ, 507, L113

Zheng, W., & O'Brien, P. T. 1990, ApJ, 353, 433

Zheng, W., et al. 2000, AJ, 120, 1607

Acknowledgments. This work was supported by NASA grant NAG 5-3234 and by the Deutsche Forschungsgemeinschaft, project SFB328 and SFB439.

Discussion

Jean-René Roy: Your supersolar metallicity is a bit puzzling. I know there are uncertainties in the models but have you made some consistency checks in the sense of calculating the total mass of metals and the amount of star formation needed to produce those metals?

Matthias Dietrich: Honestly, no. These supersolar metallicities of $\sim 8\ Z_\odot$, are derived from photoionization calculations. These photoionization models were calculated for emission line regions with densities of $n_e = 10^{10}$–10^{11} cm^{-3}. With respect to the uncertainty of the obtained abundances estimate of $8 \pm 4\ Z_\odot$, the main conclusion is that the metallicity at this high redshifts is at least solar or even higher for the gas associated with the quasars. Assuming a rapid star formation epoch this result is comparable with the chemical enrichment detected for giant ellipticals which show an overabundance of a factor of 3 times solar metallicity.

Astrophysical Ages and Time Scales
ASP Conference Series, Vol. 245, 2001
T. von Hippel, C. Simpson, N. Manset

Cosmic Flows: A Status Report

S. Courteau

Department of Physics and Astronomy, The University of British Columbia, Vancouver, British Columbia, Canada

A. Dekel

Racah Institute of Physics, The Hebrew University, Jerusalem 91904, Israel

Abstract. We give a brief review of recent developments in the study of the large-scale velocity field of galaxies since the international workshop on Cosmic Flows held in July 1999 in Victoria, B.C. Peculiar velocities (PVs) yield a tight and unique constraint on cosmological characteristics, independent of Λ and biasing, such as the cosmological matter density parameter (Ω_m) and the convergence of bulk flows on large scales. Significant progress towards incorporating non-linear dynamics and improvements of velocity field reconstruction techniques have led to a rigorous control of errors and much refined cosmic flow analyses. Current investigations favor low-amplitude ($\lesssim 250$ km s^{-1}) bulk flows on the largest scales ($\lesssim 100h^{-1}$ Mpc) probed reliably by existing redshift–distance surveys, consistent with favored ΛCDM cosmogonies. Tidal field analyses also suggest that the Shapley Concentration (SC), located behind the Great Attractor (GA), might play an important dynamical role, even at the Local Group. Low-amplitude density fluctuations on very large scales generate the overall large-scale streaming motions but massive attractors like the GA and Perseus–Pisces account for smaller scale motions, which are superposed on the large-scale flow. Likelihood analyses of galaxy PVs, in the framework of flat CDM cosmology, now provide tight constraints of $\Omega_m = 0.35 \pm 0.05$. A four-fold size increase of our data base is expected in ~ 4–5 years with the completion of next generation FP/TF surveys and automated supernovae searches within 20 000 km s^{-1}.

1. Introduction

Ever since the discovery of the microwave background dipole by Smoot, Gorenstein, & Muller (1977) and the pioneering measurements of galaxy motions by Rubin et al. (1976), the study of cosmic flows, or deviations from a smooth Hubble flow due to large-scale gravitational perturbations, has been recognized as one of the most powerful constraints to cosmological scenarios (Peebles 1980; Dekel 1994; Strauss & Willick 1995). Indeed, under the assumption that cosmic structure originated from small-amplitude density fluctuations that were amplified by gravitational instability, the peculiar velocity **v** and mass density contrast δ are together linked in the linear regime by a deceptively simple expression (from mass conservation in linear perturbation theory):

$$\nabla \cdot \mathbf{v} = -\Omega_m^{0.6}\delta. \tag{1}$$

The mean square bulk velocity on a scale R is easily calculated in Fourier space as:

$$\left\langle v^2(R) \right\rangle = \frac{\Omega_m^{1.2}}{2\pi^2} \int_0^\infty P(k)\widetilde{W}^2(kR)\mathrm{d}k, \tag{2}$$

where $P(k)$ is the mass fluctuation power spectrum and $\widetilde{W}^2(kR)$ is the Fourier transform of a top-hat window of radius R. Measurements of galaxy PVs can thus directly constrain Ω_m, the shape and amplitude of the power spectrum, and test assumptions about the gravitational instability picture (Gaussian fluctuations, etc.) statistical properties of the initial fluctuations and gravitational instability as the engine of perturbation growth.

The last major workshop on Cosmic Flows in July 1999 in Victoria, BC (Courteau, Strauss, & Willick 2000a; hereafter CFW2000) came at a time when important new data sets and critical modeling of the biasing relation between the galaxy and mass distribution were just being released. Fundamental questions debated at the conference, and central to all cosmological investigations based on cosmic flows, included[1]: (1) *What is the amplitude of bulk flows on the largest scales probed?* (2) *Can velocity analysis provide accurate estimates of Ω_m?*, and (3) *What is the value of Ω_m?* The last two years have seen significant progress providing nearly definitive answers to each of the 3 questions above, as we discuss in the remainder of this review.

Detailed information about cosmic flows can be found in the Cosmic Flows 1999 workshop proceedings (CFW2000), including the conference review by Dekel (2000). Also in Willick (1999) and Dekel (1999), as well as Willick (2000).

2. Data Sets and Bulk Flows

The radial peculiar velocity of a galaxy is derived by subtracting the Hubble velocity H_0d from the total velocity (redshift) cz in the desired frame of reference (e.g., CMB or Local Group). The distance d is inferred from a distance indicator (DI) whose accuracy dictates the range of applicability of the technique. The relative distance error of common DIs ranges from 20% (Tully-Fisher [TF], Fundamental Plane [FP], Brightest Cluster Galaxy [BCG]) down to 5-8% (Surface Brightness Fluctuations [SBF], SNIa, Kinetic Sunyaev–Zel'dovich [kSZ]). The bulk velocity $\mathbf{V}_B$ of an ensemble of galaxies within a sphere (or a shell) of radius R is computed by a least square fit of a bulk velocity model predictions $\mathbf{V}_B \cdot \hat{n}$ to the observed radial peculiar velocities, where $\hat{n}$ is a unit vector in the direction of the object. Current results are summarized in Table 1 and represented graphically in Figure 1.

---

[1] Discussions about the measurements of the small-scale velocity dispersion and the coldness of the velocity field also figured prominently in the workshop agenda but we do not offer any update below, for lack of space. The interested reader should read CFW2000.

Table 1.   Recent bulk flow measurements[†].

Survey	$R_{\rm eff}$ (km s^{-1})	$V_{\rm B}$ (km s^{-1})	Dist. Ind.
Lauer & Postman (BCG)	12500	700	BCG
Willick (LP10K)	11000	700	TF
Hudson et al. (SMAC)	8000	600	FP
Dekel et al. (POTENT/M3)	6000	350	TF,$D_{\rm n}$-σ
Tonry et al. (SBF)	3000	290	SBF
Riess et al. (SNIa)	6000	300	SN Ia
Courteau et al. (SHELLFLOW)	6000	70	TF
Dale & Giovanelli (SFI)	6500	200	TF
Colless et al. (EFAR)	10000	170	FP
Dale & Giovanelli (SCI/SCII)	14000	170	TF

† All references in CFW2000. With the exception of Lauer & Postman (1994), all results are post 1999.

The data sets can be divided into two groups which lie either exactly within or somewhat above the predictions from most (Λ)CDM families. Figure 1 shows the theoretical prediction of a ΛCDM model for the simplest statistic: the bulk-flow amplitude in a top-hat sphere. The solid line is the r.m.s. value, obtained by Equation 2. The dashed lines represent 90% cosmic scatter in the Maxwellian distribution of V, when only one random sphere is sampled. With the exception of BCG, the directions of the non-zero flow vectors are similar (they all lie within 30° of $(l, b) = (280°, 0°)$) and the velocity amplitudes can be roughly compared even though the survey geometries and inherent sample biases can differ quite appreciably. A rigorous comparison of flow analyses must however account for different window functions (Kaiser 1988; Watkins & Feldman 1995; Hudson et al. 2000). Still, the obvious interpretation of these data is that of a gradual decline of the flow amplitude, or convergence of the flow field to the rest-frame of the CMB at $\sim 100h^{-1}$ Mpc, consistent with the theoretical assumption of large-scale homogeneity.

Cosmic variance however prevents any convergence to complete rest. Some of the reported error bars are based on a careful error analysis using mock catalogs, while others are crude estimates. In most cases they represent random errors only and underestimate the systematic biases. Large error bars for surveys such as BCG, LP10, SMAC, SNIa, and Shellflow, with fewer than a thousand "test particles," are largely due to sampling errors which also increase with increasing volumes.

While present bulk flow estimates are in comforting agreement with current cosmologies, important efforts are currently underway to reduce the systematic and random errors inherent in most compilations of galaxy PVs, especially at large distance. The former is addressed by collecting homogeneous data across the entire sky, in the spirit of Lauer & Postman (1994) and Shellflow (Courteau et al. 2000b). The latter simply requires that large numbers of galaxies and

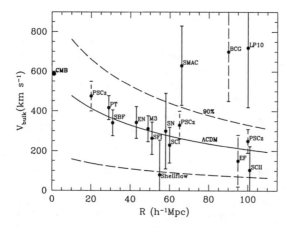

Figure 1. Amplitude of CMB bulk velocity in top-hat spheres about the LG, in comparison with theory. The curves are the predicted rms and cosmic scatter for a ΛCDM model. The measurements, based on the data listed in Table 1, are crudely translated to a top-hat bulk velocity. The error bars are random only. All the non-zero vectors (except BCG) point to $(l, b) = (280°, 0°) \pm 30°$. Shown as well are the LG dipole velocity (labeled "CMB"), and linear estimates from the PSCz redshift survey for $\beta = 0.7$. Care must be exercised when interpreting such plots since directions are not plotted and projected amplitudes (V_X, V_Y, V_Z) may differ substantially (e.g., Hudson et al. 2000).

cluster of galaxies be observed to reduce Poisson noise and systematic biases. The nominal sample size to achieve a minimum signal/noise for each spherical volume chosen must be estimated from mock catalogs based on an expected number density profile (as a function of distance or redshift from us) and sky coverage. New surveys including many thousand "test particles" and reaching out to $15\,000$ km s^{-1} should quantify the convergence of the peculiar velocity field on very large scales. These surveys include, for example, NFP200[2] for the FP measurements of ~ 4000 early-type galaxies in 100 X-ray selected clusters, 6dF[3] for the FP measurements of $\sim 15,000$ Southern hemisphere early-type galaxies, the SNfactory[4] for the serendipitous detection and subsequent follow-up of a few hundred SNe per year (G. Aldering, private communication), and the Warpfire[5] extension of Lauer & Postman's (1994) BCG analysis. These studies should be completed by 2005, if not sooner.

---

[2] astro.uwaterloo.ca/~mjhudson/nfp/

[3] msowww.anu.edu.au/colless/6dF/

[4] snfactory.lbl.gov. The detection range should actually extend out to $24\,000$ km s^{-1}.

[5] www.noao.edu/noao/staff/lauer/warpfire/

2.1.   The Large-Scale Tidal Field

The cosmological peculiar velocity field at any point can be decomposed into the sum of a divergent field due to density fluctuations inside the surveyed volume, and a tidal (shear) field, consisting of a bulk velocity and higher moments, due to the matter distribution outside the surveyed volume. This procedure was carried out by Hoffman et al. (2001), using reconstructions by POTENT (Dekel et al. 1999) or Wiener Filter (Zaroubi, Hoffman, & Dekel 1999), with respect to a sphere of radius $60h^{-1}$ Mpc about the Local Group. Their results are illustrated in figure 2 of Hoffman et al. (2001). The divergent component is dominated by the flows into the Great Attractor and Perseus–Pisces, and away from the void in between. The tidal field shows, for example, that about 50% of the velocity of the Local Group in the CMB frame is due to external density fluctuations. Their analysis suggests the non-negligible dynamical role of super-structures at distances of 100–200 h^{-1} Mpc, specifically the Shapley Concentration and two great voids. These should be taken into account when considering the convergence of bulk velocity from different surveys on different scales and of the dipole motion of the Local Group.

3.   Power Spectra and the Measurement of Ω_m

The peculiar velocities allow direct estimates of Ω_m independent of galaxy biasing and Λ. Early analyses have consistently yielded a lower bound of $\Omega_m > 0.3$ (e.g., Dekel & Rees 1994), but not a tight upper bound.

Cosmological density estimates from the confrontation of PVs and the distribution of galaxies in redshift surveys have traditionally yielded values in the range $0.3 < \Omega_m < 1$ (95% confidence). This wide span has often been attributed to nontrivial features of the biasing scheme or details of the reconstruction/likelihood method such as the choice of smoothing length. Two common approaches to measuring Ω_m are known as the *density–density* (d–d) and *velocity–velocity* (v–v) comparisons. Density–density comparisons based on POTENT-like reconstructions (e.g., Sigad et al. 1998) have produced typically large values of Ω_m, while v–v comparisons yield smaller estimates (e.g., Willick et al. 1998 [VELMOD]; Willick 2000; Branchini et al. 2001). These differences have recently been shown to be insensitive to the complexity of the biasing scheme, whether it be non-linear, stochastic, or even non-local (Berlind, Narayanan, & Weinberg 2001; see also Feldman et al. 2001). Thus, one must look for differences inherent to d–d/v–v techniques for an explanation of their apparent disagreement.

Likelihood analyses of the individual PVs (e.g., Zaroubi et al. 1997; Freudling et al. 1999; Zehavi & Dekel 1999) can be used to estimate the power spectrum of density fluctuations under the assumption that these are drawn from a Gaussian random field. In linear theory, the shape of the power spectrum $P(k)$ does not change with time and thus provides a powerful tool to estimate basic cosmological parameters. Moreover, power spectrum analyses of PVs are free of the problems that plague similar determinations from redshift surveys such as redshift distortions, triple-valued zones, and galaxy biasing, and suffer from weaker non-linear clustering effects. Likelihood methods simply require as prior a parametric functional form for $P(k)$.

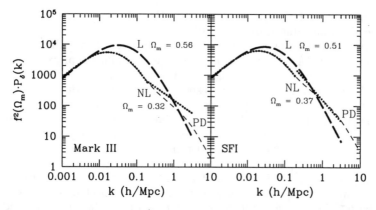

Figure 2.   The recovered power spectra by the non-linear likelihood analysis of Silberman et al. (2001) from the data of M3 (*left*) and SFI (*right*). The $P(k)$ yielded by the purely linear analysis is marked "L," and the nonlinear analysis, with a break at $k = 0.2(h^{-1} \text{ Mpc})^{-1}$, is marked "NL." The corresponding values of Ω_m are marked. Also shown for comparison is an extrapolation of the linear part of the recovered $P(k)$ into the nonlinear regime by the Peacock & Dodds (1996) approximation. The $P(k)$ is in units of $(h^{-1} \text{ Mpc})^3$.

The likelihood analysis of Silberman et al. (2001) incorporates a correction to the power spectrum for non-linear clustering effects, which has been carefully calibrated using new mock catalogs based on high-resolution simulations. The effect of this correction, shown in Figure 2, is to account for larger power on small scales and suppress the overall amplitude of $P(k)$ on larger scales where clustering is still linear. An unbiased fit of $P(k)$ in the linear regime can thus be achieved, leading to unbiased constraints on the relevant cosmological parameters. The $P(k)$ prior in their analysis assumed a flat ΛCDM cosmological model ($h = 0.65, n = 1$, COBE normalized), with only Ω_m as a free parameter. Figure 2 gives final fits based on the Mark III (Willick et al. 1997) and SFI (Haynes et al. 1999) catalogs of galaxy PVs. The Mark III catalog is more densely sampled at small distances than SFI and also includes elliptical galaxies which are absent in SFI; the correction for non-linear effects is thus stronger for Mark III. Fitted values for the Mark III data drop from $\Omega_m = 0.56 \pm 0.04$ in the earlier linear analysis to 0.32 ± 0.06 in the improved analysis, and for SFI from 0.51 ± 0.05 to 0.37 ± 0.09. These revised tight constraints from PVs represent a significant improvement in this analysis.

These results are in broad agreement with a recent v–v likelihood analysis of SFI PVs against the PSCz IRAS redshift survey by Branchini et al. (2001). Their procedure entails some assumptions about the biasing of IRAS galaxies for which PSC redshifts are measured. If linear biasing were invoked with a biasing parameter near unity, Branchini et al. (2001) would find even smaller values of the density parameter with $0.15 \leq \Omega_m \leq 0.30$. This exercise and a direct comparison with the PV-only likelihood analysis of, say, Silberman et al. (2001) is however futile without a proper prescription of galaxy biasing. The direct analysis of PVs by themselves has the advantage of being free of the complications introduced by galaxy biasing.

A χ^2 test applied by Silberman et al. (2001) to modes of a Principal Component Analysis (PCA) shows that the nonlinear procedure improves the goodness of fit and reduces a spatial gradient that was of concern in the purely linear analysis. The PCA allows to address spatial features of the data and to evaluate and fine-tune the theoretical and error models. It demonstrates in particular that the ΛCDM models used are appropriate for the cosmological parameter estimation performed. They also addressed the potential for optimal data compression using PCA, which is becoming important as the data sets are growing big.

Intriguingly, when Silberman et al. (2001) allow deviations from ΛCDM, they find an indication for a wiggle in the power spectrum: an excess near $k \sim 0.05(h^{-1} \text{ Mpc})^{-1}$ and a deficiency at $k \sim 0.1(h^{-1} \text{ Mpc})^{-1}$—a "cold flow." This may be related to a similar wiggle seen in the power spectrum from redshift surveys (Percival et al. 2001 [2dF]) and the second peak in the CMB anisotropy (e.g., Halverson et al. 2001 [DASI]).

4. The Future

Significant improvements in cosmic flow studies over the last couple of years include, for example: (1) unbiased recovery of cosmological parameters, such as Ω_m and $\sigma_8 \Omega_m^{0.6}$, via quasi-nonlinear likelihood analyses of galaxy PVs; (2) modeling of non-linear clustering effects in power spectrum analyses from PVs, and implementing tools, based on PCA, for evaluating goodness of fit; and (3) better modeling of biased galaxy formation, in order to single out biasing in the comparison of PVs with redshift surveys and to generate proper mock catalogs for calibrating PV analysis methods.

Future developments rely heavily on growth of the available data bases and on refinements of existing catalogs. The VELMOD technique has enabled improved recalibrations of the Mark III (Willick et al. 1998) and SFI (Branchini et al. 2001) catalogs using external information from IRAS redshift surveys. We are planning an improved recalibration of Mark III using as backbone the homogeneous all-sky Shellflow sample, and merging all existing catalogs of PVs of field galaxies into a new Mark IV catalog.

A number of on-going and newly envisioned surveys (NFP200, 6dF, Warpfire, SNfactory) are expected to increase the size of existing databases by a factor 4 within 2005. New wide-field surveys such as Sloan, 2MASS, and DENIS will also provide most valuable complementary data to help control distance calibration errors.

A noticeable impact to precision flow studies should come from supernovae searches whose potential to build up very large catalogs of peculiar velocities (at the rate of a few hundred detections per year) and small relative error is unparalleled by no other distance indicator. (With $\Delta d/d(\text{SNIa}) \sim 8\%$, 1 SNIa is worth ~ 6 TF or FP measurements!) If a significant fraction of the new SNe Ia can be caught at peak light and monitored to measure a light curve (yielding precise distance estimates), current TF/FP data sets will be superseded in less than 5 years. Other ambitious surveys, such as those listed above, will complement accurate SN distances with very large data bases thus enabling remarkably tight flow solutions in the near future. There are good reasons to plan a new workshop on Cosmic Flows in 2005!

References

Berlind, A. A, Narayanan, V. K., & Weinberg, D. H. 2001, ApJ, 549, 688

Branchini, E., et al. 2001, MNRAS, in press (astro-ph/0104430)

Courteau, S., Strauss, M. A., & Willick, J. A. 2000a, ed. Cosmic Flows 1999: Towards an Understanding of Large-Scale Structure (San Francisco: ASP)

Courteau, S., Willick, J. A., Strauss, M. A., Schlegel, D., & Postman, M. 2000b, ApJ, 544, 636 [Shellflow]

Dekel, A. 1994, ARA&A, 32, 271

Dekel, A. 1999, in Formation of Structure in the Universe, ed. A. Dekel & J. P. Ostriker (Cambridge: Cambridge University Press), 250

Dekel, A. 2000, in Cosmic Flows 1999: Towards an Understanding of Large-Scale Structure, ed. S. Courteau, M. A. Strauss, & J. A. Willick (San Francisco: ASP), 420

Dekel, A., Eldar, A., Kolatt, T., Yahil, A., Willick, J. A., Faber, S. M., Courteau, S., & Burstein, D. 1999, ApJ, 522, 1 [POTENT]

Dekel, A., & Rees, M. J. 1994, ApJ, 422, L1

Feldman, H. A., Frieman, J. A., Fry, J. N., Scoccimaro, R. 2001, Phys. Rev. Lett, 86, 1434

Freudling, W., et al. 1999, ApJ, 523, 1 [Linear likelihood]

Halverson, N. W., et al. 2001, ApJ, submitted (astro-ph/0104489) [DASI coll.]

Haynes, M., Giovanelli, R., Salzer, J., Wegner, G., Freudling, W., da Costa, L., Herter, T., Vogt, N. 1999, AJ, 117, 1668 [SFI]

Hoffman, Y., Eldar, A., Zaroubi, S., Dekel, A. 2001, ApJ, submitted (astro-ph/0102190)

Hudson, M. J., et al. 2000, in Cosmic Flows 1999: Towards an Understanding of Large-Scale Structure, ed. S. Courteau, M. A. Strauss, & J. A. Willick (San Francisco: ASP), 159

Kaiser, N. 1988, MNRAS, 231, 149

Lauer, T. R., & Postman, M. 1994, ApJ, 425, 418

Peacock, J. A., & Dodds, S. J. 1996, MNRAS, 280

Percival, W. J., et al. 2001, MNRAS, submitted (astro-ph/0105252) [2dF coll.]

Rubin, V. C., Thonnard, N., Ford, W. K., & Roberts, M. S. 1976, AJ, 81, 719

Sigad, Y., Eldar, A., Dekel, A., Strauss, M. A., & Yahil, A. 1998, ApJ, 495, 516

Silberman, L., Dekel, A., Eldar, A., Zehavi, I. 2001, ApJ, in press (astro-ph/0101361)

Smoot, G. F., Gorenstein, M. V., & Muller, R. A. 1977, Phys. Rev. Lett, 39, 898 (see also Corey, B. E., & Wilkinson, D. T. 1976, BAAS, 8, 351)

Strauss, M. A., & Willick, J. A. 1995, Phys. Rep., 261, 271

Watkins, R., & Feldman, H. A. 1995, ApJ, 453, L73

Willick, J. A. 1999, in Formation of Structure in the Universe, ed. A. Dekel & J. P. Ostriker (Cambridge: Cambridge University Press), 213

Willick, J. A. 2000, in Proc. XXXVth Rencontres de Moriond, Energy Densities in the Universe (astro-ph/0003232)

Willick, J. A., Courteau, S., Faber, S., Burstein, D., Dekel, A., & Strauss, M. 1997, ApJS, 109, 333 [Mark III]

Willick, J. A., Strauss, M., Dekel, A., & Kolatt, T. 1998, ApJ, 486, 629 [VELMOD]

Zaroubi, S., Hoffman, Y., & Dekel, A. 1999, ApJ, 520, 413 [Wiener]

Zaroubi, S., Zehavi, I., Dekel, A., Hoffman, Y., & Kolatt, T. 1997, ApJ, 486, 21 [Linear likelihood]

Zehavi, I., & Dekel, A. 1999, Nature, 401, 252

Acknowledgments. SC would like to thank Ted von Hippel, Chris Simpson, and the scientific organizing committee for their invitation, and for putting together a superb meeting which was so rich in content and which provided the rare opportunity to interact closely with leading (and lively!) scientists from all branches of astrophysics. The editors are also thanked for their patience while this manuscript was being written.

We remain tremendously saddened by the departure of our friend and colleague Jeff Willick who did so much for the advancement of cosmic flow studies and who touched our lives very deeply.

Discussion

Piotr Popowski: A few years ago Avishai Dekel was measuring $\Omega_m = 1$ from cosmic flows based on his POTENT method, but now the value you quote is 0.3. Can you comment on what caused this difference?

Stéphane Courteau: Please note that all previous estimates of Ω_m reported by Dekel and others based on PVs *alone* (via POTENT or other methods) actually claimed a significant lower bound of $\Omega_m > 0.3$ but no tight upper bounds.

Their results at the time were consistent with $\Omega_m \sim 1$, but they never really claimed a measurement of $\Omega_m = 1$. The claimed lower bound is still valid, but now with the addition of a significant upper bound, ruling out $\Omega_m = 1$. The main improvement came from the incorporation of nonlinear effects in the likelihood analysis, which became possible due to proper mock catalogs based on high-resolution simulations.

A wider range of estimates has indeed been obtained by comparisons of PV data with galaxy redshift surveys. For example, a density–density comparison by Sigad et al. (1998) indicated a high value for Ω_m, while velocity–velocity comparisons, such as by VELMOD (Willick et al. 1998), yielded smaller values. These analyses were contaminated by galaxy biasing and nonlinear effects, which gave rise to relatively large uncertainties.

Astrophysical Ages and Time Scales
ASP Conference Series, Vol. 245, 2001
T. von Hippel, C. Simpson, N. Manset

Type Ia Supernovae, the Hubble Constant, the Cosmological Constant, and the Age of the Universe

John L. Tonry

Institute for Astronomy, University of Hawai'i, 2680 Woodlawn Drive, Honolulu, HI 96822, USA

The High-Z Supernova Search Team

http://cfa-www.harvard.edu/cfa/oir/Research/supernova/HighZ.html

Abstract. The age of the Universe depends on both the present-day Hubble Constant and the history of cosmic expansion. For decelerating cosmologies such as $\Omega_m = 1$, the dimensionless product $H_0 t_0 < 1$ and modestly high values of the Hubble constant $H_0 > 70$ km s^{-1} Mpc^{-1} would be inconsistent with a cosmic age t_0 larger than 12 Gyr. However, if $\Omega_\Lambda > 0$, then $H_0 t_0$ can take on a range of values. Evidence from the Hubble diagram for high redshift Type Ia supernovae favors $\Omega_\Lambda \sim 0.7$ and $H_0 t_0 \sim 1$. Then, if H_0 lies in the range 65–73 km s^{-1} Mpc^{-1}, the age of the Universe, t_0, is 14 ± 1.6 Gyr.

1. It Has Been an Interesting Five Years!

Five years ago, the combination of deep seated belief in inflation, implying $\Omega = 1$, and stellar age estimates near 15 Gyr seemed to require $H_0 \sim 40$ km s^{-1} Mpc^{-1}. Measurements of $H_0 \sim 60$ km s^{-1} Mpc^{-1} and $\Omega_m \sim 0.3$ in clusters notwithstanding, Bartlett, Blanchard, Silk, and Turner (1995) wrote a provocative paper entitled, "The Case for a Hubble Constant of 30 km s^{-1} Mpc^{-1}." Persuaded by the power of theoretical reasoning, Joe Silk bet Brian Schmidt and me a case of Scotch that $H_0 < 60$ km s^{-1} Mpc^{-1}. While Joe has not yet paid up, in the past 5 years he has moved closer to the source of Scotch while the Hubble constant has moved to 60 and beyond. The new element is that supernovae have made the connection between $\Omega = 1$ and the cosmic age more flexible because of plausible evidence for cosmic acceleration.

A Danish–English team (Norgaard-Neilsen et al. 1989) initiated a program to find supernovae in clusters of galaxies at redshifts of 0.3–0.5, with the idea that they could distinguish the effects of cosmic deceleration, as expected in an $\Omega_m = 1$ universe by measuring the peak apparent magnitudes of supernova light curves. The observational problem was to find these faint ($m \sim 21$–22) and distant supernovae near the peak of their light curves. But small detectors kept this pioneering effort from yielding significant results.

The Supernova Cosmology Project (SCP) based at Lawrence Berkeley Lab forged ahead with further attempts to find distant supernovae by extending the methods of the Danes to bigger, faster telescopes. After abandoning attempts to instrument the AAT prime focus for this purpose, they used the standard large format detectors at the Kitt Peak National Observatory starting in 1992 (Perlmutter et al. 1995). By 1997, they had a preliminary result (Perlmutter et al. 1997) based on observations of seven supernovae discovered in 1994 and 1995. By comparing their supernovae with the sample at low redshift, they concluded that the evidence favored a universe with high matter density $\Omega_m = 0.88 \pm 0.6$. They argued that the supernova data at that point placed the strongest constraint on the possible value of the cosmological constant, with their best estimate being $\Omega_\Lambda = 0.06$.

Since 1995, there have been two groups pursuing evidence on cosmic deceleration using the Hubble diagram for supernovae. Our High-Z Supernova Search Team, steered by Brian Schmidt, and encompassing workers on four continents and one mid-Pacific island, found our first supernova in 1995 (Schmidt et al. 1998.) Our first results on Ω_Λ were reported at the Dark Matter conference in February of 1998 by Filippenko & Riess (1998) and published in the Astronomical Journal in September 1998 (Riess et al. 1998). In the same period, the SCP revised their analysis of earlier data (Perlmutter et al. 1998) and then independently reported evidence on Ω_Λ in June 1999 (Perlmutter et al. 1999).

2.   SN Ia Constraints on t_0

Even this first round of supernova observations, which emphasized a sample near $z \sim 0.5$, provided a good constraint on the *difference* $\Omega_m - \Omega_\Lambda$ which, since Ω_m measures deceleration, and Ω_Λ measures acceleration, translates into a surprisingly tight constraint on $H_0 t_0$.

The present samples of SN Ia published, in hand, and being reduced by the two teams provide a statistically robust measurement that $\Omega_\Lambda > 0$. The important questions now are whether the supernovae at large redshift are really the same as the supernovae nearby and whether exotic forms of grey dust might obscure both the supernovae and our understanding of cosmology (Aguirre 1999; Aguirre & Haiman 2000). The observational approach to answer these questions is to use spectra to examine the question of homogeneity (Coil et al. 2000) and multicolor observations over a wide range of wavelengths (as might be done with a superb 8-m infrared telescope at the world's best site) to constrain the properties of intergalactic dust (Riess et al. 2000). So far, although the distant supernovae could have failed these tests, they seem indistinguishable from the SN Ia nearby.

A more ambitious test for the cosmological origin of the observed effect is to extend the data set to higher redshift. While dimming due to evolution or dust would most naturally lead to larger effects at higher redshift, cosmological effects could have the opposite sign due to cosmic deceleration at early epochs ($z \sim 1.5$), followed by a transition to acceleration in the more recent past.

2.1.   What Does a Supernova Look Like?

Looking at real data helps develop an understanding of the observational issues in discovering and measuring light curves for high redshift supernovae. Here we illustrate the appearance of a $z = 0.81$ supernova, SN 1999fj, as observed in a series of images from October through December 1999. Large format detectors on telescopes at good seeing sites are the chief requirement for efficient surveys to find type Ia supernovae at $m \sim 24.5$. Follow-up observations with 8-m class telescopes enable us to construct light curves that can be used to place each SN Ia firmly on the Hubble diagram.

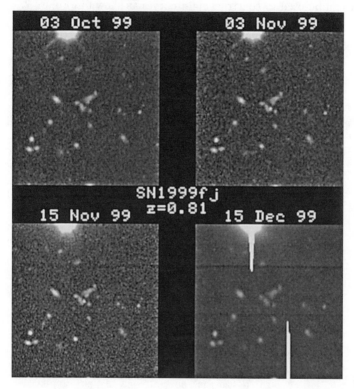

Figure 1. SN 1999fj at four epochs. Epoch 03 Oct 1999 has the supernova barely above our detection threshold using I band at the CFHT. Epoch 03 Nov 1999 is the discovery epoch. The reader is invited to find the new dot. This image covers approximately 1/3000 of the area of the detector array, and so the real search is automated. Epoch 15 Nov 1999 shows an image in 0.5″ seeing from the VLT, and Epoch 15 Dec 1999 shows an image in 0.9″ seeing from the Keck telescope.

2.2. Light Curves From Fall 1999

One of the key developments that makes SN Ia so useful as standard candles is the discovery by Mark Phillips (1993) that the luminosity of a SN Ia is related to its rate of decline after maximum light. This approach, refined by Hamuy et al. (1996) and by Riess, Press, & Kirshner (1996) allows the distance of a well-observed SN Ia with two color data to be determined to better than 10%.

The light curves for high redshift supernovae are obtained in filters which can be transformed back to rest frame B and V with good precision. Time dilation, a signature of cosmic expansion, is a powerful effect for supernovae near $z \sim 1$, transforming 40 days in the observer's frame to 20 days in the supernova's own rest frame (Leibundgut et al. 1996). Photometric calibration and scrupulous subtraction of galaxy light are serious problems for this work which become more difficult at high redshift. Nevertheless, images like those in Figure 1 can be used to construct light curves as shown in Figure 2.

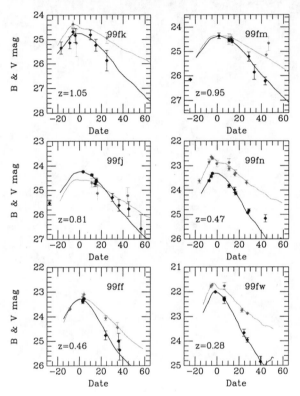

Figure 2.   Sampling of two-color light curves for 6 supernovae from fall 1999, ranging from $z = 1.05$ to $z = 0.28$.

3.   Meditations on H_0

What is the observational evidence on H_0? Many techniques have been developed for measuring extragalactic distances and the ones we love the best are SN Ia (Jha et al. 1999) and the Surface Brightness Fluctuation method, reviewed recently by Blakeslee, Ajhar, & Tonry (1999). Ajhar et al. (2001) demonstrate that these methods are internally consistent: as Figure 3 shows, distances measured by SN Ia and by SBF to the same galaxies are consistent within the quoted errors. This suggests that both methods are in good shape.

But what is H_0, so diligently sought through the decades? This depends entirely on the sample of Cepheids used and the distances assigned to those Cepheids by expert workers in that field. If you use the distances to galaxies with Cepheids as determined by various papers by the Key Project (Ferrarese

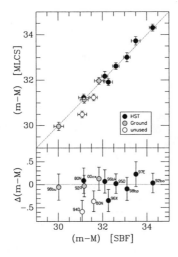

Figure 3. There is no discrepancy between the SN Ia and SBF when they get their zero points from the same set of Cepheids.

et al. 2000; Freedman et al. 2001; Gibson & Stetson 2001) to define the absolute magnitude of SN Ia and SBF, you get $H_0 = 73$, 75, and 77 for both SN Ia and SBF. If you use the distances to the same galaxies using the same Cepheid observations as reduced by Saha et al. (1997) and most recently compiled by Parodi et al. (2000), you get $H_0 = 65$ according to the methodology used by the High-Z team, or $H_0 = 58.5$ according to Parodi et al. Similarly, the distances to the SBF calibrating galaxies depend entirely on the Cepheid distances.

The Hubble constant is *not* determined by SN Ia or by SBF alone: these methods give excellent relative distances to galaxies and tie the cosmic expansion firmly to the local calibrators, but the calibration by Cepheids is now the largest uncertainty in measuring the local value of the Hubble Constant. For the rest of this paper, we will either adopt $H_0 = 73$ as a best-guess value, or else a probability distribution which consists of two Gaussians of fractional width 0.1, centered on $H_0 = 73$ and $H_0 = 65$, with a $^2/_3$:$^1/_3$ weighting. It ain't perfect, but it's our gut feeling of where H_0 really lies.

4. Differential Hubble Diagram for SBF and SN Ia

The key evidence on a value for $H_0 t_0$ comes from Figure 4. All the supernova and SBF distances are consistent at low z, and the supernova data indicate that a model with $\Omega_\Lambda = 0$ does not fit at $z \sim 0.5$. This is the case for an accelerating Universe, whether urged on by a cosmological constant or by something which varies with time. The best fit model has $\Omega_m = 0.3$ and $\Omega_\Lambda = 0.7$. None of the observers is satisfied with the current state of the statistical errors (about twice as large for each SCP supernova as for the High-Z data) or with current limits on possible systematic effects that might make distant supernovae dimmer.

Figure 4 shows that a systematic effect that just grows as the redshift is not a good fit to the data, but the most telling way to separate a systematic

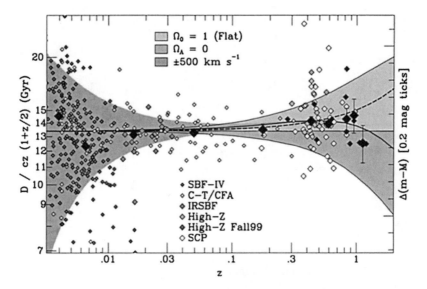

Figure 4.   Cosmological diagram for SBF and supernovae, showing observed luminosity distance (in Glyr) divided by $cz(1+z/2)$ (i.e., the luminosity distance in an empty universe) as a function of redshift. Both sets of distances are normalized to $H_0 = 73$ km s^{-1} Mpc^{-1}; the left intercept is therefore $H_0^{-1} = t_0$; the right axis is labeled in magnitudes. The darker gray region shows where cosmological models with $\Omega_\Lambda = 0$ lie, and the light and dark gray regions show where flat models ($\Omega_0 = 1$) lie. The black downturning curve shows a $(0.3, 0.7)$ cosmology, enhanced by the darkest gray region of ± 500 km s^{-1}, which corresponds to $\pm 2.5\sigma$ thermal peculiar velocities. The large black points are medians in various redshift bins, and the error bars are estimates of the uncertainty in the median judged from the scatter of the contributing points. At $z \sim 0.5$ these points lie significantly above the region permitted if $\Omega_\Lambda = 0$, and well away from $\Omega_m = 0.3$. The black dashed line illustrates how a systematic error that is proportional to z diverges from agreement with a $(0.3, 0.7)$ cosmology for $z > 0.5$.

effect that is proportional to redshift or time is to look at redshifts above 1. For a Λ-dominated cosmology, there is a transition, somewhere around $z \sim 1$, from acceleration here and now to deceleration in the distant past when the matter density, which scales as $(1 + z)^3$, would have been more important. The High-Z Team is working hard to test these ideas. Our fall 1999 data, which are being slowly beaten into photometric perfection, emphasize $z \sim 1$ objects which will provide strong evidence to distinguish cosmology from systematics. Preliminary reductions are shown here. Our fall 2000 program emphasized careful $UBVRI$ photometry in the supernova rest frame to discern the effects of not-quite-gray dust. The final reduction of those data requires observations of the galaxy *without* the supernova, obtained a year after the discovery, and will be forthcoming when these templates are in hand (Jha et al. 2001).

5.   Constraints From Ω_m–Ω_Λ and the Distribution of $H_0 t_0$

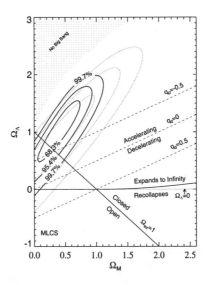

Figure 5.   Two parameter confidence contours shown for the range of cosmological parameters Ω_m and Ω_Λ, given the current SN Ia data from the High-Z collaboration. The solid and dotted contours are the constraints with and without the fall 1999 data, demonstrating the value of even a few points at $z > 0.9$.

We can employ the tools of likelihood analysis to construct contours in the Ω_m–Ω_Λ plane, as illustrated in Figure 5. Of special note is the effectiveness of just a handful of $z > 0.9$ supernovae in contracting the contours of this plot (as imagined by Goobar & Perlmutter 1995). A concise way to express the best constraint on Ω_m and Ω_Λ is $\Omega_\Lambda - 1.6\Omega_m = 0.4 \pm 0.2$. Interestingly, contours of constant age are very nearly parallel to the long axis of the error ellipse in Figure 5. This means that the competition between deceleration due to Ω_m and acceleration due to Ω_Λ is well captured by the measurement of luminosity distances, and the value of $H_0 t_0$ is well constrained even though the individual values of Ω_m and Ω_Λ are not.

Marginalizing the probabilities of Figure 5 onto the $H_0 t_0$ axis yields a probability distribution for $H_0 t_0$ as shown in Figure 6. We find $H_0 t_0 = 1.00 \pm 0.07$. The absolute value of the cosmic age depends on H_0, which, given the excellent agreement of the SBF and SN Ia distances, inherits almost all its errors from the Cepheid zero point.

6.   Summary

Precise distance estimators measured over the range from $z \sim 0$ to $z \sim 1$ provide powerful constraints on the dimensionless product $H_0 t_0 = 1.00 \pm 0.07$. Despite an ongoing struggle between deceleration and acceleration, the fortuitous result

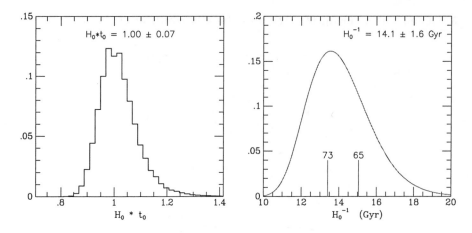

Figure 6.   *Left:* Probability density distribution for $H_0 t_0$ using our data. *Right:* Resulting distribution for t_0, using the 2:1 normalization of the SN Ia zero point to $H_0 = 73$ and 65 km s^{-1} Mpc^{-1} suggested by the HST Cepheid calibrations. The uncertainty in $H_0 t_0$ ($\sim 7\%$) is considerably smaller than the uncertainty H_0 ($\sim 12\%$), and so the uncertainty in t_0 comes mainly from the H_0 and Cepheid zero point.

is that the formulation in elementary textbooks; $t_0 = 1/H_0$ turns out to be accurate. A more formal way to express the constraint from SN Ia at $z > 1$ is that they imply $\Omega_\Lambda - 1.6\Omega_m = 0.4 \pm 0.2$. For our guess at the true value of H_0, $t_0 = 14 \pm 1.6$ Gyr. A better value for the age of the Universe hinges on nearby matters like the distance to the LMC and metallicity dependence of Cepheids!

References

Aguirre, A. 1999 ApJ, 525, 583

Aguirre, A, & Haiman, Z. 2000 ApJ, 532, 28

Ajhar, E. A., Tonry, J. L., Blakeslee, J. P., Riess, A. G., & Schmidt, B. P. 2001, ApJ, in press

Bartlett, J. G., Blanchard, A., Silk, J., & Turner, M. S. 1995, Science, 267, 980

Blakeslee, J. P., Ajhar, E. A., & Tonry, J. L. 1999, in Post-Hipparcos Cosmic Candles, ed. A. Heck & F. Caputo (Boston: Kluwer), 181

Coil, A., et al. 2000, ApJ, 544, L111

Ferrarese, L., et al. 2000, ApJ, 529, 745

Filippenko, A., & Riess, A. G. 1998, Phys. Rep., 307, 31

Freedman, W. L., et al. 2001, ApJ, in press (astro-ph/0012376)

Gibson, B. K., & Stetson, P. B. 2001, ApJ, 547, 103

Goobar, A., & Perlmutter, S. 1995, ApJ, 450, 14

Hamuy, M., Phillips, M. M., Suntzeff, N. B., Schommer, R. A., Maza, J., & Aviles, R. 1996, AJ, 112, 2398

Jha, S., et al. 1999, ApJS, 125, 73

Jha, S., et al. 2001, in New Cosmological Data and the Values of the Fundamental Parameters, ed. A. Lasenby, A. Wilkinson, & A. W. Jones (San Francisco: ASP), in press (astro-ph/0101521)

Leibundgut, B., et al. 1996, ApJ, 466, L21

Norgaard-Neilsen, H. U., Hansen, L., Jørgensen, H. E., Aragón-Salamanca, A., & Ellis, R. S. 1989, Nature, 339, 523

Parodi, B. R., Saha, A., Sandage, A., & Tammann, G. A. 2000, ApJ, 540, 634

Perlmutter, S., et al. 1995, ApJ, 440, L41

Perlmutter, S., et al. 1997, ApJ, 483, 565

Perlmutter, S., et al. 1998, Nature, 391, 51

Perlmutter, S., et al. 1999, ApJ, 517, 565

Phillips, M. M. 1993, ApJ, 413, L105

Riess, A. G., Press, W. H., & Kirshner, R. P. 1996, ApJ, 473, 88

Riess, A. G., et al. 1998, AJ, 116, 1009

Riess, A. G., et al. 2000, ApJ, 536, 62

Saha, A., Sandage, A., Labhardt, L., Tammann, G. A., Macchetto, F. D., & Panagia, N. 1997, ApJ, 486, 1

Schmidt, B. P., et al. 1998, ApJ, 507, 46

Acknowledgments. Support for this work was provided by NASA through a grant from the Space Telescope Science Institute, which is operated by the Association of Universities for Research in Astronomy, Inc. under NASA contract NAS5-26555.

Discussion

Michael Turner: Beyond the supernova data, what do you think is the strongest piece of evidence that supports the Universe accelerating?

John Tonry: I'm very impressed by the CMB data that are saying that the Universe is flat, and I'm very impressed by the fact that we seem to be seeing $\Omega_m \sim 0.3$. And if I do the difference there I get a Universe that's accelerating as well. Apart from those two, no.

Jean-René Roy: Can you tell us what you're trying to do to exclude intergalactic grey dust? Observing more supernovae at larger distances?

John Tonry: No, just the opposite. We have 7 beautiful Type Ia supernovae at $z \sim 0.5$ that we found this fall working at CTIO and CFHT, and we've been following them with HST in five different bandpasses so if the dust is truly grey we're screwed, but for any plausible model, there's a certain amount of reddening and we ought to be sensitive enough to the reddening over that wavelength range that we can exclude the possibility that the 0.25 dimming is coming from dust.

Astrophysical Ages and Time Scales
ASP Conference Series, Vol. 245, 2001
T. von Hippel, C. Simpson, N. Manset

Time Scales of Cosmic Microwave Background Radiation: Gold Mine in the Universe

Naoshi Sugiyama

National Astronomical Observatory of Japan, 2-2-1 Osawa, Mitaka, Tokyo 181-8588, Japan

Abstract. There are at least four important time scales for the formation of Cosmic Microwave Background (CMB) anisotropies. The first is the recombination epoch on which we saw the anisotropy patterns. The second time scale is related to the sound speed. For a given wave number, sound speed determines the phase of the acoustic oscillation. The third one is the diffusion time scale to determine the cutoff scale in the power spectrum of CMB anisotropies. The final important time scale is the equality epoch of matter–radiation densities. Through these time scales, the information of various cosmological parameters is imprinted on anisotropy patterns of CMB. Theoretical work reveal that these cosmological parameters can be determined by observing CMB anisotropies.

1. Introduction

The Cosmic Microwave Background radiation (CMB) is known to provide us with information at the last scattering surface, i.e., redshift $z \sim 1000$ which is the earliest Universe (about 300 000 years after the beginning of the Universe) we can directly observe. In other word, CMB is the oldest fossil of the early Universe. It is realized that anisotropies of CMB too contain rich information although they are tiny fraction of background temperature.

After the discovery of 10^{-5} fluctuations on the CMB temperature by the Differential Microwave Radiometer (DMR) of the COsmic Background Explore (COBE) in 1992 (Smoot et al. 1992), the number of observational data of temperature fluctuations has been rapidly increasing together with numerous works on theoretical understanding of CMB anisotropies.

Besides foreground contamination, there are two types of CMB anisotropies. The first one is the spectrum distortion from the back body shape which is caused by energy transfer from external sources to CMB radiation. COBE/FIRAS sets very stringent limits on both y (Compton y-parameter) and μ (chemical potential) distortions as $y < 1.5 \times 10^{-5}$ and $|\mu| < 9 \times 10^{-5}$ (Fixsen et al. 1996). These upper limits strongly constrain possible drastic events in thermal history of the universe, such as explosions of population III stars, or radiative decay of massive particles. On the other hand, the Sunyaev–Zeldovich (SZ) effect (Zeldovich & Sunyaev 1969; Sunyaev & Zeldovich 1970) caused by a cluster of galaxies has been actually observed and informs us about the real physical size of the cluster once we know the temperature of the cluster from X-ray data (see a recent review by Birkinshaw 1999). Assuming spherical symmetry of the cluster, we can thus measure the Hubble parameter.

The second type of CMB anisotropy is the primary spatial anisotropies. These anisotropies are mostly formed before recombination. There are several observational quantities. First one is the two dimensional temperature power spectrum, so-called C_ℓ where ℓ is the mutipole component and roughly corresponds to π/θ with θ being the angular separation on the sky map. As we will see later, C_ℓ contains information of cosmological parameters. Second quantity is the phase of fluctuations. This informs us on (1) the validity of the Gaussianity of the fluctuations, (2) agreement between the large scale structure of the universe and the CMB map, and (3) global topology of the Universe. Finally, polarization is the other important quantity of CMB anisotropies from which we could get information about the thermal history of the Universe and the type of perturbations, namely, scalar, vector or tensor types. The polarization has been recognized to provide complementary information to the temperature power spectrum C_ℓ. Because of the page limit, we only concentrate on the power spectrum of the primary anisotropies here.

2. Physical Processes and Their Time Scales

There are several physical effects working on the evolution of CMB anisotropies (see e.g., Hu & Sugiyama 1995a, 1995b; Hu, Sugiyama, & Silk 1997).

- Acoustic oscillations: Before recombination, photons and baryons are tightly coupled to each other. Once fluctuations cross the sound horizon, fluctuations of this tight coupled fluid oscillate as acoustic waves. These acoustic oscillations create peaks and wiggles on the power spectrum. Obviously the sound speed of the photon–baryon fluid is relevant to the time scale of acoustic oscillations.

- Diffusion Damping: Random walk of photons causes the exponential damping on fluctuations (Silk 1968). The time scale of the diffusion can be written by the photon mean free time times the square root of the number of scatterings, i.e., the age overs the photon mean free time.

- Gravitational redshift: After recombination, photon perturbations are only disturbed by the gravitational potential. Climbing up the gravitational well, photons lose energy and are redshifted. This redshift effect caused by the gravitational potential at recombination (last scattering surface) is known as the Sachs–Wolfe (SW) effect (Sachs & Wolfe 1967). For adiabatic perturbations, the gravitational potential stays constant during pure matter or radiation dominated regime in a flat $\Omega_0 = 1$ universe. However, it decays if one of these assumptions is broken. The decay of the gravitational potential in fact causes the redshift (or rather blueshift) of photons. Here we refer to this effect as the Integrated Sachs–Wolfe (ISW) effect. The ISW contribution is separated into two parts. Right after recombination, if the Universe is still not purely matter dominated, the potential decays. We call this the early ISW effect. In the case of a low density universe, the potential starts to decay very near the present epoch when the curvature or the cosmological constant starts to become the dominant component. We refer to this effect as the late ISW effect. Roughly speaking, the time scale of the ISW effect is determined by the age of the Universe when the decay of the potential takes place.

- Doppler effect: If the baryon velocity is different from the photons', the difference between the two velocities induces temperature fluctuations via Compton drag. During the tight coupling regime, baryon and photon velocities are identical. If the Universe becomes transparent after recombination, there is no interaction between photons and baryons, thus no Compton drag. Therefore this effect only appears to be important in late time reionized models. The time scale of the Doppler effect is relevant to the bulk velocity of ionized regions.

The acoustic oscillation of baryon–photon fluid is described by the analogy of a harmonic oscillator: Balls connected by a spring are set inside the gravitational well. The deviation of the balls' location from the center of the oscillation corresponds to the amplitude of the fluctuations. The spring represents the pressure. The initial location is specified by the initial condition, i.e., adiabatic or isocurvature conditions. Balls are kept at this initial location until fluctuations cross the sound horizon. Once crossing the sound horizon, they start to oscillate. At the recombination epoch, all oscillations are frozen out. They climb up the potential well to transfer to observers at present. If fluctuations of wavelength larger than the sound horizon at last scattering surface are considered, the balls stay at the initial location until the recombination epoch, which leads to the Sachs–Wolfe effect. Next let us consider shorter wavelength fluctuations. The recombination occurs after oscillation starts. A first compression mode at the last scattering surface corresponds to the first peak, a first depression mode corresponds to the second peak and so on.

3. Power Spectrum

Using our simple analytic picture, we can easily explain the dependence of the CMB spectrum on various cosmological parameters. First, we investigate the height of the peaks. General statements can be made as follows. The pressure of the photon–baryon fluid prevents the growth of fluctuations; hence less pressure (which corresponds to heavier balls in our analogy) means higher peaks. On the other hand, the gravitational potential induces adiabatic growth, which causes blueshift. A deeper potential seems to generate higher peaks; however, a deeper potential well means a bigger SW effect which causes redshift. Therefore a deeper potential doesn't necessarily produce higher peaks. It is known that the gravitational potential decays if fluctuations cross the horizon during the radiation dominated era. This decay boosts acoustic oscillations. The first reason for this boost is resonant oscillation since the decay of potential approximately synchronizes with the cycle of acoustic oscillations. The second reason is the time dilation effect. Gravitational potential stretches the geometry and causes time delay, i.e., redshift on CMB anisotropies. As the potential decays, photons get blueshifted. Eventually, the shallower (deeper) potential makes peaks higher (lower).

Let us now investigate the dependence of each cosmological parameters. If we consider $\Omega_B = 0$ limit, the sound speed is $c/\sqrt{3}$ where c is the speed of light. In this case, the amplitudes of density and velocity peaks are identical. Therefore there is no peak in C_ℓ simply because density and velocity perturbations are $\pi/2$ out of phase. Let us consider a more realistic case. The sound speed is always smaller than $c/\sqrt{3}$ if we take into account the baryon density. Increasing $\Omega_b h^2$ decreases the sound speed. Therefore, we expect a larger amplitude of the acoustic oscillation for lower baryon density. Since the initial condition fixes the initial point of the oscillation, the above effect shifts the zero point of the oscillation. Hence we can expect higher peaks for the odd peaks which correspond to the compression modes. Increasing $\Omega_0 h^2$ pushes the matter–radiation equality to earlier times. This causes potential to be deeper and the peaks to be lower.

We can explain most of parameter dependence from above arguments. For example, if we increase Ω_B while fixing Ω_0 and h, the sound speed becomes smaller. Therefore we get larger peaks as shown in Figure 1(a). Because above two effects compensate each other, the h dependence at fixed Ω_B and Ω_0 is complicated. However, if we fix $\Omega_B h^2$ which is determined by BBN, increasing h simply provides lower peaks (see Figure 1(b)).

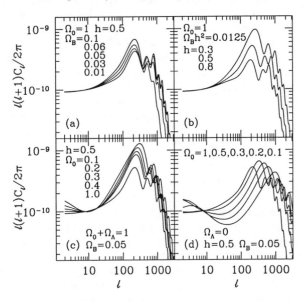

Figure 1.   Cosmological parameter dependence of CMB power spectrum C_ℓ' as function of multipole ℓ. The C_ℓs are normalized to COBE-DMR 2 year data. Panel (a) shows Ω_B dependence at fixed Ω_0 and h. From the top to the bottom, $\Omega_b = 0.1, 0.06, 0.05, 0.03$, and 0.01. Panel ($b$) shows h dependence with fixed $\Omega_B h^2$. From the top to the bottom, $h = 0.3, 0.5$, and 0.8. Panel (c) and (d) show Ω_0 dependence for flat cosmological constant dominated models and open models, respectively. From the top to the bottom of panel (c), $\Omega_0 = 0.1, 0.2, 0.3, 0.4$, and 1.0. From the left to the right of the first peak location of panel (d), $\Omega_0 = 1.0, 0.5, 0.3, 0.2$, and 0.1.

Let us next consider the dependence on the cosmological parameters of the peak location. Because the sound horizon is a weak function of $\Omega_B h^2$, the peak location is almost independent of Ω_B and h. On the other hand, it strongly depends on Ω_0, since a low density universe has a longer age, and the last scattering surface is further than in a high density universe. This effect makes the sound horizon correspond to a smaller angular scale. Moreover, in an open universe, there is a geodesic effect which causes a similar but bigger effect (Kamionkowski, Spergel, & Sugiyama 1994). These effects can be seen in Figure 1(c) and (d).

Finally, we would like to mention the peculiar behavior of CMB power spectrum on large scales (small ℓ) for low density models which is shown in Figure 1(c) and (d). On large scales, the SW effect which produces the flat tail makes the dominant contribution for the $\Omega_0 = 1$ model. On the other hand, there is the late ISW contribution for low density models. As for the Λ model, the ISW effect dominates on large scales. Because this has a damping effect caused by the finite thickness of *gravitational last scattering* (Hu & Sugiyama 1994), it is only significant on very large scales (small ℓ). For the open model, there is another effect, i.e., the cutoff at the curvature scale (Wilson 1983). Because the fluctuations outside the curvature scale do not contribute on C_ℓ, there is a cutoff on very large scales.

4. Implication From Recent Observations

Two very important observations of CMB anisotropies announced their results in 2000. First one is the long duration balloon-borne experiment in South Antarctica, BOOMERANG (de Bernardis et al. 2000; Lange et al. 2000). The angular resolution is about 10 arcminutes and depends weakly on the frequency. Analysing some part of total 200 hours data, they conclude the first peak location of the CMB angular power spectrum is $\ell = 197 \pm 6$ with an amplitude $\delta T_{\ell=200} = (68 \pm 8)$ μK. This peak location strongly rules out an open geometry for the Universe. The second observation is from MAXIMA, which is a balloon-borne experiment in North America (Hanany et al. 2000). Although MAXIMA observed a different part of the sky, their result is pretty much consistent with BOOMERANG. More or less, these two observations are consistent with the flat cold dark matter model (Jaffe et al. 2001; see Figure 2). Together with resent distant SNe surveys, the Universe appears to be almost flat, and cosmological constant-dominated.

If we take the observational results rigorously, two problems arise. First, the first peak location may suggest a closed model instead of the flat model since the peak location of the flat CDM model is about $l = 220$. A relatively low second (= even) peak may suggest a high baryon density which may not be consistent with BBN. However it is still premature to obtain any robust conclusions. For example, a recent new interferometer experiment by the Caltech group (Padin et al. 2001) shows a relatively high second peak which is not fully consistent (difference is more than 1σ but less than 2σ) with the BOOMERANG and MAXIMA results. We need to wait for whole sky measurement of CMB anisotropies with fine resolution.

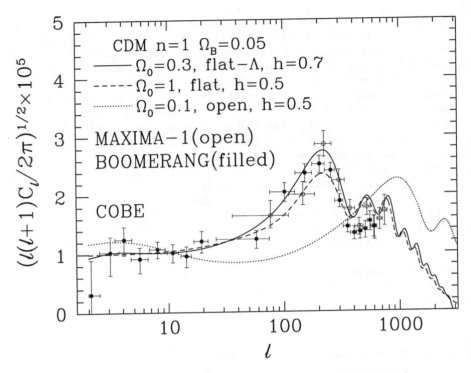

Figure 2.   Recent BOOMERANG (filled circles) and MAXIMA (open circles) observations with COBE/DMR ($\ell \leq 20$) and theoretical power spectra. We employ CDM models with flat $\Omega_0 = 0.3$ with cosmological constant (solid), flat $\Omega_0 = 1$ (dashed), and open $\Omega_0 = 0.1$ (dotted).

5.   Conclusions and the Future

It has been revealed that CMB anisotropies are gold mines in the universe and they contain almost all the information we need. The evolution of anisotropies is influenced by several physical processes such as the acoustic oscillations, the gravitational redshift, the diffusion damping, and the Doppler effect. Each physical process is controlled by the specific time scale which is a function of cosmological parameters such as $\Omega_0 h^2$ and $\Omega_B h^2$. Therefore by observing CMB anisotropies, we can determine cosmological parameters. Now we are eagerly waiting for post-COBE satellites, MAP (http://map.gsfc.nasa.gov/) which will be launched in June 2001, and PLANCK (http://astro.estec.esa.nl/SA-general/Projects/Planck/) which is scheduled for 2007. We are now standing just in front of the age of the discovery and the era of the precise cosmology. Let us see what will happen in the next decade.

References

de Bernardis, P., et al. 2000, Nature, 404, 955

Birkinshaw, M. 1999, Phys. Rep., 310, 97

Fixsen, D. J., Cheng, E. S., Gales, J. M., Mather, J. C., Shafer, R. A., & Wright, E. L. 1996. ApJ, 473, 576

Hanany, S., et al. 2000, ApJ, 545, L5

Hu, W., & Sugiyama, N. 1994, Phys. Rev. D, 50, 627

Hu, W., & Sugiyama, N. 1995a, ApJ, 444, 489

Hu, W., & Sugiyama, N. 1995b, Phys. Rev. D, 51, 2599

Hu, W., Sugiyama, N., & Silk, J., 1997, Nature 386, 37

Jaffe, A. H., et al. 2001, Phys. Rev. Lett., 86, 3475

Kamionkowski, M., Spergel, D. N., & Sugiyama, N. 1994, ApJ, 426, L57

Lange, A. E., et al. 2001, Phys. Rev. D, 63, 042001

Padin, S., et al. 2001, ApJ, 549, L1

Sachs, R. K., & Wolfe, A. M. 1967, ApJ, 147, 73

Silk, J. 1968, ApJ, 151, 459

Smoot, G., et al. 1992, ApJ. 396, L1

Sunyaev, R. A., & Zeldovich, Ya. B. 1970, Ap&SS, 9, 378

Wilson, M. L. 1983, ApJ, 273, 2

Zeldovich, Ya. B., & Sunyaev, R. A. 1969, Ap&SS, 4, 301

Discussion

Alan Guth: One of the areas that one hears a lot of talk about is the measurement of polarization in the microwave background. I don't think you said very much about that during your talk; would you like to comment on it?

Naoshi Sugiyama: Oh yeah, sure, I didn't have time. With the CMB anisotropy alone we can measure some cosmological parameters but using the polarization we can determine the cosmological paramters a bit more precisely. But not only that, because the polarization has two modes, E-mode and B-mode, and B-mode is only produced by tensor fluctuations, namely gravity waves, so it's a very good indicator of the tensor perturbations which may be induced during the inflation era. And the other interesting thing is the reionization of the Universe which I originally planned to concentrate on in this talk, but we know that the Universe got reionized around z of 10 or 20…who knows…? but by $z \sim 6$ it certainly got reionized due to galaxy formation or quasars or something like that. Then polarization is very very sensitive to the reionization process because it's just produced by the last scattering of the photons so to measure the polarization is very very important.

Wendy Freedman: Could you comment on the current status of plans to measure the polarization?

Naoshi Sugiyama: There are of course several ongoing experiments, projects like POLAR, but the sensitivity is not high enough to really measure the polarization, but the Taiwanese now plan to build a new instrument called AMIBA. And also one of the primary targets of the Planck satellite is polarization, and people have said that MAP *may* detect the polarisation, but the sensitivity to polarization for MAP is something like the CMB anisotropies for COBE, so maybe they can confirm the detection but it's hard to get any precise measurement for polarization, so maybe we can wait another 7 or 8 years, to get a really precise measurement. But there are some experiments ongoing.

Ofer Lahav: You mentioned the predictions for isocurvature and adiabatic perturbations. It's probably fair to say that most people in this field use just the predictions of adiabatic perturbations, and I wonder if you could comment on the success or lack of success of the isocurvature model?

Naoshi Sugiyama: Well maybe you ask wrong person. I have open mind so I'm really now working on isocurvature perturbation or something. So what I skipped is we have the first peak's location, ℓ of 200, from observations. So if we have adiabatic perturbations that indicates that $\Omega_0 + \Lambda = 1$, but if the perturbation is isocurvature we can have an open Universe by shifting that first peak which is off from $\ell \sim 100$ to the right or a closed Universe by shifting the second peak to $\ell \sim 200$. The open Universe model seems to be interesting so I'm now working on that but maybe the problem is there are not many natural inflation-motivated models to produce pure isocurvature mode.

Astrophysical Ages and Time Scales
ASP Conference Series, Vol. 245, 2001
T. von Hippel, C. Simpson, N. Manset

Cosmic Age in a Lepton-Asymmetric Universe: New Constraints From Primordial Nucleosynthesis and Cosmic Microwave Background Fluctuations

T. Kajino, M. Orito, K. Ichiki, S. Kawanomoto, H. Ando

National Astronomical Observatory and Graduate University for Advanced Studies, Mitaka, Tokyo 181-8588, and Department of Astronomy, University of Tokyo, Bunkyo-ku, Tokyo 113-0033, Japan

G. J. Mathews

Department of Physics and Center for Astrophysics, University of Notre Dame, Notre Dame, IN 46556, USA

R. N. Boyd

Department of Physics and Department of Astronomy, Ohio State University, Columbus, OH 43210, USA

Abstract. We study the cosmic age problem in the presence of a net lepton asymmetry as well as a baryon asymmetry. We explore a previously unnoted region of the parameter space in which very large baryon densities $0.1 \leq \Omega_b \leq 1$ can be accommodated within the light-element constraints from primordial nucleosynthesis. This parameter space consists of ν_μ and ν_τ degeneracies with a small ν_e degeneracy. We also discuss constraints from cosmic microwave background fluctuations.

1. Introduction

Recent progress in cosmological deep survey has clarified progressively the origin and distribution of matter and evolution of galaxies in the Universe. The origin of the light elements among them has been a topic of broad interest for its significance in constraining the dark matter component in the Universe and also in seeking the cosmological model which best fits the recent data of cosmic microwave background (CMB) fluctuations. This paper is concerned with neutrinos during Big Bang Nucleosynthesis (BBN). In particular, we consider new insights into the possible role which degenerate neutrinos may have played in the early Universe. There have been many important contributions toward constraining neutrino physics. Hence, a discussion of neutrinos and BBN is even essential in particle physics as well as cosmology.

There is no observational reason to insist that the universal lepton number is zero. It is possible, for example, for the individual lepton numbers to be large compared to the baryon number of the Universe, while the net total lepton number is small $L \sim B$. It has been proposed recently (Casas, Cheng, & Gelmini 1999) that models based upon the Affleck–Dine scenario of baryogenesis might naturally generate a lepton number asymmetry which is seven to ten orders of magnitude larger than the baryon number asymmetry. Neutrinos with large lepton asymmetry and masses ~ 0.07 eV might even explain the existence of cosmic rays with energies in excess of the Greisen–Zatsepin–Kuzmin cutoff (Gelmini & Kusenko 1999). It is, therefore, important for both particle physics and cosmology to carefully scrutinize the limits which cosmology places on the allowed range of both the lepton and baryon asymmetries.

2. Primordial Nucleosynthesis

The CMB power spectrum is expected to provide a precise value of the universal baryon-mass density parameter Ω_b along with the other cosmological parameters. It is therefore a critical test if BBN can predict a consistent Ω_b value.

There is a potential difficulty in the determination of Ω_b from primordial nucleosynthesis, which has been imposed by recent detections of a low deuterium abundance, $2.9 \times 10^{-5} \leq D/H \leq 4.0 \times 10^{-5}$, in Lyman-$\alpha$ clouds along the line of sight to high redshift quasars (Burles & Tytler 1998a, 1998b). Primordial abundance of ^{7}Li is constrained from the observed "Spite plateau," $0.91 \times 10^{-10} \leq^7 Li/H \leq 1.91 \times 10^{-10}$ (Ryan et al. 2000a, 2000b), and the ^{4}He abundance by mass, $0.226 \leq Y_p \leq 0.247$ (Olive, Steigman, & Walker 2000), from the observations in the H II regions. In order to satisfy these abundance constraints by a single Ω_b value, one has to assume an appreciable depletion in the observed abundance of ^{7}Li, which is still controversial both theoretically and observationally.

It depends on how accurately the nuclear reaction rates for the production of ^{7}Li are known. The ^{7}Li abundance is subject to large error bars associated with the measured cross sections for ^{4}He$(^3$H,$\gamma)^7$Li at $\eta \lesssim 2 \times 10^{-10}$ and ^{4}He$(^3$He,$\gamma)^7$Be at $3 \times 10^{-10} \lesssim \eta$. We studied these two reactions in quantum mechanics very carefully and concluded that the proper 2σ error bars could be $\sim 1/4$–$1/3$ of the previous ones. This improvement is mostly due to, first, the new precise measurement (Brune, Kavanagh, & Rolfs 1994) of the cross sections for ^{4}He$(^3$H,$\gamma)^7$Li and, second, the systematic theoretical studies (T. Kajino, M. Orito, & K. Ichiki, in preparation) of both reaction dynamics and quantum nuclear structures of ^{7}Li and ^{7}Be, whose validity is critically tested by electromagnetic form factors measured by high-energy electron scattering experiments. When our recommended error estimate is applied to the determination of Ω_b, we lose Ω_b value to explain both D/H and ^{7}Li/H simultaneously.

In order to better estimate the Ω_b value, we propose a new method to determine the primordial ^{7}Li by the use of isotopic abundance ratio ^{7}Li/^{6}Li in the interstellar medium which exhibits the minimum effects of the stellar processes including depletion effect. Details are reported elsewhere (Kajino et al. 2000; S. Kawanomoto et al., in preparation).

3. Neutrino Decoupling in Lepton Asymmetric Cosmology

Although lepton-asymmetric BBN has been studied in many papers (Kang & Steigman 1992, and references therein), there are several differences in the present work: For one, we have included finite temperature corrections to the

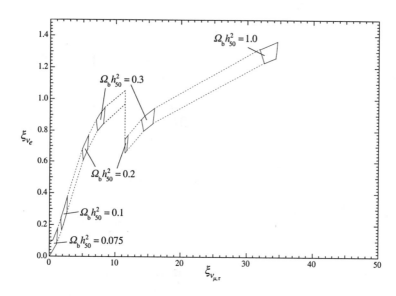

Figure 1.   Allowed values of ξ_{ν_e} and $\xi_{\nu_{\mu,\tau}}$ for which the constraints from light element abundances are satisfied for values of $\Omega_b h_{50}^2 = 0.075$, 0.1, 0.2, 0.3, and 1.0 as indicated.

mass of the electron and photon (Forengo, Kim, & Song 1997). Another is that we have calculated the neutrino annihilation rate in the cosmic comoving frame, in which the Møller velocity instead of the relative velocity is to be used for the integration of the collision term in the Boltzmann equations (Gondolo & Gelmini 1991; Enqvist, Kainulainen, & Semikoz 1992).

Neutrinos and anti-neutrinos drop out of thermal equilibrium with the background thermal plasma when the weak reaction rate becomes slower than the universal expansion rate. If the neutrinos decouple early, they are not heated as the particle degrees of freedom change. Hence, the ratio of the neutrino to photon temperatures, T_ν/T_γ, is reduced. The biggest drop in temperature for all three neutrino flavors occurs for a neutrino degeneracy parameter $\xi_\nu = \mu_\nu/T_\nu \sim 10$, where μ_ν is the neutrino chemical potential. This corresponds to a decoupling temperature above the cosmic QCD phase transition.

Non-zero lepton numbers affect nucleosynthesis in two ways. First, neutrino degeneracy increases the expansion rate, increasing the ^{4}He production. Secondly, the equilibrium n/p ratio is affected by the electron neutrino chemical potential, n/p $= \exp\{-(\Delta M/T_{n\leftrightarrow p}) - \xi_{\nu_e}\}$, where ΔM is the neutron–proton mass difference and $T_{n\leftrightarrow p}$ is the freeze-out temperature for the relevant weak reactions. This effect either increases or decreases ^{4}He production, depending upon the sign of ξ_{ν_e}.

A third effect emphasized in this paper is that T_ν/T_γ can be reduced if the neutrinos decouple early. This lower temperature reduces the energy density of neutrinos during BBN, and slows the expansion of the Universe. This decreases ^{4}He production.

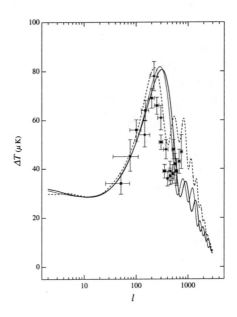

Figure 2.   CMB power spectrum from MAXIMA-1 (Hanany et al. 2000; circles) and BOOMERANG (de Bernardis et al. 2000; squares) binned data compared with calculated $\Omega = 1$ models.

Figure 1 highlights the main result of this study (Orito et al. 2001), where we take $\xi_{\nu_\mu} = -\xi_{\nu_\tau}$. For low $\Omega_b h_{50}^2$ models, only the usual low values for ξ_{ν_e} and $\xi_{\nu_{\mu,\tau}}$ are allowed. For $0.188 \lesssim \Omega_b h_{50}^2 \lesssim 0.3$, however, more than one allowed region emerges. For $\Omega_b h_{50}^2 \gtrsim 0.4$ only the large degeneracy solution is allowed. Neutrino degeneracy can even allow baryonic densities up to $\Omega_b h_{50}^2 = 1$.

4.   Constraints From Cosmic Microwave Background

Several recent works (Kinney & Riotto 1999; Lesgourges & Pastor 1999; Hannestad 2000) have shown that neutrino degeneracy can dramatically alter the power spectrum of the CMB. However, only small degeneracy parameters with the standard relic neutrino temperatures have been utilized. Here, we have calculated the CMB power spectrum to investigate effects of a diminished relic neutrino temperature.

    The solid line in Figure 2 shows a $\Omega_\Lambda = 0.4$ model for which the power-law index of the primordial density fluctuations $n = 0.78$. This fit is marginally consistent with the data at a level of 5.2σ. The dotted line in Figure 2 shows the matter-dominated $\Omega_\Lambda = 0$ best-fit model with $n = 0.83$, which is consistent with the data at the level of 3σ. The main differences in the fits between the large degeneracy models and our adopted benchmark model are that the first peak is shifted to slightly higher l value and the second peak is suppressed. One can clearly see that the suppression of the second acoustic peak is consistent with our derived neutrino-degenerate models. In particular, the MAXIMA-1 results are in very good agreement with the predictions of our neutrino-degenerate cosmological models (Orito et al. 2001; Mathews et al. 2001). It is clear that these new data sets substantially improve the goodness of fit for the neutrino-degenerate models (Lesgourges & Pastor 1999). Moreover, both data sets seem to require an increase in the baryonic contribution to the closure density as allowed in our neutrino-degenerate models.

5. Cosmic Age

There are several important implications of the neutrino degenerate Universe models. One of them is on the cosmic age problem. Recent balloon experiments to detect the CMB anisotropy have indicated that a flat cosmology is more likely. Combining this with the result from high-redshift supernova search, one may deduce a finite cosmological constant $\Omega_\Lambda \sim 0.6$, leading to a cosmic age ~ 15 Gyr. If this were the case, a potential difficulty that the cosmic age is likely to be shorter than the age of the Milky Way might be resolved. However, CMB anisotropy data provide with more details of several cosmological parameters which may not necessarily accept this simplified interpretation.

In our neutrino degenerate Universe models with $\Omega = 1$, $\Omega_\Lambda = 0.4$, and $\Omega_b h_{50}^2 = 0.1$, the neutrino mass $m_{\nu_{\mu,\tau}}$ is constrained to be less than 0.3 eV if $\Omega_\nu \leq 0.5$ (Orito et al. 2001; Mathews et al. 2001). Even should the mass be 0.3 eV, our conclusion on the primordial nucleosynthesis does not change at all. Therefore, we assumed massless neutrino. With this possible choice of the parameters in cosmology and particle physics, we can estimate the cosmic expansion age ≈ 12–13 Gyr. So the cosmic age problem seems to still remain. Further careful studies of the age problem and also the nature of cosmological constant (Yahiro et al. 2001) are highly desirable.

References

Brune, C. R., Kavanagh, R. W., & Rolfs, C. 1994, Phys. Rev. C, 50, 2205

Burles, S., & Tytler, D. 1998a, ApJ, 499, 699

Burles, S., & Tytler, D. 1998b, ApJ, 507, 732

Casas, A., Cheng, W. Y., & Gelmini, G. 1999, Nucl. Phys. B, 538, 297

de Bernardis, P., et al. 2000, Nature, 404, 955

Enqvist, K., Kainulainen, K., & Semikoz, V. 1992, Nucl. Phys. B, 374, 392

Fornengo, N., Kim, C. W., & Song, J. 1997, Phys. Rev. D, 56, 5123

Gelmini, G., & Kusenko, A. 1999, Phys. Rev. Lett, 82, 5202

Gondolo, P., & Gelmini, G. 1991, Nucl. Phys. B, 360, 145

Hanany, S., et al. 2000, ApJ, 545, L5

Hannestad, S. 2000, Phys. Rev. Lett, 85, 4203

Kajino, T., Suzuki, T.-K., Kawanomoto, S., & Ando, H. 2000, in IAU Symp. 198, The Light Elements and Their Evolution, ed. L. da Silva, M. Spite, & J. R. de Medeiros (San Francisco: ASP), 344

Kang, H., & Steigman, G. 1992, Nucl. Phys. B, 372, 494

Kinney, W. K., & Riotto, A. 1999, Phys. Rev. Lett, 83, 3366

Lesgourges, J., & Pastor, S. 1999, Phys. Rev. D, 60, 103521

Mathews, G. J., Orito, M., Kajino, T., & Wang, Y. 2001, Phys. Rev. D, submitted

Olive, K., Steigman, G., & Walker, T. 2000, Phys. Rep., 333, 389

Orito, M., Kajino, T., Mathews, G. J., & Boyd, R. N. 2001, ApJ, submitted (astro-ph/0005446)

Ryan, S., Beers, T., Olive, K., Fields, B., & Norris, J. 2000a, ApJ, 530, L57

Ryan, S., Kajino, T., Beers, T. C., Suzuki, T.-K., Romano, D., Matteucci, F., & Rosolankova, K. 2000b, ApJ, 459, 55

Yahiro, M., Mathews, G. J., Ichiki, K., Kajino, T., & Orito, M. 2001, Phys. Rev. D, submitted

Discussion

Gavin Dalton: You get $\Omega_\Lambda = 0.4$ and we measure $\Omega_m \leq 0.4$, so this sounds like a discrepancy with $\Omega_{total} = 1$.

Taka Kajino: We have recently communicated with Garnavich and other supernova guys through my collaborator Grant Mathews. Since they claim that $\Omega_\Lambda = 0.5$ is probably acceptable at the 3σ level, we explored wide possibilities of having different Ω_Λ values. After changing Ω_Λ from 0.6 to 0.0, we still have a χ^2 minimum at finite ξ for these Ω_Λ values at the same χ^2 level, which means that lepton-asymmetric Universe models are as good as the lepton-symmetric model with $\xi = 0$. But, there is such a strong degeneracy among flat Universe models with various Ω_Λs that we can hardly say which is the most favorable Ω_Λ value. This clearly shows that we need more precise data of CMB power spectrum. Let me stress again that this neutrino degeneracy parameter ξ 1-to-1 corresponds to Ω_b and also finite Ω_ν. Namely, in our Universe models Ω_b and Ω_ν are not independent adjustable parameters. This says that all three abundance constraints on the light elements, this lower limit to ^{7}Li, lower limit to deuterium, and upper limit to ^{4}He, are satisfied only when the finite lepton asymmetry is introduced.

Alan Guth: How large a lepton asymmetry are you actually talking about, say in terms of lepton number to photons?

Taka Kajino: That's a good question. It's very huge. In our present models we set out $\xi_{\nu_\mu} = -\xi_{\nu_\tau}$ so that these two counterbalance with each other to zero. However, we introduced a small $\xi_{\nu_e} \sim 0.4$, so $L \sim 10^{-1}$. Still 7 orders of magnitude bigger than B, so we are now looking for the fundamental field theory to allow $L \neq B$.

Astrophysical Ages and Time Scales
ASP Conference Series, Vol. 245, 2001
T. von Hippel, C. Simpson, N. Manset

The Age of the Universe From Joint Analysis of Cosmological Probes

Ofer Lahav

Institute of Astronomy, Madingley Road, Cambridge CB3 0HA, UK

Abstract. Analyses of various cosmological probes, including the latest Cosmic Microwave Background anisotropies, the 2dF Galaxy Redshift Survey, and Cepheid-calibrated distance indicators suggest that the age of expansion is 13 ± 3 Gyr. We discuss some statistical aspects of this estimation, and we also present results for joint analysis of the latest CMB and Cepheid data by utilizing Hyper-Parameters. The deduced age of expansion might be uncomfortably close to the age of the oldest globular clusters, in particular, if they formed relatively recently.

1. Introduction

Estimating the age of the Universe in the framework of the Big Bang model is an old problem. The rapid progress in observational cosmology in recent years has led to more accurate values of the fundamental cosmological parameters, including the age of the Universe. We summarize the basics in Section 2, we point out possible problems in joint analysis of cosmological probes in section 3, and we summarize some recent results in Section 4. In Section 5 we introduce a new method of Hyper-Parameters for combining different data sets, and we apply it in sections 6 and 7 to the latest Cepheid and CMB data. In section 8 we contrast the age of expansion with the ages of globular clusters.

2. The Age of Expansion

In the standard Big Bang model the age of the Universe is found by integrating $dt = H^{-1}da/a$, where a is the scale factor and $H \equiv \dot{a}/a$ is the Hubble parameter as given by Einstein's equations. This gives the present age in terms of three present-epoch parameters, $H_0 = 100h$ km s^{-1} Mpc^{-1} = $(9.78$ Gyr$)^{-1}h$, the mass density parameter $\Omega_{\rm m}$, and the scaled cosmological constant $\Omega_\Lambda \equiv \Lambda/(3H_0^2)$:

$$t_0 = H_0^{-1} \int_0^1 da\, a^{1/2}\, [\Omega_\Lambda a^3 + (1 - \Omega_{\rm m} - \Omega_\Lambda)a + \Omega_{\rm m}]^{-1/2} \ . \tag{1}$$

The difficulty is that in practice the parameters $(H_0, \Omega_{\rm m}, \Omega_\Lambda)$, when estimated from various cosmic probes, are commonly correlated with each other. For a flat universe $(\Omega_{\rm m} + \Omega_\Lambda = 1)$, supported by the recent Cosmic Microwave Background (CMB) experiments, this integral has an analytic solution in terms of only two free parameters,

$$t_0 = \frac{2}{3}H_0^{-1}\Omega_\Lambda^{-1/2}\ln[(1 + \sqrt{\Omega_\Lambda})(1 - \Omega_\Lambda)^{-1/2}] \qquad (2)$$

which is well approximated (e.g., Peacock 1999) by

$$t_0 \approx \frac{2}{3}H_0^{-1}\Omega_m^{-0.3}. \qquad (3)$$

This gives an insight to the way the errors propagate in the determination of the cosmic age:

$$\frac{\Delta t_0}{t_0} \approx \frac{\Delta H_0}{H_0} + 0.3\frac{\Delta\Omega_m}{\Omega_m}. \qquad (4)$$

This shows that the fractional error in H_0 is about three times more important than the fractional error in Ω_m. Typically the quoted error on H_0 is 10% (e.g., Freedman et al. 2001). The range of recent quoted values for the density parameter suggests $\Omega_m \sim 0.3 \pm 50\%$, so the expected fractional error in age is about 25% (e.g., for $t_0 \approx 13$ Gyr, $\Delta t_0 \approx 3$ Gyr). Again, Ω_m and H_0 are not always measured independently. For example, redshift surveys constrain the shape of the Cold Dark Matter (CDM) power-spectrum via the product $\Gamma \equiv \Omega_m h$, while the CMB angular power-spectrum constrains $\omega_m = \Omega_m h^2$.

3.   Cosmological Parameters From Joint Analysis: Cosmic Harmony?

A simultaneous analysis of the constraints placed on cosmological parameters by different kinds of data is essential because each probe (e.g., CMB, SNe Ia, redshift surveys, cluster abundance, and peculiar velocities) typically constrains a different combination of parameters. By performing joint likelihood analyses, one can overcome intrinsic degeneracies inherent in any single analysis and so estimate fundamental parameters much more accurately. The comparison of constraints can also provide a test for the validity of the assumed cosmological model or, alternatively, a revised evaluation of the systematic errors in one or all of the data sets. Recent papers that combine information from several data sets simultaneously include Webster et al. (1998); Lineweaver (1998); Gawiser & Silk (1998), Bridle et al. (1999, 2001), Eisenstein, Hu, & Tegmark (1999); Efstathiou et al. (1999); and Bahcall et al. (1999).

While joint Likelihood analyses employing both CMB and LSS data allow more accurate estimates of cosmological parameters, they involve various subtle statistical issues:

- There is the uncertainty that a sample does not represent a typical patch of the FRW Universe to yield reliable global cosmological parameters.

- The choice of the model parameter space is somewhat arbitrary.

- One commonly solves for the probability for the data given a model (e.g., using a likelihood function), while in the Bayesian framework this should be modified by the prior for the model and its parameters.

- If one is interested in a small set of parameters, should one marginalize over all the remaining parameters, rather than fix them at certain (somewhat ad hoc) values?

- The 'topology' of the likelihood contours may not be simple. It is helpful when the likelihood contours of different probes 'cross' each other to yield a global maximum (e.g., in the case of CMB and SNe), but in other cases they may yield distinct separate 'mountains', and the joint maximum likelihood may lie in a 'valley'.

- Different probes might be spatially correlated, i.e., not necessarily independent.

- What weight should one give to each data set?

In a long term collaboration in Cambridge (Bridle et al. 1999, 2001; Efstathiou et al. 1999; Lahav et al. 2000) we have compared and combined in a self-consistent way the most powerful cosmic probes: CMB, galaxy redshift surveys, galaxy cluster number counts, type Ia Supernovae, and galaxy peculiar velocities. Our analysis suggests, in agreement with studies by other groups, that we live in a flat accelerating Universe, with comparable amounts of dark matter and 'vacuum energy' (the cosmological constant Λ).

4. Some Recent Best Fit Cosmological Parameters

To give the flavor of favoured parameters we quote below two recent studies. These and numerous other studies support a Λ-CDM model with $\Omega_m = 1 - \Omega_\Lambda \sim 0.3$ and $h \sim 0.75$, which corresponds to an expansion age of $t_0 \sim 12.6$ Gyr (and $H_0 t_0 = 0.96$).

4.1. Combining CMB, Supernovae Ia, and Peculiar Velocities

A recent study (Bridle et al. 2001) is an example of combining 3 different data sets. We compared and combined likelihood functions for the matter density parameter Ω_m, the Hubble constant h, and the normalization σ_8 (in terms of the variance in the mass density field measured in an $8h^{-1}$ Mpc radius sphere) from peculiar velocities, CMB (including the earlier BOOMERANG and MAXIMA data), and type Ia Supernovae. These three data sets directly probe the mass in the Universe, without the need to relate the galaxy distribution to the underlying mass via a biasing relation.

Our analysis assumes a flat Λ-CDM cosmology with a scale-invariant adiabatic initial power spectrum and baryonic fraction as inferred from Big Bang Nucleosynthesis. We find that all three data sets agree well, overlapping significantly at the 2σ level. This therefore justifies a joint analysis, in which we find a best fit model and 95% confidence limits of $\Omega_m = 0.28(0.17, 0.39)$, $h = 0.74(0.64, 0.86)$, and $\sigma_8 = 1.17(0.98, 1.37)$. In terms of the natural parameter combinations for these data $\sigma_8 \Omega_m^{0.6} = 0.54(0.40, 0.73)$, $\Omega_m h = 0.21(0.16, 0.27)$. Also for the best fit point, $Q_{rms} = 19.7$ μK and the age of the Universe is 13.0 Gyr.

4.2.   The 2dF Galaxy Redshift Survey

The 2dF Galaxy Redshift Survey (2dFGRS) has now measured in excess of 160,000 galaxy redshifts and is the largest existing galaxy redshift survey. A sample of this size allows large-scale structure statistics to be measured with very small random errors. An initial analysis of the power-spectrum of the 2dFGRS (Percival et al. 2001) yields 68% confidence limits on the total matter density times the Hubble parameter $\Omega_m h = 0.20 \pm 0.03$, and the baryon fraction $\Omega_b/\Omega_m = 0.15 \pm 0.07$, assuming scale-invariant primordial fluctuations and a prior on the Hubble constant ($h = 0.7 \pm 10\%$). Although the Λ-CDM model with comparable amounts of dark matter and dark energy is not so elegant, it is remarkable that various measurements show such good consistency.

5.   Hyper-Parameters

We have addressed recently (Lahav et al. 2000; Lahav 2001) the issue of combining different data sets, which may suffer different systematic and random errors. We generalized the standard procedure of combining likelihood functions by allowing freedom in the relative weights of various probes. This is done by including in the joint likelihood function a set of 'Hyper-Parameters', which are dealt with using Bayesian considerations. The resulting algorithm, which assumes uniform priors on the logarithm of the Hyper-Parameters, is simple to implement. Here we show some examples of and results from the joint analysis of the latest CMB and Cepheid data sets.

Assume that we have two independent data sets, D_A and D_B (with N_A and N_B data points respectively) and that we wish to determine a vector of free parameters $\mathbf{w}$ (such as the density parameter Ω_m, the Hubble constant H_0 etc.). This is commonly done by minimizing

$$\chi^2_{\text{joint}} = \chi^2_A + \chi^2_B, \qquad (5)$$

(or, more generally, maximizing the product of likelihood functions).

Such procedures assume that the quoted observational random errors can be trusted, and that the two (or more) χ^2s have equal weights. However, when combining "apples and oranges" one may wish to allow freedom in the relative weights. One possible approach is to generalize Equation 5 to be

$$\chi^2_{\text{joint}} = \alpha\chi^2_A + \beta\chi^2_B, \qquad (6)$$

where α and β are 'Hyper-Parameters', which are to be dealt with in a Bayesian way. There are a number of ways to interpret the meaning of the HPs. One way is to understand α and β as controlling the relative weight of the two data sets. It is not uncommon that astronomers accept and discard measurements (e.g., by assigning $\alpha = 1$ and $\beta = 0$) in an ad hoc way. The HP procedure gives an objective diagnostic as to which measurements are problematic and deserve further understanding of systematic or random errors.

How do we eliminate the unknown HPs α and β? This is done by marginalization over α and β with Jeffreys' uniform priors in the log, $P(\ln \alpha) = P(\ln \beta) = 1$. We can then get the probability for the parameters $\mathbf{w}$ given the data sets:

$$-2\ln P(\mathbf{w}|D_A, D_B) = N_A \ln(\chi^2_A) + N_B \ln(\chi^2_B). \qquad (7)$$

To find the best fit parameters $\mathbf{w}$ requires us to minimize the above probability in the $\mathbf{w}$ space. It is as easy to calculate this statistic as the standard χ^2, and it can be generalized for any number of data sets.

Since α and β have been eliminated from the analysis by marginalization they do not have particular values that can be quoted. Rather, each value of α and β has been considered and weighted according to the probability of the data given the model. It can be shown that the 'weights' are $\alpha_{\text{eff}} = \frac{N_A}{\chi_A^2}$ and $\beta_{\text{eff}} = \frac{N_B}{\chi_B^2}$, both evaluated at the joint peak.

6. H_0 From Cepheids

One of the most important results for the Hubble constant comes from the Hubble Space Telescope Key Project (Freedman et al. 2001). The method is based on period–luminosity Cepheid calibration of several secondary distance indicators measured over distances of 400 to 600 Mpc. Freedman et al. (2001; see also these proceedings) combined the different measurements by several statistical methods and derived as the final result $H_0 = 72 \pm (3)_r \pm (7)_s$ km s^{-1} Mpc^{-1} (1σ random and systematic errors).

Given the importance of this work, we have attempted to combine the data by a different method, using the HPs. We used the raw data given by Freedman et al. (2001) for Surface Brightness Fluctuations (SBF), Supernovae Ia (SNIa), Tully–Fisher (TF), and Fundamental Plane (FP). To the random errors given in the tables we added (in quadrature) the quoted systematic errors.

The results are shown on the right top and bottom panels in Figure 1. We see that the four methods give a range of values for H_0, with the most discrepant result being the FP ($H_0 = 88$ km s^{-1} Mpc^{-1}). However, using the HPs, the most probable result, $H_0 = 73$ km s^{-1} Mpc^{-1}, agrees well with the result of Freedman et al. (2001). We also see that in this case the standard joint χ^2 and the HPs give a very similar answer. The resulting HPs (weights) for SBF, TF, SN, and FP are 3.4, 2.7, 1.9, and 0.5, respectively.

7. Combining Cepheids and CMB Data

The latest BOOMERANG (Netterfield et al. 2001; de Bernardis et al. 2001), MAXIMA (Stompor et al. 2001) and DASI (Pryke et al. 2001) CMB anisotropy measurements indicate 3 acoustic peaks. Parameter fitting to a Λ-CDM model suggests consistency between the different experiments, and a best-fit universe with zero curvature, and an initial spectrum with spectral index $n = 1$ (e.g., Wang, Tegmark, & Zaldarriaga 2001 and references therein). Unlike the earlier BOOMERANG and MAXIMA results, the new data also show that the baryon contribution is consistent with the Big Bang Nucleosynthesis value $\Omega_b h^2 \sim 0.02$ (O'Meara et al. 2001).

Several authors (e.g., Wang et al. 2001) have pointed out that the Hubble constant h itself cannot be determined accurately from CMB data alone. As the CMB constrains well the combination $\omega_m \equiv \Omega_m h^2$, the curvature Ω_k and Ω_Λ,

622   *Lahav*

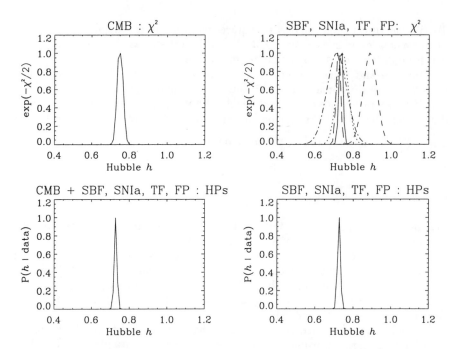

Figure 1.   Probabilities for the Hubble constant. *Top right:* χ^2 statistic for four Cepheid-calibrated distance indicators from Freedman et al. (2001): SBF (dashed-dotted line), SNIa (long-dashed), TF (dotted), and FP (dashed). The joint χ^2 for the Cepheid data is shown by the solid line, centered on $h = 0.74$. *Bottom right:* The probability based on the Hyper-Parameters approach for the Hubble constant given the four Cepheid-calibrated data sets. The maximum probability is at $h = 0.73$. *Top left:* χ^2 statistic derived for a compilation of the latest CMB data including BOOMERANG, MAXIMA, and DASI from Wang et al. (2001), and a grid of CMB models for an assumed Λ-CDM model with $n = 1, \Omega_m = 1 - \Omega_\Lambda = 0.3, \Omega_b h^2 = 0.02$ (BBN value), and $Q_{rms} = 18 \ \mu$K (COBE normalization). The maximum probability is at $h = 0.75$ (see also Figure 2). The distribution will be wider if some of the above parameters are kept free. *Bottom left:* The probability based on the Hyper-Parameters for the four Cepheid-calibrated data sets and the CMB compilation. The maximum probability is at $h = 0.73$.

the Hubble constant can be derived from

$$h = \sqrt{\omega_m/(1 - \Omega_k - \Omega_\Lambda)}. \tag{8}$$

It is not surprising therefore that estimates of h from the latest CMB data strongly depend on the assumed set of free parameters, and on the assumed priors from other probes. For example, from the latest BOOMERANG data, Netterfield et al. (2001) derive for weak priors $h = 0.56 \pm 0.11$ and $t_0 = 15.4 \pm 2.1$ Gyr, while with strong priors $h = 0.66 \pm 0.05$ and $t_0 = 14.0 \pm 0.6$ Gyr).

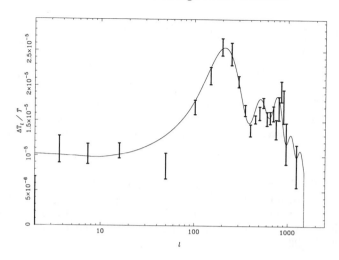

Figure 2. Compilation of the latest CMB $\frac{\Delta T}{T}$ data points against spherical harmonic l (from Wang et al. 2001). The line shows the predicted angular power-spectrum for a Λ-CDM model with $n = 1$, $\Omega_m = 1 - \Omega_\Lambda = 0.3$, $\Omega_b h^2 = 0.02$ (BBN value), $Q_{rms} = 18\ \mu K$ (COBE normalization), and $h = 0.75$. A similar model (with $h = 0.7$) is also the best fit to the the 2dF galaxy power-spectrum (Percival et al. 2001). Hence, we see good agreement of two entirely different data sets.

Wang et al. (2001) find, e.g., ($h = 0.42 \pm 0.23$; $t_0 = 20.5 \pm 9.0$ Gyr) from CMB alone, and ($h = 0.57 \pm 0.30$; $t_0 = 14.2 \pm 4.3$ Gyr) by combining the CMB with the IRAS PSCz data.

Here, for simplicity, we take the approach of fixing all the other parameters, apart from h. We assume that CMB fluctuations arise from adiabatic initial conditions with Cold Dark Matter and negligible tensor component, in a flat Universe with $\Omega_m = 1 - \Omega_\Lambda = 0.3$, $n = 1$, $Q_{rms} = 18\ \mu K$ (COBE normalization) and $\Omega_b h^2 = 0.02$. This choice is motivated by numerous other studies which combined CMB data with other cosmological probes (e.g., Bridle et al. 2001; Hu et al. 2001; Wang et al. 2001; Section 4 above). Of course, one may keep more free parameters, and marginalize over some of them, as done in numerous other studies. We obtain theoretical CMB power-spectra using the CMBFAST and CAMB codes (Slejak & Zaldarriaga 1996; Lewis, Challinor, & Lasenby 2000). Increasing h decreases the height of the first acoustic peak, and makes few other significant changes to the angular power spectrum (e.g., Hu et al. 2001). The range in h investigated here is ($0.5 < h < 1.1$).

Different CMB data sets can be combined in different ways (e.g., Jaffe et al. 2000; Lahav et al. 2000; Lahav 2001). For simplicity we use here a compilation of 24 $\Delta T/T$ data points from Wang et al. (2001), which is based on 105 band-power measurements (including the latest from BOOMERANG, MAXIMA, and DASI). The left top panel of Figure 1 shows the CMB likelihood function (with the correlation matrix not yet taken into account) when the only free parameter is the Hubble constant. It favours a value of $h \sim 0.75$. Figure 2 shows the

CMB data points, and we also projected our best-fit model (for the above set of assumptions). A similar model (with $h = 0.7$) also fits well other cosmological measurements, e.g., the 2dF galaxy power-spectrum (Percival et al. 2001). The left bottom panel of Figure 1 shows the Hyper-Parameters joint probability for the CMB and the four Cepheid-calibrated data sets. The maximum probability is at $h \sim 0.73$. We note that if more cosmological parameters are left free and then marginalized over, the error in h would typically be much larger. We also note that in the context of CDM models we can estimate the Hubble constant from the ratio

$$h = \omega_{\mathrm{m}}/\Gamma \qquad (9)$$

(see Section 2). The recent CMB data suggest $\omega_{\mathrm{m}} \sim 0.15$ (e.g., Netterfield et al. 2001) and from various redshift surveys $\Gamma \sim 0.2$, so $h \sim 0.75$, in good agreement with our derived value. Of course h and other parameters can be derived more quantitatively by joint likelihood analysis of CMB and redshift surveys (e.g., Webster et al. 1998).

8. Discussion

Joint analyses of cosmic probes suggests a flat Universe with $\Omega_{\mathrm{m}} = 1 - \Omega_{\Lambda} \approx 0.3$. The measurement of the Hubble constant from Cepheids and from the CMB suggests $H_0 \approx 75$ km s^{-1} Mpc^{-1}. This set of best-fit parameters yields an expansion age $t_0 = 13 \pm 3$ Gyr. The error estimate is due to errors in both H_0 and Ω_{m} (based on the range of recently quoted values). While this is currently the most popular model there are potential problems with this set of parameters:

(1) There is no simple theoretical explanation why the present epoch contributions to matter Ω_{m} and dark energy (Ω_{Λ}) are nearly equal.

(2) The age of expansion is commonly compared with the age t_{GC} of globular clusters (GCs) and other old objects. More precisely, one requires

$$t_0 = t_{\mathrm{f}} + t_{\mathrm{GC}}, \qquad (10)$$

where t_{f} is the time scale of formation.

The age of the oldest GC was estimated to be $t_{\mathrm{GC}} = 11.5 \pm 1.3$ Gyr by Chaboyer (1998), but he revised it upwards to $t_{\mathrm{GC}} = 13.2 \pm 1.5$ Gyr (Chaboyer, these proceedings). A recent radioactive dating (using ^{238}U, with half-life of 4.5 Gyr) of a very metal-poor star in the Galaxy, gives an age of 12.5 ± 3 Gyr (Cayrel et al. 2001).

We see that the ages of old objects might be uncomfortably close to the age of expansion (despite having a non-vanishing cosmological constant, which tends to stretch the age of the Universe). Furthermore, it is commonly assumed that $t_{\mathrm{f}} \sim 0.5\text{--}2$ Gyr (e.g., Chaboyer 1998), and hence t_{f} is neglected in Equation 10. However, we point out that t_{f} is model dependent, and it can span a wide range of values. For example Peebles & Dicke (1968) suggested (before dark matter was recognized as a major component) that GCs can be identified with the Jeans' mass after recombination, i.e., $z_{\mathrm{f}} \sim 100\text{--}1000$ and indeed in this case t_{f} is negligible.

On the other hand, the model of Fall & Rees (1985) for the formation of GCs in the Galactic halo as a result of thermal instability suggests $z_{\mathrm{f}} \sim 1\text{--}3$, i.e., $t_{\mathrm{f}} \sim$

2–5 Gyr for the above world model. These and other models will be discussed elsewhere (N. Y. Gnedin, O. Lahav, & M. J. Rees, in preparation). Obviously, late formation of the oldest GC means that the derived age of expansion t_0 is too short compared with $t_f + t_{GC}$. This possible age crisis might indicate a potential problem for either the cosmological model or the age estimation of GCs.

The age of the Universe and other cosmological parameters will be revisited soon with larger and more accurate data sets such as the big redshift surveys (2dF, SDSS) and CMB (MAP, Planck) data. Other relevant probes for the Hubble constant are the Sunyaev–Zeldovich effect (e.g., Mason, Myers, & Readhead 2001), the gravitational lensing time delay (e.g., Williams & Saha 2000) and the baryon fraction in clusters (e.g., Ettori 2001; Douspis et al. 2001). New high-quality data sets will allow us to study a wider range of models and parameters.

References

Bahcall, N. A., Ostriker, J. P., Perlmutter, S., & Steinhardt, P. J. 1999, Science, 284, 148

Bridle, S. L., Eke, V. R., Lahav, O., Lasenby, A. N., Hobson, M. P., Cole, S., Frenk, C. S., & Henry, J. P. 1999, MNRAS, 310, 565

Bridle, S. L., Zehavi, I., Dekel, A., Lahav, O., Hobson, M. P., & Lasenby, A. N. 2001, MNRAS, 321, 333

Cayrel, R., et al. 2001, Nature, 409, 691

Chaboyer, B. 1998, Phys. Rep., 307, 23

de Bernardis, P., et al. 2000, Nature, 404, 955

Douspis, M., et al. 2001 (astro-ph/0105129)

Efstathiou, G., Bridle, S. L., Lasenby, A. N., Hobson, M. P., & Ellis, R. S. 1999, MNRAS, 303, L47

Eisenstein, D. J., Hu, W., & Tegmark, M. 1999, ApJ, 518, 2

Ettori, S. 2001, MNRAS, 323, 1

Fall, S. M., & Rees, M. J. 1985, ApJ, 298, 18

Freedman, W. L., et al. 2001, ApJ, in press (astro-ph/0012376)

Gawiser, E., & Silk, J. 1998, Science, 280, 1405

Hu, W., Fukugita, M., Zaldarriaga, M., & Tegmark, M. 2001, ApJ, 549, 669

Jaffe, A., et al. 2000, Phys. Rev. Lett, 86, 3475

Lahav, O. 2001, in New Cosmological Data and the Values of the Fundamental Parameters, ed. A. Lasenby & A. Wilkinson (San Francisco: ASP), in press (astro-ph/0012475)

Lahav, O., Bridle, S. L., Hobson, M. P., Lasenby, A. L., & Sodré, L. 2000, MNRAS, 315, L45

Lewis, A., Challinor, A., & Lasenby, A. 2000, ApJ, 538, 473

Lineweaver, C. H. 1998, ApJ, 505, L69

Mason, B. S., Myers, S. T., & Readhead, A. C. S. 2001, Proc. 9th Marcel Grossman Meeting, ed. V. G. Gurzadyan, R. Jantzen, & R. Ruffini (Singapore: World Scientific), in press (astro-ph/0101170)

Netterfield, C. B., et al. 2001, ApJ, submitted (astro-ph/0104460)

O'Meara, J. M., Tytler, D., Kirkman, D., Suzuki, N., Prochaska, J. X., Lubin, D., & Wolfe, A. M. 2001, ApJ, 552, 718

Peacock, J. A. 1999, Cosmological Physics (Cambridge: Cambridge University Press)

Peebles, P. J. E., & Dicke, R. H. 1968, ApJ, 154, 891

Percival, W. J., et al. 2001, MNRAS, submitted (astro-ph/0105252)

Pryke, C., Halverson, N. W., Leitch, E. M., Kovac, J., Carlstrom, J. E., Holzapfel, W. L., & Dragovan, M. 2001, ApJ, submitted (astro-ph/0104490)

Seljak, U., & Zaldarriaga, M. 1996, ApJ, 469, 437

Stompor, R., et al. 2001, ApJ, submitted (astro-ph/0105062)

Wang, X., Tegmark, M., & Zaldarriaga, M. 2001, Phys. Rev. D, submitted (astro-ph/01050910)

Webster, M., Bridle, S. L., Hobson, M. P., Lasenby, A. N., Lahav, O., & Rocha, G. 1998, ApJ, 509, L65

Williams, L. L. R., & Saha, P. 2000, AJ, 119, 439

Acknowledgments. I thank Sarah Bridle, Pirin Erdogdu, Carolina Ödman, and the 2dFGRS team for their contribution to the work presented here, and Wendy Freedman, Oleg Gnedin, Jeremy Mould, and Martin Rees for helpful discussions. I also thank Ted von Hippel and the other conference organizers for the hospitality in Hawai'i.

Discussion

Brent Tully: I wanted to take exception to the statement you made a little earlier that a decade ago there was commonality similar to what we're seeing right now. I was there back then and I remember that it was rather different. It may have been true that there was a standard model back then, as you said, but there were clearly things wrong with that standard model, and I'll just say the Hubble constant was one of the most obvious ones. And the thing that's striking right now is how everyone is agreeing, and that's just scary.

Astrophysical Ages and Time Scales
ASP Conference Series, Vol. 245, 2001
T. von Hippel, C. Simpson, N. Manset

The Static Universe Hypothesis and Observational Proof

Thomas B. Andrews

3828 Atlantic Avenue, Brooklyn, NY 11224, USA

Abstract. From the axiom of the unrestricted repeatability of all experiments, Bondi and Gold argued that the Universe is in a stable, self-perpetuating equilibrium state. Their reasoning extended the usual cosmological principle to the perfect cosmological principle (PCP) in which the Universe looks the same from any location at any time. By itself, the PCP predicts a static universe since the Universe must be in an equilibrium state.

However, Bondi and Gold rejected this prediction for two reasons: First, they believed the Universe was expanding on the basis of the Hubble redshift. Second, the Universe appeared to be far from thermodynamic equilibrium. Thus, they hypothesized the steady state universe an expanding universe model in which matter is created to maintain the Universe in a stationary state.

Instead, based on the PCP and a new process for the Hubble redshift in a static universe, I hypothesize that the Universe is flat, static, and non-evolving. Then, applying the well-known scientific method, I test deductions from this hypothesis against current observational data.

Using three global tests of the space-time metric, I find that the observational data consistently fits the static universe hypothesis. In particular, the tests show (1) the luminosity and physical sizes of galaxies have not evolved over the last 10 to 20 billion years, and (2) the luminosity of Type Ia supernovae do not decrease at high z in the static universe.

Because the static universe hypothesis is a simple and logical deduction from the PCP and the hypothesis is amply confirmed by the observational data, I conclude that the Universe is static and non-evolving.

Astrophysical Ages and Time Scales
ASP Conference Series, Vol. 245, 2001
T. von Hippel, C. Simpson, N. Manset

Lengths of the First Days of the Universe

Moshe Carmeli

Department of Physics, Ben Gurion University, Beer Sheva 84105,
Israel

Abstract. The early stage of the Universe is discussed and the time lengths of its first days are given. If we denote the Hubble time in the zero-gravity limit by τ (approximately 12.16 billion years), and T_n denotes the length of the n-th day, then we have the very simple relation $T_n = \tau/(2n - 1)$. Hence, we obtain for the first days the following lengths of time: $T_1 = \tau$, $T_2 = \tau/3$, $T_3 = \tau/5$, etc.

In this Note we calculate the lengths of days of the early Universe, day by day, from the first day after the Big Bang on up to our present time. We find that the first day actually lasted the Hubble time in the limit of zero gravity. If we denote the Hubble time in the zero-gravity limit by τ which equals about 12.16 billion years and T_n denotes the length of the n-th day in units of times of the early Universe, then we have a very simple relation

$$T_n = \frac{\tau}{2n - 1}. \tag{1}$$

Hence we obtain for the first few days the following lengths of time:

$$T_1 = \tau, \quad T_2 = \frac{\tau}{3}, \quad T_3 = \frac{\tau}{5}, \quad T_4 = \frac{\tau}{7}, \quad T_5 = \frac{\tau}{9}, \quad T_6 = \frac{\tau}{11}. \tag{2}$$

It also follows that the accumulation of time from the first day to the second, third, fourth, etc., up to now is just exactly the Hubble time. The Hubble time in the limit of zero gravity is the maximum time allowed in nature.

Using Cosmological Special Relativity (Carmeli 1995, 1996, 1997a, 1997b), the calculation is very simple. We assume that the Big Bang time with respect to us now was $t_0 = \tau$, the time of the first day after that was t_1, the time of the second day was t_2, and so on. In this way the time scale is progressing in units of one day (24 hours) in our units of present time. The time difference between t_0 and t_1, denoted by T_1, is the time as measured at the early Universe and is by no means equal to one day of our time. In this way we denote the times elapsed from the Big Bang to the end of the first day t_1 by T_1, between the first day t_1 and the second day t_2 by T_2 and so on. According to the rule of the addition of cosmic times one has, for example,

$$t_6 + 1(\text{day}) = \frac{t_6 + T_6}{1 + t_6 T_6/\tau^2}. \tag{3}$$

A straightforward calculation then shows that

$$T_6 = \frac{\tau^2}{\tau^2 - (\tau - 6)(\tau - 5)} = \frac{\tau^2}{11\tau - 30}. \tag{4}$$

628

In general one finds that

$$T_n = \frac{\tau^2}{\tau^2 - (\tau - n)(\tau - n + 1)}, \tag{5}$$

or

$$T_n = \frac{\tau}{n + (n-1) - n(n-1)/\tau}. \tag{6}$$

As is seen from the last formula one can neglect the last term in the denominator in the first approximation and we get the simple Eq. (1).

From the above one reaches the conclusion that the age of the Universe exactly equals the Hubble time in vacuum τ, i.e., 12.16 billion years, and it is a universal constant (Carmeli & Kuzmenko 2001). This means that the age of the Universe tomorrow will be the same as it was yesterday or today.

But this might not go along with our intuition since we usually deal with short periods of times in our daily life, and the unexperienced person will reject such a conclusion. Physics, however, deals with measurements.

In fact we have exactly a similar situation with respect to the speed of light c. When measured in vacuum, it is $300\,000$ km s^{-1}. If the person doing the measurement tries to decrease or increase this number by moving with a very high speed in the direction or against the direction of the propagation of light, he will find that this is impossible and he will measure the same number as before. The measurement instruments adjust themselves in such a way that the final result remains the same. In this sense the speed of light in vacuum c and the Hubble time in vacuum τ behave the same way and are both universal constants.

The similarity of the behavior of velocities of objects and those of cosmic times can also be demonstrated as follows. Suppose a rocket moves with the speed V_1 with respect to an observer on the Earth. We would like to increase that speed to V_2 as measured by the observer on the Earth. In order to achieve this, the rocket has to increase its speed not by the difference $V_2 - V_1$, but by

$$\Delta V = \frac{V_2 - V_1}{1 - V_1 V_2/c^2}. \tag{7}$$

As can easily be seen ΔV is much larger than $V_2 - V_1$ for velocities V_1 and V_2 close to that of light c. This result follows from the rule for the addition of velocities,

$$V_{1+2} = \frac{V_1 + V_2}{1 + V_1 V_2/c^2}, \tag{8}$$

a consequence of Einstein's (1905) famous Special Relativity Theory. In cosmology, we have the analogous formula

$$T_{1+2} = \frac{T_1 + T_2}{1 + T_1 T_2/\tau^2} \tag{9}$$

for the cosmic times.

References

Carmeli, M. 1995, Found. Phys., 25, 1029

Carmeli, M. 1996, Found. Phys., 26, 413

Carmeli, M. 1997a, Inter. J. Theor. Phys., 36, 757

Carmeli, M. 1997b, Cosmological Special Relativity: The Large-Scale Structure of Space, Time and Velocity (World Scientific)

Carmeli, M., & Kuzmenko, T. 2001, in Proceedings of the 20th Texas Symposium on Relativistic Astrophysics, ed. J. C. Wheeler, in press

Einstein, A. 1905, Annalen der Physik, 17, 891; English translation: Einstein, A., Lorentz, H. A., Minkowski, H., & Weyl, H. 1923, The Principle of Relativity (Dover Publications), 35

Astrophysical Ages and Time Scales
ASP Conference Series, Vol. 245, 2001
T. von Hippel, C. Simpson, N. Manset

The Scale Expanding Cosmos Theory and a New Conception of the Progression of Time

C. Johan Masreliez

The EST Foundation, USA

Abstract. A new cosmological theory based on the assumption that all four spacetime metrics expand, predicts that the planets slowly spiral toward the Sun while accelerating. This would result in a difference between dynamic time as defined by the planetary motions and atomic time. The Expanding Spacetime Theory is presented together with observational evidence for this effect.

1. The Expanding Spacetime Theory (EST)

General Relativity (GR) is scale invariant, which means that the GR equations do not change when multiplying the metrics in the line element by a constant scale factor. The EST theory, which proposes that the Universe evolves by incrementally changing its scale, is based on the following assumptions:

(1) The scale of material objects is determined by the spacetime metrics. Vice versa, matter defines the metrics of spacetime.

(2) The Universe expands by changing the scale, i.e., the metrics of both space and time.

(3) All spatial locations and epochs are strongly equivalent.

1.1. Scale Expansion by Gauge

A changing cosmological scale suggests a with time increasing gauge $a(t)$:

$$\mathrm{d}s^2 = a^2(t) \cdot g_{ij}(t, x)\mathrm{d}x_i\mathrm{d}x_j. \tag{1}$$

However, different epochs are not covariant with this line element since generally no (continuous) variable transformation relating different metrics $\mathrm{d}s^2(t_1)$ and $\mathrm{d}s^2(t_2)$ exists. They are therefore not strongly equivalent in the Einstein sense.

1.2. Scaled Spacetimes

Two line elements $\mathrm{d}s^2 = g_{ij}(t, x)\mathrm{d}x_i\mathrm{d}x_j$ and $\mathrm{d}s'^2 = C \cdot g_{ij}(t, x)\mathrm{d}x_i\mathrm{d}x_j$ differing by a constant scale factor C are termed "scaled line elements" and corresponding spacetimes "scaled spacetimes." They satisfy Einstein's GR relations identically—they are "spacetime equivalent." Since the energy–momentum tensor is the same for these line elements the energy density is the same, which supports the proposition that the scale of material objects depends on the scale of spacetime. An imaginary universe in which the metrics of space and time differ from our Universe by a constant factor should appear identical in all respects.

1.3.   The EST Line Element

Consider the line element:

$$ds^2 = \exp(2t/T) \cdot (dt^2 - dx^2 - dy^2 - dz^2), \qquad (2)$$

where T is the Hubble time. This line element satisfies the GR equations identically for an arbitrary discrete time translation $t \mapsto t + t_0$, that generates a scale change. Furthermore, the corresponding scaled spacetimes are related via a simple variable transformation $t' = t + t_0$, which implies that they are covariant and strongly equivalent in Einstein's sense. Thus, the EST line element is spacetime equivalent under discrete time translations. The EST expansion mode might be modeled by the following cyclic iteration:

(1) The scale expands continuously from $\exp(t/T)$ to $\exp((t + t_0)/T)$.

(2) A $t + t_0$ the pace of time decreases by changing the pace of proper time $ds \mapsto ds \exp(t_0/T)$. This eliminates the scale factor $\exp(t_0/T)$ and restores the EST line element.

(3) Returning to step 1 above activates the iteration.

This means that the pace of time decreases in discrete increments in the Expanding Spacetime.

With this expansion mode all locations in space and time are equivalent and the line element satisfies the GR relations identically. Although the Universe evolves, spacetime is invariant to translations in space or time. This is the Expanding Spacetime model.

This discrete cosmological expansion mode might provide a clue to the progression of time, which might not be a continuous process to be modeled by GR. Since GR is based on Riemann geometry, any cosmological model using GR describes both the past and the future as a continuous manifold but it offers no explanation to the progression of time. In the EST model the progression of time projects the present into the future via scale expansion checked by discrete reduction in the pace of time that preserves spacetime equivalence. In spite of the cosmological expansion all epochs are equivalent and the line element does not change. The duration of the (proper) second increases in a stepwise manner in order to accommodate the increasing spatial scale. This preserves the relation between space and time. The spacetime geometry remains the same and the Perfect Cosmological Principle applies.

2.   A Few Properties of the EST Model

(1) The horizon problem disappears. All locations in the Universe communicate and have always communicated.

(2) There is no particle horizon and no event horizon.

(3) The Hubble time, T, which is believed to be related to the age of the Universe, is not the aging time but a time constant related to the expansion. It is also the age of the Universe measured with the current temporal metric. This is true for all epochs. However, since the length of the (atomic) second increases with time, the aging time could be infinite. This resolves the aging enigma.

(4) Instead of interpreting the redshift as a Doppler effect the new theory suggests that the redshift be Tired Light caused by the scale expansion. Tired Light

is known to agree better with observations than the Doppler shift mechanism. The Tired Light effect follows directly from the EST line element.

(5) The black body spectrum is preserved in the EST. In this respect the EST behaves just like a classical black body cavity. The electromagnetic energy added from radiating sources is balanced by the energy lost due to Tired Light resulting in the CMB equilibrium temperature, 2.73 K.

(6) The scale expansion generates a vacuum energy–momentum tensor with non-zero components, the Cosmic Energy Tensor. The T_{00} component corresponding to mass density equals the critical density. However, this positive energy is balanced by negative pressure so that the net gravitating energy is zero. This suggests that the missing dark matter is spacetime energy generated by the cosmological expansion rather than consisting of exotic particles.

(7) The Ω-enigma disappears. The EST expansion does not change the relative position between galaxies; these distances always remain the same measured with the expanding metric. There is no net motion or cosmological mass flow due to the expansion; the cosmological mass density is always constant. Also, there is no creation of matter as in the Steady State theories of Hoyle, Gold and Bondi.

(8) The Cosmic Drag effect generates a cosmological reference frame.

(9) The discrete cosmological expansion mode implied by the EST theory also provides a possible explanation for the quantum world.

3.   Cosmic Drag: A New Testable Property Predicted by the EST Theory

From the geodesic equations for the EST line element the (relative) velocity and angular momentum of a freely moving object diminishes over time. This effect generates a cosmological reference frame toward which all freely moving bodies converge and explains the relatively low galaxy velocities (less than 1% of c). It also explains the spiral shape of galaxies and their loss of angular momentum, which cannot be explained by standard physics. The majority of observed pulsars slow down with a time constant close to the Hubble time, about fourteen billion years. There is no physical explanation for this loss of angular momentum other than Cosmic Drag. Cosmic Drag also influences planetary motions in the Solar System, and the secular accelerations of Mercury, Venus and the Earth predicted by the EST theory have recently been confirmed by Kolesnik (these proceedings). The from observations estimated secular accelerations agree excellently with the EST theory's prediction if the Hubble time $T = 14$ Gyr:

$$dw/dt = 3w/T, \qquad (3)$$

where w is the angular velocity.

The EST theory also agrees excellently with cosmological tests, for example the Number Count Test and the Angular Size Test. Further information on the EST theory may be found in Masreliez (1999) and at http://www.estfound.org.

References

Masreliez, C. J. 1999, Ap&SS, 266, 399

Astrophysical Ages and Time Scales
ASP Conference Series, Vol. 245, 2001
T. von Hippel, C. Simpson, N. Manset

JHK Spectroscopy of the $z = 2.39$ Radio Galaxy 53W002 and Its Companions

K. Motohara

Subaru Telescope, National Astronomical Observatory of Japan, 650 North A'ohōkū Place, Hilo, HI 96720, USA

T. Yamada

National Astronomical Observatory of Japan, Mitaka, Tokyo 181-8588, Japan

F. Iwamuro, T. Maihara

Kyoto University, Kitashirakawa, Kyoto 606-8502, Japan

Abstract. We present low-resolution, near-IR *JHK* spectra of the weak $z = 2.39$ radio galaxy 53W002 and its companion objects #18 and #19, obtained with OHS/CISCO on the Subaru Telescope. They cover rest-frame wavelengths of 3400–7200 Å, and many rest-optical emission lines are detected. Contributions to the broad-band flux from these emission lines are found to be very large, up to 40% in the H and K'-bands and 30% in the J-band.

1. Introduction

Recent narrow-band imaging has revealed the existence of a cluster of Lyα emitters around the $z = 2.39$ radio galaxy 53W002 (Pascarelle et al. 1996a, 1996b; Pascarelle, Windhorst, & Keel 1998; Keel et al. 1999). These emitters have sub-galactic sizes, and are thought to be building blocks which will merge into a luminous galaxy at the present epoch.

To investigate the rest-frame optical nature of these objects, we carried out low-dispersion *JHK* spectroscopy of 53W002, Object #18, and #19 using the newly commissioned instrument of the Subaru telescope, OH-airglow suppression spectrograph (OHS; Iwamuro et al. 2001) and Cooled Infrared Spectrograph and Camera for OHS (CISCO; Motohara et al. 1998).

2. Detection of Strong Rest-Frame Optical Emission Lines

The resulting spectra of these objects are shown in Figure 1. Both 53W002 and Object #18 show very strong [O III] and Hα+[N II] lines. The Balmer jump is also detected in the continuum of 53W002. Object #19, which is known to be a quasar, shows a power-law continuum and a broad (8000 km s^{-1}), strong

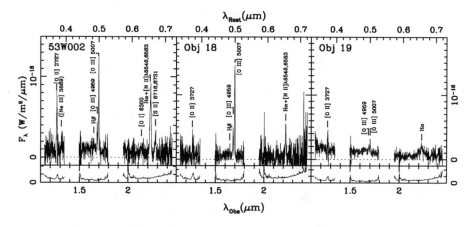

Figure 1.   *JHK* spectra. Lower box shows the 1σ noise level calculated from the background level.

(3×10^{43} erg s^{-1}; $H_0 = 65$ km s^{-1} Mpc^{-1}, $q_0 = 0.1$) Hα line. The contribution
of these lines to the broad-band flux is as high as 45%, as shown in Table 1.

3.   Emission Line Diagnosis

Because the wavelength resolution is low ($\lambda/\Delta\lambda = 200$–$400$), we deconvolved
the blended Hα+[N II] lines by multiple Gaussian fitting. We then estimated
the dust extinction from Hα/Hβ ratio, assuming the SMC extinction curve. The
estimate of $E(B - V)$ is 0.14 for 53W002 and 0.50 for Object #18.

In Figure 2, two diagrams of reddening corrected emission-line ratios are
presented. One shows [N II]λ6583/Hα versus [O III]λ5007/Hβ and the other
[O II]λ3727/[O III]λ5007 versus [O III]λ5007/Hβ.

We next carried out photoionization calculation using CLOUDY94 (Ferland
2000), and over-plotted the results on Figure 2. These results show that the
emission line ratios of 53W002 are well reproduced by a cloud with electron
density $n_e = 1 \times 10^{3-4}$ cm^{-3} and solar metallicity, ionized by an $\alpha = -0.7$
power-law continuum. Object #18 seems to be not a star-forming galaxy but a
type 2 AGN, and its line ratios are reproduced by a cloud of solar metallicity,
ionized by an $\alpha = -1.5$ power-law continuum.

Table 1.   Contribution of the emission lines to the broad-band flux.

Object	K'	H	J
53W002	42%	30	22
Object #18	32	45	11
Object #19	29	6	4

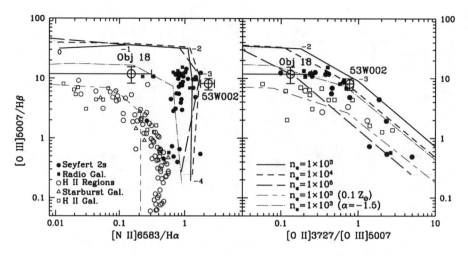

Figure 2.   Reddening corrected line ratios of 53W002 and Object #18
(open circles with error bars). Spectral index is assumed to be $\alpha = -0.7$
except for one case of $\alpha = -1.5$ shown by thin dot-dashed line. The
metallicity is set to be $Z = 1.0\ Z_\odot$, except for one case of $Z = 0.1\ Z_\odot$
shown by thin dash-long-dashed line. Ionization parameter varies along
each curve, and representative points are labeled by their powers.

Both 53W002 and Object #18 show high metallicity (solar abundances).
We suggest that they are produced by starburst activity during merger events
with surrounding objects, for which we find evidence in our spectrum of 53W002
in the form of the Balmer discontinuity at 4000 Å.

References

Ferland, G. J. 2000, RMxAC, 9, 153

Iwamuro, F., Motohara, K., Maihara, T., Hata, R., Harashima, T., & Sekiguchi,
K. 2001, PASJ, in press.

Keel, W. C., Cohen, S. H., Windhorst, R. A., & Waddington, I. 1999, AJ, 118,
2547

Motohara, K., et al. 1998, in Proc. SPIE 3354, Infrared Astronomical Instrumentation, ed. A. M. Fowler (Bellingham: SPIE), 659

Pascarelle, S. M., Windhorst, R. A., Driver, S. P., Ostrander, E. J., & Keel, W.
C. 1996a, ApJ, 456, L21

Pascarelle, S. M., Windhorst, R. A., & Keel, W. C. 1998, AJ, 116, 2659

Pascarelle, S. M., Windhorst, R. A., Keel, W. C., & Odewahn, S. C. 1996b,
Nature, 383, 45

Author index

ASTRONOMICAL SOCIETY OF THE PACIFIC CONFERENCE SERIES

and

INTERNATIONAL ASTRONOMICAL UNION VOLUMES

Published
by

The Astronomical Society of the Pacific
(ASP)

ASP CONFERENCE SERIES VOLUMES
Published by the Astronomical Society of the Pacific

PUBLISHED: 1988 (* asterisk means OUT OF STOCK)

Vol. CS -1 PROGRESS AND OPPORTUNITIES IN SOUTHERN HEMISPHERE
OPTICAL ASTRONOMY: CTIO 25TH Anniversary Symposium
eds. V. M. Blanco and M. M. Phillips
ISBN 0-937707-18-X

Vol. CS-2 PROCEEDINGS OF A WORKSHOP ON OPTICAL SURVEYS FOR QUASARS
eds. Patrick S. Osmer, Alain C. Porter, Richard F. Green, and Craig B. Foltz
ISBN 0-937707-19-8

Vol. CS-3 FIBER OPTICS IN ASTRONOMY
ed. Samuel C. Barden
ISBN 0-937707-20-1

Vol. CS-4 THE EXTRAGALACTIC DISTANCE SCALE:
Proceedings of the ASP 100th Anniversary Symposium
eds. Sidney van den Bergh and Christopher J. Pritchet
ISBN 0-937707-21-X

Vol. CS-5 THE MINNESOTA LECTURES ON CLUSTERS OF GALAXIES
AND LARGE-SCALE STRUCTURE
ed. John M. Dickey
ISBN 0-937707-22-8

PUBLISHED: 1989

Vol. CS-6 SYNTHESIS IMAGING IN RADIO ASTRONOMY: A Collection of Lectures
from the Third NRAO Synthesis Imaging Summer School
eds. Richard A. Perley, Frederic R. Schwab, and Alan H. Bridle
ISBN 0-937707-23-6

PUBLISHED: 1990

Vol. CS-7 PROPERTIES OF HOT LUMINOUS STARS: Boulder-Munich Workshop
ed. Catharine D. Garmany
ISBN 0-937707-24-4

Vol. CS-8* CCDs IN ASTRONOMY
ed. George H. Jacoby
ISBN 0-937707-25-2

Vol. CS-9 COOL STARS, STELLAR SYSTEMS, AND THE SUN: Sixth Cambridge Workshop
ed. George Wallerstein
ISBN 0-937707-27-9

Vol. CS-10* EVOLUTION OF THE UNIVERSE OF GALAXIES:
Edwin Hubble Centennial Symposium
ed. Richard G. Kron
ISBN 0-937707-28-7

Vol. CS-11 CONFRONTATION BETWEEN STELLAR PULSATION AND EVOLUTION
eds. Carla Cacciari and Gisella Clementini
ISBN 0-937707-30-9

Vol. CS-12 THE EVOLUTION OF THE INTERSTELLAR MEDIUM
ed. Leo Blitz
ISBN 0-937707-31-7

PUBLISHED: 1991

Vol. CS-13 THE FORMATION AND EVOLUTION OF STAR CLUSTERS
ed. Kenneth Janes
ISBN 0-937707-32-5

ASP CONFERENCE SERIES VOLUMES
Published by the Astronomical Society of the Pacific

PUBLISHED: 1991 (* asterisk means OUT OF STOCK)

Vol. CS-14   ASTROPHYSICS WITH INFRARED ARRAYS
ed. Richard Elston
ISBN 0-937707-33-3

Vol. CS-15   LARGE-SCALE STRUCTURES AND PECULIAR MOTIONS IN THE UNIVERSE
eds. David W. Latham and L. A. Nicolaci da Costa
ISBN 0-937707-34-1

Vol. CS-16   Proceedings of the 3rd Haystack Observatory Conference on ATOMS, IONS, AND MOLECULES: NEW RESULTS IN SPECTRAL LINE ASTROPHYSICS
eds. Aubrey D. Haschick and Paul T. P. Ho
ISBN 0-937707-35-X

Vol. CS-17   LIGHT POLLUTION, RADIO INTERFERENCE, AND SPACE DEBRIS
ed. David L. Crawford
ISBN 0-937707-36-8

Vol. CS-18   THE INTERPRETATION OF MODERN SYNTHESIS OBSERVATIONS OF SPIRAL GALAXIES
eds. Nebojsa Duric and Patrick C. Crane
ISBN 0-937707-37-6

Vol. CS-19   RADIO INTERFEROMETRY: THEORY, TECHNIQUES, AND APPLICATIONS, IAU Colloquium 131
eds. T. J. Cornwell and R. A. Perley
ISBN 0-937707-38-4

Vol. CS-20   FRONTIERS OF STELLAR EVOLUTION:
50th Anniversary McDonald Observatory (1939-1989)
ed. David L. Lambert
ISBN 0-937707-39-2

Vol. CS-21   THE SPACE DISTRIBUTION OF QUASARS
ed . David Crampton
ISBN 0-937707-40-6

PUBLISHED: 1992

Vol. CS-22   NONISOTROPIC AND VARIABLE OUTFLOWS FROM STARS
eds. Laurent Drissen, Claus Leitherer, and Antonella Nota
ISBN 0-937707-41-4

Vol CS-23   ASTRONOMICAL CCD OBSERVING AND REDUCTION TECHNIQUES
ed. Steve B. Howell
ISBN 0-937707-42-4

Vol. CS-24   COSMOLOGY AND LARGE-SCALE STRUCTURE IN THE UNIVERSE
ed. Reinaldo R. de Carvalho
ISBN 0-937707-43-0

Vol. CS-25   ASTRONOMICAL DATA ANALYSIS, SOFTWARE AND SYSTEMS I - (ADASS I)
eds. Diana M. Worrall, Chris Biemesderfer, and Jeannette Barnes
ISBN 0-937707-44-9

Vol. CS-26   COOL STARS, STELLAR SYSTEMS, AND THE SUN:
Seventh Cambridge Workshop
eds. Mark S. Giampapa and Jay A. Bookbinder
ISBN 0-937707-45-7

Vol. CS-27   THE SOLAR CYCLE: Proceedings of the
National Solar Observatory/Sacramento Peak 12th Summer Workshop
ed. Karen L. Harvey
ISBN 0-937707-46-5

ASP CONFERENCE SERIES VOLUMES
Published by the Astronomical Society of the Pacific

ASP CONFERENCE SERIES VOLUMES
Published by the Astronomical Society of the Pacific

ASP CONFERENCE SERIES VOLUMES
Published by the Astronomical Society of the Pacific

ASP CONFERENCE SERIES VOLUMES
Published by the Astronomical Society of the Pacific

PUBLISHED: 1995 (* asterisk means OUT OF STOCK)

Vol. CS-71　　TRIDIMENSIONAL OPTICAL SPECTROSCOPIC METHODS IN ASTROPHYSICS,
　　　　　　　IAU Colloquium 149
　　　　　　　eds. Georges Comte and Michel Marcelin
　　　　　　　ISBN 0-937707-90-2

Vol. CS-72　　MILLISECOND PULSARS: A DECADE OF SURPRISE
　　　　　　　eds. A. S Fruchter, M. Tavani, and D. C. Backer
　　　　　　　ISBN 0-937707-91-0

Vol. CS-73　　AIRBORNE ASTRONOMY SYMPOSIUM ON THE GALACTIC ECOSYSTEM:
　　　　　　　FROM GAS TO DUST
　　　　　　　eds. Michael R. Haas, Jacqueline A. Davidson, and Edwin F. Erickson
　　　　　　　ISBN 0-937707-92-9

Vol. CS-74　　PROGRESS IN THE SEARCH FOR EXTRATERRESTRIAL LIFE:
　　　　　　　1993 Bioastronomy Symposium
　　　　　　　ed. G. Seth Shostak
　　　　　　　ISBN 0-937707-93-7

Vol. CS-75　　MULTI-FEED SYSTEMS FOR RADIO TELESCOPES
　　　　　　　eds. Darrel T. Emerson and John M. Payne
　　　　　　　ISBN 0-937707-94-5

Vol. CS-76　　GONG '94: HELIO- AND ASTERO-SEISMOLOGY FROM THE EARTH
　　　　　　　AND SPACE
　　　　　　　eds. Roger K. Ulrich, Edward J. Rhodes, Jr., and Werner Däppen
　　　　　　　ISBN 0-937707-95-3

Vol. CS-77　　ASTRONOMICAL DATA ANALYSIS SOFTWARE AND SYSTEMS IV - (ADASS IV)
　　　　　　　eds. R. A. Shaw, H. E. Payne, and J. J. E. Hayes
　　　　　　　ISBN 0-937707-96-1

Vol. CS-78　　ASTROPHYSICAL APPLICATIONS OF POWERFUL NEW DATABASES:
　　　　　　　Joint Discussion No. 16 of the 22nd General Assembly of the IAU
　　　　　　　eds. S. J. Adelman and W. L. Wiese
　　　　　　　ISBN 0-937707-97-X

Vol. CS-79*　ROBOTIC TELESCOPES: CURRENT CAPABILITIES, PRESENT
　　　　　　　DEVELOPMENTS, AND FUTURE PROSPECTS
　　　　　　　FOR AUTOMATED ASTRONOMY
　　　　　　　eds. Gregory W. Henry and Joel A. Eaton
　　　　　　　ISBN 0-937707-98-8

Vol. CS-80*　THE PHYSICS OF THE INTERSTELLAR MEDIUM
　　　　　　　AND INTERGALACTIC MEDIUM
　　　　　　　eds. A. Ferrara, C. F. McKee, C. Heiles, and P. R. Shapiro
　　　　　　　ISBN 0-937707-99-6

Vol. CS-81　　LABORATORY AND ASTRONOMICAL HIGH RESOLUTION SPECTRA
　　　　　　　eds. A. J. Sauval, R. Blomme, and N. Grevesse
　　　　　　　ISBN 1-886733-01-5

Vol. CS-82*　VERY LONG BASELINE INTERFEROMETRY AND THE VLBA
　　　　　　　eds. J. A. Zensus, P. J. Diamond, and P. J. Napier
　　　　　　　ISBN 1-886733-02-3

Vol. CS-83*　ASTROPHYSICAL APPLICATIONS OF STELLAR PULSATION,
　　　　　　　IAU Colloquium 155
　　　　　　　eds. R. S. Stobie and P. A. Whitelock
　　　　　　　ISBN 1-886733-03-1

ATLAS　　　　INFRARED ATLAS OF THE ARCTURUS SPECTRUM, 0.9 - 5.3 μm
　　　　　　　eds. Kenneth Hinkle, Lloyd Wallace, and William Livingston
　　　　　　　ISBN: 1-886733-04-X

ASP CONFERENCE SERIES VOLUMES
Published by the Astronomical Society of the Pacific

PUBLISHED: 1995 (* asterisk means OUT OF STOCK)

Vol. CS-84   THE FUTURE UTILIZATION OF SCHMIDT TELESCOPES, IAU Colloquium 148
eds. Jessica Chapman, Russell Cannon, Sandra Harrison, and Bambang Hidayat
ISBN 1-886733-05-8

Vol. CS-85*   CAPE WORKSHOP ON MAGNETIC CATACLYSMIC VARIABLES
eds. D. A. H. Buckley and B. Warner
ISBN 1-886733-06-6

Vol. CS-86   FRESH VIEWS OF ELLIPTICAL GALAXIES
eds. Alberto Buzzoni, Alvio Renzini, and Alfonso Serrano
ISBN 1-886733-07-4

PUBLISHED: 1996

Vol. CS-87   NEW OBSERVING MODES FOR THE NEXT CENTURY
eds. Todd Boroson, John Davies, and Ian Robson
ISBN 1-886733-08-2

Vol. CS-88*   CLUSTERS, LENSING, AND THE FUTURE OF THE UNIVERSE
eds. Virginia Trimble and Andreas Reisenegger
ISBN 1-886733-09-0

Vol. CS-89   ASTRONOMY EDUCATION: CURRENT DEVELOPMENTS,
FUTURE COORDINATION
ed. John R. Percy
ISBN 1-886733-10-4

Vol. CS-90   THE ORIGINS, EVOLUTION, AND DESTINIES OF BINARY STARS
IN CLUSTERS
eds. E. F. Milone and J. -C. Mermilliod
ISBN 1-886733-11-2

Vol. CS-91   BARRED GALAXIES, IAU Colloquium 157
eds. R. Buta, D. A. Crocker, and B. G. Elmegreen
ISBN 1-886733-12-0

Vol. CS-92*   FORMATION OF THE GALACTIC HALO .... INSIDE AND OUT
eds. Heather L. Morrison and Ata Sarajedini
ISBN 1-886733-13-9

Vol. CS-93   RADIO EMISSION FROM THE STARS AND THE SUN
eds. A. R. Taylor and J. M. Paredes
ISBN 1-886733-14-7

Vol. CS-94   MAPPING, MEASURING, AND MODELING THE UNIVERSE
eds. Peter Coles, Vicent J. Martinez, and Maria-Jesus Pons-Borderia
ISBN 1-886733-15-5

Vol. CS-95   SOLAR DRIVERS OF INTERPLANETARY AND TERRESTRIAL DISTURBANCES:
Proceedings of 16th International Workshop National Solar
Observatory/Sacramento Peak
eds. K. S. Balasubramaniam, Stephen L. Keil, and Raymond N. Smartt
ISBN 1-886733-16-3

Vol. CS-96   HYDROGEN-DEFICIENT STARS
eds. C. S. Jeffery and U. Heber
ISBN 1-886733-17-1

Vol. CS-97   POLARIMETRY OF THE INTERSTELLAR MEDIUM
eds. W. G. Roberge and D. C. B. Whittet
ISBN 1-886733-18-X

ASP CONFERENCE SERIES VOLUMES
Published by the Astronomical Society of the Pacific

PUBLISHED: 1996 (* asterisk means OUT OF STOCK)

Vol. CS-98   FROM STARS TO GALAXIES: THE IMPACT OF STELLAR PHYSICS
ON GALAXY EVOLUTION
eds. Claus Leitherer, Uta Fritze-von Alvensleben, and John Huchra
ISBN 1-886733-19-8

Vol. CS-99   COSMIC ABUNDANCES:
Proceedings of the 6th Annual October Astrophysics Conference
eds. Stephen S. Holt and George Sonneborn
ISBN 1-886733-20-1

Vol. CS-100   ENERGY TRANSPORT IN RADIO GALAXIES AND QUASARS
eds. P. E. Hardee, A. H. Bridle, and J. A. Zensus
ISBN 1-886733-21-X

Vol. CS-101   ASTRONOMICAL DATA ANALYSIS SOFTWARE AND SYSTEMS V – (ADASS V)
eds. George H. Jacoby and Jeannette Barnes
ISBN 1080-7926

Vol. CS-102   THE GALACTIC CENTER, 4th ESO/CTIO Workshop
ed. Roland Gredel
ISBN 1-886733-22-8

Vol. CS-103   THE PHYSICS OF LINERS IN VIEW OF RECENT OBSERVATIONS
eds. M. Eracleous, A. Koratkar, C. Leitherer, and L. Ho
ISBN 1-886733-23-6

Vol. CS-104   PHYSICS, CHEMISTRY, AND DYNAMICS OF INTERPLANETARY DUST,
IAU Colloquium 150
eds. Bo Å. S. Gustafson and Martha S. Hanner
ISBN 1-886733-24-4

Vol. CS-105   PULSARS: PROBLEMS AND PROGRESS, IAU Colloquium 160
ed. S. Johnston, M. A. Walker, and M. Bailes
ISBN 1-886733-25-2

Vol. CS-106   THE MINNESOTA LECTURES ON EXTRAGALACTIC NEUTRAL HYDROGEN
ed. Evan D. Skillman
ISBN 1-886733-26-0

Vol. CS-107   COMPLETING THE INVENTORY OF THE SOLAR SYSTEM:
A Symposium held in conjunction with the 106th Annual Meeting of the ASP
eds. Terrence W. Rettig and Joseph M. Hahn
ISBN 1-886733-27-9

Vol. CS-108   M.A.S.S. -- MODEL ATMOSPHERES AND SPECTRUM SYNTHESIS:
5th Vienna - Workshop
eds. Saul J. Adelman, Friedrich Kupka, and Werner W. Weiss
ISBN 1-886733-28-7

Vol. CS-109   COOL STARS, STELLAR SYSTEMS, AND THE SUN: Ninth Cambridge Workshop
eds. Roberto Pallavicini and Andrea K. Dupree
ISBN 1-886733-29-5

Vol. CS-110   BLAZAR CONTINUUM VARIABILITY
eds. H. R. Miller, J. R. Webb, and J. C. Noble
ISBN 1-886733-30-9

Vol. CS-111   MAGNETIC RECONNECTION IN THE SOLAR ATMOSPHERE:
Proceedings of a Yohkoh Conference
eds. R. D. Bentley and J. T. Mariska
ISBN 1-886733-31-7

ASP CONFERENCE SERIES VOLUMES
Published by the Astronomical Society of the Pacific

PUBLISHED: 1996 (* asterisk means OUT OF STOCK)

Vol. CS-112   THE HISTORY OF THE MILKY WAY AND ITS SATELLITE SYSTEM
eds. Andreas Burkert, Dieter H. Hartmann, and Steven R. Majewski
ISBN 1-886733-32-5

PUBLISHED: 1997

Vol. CS-113   EMISSION LINES IN ACTIVE GALAXIES: NEW METHODS AND TECHNIQUES,
IAU Colloquium 159
eds. B. M. Peterson, F.-Z. Cheng, and A. S. Wilson
ISBN 1-886733-33-3

Vol. CS-114   YOUNG GALAXIES AND QSO ABSORPTION-LINE SYSTEMS
eds. Sueli M. Viegas, Ruth Gruenwald, and Reinaldo R. de Carvalho
ISBN 1-886733-34-1

Vol. CS-115   GALACTIC CLUSTER COOLING FLOWS
ed. Noam Soker
ISBN 1-886733-35-X

Vol. CS-116   THE SECOND STROMLO SYMPOSIUM:
THE NATURE OF ELLIPTICAL GALAXIES
eds. M. Arnaboldi, G. S. Da Costa, and P. Saha
ISBN 1-886733-36-8

Vol. CS-117   DARK AND VISIBLE MATTER IN GALAXIES
eds. Massimo Persic and Paolo Salucci
ISBN-1-886733-37-6

Vol. CS-118   FIRST ADVANCES IN SOLAR PHYSICS EUROCONFERENCE:
ADVANCES IN THE PHYSICS OF SUNSPOTS
eds. B. Schmieder. J. C. del Toro Iniesta, and M. Vázquez
ISBN 1-886733-38-4

Vol. CS-119   PLANETS BEYOND THE SOLAR SYSTEM
AND THE NEXT GENERATION OF SPACE MISSIONS
ed. David R. Soderblom
ISBN 1-886733-39-2

Vol. CS-120   LUMINOUS BLUE VARIABLES: MASSIVE STARS IN TRANSITION
eds. Antonella Nota and Henny J. G. L. M. Lamers
ISBN 1-886733-40-6

Vol. CS-121   ACCRETION PHENOMENA AND RELATED OUTFLOWS, IAU Colloquium 163
eds. D. T. Wickramasinghe, G. V. Bicknell, and L. Ferrario
ISBN 1-886733-41-4

Vol. CS-122   FROM STARDUST TO PLANETESIMALS:
Symposium held as part of the 108th Annual Meeting of the ASP
eds. Yvonne J. Pendleton and A. G. G. M. Tielens
ISBN 1-886733-42-2

Vol. CS-123   THE 12th 'KINGSTON MEETING': COMPUTATIONAL ASTROPHYSICS
eds. David A. Clarke and Michael J. West
ISBN 1-886733-43-0

Vol. CS-124   DIFFUSE INFRARED RADIATION AND THE IRTS
eds. Haruyuki Okuda, Toshio Matsumoto, and Thomas Roellig
ISBN 1-886733-44-9

Vol. CS-125   ASTRONOMICAL DATA ANALYSIS SOFTWARE AND SYSTEMS VI
eds. Gareth Hunt and H. E. Payne
ISBN 1-886733-45-7

ASP CONFERENCE SERIES VOLUMES
Published by the Astronomical Society of the Pacific

ASP CONFERENCE SERIES VOLUMES
Published by the Astronomical Society of the Pacific

PUBLISHED: 1998 (* asterisk means OUT OF STOCK)

Vol. CS-140 SYNOPTIC SOLAR PHYSICS --18th NSO/Sacramento Peak Summer Workshop
eds. K. S. Balasubramaniam, J. W. Harvey, and D. M. Rabin
ISBN 1-886733-60-0

Vol. CS-141 ASTROPHYSICS FROM ANTARCTICA:
A Symposium held as a part of the 109th Annual Meeting of the ASP
eds. Giles Novak and Randall H. Landsberg
ISBN 1-886733-61-9

Vol. CS-142 THE STELLAR INITIAL MASS FUNCTION: 38th Herstmonceux Conference
eds. Gerry Gilmore and Debbie Howell
ISBN 1-886733-62-7

Vol. CS-143* THE SCIENTIFIC IMPACT OF THE GODDARD HIGH RESOLUTION
SPECTROGRAPH (GHRS)
eds. John C. Brandt, Thomas B. Ake III, and Carolyn Collins Petersen
ISBN 1-886733-63-5

Vol. CS-144 RADIO EMISSION FROM GALACTIC AND EXTRAGALACTIC COMPACT
SOURCES, IAU Colloquium 164
eds. J. Anton Zensus, G. B. Taylor, and J. M. Wrobel
ISBN 1-886733-64-3

Vol. CS-145 ASTRONOMICAL DATA ANALYSIS SOFTWARE AND SYSTEMS VII – (ADASS VII)
Eds. Rudolf Albrecht, Richard N. Hook, and Howard A. Bushouse
ISBN 1-886733-65-1

Vol. CS-146 THE YOUNG UNIVERSE GALAXY FORMATION
AND EVOLUTION AT INTERMEDIATE AND HIGH REDSHIFT
eds. S. D'Odorico, A. Fontana, and E. Giallongo
ISBN 1-886733-66-X

Vol. CS-147 ABUNDANCE PROFILES: DIAGNOSTIC TOOLS FOR GALAXY HISTORY
eds. Daniel Friedli, Mike Edmunds, Carmelle Robert, and Laurent Drissen
ISBN 1-886733-67-8

Vol. CS-148 ORIGINS
eds. Charles E. Woodward, J. Michael Shull, and Harley A. Thronson, Jr.
ISBN 1-886733-68-6

Vol. CS-149 SOLAR SYSTEM FORMATION AND EVOLUTION
eds. D. Lazzaro, R. Vieira Martins, S. Ferraz-Mello, J. Fernández, and C. Beaugé
ISBN 1-886733-69-4

Vol. CS-150 NEW PERSPECTIVES ON SOLAR PROMINENCES, IAU Colloquium 167
eds. David Webb, David Rust, and Brigitte Schmieder
ISBN 1-886733-70-8

Vol. CS-151 COSMIC MICROWAVE BACKGROUND
AND LARGE SCALE STRUCTURES OF THE UNIVERSE
eds. Yong-Ik Byun and Kin-Wang Ng
ISBN 1-886733-71-6

Vol. CS-152 FIBER OPTICS IN ASTRONOMY III
eds. S. Arribas, E. Mediavilla, and F. Watson
ISBN 1-886733-72-4

Vol. CS-153 LIBRARY AND INFORMATION SERVICES IN ASTRONOMY III -- (LISA III)
eds. Uta Grothkopf, Heinz Andernach, Sarah Stevens-Rayburn,
and Monique Gomez
ISBN 1-886733-73-2

ASP CONFERENCE SERIES VOLUMES
Published by the Astronomical Society of the Pacific

ASP CONFERENCE SERIES VOLUMES
Published by the Astronomical Society of the Pacific

PUBLISHED: 1999 (* asterisk means OUT OF STOCK)

Vol. CS-169   11th EUROPEAN WORKSHOP ON WHITE DWARFS
eds. J.-E. Solheim and E. G. Meištas
ISBN 1-886733-91-0

Vol. CS-170   THE LOW SURFACE BRIGHTNESS UNIVERSE, IAU Colloquium 171
eds. J. I. Davies, C. Impey, and S. Phillipps
ISBN 1-886733-92-9

Vol. CS-171   LiBeB, COSMIC RAYS, AND RELATED X- AND GAMMA-RAYS
eds. Reuven Ramaty, Elisabeth Vangioni-Flam, Michel Cassé, and Keith Olive
ISBN 1-886733-93-7

Vol. CS-172   ASTRONOMICAL DATA ANALYSIS SOFTWARE AND SYSTEMS VIII
eds. David M. Mehringer, Raymond L. Plante, and Douglas A. Roberts
ISBN 1-886733-94-5

Vol. CS-173   THEORY AND TESTS OF CONVECTION IN STELLAR STRUCTURE:
First Granada Workshop
ed. Álvaro Giménez, Edward F. Guinan, and Benjamín Montesinos
ISBN 1-886733-95-3

Vol. CS-174   CATCHING THE PERFECT WAVE: ADAPTIVE OPTICS AND
INTERFEROMETRY IN THE 21st CENTURY,
A Symposium held as a part of the 110th Annual Meeting of the ASP
eds. Sergio R. Restaino, William Junor, and Nebojsa Duric
ISBN 1-886733-96-1

Vol. CS-175   STRUCTURE AND KINEMATICS OF QUASAR BROAD LINE REGIONS
eds. C. M. Gaskell, W. N. Brandt, M. Dietrich, D. Dultzin-Hacyan,
and M. Eracleous
ISBN 1-886733-97-X

Vol. CS-176   OBSERVATIONAL COSMOLOGY: THE DEVELOPMENT OF GALAXY SYSTEMS
eds. Giuliano Giuricin, Marino Mezzetti, and Paolo Salucci
ISBN 1-58381-000-5

Vol. CS-177   ASTROPHYSICS WITH INFRARED SURVEYS: A Prelude to SIRTF
eds. Michael D. Bicay, Chas A. Beichman, Roc M. Cutri, and Barry F. Madore
ISBN 1-58381-001-3

Vol. CS-178   STELLAR DYNAMOS: NONLINEARITY AND CHAOTIC FLOWS
eds. Manuel Núñez and Antonio Ferriz-Mas
ISBN 1-58381-002-1

Vol. CS-179   ETA CARINAE AT THE MILLENNIUM
eds. Jon A. Morse, Roberta M. Humphreys, and Augusto Damineli
ISBN 1-58381-003-X

Vol. CS-180   SYNTHESIS IMAGING IN RADIO ASTRONOMY II
eds. G. B. Taylor, C. L. Carilli, and R. A. Perley
ISBN 1-58381-005-6

Vol. CS-181   MICROWAVE FOREGROUNDS
eds. Angelica de Oliveira-Costa and Max Tegmark
ISBN 1-58381-006-4

Vol. CS-182   GALAXY DYNAMICS: A Rutgers Symposium
eds. David Merritt, J. A. Sellwood, and Monica Valluri
ISBN 1-58381-007-2

Vol. CS-183   HIGH RESOLUTION SOLAR PHYSICS: THEORY, OBSERVATIONS,
AND TECHNIQUES
eds. T. R. Rimmele, K. S. Balasubramaniam, and R. R. Radick
ISBN 1-58381-009-9

ASP CONFERENCE SERIES VOLUMES
Published by the Astronomical Society of the Pacific

PUBLISHED: 1999 (* asterisk means OUT OF STOCK)

Vol. CS-184   THIRD ADVANCES IN SOLAR PHYSICS EUROCONFERENCE:
MAGNETIC FIELDS AND OSCILLATIONS
eds. B. Schmieder, A. Hofmann, and J. Staude
ISBN 1-58381-010-2

Vol. CS-185   PRECISE STELLAR RADIAL VELOCITIES, IAU Colloquium 170
eds. J. B. Hearnshaw and C. D. Scarfe
ISBN 1-58381-011-0

Vol. CS-186   THE CENTRAL PARSECS OF THE GALAXY
eds. Heino Falcke, Angela Cotera, Wolfgang J. Duschl, Fulvio Melia,
and Marcia J. Rieke
ISBN 1-58381-012-9

Vol. CS-187   THE EVOLUTION OF GALAXIES ON COSMOLOGICAL TIMESCALES
eds. J. E. Beckman and T. J. Mahoney
ISBN 1-58381-013-7

Vol. CS-188   OPTICAL AND INFRARED SPECTROSCOPY OF CIRCUMSTELLAR MATTER
eds. Eike W. Guenther, Bringfried Stecklum, and Sylvio Klose
ISBN 1-58381-014-5

Vol. CS-189   CCD PRECISION PHOTOMETRY WORKSHOP
eds. Eric R. Craine, Roy A. Tucker, and Jeannette Barnes
ISBN 1-58381-015-3

Vol. CS-190   GAMMA-RAY BURSTS: THE FIRST THREE MINUTES
eds. Juri Poutanen and Roland Svensson
ISBN 1-58381-016-1

Vol. CS-191   PHOTOMETRIC REDSHIFTS AND HIGH REDSHIFT GALAXIES
eds. Ray J. Weymann, Lisa J. Storrie-Lombardi, Marcin Sawicki,
and Robert J. Brunner
ISBN 1-58381-017-X

Vol. CS-192   SPECTROPHOTOMETRIC DATING OF STARS AND GALAXIES
ed. I. Hubeny, S. R. Heap, and R. H. Cornett
ISBN 1-58381-018-8

Vol. CS-193   THE HY-REDSHIFT UNIVERSE:
GALAXY FORMATION AND EVOLUTION AT HIGH REDSHIFT
eds. Andrew J. Bunker and Wil J. M. van Breugel
ISBN 1-58381-019-6

Vol. CS-194   WORKING ON THE FRINGE:
OPTICAL AND IR INTERFEROMETRY FROM GROUND AND SPACE
eds. Stephen Unwin and Robert Stachnik
ISBN 1-58381-020-X

PUBLISHED: 2000

Vol. CS-195   IMAGING THE UNIVERSE IN THREE DIMENSIONS:
Astrophysics with Advanced Multi-Wavelength Imaging Devices
eds. W. van Breugel and J. Bland-Hawthorn
ISBN 1-58381-022-6

Vol. CS-196   THERMAL EMISSION SPECTROSCOPY AND ANALYSIS OF DUST,
DISKS, AND REGOLITHS
eds. Michael L. Sitko, Ann L. Sprague, and David K. Lynch
ISBN: 1-58381-023-4

Vol. CS-197   XVth IAP MEETING DYNAMICS OF GALAXIES:
FROM THE EARLY UNIVERSE TO THE PRESENT
eds. F. Combes, G. A. Mamon, and V. Charmandaris
ISBN: 1-58381-24-2

ASP CONFERENCE SERIES VOLUMES
Published by the Astronomical Society of the Pacific

ASP CONFERENCE SERIES VOLUMES
Published by the Astronomical Society of the Pacific

PUBLISHED: 2000 (* asterisk means OUT OF STOCK)

Vol. CS-211   MASSIVE STELLAR CLUSTERS
eds. Ariane Lançon and Christian M. Boily
ISBN: 1-58381-042-0

Vol. CS-212   FROM GIANT PLANETS TO COOL STARS
eds. Caitlin A. Griffith and Mark S. Marley
ISBN: 1-58381-041-2

Vol. CS-213   BIOASTRONOMY '99: A NEW ERA IN BIOASTRONOMY
eds. Guillermo A. Lemarchand and Karen J. Meech
ISBN: 1-58381-044-7

Vol. CS-214   THE Be PHENOMENON IN EARLY-TYPE STARS, IAU Colloquium 175
eds. Myron A. Smith, Huib F. Henrichs and Juan Fabregat
ISBN: 1-58381-045-5

Vol. CS-215   COSMIC EVOLUTION AND GALAXY FORMATION:
STRUCTURE, INTERACTIONS AND FEEDBACK
The 3rd Guillermo Haro Astrophysics Conference
eds. José Franco, Elena Terlevich, Omar López-Cruz, and Itziar Aretxaga
ISBN: 1-58381-046-3

Vol. CS-216   ASTRONOMICAL DATA ANALYSIS SOFTWARE AND SYSTEMS IX
eds. Nadine Manset, Christian Veillet, and Dennis Crabtree
ISBN: 1-58381-047-1   ISSN: 1080-7926

Vol. CS-217   IMAGING AT RADIO THROUGH SUBMILLIMETER WAVELENGTHS
eds. Jeffrey G. Mangum and Simon J. E. Radford
ISBN: 1-58381-049-8

Vol. CS-218   MAPPING THE HIDDEN UNIVERSE: THE UNIVERSE BEHIND THE MILKYWAY
THE UNIVERSE IN HI
eds. Renée C. Kraan-Korteweg, Patricia A. Henning, and Heinz Andernach
ISBN: 1-58381-050-1

Vol. CS-219   DISKS, PLANETESIMALS, AND PLANETS
eds. F. Garzón, C. Eiroa, D. de Winter, and T. J. Mahoney
ISBN: 1-58381-051-X

Vol. CS-220   AMATEUR - PROFESSIONAL PARTNERSHIPS IN ASTRONOMY:
The 111th Annual Meeting of the ASP
eds. John R. Percy and Joseph B. Wilson
ISBN: 1-58381-052-8

Vol. CS-221   STARS, GAS AND DUST IN GALAXIES: EXPLORING THE LINKS
eds. Danielle Alloin, Knut Olsen, and Gaspar Galaz
ISBN: 1-58381-053-6

PUBLISHED: 2001

Vol. CS-222   THE PHYSICS OF GALAXY FORMATION
eds. M. Umemura and H. Susa
ISBN: 1-58381-054-4

Vol. CS-223   COOL STARS, STELLAR SYSTEMS AND THE SUN:
Eleventh Cambridge Workshop
eds. Ramón J. García López, Rafael Rebolo, and María Zapatero Osorio
ISBN: 1-58381-056-0

Vol. CS-224   PROBING THE PHYSICS OF ACTIVE GALACTIC NUCLEI
BY MULTIWAVELENGTH MONITORING
eds. Bradley M. Peterson, Ronald S. Polidan, and Richard W. Pogge
ISBN: 1-58381-055-2

ASP CONFERENCE SERIES VOLUMES
Published by the Astronomical Society of the Pacific

ASP CONFERENCE SERIES VOLUMES
Published by the Astronomical Society of the Pacific

All book orders or inquiries concerning ASP or IAU volumes listed should be directed to the:

The Astronomical Society of the Pacific Conference Series
390 Ashton Avenue
San Francisco CA 94112-1722 USA

Phone: 415-337-2126
Fax: 415-337-5205

E-mail: catalog@astrosociety.org
Web Site: http://www.astrosociety.org

IAU VOLUMES CAN BE FOUND ON THE NEXT PAGE

INTERNATIONAL ASTRONOMICAL UNION (IAU) VOLUMES
Published by the Astronomical Society of the Pacific

PUBLISHED: 1999

Vol. No. 190　　NEW VIEWS OF THE MAGELLANIC CLOUDS
eds. You-Hua Chu, Nicholas B. Suntzeff, James E. Hesser,
and David A. Bohlender
ISBN: 1-58381-021-8

Vol. No. 191　　ASYMPTOTIC GIANT BRANCH STARS
eds. T. Le Bertre, A. Lèbre, and C. Waelkens
ISBN: 1-886733-90-2

Vol. No. 192　　THE STELLAR CONTENT OF LOCAL GROUP GALAXIES
eds. Patricia Whitelock and Russell Cannon
ISBN: 1-886733-82-1

Vol. No. 193　　WOLF-RAYET PHENOMENA IN MASSIVE STARS AND STARBURST GALAXIES
eds. Karel A. van der Hucht, Gloria Koenigsberger, and Philippe R. J. Eenens
ISBN: 1-58381-004-8

Vol. No. 194　　ACTIVE GALACTIC NUCLEI AND RELATED PHENOMENA
eds. Yervant Terzian, Daniel Weedman, and Edward Khachikian
ISBN: 1-58381-008-0

PUBLISHED: 2000

Vol. XXIVA　　TRANSACTIONS OF THE INTERNATIONAL ASTRONOMICAL UNION
REPORTS ON ASTRONOMY 1996-1999
ed. Johannes Andersen
ISBN: 1-58381-035-8

Vol. No. 195　　HIGHLY ENERGETIC PHYSICAL PROCESSES AND MECHANISMS FOR
EMISSION FROM ASTROPHYSICAL PLASMAS
eds. P. C. H. Martens, S. Tsuruta, and M. A. Weber
ISBN: 1-58381-038-2

Vol. No. 197　　ASTROCHEMISTRY: FROM MOLECULAR CLOUDS TO PLANETARY SYSTEMS
eds. Y. C. Minh and E. F. van Dishoeck
ISBN: 1-58381-034-X

Vol. No. 198　　THE LIGHT ELEMENTS AND THEIR EVOLUTION
eds. L. da Silva, M. Spite, and J. R. de Medeiros
ISBN: 1-58381-048-X

PUBLISHED: 2001

IAU SPS　　ASTRONOMY FOR DEVELOPING COUNTRIES
Special Session of the XXIV General Assembly of the IAU
ed. Alan H. Batten
ISBN: 1-58381-067-6

Vol. No. 196　　PRESERVING THE ASTRONOMICAL SKY
eds. R. J. Cohen and W. T. Sullivan, III
ISBN: 1-58381-078-1

Vol. No. 200　　THE FORMATION OF BINARY STARS
eds. Hans Zinnecker and Robert D. Mathieu
ISBN: 1-58381-068-4

Vol. No. 203　　RECENT INSIGHTS INTO THE PHYSICS 0F THE SUN AND HELIOSPHERE:
HIGHLIGHTS FROM SOHO AND OTHER SPACE MISSIONS
eds. Pål Brekke, Bernhard Fleck, and Joseph B. Gurman
ISBN: 1-58381-069-2

INTERNATIONAL ASTRONOMICAL UNION (IAU) VOLUMES
Published by the Astronomical Society of the Pacific

PUBLISHED: 2001

Vol. No. 204   THE EXTRAGALACTIC INFRARED BACKGROUND AND ITS COSMOLOGICAL
IMPLICATIONS
eds. Martin Harwit and Michael G. Hauser
ISBN: 1-58381-062-5

Vol. No. 205   GALAXIES AND THEIR CONSTITUENTS
AT THE HIGHEST ANGULAR RESOLUTIONS
eds. Richard T. Schilizzi, Stuart N. Vogel, Francesco Paresce, and Martin S. Elvis
ISBN: 1-58381-066-8

Complete lists of proceedings of past IAU Meetings are maintained at the
IAU Web site at the URL: http://www.iau.org/publicat.html

Volumes 32 - 189 in the IAU Symposia Series may be ordered from
Kluwer Academic Publishers
P. O. Box 117
NL 3300 AA Dordrecht
The Netherlands